Polymer Nanocomposites

Sachin N. Kothawade • Vishal Pande
Surendra Gattani • Rajashri Sumbe
Editors

Polymer Nanocomposites

Advances in Design, Synthesis, and Applications

Springer

Editors
Sachin N. Kothawade
Department of Pharmaceutics
SCSSS's Sitabai Thite College of Pharmacy
Shirur, Maharashtra, India

Vishal Pande
Department of Pharmaceutics
RSM's N. N. Sattha College of Pharmacy, Ahmednagar
Ahilyanagar, Maharashtra, India

Surendra Gattani
Department of Pharmaceutics
School of Pharmacy, S.R.T.M.University, Nanded
Nanded, Maharashtra, India

Rajashri Sumbe
Department of Pharmaceutical Chemistry
RSM's N. N. Sattha College of Pharmacy, Ahmednagar
Ahilyanagar, Maharashtra, India

ISBN 978-3-032-00364-5 ISBN 978-3-032-00365-2 (eBook)
https://doi.org/10.1007/978-3-032-00365-2

This Springer imprint is published by the registered company Springer Nature Switzerland AG
The registered company address is: Gewerbestrasse 11, 6330 Cham, Switzerland

Foreword

It is with great pleasure that I present this edited volume, which offers a profound and up-to-date examination of the dynamic and ever-evolving field of polymer nanocomposites. This compilation represents a significant academic endeavor, bringing together distinguished contributions from researchers and practitioners who are shaping the future of material science and nanotechnology.

Polymer nanocomposites, by virtue of their unique structural and functional properties, have revolutionized the landscape of advanced materials. Through the synergistic integration of nanoscale reinforcements into polymer matrices, these systems demonstrate enhanced mechanical, thermal, electrical, and barrier properties—making them indispensable in diverse domains, such as environmental remediation, energy systems, pharmaceuticals, and biomedical engineering.

This volume serves not only as a scholarly resource but also as a strategic guide for future innovation. The chapters encapsulate cutting-edge research and reflect multidisciplinary perspectives, covering synthesis strategies, structure–property relationships, characterization techniques, and application-driven insights. The content is both foundational and forward-looking, ensuring relevance to academic researchers, postgraduate students, and industry professionals alike.

As an academic deeply engaged in materials research, I recognize the critical importance of such a comprehensive reference. I commend the editorial team and all contributing authors for their dedication, academic integrity, and clarity of vision. Their collaborative effort has resulted in a volume that does not merely disseminate knowledge but also inspires continued exploration and advancement in the field of polymer nanocomposites.

It is my hope that this book will serve as a valuable tool for learning, teaching, and innovation across institutions and industries and will continue to spark intellectual curiosity and scientific breakthroughs in the years to come.

PES's Modern College of Pharmacy, Nigdi
Pune, Maharashtra, India
P. D. Chaudhari

Preface

Polymer nanocomposites represent a dynamic and rapidly evolving field that integrates the disciplines of materials science, nanotechnology, chemistry, and engineering. These advanced materials have attracted significant attention due to their remarkable ability to enhance mechanical strength, thermal stability, chemical resistance, and multifunctionality through the incorporation of nanoscale fillers into polymer matrices. Their widespread utility spans diverse domains, such as water purification, drug delivery systems, advanced packaging materials, energy storage devices, and structural reinforcements.

This edited volume aims to provide a comprehensive overview of recent developments in the field of polymer nanocomposites, reflecting the interdisciplinary nature and technological potential of these materials. The chapters collectively explore critical aspects, such as novel design strategies, state-of-the-art synthesis methods, advanced characterization tools, reinforcement mechanisms, and future-oriented applications. Special attention has been given to both the fundamental principles and the innovative approaches that drive this field forward.

Each contribution has been carefully selected and peer-reviewed to ensure scholarly rigor and coherence with the overarching themes of this book. While individual chapters delve into specific subject areas, they are interconnected through a shared vision of advancing material performance and functionality at the nanoscale. The compilation seeks to bridge the gap between academic research and industrial application by highlighting translational prospects and real-world implementation.

This book is intended to serve as a valuable reference for a broad audience—including academic researchers, industrial scientists, postgraduate students, and professionals—who are engaged in the exploration, development, and application of polymer nanocomposite technologies. We trust that the insights and knowledge shared within these pages will stimulate further innovation and collaboration in this exciting domain of materials science.

We extend our sincere gratitude to all contributing authors and reviewers for their commitment to excellence, and we hope that this work proves to be an inspiring and informative resource for the global scientific community.

Shirur, Maharashtra, India — Sachin N. Kothawade
Ahilyanagar, Maharashtra, India — Vishal Pande
Nanded, Maharashtra, India — Surendra Gattani
Ahilyanagar, Maharashtra, India — Rajashri Sumbe

Acknowledgments

The editorial team sincerely expresses its deepest appreciation to Dr. Rajendra N. Thite, President of Shri Chhatrapati Sambhaji Shikshan Sanstha, Shirur, for his steadfast support, visionary leadership, and continued encouragement throughout the development of this book. His commitment to fostering academic excellence has been a guiding force behind the successful completion of this endeavor.

We are profoundly grateful to all the contributing authors who shared their valuable insights, conducted rigorous research, and demonstrated remarkable academic integrity. Their contributions have not only enriched the content but also elevated the overall quality of this volume.

Special thanks are also due to the peer reviewers, whose critical evaluations and constructive feedback helped ensure the scholarly standard of each chapter. Their time and dedication significantly contributed to the refinement and credibility of the work.

Our heartfelt thanks extend to the administrative and editorial staff, as well as the production and publishing teams, whose coordinated efforts, technical expertise, and unwavering professionalism facilitated the seamless execution and publication of this book.

Without the collaborative spirit, dedication, and tireless efforts of each individual involved, this volume would not have been possible. We remain truly thankful to all.

Contents

Editors and Contributors

About the Editors

Sachin N. Kothawade Dr. Sachin N. Kothawade is a Principal and Associate Professor in Pharmaceutics at Shri Chhatrapati Sambhaji Shikshan Sanstha's Sitabai Thite College of Pharmacy, Shirur. He holds a post-graduation degree from Poona College of Pharmacy, Pune, and a Ph.D. from Savitribai Phule Pune University. Specializing in novel drug delivery systems, nanoparticles, and solubility enhancement, he has 18 years of teaching experience and a robust academic background. Dr. Kothawade has authored over 70 research papers in national and international journals, published 15 textbooks, 35 chapters in Scopus-indexed books, and holds credit for 4 Indian patents. He has also served as editor for six edited books, with one published by De Gruyter in Germany and the others by Bentham Science, Elsevier, Springer Nature, and CRC Press. Acknowledged for his research excellence, he has received grants from SPPU, Pune, and DBATU, Lonere. He is a life member of the Association of Pharmaceutical Teachers in India, showcasing his active engagement in the academic community.

Vishal Pande Dr. Vishal Vijay Pande is a Principal and Professor in Pharmaceutics at Rashtriya Shikshan Mandal's N.N. Sattha College of Pharmacy, Ahmednagar. He has 18 years of teaching and 1.2 years of industrial experience. He has also published more than 150 research papers in national and international peer-reviewed journals, authored five textbooks on pharmaceutical engineering, and has filed five Indian patents while being awarded two international patents. He is an approved PG and PhD guide for various universities in India, and he has guided more than 70 M.Pharm and 3 Ph.D. students for their Research work. He is a life member of the Association of Pharmaceutical Teachers in India, the Indian Pharmaceutical Association, the National Environmentalist Association (Ranchi), the Institution of Chemists (Kolkata), the Indian Society for Technical Education (Mumbai), the Maharashtra State Pharmacy Council (Mumbai), and the International Association of Advanced Materials (Sweden).

Surendra Gattani Dr. Surendra Ganeshlal Gattani is a Professor and Head of the Department of Pharmaceutics at the School of Pharmacy, Swami Ramanand Teerth Marathwada University, Nanded. He holds a Ph.D. from Rajiv Gandhi Technological University, Bhopal, and a postgraduate degree from Shivaji University, Kolhapur. With over 21 years of academic experience, including 17 years at the postgraduate level, Dr. Gattani has made substantial contributions to the fields of novel drug delivery systems, material science, nanotechnology, and community health. He has published 97 research papers in reputed journals, authored 10 books (including 2 e-books), contributed 2 book chapters, and holds 2 patents. A committed research mentor, he has guided 12 Ph.D. and 71 M.Pharm scholars. Dr. Gattani has delivered 41 invited talks at national forums and served in various prestigious academic and administrative roles, such as director, dean, chairman BOS, and principal. He has completed four research projects with a total funding of ₹60 lakhs. As an active member of APTI, PCI, and AICTE committees, he also contributes to policy evaluation and quality assurance.

Rajashri Sumbe Ms. Rajashri Balasaheb Sumbe is an Assistant Professor in Pharmaceutics at N. N. Sattha College of Pharmacy, Ahmednagar, affiliated with MSBTE and DBATU. She holds a Master's degree in Pharmacy (QAT) and is currently pursuing her Ph.D. in the domain of herbal dermaceuticals. With over 7 years of teaching experience, she has made notable contributions to pharmaceutical research, particularly in the area of herbal formulations and nanotechnology-based drug delivery systems. Ms. Sumbe has published nine research papers in reputed national and international journals and is credited with two published patents (including a German patent) and one Indian patent filed. She has presented her work at various conferences, earning accolades, including a second rank at an international conference in collaboration with AIMST University, Malaysia. Ms. Sumbe is a registered member of MSPC.

Contributors

Okello Aaron, BSc. Pharm. Sciences Department of Pharmaceutical Sciences, Faculty of Medicine, Mbarara University of Science and Technology, Mbarara, Uganda

Tanvir Ahmed Department of Pharmaceutical Sciences, School of Health and Life Sciences, North South University, Dhaka, Bangladesh

Kazi Tasnuva Alam Department of Pharmaceutical Sciences, School of Health and Life Sciences, North South University, Dhaka, Bangladesh

Sunil Vishvnath Amrutkar Department of Pharmaceutical Chemistry, GES's Sir Dr. M. S. Gosavi College of Pharmaceutical Education and Research, Nashik, Maharashtra, India

Anyole Arthur Amule, BSc., MSc. Pharm. Eng Department of Pharmaceutical Sciences, Faculty of Medicine, Mbarara University of Science and Technology, Mbarara, Uganda

AKO Life Sciences Uganda Limited, Mbarara, Uganda

Kalyani Autade RSM's N N Sattha College of Pharmacy, Ahmednagar, MH, India

Varsharani R. Avhad Arihant College of Pharmacy, Ahilyanagar, Maharashtra, India

Kagimu Braize Ben, BSc., MSc. Pharm. Analysis Department of Pharmaceutical Sciences, Faculty of Medicine, Mbarara University of Science and Technology, Mbarara, Uganda

Department of Pharmacy, Faculty of Medicine, Mbarara University of Science and Technology, Mbarara, Uganda

AKO Life Sciences Uganda Limited, Mbarara, Uganda

Unmesh Gulabrao Bhamare Department of Pharmaceutics, GES's Sir Dr. M. S. Gosavi College of Pharmaceutical Education and Research, Nashik, Maharashtra, India

Deepali Dattatray Bhandari Department of Pharmaceutical Chemistry, GES's Sir Dr. M. S. Gosavi College of Pharmaceutical Education and Research, Nashik, Maharashtra, India

Thirupataiah Boini Department of Chemistry, National Ayurveda Research Institute for Panchakarma CCRAS Ministry of AYUSH, Cheruthuruthi, Kerala, India

Mwandah Daniel Chans, BPharm., MSc., PharmD Department of Pharmacology and Toxicology, Kampala International University, Kampala, Uganda

Department of Pharmacology and Therapeutics, Faculty of Medicine, Mbarara University of Science and Technology, Mbarara, Uganda

Pallavi M. Chaudhari Department of Pharmaceutics, Dr. D. Y. Patil College of Pharmacy, Akurdi, Pune, Maharashtra, India

Odongo Denis, Dip., BSc., MSc. and Biotech Department of Pharmaceutical Sciences, Faculty of Medicine, Mbarara University of Science and Technology, Mbarara, Uganda

AKO Life Sciences Uganda Limited, Mbarara, Uganda

Delafrique Deogratias, BSc. Pharm. Sciences Department of Pharmaceutical Sciences, Faculty of Medicine, Mbarara University of Science and Technology, Mbarara, Uganda

Ujwala Desai PES Modern College of Pharmacy, Nigdi, MH, India

Kyemanta C. K. Desire, BSc., MSc. Pharm. Analysis Department of Pharmaceutical Sciences, Faculty of Medicine, Mbarara University of Science and Technology, Mbarara, Uganda

Department of Pharmacy, Faculty of Medicine, Mbarara University of Science and Technology, Mbarara, Uganda

Upendra Chandrakant Galgatte Department of Pharmaceutics, Modern College of Pharmacy, Nigdi, Pune, Maharashtra, India

Ashwini Gawade Department of Pharmaceutical Sciences, School of Health Sciences and Technology, MIT World Peace University, Pune, Maharashtra, India

Parth Ramesh Gharat Department of Pharmacognosy, Regional Ayurveda Research Institute, CCRAS, Ministry of AYUSH, Pune, Maharashtra, India

Ganesh R. Godge Department of Pharmaceutics, Dr Vithalrao Vikhe Patil Foundation's College of Pharmacy, Ahilyanagar, Maharashtra, India

Hannington Gumisiriza, BSc., MSc., PhD Chem. Department of Chemistry, Faculty of Science, Mbarara University of Science and Technology, Mbarara, Uganda

Praveen Halagali Department of Pharmaceutics, Manipal College of Pharmaceutical Sciences, Manipal Academy of Higher Education, Manipal, Karnataka, India

Ashish S. Jain Shri. D. D. Vispute College of Pharmacy and Research Center, Panvel, Maharashtra, India

Namutebi Josephine, BSc., MSc. Pharm. Analysis Department of Pharmaceutical Sciences, Faculty of Medicine, Mbarara University of Science and Technology, Mbarara, Uganda

Department of Pharmacy, Faculty of Medicine, Mbarara University of Science and Technology, Mbarara, Uganda

Mukasa Benjamin Kabuuza, BSc. Pharm. Sciences Department of Pharmaceutical Sciences, Faculty of Medicine, Mbarara University of Science and Technology, Mbarara, Uganda

Ramanlal Narayan Kachave Department of Pharmaceutical Chemistry, GES's Sir Dr. M. S. Gosavi College of Pharmaceutical Education and Research, Nashik, Maharashtra, India

Sangita Kale PES Modern College of Pharmacy, Nigdi, MH, India

Kailas Karandeb Department of Pharmaceutics, Arvind Gavali College of Pharmacy, Satara, Maharashtra, India

Sachin N. Kothawade, BPharm., MPharm., PhD Department of Pharmaceutics, SCSSS's Sitabai Thite College of Pharmacy, Shirur, Pune, Maharashtra, India

Ashwin Kuchekar Department of Pharmaceutical Sciences, School of Health Sciences and Technology, MIT World Peace University, Pune, Maharashtra, India

Vaishali P. Kudekar Department of Pharmaceutics, Shri. D. D. Vispute College of Pharmacy and Research Center, Panvel, Maharashtra, India

Ajit Kulkarnic School of Pharmacy, MIT Vishwaprayag University, Solapur, Maharashtra, India

Sunil Mahajan Department of Pharmaceutical Chemistry, Divine College of Pharmacy, Satana, Nashik, Maharashtra, India

Anitha Marimuthu Department of Pharmacology, JSS College of Pharmacy, JSS Academy of Higher Education & Research, Ooty, Nilgiris, Tamilnadu, India

Rahul Maurya Department of Pharmacy, Central Ayurveda Research CCRAS Ministry of AYUSH, Jhansi, UP, India

Lakshminarayana Mirso Department of Chemistry, National Ayurveda Research Institute for Panchakarma CCRAS Ministry of AYUSH, Cheruthuruthi, Kerala, India

Mukesh T. Mohite Department of Pharmaceutical Quality Assurance, Dr. D. Y. Patil College of Pharmacy, Akurdi, Pune, Maharashtra, India

Gershom Muganga, BSc., MSc. Pharm. Analysis Department of Pharmaceutical Sciences, Faculty of Medicine, Mbarara University of Science and Technology, Mbarara, Uganda

Department of Pharmacy, Faculty of Medicine, Mbarara University of Science and Technology, Mbarara, Uganda

Pharm-Biotechnology and Traditional Medicine Centre of Excellence, Mbarara University of Science and Technology, Mbarara, Uganda

Prakash Pannalal Muleva Department of Pharmaceutics, GES's Sir Dr. M. S. Gosavi College of Pharmaceutical Education and Research, Nashik, Maharashtra, India

Jey Kumar Pachiyappan Department of Pharmaceutics, JSS College of Pharmacy, JSS Academy of Higher Education & Research, Ooty, Nilgiris, Tamilnadu, India

Umesh S. Panchal Department of Pharmaceutics, Dr Vithalrao Vikhe Patil Foundation's College of Pharmacy, Ahilyanagar, Maharashtra, India

Vishal Pande Department of Pharmaceutics, RSM's N. N. Sattha College of Pharmacy, Ahmednagar, Ahilyanagar, Maharashtra, India

Kajal Pansare Department of Pharmacology, Divine College of Pharmacy, Satana, Nashik, Maharashtra, India

Dhrusha Patel Department of Pharmaceutics, Anand Pharmacy College, Anand, Gujarat, India

Rashmi Pathak Department of Pharmacy, Invertis University, Bareilly, UP, India

Ashwini R. Patil Department of Pharmaceutics, Shri. D. D. Vispute College of Pharmacy and Research Center, Panvel, Maharashtra, India

Bhagwat A. Patil Department of Pharmaceutics, Dr. D. Y. Patil College of Pharmacy, Akurdi, Pune, Maharashtra, India

Prashant Patil Department of Pharmaceutics, N N Sattha College of Pharmacy, Ahmednagar, Maharashtra, India

Ntale Paulo, BSc., MSc. Pharm. Analysis Department of Pharmaceutical Sciences, Faculty of Medicine, Mbarara University of Science and Technology, Mbarara, Uganda

Department of Pharmacy, Faculty of Medicine, Mbarara University of Science and Technology, Mbarara, Uganda

Hardik Rana Department of Pharmaceutics, Anand Pharmacy College, Anand, Gujarat, India

Bhagyashri B. Randhawan Arihant College of Pharmacy, Ahilyanagar, Maharashtra, India

Bhushan R. Rane Department of Pharmaceutics, Shri. D. D. Vispute College of Pharmacy and Research Center, Panvel, Maharashtra, India

Manish A. Raskar Department of Pharmaceutical Chemistry, Dr Vithalrao Vikhe Patil Foundation's College of Pharmacy, Ahilyanagar, Maharashtra, India

Mutyaba Ronald, BSc. Pharm. Sciences Department of Pharmaceutical Sciences, Faculty of Medicine, Mbarara University of Science and Technology, Mbarara, Uganda

Bole Sandesh Department of Pharmaceutics, N N Sattha College of Pharmacy, Ahmednagar, Maharashtra, India

Raagul Seenivasan Department of Pharmaceutics, Manipal College of Pharmaceutical Sciences, Manipal Academy of Higher Education, Manipal, Karnataka, India

Crispin Duncan Sessazi, BSc., MSc., MSc., PhD Department of Pharmaceutical Sciences, Faculty of Medicine, Mbarara University of Science and Technology, Mbarara, Uganda

Himanshu Sharma Teerthanker Mahaveer College of Pharmacy, Teerthanker Mahaveer University, Moradabad, UP, India

Samya Sheth Department of Pharmaceutics, Anand Pharmacy College, Anand, Gujarat, India

Dattatraya Manohar Shinkar Department of Pharmaceutics, GES's Sir Dr. M. S. Gosavi College of Pharmaceutical Education and Research, Nashik, Maharashtra, India

Anil Kumar Singh United Institute of Pharmacy, Naini Prayagraj, Uttar Pradesh, India

Sudarshan Singh Office of Research Administrations, Chiang Mai University, Chiang Mai, Thailand

Faculty of Pharmacy, Chiang Mai University, Chiang Mai, Thailand

Preethi Somanna Department of Pharmaceutics, JSS College of Pharmacy, JSS Academy of Higher Education and Research, Mysuru, Karnataka, India

Deepak Sonawane Department of Pharmaceutics, Divine College of Pharmacy, Satana, Nashik, Maharashtra, India

Ganesh Sonawane Department of Pharmaceutical Chemistry, Divine College of Pharmacy, Satana, Nashik, Maharashtra, India

Rajashri Sumbe Department of Pharmaceutical Chemistry, RSM's N. N. Sattha College of Pharmacy, Ahmednagar, Ahilyanagar, Maharashtra, India

Jayprakash Suryawanshi Department of Pharmacognosy, N N Sattha College of Pharmacy, Ahmednagar, Maharashtra, India

Dipak S. Thorat SMBT College of Pharmacy, Nashik, Maharashtra, India

Vamshi Krishna Tippavajhala Department of Pharmaceutics, Manipal College of Pharmaceutical Sciences, Manipal Academy of Higher Education, Manipal, Karnataka, India

Zarin Tasnim Tisha School of Chemistry and Chemical Sciences, University of Surrey, Guildford, UK

Yogesh V. Ushir SMBT College of Pharmacy, Nashik, Maharashtra, India

Kuldeep Vinchurkar Department of Pharmaceutics, Sandip Institute of Pharmaceutical Sciences (SIPS), Nashik, Maharashtra, India

Swathi Yempallya Center for Advanced Materials, Qatar University, Doha, Qatar

Veeranna Yempallyd School of Computing, MIT Vishwaprayag University, Solapur, Maharashtra, India

Chapter 1
Introduction to Polymer Nanocomposites

Raagul Seenivasan, Praveen Halagali, Jey Kumar Pachiyappan, Anitha Marimuthu, Himanshu Sharma, Preethi Somanna, Rashmi Pathak, Sachin N. Kothawade, and Vamshi Krishna Tippavajhala

Abstract Polymer nanocomposites are materials that introduce nanoparticles into polymer matrices to change or improve the characteristics of the underlying polymer. This family of materials' exceptional mechanical, thermal, electrical, and barrier qualities might open up new uses in the electronics, automotive, aerospace, and biomedical industries. Metal oxides, graphene, carbon nanotubes, and nanoclays are examples of high aspect ratio and surface area nanoparticles that form strong interfacial interactions with the polymer matrix and give rise to the unique characteristics of polymer nanocomposites. The types of nanoparticles employed, such as carbon-based nanomaterials, inorganic nanoparticle-modified polymers, and

R. Seenivasan · P. Halagali · V. K. Tippavajhala (✉)
Department of Pharmaceutics, Manipal College of Pharmaceutical Sciences, Manipal Academy of Higher Education, Manipal, Karnataka, India
e-mail: vamshi.krishna@manipal.edu

J. K. Pachiyappan
Department of Pharmaceutics, JSS College of Pharmacy, JSS Academy of Higher Education & Research, Ooty, Nilgiris, Tamilnadu, India

A. Marimuthu
Department of Pharmacology, JSS College of Pharmacy, JSS Academy of Higher Education & Research, Ooty, Nilgiris, Tamilnadu, India

H. Sharma
Teerthanker Mahaveer College of Pharmacy, Teerthanker Mahaveer University, Moradabad, UP, India

P. Somanna
Department of Pharmaceutics, JSS College of Pharmacy, JSS Academy of Higher Education and Research, Mysuru, Karnataka, India

R. Pathak
Department of Pharmacy, Invertis University, Bareilly, UP, India

S. N. Kothawade
Department of Pharmaceutics, SCSSS's Sitabai Thite College of Pharmacy, Shirur, Pune, Maharashtra, India

S. N. Kothawade et al. (eds.), *Polymer Nanocomposites*,
https://doi.org/10.1007/978-3-032-00365-2_1

organic nanoparticles, have been described in this chapter. Additionally included are synthesis techniques such as melt blending, solution blending, and in situ polymerization. Additionally covered the characterization techniques such as mechanical testing, thermal analysis, and structural analysis. Important performance-influencing factors are also covered, including dispersion and distribution, interfacial adhesion, and nanoparticle orientation. Additionally, challenges and solutions are described. Finally, the use of polymeric nanocomposite is discussed as well.

Keywords Polymer nanocomposites · Nanoparticles · Dispersion · Mechanical properties · Thermal properties · Interfacial interactions

Abbreviations

AFM	Atomic Force Microscopy
ASTM	American Society for Testing and Materials
BF	Bright field
CNTs	Carbon Nanotubes
DF	Darkfield
DMA	Dynamic Mechanical Analysis
DSC	Differential Scanning Calorimetry
EELS	Electron Energy Loss Spectroscopy
EMI	Electromagnetic Interference
GNPs	Graphene Nanoplatelets
HAADF	High Angle Annular Dark Field
HRTEM	High-Resolution TEM
JPs	Janus Particles
NPs	Nanoparticles
OLEDs	Organic Light-Emitting Diodes
PNCs	Polymer Nanocomposites
SAED	Selected-Area Electron Diffraction
SEM	Scanning Electron Microscopy
TEM	Transmission Electron Microscopy
VA-MWCNTs	Vertically Aligned Multiwalled Carbon Nanotubes

1.1 Introduction

Polymer nanocomposites (PNCs) are novel materials with improved characteristics that are created by incorporating nanoparticles (NPs) into both recovered waste polymers and virgin polymeric matrices. The last 20 years have seen a rapid expansion in the field of polymer nanocomposites because of their improved and adjustable optical, thermal, electrical, magnetic, and mechanical characteristics, which make them practical and well-suited for a range of uses. Since NPs have a significant surface area to volume ratio, adding them to polymer matrices in the form of nanosheets, nanotubes, nanospheres, quantum dots, etc., lessens the negative properties of polymers, such as their poor mechanical strength and lower chemical resistance. Additionally, when nanoparticles engage with the polymer chain, the polymer's water resistance rises and gas permeability decreases. PNCs have garnered a lot of theoretical attention and several practical applications across a range of scientific and technical fields because of their exceptional capabilities (Inamdar et al. 2024; Nayak et al. 2024; Ramakoti et al. 2023). Various-sized fillers are frequently used to strengthen polymers to mitigate some of their drawbacks and expand their range of uses. A radical substitute for traditional polymer composites has been created using nanoscale fillers to enhance the mechanical and physical characteristics of polymers (Reynaud et al. 1999; Young et al. 2012). The nanoscale fillers fundamentally range from isotropic to highly anisotropic sheet-like or needle-like morphologies, with at least one distinctive length scale on the order of nanometers. Innovative ways to combine nanoscale fillers with polymer materials to produce polymer nanocomposites with intriguing characteristics are made possible by nanoscience and nanotechnology.

Depending on the size of the distributed nanoscale fillers, PNCs may be broadly divided into three categories. In the first kind, polymeric matrices contain two-dimensional (2D) nanoscale fillers, such as layered silicate, graphene, or MXene (Albdiry et al. 2013; Du and Cheng 2012; Li et al. 2018; Tu et al. 2018), in the form of sheets that are hundreds to thousands of nanometers long and one to a few nanometers thick. It is possible to classify the related polymer nanocomposites within the category of layered PNCs. The second kind uses nanofibers or nanotubes, such as carbon nanofibers and nanotubes or halloysite nanotubes, as reinforcing nanofillers to create materials with remarkable qualities (Calvert 1999; Liu et al. 2014). Two of the dimensions are in the nanometer scale, and the third is larger, creating an elongated one-dimensional structure. The third category consists of nanocomposites with three-dimensional nanoscale fillers that are on the order of nanometers. These nanoscale fillers include low aspect ratio, iso-dimensional nanoparticles such as quantum dots, semiconductor nanoclusters, and spherical silica (Herron and Thorn 1998; Huang et al. 2016; Mark 1996). Schematics of many nanoscale fillers are displayed in Fig. 1.1.

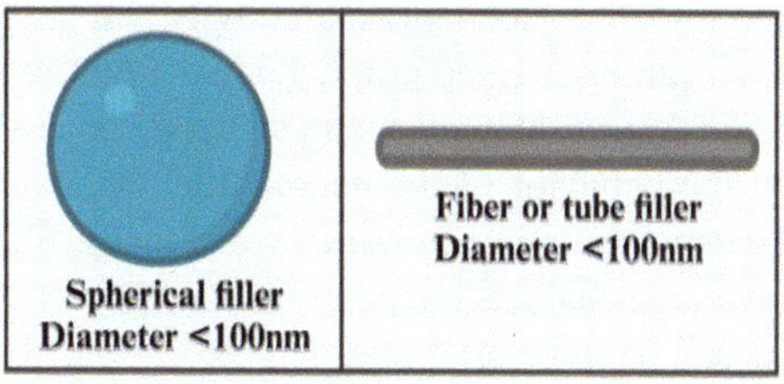

Fig. 1.1 Schematic diagram of nanoscale fillers

1.2 Types of Nanoparticles Used

1.2.1 Carbon-Based Nanomaterials

Recent years have seen a significant exponential increase in the study of carbon-based nanomaterials, including fullerenes, graphene and its derivatives, carbon nanotubes (CNTs), nanodiamonds, and other nanosized carbon allotropes. A wide range of applications have been made possible by the infinite possibilities to modify and tailor carbon nanomaterials due to their small size, which is comparable to the size of many basic biomolecules, large specific surface area, high electrical and thermal conductivity, unique optical properties, and superior mechanical properties. Graphene has been widely used in flexible electronics, fullerene derivatives have been used specifically for scavenging solar energy, carbon nanotubes have been modified to have the ability to recognize molecules, graphene quantum dots have been widely used for bio-imaging and sensing because of their photoluminescence properties, and nanodiamonds are helpful in super-resolution imaging and nanoscale temperature sensing (Díez-Pascual 2021).

The potential biological uses of carbon nanotubes (CNTs), including drug delivery, tissue engineering, diagnostics, and biosensing, have been extensively studied during the past 20 years (Simon et al. 2019). Specifically, CNTs can function as contrast agents in several imaging techniques (Gong et al. 2013). They can provide a high spatial resolution indication of the presence and location of targeted cells after functionalization and conjugation with different biomarkers. CNTs are viable options for addressing the theragnostic problems related to neurological conditions like ischemic stroke. Recently, vertically aligned multiwalled carbon nanotubes (VA-MWCNTs) were produced by Komane et al. (2020) and purified, carboxylated, acylated, and PEGylated for use in targeting investigations. The PC-12 neuronal cells, a kind of catecholamine cell that produces, stores, and releases dopamine and norepinephrine, were shown to be unaffected by the functionalized VA-MWCNTs.

In regenerative medicine, nerve regeneration using cell electrostimulation will be crucial, especially for body reconstruction, artificial organs, or neural prostheses. A nontoxic, highly electrically conductive substrate material is required for indirect electrical cell stimulation to achieve precise and efficient cell electrostimulation. Although graphene nanoplatelets (GNPs) can be used to do this, the quality of the produced coatings is hampered by their significant agglomerating propensity. Consequently, it is essential to select the right amount of surfactant to produce a coating with excellent conductivity and quality. In this context, Dybowska-Sarapuk et al. (2020) created graphene inks with varying surfactant concentrations between 2 and 20 weight percent and showed that electrostimulation had a discernible impact on the behavior of the neural stem cells contained in the graphene layer. Cellular electrostimulation may offer a remedy for neurological conditions that are now incurable, such as the potential to repair nerve connections and the spinal cord. The carbon nanomaterials are depicted in Fig. 1.2.

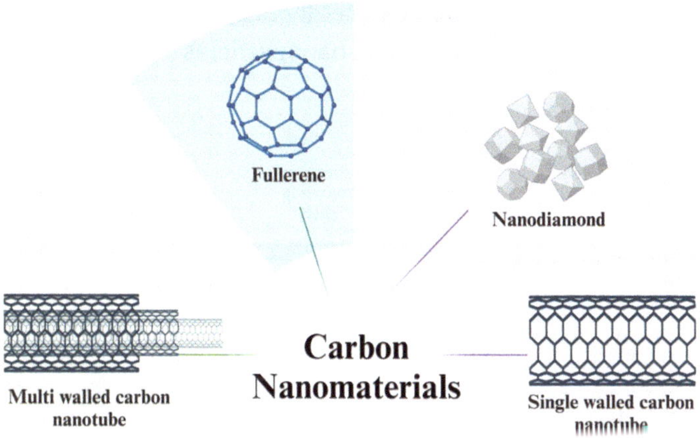

Fig. 1.2 Carbon nanomaterials

1.2.2 Inorganic-Nanoparticle Modified Polymers

Over the past several years, inorganic nanoparticle-modified polymer nanocomposites have garnered a lot of interest in the creation of materials for various uses. For use in a variety of industries, including medicine, photovoltaics, packaging, and structural applications, among others, a variety of inorganic nanoparticle types, including metal oxides, quantum dots, mesoporous silica, gold and silver nanoparticles, and ceramics, have been integrated into polymers (Díez-Pascual 2022). Improving the mechanical and thermal characteristics of polymers is one of the main goals of adding inorganic nanoparticles. Therefore, as compared to plain polymer, a small percentage of clay often results in better flame-retardant behavior, increased stiffness and strength, and improved barrier qualities against gases and vapors (Díez-Pascual and Díez-Vicente 2014, 2015; Díez-Pascual and Naffakh 2012; Kotal and Bhowmick 2015). Compared to microsized particles, nanoparticles have a larger specific surface area, which can result in superior characteristics than those obtained with traditional microfiller loadings. But certain things must be done beforehand. It can be difficult to change the surface chemistry to encourage physical or chemical interactions with polymer chains because of nanoparticle aggregation inside the matrix and poor interfacial adhesion between the nanofillers and the matrix. The chemistry of the polymer matrix, the filler-polymer affinity, the particle's shape, its degree of orientation and dispersion within the polymer, and the methods used to prepare the nanocomposite all have a direct impact on the final characteristics of the nanocomposites. The inorganic nanoparticles are given in Fig. 1.3a.

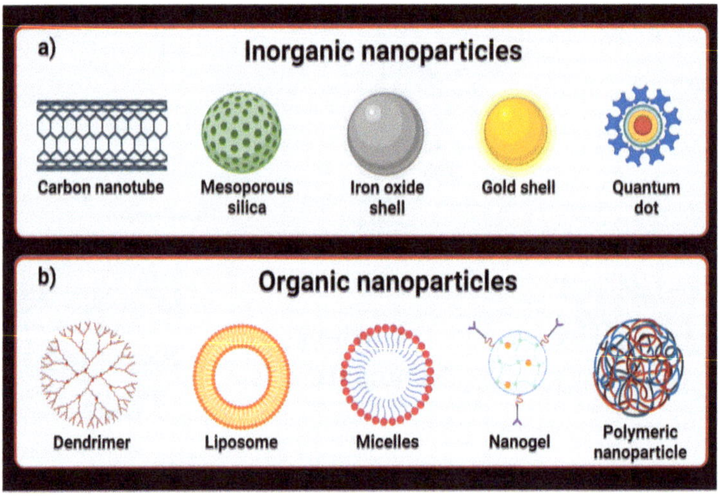

Fig. 1.3 (**a**) Inorganic nanoparticles, (**b**) Organic nanoparticles

1.2.3 Organic Nanoparticles

The use of organic nanoparticles as fillers in polymer matrices to prepare polymer nanocomposites has gained popularity because of the more effective properties they exhibit over conventional ones. There is a great deal of potential for such materials in improving mechanical, thermal, electrical, and optical properties that would suit a broad class of applications. Properties and benefits are: (a) Greater mechanical strength: organic nanoparticle incorporation stabilizes the mechanical properties and extends the life of the polymer by overcoming inherent situations like low mechanical strength, and chemical resistance. (b) Thermal stability: It is this cause that provides a certain amount of thermal stability to them for heat-resistant applications, generally attributed to the peculiarities of the nanoparticles. (c) Optical and electrical properties: The introduction of organic NPs facilitates the alteration of the optical and electrical characteristics of polymers, contributing to optoelectronic and sensor applications. (d) Reduced permeability: The use of nanoparticles can produce lower permeability against gases and increased water resistance in polymer matrices, extending the usability of materials, especially in the packaging and protective-coating applications. The organic nanoparticles are given in Fig. 1.3b (Hiremath et al. 2021).

1.3 Synthesis Techniques

1.3.1 In Situ Polymerization

A helpful technique that produces nanocomposites for drug delivery systems for targeted cancer therapy is in situ polymerization, which improves the stability, controlled release profiles, and encapsulation efficiency of the medicinal medications.

This technique combines drug molecules and polymer monomers in nanosized forms, such as dendrimers, metal or ceramic nanoparticles, carbon-based materials, etc., and starts the polymerization process while these materials are present to create a highly dispersed and architecturally integrated nanocomposite. When active pharmaceutical components and nanosized fillers are used as loading carriers, this approach usually results in a synergistic improvement in targeting efficiency and treatment efficacy (Adnan et al. 2018). When water-insoluble or difficult-to-dissolve chemotherapeutics like doxorubicin or paclitaxel are incorporated into polymeric structures, the process of in situ polymerization helps to trap them, boosting their durability and effectiveness. Furthermore, the polymerization may be enhanced by adding nanofillers with various functionalized components, such as magnetic nanoparticles or gold nanorods, to provide photosensitizing and magnetic targeting capabilities for tumor treatment. This technique optimizes the surface chemistry of the polymer as well as its architecture, allowing ligands (such as folic acid and antibodies) to be introduced for active cancer-targeting that minimizes systemic damage and off-target effects (Gagliardi et al. 2021). Additionally, stimuli-responsive nanocomposites have been effectively created using in situ polymerization, with a polymer matrix serving as the delivery mechanism that may expand and contract in response to stimuli such as pH, temperature, or enzyme activity. Enabling localized administrations with targeted drug release to the cancer locations known to have low pH or oxygen deficiencies, improves therapy effectiveness without endangering healthy cells. As a result, in situ polymerized nanocomposites offer a promising architecture for sophisticated stimuli-sensitive and tailored drug delivery systems that may be used for cancer treatment (Mittal 2014). Synthesis techniques of polymer nanocomposites are depicted in Fig. 1.4.

1.3.2 Solution Blending

This process allows for the evaporation or precipitation of the solvent to create the drug-polymer matrix after blending and dissolving metallic nanoparticles, graphene oxide, or polymer micelles into the polymer matrix to enrich the polymer solutions with medicinal and nanofiller agents. Furthermore, the technique has the advantage of protecting sensitive and/or bulky medications from hydrolysis by enclosing them in the polymer chain, which increases their availability throughout the system (Lago et al. 2016).

1.3.3 Melt Blending

Nanofibrous membranes can be embedded during hot melt blending to give thin films manufactured from amphiphilic polyethylene glycols spatial and temporal controllability as well as drug release kinetics. Without the use of solvents, medicinal drugs and nanofillers are incorporated into a polymeric matrix by the process of melt blending (Mittal 2014).

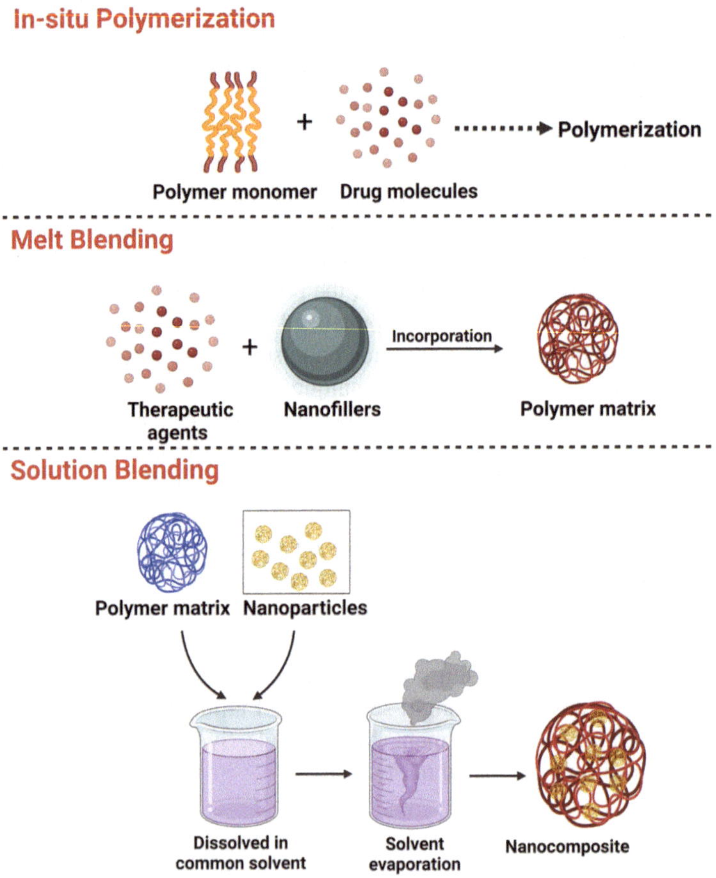

Fig. 1.4 Synthesis techniques of polymer nanocomposites

1.4 Characterization Methods

1.4.1 Structural Analysis

Scanning electron microscopy (SEM) is an imaging technique used to visualize and analyze solid materials. SEM typically provides magnification from 20–50× to 1,000,000×, allowing measurements from millimeters to nanometers, with a sensitivity of about 0.5 nm. Primary data acquisition includes secondary electron images (SE images) of surface topography affected by the surface of the sample revealed and backscattered electrons (BSE), which give rise to compositional differences. Although SEM better captures sample morphology, assessing roughness often requires cross-sectional viewing as electron emission is not always correlated with height. Furthermore, energy dispersive X-ray spectroscopy (EDS) can be used for material detection, with detection limits as low as 0.5% by weight (Abdellaoui Hind et al. 2023; Awasthi et al. 2024).

Transmission electron microscopy (TEM) is a powerful technique for investigating material properties at a very high spatial resolution, providing atomic-level insights into size, size distribution, crystal structure, stress, defects, and chemical composition. It works by electrons transferring through a thin layer of sample, typically not less than and greater than 200 nm in diameter. Different techniques enhance the capabilities of TEM: high-resolution TEM (HRTEM) produces atomic images by interference between diffuse and diffracted beams; force-dispersive TEM (EFTEM) images objects and enhances image contrast by filtering out weakly scattered electrons; energy dispersive X-ray spectroscopy (EDS) identifies chemical surfaces; selected-area electron diffraction (SAED) investigates the crystal structure and growth rate of nanowires; high angle annular dark field (HAADF) TEM examines nanoparticles with good contrast for heavy atoms, bright field (BF) TEM reveals physical structures where solid areas appear dark; darkfield (DF) TEM shows nanocrystal shapes and crystal defects; and electron energy loss spectroscopy (EELS) investigate atomic structure, chemical bonding, and surface morphology. TEM is an essential tool in the area of advanced research, including materials science and biology (Bokobza 2018).

Atomic force microscopy (AFM) is a scanning probing microscope that provides wide-ranging images of solids and analyses their mechanical and electrical properties. It works by scanning the surface of a sample with a small probe and measuring its bonds. AFM is commonly used to investigate surface properties such as thickness and roughness, especially silicon probes with a cantilever beam and sharp tip. Tapping mode is a standard operating mode in which the cantilever is vibrated vertically, "tapping" the sample to move it and maintain a constant intensity through the feedback loop, which generates AFM height maps that reflect surface topography. Still, these maps can be misleading because they rely on continuous intensity rather than maintained direct height measurements. The shape of AFM images is affected by probe sharpness, and although it can detect sample roughness accurately, using AFM tips can distort data. Disposables will, over time, deteriorate; their structure is appropriate to the specific data required, and their properties are analyzed (Batool et al. 2022; Gobena and Woldeyonnes 2024).

1.4.2 Thermal Analysis

Differential scanning calorimetry (DSC) is an essential technique for investigating the thermal properties of polymer nanocomposites containing carbon nanotubes and metal nanoparticles. This information helps to calculate thermal stability, permeability, and temperature gradient. Various studies have demonstrated that the presence of carbon nanotubes can significantly affect the glass transition temperature of polymeric matrices. For example, single-walled carbon nanotubes increase the glass transition temperature, while condensed carbon nanotubes decrease it, indicating their different roles in heat absorption and energy exchange. DSC can also measure the onset and rise temperatures of materials involved in metallic deposits. Less quantities are required for DSC analysis (Sarma and Rao 2024).

Thermogravimetric analysis is also a thermal analysis technique used to evaluate the thermal stability of polymer nanocomposites by measuring the effect of decreasing temperature under controlled conditions. Nanotubes have slow decomposition and may increase thermal temperature. Still, precautions are required while interpreting mass reduction as an indication of decomposition, and TGA is not suitable for samples with no homogeneity or weight loss (Corcione and Frigione 2012).

Dynamic mechanical thermal analysis (DMTA) combines both dynamic mechanical analysis (DMA) and thermal analysis techniques to analyze the Visco critical thermal properties of polymer nanocomposites. In the DMTA test, a sinusoidal pressure or strain is applied to the sample, and the response obtained is used to measure temperature or time. This analysis yields basic parameters such as storage parameter (E'), which indicates the ability of the material to store mechanical energy; loss parameter (E''), which means energy dissipation; and damping factor (tan δ). As Hajializadeh et al. found that CNT-water polyurethane composites have higher Tg and E' with increasing multi-walled carbon nanotube (MWCNT) content, indicating increased toughness due to stronger interactions between CNTs and polymer matrix, this should be noted that Tg values from DMTA generally differential scanning are higher than those obtained from calorimetry (DSC). In addition, the 3ω method is effective for investigating the thermal properties of PNCs by applying an AC to a metal heater, which induces temperature changes, which can be analyzed to determine thermal conductivity and diffusivity even at initial measurements the CNT thermal conductivity using this method is then lower than theoretical values Results, including the development of standalone sensors for however, if the sample surface is rough and ambient, the reliability of these methods may be reduced (Sui et al. 2022).

1.4.3 Mechanical Testing

Tensile tests were conducted to determine the elastic parameter, ultimate stress, and ultimate strain for six systems using a "dog-bone" shaped specimen. The specimen was placed in a screw-driven machine with movable and static grips that slipped to failure; the typical grips in this test measured the applied load against the extension; a weight cell was needed for accurate weight measurement and a width gauge for detecting small widths. Output data, including weight and width, are written to a text file, and widths are converted from volts to millimeters. The mechanical properties were obtained from stress-strain plots with load and elongation data. Tensile testing is a destructive method, and valid American Society for Testing and Materials (ASTM) standards include D638 for plastics and D3039 for polymeric matrix composites (Veigel et al. 2023).

Dynamic mechanical analysis (DMA) was used to obtain elastic parameter data for six different systems, with measured elastic parameters, loss parameters, and inhibition coefficients as a function of temperature, frequency, or time. The method applies torsional oscillations in the specimen as it traverses a specific temperature range, with input parameters including frequency, oscillation amplitude, static load, and temperature range. The results are usually expressed as graphs of elastic parameters, loss measures, and damping coefficients versus temperature. Like tensile testing, DMA is a destructive method. Appropriate ASTM standards for DMA include D4065 for the identification of dynamic mechanical properties and D4092 for terminology related to dynamic mechanical plastics (Qiang et al. 2020).

Nanoindentation was used to obtain comparative data on elastic parameters and hardness for the six systems. This method involves pressing the detected material into a sample to check its mechanical properties, where nanoindentation typically measures penetration in nanometers. Thin film samples are ideal for this method due to their smaller surface and inner surface area for the sake of love. In nanoindentation, the contact area of the indenter and the sample is critical. While traditional indentation testing estimates this area from memory theory, nanoindentation measures penetration depth. It uses the geometry of the indenter to calculate load-displacement data collected during the process, allowing the determination of modulus and hardness based on maximum penetration depth and slope of the unloading curve. There may be creep during a given loading, affecting the indentation depth, which should be carefully interpreted. Polymers such as polystyrene (PS), poly(methyl methacrylate) (PMMA), and others have shown that nanoindentation often gives higher modulus values than parameter values from tensile testing or DMA. Despite these differences, this study assumes that nanoindentation provides reliable comparative data as tested systems are all observed under constant parameters and are generally considered non-destructive. ASTM Working Group E28.06.11 is currently developing a standard test method for indentation testing (Liu et al. 2011).

1.5 Key Factors Influencing Performance

1.5.1 Dispersion and Distribution

When fabricating polymer nanocomposites, achieving stable dispersion of nanoparticles can be challenging due to their high surface energy, which promotes aggregation. This review examines the effects of nanoparticle size, surface chemistry, and thermodynamic parameters on nanoparticle dispersion in PNCs. Ideal dispersion occurs when nanoparticles are smaller than the rotation radius of polymer chains and exhibit strong compatibility with the polymer matrix. In polymer solutions, establishing appropriate thermodynamic conditions is essential for balancing interactions among nanoparticles, solvents, and polymers to create well-structured

PNCs. However, even with favorable thermodynamics, effective dispersion may not be achieved without suitable kinetic parameters. This highlights the need for further research into both the thermodynamics and kinetics of nanoparticle dispersion processes (Müller et al. 2017).

1.5.2 Interfacial Adhesion

Nanofillers are increasingly popular in various technological applications, including biomedical devices and automotive products, due to their advantageous properties, such as large surface areas, high surface energy, and reduced structural imperfections compared to microfillers. In adhesive science, incorporating nanofillers into adhesive matrices enhances properties like wettability and rheology while also providing functional benefits such as improved electrical and thermal conductivity. These enhancements are typically more challenging to achieve with microfillers in adhesives (Gou et al. 2012).

1.5.3 Nanoparticle Orientation

Achieving unique mechanical properties in energy-saving lightweight polymer composites is challenging. Reinforced polymer composites and nanofillers have gained attention due to their large interface area, which can be more efficient and reinforced at lower loads compared to conventional microfillers. Appropriate width is associated with uniform stress distribution. On the contrary, the formation of connected networks that enhance load transfer and the formation of interphases around nanoparticles exhibit comparatively better mechanical properties around the bulk polymers; poor interfacial adhesion can cause slippage or debonding, reducing the effective load-bearing area. Recent advances highlight the importance of controlling nanoparticle orientation at the nanoscale to improve the effectiveness of composite materials. Unidirectional alignment of anisotropic complements has been shown to increase the longitudinal parameter compared to random orientations. Matching insertions can also be a direct means of increasing thermoelectric conductivity. In dielectric nanocomposites, fillers oriented perpendicular to the electric field improve separation strength by forming circular paths that hinder charge diffusion. Although conventional liquid mixing is a cost-effective way to incorporate nanofillers, it does not follow their direction correctly. Often, other methods, such as solution-based methods or the use of external forces, are used. For example, the migration of the elastomer into compatible carbon nanotubes by applying capillary action before treatment can be excessive (Kamal et al. 2022).

1.6 Challenges and Solutions

1.6.1 Agglomeration Issues

The growing interest in PNCs in the last few decades is due to their improved mechanical, thermal, chemical, and barrier properties, which make them useful in various industrial applications. Higher levels of nanoparticles also result in stronger tensile forces that can lead to aggregation or agglomeration, which adversely affects the mechanical properties of nanoparticles. Research has shown that CNT aggregation limits the reinforcement capacity, leading to poor device performance. Therefore, it is essential to control the filler aggregation to optimize the mechanical behavior of PNCs. Methods such as surface treatment and compatibilization have been proposed to reduce these aggregation issues. Beyond aggregation, the interphase conditions—in the region around the nanoparticles—significantly influence the PNC properties. Compared to the polymer matrix and nanofillers, the interphase exhibits different properties. Interface adhesion is difficult to measure and can be quantitatively evaluated using models of mechanical properties such as yield strength or Young's modulus. Aggregation plays a vital role in influencing interface properties while reducing the effective interface area between polymer and nanoparticles (Verma et al. 2020).

1.6.2 Interfacial Compatibility

Polymer materials have received much attention due to their excellent properties. However, the increasing demand for high-performance multifunctional polymers has highlighted the limitations of single-layer systems. Blending technology, in which two or more polymers are physically mixed to form a homogeneous "polymer alloy," is a versatile and cost-effective method for producing flexible and limited advanced polymer materials in terms of mechanical and functional improvements expected from blending. To address this issue, introducing compatibilizers in immiscible blends may well tailor phase morphologies and improve interface adhesion. Common compatibilizers are amphiphilic block copolymers with micro-or nanofillers. However, conventional compatibilizers often struggle to migrate quickly to the interface while maintaining stability, lacking the dual advantages of compatibilizers and functionalization. Recently, Janus particles (JPs) have emerged as new compatibilizers capable of enhancing phase morphology and interface properties types developed in polymer blends. Janus particles that can migrate to the interface and effectively stabilize can optimize the interface characteristics by refining the dispersed phase size, thereby improving the mechanical properties and performance of the resulting mixtures (Uthale et al. 2021).

1.6.3 Processing Techniques

The properties of materials, such as mechanical strength, electrical conductivity, and chemical stability, are essential for their performance and applications, especially in porous polymer nanocomposites, which can lead to electrical conductivity but may also suffer from defects due to porosity introduction, resulting in reduced machinability. Enhancing the properties of the fillers in the polymer matrix is necessary to obtain an even dispersion. At low filler concentrations, voids can be filled effectively without significant aggregation, preserving internal porosity and strengthening the material. Still, increasing filler volume can lead to undesired porosity due to aggregation and, thus, mechanical integrity. Strategies play an essential role in addressing these challenges. For example, Tay et al. revealed that self-assembly by immersion methods is more effective than dip coating for the preparation of polymer/metal nanocomposites used in photovoltaic applications, resulting in higher power conversion efficiency due to improved homogeneity of nanocomposite coating on the silicon surface approx porosity by careful selection of techniques that take into account the factors affecting the external porosity and understand its significance for polymer optimizing the performance of nanocomposites in various applications (Okamoto 2023).

1.7 Application of Polymeric Nanocomposite

Nanocomposites are advanced materials that can be incorporated into the polymer matrix at the nanoscale. The nanocomposites represent a diverse range of materials, such as metallic components, fibers, ceramics, and polymers, that are inoculated to enhance the basic material properties. This modified base material will have a unique characteristic that can be tailored for various applications, from aerospace to biomedical healthcare (Darwish et al. 2022). This section will focus on some key applications of PNC materials.

1.7.1 Automotive and Aerospace

Polymeric nanocomposites have emerged as a prime advanced technology in the automotive industry. Their unique properties facilitate the development of strong, safe, lightweight, and sustainable solutions that meet the necessity for modern vehicles' evolution (Shah et al. 2022). Following the traditional material, this polymeric nanocomposite material is rewarded for its excellent thermal and mechanical strength (Chandgude and Salunkhe 2021), corrosion resistance, fuel efficiency, high strength-to-weight ratio, and electrical and barrier properties (Harun-Ur-Rashid et al. 2022). The PNCs have received tremendous applications in automotive manufacturing, such as display LEDs, scratch-resistant paint, dirt-repellent coating, antireflective coating, antifogging coating (Chemin et al. 2018), water-resistant tires (Asha Krishnan 2024), batteries (Chowdhury et al. 2022), fuel cell catalysts (Chen et al. 2022), enhanced liquid cooling, and heat transfer.

Polymeric nanocomposites in aerospace have a diverse range of applications in weight reduction, thermal stability, mechanical strength, coating, and surface protection, electromagnetic interference shielding, and environmental resistance. Polymeric matrix impregnated with carbon tubes or graphene will increase the mechanical properties of the fuselage structure (Guzzi et al. 2021). The study found that 3% of carbon nanotubes (CNTs) in epoxy matrix resulted in a significant increase in tensile strength of modulus. Ali MR et al. investigated the graphene-reinforced epoxy nanocomposite material for its anticorrosion and multiphysical properties. The reinforced 3–5% of the graphene synthesized by the electrochemical exfoliation process has better coating properties in the case of aerospace and electronic applications than single coating materials (Ali et al. 2024).

1.7.2 Electronics and Energy

Some of the nanofillers like carbon nanotubes, graphene, the metal oxide, will have unique characteristics in the materials with improved thermal, mechanical, electrical, and optical properties (Arunachalam 2018). Conductive fibers such as graphene or carbon nanotubes are used in flexible electronic devices in the organic light-emitting diodes (OLEDs). Additionally, the use of gold or silver nanoparticles enhances the signal transduction for the biosensor application for their selectivity and sensitivity. Further, the use of PNCs in organic photovoltaic cells will improve the charge transport properties and the energy storage data with the use of supercapacitors due to their high surface areas and electrical conductivity (Yang et al. 2015). PNCs electrodes used in lithium-ion batteries can increase the structural integrity of the electrodes, electrical conductivity, high energy density, and cycle stability. Scientists focused on the perovskite solar cells, where the nanocomposite material improves the stability and transport of charge carriers. Piezoelectric is another nanocomposite material that can be used to harvest the energy from mechanical and moment vibrations to convert them into electrical energy (Asrafuzzaman Amin et al. 2022). Barman Jaydev et al. reported the use of Al_2O_3 with the reduced graphene oxide introduced into polyvinylidene fluoride nanocomposite. This film was synthesized by using the hydrothermal process with the piezoelectric nanogenerator to form the nanosheets with the incorporated nanocomposite material. This device reported an open-circuit voltage of around 42 V with an output of about 8.25 mW/cm^3 (Barman et al. 2024).

1.7.3 Biomedical and Healthcare

PNCs have emerged as a versatile and promising class of nanomaterials that have significant attention on healthcare and biomedical applications. This is due to their unique tailored functionalities and physicochemical properties to tackle the clinical requirement (Li et al. 2021). There are two types of polymeric nanocomposite used in the medical and healthcare system such as natural and synthetic. This can be used in the generation of tissue engineering, wound healing, medical implants, and drug delivery systems (Sofini et al. 2024). Nanofillers such as hydroxyapatite or silica nanoparticles introduced into the polymer matrix can result in bone regeneration and repair. As well as the use of polymeric nanocomposite in the drug delivery system which acts as the therapeutic agent in the controlled release that enhances the therapeutic efficacy and bioavailability (Omar et al. 2024). Geyik Gulcan et al. investigated κ-carrageenan with the poly (acrylic acid/dimethylamino ethyl methacrylate) with the microwave-assisted coprecipitation method. This nanosphere acts as a triple-modal-triggered properties by temperature, pH, and magnetic field stimulation. Along with that, it has a sustained release pattern with the generation of nanostructure for biomedical applications like targeted and controlled drug delivery (Geyik and Işıklan 2021).

1.8 Future Trends and Opportunities

1.8.1 Smart and Functional Materials

In recent years, the polymeric nanocomposite has been rapidly advancing in the research and development of innovative and sustainable goals with diverse applications across various industries. Among these, it has gained distinctive structural and functional characteristics to make it novel and enhanced synergistic properties (Ganesh et al. 2014). The smart or functional material will be designed to respond to some of the external environmental factors or stimuli including pH, temperature, light, and electric field (Wang et al. 2020). Triggering these nanomaterials leads to the development of a controlled release mechanism in the drug delivery system. Additionally, the integration of 3D printing of this nanomaterial will customize the material property by enhancing the lightweight to higher performance and a reliable source to the end-user.

The use of smart materials creates new avenues in the field of biomedical applications with the development of biosensors, tissue engineering, and drug delivery systems (Arif et al. 2022). Materials like graphene, metal oxide, and carbon nanotubes remarkably increase the mechanical strength and durability of bone repair implants and dental applications (Zhang et al. 2023). Further, this polymeric nanocomposite is utilized not only to protect the wound but also to promote antibacterial activity and moisture retention.

1.8.2 Sustainability and Environmental Impact

The polymeric nanocomposite is one of the key factors in developing biodegradable plastic for sustainable and environmental solutions. The use of renewable sources like adipate-co-terephthalate (PBAT), cellulose, and other nanomaterials is a recent advancement focusing on biodegradable production (Tamjid et al. 2024). It has also been developed for agricultural applications including mulching films and controlled-release fertilizer. It not only reduces the plastic waste in the soil health but also decreases the hazardous chemical fertilizers by promoting sustainable practices. It also promises environmental application in the sense of pollutant sensing, degradation, and detection (Darwish et al. 2022).

Additionally, it represents a promising solution for addressing environmental challenges through the remediation process. The enhanced materials are used for adsorbing the hydrophobic substances such as heavy metals, dyes, and organic pollutants which are harmful to the aquatic marine ecosystem (Agboola et al. 2021). Further, polymeric nanocomposite can be employed for the air filtration system to decontaminate or capture the particulate matter, fibrous materials, and organic compounds. Some of the nanocomposite materials potentially facilitate the breakdown of harmful substances into nontoxic forms. It addresses global challenges influenced by advanced technologies and emerging trends in different research sectors. AI has developed a transforming model for the vast dataset by developing the predictive model and more efficient experimental design (Imoisili et al. 2023). It also advances in biotechnology and genomics therapeutic areas by tailoring the research toward the individual genetic profile (personalized medicine).

1.9 Conclusion

This chapter highlights the various nanocomposites and their application in various domains of nanotechnology. These nanocomposites are versatile and can be fabricated based on their characteristic use. With advanced synthesis techniques and precise characterization methods, there is a better understanding of these materials. These nanocomposites have a wide range of applications in many fields such as medicine, energy storage, pharmaceutical, biomedical, aeronautical, and environmental remediation. Despite their advantages, there are several limitations such as toxicities, scalability, cost-effectiveness, and stability issues. These limitations can be addressed by green synthesis or biogenic or ecofriendly-based nanocomposite production techniques, real-time monitoring, and other environmental approaches. Interdisciplinary research in this area could open many avenues in the future and thus address many research problems.

References

Abdellaoui Hind MR, Sanjay, Suchart S (2023) Mechanics of nanomaterials and polymer nanocomposites. Springer

Adnan MM, Dalod ARM, Balci MH, Glaum J, Einarsrud MA (2018) In situ synthesis of hybrid inorganic-polymer nanocomposites. Polymers 10(10):1129. https://doi.org/10.3390/polym10101129

Agboola O, Fayomi OSI, Ayodeji A, Ayeni AO, Alagbe EE, Sanni SE, Okoro EE, Moropeng L, Sadiku R, Kupolati KW, Oni BA (2021) A review on polymer nanocomposites and their effective applications in membranes and adsorbents for water treatment and gas separation. Membranes 11(2):1–33. https://doi.org/10.3390/membranes11020139

Albdiry MT, Yousif BF, Ku H, Lau KT (2013) A critical review on the manufacturing processes in relation to the properties of nanoclay/polymer composites. J Compos Mater 47(9):1093–1115. https://doi.org/10.1177/0021998312445592

Ali MR, Chowdhury MA, Shahin M, Rahman MM, Ali MO, Gafur MA (2024) Multi-physical and anti-corrosion properties of graphene-reinforced epoxy nanocomposite coatings for industrial applications. Arab J Chem 17(1):105424. https://doi.org/10.1016/j.arabjc.2023.105424

Arif ZU, Khalid MY, Zolfagharian A, Bodaghi M (2022) 4D bioprinting of smart polymers for biomedical applications: recent progress, challenges, and future perspectives. React Funct Polym 179:105374. https://doi.org/10.1016/j.reactfunctpolym.2022.105374

Arunachalam P (2018) Polymer-based nanocomposites for energy and environmental applications. In: Polymer-based nanocomposites for energy and environmental applications: a volume in Woodhead Publishing series in composites science and engineering. Woodhead Publishing, pp 186–203. https://doi.org/10.1016/B978-0-08-102262-7.00006-4

Asha Krishnan K (2024) Polymer/clay nanocomposites for tire applications. In: Functional nanocomposites and their applications. Apple Academic Press, pp 49–73

Asrafuzzaman Amin KF, Sen A, Hoque ME (2022) Polymer nanocomposites for energy. In: Advanced polymer nanocomposites: science, technology and applications. Elsevier, pp 335–372. https://doi.org/10.1016/B978-0-12-824492-0.00007-6

Awasthi S, De S, Pandey SK (2024) Advancement in fabrication and characterization techniques of nanocomposites. ACS Omega 9(18):19756–19769. https://doi.org/10.1021/acsomega.3c09981

Barman J, Sarkar M, Mukherjee N, Das S, Bag R, Bose N (2024) Al2O3-decorated reduced graphene oxide composite for flexible piezoelectric energy-harvesting applications. J Electron Mater 53:7621. https://doi.org/10.1007/s11664-024-11475-1

Batool M, Haider MN, Javed T (2022) Applications of spectroscopic techniques for characterization of polymer nanocomposite: a review. J Inorg Organomet Polym Mater 32(12):4478–4503. https://doi.org/10.1007/s10904-022-02461-3

Bokobza L (2018) Spectroscopic techniques for the characterization of polymer nanocomposites: a review. Polymers 10(1):7. https://doi.org/10.3390/polym10010007

Calvert P (1999) Nanotube composites: a recipe for strength. Nature 399(6733):210–211. https://www.nature.com/articles/20326

Chandgude S, Salunkhe S (2021) In state of art: mechanical behavior of natural fiber-based hybrid polymeric composites for application of automobile components. Polym Compos 42(6):2678–2703. https://doi.org/10.1002/pc.26045

Chemin JB, Bulou S, Baba K, Fontaine C, Sindzingre T, Boscher ND, Choquet P (2018) Transparent anti-fogging and self-cleaning TiO2/SiO2 thin films on polymer substrates using atmospheric plasma. Sci Rep 8(1):9603. https://doi.org/10.1038/s41598-018-27526-7

Chen TW, Kalimuthu P, Veerakumar P, Lin KC, Chen SM, Ramachandran R, Mariyappan V, Chitra S (2022) Recent developments in carbon-based nanocomposites for fuel cell applications: a review. Molecules 27(3):761. https://doi.org/10.3390/molecules27030761

Chowdhury MIS, Autul YS, Rahman S, Hoque ME (2022) Polymer nanocomposites for automotive applications. In: Advanced polymer nanocomposites: science, technology and applications. Elsevier, pp 267–317. https://doi.org/10.1016/B978-0-12-824492-0.00010-6

Corcione CE, Frigione M (2012) Characterization of nanocomposites by thermal analysis. Materials 5(12):2960–2980. https://doi.org/10.3390/ma5122960

Darwish MSA, Mostafa MH, Al-Harbi LM (2022) Polymeric nanocomposites for environmental and industrial applications. Int J Mol Sci 23(3):1023. https://doi.org/10.3390/ijms23031023

Díez-Pascual AM (2021) Carbon-based nanomaterials. Int J Mol Sci 22(14):7726. https://doi.org/10.3390/ijms22147726

Díez-Pascual AM (2022) Inorganic-nanoparticle modified polymers. Polymers 14(10):1979. https://doi.org/10.3390/polym14101979

Díez-Pascual AM, Díez-Vicente AL (2014) ZnO-reinforced poly(3-hydroxybutyrate-co-3-hydroxyvalerate) bionanocomposites with antimicrobial function for food packaging. ACS Appl Mater Interfaces 6(12):9822–9834. https://doi.org/10.1021/am502261e

Díez-Pascual AM, Díez-Vicente AL (2015) Nano-TiO2 reinforced PEEK/PEI blends as biomaterials for load-bearing implant applications. ACS Appl Mater Interfaces 7(9):5561–5573. https://doi.org/10.1021/acsami.5b00210

Díez-Pascual AM, Naffakh M (2012) Towards the development of poly(phenylene sulphide) based nanocomposites with enhanced mechanical, electrical and tribological properties. Mater Chem Phys 135(2–3):348–357. https://doi.org/10.1016/j.matchemphys.2012.04.057

Du J, Cheng HM (2012) The fabrication, properties, and uses of graphene/polymer composites. Macromol Chem Phys 213(10–11):1060–1077. https://doi.org/10.1002/macp.201200029

Dybowska-Sarapuk L, Sosnowicz W, Krzeminski J, Grzeczkowicz A, Granicka LH, Kotela A, Jakubowska M (2020) Printed graphene layer as a base for cell electrostimulation—preliminary results. Int J Mol Sci 21(21):1–13. https://doi.org/10.3390/ijms21217865

Gagliardi A, Giuliano E, Venkateswararao E, Fresta M, Bulotta S, Awasthi V, Cosco D (2021) Biodegradable polymeric nanoparticles for drug delivery to solid tumors. Front Pharmacol 12:1. https://doi.org/10.3389/fphar.2021.601626

Ganesh VA, Baji A, Ramakrishna S (2014) Smart functional polymers—a new route towards creating a sustainable environment. RSC Adv 4(95):53352–53364. https://doi.org/10.1039/c4ra10631h

Geyik G, Işıklan N (2021) Design and fabrication of hybrid triple-responsive κ-carrageenan-based nanospheres for controlled drug delivery. Int J Biol Macromol 192:701–715. https://doi.org/10.1016/j.ijbiomac.2021.10.007

Gobena ST, Woldeyonnes AD (2024) A review of synthesis methods, and characterization techniques of polymer nanocomposites for diverse applications. Discov Mater 4(1):52. https://doi.org/10.1007/s43939-024-00119-0

Gong H, Peng R, Liu Z (2013) Carbon nanotubes for biomedical imaging: the recent advances. Adv Drug Deliv Rev 65(15):1951–1963. https://doi.org/10.1016/j.addr.2013.10.002

Gou J, Zhuge J, Liang F (2012) Processing of polymer nanocomposites. In: Manufacturing techniques for polymer matrix composites (PMCs). Springer, Cham, pp 95–119. https://doi.org/10.1533/9780857096258.1.95

Guzzi R, Marra F, Pulci G (2021) Nanocomposite and micro-nanostructured materials with applications of MEMS/Nano devices for aerospace. In: Nanotechnology in space. Jenny Stanford Publishing, pp 23–70. https://doi.org/10.1201/9781003131915-2

Harun-Ur-Rashid M, Imran AB, Susan MABH (2022) Green polymer nanocomposites in automotive and packaging industries. Curr Pharm Biotechnol 24(1):145–163. https://doi.org/10.2174/1389201023666220506111027

Herron N, Thorn DL (1998) Nanoparticles: uses and relationships to molecular cluster compounds. Adv Mater 10(15):1173–1184. https://doi.org/10.1002/(SICI)1521-4095(199810)10:15<1173::AID-ADMA1173>3.0.CO;2-6

Hiremath A, Murthy AA, Thipperudrappa S, Bharath KN (2021) Nanoparticles filled polymer nanocomposites: a technological review. Cogent Eng 8(1):1991229. https://doi.org/10.1080/23311916.2021.1991229

Huang P, Shi HQ, Fu SY, Xiao HM, Hu N, Li YQ (2016) Greatly decreased redshift and largely enhanced refractive index of mono-dispersed ZnO-QD/silicone nanocomposites. J Mater Chem C 4(37):8663–8669. https://doi.org/10.1039/c6tc02047j

Imoisili PE, Adeleke O, Makhatha ME, Jen TC (2023) Response surface methodology (RSM)-artificial neural networks (ANN) aided prediction of the impact strength of natural fibre/carbon nanotubes hybrid reinforced polymer nanocomposite. Eng Sci 23:10.30919/es8d852

Inamdar A, Gurupadayya B, Halagali P, Tippavajhala VK, Khan F, Pathak R, Sharma H (2024) Unraveling neurological drug delivery: polymeric Nanocarriers for enhanced blood-brain barrier penetration. Curr Drug Targets 26:243. https://doi.org/10.2174/0113894501339455241101065040

Kamal A, Ashmawy M, Shanmugan S, Algazzar AM, Elsheikh AH (2022) Fabrication techniques of polymeric nanocomposites: A comprehensive review. Proc Inst Mech Eng C J Mech Eng Sci 236(9):4843–4861. https://doi.org/10.1177/09544062211055662

Komane PP, Kumar P, Choonara YE, Pillay V (2020) Functionalized, vertically super-aligned multiwalled carbon nanotubes for potential biomedical applications. Int J Mol Sci 21(7):2276. https://doi.org/10.3390/ijms21072276

Kotal M, Bhowmick AK (2015) Polymer nanocomposites from modified clays: recent advances and challenges. Prog Polym Sci 51:127–187. https://doi.org/10.1016/j.progpolymsci.2015.10.001

Lago E, Toth PS, Pugliese G, Pellegrini V, Bonaccorso F (2016) Solution blending preparation of polycarbonate/graphene composite: boosting the mechanical and electrical properties. RSC Adv 6(100):97931–97940. https://doi.org/10.1039/c6ra21962d

Li X, Wang C, Cao Y, Wang G (2018) Functional MXene materials: Progress of their applications. Chem Asian J 13(19):2742–2757. https://doi.org/10.1002/asia.201800543

Li X, Xu W, Xin Y, Yuan J, Ji Y, Chu S, Liu J, Luo Q (2021) Supramolecular polymer nanocomposites for biomedical applications. Polymers 13(4):1–17. https://doi.org/10.3390/polym13040513

Liu J, Gao Y, Cao D, Zhang L, Guo Z (2011) Nanoparticle dispersion and aggregation in polymer nanocomposites: insights from molecular dynamics simulation. Langmuir 27(12):7926–7933. https://doi.org/10.1021/la201073m

Liu M, Jia Z, Jia D, Zhou C (2014) Recent advance in research on halloysite nanotubes-polymer nanocomposite. Prog Polym Sci 39(8):1498–1525. https://doi.org/10.1016/j.progpolymsci.2014.04.004

Mark JE (1996) Ceramic-reinforced polymers and polymer-modified ceramics. Polym Eng Sci 36(24):2905–2920. https://doi.org/10.1002/pen.10692

Mittal V (2014) Synthesis techniques for polymer nanocomposites. In: Synthesis techniques for polymer nanocomposites, vol 1–297. Wiley. https://doi.org/10.1002/9783527670307

Müller K, Bugnicourt E, Latorre M, Jorda M, Sanz YE, Lagaron JM, Miesbauer O, Bianchin A, Hankin S, Bölz U, Pérez G, Jesdinszki M, Lindner M, Scheuerer Z, Castelló S, Schmid M (2017) Review on the processing and properties of polymer nanocomposites and nanocoatings and their applications in the packaging, automotive and solar energy fields. Nanomaterials 7(4):74. https://doi.org/10.3390/nano7040074

Nayak U, Halagali P, Panchal K, Tippavajhala VK, Mudgal J, Radhakrishnan R, Manikkath J (2024) Nanoparticles in CNS therapeutics: pioneering drug delivery advancements. Curr Pharm Des 31:443. https://doi.org/10.2174/0113816128328722240828184410

Okamoto M (2023) Polymer nanocomposites. Eng 4(1):457–479. https://doi.org/10.3390/eng4010028

Omar AM, Hassan MH, Daskalakis E, Smith A, Dooghue J, Mirihanage W, Bartolo PJDS (2024) Biomimetic dual sensing polymer nanocomposite for biomedical applications. Front Bioeng Biotechnol 12:1322753. https://doi.org/10.3389/fbioe.2024.1322753

Qiang D, Wang Y, Alhabill FN, Chen G, Andritsch T (2020) Study on quantification method for dispersion and distribution of sphere-like particles and relationship with AC/DC breakdown strength in polymer nanocomposites. IEEE Trans Dielectr Electr Insul 27(2):343–351. https://doi.org/10.1109/TDEI.2019.008242

Ramakoti IS, Panda AK, Gouda N (2023) A brief review on polymer nanocomposites: current trends and prospects. J Polym Eng 43(8):651–679. https://doi.org/10.1515/polyeng-2023-0103

Reynaud E, Gauthier C, Perez J (1999) Nanophases in polymers. Revue de Metallurgie Cahiers D'Informations Techniques 96(2):169–176. https://doi.org/10.1051/metal/199996020169

Sarma S, Rao VR (2024) Emerging synthesis and characterization techniques for hybrid polymer nanocomposites. Nanotechnology 35(1):012002. https://doi.org/10.1088/1361-6528/acfef8

Shah V, Bhaliya J, Patel GM, Deshmukh K (2022) Advances in polymeric nanocomposites for automotive applications: a review. Polym Adv Technol 33(10):3023–3048. https://doi.org/10.1002/pat.5771

Simon J, Flahaut E, Golzio M (2019) Overview of carbon nanotubes for biomedical applications. Materials 12(4):624. https://doi.org/10.3390/ma12040624

Sofini AH, Balasubramanian SPS, Girigoswami DA, Girigoswami K (2024) Biomedical applications of natural and synthetic polymer based nanocomposites. J Biomater Sci Polym Ed 35(2):269–294. https://doi.org/10.1080/09205063.2023.2283910

Sui Y, Cui Y, Meng X, Zhou Q (2022) Research progress on the correlation between properties of nanoparticles and their dispersion states in polymer matrix. J Appl Polym Sci 139(19):1 https://doi.org/10.1002/app.52096

Tamjid E, Najafi P, Khalili MA, Shokouhnejad N, Karimi M, Sepahdoost N (2024) Review of sustainable, eco-friendly, and conductive polymer nanocomposites for electronic and thermal applications: current status and future prospects. Discov Nano 19(1):29. https://doi.org/10.1186/s11671-024-03965-2

Tu S, Jiang Q, Zhang X, Alshareef HN (2018) Large dielectric constant enhancement in MXene Percolative polymer composites. ACS Nano 12(4):3369–3377. https://doi.org/10.1021/acsnano.7b08895

Uthale SA, Dhamal NA, Shinde DK, Kelkar AD (2021) Polymeric hybrid nanocomposites processing and finite element modeling: an overview. Sci Prog 104(3):00368504211029471. https://doi.org/10.1177/00368504211029471

Veigel D, Rishi K, Okoli U, Beaucage G, Galloway JA, Campanelli H, Ilavsky J, Kuzmenko I, Fickenscher M (2023) Comparison of nanocomposite dispersion and distribution for several melt mixers. Polymer 269:125735. https://doi.org/10.1016/j.polymer.2023.125735

Verma R, Rathod MJ, Goyal RK (2020) A review on the effect of processing techniques and functionalization of filler on mechanical properties of polymer nanocomposites. IOP Conf Ser Mater Sci Eng 798(1):012031. https://doi.org/10.1088/1757-899X/798/1/012031

Wang K, Amin K, An Z, Cai Z, Chen H, Chen H, Dong Y, Feng X, Fu W, Gu J, Han Y, Hu D, Hu R, Huang D, Huang F, Huang F, Huang Y, Jin J, Jin X, Tang BZ (2020) Advanced functional polymer materials. Mater Chem Front 4(7):1803–1915. https://doi.org/10.1039/d0qm00025f

Yang C, Wei H, Guan L, Guo J, Wang Y, Yan X, Zhang X, Wei S, Guo Z (2015) Polymer nanocomposites for energy storage, energy saving, and anticorrosion. J Mater Chem A 3(29):14929–14941. https://doi.org/10.1039/c5ta02707a

Young RJ, Kinloch IA, Gong L, Novoselov KS (2012) The mechanics of graphene nanocomposites: a review. Compos Sci Technol 72(12):1459–1476. https://doi.org/10.1016/j.compscitech.2012.05.005

Zhang J, Zhuang Y, Sheng R, Tomás H, Rodrigues J, Yuan G, Wang X, Lin K (2023) Smart stimuli-responsive strategies for titanium implant functionalization in bone regeneration and therapeutics. Mater Horizons 11(1):12–36. https://doi.org/10.1039/d3mh01260c

Chapter 2
Types of Nanofillers and Matrices

Prashant Patil, Ganesh Sonawane, Kajal Pansare, Bole Sandesh, Jayprakash Suryawanshi, Vishal Pande, and Sachin N. Kothawade

Abstract This chapter explores the types of nanofillers and matrices used in polymer nanocomposites, emphasizing their essential role in enhancing material properties for high-performance applications. Beginning with an introduction to polymer nanocomposites, it highlights the significance of nanoscale fillers in improving mechanical, thermal, electrical, and barrier characteristics. The first section categorizes nanofillers, examining their structural and functional features. Inorganic nanofillers—such as layered silicates, carbon-based materials (carbon nanotubes, graphene, and fullerenes), and metal or ceramic nanoparticles—offer reinforcement, conductivity, and stability. Organic nanofillers, including cellulose nanocrystals, dendrimers, and polyhedral oligomeric silsesquioxane (POSS), are noted for biodegradability and potential in sustainable composites. Hybrid nanofillers like core-shell nanoparticles and multicomponent systems provide multifunctionality. The chapter then investigates polymer matrices, discussing thermoplastics, thermosets, elastomers, and biodegradable polymers, with emphasis on compatibility with nanofillers for specific property enhance-

P. Patil (✉) · B. Sandesh
Department of Pharmaceutics, N N Sattha College of Pharmacy, Ahmednagar, Maharashtra, India

G. Sonawane
Department of Pharmaceutical Chemistry, Divine College of Pharmacy, Satana, Nashik, Maharashtra, India

K. Pansare
Department of Pharmacology, Divine College of Pharmacy, Satana, Nashik, Maharashtra, India

J. Suryawanshi
Department of Pharmacognosy, N N Sattha College of Pharmacy, Ahmednagar, Maharashtra, India

V. Pande
Department of Pharmaceutics, RSM's N. N. Sattha College of Pharmacy, Ahmednagar, Ahilyanagar, Maharashtra, India

S. N. Kothawade
Department of Pharmaceutics, SCSSS's Sitabai Thite College of Pharmacy, Shirur, Pune, Maharashtra, India

S. N. Kothawade et al. (eds.), *Polymer Nanocomposites*,
https://doi.org/10.1007/978-3-032-00365-2_2

ments. Key filler-matrix interactions are explored, focusing on surface functionalization to improve bonding and overall performance. A review of synthesis and processing techniques, including in situ polymerization, melt compounding, and solution blending, reveals their influence on filler dispersion and composite quality. Performance attributes such as strength, thermal stability, and conductivity are examined in relation to filler and matrix selection. Applications in automotive, aerospace, electronics, biomedicine, and sustainable packaging are highlighted, alongside challenges like dispersion, scalability, and environmental impact. The chapter concludes with future directions in sustainable nanofillers and biodegradable matrices, offering a foundation for advancing polymer nanocomposite design and applications across industries.

Keywords Polymer nanocomposites · Nanofillers · Polymer matrices · Mechanical reinforcement · Surface functionalization · Sustainable materials

2.1 Introduction

2.1.1 Overview of Polymer Nanocomposites and Their Relevance in Material Science

Polymer nanocomposites are advanced materials that combine a polymer matrix with nanoscale fillers (nanofillers) to significantly enhance the composite's physical, chemical, and mechanical properties. These materials have garnered substantial interest in material science due to their unique ability to bridge the gap between conventional composite materials and nano-engineered systems. Nanocomposites leverage the distinct characteristics of nanofillers, such as high surface area, nanoscale size, and exceptional physical properties, to improve the performance of polymer matrices (Hussain et al. 2006a; Alexandre and Dubois 2000).

The inclusion of nanofillers transforms the inherent limitations of pure polymers, such as low mechanical strength, thermal stability, and poor barrier properties. This makes polymer nanocomposites suitable for diverse applications in automotive, aerospace, biomedical, and packaging industries (Paul and Robeson 2008).

Moreover, the nanoscale dispersion of fillers ensures a significant improvement in the interfacial interaction between the polymer and the filler, which is crucial for transferring stress and enhancing overall material performance (Fischer 2003). The advancements in nanocomposite technology align with the growing demand for high-performance materials in various engineering and scientific fields.

2.1.2 Key Advantages of Integrating Nanofillers into Polymer Matrices

- *Enhanced mechanical* properties: The inclusion of nanofillers such as carbon nanotubes, graphene, and clay nanoparticles improves the tensile strength, modulus, and impact resistance of polymer matrices. For instance, the high aspect ratio of carbon nanotubes contributes to exceptional load transfer and mechanical reinforcement (Moniruzzaman and Winey 2006; Coleman et al. 2006).

- *Improved thermal stability*: Nanofillers such as titanium dioxide and silica enhance the thermal stability of polymers by acting as thermal barriers, slowing down degradation processes during high-temperature applications (Zeng et al. 2008a).
- *Superior barrier properties:* Layered nanofillers like montmorillonite clay create tortuous pathways that significantly reduce the permeability of gases and liquids through the polymer matrix, making them ideal for food packaging and protective coatings (Ray and Okamoto 2003a).
- *Electrical and optical properties*: Nanocomposites incorporating graphene or carbon nanotubes exhibit excellent electrical conductivity, making them suitable for electronic devices, sensors, and energy storage systems. Additionally, optical nanofillers enable the development of materials with tailored optical properties (Kaminska et al. 2012).
- *Lightweight and multifunctional characteristics*: Compared to traditional composites, polymer nanocomposites maintain lightweight characteristics while offering multifunctionality. For example, they combine electrical conductivity, thermal resistance, and mechanical strength in a single material (Krishnamoorti and Vaia 2007).
- *Eco-friendly and sustainable options:* With the incorporation of bio-based matrices and biodegradable nanofillers such as cellulose nanocrystals, polymer nanocomposites are increasingly aligning with the principles of sustainability and environmental conservation (Habibi et al. 2010a).

2.2 Types of Nanofillers in Polymer Nanocomposites

Nanofillers are integral components of polymer nanocomposites, providing tailored properties to meet specific application requirements. They are classified into inorganic, organic, and hybrid types, each offering distinct advantages based on their structural and chemical characteristics. Table 2.1 summarizes the key features and applications of these nanofillers, highlighting their diverse utility in enhancing the performance of polymer matrices.

2.2.1 Inorganic Nanofillers

- *Layered silicates (clay nanoparticles)*: Layered silicates, commonly known as clay nanoparticles, include structures such as montmorillonite, hectorite, and kaolinite. These nanofillers consist of stacked silicate layers separated by an interlayer space. Montmorillonite is the most widely used type due to its high aspect ratio and large surface area. It has a 2:1 layer structure, with one alumina octahedral sheet sandwiched between two silica tetrahedral sheets (Alexandre

Table 2.1 Types of nanofillers in polymer nanocomposites

Category	Type	Key features	Applications
Inorganic nanofillers	Layered silicates (e.g., montmorillonite)	Plate-like structure, high aspect ratio, improves mechanical and barrier properties	Packaging, coatings, and structural materials Alexandre and Dubois (2000), Ray and Okamoto (2003a)
	Carbon-based nanofillers		
	– Carbon nanotubes (CNTs)	Single-walled (SWCNTs) and multi-walled (MWCNTs); exceptional electrical and mechanical properties	Electronics, sensors, and structural reinforcement Iijima (1991), Coleman et al. (2006)
	– Graphene and graphene oxide	High thermal stability, electrical conductivity, and mechanical strength	Conductive films, energy storage, and composites Novoselov et al. (2004), Zhang et al. (2011)
	– Fullerenes	Spherical structure; excellent electrical and energy storage properties	Photovoltaics, energy devices, and nanomedicine Kroto et al. (1985)
	Metal and metal oxide nanoparticles	Examples: Silver, gold, titanium dioxide, zinc oxide; antimicrobial properties, photocatalysis	Medical devices, coatings, and environmental cleanup Morones et al. (2005), Fujishima and Honda (1972)
	Ceramic nanoparticles	High thermal stability, enhanced mechanical strength	Aerospace, electronics, and thermal insulation Koo (2006)
Organic nanofillers	Cellulose nanocrystals (CNCs) and nanofibers (CNFs)	Renewable, biodegradable, lightweight; derived from natural sources	Sustainable composites and packaging materials Habibi et al. (2010a), Moon et al. (2011)
	Dendrimers and hyperbranched polymers	Highly branched structures; potential for controlled drug delivery	Drug delivery systems and nanomedicine Frechet (1994)
	Polyhedral oligomeric silsesquioxane (POSS)	Nano-sized cage structure; enhances thermal stability and chemical resistance	High-performance coatings and electronics Lichtenhan et al. (1993)
Hybrid nanofillers	Core-shell nanoparticles	Multifunctional; customizable core and shell compositions	Catalysis, drug delivery, and optical devices Caruso et al. (1998)
	Multicomponent systems	Combines properties of different fillers for tailored performance	High-performance and multifunctional composites Mittal (2009a)

and Dubois 2000). Clay nanoparticles enhance mechanical strength, thermal stability, and gas barrier properties of polymer matrices. The intercalation or exfoliation of these nanoparticles into the polymer matrix creates tortuous paths, significantly reducing gas permeability (Ray and Okamoto 2003a).

- **Carbon-Based Nanofillers**
  - *Carbon nanotubes (CNTs):* CNTs exist as single-walled (SWCNTs) or multi-walled (MWCNTs) structures. Their unique tubular structure, high aspect ratio, and exceptional electrical and mechanical properties make them ideal for reinforcing polymers (Iijima 1991). CNTs provide high tensile strength and electrical conductivity. They are used in electronic devices, energy storage systems, and high-strength composites (Coleman et al. 2006).
  - *Graphene and graphene oxide (GO)*: Graphene is a single layer of carbon atoms arranged in a hexagonal lattice, while graphene oxide (GO) includes oxygen-containing functional groups. Graphene offers high thermal conductivity and electrical properties, while GO facilitates better dispersion due to its hydrophilic nature (Novoselov et al. 2004). Graphene is widely used in thermal management systems, conductive coatings, and sensors (Zhang et al. 2011).
  - *Fullerenes*: These are spherical carbon molecules with unique properties such as electron acceptor ability. Fullerenes are employed in organic photovoltaics, energy storage, and electronic devices (Kroto et al. 1985).

- *Metal and metal oxide nanoparticles*: Metal nanoparticles (e.g., silver, gold) and metal oxide nanoparticles (e.g., titanium dioxide, zinc oxide) have gained prominence due to their multifunctional properties. These nanoparticles exhibit antimicrobial, photocatalytic, and UV-blocking properties. They are used in coatings, medical devices, and photocatalysis. For example, silver nanoparticles are known for their antimicrobial efficacy, while titanium dioxide is used in sunscreens and self-cleaning surfaces (Morones et al. 2005; Fujishima and Honda 1972).
- *Ceramic nanoparticles*: Ceramic nanofillers include materials like alumina, silica, and zirconia. They provide excellent thermal stability, wear resistance, and hardness. They are used in thermal barrier coatings, cutting tools, and high-temperature applications (Koo 2006).

2.2.2 Organic Nanofillers

- *Cellulose nanocrystals (CNCs) and nanofibers (CNFs)*: CNCs and CNFs are derived from natural cellulose sources such as wood, cotton, and bacterial cellulose. They are prepared via chemical treatments like acid hydrolysis or mechanical disintegration (Habibi et al. 2010a). These nanofillers are environmentally friendly, offering sustainability in polymer nanocomposites and used in packaging, medical implants, and biodegradable composites (Moon et al. 2011).
- *Dendrimers and hyperbranched polymers*: Dendrimers are highly branched, tree-like structures with controlled molecular weight and functional groups. Primarily used in drug delivery systems and as additives to enhance polymer properties (Frechet 1994).

- *Polyhedral oligomeric silsesquioxane (POSS)*: POSS molecules consist of a silica core surrounded by organic functional groups, offering hybrid organic-inorganic characteristics. Used to enhance thermal stability, chemical resistance, and mechanical properties of polymers (Lichtenhan et al. 1993).

2.2.3 *Hybrid Nanofillers*

- *Core-shell nanoparticles*: These consist of a core material (e.g., metal or silica) surrounded by a shell (e.g., polymer or another material). They are synthesized using techniques like sol-gel or layer-by-layer deposition. Core-shell structures offer multifunctionality, including optical, thermal, and mechanical enhancements (Caruso et al. 1998).
- *Multicomponent systems*: Combining multiple nanofiller types (e.g., CNTs with clay) tailors the properties of nanocomposites, achieving synergy between different filler characteristics. Used in high-performance materials for aerospace, automotive, and biomedical applications (Mittal 2009a).

2.3 Polymer Matrices in Nanocomposites

Polymer matrices are the backbone of nanocomposites, providing structural integrity and a medium for dispersing nanofillers. The choice of polymer matrix influences the final properties of nanocomposites, including mechanical strength, thermal stability, and functional attributes. Figure 2.1 illustrates the types of polymer matrices used in nanocomposites.

2.3.1 *Thermoplastic Matrices*

Thermoplastics are linear or branched polymers that soften upon heating and solidify upon cooling, making them widely used in nanocomposites. Common thermoplastic matrices include polyolefins, polyesters, and polyamides.

- *Polyolefins*: Polyolefins offer excellent chemical resistance, low cost, and processability. However, their non-polar nature can make compatibility with polar nanofillers challenging (Paul and Robeson 2008). Used in packaging, automotive parts, and consumer goods when combined with nanofillers like clay or carbon nanotubes for enhanced barrier and mechanical properties (Ray and Okamoto 2003b). *Examples*: Polyethylene (PE) and polypropylene (PP).
- *Polyesters*: Polyesters are polar polymers with high mechanical strength and thermal stability, making them compatible with a wide range of nanofillers

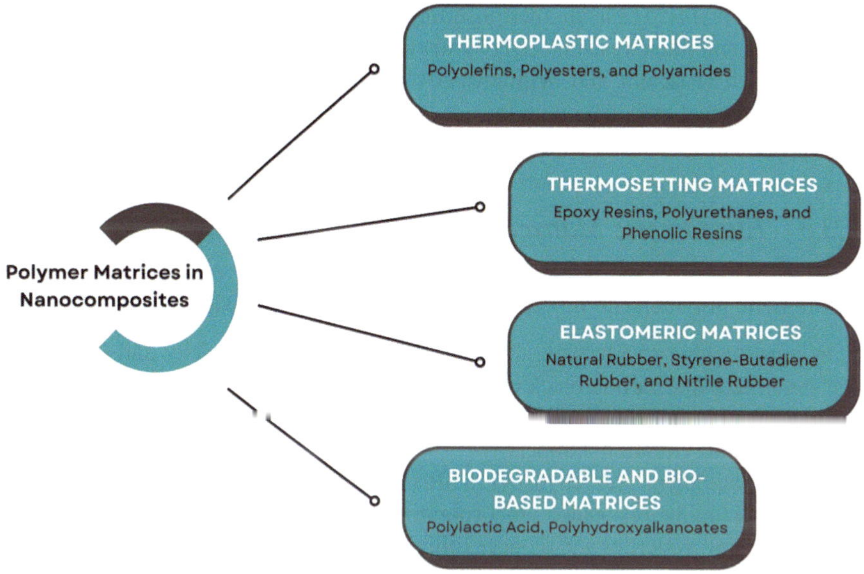

Fig. 2.1 Types of polymer matrices in nanocomposites

(Sánchez et al. 2005). Widely used in textiles, electronics, and structural components. *Examples*: polyethylene terephthalate (PET) and polybutylene terephthalate (PBT).

- *Polyamides*: Polyamides are known for their toughness, thermal resistance, and adhesion to nanofillers such as clay and graphene (Gorrasi and Sorrentino 2015). Used in engineering plastics, gears, and automotive components. *Examples*: Nylon-6 and Nylon-66.

2.3.2 *Thermosetting Matrices*

Thermosetting polymers form irreversible networks upon curing, offering superior thermal stability and mechanical strength.

- *Epoxy resins*: Epoxies are highly versatile due to their excellent adhesive properties, low shrinkage, and resistance to chemicals. They are compatible with a range of nanofillers, including carbon nanotubes and silica nanoparticles (Kim et al. 2010). Commonly used in coatings, adhesives, and aerospace composites.
- *Polyurethanes*: Polyurethanes offer flexibility, abrasion resistance, and good adhesion to nanofillers. Their properties can be tailored by varying the ratio of hard and soft segments (Petrovic and Ferguson 1991). Used in foams, elastomers, and medical devices.

- *Phenolic resins*: These are thermosetting polymers known for their high thermal stability and flame resistance. They work well with nanofillers like carbon black and silica (Pascault and Williams 2013). Used in electrical insulation, brake pads, and heat-resistant components.

2.3.3 Elastomeric Matrices

Elastomeric polymers are rubber-like materials with high flexibility and resilience.

- *Natural rubber (NR)*: Natural rubber offers excellent elasticity and toughness. Nanofillers like carbon black or graphene are often incorporated to improve tensile strength and abrasion resistance (Barbin and Rodgers 1994). Used in tires, gaskets, and vibration dampers.
- *Styrene-butadiene rubber (SBR)*: SBR is a synthetic elastomer with good abrasion and aging resistance. Incorporating nanofillers like silica enhances its mechanical properties and reduces rolling resistance (Bokobza 2004). Commonly used in automotive tires and industrial goods.
- *Nitrile rubber (NBR):* NBR is resistant to oils and chemicals, making it ideal for nanocomposites used in harsh environments. Nanofillers improve its thermal and mechanical properties (Franta 2012). Used in seals, hoses, and gloves.

2.3.4 Biodegradable and Bio-Based Matrices

The focus on sustainability has led to increased use of biodegradable and bio-based polymers in nanocomposites.

- *Polylactic acid (PLA)*: PLA is a biodegradable thermoplastic derived from renewable resources like corn starch. It offers good stiffness and transparency but requires nanofillers like clay or CNCs to enhance its toughness and thermal stability (Auras et al. 2004). Used in packaging, biomedical devices, and disposable products.
- *Polyhydroxyalkanoates (PHA):* PHAs are biodegradable polymers produced by microbial fermentation. They exhibit good biodegradability and compatibility with various nanofillers (Philip et al. 2007). Used in agricultural films, medical implants, and packaging materials.
- *Other bio-based polymers:* These materials reduce dependency on petroleum-based resources and can be combined with nanofillers for improved mechanical and barrier properties (Keshavarz and Roy 2010). *Examples*: polycaprolactone (PCL) and starch-based polymers.

2.4 Nanofiller-Matrix Interactions

The interactions between nanofillers and the polymer matrix play a critical role in determining the overall properties of polymer nanocomposites. The effectiveness of these interactions depends on surface chemistry, compatibility, and the extent of bonding between the nanofillers and the matrix.

2.4.1 *Importance of Surface Chemistry and Compatibility*

- *Surface chemistry*: The surface characteristics of nanofillers, such as surface area, charge, and functional groups, influence their dispersion within the matrix. Poor compatibility often leads to agglomeration, reducing the effectiveness of the nanofillers (Ray and Okamoto 2003b). Nanofillers with reactive surface groups, such as hydroxyl or carboxyl groups, can form stronger bonds with polar polymer matrices like epoxy or polyesters (Mittal 2009b).
- *Compatibility*: Compatibility between the polymer matrix and nanofillers ensures uniform dispersion, maximizing the interfacial area for stress transfer. For instance, polar nanofillers like layered silicates are more compatible with polar polymers, while nonpolar matrices require surface modifications of nanofillers (Paul and Robeson 2008). Enhancing compatibility improves the mechanical, thermal, and electrical properties of the nanocomposite. For example, the incorporation of silane-treated silica nanoparticles into rubber matrices significantly improves tensile strength and elasticity (Gorrasi and Sorrentino 2015).

2.4.2 *Functionalization Techniques to Enhance Filler-Matrix Bonding*

2.4.2.1 Surface Functionalization

- *Chemical grafting*: Chemical grafting is a widely used technique for enhancing the compatibility between nanofillers and polymer matrices by chemically modifying the nanofiller surface. The principle of chemical grafting involves the covalent attachment of reactive functional groups to the nanofiller surface. These groups, such as carboxylic acid (-COOH), amine (-NH_2), or silane groups, are introduced via specific chemical reactions tailored to the nature of the nanofiller and the desired compatibility with the polymer matrix. This process improves the interaction between the filler and the matrix by enabling strong chemical bonding, which is critical for effective load transfer and uniform stress distribution in composite materials.

- For instance, CNTs are frequently functionalized with carboxylic or amine groups. These functional groups improve the dispersion of CNTs in polar polymer matrices such as polyamides or epoxies, which are otherwise challenging due to CNTs' intrinsic hydrophobicity. The enhanced dispersion not only prevents agglomeration but also establishes robust chemical interactions with the polymer chains, resulting in a uniform stress transfer at the filler-matrix interface. This leads to improved mechanical properties, such as tensile strength and stiffness, in the resulting composite material. The advantages of chemical grafting are significant, as it forms strong covalent bonds that ensure a durable and stable interface between the nanofiller and the matrix. However, the process is not without challenges. It can be time-consuming, often requiring multiple steps to achieve the desired level of functionalization. Additionally, the chemical modification might alter the intrinsic properties of the nanofillers, such as their electrical conductivity or thermal stability, depending on the functional groups and reaction conditions used (Kim et al. 2010).
- *Plasma treatment*: Plasma treatment is an advanced surface modification technique widely used to enhance the compatibility of nanofillers with polymer matrices. The process involves exposing nanofillers to ionized gas (plasma), which consists of charged particles, radicals, and photons. This interaction modifies the nanofiller surface by introducing functional groups such as hydroxyl (-OH) or carboxyl (-COOH) or by altering the surface energy. These modifications improve the adhesion between the nanofillers and the polymer matrix, facilitating better load transfer and uniform dispersion.

- A prominent example of this technique is the oxygen plasma treatment of graphene. This treatment introduces polar functional groups onto the graphene surface, making it more hydrophilic. The modified graphene exhibits significantly enhanced bonding with hydrophilic polymer matrices such as polyvinyl alcohol or epoxy. This improved interfacial bonding leads to better mechanical, thermal, and barrier properties in the resulting composite materials. The advantages of plasma treatment include its eco-friendly nature, as it does not involve harmful solvents or chemicals. It is also a rapid process and is highly effective across a wide range of nanofillers, making it a versatile solution for enhancing composite performance. However, the technique has limitations. It requires specialized and often expensive equipment, which may not be accessible for all applications. Additionally, excessive exposure to plasma can degrade the nanofiller surface, potentially diminishing its intrinsic properties such as electrical conductivity or structural integrity (Morent et al. 2011).

2.4.2.2 Physical Treatments

Ultrasonication and high-shear mixing are widely used physical techniques for dispersing nanofillers in polymer matrices, particularly when dealing with materials prone to agglomeration. These methods rely on mechanical forces to break down clusters of nanofillers, ensuring uniform dispersion throughout the matrix.

Ultrasonication uses high-frequency sound waves to create cavitation forces in a liquid medium. The rapid formation and collapse of bubbles generate intense localized energy, which effectively separates aggregated nanofillers. In contrast, high-shear mixing applies intense mechanical forces through rotating blades or other mixing elements, promoting the even distribution of nanofillers within viscous matrices.

For example, ultrasonication is often employed to disperse CNTs in polymer solutions, addressing their natural tendency to form bundles due to van der Waals forces. This results in composites with enhanced electrical and mechanical properties. High-shear mixing, on the other hand, is particularly effective for dispersing silica nanoparticles in thermosetting resins, improving the thermal and mechanical performance of the final product.

The advantages of these techniques are significant. Both methods are relatively simple to implement, scalable for industrial applications, and highly effective in reducing filler agglomeration. The uniform dispersion achieved enhances the mechanical strength, thermal stability, and overall performance of nanocomposites. However, these techniques are not without limitations. Prolonged ultrasonication can lead to the fragmentation or functional alteration of nanofillers, potentially affecting their structural integrity and intrinsic properties. Similarly, excessive shear forces during high-shear mixing can introduce defects or alter the morphology of delicate nanofillers (Potschke et al. 2004a).

2.4.2.3 Polymer Coating

Polymer coating is a functionalization technique that involves applying a thin polymer layer to the surface of nanofillers. This process enhances the compatibility of nanofillers with the surrounding polymer matrix, improving their dispersion and interaction. The polymer coating can be achieved through methods such as solution mixing, in situ polymerization, or adsorption. The coating serves as a bridge between the nanofiller and the matrix, facilitating better adhesion and load transfer, which are crucial for optimizing the properties of composite materials.

An example of this technique is the coating of graphene oxide with epoxy resin or thermoplastic polymers. This coating not only improves the dispersion of graphene oxide in the matrix but also enhances bonding and stress transfer when embedded in thermosetting matrices like epoxy systems. As a result, the composite exhibits superior mechanical properties, such as increased tensile strength and toughness, along with improved thermal stability.

The advantages of polymer coating are noteworthy. It significantly enhances the dispersion of nanofillers, reduces agglomeration, and improves filler-matrix adhesion. Additionally, this technique provides a versatile solution to compatibility challenges, as the choice of coating polymer can be tailored to suit specific matrix materials. However, the process has limitations. It involves additional processing steps, which can increase production costs and complexity. Furthermore, the thickness and uniformity of the coating must be carefully controlled to avoid altering the desired properties of the nanofillers or the matrix (Ramanathan et al. 2008).

2.4.2.4 Surfactant Modification

Surfactant modification is a widely used technique to improve the dispersion of nanofillers in polymer matrices by altering their surface properties. Surfactants consist of hydrophobic and hydrophilic segments that adsorb onto the nanofiller surface. This adsorption reduces the surface energy of the nanofillers, making them more compatible with the matrix and preventing aggregation. By facilitating better interaction between the nanofillers and the matrix, surfactant modification ensures uniform distribution, which is essential for optimizing the mechanical, thermal, and other functional properties of composites.

A common example of this technique is the use of non-ionic surfactants like Tween or Triton X to disperse clay or silica nanoparticles in polar polymer matrices such as epoxy or polyester. These surfactants effectively stabilize the nanofillers within the matrix, enhancing the composite's mechanical strength, thermal conductivity, and durability.

The advantages of surfactant modification include its cost-effectiveness and ease of implementation. It is a straightforward process that can be applied to a wide range of nanofillers, making it a versatile solution for improving composite properties. However, surfactant modification has limitations. Over time, surfactants may migrate within the composite material, which can lead to a reduction in long-term stability and performance. This migration might also impact the uniformity of the nanofiller dispersion or alter the composite's surface characteristics (Sinha Ray et al. 2002).

2.4.3 Impact of Interactions on Mechanical, Thermal, and Electrical Properties

2.4.3.1 Mechanical Properties

Strong filler-matrix interactions enhance tensile strength, Young's modulus, and impact resistance. For example, functionalized graphene oxide in polyolefin matrices improves load transfer at the interface, resulting in superior mechanical performance (Rafiee et al. 2009a).

2.4.3.2 Thermal Properties

Enhanced interactions improve thermal stability and resistance to degradation. Clay nanofillers, when uniformly dispersed, form heat barriers, delaying thermal degradation (Hussain et al. 2006b). Nanofillers like metal oxides improve thermal conductivity, making them suitable for applications requiring heat dissipation (Yu et al. 2007a).

2.4.3.3 Electrical Properties

Functionalization of conductive nanofillers, such as CNTs and graphene, ensures uniform percolation networks within the matrix. This significantly enhances electrical conductivity and electromagnetic shielding capabilities (Bauhofer and Kovacs 2009).

For example, incorporating functionalized CNTs into epoxy matrices results in composites with excellent conductivity for electronic applications (Spitalsky et al. 2010).

2.5 Synthesis and Processing Techniques for Polymer Nanocomposites

The synthesis and processing of polymer nanocomposites are vital to achieving uniform nanofiller dispersion, which directly impacts the mechanical, thermal, and electrical properties of the final material. Proper dispersion enhances interfacial interactions between the nanofillers and the polymer matrix, ensuring superior performance in various applications. The three commonly used techniques are in situ polymerization, melt compounding, and solution blending, each with distinct processes, advantages, challenges, and applications.

2.5.1 *In Situ Polymerization*

In situ polymerization is a synthesis method where nanofillers are integrated into a polymer matrix during the polymerization process, ensuring they are uniformly embedded in the growing polymer chains. The process begins with dispersing nanofillers such as layered silicates or CNTs in a monomer. Polymerization is then initiated using heat, radiation, or chemical catalysts. As the polymer chains grow, they envelop the nanofillers, resulting in a composite material with excellent dispersion and strong interfacial interactions. This technique is widely applied in the production of epoxy-based nanocomposites, which find use in aerospace, electronics, and structural applications due to their enhanced mechanical properties, such as improved tensile strength and modulus. The method offers significant advantages, including high compatibility between nanofillers and the polymer matrix and uniform dispersion even at high nanofiller concentrations. However, the process requires meticulous control of reaction conditions to avoid filler aggregation and can be complex and time-consuming compared to other fabrication methods (Alexandre and Dubois 2000; Ray and Okamoto 2003b).

2.5.2 Melt Compounding

Melt compounding is a solvent-free processing method where nanofillers are mechanically mixed into a molten polymer to create nanocomposites. The process involves adding nanofillers directly into the molten polymer using an extruder or internal mixer. High shear forces are applied during the mixing to break apart filler agglomerates and achieve a uniform distribution of nanofillers throughout the polymer matrix. This technique is widely used for thermoplastics like polyethylene, polypropylene, and polystyrene, with applications in automotive components, packaging materials, and consumer goods. Melt compounding offers several advantages, including being environmentally friendly due to the absence of solvents, cost-effective, and scalable for industrial production. However, it also presents challenges, such as difficulties in ensuring uniform dispersion for highly agglomerated fillers and the potential degradation of thermally sensitive nanofillers under high processing temperatures, which can compromise the composite's performance (Sullivan et al. 2016; Zeng et al. 2008b).

2.5.3 Solution Blending

Solution blending is a technique where the polymer is dissolved along with nanofillers in a common solvent, and the solvent is later evaporated to form a nanocomposite material. The process involves dissolving the polymer and dispersing nanofillers, such as graphene oxide, in solvents like chloroform or tetrahydrofuran. The mixture is then stirred or ultrasonicated to achieve a uniform dispersion of the nanofillers. Afterward, the solvent is evaporated, leaving behind a homogeneous composite material. Solution blending is especially useful for creating polymer/graphene oxide composites in electronic devices, sensors, and barrier applications due to its ability to precisely control the dispersion and alignment of nanofillers. This method offers the advantage of excellent nanofiller dispersion, especially for polar polymers, and allows for fine control over filler distribution. However, challenges include the use of organic solvents, which pose environmental and health risks, as well as the additional costs and complexity of solvent recovery and drying, which limit its scalability for large-scale production (Mittal 2010; Stankovich et al. 2006).

2.5.4 Influence of Processing Techniques on Dispersion and Performance

- *Dispersion quality*: The method of nanocomposite synthesis directly impacts the dispersion of nanofillers, which in turn affects properties such as mechanical strength, electrical conductivity, and thermal stability. Uniform dispersion is essential for effective stress transfer at the filler-matrix interface (Hussain et al. 2006b).

- *Mechanical properties*: Techniques like melt compounding often yield composites with high mechanical properties due to the alignment of nanofillers under shear forces during processing (Potschke et al. 2004b).
- *Thermal properties:* In situ polymerization produces composites with superior thermal stability by enabling strong interactions between the matrix and nanofillers (Yu et al. 2007b).
- *Electrical properties*: Solution blending is particularly effective for ensuring the formation of conductive networks in composites containing graphene or CNTs, enhancing electrical conductivity (Rafiee et al. 2009b).

2.6 Properties and Performance of Polymer Nanocomposites

Polymer nanocomposites have emerged as a prominent class of materials with enhanced mechanical, thermal, electrical, and optical properties compared to traditional composites. The inclusion of nanofillers into polymer matrices enables significant improvements in various performance aspects, driven by the high surface area and unique properties of the nanofillers.

2.6.1 *Mechanical Properties*

- *Strength and modulus*: Polymer nanocomposites exhibit significantly improved tensile strength and modulus due to effective stress transfer between the polymer matrix and nanofillers. For instance, adding 5% montmorillonite clay to nylon-6 increases its tensile strength by up to 40% (Alexandre and Dubois 2000). CNTs are particularly effective in enhancing modulus due to their exceptional intrinsic strength (~1 TPa) (Coleman et al. 2006).
- *Impact resistance*: Nanofillers such as graphene and carbon nanotubes improve the fracture toughness of polymers by impeding crack propagation through mechanisms like filler pull-out and crack bridging (Rafiee et al. 2010). Elastomeric matrices with cellulose nanofibers (CNFs) offer enhanced impact resistance while maintaining flexibility (Habibi et al. 2010a).

2.6.2 *Thermal Stability and Barrier Properties*

- *Thermal stability*: Nanocomposites often show higher thermal degradation temperatures due to the heat dissipation capability of nanofillers. Metal oxide nanoparticles like TiO_2 and ZnO improve the thermal stability of polymers by acting as thermal insulators (Jancar et al. 2010).

- *Barrier properties*: Layered silicates and graphene create tortuous paths for gas diffusion, significantly reducing gas permeability in nanocomposites (Ray and Okamoto 2003b). This property makes polymer nanocomposites suitable for packaging applications where oxygen or moisture barrier is critical.

2.6.3 iElectrical and Optical Properties

- *Electrical properties*: Carbon-based nanofillers, such as CNTs and graphene, impart electrical conductivity to polymer matrices, transforming insulating polymers into conductive composites. Applications include electromagnetic interference (EMI) shielding and conductive coatings (Al-Saleh and Sundararaj 2009a).
- *Optical properties*: Nanofillers like quantum dots and metal nanoparticles enhance the optical transparency and luminescence of nanocomposites. These materials are used in optoelectronic devices and light-emitting diodes (LEDs) (Pumera et al. 2009).

2.6.4 Effect of Nanofiller Types and Matrix Selection on Overall Performance

The effect of nanofiller types and matrix selection on the overall performance of polymer nanocomposites is critical in determining the final properties of the material, including mechanical, thermal, electrical, and optical characteristics. Both the type of nanofiller used and the selection of the polymer matrix significantly influence the behavior and application of the composite material.

2.6.4.1 Impact of Nanofiller Types

- *Layered silicates*: These nanofillers, including montmorillonite and bentonite, are known for enhancing barrier properties (such as resistance to gases and liquids) and mechanical strength. They are particularly useful in applications requiring resistance to environmental factors. However, they generally have low compatibility with nonpolar matrices like polyethylene or polypropylene. This necessitates surface modification techniques, such as silanization or grafting, to improve the interaction between the nanofiller and the polymer matrix, ensuring better dispersion and bonding (Tjong 2006).

- *Carbon-based nanofillers*: CNTs and graphene are highly regarded for their exceptional mechanical, thermal, and electrical properties. CNTs provide excellent tensile strength, while graphene has remarkable electrical conductivity, making these fillers ideal for applications in electronics, sensors, and conductive coatings. However, their cost can be prohibitive, and they are often challenging to disperse uniformly in the matrix, which can lead to agglomeration and poor filler-matrix interaction. To mitigate this, surface functionalization or dispersion techniques like ultrasonication or melt compounding are often employed (Feller et al. 2008).
- *Metal and metal oxide nanoparticles*: Metal nanoparticles, such as silver and copper, and metal oxide nanoparticles, like titanium dioxide (TiO_2) and zinc oxide (ZnO), offer thermal stability, antimicrobial properties, and UV protection. These fillers are often used in coatings, medical devices, and packaging materials. However, they can negatively affect the optical transparency of the composite, making them unsuitable for applications that require clear or transparent materials, such as optical films or lenses (Huang et al. 2006).

2.6.4.2 Matrix Selection

- *Thermoplastics*: Thermoplastic polymers, such as polyethylene, polypropylene, and polystyrene, are known for their lightweight and flexible properties, making them ideal for packaging, automotive, and consumer goods applications. These materials can be easily processed using techniques like melt compounding or extrusion, which allows for scalable production. However, thermoplastics may have lower thermal stability and mechanical properties compared to thermosets.
- *Thermosets*: Thermoset polymers, such as epoxy, polyester, and phenolic resins, offer superior thermal and mechanical properties, making them suitable for high-performance applications like aerospace, automotive structural parts, and electronics. Thermosets undergo a curing process that crosslinks the polymer chains, resulting in a material that is rigid and has excellent dimensional stability. However, once cured, thermosets cannot be reprocessed, which limits their recyclability (Schmidt et al. 2002).
- *Elastomers*: Elastomers are flexible, rubber-like materials that can be combined with nanofillers such as cellulose nanofibers (CNFs) to enhance their strength while maintaining flexibility. These nanocomposites are ideal for applications requiring both elasticity and enhanced mechanical properties, such as biomedical devices (e.g., implants, prosthetics), sports equipment, and seals. The addition of nanofillers can significantly improve the strength, durability, and wear resistance of elastomers, making them more versatile in demanding applications (Liu et al. 2008).

2.7 Applications of Polymer Nanocomposites with Specific Nanofillers and Matrices

Polymer nanocomposites have found numerous applications across a wide range of industries due to their remarkable mechanical, thermal, electrical, and barrier properties. The synergy between specific nanofillers and polymer matrices allows for the customization of materials to meet the specific requirements of high-performance applications. These nanocomposites enhance the overall performance of materials, making them suitable for use in industries such as automotive, aerospace, electronics, energy storage, biomedical fields, and environmental applications.

2.7.1 Automotive and Aerospace

- **Lightweight and High-Strength Components.**
- In the automotive and aerospace industries, reducing the weight of components is critical for enhancing fuel efficiency, performance, and reducing the environmental impact of vehicles and aircraft. Polymer nanocomposites, which combine nanofillers with specific polymer matrices, provide a solution to this challenge by offering a balance of lightweight properties and high strength. Nanofillers such as CNTs, graphene, and layered silicates are incorporated into the polymer matrix to significantly enhance the mechanical properties of the composites. These nanofillers contribute to the increased stiffness, strength, and impact resistance of the material without adding significant weight. For instance, CNTs are known for their exceptional strength-to-weight ratio, making them ideal for applications where both lightness and durability are required, such as in aerospace. In this context, CNT-epoxy composites are widely used in the structural components of aircraft, such as wings and fuselages, to reduce weight while maintaining or improving structural integrity.
- On the other hand, the matrices used to form these nanocomposites are also crucial in determining the final material properties. Thermosetting resins like epoxy and phenolic resins, along with thermoplastics like polyamides and polyesters, are commonly used in automotive and aerospace applications. These matrices provide the required thermal stability, chemical resistance, and molding capability for creating durable components.
- In addition to improving the strength-to-weight ratio, these nanocomposites also offer enhanced barrier properties, which are vital in specific applications such as automotive fuel tanks. For example, clay-polymer nanocomposites are employed

to improve the barrier resistance of polymer materials to fuel permeation, ensuring the safety, durability, and long-term performance of fuel tanks in vehicles. These advanced nanocomposites contribute not only to the mechanical performance but also to safety and sustainability by reducing fuel loss and improving the overall efficiency of automotive systems. Thus, the integration of nanofillers with polymer matrices in the automotive and aerospace sectors leads to lightweight, high-performance materials that contribute to energy savings, reduced emissions, and enhanced safety standards (Mittal 2009c; Alexandre and Dubois 2000).

- **Thermal and Flame Resistance**

- The incorporation of nanofillers into polymer composites significantly improves their thermal stability and flame-retardant properties, which is particularly critical in industries like automotive and aerospace where fire safety and heat resistance are paramount. Metal oxides, such as aluminum oxide (Al_2O_3) and zinc oxide (ZnO), are commonly used as nanofillers because of their inherent ability to withstand high temperatures and act as effective flame retardants. Additionally, layered silicates are also employed to enhance the thermal resistance of the polymer matrix. These nanofillers improve the heat resistance of the composite materials, enabling them to maintain their mechanical integrity even under extreme conditions.
- The matrix materials typically used in these composites include polycarbonate and polyethylene, which are well-known for their versatility, impact resistance, and ease of processing. When combined with metal oxides or layered silicates, these matrices significantly enhance the thermal stability and flame retardancy of the composite material. For example, polycarbonate, with its high thermal stability, becomes even more heat-resistant when reinforced with metal oxide nanofillers. This makes the composite suitable for use in environments where exposure to high temperatures or fire hazards is a concern.

- In practical applications, such as in vehicles and aircraft, these thermal and flame-resistant composites are crucial for ensuring the safety of the occupants. They are commonly used in interior components such as dashboards, seating, and insulation materials, where the risk of fire is elevated. The metal oxide nanofillers play a vital role in preventing the spread of flames and improving the fire resistance of the polymer composites. In the event of a fire, these composites can withstand higher temperatures for longer periods, preventing the rapid spread of flames and providing critical time for evacuation. As a result, the use of nanofiller-enhanced composites in automotive and aerospace applications is essential for improving fire safety, ensuring that materials used in these industries are not only durable and lightweight but also capable of withstanding extreme fire and heat conditions (Morgan and Wilkie 2007).

2.7.2 *Electronics and Energy Storage*

- **Conductive and Electromagnetic Shielding Materials**
- Conductive and electromagnetic shielding materials are essential in modern electronics, where controlling electrical properties and protecting sensitive circuits from external interference are crucial. Nanofillers, such as CNTs, graphene, and silver nanoparticles, are often incorporated into polymer matrices to enhance the electrical conductivity and electromagnetic shielding capabilities of the resulting composites. CNTs are particularly prized for their remarkable electrical conductivity, enabling the creation of high-performance conductive materials. Likewise, graphene, known for its exceptional electrical properties, provides significant benefits in terms of conductivity and can also contribute to mechanical reinforcement in composites. Silver nanoparticles, due to their excellent conductivity and high surface area, are also frequently used to boost the conductive properties of the matrix.
- The matrices typically used in conductive and electromagnetic shielding composites include polyimides, polyaniline, and various thermoplastics such as polyethylene and polystyrene. Polyimides are particularly favored in flexible electronic applications due to their excellent thermal stability, mechanical properties, and flexibility. When combined with CNTs or graphene, these matrices form nanocomposites with enhanced electrical conductivity, making them ideal for use in flexible circuits, antennas, and sensor systems. Polyaniline, a conductive polymer, also works well with nanofillers to provide electromagnetic shielding while maintaining the overall flexibility and processability of the composite.
- In practical applications, CNT-polyimide nanocomposites are commonly used in flexible electronics, such as printed circuits and wearable devices, where high conductivity is essential for efficient operation. The CNTs impart a high degree of electrical conductivity, while the polyimide matrix ensures that the material can bend without losing performance. In contrast, graphene-based composites are widely used in electromagnetic interference (EMI) shielding applications. EMI shielding is crucial in protecting sensitive electronic circuits from the damaging effects of electromagnetic radiation, which can disrupt the functionality of devices such as smartphones, computers, and medical equipment. The graphene-polymer composites provide an effective barrier against electromagnetic waves, ensuring that EMI shielding is robust and reliable. These composites can be used in mobile phone casings, laptop shields, and circuit boards to prevent unwanted interference from external sources and ensure the proper functioning of sensitive electronic components.
- By integrating conductive nanofillers like CNTs, graphene, and silver nanoparticles into polymer matrices, manufacturers can create advanced materials that meet the growing demand for lightweight, flexible, and electrically conductive materials, while also providing protection from electromagnetic radiation. This

makes them indispensable in the development of next-generation electronic devices (Al-Saleh and Sundararaj 2009b; Kim et al. 2010).

- **Energy Storage Devices.**

- Energy storage devices, such as batteries and supercapacitors, are integral to modern technologies, and the integration of nanocomposites into their design is significantly improving performance. The use of nanofillers like graphene, fullerenes, and metal oxide nanoparticles is at the forefront of enhancing the energy density, charge-discharge efficiency, and overall performance of these devices. Graphene is particularly valuable in energy storage applications due to its exceptional electrical conductivity, high surface area, and mechanical strength, which contribute to faster charge and discharge cycles and longer device lifespans. Similarly, fullerenes, with their spherical structure, are utilized for their ability to improve the electrochemical performance and stability of energy storage systems. Metal oxide nanoparticles, such as titanium oxide (TiO_2) and zinc oxide (ZnO), also play a critical role in enhancing the charge capacity and thermal stability of energy storage devices, contributing to greater cycle stability.
- The matrices used to form these nanocomposites are crucial in determining the material's overall performance. Polyvinylidene fluoride (PVDF) and polyethylene oxide (PEO) are often used as matrix materials in energy storage devices. PVDF is known for its high dielectric constant, chemical stability, and ionic conductivity, making it ideal for use in batteries and supercapacitors. It enhances the electrochemical properties of the composite and ensures that the nanofillers are well-dispersed within the polymer matrix. PEO, on the other hand, is commonly used in solid-state batteries and supercapacitors due to its excellent ionic conductivity and flexibility, which aids in the effective movement of ions within the energy storage device.

- The combination of graphene-based nanocomposites and battery electrodes, particularly in lithium-ion batteries, demonstrates a significant improvement in capacity, charge rate, and cycle stability. Graphene-enhanced electrodes allow for faster electron and ion transport, leading to quick charging and discharging capabilities, which is crucial for modern energy storage applications, such as in electric vehicles and portable electronics. These graphene-based composites also offer improved cycle life, reducing the degradation of materials after repeated charge and discharge cycles, which is a common limitation in traditional battery systems.

- In supercapacitors, which rely on the electrostatic storage of energy, the inclusion of graphene and fullerenes results in increased charge storage and faster charge-discharge rates compared to conventional materials. The nanofillers contribute to enhanced surface area, enabling more efficient storage of energy in the form of electrostatic charge. This makes the supercapacitors more efficient for applications requiring quick bursts of power, such as in energy harvesting, regenerative braking systems, and power backups (**Zhan et al.** 2011).

2.7.3 Biomedical Applications

- **Drug Delivery Systems**
- In biomedical applications, particularly in drug delivery systems, the integration of nanocomposites significantly enhances the efficacy and precision of therapeutic treatments. Nanofillers such as dendrimers, cellulose nanocrystals (CNCs), and polyhedral oligomeric silsesquioxane (POSS) are widely used due to their unique properties that allow for controlled release and targeted delivery of drugs. Dendrimers, with their highly branched, tree-like structure, offer high surface areas and functional groups that enable them to encapsulate drugs and release them in a controlled manner over time. Cellulose nanocrystals (CNCs), derived from natural sources, provide biocompatibility and the ability to modulate the release rate of drugs. POSS, a silica-based nanofiller, is particularly valuable in drug delivery due to its biocompatibility, nanostructured nature, and ability to stabilize drugs, ensuring they are released only under specific conditions.
- To optimize the delivery of these drugs, the matrix materials used in conjunction with the nanofillers are typically biodegradable polymers such as polylactic acid (PLA) and polycaprolactone (PCL). These polymers are preferred because they degrade naturally within the body, reducing the risk of accumulation and toxicity. PLA is often used due to its ability to slowly degrade and release the drug over an extended period, making it ideal for applications that require sustained therapeutic effects. Similarly, PCL is a biodegradable polymer that offers flexibility and slow degradation, allowing for controlled release over extended periods, which is beneficial in chronic disease management.
- One of the most impactful applications of these nanocomposites is in targeted drug delivery, particularly in cancer treatment. For instance, POSS-PLA systems are used to encapsulate chemotherapy drugs, providing targeted delivery directly to cancer cells while minimizing side effects. This is achieved by modifying the nanocomposite to recognize and bind to specific biomarkers or receptors expressed on cancer cells, ensuring the drug is released only at the site of action, thereby increasing its effectiveness while reducing its impact on healthy cells. This controlled release not only improves the efficiency of the drug but also significantly reduces toxicity and adverse effects, leading to better patient outcomes (Bharali et al. 2005).
- **Tissue Engineering**
- In tissue engineering, particularly for bone regeneration, the combination of nanofillers and biodegradable matrices plays a crucial role in creating scaffolds that closely mimic the natural bone structure. Hydroxyapatite nanoparticles, which are the primary mineral component of bone, are commonly used as nanofillers in these nanocomposites. When incorporated into polymers, such as collagen and polyhydroxyalkanoates (PHAs), hydroxyapatite nanoparticles help replicate the mineral content of bone, providing the necessary structural and biochemical cues to promote osteogenesis (bone formation).

- Collagen, a natural protein found in the extracellular matrix of bones, provides excellent biocompatibility and promotes cell adhesion, which is essential for tissue integration and regeneration. PHAs, which are biodegradable and biocompatible, serve as an effective matrix for supporting cell growth and tissue formation. These polymers not only offer structural support but also allow for controlled degradation, ensuring that the scaffold gradually dissolves as the new bone tissue forms, without causing adverse effects.
- The hydroxyapatite-polymer nanocomposites used in bone regeneration scaffolds are particularly effective because they create an environment that promotes osteoblast differentiation and the formation of bone-like mineralized tissue. The nanoparticles' size and surface characteristics also enhance the interaction with bone cells, leading to better integration and functionality. For example, in bone tissue engineering, these composites are used to repair fractures, defects, or damaged bone, offering a solution that supports the natural healing process. The incorporation of hydroxyapatite into the scaffold not only mimics the bone's mineral content but also encourages the formation of new bone tissue, making these nanocomposites ideal for bone regeneration applications. Ultimately, they provide a highly effective platform for repairing or regenerating bone tissue, especially in cases of fractures or bone defects, where promoting the natural formation of bone is crucial for healing (Rezwan et al. 2006).
- **Antimicrobial Coatings**
- Antimicrobial coatings incorporating nanofillers such as silver nanoparticles and zinc oxide are gaining significant importance in medical applications due to their powerful antimicrobial properties. These nanofillers exhibit strong bactericidal and fungicidal activities, effectively inhibiting the growth of a wide range of harmful microorganisms. Silver nanoparticles are known for their ability to disrupt microbial cell membranes and interfere with cellular functions, while zinc oxide offers antimicrobial activity by generating reactive oxygen species that damage the cell structure of bacteria.
- These nanofillers are typically embedded in matrices like polyurethane and polyethylene, which are commonly used in medical devices due to their biocompatibility, flexibility, and ease of processing. Polyurethane provides a durable and flexible substrate, making it ideal for applications such as wound dressings and catheters, while polyethylene is often chosen for its strength and resistance to environmental factors, making it suitable for implants.
- The application of silver nanoparticle or zinc oxide-loaded nanocomposites to medical devices such as catheters, implants, and wound dressings serves as a preventive measure against bacterial infections. These coatings create an antimicrobial barrier on the surface of the medical devices, reducing the risk of infection and improving patient outcomes. For example, in catheters and implants, where the risk of microbial colonization is high, these nanocomposites prevent the adhesion and growth of bacteria, thereby significantly lowering the chances of hospital-acquired infections (HAIs). In wound dressings, they help to maintain

a sterile environment, accelerating healing and minimizing the risk of secondary infections (Rai et al. 2009).

2.7.4 Packaging and Environmental Applications

In packaging applications, nanofillers like clay nanoparticles and silver nanoparticles are incorporated into polymers to significantly enhance the barrier properties and provide antimicrobial activity. Clay nanoparticles are known for their ability to improve the oxygen and moisture barrier properties of polymers, which are essential factors in preserving the quality and freshness of packaged goods. By incorporating clay nanoparticles into packaging materials, the rate of oxygen permeation is reduced, which slows down the oxidation of food products and prolongs their shelf life. This makes clay-polymer nanocomposites, such as clay-PLA (polylactic acid) blends, highly effective for food packaging, where maintaining freshness is critical.

Similarly, silver nanoparticles are added to packaging materials for their antimicrobial properties. Silver is well-known for its ability to kill or inhibit the growth of a wide range of bacteria, fungi, and other microorganisms. When embedded in polymer films, silver nanoparticles provide a constant antimicrobial action, preventing microbial contamination that can spoil food. This is particularly useful in packaging perishable food items or products that are susceptible to bacterial or fungal growth. Silver-embedded films thus help maintain food safety and quality by reducing the risk of contamination, which in turn helps extend the product's shelf life.

For these reasons, clay-PLA and silver-containing nanocomposites are becoming increasingly popular in the food packaging industry. They combine the environmentally friendly nature of biodegradable polymers like PLA with the functional advantages of nanofillers, such as improved barrier properties and antimicrobial action, ensuring better food preservation and reduced environmental impact. This technology not only improves the quality and safety of packaged food products but also contributes to sustainable packaging solutions (Ray and Bousmina 2005; Rhim et al. 2006).

- **Sustainable and Biodegradable Solutions**
- Sustainable and biodegradable solutions are gaining significant traction in the packaging industry as a response to the growing concerns over plastic waste and environmental sustainability. Nanofillers such as cellulose nanocrystals (CNCs) and graphene oxide are increasingly being used in these applications due to their environmental benefits. CNCs, derived from renewable sources like wood and plants, are not only biodegradable but also offer exceptional mechanical strength and rigidity, making them ideal for enhancing the properties of bio-based polymers. Graphene oxide, though typically more associated with enhancing electrical conductivity, can also contribute to improving the barrier properties of packaging materials, making them more effective in extending the shelf life of products while still being environmentally friendly.

- When combined with bio-based polymers like polylactic acid (PLA) and polyhydroxyalkanoates (PHAs), these nanofillers create biodegradable nanocomposites that are strong, functional, and sustainable alternatives to conventional plastics. PLA and PHAs are biodegradable polymers made from renewable resources such as corn starch or sugarcane, and they decompose naturally, reducing the environmental impact compared to petroleum-based plastics. These bio-based matrices, when reinforced with CNCs or graphene oxide, result in packaging materials that are not only strong and effective in their role but also have the potential to break down without contributing to plastic pollution.
- The growing use of biodegradable nanocomposites in single-use products like food containers, wrappers, and packaging films aligns with the push for more sustainable packaging solutions. These materials help reduce the environmental footprint associated with traditional plastic packaging by offering a biodegradable option that naturally breaks down over time. As a result, they contribute to reducing plastic waste and its harmful effects on ecosystems, particularly in marine environments. Thus, biodegradable nanocomposites represent an essential step towards creating more sustainable, eco-friendly packaging solutions that meet both functional and environmental needs (Habibi et al. 2010b).
- **Water Purification.**
- Water purification is one of the most vital applications of nanocomposite materials, particularly in providing safe drinking water in areas affected by contamination or scarcity. Nanofillers such as metal oxide nanoparticles like titanium dioxide (TiO_2) and zinc oxide (ZnO), along with carbon nanofibers, are utilized for their exceptional ability to enhance filtration processes. These nanofillers have unique properties, such as high surface area, photocatalytic activity, and antimicrobial effects, making them highly effective at removing various contaminants. TiO_2 and ZnO nanoparticles, for example, can degrade organic pollutants and kill bacteria under UV light, while carbon nanofibers improve the mechanical strength and permeability of filtration membranes.
- The nanofillers are typically incorporated into matrices like polyethersulfone and polyaniline, which are commonly used to create filtration membranes. Polyethersulfone is a durable, hydrophilic polymer known for its excellent mechanical properties and resistance to fouling, which is crucial in ensuring long-term filtration efficiency. Polyaniline, a conducting polymer, contributes to the enhanced performance of filtration membranes, especially in removing organic pollutants and heavy metals.
- In water purification systems, these nanocomposite membranes play a critical role in improving filtration efficiency, enabling the removal of contaminants such as bacteria, heavy metals, and organic pollutants. The advanced filtration capabilities of these membranes ensure that water is purified to a level suitable for drinking, making them indispensable in desalination, wastewater treatment, and point-of-use filtration systems. This technology is particularly valuable in regions with limited access to clean water, where these nanocomposite membranes can

help provide safe drinking water, reduce the spread of waterborne diseases, and contribute to environmental sustainability by offering more efficient and cost-effective filtration solutions (Thakur et al. 2013).

2.8 Challenges and Future Directions in Polymer Nanocomposites

Despite the remarkable advantages of polymer nanocomposites, their practical applications face numerous challenges. Addressing these issues will pave the way for broader industrial adoption and innovative solutions.

2.8.1 Dispersion Issues and Scaling Up for Industrial Applications

- *Nanofiller dispersion*: Uniform dispersion of nanofillers within polymer matrices is critical for achieving optimal properties. Agglomeration and poor compatibility between fillers and matrices reduce mechanical, thermal, and electrical performance (Hussain et al. 2006b). Techniques such as surface functionalization and the use of compatibilizers have been developed to enhance filler dispersion. However, these methods add complexity and cost to the manufacturing process (Balazs et al. 2006).
- *Challenges in scaling up*: Laboratory-scale synthesis methods like solution blending and in situ polymerization are often unsuitable for industrial production due to high costs and processing times. Melt compounding, while scalable, requires precise control of parameters to prevent filler degradation and ensure uniform properties (Paul and Robeson 2008). Developing scalable, cost-effective, and environmentally friendly production techniques remains a key challenge (Winey and Vaia 2007).

2.8.2 Environmental and Toxicity Considerations

- *Nanofiller toxicity*: Certain nanofillers, such as CNTs and metal oxide nanoparticles, pose health risks due to their potential to cause cellular damage or respiratory issues (Poland et al. 2008). Proper risk assessments and safer handling protocols during production are necessary to mitigate these risks.
- *Environmental impact*: Non-biodegradable polymer matrices combined with persistent nanofillers contribute to long-term environmental pollution (Chen

et al. 2005). Recycling polymer nanocomposites is challenging due to the difficulty in separating fillers from matrices and the degradation of filler properties during processing (Ragaert et al. 2017).
- *Regulatory and safety concerns*: The lack of comprehensive guidelines for the safe disposal and recycling of polymer nanocomposites hinders their widespread adoption (OECD 2020). Future advancements must include strategies for environmentally safe production and lifecycle management.

2.8.3 Innovations in Sustainable Nanofillers and Biodegradable Matrices

- *Development of green nanofillers*: Sustainable nanofillers like cellulose nanocrystals (CNCs), biochar, and natural clay offer renewable and eco-friendly alternatives to traditional fillers (Moon et al. 2011). CNCs derived from plant biomass provide high strength, biodegradability, and compatibility with bio-based polymers, making them ideal for green composites (Habibi et al. 2010c).
- *Advancements in biodegradable matrices*: Polylactic acid (PLA), polyhydroxyalkanoates (PHA), and other biodegradable polymers are gaining traction as matrices for sustainable nanocomposites. Innovations in polymer chemistry are improving the thermal and mechanical properties of biodegradable matrices, enabling their use in high-performance applications (Reddy et al. 2013).
- *Circular economy approaches*: Incorporating circular economy principles, such as recycling, reuse, and upcycling, can minimize waste and environmental impact. Strategies like chemical recycling of matrices and recovery of nanofillers are being explored for sustainable lifecycle management (Geyer et al. 2017).

2.8.4 Future Directions

- *Functional nanocomposites*: The focus is shifting towards multifunctional nanocomposites that combine mechanical strength with advanced properties like self-healing, shape memory, or stimuli responsiveness (Qiu et al. 2019).
- *Integration with emerging technologies*: Combining polymer nanocomposites with artificial intelligence and additive manufacturing techniques (e.g., 3D printing) can revolutionize their design and production processes (Fortunati et al. 2013).
- *Regulatory frameworks and standardization*: Establishing globally accepted standards and regulatory frameworks for production, safety, and environmental management is essential for accelerating industrial adoption (Mech et al. 2019).

2.9 Conclusion

In conclusion, polymer nanocomposites represent a promising frontier in material science, offering a unique combination of enhanced mechanical, thermal, electrical, and environmental properties. The integration of nanofillers such as inorganic materials (e.g., clay nanoparticles, carbon nanotubes, and metal oxides) and organic materials (e.g., cellulose nanocrystals and dendrimers) into polymer matrices has proven to significantly improve the performance of composites for various applications. These advancements provide valuable solutions for industries ranging from automotive and aerospace to electronics, biomedical devices, and sustainable packaging. The choice of nanofillers and matrices, whether thermoplastic, thermosetting, elastomeric, or biodegradable, plays a crucial role in determining the composite's final properties. Inorganic fillers like CNTs and graphene enhance electrical conductivity and mechanical strength, while organic fillers like CNCs contribute to sustainability and biodegradability. Furthermore, hybrid systems combining different types of nanofillers offer the potential to tailor composites to meet specific performance requirements, broadening their application scope. However, significant challenges remain in the commercial scale-up of polymer nanocomposites, including issues related to nanofiller dispersion, cost-effective processing, and potential environmental and toxicity concerns. Innovations in sustainable nanofillers and biodegradable matrices hold promise in addressing these challenges, paving the way for eco-friendly and high-performance materials in the future. Emerging trends in the field include the development of multifunctional nanocomposites with integrated properties such as self-healing, responsive behaviors, and smart functionalities. Advances in processing techniques, like 3D printing and advanced blending methods, also hold great potential for producing highly specialized composites with tailored properties. As research continues to evolve, the interdisciplinary nature of polymer nanocomposites promises exciting new applications in various technological and industrial sectors. In conclusion, the future of polymer nanocomposites lies in overcoming current limitations through innovative solutions that emphasize sustainability, performance optimization, and safe processing techniques. This vibrant area of research is expected to play a transformative role in the development of next-generation materials, driving advancements in both traditional and emerging industries.

References

Alexandre M, Dubois P (2000) Polymer-layered silicate nanocomposites: Preparation, properties, and uses of a new class of materials. Mater Sci Eng R Rep 28(1):1–63

Al-Saleh MH, Sundararaj U (2009a) Review of the electrical properties of polymer nanocomposites containing carbon nanotubes. Compos A: Appl Sci Manuf 40(11):1732–1741

Al-Saleh MH, Sundararaj U (2009b) EMI shielding mechanisms of CNT/polymer composites. Polymer 50(5):1103–1111

Auras R et al (2004) An overview of polylactides as packaging materials. Macromol Biosci 4(9):835–864
Balazs AC et al (2006) Modeling the interactions between polymers and clay surfaces. Science 314(5802):1107–1110
Barbin WW, Rodgers MB (1994) The science of rubber compounding. In: Science and technology of rubber. Academic Press, pp 419–469
Bauhofer W, Kovacs JZ (2009) A review and analysis of electrical percolation in carbon nanotube polymer composites. Compos Sci Technol 69(10):1486–1498
Bharali DJ et al (2005) Nanoparticles and dendrimers in cancer drug delivery. J Nanosci Nanotechnol 5(5):892–901
Bokobza L (2004) Reinforcement of natural rubber. Compos Sci Technol 64(11):1497–1513
Caruso F et al (1998) Hollow capsule preparation through the layer-by-layer assembly of polyelectrolytes. Science 282(5391):1111–1114
Chen GX et al (2005) Nanocomposites and environmental concerns. Environ Sci Technol 39(14):5520–5526
Coleman JN, Khan U, Blau WJ, Gun'ko YK (2006) Small but strong: a review of the mechanical properties of carbon nanotube–polymer composites. Carbon 44(9):1624–1652
Feller JF et al (2008) Polymer nanocomposites for electrical conductivity applications. Mater Sci Eng B 151(1):83–91
Fischer H (2003) Polymer nanocomposites: from fundamental research to specific applications. Mater Sci Eng C 23(6–8):763–772
Fortunati E et al (2013) Multifunctional nanocomposites for environmental and biomedical applications. Polym Degrad Stab 98(11):2093–2100
Franta I (ed) (2012) Elastomers and rubber compounding materials, vol 1. Elsevier
Frechet JMJ (1994) Functional polymers and dendrimers: reactivity, molecular architecture, and interfacial energy. Science 263(5154):1710–1715
Fujishima A, Honda K (1972) Electrochemical photolysis of water at a semiconductor electrode. Nature 238(5358):37–38
Geyer R et al (2017) Production, use, and fate of all plastics ever made. Sci Adv 3(7):e1700782
Gorrasi G, Sorrentino A (2015) Mechanical properties and barrier performance of polymer/clay nanocomposites. J Polym Sci B Polym Phys 53(12):783–792
Habibi Y, Lucia LA, Rojas OJ (2010a) Cellulose nanocrystals: chemistry, self-assembly, and applications. Chem Rev 110(6):3479–3500
Habibi Y et al (2010b) Cellulose nanocrystals and their applications. Chem Rev 110(6):3479–3500
Habibi Y et al (2010c) Nanocomposites from cellulose. Chem Rev 110(6):3479–3500
Huang YY et al (2006) Multifunctional properties of metal oxide–polymer nanocomposites. J Nanomater 2006:1–7
Hussain F, Hojjati M, Okamoto M, Gorga RE (2006a) Review article: polymer-matrix nanocomposites, processing, manufacturing, and application: an overview. J Compos Mater 40(17):1511–1575
Hussain F et al (2006b) Polymer-matrix nanocomposites, processing, manufacturing, and application: an overview. J Compos Mater 40(17):1511–1575
Iijima S (1991) Helical microtubules of graphitic carbon. Nature 354(6348):56–58
Jancar J et al (2010) Polymer-matrix nanocomposites, processing, manufacturing, and application: An overview. Polym Eng Sci 50(11):2087–2106
Kaminska I, Das MR, Coffinier Y, Niedziolka-Jonsson J, Sobczak J, Woisel P, Szunerits S (2012) Reduction and functionalization of graphene oxide sheets using biomimetic dopamine derivatives in one step. ACS Appl Mater Interfaces 4(2):1016–1020
Keshavarz T, Roy I (2010) Polyhydroxyalkanoates: bioplastics with a green agenda. Curr Opin Microbiol 13(3):321–326
Kim H et al (2010) Graphene/polymer nanocomposites. Macromolecules 43(16):6515–6530
Koo JH (2006) Polymer nanocomposites: Processing, characterization, and applications. McGraw-Hill

Krishnamoorti R, Vaia RA (2007) Polymer nanocomposites. J Polym Sci B Polym Phys 45(24):3252–3256
Kroto HW et al (1985) C60: Buckminsterfullerene. Nature 318(6042):162–163
Lichtenhan JD et al (1993) Polyhedral oligomeric silsesquioxane (POSS) molecules as model silica surfaces. Mater Res Soc Symp Proc 274:33–38
Liu Z et al (2008) Elastic and viscoelastic properties of polymer nanocomposites. Adv Mater Res 47–50:391–394
Mech A, Rasmussen K, Jantunen P, Aicher L, Alessandrelli M, Bernauer U, Bleeker EA, Bouillard J, Di Prospero FP, Draisci R, Dusinska M (2019) Insights into possibilities for grouping and read-across for nanomaterials in EU chemicals legislation. Nanotoxicology 13(1):119–141
Mittal V (ed) (2009a) Polymer nanocomposites: Advances in filler surface modification techniques. Nova Science Publishers, Incorporated
Mittal V (2009b) Polymer layered silicate nanocomposites: a review. Materials 2(3):992–1057
Mittal V (2009c) Polymer nanotube nanocomposites: synthesis, properties, and applications. Macromol Mater Eng 294(1):87–95
Mittal V (2010) Solution blending for polymer nanocomposites: Opportunities and challenges. Macromol Mater Eng 295(5):391–408
Moniruzzaman M, Winey KI (2006) Polymer nanocomposites containing carbon nanotubes. Macromolecules 39(16):5194–5205
Moon RJ et al (2011) Cellulose nanomaterials review: structure, properties, and nanocomposites. Chem Soc Rev 40(7):3941–3994
Morent R, De Geyter N, Desmet T, Dubruel P, Leys C (2011) Plasma surface modification of biodegradable polymers: a review. Plasma Process Polymers 8(3):171–190
Morgan AB, Wilkie CA (2007) Flame Retardant Polymer Nanocomposites. Wiley
Morones JR et al (2005) The bactericidal effect of silver nanoparticles. Nanotechnology 16(10):2346
Novoselov KS et al (2004) Electric field effect in atomically thin carbon films. Science 306(5696):666–669
OECD. (2020). Nanotechnology: Safety and Regulatory Issues
Pascault JP, Williams RJ (2013) Thermosetting polymers. In: Handbook of polymer synthesis, characterization, and processing. Wiley, pp 519–533
Paul DR, Robeson LM (2008) Polymer nanotechnology: Nanocomposites. Polymer 49(15):3187–3204
Petrovic ZS, Ferguson J (1991) Polyurethane elastomers. Prog Polym Sci 16(5):695–836
Philip S et al (2007) Polyhydroxyalkanoates: biodegradable plastics with a range of applications. J Chem Technol Biotechnol 82(3):233–247
Poland CA et al (2008) Carbon nanotubes introduced into the abdominal cavity of mice show asbestos-like pathogenicity in a pilot study. Nat Nanotechnol 3(7):423–428
Potschke P et al (2004a) Dispersion of carbon nanotubes in a polymer matrix. Polymer 45(6):1990–2001
Potschke P et al (2004b) Melt mixing of multiwalled carbon nanotubes with polycarbonate. Polymer 45(6):1990–2001
Pumera M et al (2009) Nanomaterials meet microfluidics. Chem Soc Rev 38(6):1934–1945
Qiu J et al (2019) Multifunctional polymer nanocomposites with tailored properties. Prog Polym Sci 92:1–41
Rafiee MA et al (2009a) Graphene oxide and its polymer nanocomposites. Nat Mater 8(10):837–843
Rafiee MA et al (2009b) Solution blending for high-performance polymer/graphene nanocomposites. Mater Sci Eng A 504(1–2):47–52
Rafiee MA et al (2010) Fracture and fatigue in graphene nanocomposites. Small 6(2):179–183
Ragaert K et al (2017) Mechanical and chemical recycling of solid plastic waste. Waste Manag 69:24–58
Rai M et al (2009) Silver nanoparticles as antimicrobial agents. Appl Microbiol Biotechnol 85(4):1115–1122

Ramanathan T et al (2008) Functionalized graphene sheets for polymer nanocomposites. Nat Nanotechnol 3(6):327–331
Ray SS, Bousmina M (2005) Biodegradable polymers and their layered silicate nanocomposites. Macromol Rapid Commun 26(19):1639–1647
Ray SS, Okamoto M (2003a) Polymer/layered silicate nanocomposites: A review from preparation to processing. Prog Polym Sci 28(11):1539–1641
Ray SS, Okamoto M (2003b) Polymer/layered silicate nanocomposites: A review. Prog Polym Sci 28(11):1539–1641
Reddy RL et al (2013) Biodegradable composites: a review. Mater Sci Eng C 33(4):1817–1831
Rezwan K et al (2006) Biodegradable and bioactive porous polymer/inorganic composite scaffolds. Biomaterials 27(18):3413–3431
Rhim JW et al (2006) Preparation and characterization of silver-polymer nanocomposite films. J Appl Polym Sci 101(1):1–8
Sánchez C et al (2005) Hybrid organic–inorganic materials–a land of multidimensionality. Nat Mater 4(4):277–288
Schmidt D et al (2002) Thermoset polymer nanocomposites. Polymer Sci Eng [illegible]
Sinha Ray S et al (2002) Clay-based nanocomposites: properties and applications. Adv Polymer Sci 155:1–78
Spitalsky Z et al (2010) Carbon nanotube–polymer composites: chemistry, processing, mechanical and electrical properties. Prog Polym Sci 35(3):357–401
Stankovich S et al (2006) Graphene-based polymer nanocomposites. Nat Nanotechnol 1(2):133–137
Sullivan EM, Karimineghlani P, Naraghi M, Gerhardt RA, Kalaitzidou K (2016) The effect of nanofiller geometry and compounding method on polylactic acid nanocomposite films. Euro Polymer J 77:31–42
Thakur VK et al (2013) Green nanocomposites from cellulose and chitosan. Carbohydr Polym 91(2):645–650
Tjong SC (2006) Structural and mechanical properties of polymer nanocomposites. Mater Sci Eng R Rep 53(3–4):73–197
Winey KI, Vaia RA (2007) Polymer nanocomposites. MRS Bulletin 32(4):314–322
Yu A et al (2007a) Enhanced thermal conductivity in a hybrid graphite nanoplatelet–carbon nanotube filler for epoxy composites. Adv Mater 19(17):2181–2184
Yu A et al (2007b) Enhanced thermal properties in polymer nanocomposites. Adv Mater 19(17):2181–2184
Zeng QH, Yu AB, Lu GQ, Paul DR (2008a) Clay-based polymer nanocomposites: Research and commercial development. J Nanosci Nanotechnol 8(4):1598–1616
Zeng QH et al (2008b) Polymer nanocomposites: the dispersion state of fillers and mechanical properties. J Mater Chem 18(15):1766–1770
Zhan C et al (2011) Carbon nanomaterials for lithium-ion batteries. Chem Rev 111(7):3572–3614
Zhang Y et al (2011) Graphene-based composites. Chem Soc Rev 40(5):2226–2232

Chapter 3
Organic and Inorganic Nanofillers: Properties and Applications

Sangita Kale, Ashwin Kuchekar, Kalyani Autade, Rajashri Sumbe, Ujwala Desai, and Ashwini Gawade

Abstract The potential of both organic and inorganic nanofillers to improve the characteristics of composites in a range of industrial applications is drawing a lot of interest. Because of their special qualities, such as flexibility, low weight, and functionalization capabilities, organic nanofillers such as carbon nanotubes, graphene oxide, and organic polymers are perfect for use in sensors, electronics, and medicine. The strength, thermal stability, and conductivity of inorganic nanofillers, such as silica, metal nanoparticles, and clay minerals, on the other hand, make them appropriate for use in environmental, automotive, and construction engineering. The features, synthesis processes, and functionalization strategies of both organic and inorganic nanofillers are examined in this chapter, with an emphasis on how they can enhance the mechanical, thermal, and electrical characteristics of nanocomposites. Additionally, the study examines the several applications of nanofillers, which improve the functionality and effectiveness of products in a variety of industries, such as energy storage, biology, and materials research. This will emphasize how crucial these nanofillers are becoming to next-generation technologies.

Keywords Inorganic nanocomposite · Nanofillers · Organic · Polymer

---

S. Kale · U. Desai
PES Modern College of Pharmacy, Nigdi, MH, India

A. Kuchekar · A. Gawade (✉)
Department of Pharmaceutical Sciences, School of Health Sciences and Technology, MIT World Peace University, Pune, Maharashtra, India

K. Autade
RSM's N N Sattha College of Pharmacy, Ahmednagar, MH, India

R. Sumbe
Department of Pharmaceutical Chemistry, RSM's N. N. Sattha College of Pharmacy, Ahmednagar, Ahilyanagar, Maharashtra, India

S. N. Kothawade et al. (eds.), *Polymer Nanocomposites*,
https://doi.org/10.1007/978-3-032-00365-2_3

3.1 Introduction

3.1.1 *Definition of Nanofillers*

Nanofillers are materials with a size of nanometers that are integrated into coatings, drug delivery systems, pharmaceutical formulations, and biomedical devices to improve their mechanical, functional, and physical characteristics. These fillers, typically ranging from 1 to 100 nm in size, play a crucial role in improving drug stability, bioavailability, targeted delivery, and controlled release. They enhance pharmaceutical materials by modifying solubility, permeability, mechanical strength, and biocompatibility, leading to more efficient and effective therapeutic outcomes.

3.1.2 *Classification of Nanofillers in Pharmaceuticals*

Nanofillers in pharmaceuticals can be classified based on their dimensional characteristics and composition:

1. **Based on Dimensionality:**
   (a) *Zero-dimensional (0D)*: Nanoparticles (e.g., silica, titanium dioxide, gold nanoparticles, silver nanoparticles)
   (b) *One-dimensional (1D):* Nanotubes and nanorods (e.g., halloysite nanotubes, carbon nanotubes (CNTs), lipid-based nanorods)
   (c) *Two-dimensional (2D):* Nanoplates and nanosheets (e.g., graphene oxide, montmorillonite clay, layered silicates)
   (d) *Three-dimensional (3D):* Nanostructured networks (e.g., dendrimer-based nanogels, metal–organic frameworks for drug carriers)
2. **Based on Chemical Composition:**
   (a) *Carbon-based:* Graphene oxide, carbon nanotubes for drug loading and targeted therapy.
   (b) *Inorganic-based:* Silica, titanium dioxide, calcium phosphate for controlled drug release and stability.
   (c) *Polymer-based:* Dendrimers, nanogels, hyperbranched polymers for drug encapsulation and sustained release.
   (d) *Natural-based:* Chitosan, cellulose nanofibers, montmorillonite clay for biodegradable drug carriers and bioadhesive formulations.

3.1.2.1 Relevance of Nanofillers in Pharmaceutical Science

Nanofillers are crucial in pharmaceutical research and development, offering significant improvements in drug formulations and medical materials:

- *Enhanced* ***drug stability***: Nanofillers protect active pharmaceutical ingredients (APIs) from degradation caused by environmental factors like moisture, heat, and light.
- *Improved* ***bioavailability***: By increasing solubility and permeability, nanofillers enhance the absorption of poorly water-soluble drugs.
- *Controlled* ***drug release***: Nanostructured carriers allow for sustained, delayed, o or drug release at targeted site, enhancement of therapeutic efficacy and reducing side effects.
- *Biocompatibility and* ***safety***: Natural and polymer-based nanofillers improve the safety profile of drug formulations by reducing toxicity.
- *Mechanical* ***strength in biomedical devices***: They reinforce implants, prosthetics, and medical coatings for durability and enhanced performance.
- *Antimicrobial* ***properties***: Silver and zinc oxide nanofillers provide antimicrobial effects in pharmaceutical coatings and wound dressings.

3.1.3 Applications of Nanofillers in the Pharmaceutical Industry

3.1.3.1 Drug Delivery Systems

- *Targeted* ***drug delivery***: Nanocarriers such as graphene oxide and carbon nanotubes enable precise drug delivery to specific tissues, reducing side effects.
- *Sustained* ***release formulations***: Nanogels and polymeric nanofillers help control drug release rates for long-term therapeutic effects.
- *Oral* ***bioavailability enhancement***: Nanoclay-based drug carriers enhance the solubility and absorption of poorly soluble drugs.

3.1.3.2 Biomedical Devices and Implants

- *Dental* ***composites***: Nanofillers like silica and calcium phosphate improve strength, wear resistance, and antibacterial properties in dental materials.
- *Orthopedic* ***implants***: Carbon and ceramic nanofillers enhance bone integration and durability in joint and bone implants.
- *Wound* ***dressings***: Silver nanoparticles and chitosan nanofillers provide antimicrobial protection and promote healing in wound care products.

3.1.3.3 Pharmaceutical Coatings

- *Tablet and* ***capsule coatings***: Nanofillers improve film integrity, stability, and controlled dissolution rates for oral dosage forms.
- *Antimicrobial* ***coatings***: Silver and zinc oxide nanoparticles enhance the sterility of medical devices and hospital surfaces.

- *UV-**protective coatings***: Titanium dioxide nanofillers protect light-sensitive drugs from degradation.

3.1.3.4 Tissue Engineering and Regenerative medicine

- *Scaffolds for **tissue repair***: Biodegradable nanofillers like collagen and chitosan support cell growth in regenerative therapies.
- *Bone regeneration*: Hydroxyapatite and silica-based nanofillers promote bone healing and mineralization.

3.1.3.5 Vaccine and Biopharmaceutical Formulations

- *Adjuvants for vaccines*: Nanoparticles act as immune enhancers, improving vaccine efficacy.
- *Protein and **peptide stability***: Nanofillers help stabilize biologics and prevent protein aggregation.

3.1.3.6 Diagnostic and Imaging Applications

- *Contrast **agents in imaging***: Gold and iron oxide nanoparticles enhance contrast in MRI and CT scans for better disease diagnosis.
- *Biosensors for **disease detection***: Nanofiller-based biosensors improve sensitivity and accuracy in detecting biomarkers for diseases like cancer and infections.

3.1.4 Future Prospects of Nanofillers in Pharmaceuticals

The advancement of nanofillers continues to revolutionize the pharmaceutical industry by enabling precision medicine, enhancing drug formulations, and improving biomedical materials. Ongoing research focuses on biodegradable and biocompatible nanofillers to ensure safe and effective applications in drug delivery and healthcare. Their integration into pharmaceuticals holds immense potential for personalized therapies, regenerative medicine, and next-generation medical technologies.

3.2 Organic Nanofillers

Organic nanofillers are a class of nanomaterials that are derived from organic sources, specifically designed to modify and enhance the properties of composite materials. These fillers, typically ranging in size from 1 to 100 nm, interact with the polymer matrices or other substrates at a molecular level to improve mechanical, thermal, electrical, and barrier characteristics. Their small size grants them an unusually large surface area relative to their volume, which is crucial for improving the overall performance of the material. This high surface-area-to-volume ratio contributes to an increased number of interfacial interactions between the filler and the matrix, which can lead to significant improvements in the composite's functionality.

The incorporation of organic nanofillers into a polymer matrix or other composite structures results in a synergistic effect, wherein the properties of the composite material are significantly enhanced due to the interactions between the filler and the matrix. The interaction between organic nanofillers and polymers is particularly important because it can lead to a change in the material's microstructure, thus improving its overall stability and performance. For example, the addition of organic nanofillers such as carbon-based nanomaterials like carbon nanotubes or graphene can improve the mechanical strength of a polymer, making it stronger and more durable. Additionally, the increased surface area provided by these nanofillers facilitates the formation of stronger interfacial bonds, leading to improved dispersion and compatibility within the composite material.

Furthermore, organic nanofillers offer a unique advantage when it comes to enhancing thermal stability. By incorporating fillers such as carbon nanotubes or polymeric nanoparticles, composites can exhibit improved heat resistance, making them suitable for applications that require the material to withstand high temperatures. This makes organic nanofillers particularly useful in the automotive, aerospace, and electronics industries, where materials are subjected to extreme conditions. In addition to thermal properties, the electrical conductivity of composites can also be significantly enhanced. For example, carbon-based organic nanofillers are known for their exceptional electrical conductivity, and their inclusion into a polymer matrix can turn an insulating polymer into a conductive material. This opens up possibilities for organic nanofillers in the development of electronic devices, sensors, and energy storage applications.

In the context of biomedicine, organic nanofillers are increasingly being used in drug delivery systems, tissue engineering, and diagnostic applications. The ability of organic nanofillers to modify the physical and chemical properties of composites makes them highly valuable in the creation of materials with controlled release profiles, enhanced biocompatibility, and improved tissue integration. For instance, organic nanofillers like polymeric nanoparticles are often employed to encapsulate drugs or bioactive molecules, ensuring their sustained release over a prolonged period. Moreover, the surface characteristics of organic nanofillers can be modified to improve their interaction with biological systems, such as targeting specific cells or tissues, which is a critical aspect of targeted drug delivery.

The development of organic nanofillers has also led to the creation of materials that can effectively address environmental challenges. Organic nanofillers can be used to produce composite materials with enhanced barrier properties, making them more resistant to moisture, gases, and other external factors. This is particularly beneficial in applications such as food packaging, where preserving the freshness and extending the shelf-life of products is of paramount importance. In addition to food packaging, these advanced materials can be utilized in various other sectors, including agriculture, construction, and environmental protection, where durable and sustainable materials are in demand.

3.2.1 Characteristics of Organic Nanofillers

Organic nanofillers are distinguished by their remarkable characteristics, primarily driven by their small size and the associated increase in surface area-to-volume ratio. The nanoscale dimensions, typically between 1 and 100 nm, endow these materials with unique physicochemical properties that are absent in their bulk counterparts. These properties enable organic nanofillers to significantly enhance the performance of composite materials in terms of mechanical, thermal, and electrical behavior. At this scale, the behavior of these nanomaterials is dominated by surface effects, which govern their interactions with other substances, including polymer matrices, metals, ceramics, and other nanomaterials.

A primary characteristic of organic nanofillers is their high surface energy, which allows for the formation of strong interfacial bonds with the surrounding polymer matrix. The increased surface area and reactivity of organic nanofillers are key factors in ensuring effective bonding between the filler and the matrix, resulting in composites with improved physical properties. These strong interfacial bonds help distribute the applied forces more effectively across the material, leading to enhanced mechanical strength and overall structural integrity. The interfacial properties of organic nanofillers are essential for promoting good dispersion within the composite material, preventing agglomeration, and ensuring uniformity in the final product.

The molecular structure of organic nanofillers also plays a significant role in determining their characteristics. The interactions between the nanofillers and the host material are influenced by factors such as the functional groups on the surface of the filler particles, the polarity of the filler, and the compatibility between the nanofiller and the polymer matrix. These interactions are essential for tailoring the properties of the composite material to meet the specific requirements of various applications. For example, by modifying the surface of organic nanofillers with certain chemical groups or functionalizing the nanofiller surface, their compatibility with different polymer matrices can be improved, thus enhancing the overall mechanical and thermal properties of the composite.

In terms of mechanical strength, organic nanofillers such as carbon-based nanomaterials (e.g., carbon nanotubes, graphene) have been shown to significantly enhance the tensile strength, elasticity, and fracture toughness of polymer composites. These organic nanofillers can provide a reinforcing effect due to their high intrinsic strength and stiffness, which is translated to the composite material. For example, carbon nanotubes, due to their exceptional mechanical properties, can impart outstanding strength and toughness to polymers, resulting in composites that are both strong and lightweight, ideal for applications in aerospace, automotive, and electronics industries. Furthermore, the flexibility of certain organic nanofillers, such as some polymeric nanofillers, can contribute to the overall flexibility of the composite material, which is beneficial for applications requiring materials that can withstand deformation without failure.

The thermal stability of organic nanofillers is another important characteristic. The presence of nanofillers in a polymer matrix can improve the thermal resistance of the material, making it more suitable for high-temperature applications. For example, organic nanofillers like carbon nanotubes and graphene can improve the thermal conductivity of the composite, helping to dissipate heat more efficiently. Additionally, these fillers can prevent the polymer from degrading at elevated temperatures by forming a protective network within the composite. This enhanced thermal stability is crucial in many industrial applications where materials are exposed to varying temperatures, such as in electronics, automotive parts, and high-performance coatings.

One of the unique characteristics of organic nanofillers is their ability to modify the optical properties of composites. Organic nanofillers, particularly those based on carbon and polymers, can significantly influence the optical behavior of the material, including light absorption, transmission, and reflection. This property is particularly useful in applications such as optical sensors, coatings, and solar cells. For instance, graphene-based nanofillers can enhance the optical transparency of composites while maintaining their strength, making them suitable for use in transparent conductive films and displays.

In addition to these properties, organic nanofillers also offer advantages in terms of their versatility and functionalization. Their surface chemistry allows for easy modification, enabling researchers to design nanofillers with specific functional groups that can enhance properties like biocompatibility, conductivity, or hydrophobicity. This makes organic nanofillers highly adaptable for a wide range of applications, from electronics to biomedical devices. For instance, by functionalizing the surface of organic nanofillers, they can be tailored for use in targeted drug delivery systems, where their surface properties can help with the controlled release of therapeutic agents.

Overall, the characteristics of organic nanofillers, including their high surface energy, molecular structure, mechanical strength, flexibility, thermal stability, and optical properties, make them ideal candidates for enhancing the properties of composite materials. These unique behaviors enable organic nanofillers to improve the performance of materials in various fields, including nanocomposites, electronics, energy storage, and biomedicine. By carefully selecting and functionalizing organic nanofillers, materials scientists and engineers can tailor composites to meet the specific needs of a wide range of industrial applications.

3.2.2 Types of Organic Nanofillers

Organic nanofillers can be broadly categorized into two primary types: carbon-based nanofillers and polymeric nanofillers. Both types of nanofillers possess unique properties that make them highly valuable in modifying and enhancing the performance of composite materials. While polymeric nanofillers are often used for flexibility and biocompatibility, carbon-based nanofillers are particularly renowned for their exceptional mechanical, electrical, and thermal properties. These distinctive characteristics make carbon-based nanofillers ideal for high-performance applications in fields ranging from electronics to aerospace. Below, we explore carbon-based organic nanofillers in detail, focusing on their properties, types, and applications.

3.2.3 Carbon-Based Nanofillers

Carbon-based organic nanofillers are among the most extensively studied and utilized types of nanomaterials due to their extraordinary mechanical, electrical, and thermal properties. These nanofillers are primarily composed of carbon atoms arranged in various nanostructures that impart unique performance characteristics to the composites they are incorporated into. Among the most prominent carbon-based nanofillers are carbon nanotubes (CNTs), graphene, and carbon nanofibers (CNFs). Each of these nanofillers has distinct attributes that make them suitable for different applications.

3.2.4 Carbon Nanotubes (CNTs)

Carbon nanotubes are one of the most widely researched and commercially utilized carbon-based nanofillers due to their mechanical strength, electrical conductivity, and high surface area with diameters typically ranging from 1 to 100 nm. These tubes can vary in length from micrometers to several millimeters, depending on their synthesis method. The remarkable strength of CNTs is largely attributed to the strong covalent bonds between the carbon atoms, which provide them with tensile strengths far superior to conventional materials like steel. As a result, CNTs can significantly improve the mechanical properties of composite materials by providing reinforcement without significantly increasing the weight of the material.

CNTs also exhibit excellent electrical conductivity, which makes them ideal for use in composite materials intended for electronic and conductive applications. Their high electrical conductivity is particularly advantageous in the development of lightweight, flexible, and high-performance conductors. Additionally, CNTs possess superior thermal conductivity, making them useful in applications that require efficient heat dissipation, such as in electronics and high-performance coatings. When incorporated into polymers, CNTs create conductive pathways that enhance the overall electrical properties of the material, enabling the development of composites for advanced electronic devices, sensors, and energy storage applications.

3.2.5 Graphene

Graphene is another prominent carbon-based nanomaterial that has garnered significant attention due to its outstanding properties. Graphene's mechanical properties are extraordinary; it is around 200 times stronger than steel by weight, while remaining extremely flexible and lightweight.

The electrical conductivity of graphene is one of its defining features. It allows electrons to flow freely across its surface, making it an excellent material for enhancing the electrical properties of composite materials. This high conductivity makes graphene an ideal filler for applications in electronics, energy storage, and supercapacitors. Graphene-based composites have been shown to exhibit significant improvements in the performance of batteries and supercapacitors, including higher energy densities, faster charging times, and longer cycle lives. In addition to its electrical properties, graphene's thermal conductivity is superior to many other materials, which makes it valuable for applications that require high thermal stability and efficient heat management, such as in thermal interface materials or heat sinks in electronic devices.

Furthermore, graphene is also beneficial for enhancing the barrier properties of composites. Its impermeability to gases and liquids, coupled with its ability to form strong interfacial bonds with polymers, allows graphene-based composites to be used in packaging materials, where they can provide improved moisture and oxygen barrier properties. This has significant implications in industries such as food packaging and medical device manufacturing.

3.2.6 Carbon Nanofibers (CNFs)

Carbon nanofibers are another class of carbon-based nanofillers that share some similarities with CNTs but differ in their structure and properties. CNFs are typically composed of hollow or solid cylindrical filaments with diameters ranging from a few nanometers to several micrometers, and they have a higher aspect ratio than other carbon nanofillers. CNFs exhibit excellent mechanical strength, chemical stability, and electrical conductivity, though their properties are generally not as extreme as those of CNTs and graphene. However, CNFs can still offer substantial improvements in the strength, stiffness, and electrical properties of polymer composites.

One of the key advantages of CNFs is their relatively simple and cost-effective production process compared to CNTs and graphene. This makes CNFs a more accessible option for large-scale manufacturing of composite materials. CNFs have been used in a variety of applications, including conductive polymers, structural reinforcement, and thermal management. In particular, CNFs are beneficial in automotive and aerospace industries, where their lightweight and high-strength properties are valuable for creating durable and energy-efficient materials. Additionally, CNFs are being explored in the development of energy storage devices such as batteries and supercapacitors, where their high surface area and conductivity can significantly enhance performance.

3.2.7 Applications of Carbon-Based Nanofillers

The exceptional mechanical, electrical, and thermal properties of carbon-based organic nanofillers make them highly suitable for a wide range of applications. In the field of electronics, carbon nanotubes, graphene, and carbon nanofibers are incorporated into polymer matrices to produce conductive composites for flexible electronics, sensors, and wearable devices. The ability of carbon nanofillers to withstand high temperatures and mechanical stresses also makes them ideal for high-performance applications, such as in turbine engines, heat shields, and structural components.

Lastly, carbon-based nanofillers are increasingly being explored in biomedical applications, including drug delivery systems, tissue engineering, and diagnostic devices. The biocompatibility and functionalization potential of carbon nanotubes, graphene, and carbon nanofibers enable the development of advanced biomaterials with controlled release capabilities, targeted therapies, and improved interactions with biological systems.

3.2.8 Polymeric Nanofillers

Polymeric nanofillers are a class of organic nanofillers made from polymeric materials, such as polystyrene, polyolefins, and polyesters, that are engineered to achieve nanoscale dimensions. These fillers are primarily utilized to enhance various properties of composite materials, including mechanical, optical, and barrier characteristics. Polymeric nanofillers are highly valued for their ability to improve the overall performance of polymer composites, while maintaining ease of processing and compatibility with the matrix materials. Their unique ability to modify the properties of the matrix makes them crucial in a variety of applications, ranging from packaging and biomedical to automotive industries.

3.2.9 Synthesis and Structure of Polymeric Nanofillers

Polymeric nanofillers are typically synthesized by techniques such as solution polymerization, emulsion polymerization, or by employing mechanical processes like extrusion and grinding. In these methods, the polymeric material is broken down to nanoscale dimensions, often ranging from 1 to 100 nm. The process ensures that the polymer chains are finely dispersed within the composite material, allowing them to interact effectively with the polymer matrix. These nanofillers may take on various shapes and forms, including nanoparticles, nanospheres, nanorods, and nanofibers, each with unique surface characteristics that influence their interaction with the host material.

One of the key benefits of polymeric nanofillers is their ability to form strong interactions with polymer matrices, leading to improved compatibility and dispersion within the composite material. This is primarily due to the similar chemical nature of the polymeric nanofillers and the matrices, which allows for a more uniform distribution of the filler within the polymer. The improved dispersion of the nanofillers ensures that their properties are efficiently transferred to the composite, contributing to significant enhancements in the overall material performance.

3.2.10 Properties of Polymeric Nanofillers

Polymeric nanofillers are recognized for their unique set of properties that stem from their nanoscale dimensions and molecular structure. These properties can vary depending on the specific type of polymer used and the processing method, but common characteristics include:

1. *Mechanical properties*: Polymeric nanofillers significantly enhance the mechanical properties of the composite, such as tensile strength, stiffness, and impact resistance. By dispersing uniformly within the matrix, these nanofillers reinforce the polymer network, improving its overall structural integrity without compromising flexibility or elongation. This makes polymeric nanofillers ideal for applications where both strength and flexibility are required.
2. *Optical properties*: Some polymeric nanofillers, such as those made from polyesters or other optically active polymers, can enhance the optical properties of composites. They can improve the transparency, gloss, and color stability of the material, making them suitable for applications in coatings, films, and packaging materials. The ability to tune the optical properties through the incorporation of polymeric nanofillers is especially valuable in industries like packaging, where visual appeal and product protection are important.
3. *Barrier properties*: Polymeric nanofillers can enhance the barrier properties of composites, especially in terms of moisture and gas permeability. These improve-

ments are crucial in applications such as food packaging, where protection against moisture, oxygen, and other environmental factors is essential for preserving product quality. The nanofillers reduce the rate at which gases or liquids permeate through the material, increasing the shelf life of packaged goods.
4. *Moisture resistance and hydrophobicity*: Polymeric nanofillers are known to improve the moisture resistance and hydrophobicity of composite materials. The incorporation of hydrophobic polymeric fillers, such as certain polyolefins, can enhance the composite's ability to repel water and resist environmental degradation caused by moisture absorption. This is especially beneficial in applications exposed to harsh environmental conditions, such as automotive parts and outdoor equipment.

3.2.11 Advantages of Polymeric Nanofillers

Polymeric nanofillers offer several advantages over other types of nanofillers, particularly in terms of ease of processing, cost-effectiveness, and compatibility with polymer matrices. These advantages include:

1. *Ease of processing*: Compared to carbon-based nanofillers, polymeric nanofillers are typically easier to process and handle. This is primarily because they are derived from organic polymers, which are generally more compatible with common polymer processing techniques such as extrusion, injection molding, and casting. Their nanoscale dimensions also allow for simpler dispersion into the matrix, resulting in homogeneous composites.
2. *High compatibility*: One of the most notable advantages of polymeric nanofillers is their high degree of compatibility with polymer matrices. Since both the nanofillers and the matrix are typically organic materials, there is a natural affinity between the two, leading to better interfacial interactions and improved dispersion within the composite. This enhances the performance of the material, as the filler's properties are more effectively transferred to the composite.
3. *Cost-effectiveness*: Polymeric nanofillers are often less expensive than carbon-based alternatives, such as CNTs or graphene. Their lower cost makes them an attractive option for large-scale manufacturing processes, particularly in industries like packaging, automotive, and consumer goods, where cost efficiency is essential. Additionally, the ease of processing polymeric nanofillers often results in lower manufacturing costs compared to the more complex processes required for carbon-based nanofillers.
4. *Improved toughness and flexibility*: Polymeric nanofillers can improve the toughness and flexibility of composites, which is crucial for applications that require materials that can withstand impact or deformation. Unlike rigid fillers like carbon nanotubes, polymeric nanofillers can enhance the ability of the composite to absorb and dissipate energy, making them ideal for applications in automotive

parts, protective coatings, and biomedical devices where flexibility and toughness are required.

3.2.12 Applications of Polymeric Nanofillers

The unique properties of polymeric nanofillers make them suitable for a wide range of applications across various industries. Some of the key areas of application include:

1. *Packaging*: Polymeric nanofillers are widely used in packaging materials, particularly in food and beverage packaging, where they enhance the barrier properties, moisture resistance, and mechanical strength of the material. These nanofillers are also employed in creating biodegradable or eco-friendly packaging solutions that meet the growing demand for sustainable materials in the packaging industry.
2. *Biomedical applications*: In the biomedical field, polymeric nanofillers are incorporated into drug delivery systems, wound dressings, and tissue engineering scaffolds. Their biocompatibility and ability to be functionalized for specific applications make them suitable for controlled release formulations and implantable devices. Polymeric nanofillers can also improve the mechanical properties of biopolymers used in medical devices, ensuring durability while maintaining flexibility.
3. *Electronics*: In the electronics industry, polymeric nanofillers are used to enhance the properties of flexible electronics, sensors, and capacitors. Their ability to improve the mechanical and electrical properties of composites is beneficial for producing lightweight, flexible, and durable electronic devices. The optical properties of polymeric nanofillers are also utilized in the development of displays, lighting systems, and other electronic components.

3.2.13 Properties of Organic Nanofillers

The unique properties of organic nanofillers arise primarily from their nanoscale dimensions, which impart them with exceptional surface area-to-volume ratios. These properties, including mechanical, thermal, chemical, and electrical characteristics, significantly enhance the performance of composite materials when organic nanofillers are incorporated. The small size of these fillers allows them to interact at the molecular level with the matrix material, leading to improvements that are not possible with bulk materials. This makes organic nanofillers particularly valuable in the development of high-performance composite materials across a wide range of applications, from electronics and sensors to automotive and biomedical fields.

3.2.14 Size-Dependent Effects

One of the most significant factors that contributes to the unique properties of organic nanofillers is their size-dependent effects. At the nanoscale, materials behave differently compared to their bulk counterparts due to the increased surface area and the corresponding increase in surface energy. This results in several important phenomena, including enhanced strength, flexibility, and chemical reactivity. The size reduction of organic nanofillers to the nanoscale allows for a high surface area to volume ratio, which increases the number of active sites available for interaction with the matrix material.

3.2.15 Surface Chemistry and Functionalization

The surface chemistry of organic nanofillers plays a crucial role in determining their interactions with the polymer matrix. The functional groups present on the surface of the nanofillers can significantly influence their dispersion, compatibility, and overall performance in the composite material. Surface functionalization, which involves attaching specific chemical groups or molecules to the surface of the nanofillers, can tailor the interactions between the filler and the matrix. This allows for the optimization of the filler-matrix interface, which is critical for achieving superior material properties.

For example, surface functionalization can be used to improve the compatibility of organic nanofillers with polar or non-polar polymer matrices. In the case of carbon nanotubes or graphene, surface modification with functional groups such as carboxyl, amine, or hydroxyl groups can enhance their dispersion within the matrix and improve interfacial bonding. The functionalized surface groups can form stronger chemical bonds with the polymer chains, leading to improved mechanical, thermal, and electrical properties. This makes surface functionalization an essential tool for tailoring the properties of composite materials and ensuring the efficient use of organic nanofillers.

3.2.16 Mechanical Properties Enhancement

Organic nanofillers are known for their remarkable ability to enhance the mechanical properties of composite materials. The incorporation of materials such as carbon nanotubes, graphene, and other organic nanofillers into polymers has been shown to significantly improve properties like tensile strength, stiffness, and toughness. The presence of nanofillers within the polymer matrix helps to reinforce the material at the nanoscale, leading to improved structural integrity.

For example, carbon nanotubes, due to their high aspect ratio and strength, have been widely studied for their ability to reinforce polymer composites. When added to a polymer matrix, carbon nanotubes provide additional reinforcement, enhancing the composite's resistance to mechanical stress. Similarly, graphene, with its two-dimensional lattice structure, imparts extraordinary strength and flexibility to the composite material. These improvements are particularly beneficial in applications where high strength-to-weight ratios are required, such as in aerospace and automotive industries.

The mechanical properties of organic nanofillers also extend to enhancing the toughness of the composite material. The inclusion of nanofillers can help prevent the formation and propagation of cracks within the matrix, thereby increasing the material's resistance to fracture. This improvement in toughness, combined with enhanced strength and stiffness, makes organic nanofillers ideal for a wide range of structural applications.

3.2.17 Thermal Properties

Organic nanofillers can also significantly enhance the thermal properties of composite materials. The high surface area and unique molecular structure of organic nanofillers allow for better heat transfer and distribution within the composite material. For instance, the inclusion of carbon nanotubes or graphene in a polymer matrix can improve the material's thermal conductivity, enabling it to dissipate heat more effectively. This is particularly useful in applications where heat management is critical, such as in electronics, batteries, and energy storage systems.

The incorporation of organic nanofillers can also improve the material's thermal stability, which is the ability of the composite to maintain its properties under elevated temperatures. Nanofillers, especially carbon-based ones, can help increase the thermal degradation temperature of polymers, thereby improving the material's resistance to high-temperature environments. This is particularly advantageous in applications where the material is subjected to extreme temperatures, such as in automotive components, aerospace materials, and industrial machinery.

3.2.18 Electrical Properties

The electrical properties of organic nanofillers are one of the most attractive features of these materials. Carbon-based nanofillers, in particular, exhibit exceptional electrical conductivity, which can be leveraged to enhance the electrical properties of polymer composites. For example, carbon nanotubes and graphene are excellent conductors of electricity, and their incorporation into a polymer matrix can result in composites that exhibit significantly improved electrical conductivity. These composites are ideal for use in electronic devices, sensors, and conductive coatings.

3.2.19 Chemical Properties and Reactivity

The chemical properties of organic nanofillers are closely tied to their surface chemistry and functionalization. The increased surface area and reactivity of these nanofillers allow them to interact more effectively with other chemicals, improving the overall chemical resistance of the composite material. For instance, the incorporation of organic nanofillers into polymer matrices can enhance the material's resistance to chemical degradation, such as corrosion, oxidation, or hydrolysis. This makes composites with organic nanofillers suitable for use in harsh chemical environments, including industrial applications where exposure to corrosive substances is common.

The chemical reactivity of organic nanofillers also opens up opportunities for developing smart materials with responsive properties. For example, nanofillers can be functionalized with specific groups that react to environmental stimuli such as pH, temperature, or light, enabling the creation of materials that can change their properties in response to these factors. This makes organic nanofillers ideal for applications in areas such as sensors, drug delivery systems, and environmental monitoring.

3.2.20 Overall Performance Improvement

In summary, the properties of organic nanofillers make them an invaluable tool for enhancing the performance of composite materials. Their nanoscale size, high surface area, and unique chemical and physical characteristics allow them to interact at the molecular level with the polymer matrix, leading to significant improvements in mechanical, thermal, electrical, and chemical properties. The ability to tailor the surface chemistry of organic nanofillers further expands their potential applications, enabling the development of advanced composite materials with superior performance in a wide range of industries.

3.2.21 Improvement in Strength and Mechanical Properties

The integration of organic nanofillers into composite materials leads to profound improvements in their mechanical properties, particularly in terms of strength, stiffness, flexibility, and impact resistance. Organic nanofillers, such as carbon nanotubes (CNTs) and graphene, are well-known for their extraordinary mechanical properties, and their inclusion in polymeric matrices significantly enhances the overall performance of the composite material. The fundamental mechanisms responsible for this enhancement stem from the unique characteristics of nanofillers and their ability to interact at the molecular level with the polymer matrix.

3.2.22 Enhancement in Tensile Strength and Stiffness

The inclusion of organic nanofillers like carbon nanotubes or graphene significantly enhances the tensile strength and stiffness of composite materials. These nanofillers can improve the ability of the composite to withstand tensile forces without elongating or breaking. In particular, the high aspect ratio of carbon nanotubes, which can span large distances within the polymer matrix, makes them ideal for reinforcing composites. As a result, the overall tensile strength of the composite is increased, allowing it to bear greater loads before failure.

The stiffness of a composite is also directly influenced by the addition of nanofillers. Nanofillers with high stiffness, such as graphene, can increase the rigidity of the composite material, reducing its tendency to deform under stress. This improvement in stiffness is crucial in applications that require high structural integrity, such as in aerospace, automotive, and construction materials. The enhancement of both tensile strength and stiffness by organic nanofillers makes composites more reliable and effective in withstanding mechanical forces over extended periods.

3.2.23 Impact Resistance and Toughness

In addition to improving strength and stiffness, organic nanofillers can also enhance the impact resistance and toughness of composite materials. The introduction of nanofillers helps to prevent the propagation of cracks and fractures within the material, improving its ability to absorb energy and resist impact without breaking. This is particularly important in applications where composites are exposed to sudden mechanical forces, such as in the automotive and aerospace industries, where materials must withstand collisions or other high-impact events.

Nanofillers such as carbon nanotubes and graphene play a role in toughening the composite material by reinforcing the polymer matrix at the molecular level. These nanofillers can help to distribute the applied stress more evenly across the material, preventing localized damage that could lead to catastrophic failure. The high tensile strength of carbon nanotubes, coupled with their ability to form strong bonds with the polymer matrix, helps to prevent crack formation and propagation, improving the overall toughness of the composite.

3.2.24 Flexibility and Ductility

Although the incorporation of nanofillers is often associated with increased stiffness and strength, the flexibility and ductility of composite materials can also be improved, provided that the nanofillers are carefully selected and dispersed within the matrix. Organic nanofillers like polymeric nanofillers and functionalized carbon nanotubes can enhance the flexibility of the composite material without compromising its strength. For instance, the inclusion of polymeric nanofillers such as polystyrene or polyolefins into a composite matrix can improve the material's ability to bend and stretch without breaking, thus enhancing its overall mechanical performance in applications that require flexibility, such as in packaging, textiles, and flexible electronics.

3.2.25 Synergistic Effect with Polymer Matrix

The mechanical properties of organic nanofillers are not only dependent on their intrinsic properties but also on their interaction with the polymer matrix. The combination of nanofillers and polymer matrices results in a synergistic effect, where the properties of the composite material are greater than the sum of the individual components. The unique interfacial interactions between the nanofillers and the polymer chains lead to the formation of a stronger and more resilient composite.

3.2.26 Applications of Organic Nanofillers

Organic nanofillers, due to their remarkable properties, find wide-ranging applications across various industries. Their ability to enhance the properties of composite materials makes them invaluable in numerous sectors. The most prominent applications of organic nanofillers include industries such as automotive, electronics, biomedical, and construction, as well as energy-efficient materials.

3.2.27 Automotive Industry

In the automotive industry, organic nanofillers are used to produce lightweight, durable composite materials. The reduction in material weight is crucial for improving fuel efficiency and lowering carbon emissions. For instance, the incorporation of organic nanofillers like carbon nanotubes (CNTs) or graphene into polymeric composites results in stronger yet lighter materials. These materials not only reduce the overall weight of vehicles but also enhance their strength and safety. The high mechanical strength and toughness of the composites make them ideal for critical automotive components such as body panels, bumpers, and interior parts, which must withstand impacts and provide structural integrity.

3.2.28 Electronics Industry

Organic nanofillers are increasingly used in the electronics industry to fabricate high-performance materials. In particular, they are utilized to create conductive polymers, flexible electronics, and sensors. The incorporation of nanofillers like graphene or carbon nanotubes significantly improves the electrical conductivity of composite materials, which is essential for applications in sensors, transistors, and various electronic devices. Furthermore, the flexibility of organic nanofillers makes them suitable for the production of flexible electronic devices, such as wearable electronics, flexible displays, and printed circuit boards. The combination of high conductivity and flexibility is a game-changer in the development of new-generation electronics.

3.2.29 Biomedical Applications

Organic nanofillers have proven to be extremely beneficial in the biomedical field, where their biocompatibility and functionalization capabilities make them ideal for use in drug delivery systems, tissue engineering, and medical devices. For example, carbon-based nanofillers like carbon nanotubes and graphene are used to develop drug carriers that improve the bioavailability and targeted delivery of therapeutic agents. These nanofillers can be functionalized with specific molecules to enhance their interaction with biological systems, making them suitable for applications in personalized medicine and regenerative medicine. Additionally, the mechanical properties of organic nanofillers make them useful in tissue engineering, where they can be incorporated into scaffolds that mimic the natural extracellular matrix, facilitating tissue regeneration.

3.2.30 Challenges in Processing Organic Nanofillers

While organic nanofillers offer significant improvements in material properties, their processing and incorporation into composite matrices pose several challenges. These challenges need to be addressed to fully harness the potential of organic nanofillers in various applications.

3.2.31 Poor Dispersion and Agglomeration

One of the major challenges in incorporating organic nanofillers into composites is achieving uniform dispersion. Nanofillers, due to their small size and high surface energy, tend to agglomerate, which can hinder their distribution within the matrix. Poor dispersion leads to the formation of clusters, reducing the effectiveness of the nanofillers in enhancing the properties of the composite. Agglomeration can also create weak spots within the material, leading to reduced mechanical performance.

To overcome this issue, various processing techniques are employed to ensure optimal dispersion of the nanofillers. These techniques include melt blending, solution mixing, and electrospinning. By using these methods, the nanofillers can be evenly distributed throughout the matrix, maximizing their reinforcing effects. Achieving good dispersion is essential to ensure that the desired properties of the composite, such as strength, flexibility, and electrical conductivity, are realized.

3.2.32 Compatibility with Polymer Matrix

Another significant challenge in processing organic nanofillers is ensuring their compatibility with the polymer matrix. The polymer matrix must provide adequate interaction with the nanofillers to achieve proper bonding and enhance the overall properties of the composite. Incompatibility between the nanofillers and the polymer matrix can lead to poor dispersion, weak interfacial bonding, and reduced performance.

The compatibility of organic nanofillers with the polymer matrix can be influenced by factors such as the chemical structure of the nanofillers and the matrix, as well as the physical properties of the materials. Surface modification or functionalization is often employed to improve the compatibility between the nanofillers and the matrix. This helps in achieving a better interaction between the nanofillers and the polymer chains, leading to improved material properties.

3.2.33 Surface Functionalization of Organic Nanofillers

Surface functionalization is a critical technique used to enhance the performance of organic nanofillers in composite materials. By modifying the surface of the nanofillers with specific functional groups, their interaction with the polymer matrix can be improved, leading to better dispersion, bonding, and overall material properties.

3.2.34 Tailored Interactions with the Matrix

The primary objective of surface functionalization is to improve the interaction between the nanofillers and the polymer matrix. By introducing functional groups such as amines, carboxylates, or hydroxyls to the surface of the nanofillers, their chemical reactivity can be increased, enhancing their compatibility with the matrix. For instance, amine-functionalized nanofillers can form stronger hydrogen bonds or covalent bonds with the polymer matrix, leading to improved dispersion and mechanical properties.

Surface functionalization also allows for the customization of the nanofillers for specific applications. For example, the introduction of hydrophobic groups to the surface of organic nanofillers can make them more suitable for use in applications that require water resistance, such as waterproof coatings and packaging materials. Functionalization also enables the incorporation of bioactive molecules onto the surface of the nanofillers, which is particularly useful in biomedical applications like drug delivery and tissue engineering.

3.2.35 Enhanced Dispersion and Bonding

Surface functionalization plays a vital role in improving the dispersion of organic nanofillers within the polymer matrix. By modifying the surface of the nanofillers, it is possible to reduce the tendency for agglomeration, leading to better dispersion and uniform distribution within the matrix. Functionalized nanofillers are more likely to interact effectively with the polymer chains, which helps to form stronger and more stable bonds between the filler and the matrix. This enhances the overall mechanical, thermal, and electrical properties of the composite material.

3.2.36 Inorganic Nanofillers

A particulate inorganic material with a distinctive size in the range of nanometers is called an inorganic nanofiller. Its dimensional property offers several advantages over traditional fillers, including: (a) a very wide interfacial area with a polymer matrix, (c) a decrease in optical wavelength scattering efficiency, and (b) size-dependent physical characteristics. The nanofillers are inorganic substances with superior mechanical and thermal characteristics. At very low compositions (less than 10 weight percent), they are employed to strengthen the polymer matrix, improving the polymer's characteristics without changing its density.

Many nanofillers, including silica, carbon nanotubes, clay, graphene, fullerene, and boron nitride, are currently accessible and can be utilized to strengthen polymers. In order to create composite materials, “filler” is a type of substance, usually particulate, that is placed in a polymeric matrix. In order to enhance or add different qualities, such as dielectric, magnetic, mechanical, thermal, optical, electric, etc., fillers are typically added to composite materials. The chemical makeup and quantity of the fillers are only two of the variables that affect the composites’ characteristics.

3.2.37 General Features

3.2.37.1 Size and Shape

Nanofillers usually have a size range of 1–100 nm. The dispersion inside the matrix and the interfacial interactions can be strongly impacted by the particular size. The nanocomposite’s mechanical, thermal, and electrical characteristics are influenced by its size. [9] Spherical, tubular, plate-like, or fibrous structures are examples of common shapes (Nayak 2023).

3.2.37.2 Large Surface Area

Better composite properties are the outcome of their huge surface area, which improves interaction with the matrix, for example, in polymer/graphene nanocomposites, a higher elastic modulus, which indicates a stiffer material, is correlated with an increase in the effective specific surface area of the nanofiller.

3.2.37.3 Surface Chemistry

To enhance dispersion and compatibility with various matrices, it can be functionalized or altered. In order to improve compatibility with the polymer matrix, surface functionalization entails altering the surface of nanofillers. This procedure can enhance interfacial bonding and dispersion, giving the nanocomposite better mechanical and thermal properties (Zhang et al. 2020).

3.2.37.4 Mechanical Properties

Reinforcement is aided by high stiffness, strength, and toughness. Significant increases in stiffness and tensile strength can result from the use of nanofillers (Moskalyuk and Yudin 2020).

3.2.37.5 Thermal Properties

They frequently improve resistance to deterioration and thermal stability. By strengthening their resistance to heat degradation, the thermal stability of polymer composites can be enhanced using nanofillers. The result depends on the filler's kind and concentration, as well as the interfacial bonding between the filler and matrix.

3.2.37.6 Electrical and Optical Properties

A lot of nanofillers have special optical or electrical conductivity, such UV absorption or fluorescence. Dielectric characteristics: polymers' capacitance and capacity to store energy can be impacted by nanofillers' effects on their dielectric characteristics (Moshkalyuk and Yudin 2020). It has been demonstrated that adding nanofillers, such as $CuCoO_4$ nanoparticles, to PEO/PVA blends improves electrical conductivity. The development of new energy states inside the polymer matrix, which promotes electron migration and narrows the band gap, is responsible for this improvement (Alghamdi and Rajeh 2024).

3.2.37.7 Barrier Properties

When added to materials like polymers, they increase resistance to gases and moisture. By improving interfacial contacts and increasing crystallinity, nanofillers dramatically improve the barrier qualities of polymers (Jali 2023).

Metal Oxide Nanofillers: Metal oxide nanofillers are tiny particles made of metal oxides, including aluminum oxide (AlO_3), titanium dioxide (TiO_2), and zinc oxide (ZnO). Owing to their special qualities, these nanofillers are widely utilized to improve the functionality of materials, such as composites, ceramics, and polymers.

The general features of metal oxide nanofillers include:

Size and Shape: The average size range of metal oxide nanofillers is 1–100 nm. Their dispersion and interaction within the matrix are influenced by their shape, which can range from spherical to rod-like or plate-like structures.

Surface Area: These nanofillers have a greater specific surface area because of their nanoscale size, which improves their reactivity and interaction with the surrounding matrix.

Crystallinity: The mechanical and thermal properties of metal oxide nanofillers are affected by their crystallinity. For example, ZnO nanostructures frequently display a hexagonal wurtzite structure, which may affect how they behave optically and electrically.

Optical Activity: Under UV light exposure, some metal oxide nanofillers, such TiO_2, have photocatalytic qualities that allow them to break down organic contaminants.

Electrical Characteristics.

Semiconducting Behavior: A variety of metal oxide nanofillers, such as ZnO and TiO_2, have semiconducting characteristics that make them suitable for use in opto-electronic and electrical devices.

Conductivity Enhancement: Polymer matrices can be made more electrically conductive by adding metal oxide nanofillers, which makes them appropriate for uses such as conductive coatings and sensors.

Mechanical Properties-Reinforcement: Metal oxide nanofillers can enhance mechanical characteristics like impact resistance, stiffness, and tensile strength when added to polymers.

Thermal Stability: By adding metal oxide nanofillers, polymer composites' capacity to tolerate greater temperatures without degrading can be improved.

Surface Functionalization: To enhance compatibility with diverse matrices and impart certain functionalities, metal oxide nanofillers' surfaces may have different chemical groups added to it to make it useful.

Dispersion: Reaching consistent dispersion in the matrix is made possible by adequate surface treatment, and this is necessary for maximizing the composite material's improved qualities. (Schmidt-Mende 2007).

Applications in the biomedical field include:

- Drug delivery systems.
- Tissue engineering.
- Imaging and diagnostics.
- Wound healing.
- Cancer treatment.
- Antimicrobial agents (Sharifi 2024).

Types of Drug Delivery Metal Oxide Nanoparticles

1. Nanoparticles of zinc oxide (ZnO):

   ZnO nanoparticles have been investigated for the delivery of cancer drugs and are biodegradable. For example, a doxorubicin-ZnO nanocomplex has shown efficacy against hepatocarcinoma cells by raising doxorubicin's intracellular concentration, which improves the effectiveness of chemotherapy (Nikolova and Chavali 2020).
2. Iron oxide nanoparticles:

   Targeted drug delivery under an external magnetic field is made possible by the magnetic characteristics of superparamagnetic iron oxide nanoparticles (SPIONs). To increase circulation time and decrease toxicity, they might be coated with biocompatible substances like polyethylene glycol (PEG).
3. Magnesium oxide (MgO) nanoparticles:

   These particles are thought to be non-toxic and may be used as drug delivery systems in antibacterial and anticancer therapies. They are good options for delivering therapeutic drugs because of their biocompatibility and functionalization potential (Negrescu and Killian 2022).

Metal Oxide Nanoparticles' Benefits for Drug Delivery

1. Improved stability and half-life: By extending the stability and circulation half-life of drug carriers, metal oxide nanoparticles can guarantee that therapeutic medicines stay active in the bloodstream for prolonged periods of time (Chandrakala 2022).
2. Targeted delivery: Sick tissues or cells can be actively targeted by changing the surface of these nanoparticles with specific ligands or antibodies, reducing off-target effects and improving treatment efficacy (Chandrakala 2022).
3. Regulated release: Because metal oxide nanoparticles are porous, medications can be encapsulated and released under regulated conditions in response to particular stimuli like temperature changes, pH shifts, or magnetic fields.

Despite their advantages, the use of metal oxide nanoparticles in drug delivery systems faces challenges, including potential toxicity, biodegradability, and the requirement for exact particle size control and surface characteristics (Chandrakala 2022).

Tissue Engineering Because of their special qualities that improve scaffold performance and encourage tissue regeneration, metal oxide nanofillers have attracted a lot of interest in tissue engineering.

Metal oxide nanofiller types:

The magnetic characteristics of iron oxide nanoparticles (IONPs), which enable targeted medication administration and imaging, have been the subject of much research. In tissue engineering, they enhance the mechanical strength of scaffolds and encourage cell division and proliferation (Friedrich 2021).

Nanoparticles of Titanium Dioxide (TiO_2): TiO_2 nanoparticles are well known for having a large surface area and being biocompatible. They are perfect for bone tissue engineering because, when used as nanofillers, they enhance the mechanical characteristics of scaffolds and encourage osteogenic differentiation. (Jamroz 2019).

Zinc Oxide (ZnO) Nanoparticles: ZnO nanoparticles can improve scaffold bioactivity and have antibacterial qualities. It has been demonstrated that their integration into scaffolds promotes cell adhesion and proliferation, which is advantageous for tissue regeneration. (Jamroz 2019).

Tissue engineering applications:

1. Bone tissue engineering:

   IONPs and TiO_2 nanoparticles are examples of metal oxide nanofillers that are used in scaffolds to replicate the natural bone environment in bone tissue engineering. In order to promote bone regeneration, they increase mechanical strength and aid in the differentiation of stem cells into osteoblasts.
2. Cartilage regeneration:

   The mechanical characteristics of scaffolds used in cartilage regeneration are enhanced by nanofillers. Metal oxide nanoparticles, for example, can promote chondrocyte proliferation and increase scaffold rigidity (Wu and Huang 2009).

3. Wound healing:

   ZnO nanoparticles are utilized in scaffolds for skin tissue engineering because of their antibacterial qualities. They contribute to wound healing by preventing infection and encouraging cell growth (Jamroz 2019).
4. Problems and considerations:

   Although metal oxide nanofillers have several advantages, problems such long-term stability, aggregation in the scaffold matrix, and possible cytotoxicity must be resolved. To reduce side effects and attain the appropriate scaffold qualities, it is essential to manage the concentration of nanoparticles and ensure their uniform dispersion (Jamroz 2019).

Imaging and Diagnostics Because of their distinct physicochemical properties, such as their magnetic, optical, and electronic qualities, metal oxide nanoparticles (MONPs) have attracted a lot of interest in biomedical imaging and diagnostics. These characteristics allow them to be used in a variety of imaging modalities, improving therapeutic results and diagnostic precision.

1. Magnetic resonance imaging (MRI): Superparamagnetic iron oxide nanoparticles, or SPIONs, are frequently employed as contrast agents. Because of their superparamagnetic nature, signals are significantly enhanced, which improves the detection of vascular anomalies, inflammation, and malignancies. Because SPIONs produce high-resolution images that aid in early detection, they have been used to examine the liver, lymph nodes, and inflammatory diseases (Dadfar 2019)
2. Computed tomography (CT): As contrast agents in CT imaging, metal oxide nanoparticles, including those based on cerium oxide, have been investigated. Their high electron density helps to improve contrast, which makes it easier to see tumor borders and vascular architecture. Furthermore, cerium oxide nanoparticles' antioxidant qualities may have therapeutic uses.
3. Optical imaging: Metal oxide-based quantum dots, such as zinc oxide, have special optical characteristics, such as size-tunable fluorescence. In optical imaging, these nanoparticles function as fluorescent probes, allowing for highly sensitive viewing of cellular and molecular processes. They can be used for both in vivo imaging of malignant tissues and in vitro tests.
4. Nuclear imaging: For usage in single-photon emission computed tomography (SPECT) and positron emission tomography (PET), metal oxide nanoparticles can be radiolabeled with diagnostic radionuclides. Titanium dioxide nanoparticles, for example, have been functionalized to enable targeted imaging of particular tissues, producing high-resolution images and aiding in the early diagnosis of disease (Janska 2024)
5. Ultrasound imaging: Although metal oxide nanoparticles are not as frequently employed in this field, studies have looked into how they can improve contrast. Iron oxide nanoparticles, for instance, have been studied for their potential to enhance tumor ultrasound imaging, which could help in tumor boundary delineation.

Wound Healing Because of their special qualities, which include antibacterial action, cell proliferation encouragement, and tissue regeneration facilitation, metal oxide nanofillers have attracted a lot of interest in wound healing applications.

1. Zinc oxide nanoparticles, or ZnO NPs, are well known for their regenerative and biocompatible properties. They encourage collagen deposition, keratinocyte migration, and re-epithelialization. Furthermore, ZnO NPs are efficient against both Gram-negative and Gram-positive bacteria due to their antibacterial qualities. Improved wound closure and decreased infection rates have been demonstrated by their inclusion into hydrogels and scaffolds (Mendes 2022).
2. Titanium dioxide nanoparticles (TiO_2 NPs): TiO_2 NPs have anti-inflammatory and antibacterial qualities. They improve mechanical strength, hydrophilicity, and biocompatibility when used into wound dressings. TiO_2 NPs have been shown in studies to accelerate wound healing by inducing the synthesis of vital molecules for tissue regeneration (Dam 2023).
3. MgO NPs, or magnesium oxide nanoparticles, are prized for their low toxicity, thermal stability, and biocompatibility. According to studies, MgO NPs enhance the mechanical characteristics, biocompatibility, and antibacterial activity of wound dressings, which speeds up the healing process in models of skin defects (Dam 2023)
4. Copper oxide nanoparticles, or CuO NPs, have fungicidal and antibacterial properties. By stimulating angiogenesis and vascular endothelial growth factor (VEGF), they aid in tissue healing. Controlling their concentration is essential, though, because too much copper can produce free radicals, which can result in lipid peroxidation and cell death. [11]
5. Aluminum oxide nanoparticles (AlO_3 NPs): These nanoparticles have been utilized in biodegradable hydrogels for wound healing applications and exhibit antibacterial properties. Al_2O_3 NPs have been shown to improve mechanical qualities, lower the number of bacteria, lower the expression of inflammatory genes, and raise the expression of collagen, fibroblasts, and proliferative genes, all of which aid in wound healing (Dam 2023)

Cancer Treatment Because of its special qualities, such as their enhanced surface area, magnetic behavior, and capacity to produce reactive oxygen species (ROS), metal oxide nanoparticles (MONPs) have created a lot of interest in the treatment of cancer. These qualities qualify them for medicinal, imaging, and targeted drug delivery applications.

(a) Targeted Drug Delivery and Imaging

It is possible to create MONPs like iron oxide, zinc oxide, and copper oxide to improve imaging and drug administration in cancer treatment. Their magnetic characteristics make it possible to create nanocarriers that can be guided by external magnetic fields to tumor locations, resulting in targeted medication release and better treatment results. For example, iron oxide nanoparticles have been used in heat/ROS-mediated treatment and targeted administration to increase the therapeutic efficacy of anticancer medications (Poonia 2024).

(b) Antitumor activity: Some MONPs are intrinsically cytotoxic to cancer cells but do not harm healthy cells. The production of ROS frequently mediates this selective toxicity, which causes tumor cells to undergo necrosis or death. For instance, zinc oxide nanoparticles have demonstrated anticancer properties both on their own and when combined with medications like as doxorubicin. Cobalt oxide, iron oxide, and copper oxide are among the other nanoparticles that have demonstrated lethal effects on cancer cells (Vinardell 2015).
(c) MONPs' clinical use is beset by issues such production scalability, regulatory barriers, and guaranteeing safety and biocompatibility. In order for MONPs to be successfully included into cancer treatment, several problems must be resolved (Yassin 2024). The goal of these developments is to reduce side effects while increasing the specificity and effectiveness of cancer treatments (Hheidari 2024).

Antimicrobial Agents Because of their unique physicochemical properties, metal oxide nanofillers have attracted a lot of interest as antimicrobial agents. Their unique electrical structures and greater surface area-to-volume ratio make it easier for them to interact with microbial cells, which results in potent antibacterial action.

Antimicrobial Action Mechanisms:

1. Production of reactive oxygen species (ROS): Under some circumstances, metal oxide nanoparticles (MeO-NPs) can generate ROS, superoxide anions, and hydroxyl radicals. These ROS cause oxidative stress in microbial cells, which damages DNA, proteins, and lipids and finally results in cell death.
2. Cell membrane disruption: Specific MeO-NPs have the ability to interact with microbial cell membranes, resulting in structural alterations that raise permeability. Cell lysis and the release of vital intracellular substances result from this.
3. Metal ion release: Certain metal oxide nanoparticles release metal ions that can interact with proteins and enzymes in microorganisms, preventing them from growing normally.

Metal Oxide Nanofiller Examples:

1. Zinc Oxide (ZnO) Nanoparticles: Due to the production of ROS and the release of Zn2+ ions, ZnO nanoparticles have potent antibacterial action. They have been shown to be effective against a range of bacterial strains.
2. Copper Oxide (CuO) Nanoparticles: By producing reactive oxygen species (ROS) and disrupting membranes, CuO nanoparticles have shown antibacterial characteristics. Both Gram-negative bacteria and Gram-positive can be effectively combatted by them.
3. Iron Oxide (FeO_4) Nanoparticles: These nanoparticles have demonstrated potential in the fight against microbial illnesses, particularly when doped with additional metals. In medicinal applications, their magnetic qualities also enable targeted distribution (Tasnim 2023).

3.2.38 *Metal-Based Nanofillers*

Renowned Nobel laureate Richard P. Feynman outlined the basic ideas of nanotechnology and nanoscience. His well-known speech, "There's Plenty of Room at the Bottom," was delivered on December 29, 1959, at the California Institute of Technology (Caltech) (Bhushan 2017). Since then, a lot of research has been done on the synthesis of nanomaterials. The potential uses of nanoparticles are numerous and are determined by a variety of implementation factors (Pattanayak et al. 2020–2023). These modernized and engineered material nanofillers blend in perfectly with the host matrix, bringing forth a paradigm shift of enhanced and all-encompassing material properties (Yadav et al. 2024).

Two categories have been created in order to recognize the nanofillers and comprehend how they are incorporated: materialistic (Jagadeesh et al. 2021) and morphological categories (Basavegowda et al. 2021). Nanofillers have a variety of materialistic traits that add to their special qualities and uses. Organic, inorganic, carbon-based, and clay-based nanofillers are some of the most widely used types of nanofillers (Jagadeesh et al. 2021). As conductive fillers in polymer composites, carbonaceous nanofillers—such as carbon nanotubes (CNTs), graphene, and carbon black (CB)—have grown in popularity. They are very sought-after in many applications due to their extensive availability, lightweight nature, low cost, big aspect ratio, superior conductivity, and reinforcing qualities (Kuang et al. 2023; Vu et al. 2023; Pattanayak et al. 2020–2023).

In composites, one or more minor components serve as fibers, sometimes referred to as reinforcing materials, while the major components are typically referred to as matrix. Usually, the reinforcement material gives the matrix its unique properties or improvements. In some cases, synergism is observed, when the extraordinary features of the reinforcing material and matrix complement one another. A substance is referred to as a nanocomposite if one or more of its constituent parts have dimensions less than 100 nm. Greater surface-to-volume ratios, aspect ratios, or combinations of matrix and reinforcing materials are the results of improving qualities of interest in nanocomposites. Because the components in nanocomposites are so minute, even a tiny quantity of reinforcing material has a big impact on the macroscopic characteristics, sometimes exhibiting multifunctional behavior and unexpected flexibility. Furthermore, because filler material is often larger in size, it has a greater density, which results in heavier composites and restricts its use in a number of applications (Basavegowda et al. 2021).

3.2.39 *Carbon Nanofillers*

3.2.39.1 Carbon Nanotube

In addition to its many uses, carbon is the basic component of all life on Earth, including proteins, lipids, genetic components, and more. Diamond, fullerenes, graphite, and carbon nanotubes (CNT) are allotropic forms of carbon. Every carbon allotrope has a unique set of physical characteristics.

We focus primarily on carbon nanotubes (CNTs) among the different types of carbon in the nanoscale. Single-walled carbon nanotubes (SWCNTs) and multi-walled carbon nanotubes (MWCNTs) are the two types of CNT (Iijima and Bethune 1993). While MWCNT consists of concentric graphene cylinders, a single sheet of graphene folded into a cylinder is known as SWCNT. SWCNT is divided into three varieties according to its crystallographic configurations: chiral, armchair, and zig-zag (Iijima and Bethune 1993). The way the graphene sheet is folded to create the SWCNT determines this categorization: arm-chair conformation happens when the opposing c-c bonds are perpendicular to the axis of the cylinder, whereas zigzag confirmation happens when the opposite carbons come together to form a cylinder and the c-c bonds are in the axis of the cylinder.

3.2.40 *Polymer Nanocomposites Fortified CNT*

3.2.40.1 Polyethylene CNT Nanocomposite

Concerns have been raised about CNTs' potential to be cytotoxic and/or incompatible with human cells in biomedical and pharmacological applications. It is generally accepted that CNTs, both functionalized and unfunctionalized, are not optimal, regardless of size (Alshehri et al. 2013–2016). The primary reasons why CNTs are not appropriate for these types of applications are their size, aggregation, and impurities.

The superior physicochemical characteristics of ultra-high molecular weight polyethylene (UHPE) make it a popular choice for biological applications. Although UHPE in long-chain form is preferred for biomedical applications, some applications are limited by its low strength. In order to get around this problem, Mamidi et al. used human fibroblast cells to test the cytotoxicity of UHPE reinforced with MWCNT (Mamidi et al. 2017). The most common use of UHPE polymer is in joint replacement applications (implants) that interact with ceramic and metal surfaces (Kanagaraj and Liu 2011–2013). These applications are common because of their exceptional biocompatibility, impact strength, low friction, and resistance to abrasive sliding wear, and outstanding toughness (Bakshi et al. 2007). The longevity of implants made of these materials is significantly impacted by their tribological behavior, which includes critical wear and low yield strength that leads to early failure (Xue et al. 2016). Furthermore, wear debris produced by UHPE implants moving relative to metal or ceramic material moves to tissues, causing osteolysis (Amstutz et al. 1993). The use of carbon nanotubes (CNTs) to reinforce UHPE can help to alleviate its inadequacies for these kinds of applications.

3.2.40.2 Polystyrene CNT Nanocomposites

Using atomistic models of polystyrene and polystyrene-modified SWCNTs, Gul et al. investigated how they interacted with a model lipid bilayer that contained or did not include the medication ibuprofen. The use of polystyrene as a functional group for improved and safe drug delivery is particularly promising (Gulsah et al. 2022).

3.2.40.3 Polylactide (PLA) Nanocomposite

Different industries, including biomedicine, packaging and agriculture, use biodegradable polymers because of their superior mechanical qualities, ease of processing affordability and biocompatibility. Polylactide nanocomposite is widely used in tissue engineering, as synthetic bone substitute and in drug delivery systems (Drumright et al. 2000).

3.2.40.4 Graphene Nanosheets

Graphene exhibits numerous filler qualities, making it a viable material to enhance the electrical, mechanical, and thermal properties of polymers. Crucially, in addition to the uniform suspension of graphene in the polymer, which improves the same load transfer via the polymer matrix, the composites' characteristics are controlled by the strong connection between polymer matrix and graphene. Adding graphene functionalization is one method of enhancing graphene's dispersion capacity. Good examples of functionalized polymers used as fillers in polymer composites include reduced graphene oxide (rGO) and graphene oxide (GO). The primary cause of this is that GO has oxygen functional groups that readily react with the polymer matrix, including hydroxyl, epoxy, and carboxylic groups. Additionally, many of these residual functional groups are present in the rGO molecule, which is created when GO oxide is reduced. This results in a greater interlayer separation and a decrease in the van der Waals force contact effects in graphene interlayers.

3.2.40.5 Fullerene

Fullerenes have been widely utilized in nanocarbon materials since their early discovery. A spherical π-conjugated structure with a significant pyramidalization angle is one of the many intriguing characteristics of fullerenes. This structure makes it easier to chemically functionalize than carbon nanotubes which results in elevated solubility in a variety of aqueous and organic solvents. Carbon fiber-reinforced composites are among the most significant uses of polymers and fullerenes.

3.2.40.6 Graphite

The most solid form of carbon in most conditions is graphite, which is stronger than diamond. Each layer of its planar, layered structure is organized in a hexagonal lattice. Solution blending, gentle blending, in situ polymerization, and a variety of other techniques were used to create the polymer/graphite nanocomposites.

3.2.41 Metal-Based Nanofillers

Metal-based nanofillers are a fascinating area of materials science! They are typically used to improvise the characteristics of composites, such as strength, conductivity, and thermal stability. Here are some common types of metal-based nanofillers:

3.2.42 Metal Nanoparticles

- *Gold nanoparticles*: They are employed in drug administration, cancer treatment, and diagnostic imaging because of their distinctive optical qualities and biocompatibility.
- *Silver nanoparticles*: Highly valued for their antimicrobial properties, they are incorporated into wound dressings, coatings, and medical devices to prevent infections.
- *Platinum nanoparticles*: Used in catalysis, cancer treatment, and as antioxidant agents due to their greater reactivity and stability.
- *Palladium nanoparticles*: Commonly used in catalysis and hydrogen storage due to their excellent catalytic properties.

3.2.43 Applications

Metal-based nanofillers have several promising applications in the field of pharmacy and pharmaceuticals. Here are some key areas:

1. *Drug delivery systems*: Targeted drug delivery systems can be made with metal nanoparticles, including those made of silver and gold. By lowering side effects and increasing therapeutic results, these systems can increase the effectiveness and specificity of medication delivery to sick cells.
2. *Antimicrobial agents*: Silver nanoparticles are widely used for their antimicrobial properties. They can be incorporated into wound dressings, coatings for medical devices, and packaging materials to prevent infections.
3. *Cancer therapy*: Metal nanoparticles are used in cancer therapy for both diagnosis and treatment. Gold nanoparticles, e.g., can be used for targeted drug delivery and photothermal therapy, where they help to destroy cancer cells by converting light into heat.
4. *Medical imaging*: Metal nanoparticles, such as gold and iron oxide, enhance the contrast in imaging techniques like MRI and CT scans, improving the accuracy of diagnoses.[155]
5. *Tissue engineering*: Scaffolds for tissue engineering can employ metal-based nanofillers to encourage tissue regeneration and cell growth (Ranjan and Dave 2024).

6. *Diagnostics*: Metal nanoparticles are used in biosensors to detect biomolecules and pathogens with greater specificity and sensitivity.

3.2.44 Nanoclay

Nanoclay is a kind of nanomaterial made up of layered crystalline silicates with nanometer-sized particles, usually less than 100 nm in at least one dimension. Nanoclays are employed as fillers and reinforcements in polymer nanocomposites, among other uses, because of their special structure and characteristics. Due to their extensive availability, affordability, and minimal environmental impact, these materials are highly used in research and development for a range of uses.

Nanoclays have garnered a lot of interest because of their special qualities and adaptability for cutting-edge functional applications. These applications include polymer composites, nanoelectronics, biomedical engineering, wastewater treatment, and energy storage. Nanoclay has gained popularity due to its elevated surface area, large aspect ratio, adjustable characteristics, and capacity to enhance mechanical and barrier qualities. By stacking layered structural units, layered mineral silicates known as nanoclays create intricate clay crystallites. The composition of the individual units of nanoclays, which are made up of octahedral and/or tetrahedral sheets, affects their distinctive characteristics. The interconnecting silicon-oxygen tetrahedra that make up the tetrahedral sheets of this material share three corners with their surrounding tetrahedra. A neighboring octahedral sheet is covalently bonded to the fourth vertex of each tetrahedral sheet.

High-performance polymer nanocomposites with improved stiffness, strength, toughness, and thermal stability are produced when nanoclay is incorporated into polymer matrices. While nanoclays have numerous potential advantages, they also have many disadvantages, including limited availability, processing challenges due to their small particle size, potential health concerns from ingesting or inhalation, and environmental effects from raw material extraction and processing.

Nanoclay is employed as a drug delivery system, a scaffold for tissue engineering, and a wound healing agent. The unique characteristics of nanoclay, including its high biocompatibility and capability for controlled drug release, render it an excellent candidate for biomedical engineering applications.

3.2.45 Types of Nanoclay

Three primary categories are used to classify nanoclays according to their electrochemical charge. Every major category has a subcategory. The three main types of nanoclays are given as follows:

1. Cationic nanoclays
2. Anionic nanoclays
3. Neutral nanoclays

3.2.45.1 Cationic Nanoclays

Groups such as smectite, vermiculite, mica, brittle mica, and chlorite are examples of cationic nanoclays. They have a 2:1 or 2:1:1 layer structure and a substantial negative charge between 0.2 and 2. Because of isomorphous replacements inside the tetrahedral or octahedral sheets, such as Mg^{2+} replacing Al^{3+} or Al^{3+} replacing Si^{4+}, these nanoclays have a negative charge. The cations in the interlayer spaces balance out the charge imbalance caused by this replacement, improving their cation exchange capacity and swelling behavior. These characteristics make cationic nanoclays very useful in processes where high surface area and reactivity are advantageous, like catalysis, adsorption, and as reinforcement agents in polymer nanocomposites.

Montmorillonite

Montmorillonite is a layered nanoclay mineral that is a member of the smectite group. The chemical formula is $(NaCa)_{0.33}(AlMg)_2(Si_4O_{10})(OH)_2{\cdot}nH_2O$, where n is water molecules number in the interlayer gap. The mineral is able to exchange ions with the environment due to greater cation exchange capacity.

Montmorillonite is a highly sought-after mineral in many industries, including plastics, rubbers, coatings, drilling fluids, and cosmetics, due to its outstanding qualities. In composites, nanoclay acts as a reinforcing ingredient, improving the polymer matrix's mechanical qualities while also lowering the end products' weight and cost.

Montmorillonite's remarkable adsorption ability makes it useful for environmental cleanup as well. It can efficiently remove heavy metals and other impurities from contaminated water in this situation. This multipurpose nanoclay has a layered structure, a high capacity for exchanging cations, and a number of uses. In composite materials, montmorillonite can be used to increase mechanical and barrier properties, reduce weight and cost. Furthermore, because of its high surface area, porosity, and cation exchange capability, montmorillonite can be a great catalytic support material. Montmorillonite-based catalysts have been prepared using a variety of techniques, such as ion exchange, impregnation, and sol–gel processes.

Montmorillonite does have several drawbacks, though, such as limited stability in certain situations, such as high temperatures, acidic or alkaline environments, and organic solvents, which can cause it to lose its qualities. Despite obstacles, montmorillonite's remarkable versatility and unique properties continue to make it a promising material for a range of industries and uses.

Numerous scientific studies have recently looked into many uses of montmorillonite in various disciplines. Zango et al. researched the potential of montmorillonite as a catalyst support material and highlighted its exceptional qualities, including high surface area, porosity, and cation exchange capacity, which make it an excellent catalyst support material. In the meanwhile, Saba et al. investigated the advantages of adding montmorillonite to composites and found that it decreased weight and cost while improving mechanical, thermal, and barrier qualities. In polymer nanocomposites, Mllakpour et al. showed that adding montmorillonite as a filler improved the material's mechanical, thermal, and barrier qualities and biodegradability.

Hectorite

Hectorite is a layered silicate mineral that is a member of the clay mineral smectite group. It has the chemical formula $(Mg, Li)_3Si_4O_{10}(OH)_2 \cdot 4H_2O$ and, like other smectite minerals such as montmorillonite, it is made of two tetrahedral sheets that are sandwiched by a central octahedral sheet. The layered silicate mineral hectorite, which is a candidate of the smectite group of clay minerals, has a special structure because of its high lithium content. As a result, the octahedral sheet is deformed and the interlayer structure is stiffer than in other smectite minerals.

One of the most promising uses of hectorite is as a nanofiller in polymer composites, where it can improve mechanical characteristics, thermal stability. It serves as a thickening and stabilizer in cosmetics, and a medication delivery vehicle in medicines.

According to Yan et al., stiffness and an increased aspect ratio are fortunate traits that enhance the thermal and mechanical properties of polymer composites. According to Thompson et al., hectorite's high porosity and surface area make it a perfect option for drug delivery vehicles. In the meanwhile, Fernando et al. provide a summary of composites based on hectorite nanoclay, emphasizing their high aspect ratio, cation exchange capacity. The authors describe different processing methods for adding hectorite to polymer matrices and how they affect the composite's ultimate characteristics.

Hectorite nanoclay's thermal stability has been shown to be less than ideal, nevertheless, because the substance degrades easily at high temperatures. This thermal instability is thought to be caused mostly by the presence of water molecules and other organic compounds that have been absorbed on the surface of the nanoclay particles. Furthermore, when hectorite nanoclay is distributed in particular solvents or polymer matrices, it has a tendency to clump or agglomerate, which results in poor separation of the nanoclay particles and a decrease in the composite material's mechanical properties.

Additionally, adding hectorite nanoclay can have a negative impact on polymer composites' rheological characteristics, increasing their viscosity and decreasing their processability. The greater aspect ratio and elevation surface area of the nanoclay particles are responsible for this, as they produce a high degree of interparticle contacts and hinder the composite material's flow.

Laponite

Laponite is a synthetic layered silicate mineral belonging to a member of the clay mineral smectite group. It has the chemical formula $Na_{0.7}(Mg_5.5Li0.3Si_8O_{20})(OH)_4Na_{0.7}(CO_3)_{0.3}$ nH_2O. Like other smectite minerals including montmorillonite and hectorite, it has a structure with two tetrahedral sheets around a center octahedral sheet.

Mostly, laponite is utilized as a rheological addition. It is also used in ceramics, coatings, and personal hygiene items. It is a powerful thickening and stabilizer because of its special structure, which enables it to create stable colloidal suspensions. Laponite is also utilized as a nanofiller to increase the mechanical and thermal qualities of polymer composites.

Laponite has been investigated for a number of additional uses, such as medication administration and catalysis, in addition to its use in composites and rheological additives. Laponite is a potential material for drug delivery vehicles and a support material for catalytic metals in the chemical industry because of its large surface area and cation exchange capacity.

Ruiz-Hitzky et al. emphasize on Laponite's special qualities, such as its high surface area, cation exchange capacity, and uniform and regular structure. Additionally, they look into different synthesis procedures, like sol–gel and precipitation methods, as well as how synthesis conditions affect the material's final qualities.

Buruga et al. concentrate on the versatility of Laponite-based nanocomposites. The authors talk about how Laponite's high surface area, homogeneous structure, and colloidal stability make it a viable nanofiller for enhancing the mechanical and thermal characteristics of polymer composites. Additionally, the authors emphasize the applications of composites in biomedical devices and packaging materials.

Laponite has demonstrated promise in a number of applications, but it also has a number of disadvantages. Laponite particles have been demonstrated to induce oxidative stress and DNA damage in human lung cells, and inhalation of these particles is associated with respiratory issues and lung inflammation.

Furthermore, it may be difficult for laponite to disperse evenly in a matrix due to its large surface area and propensity to agglomerate, which could result in nonuniformity in the final product.

Smectite

An unusual layered structure composed of 2 tetrahedral sheets encircling a core octahedral sheet joined by weak Van der Waals interactions characterizes smectite, a fascinating member of the clay mineral family. Its structure allows it to be readily separated and to inflate when water is present.

Smectite nanoclays are perfect for a various sectors due to their elevated surface area, high aspect ratio, and superior to montmorillonite in polymers. For instance, they can be utilized in polymer composites as thickeners, emulsion stabilizers, rheological modifiers, and nanofillers.

Smectite nanoclays have been employed as nanofillers in polymer composites by Mine et al. The authors discussed the critical impact of smectite concentration and dispersion on the properties of the final composites by thoroughly examining processing methods like melt blending and solution casting. Glotov et al. continue the study by exploring the fascinating area of layered silicate nanoplatelets, which contain smectite nanoclays.

The use of smectite in high-temperature settings and processes is, however, restricted by certain issues. For example, at high temperatures, it may experience permanent structural changes that result in the loss of its adsorption and rheological characteristics. Furthermore, smectite's mechanical qualities and structural integrity may be impacted by its vulnerability to swelling and hydration, which could impact how effectively it works as a stabilizing or reinforcing agent.

The inherent heterogeneity in the composition and characteristics of smectite deposits, which impacts their performance and consistency, is another major obstacle. The adsorption capacity and rheological behavior of smectite can be impacted by impurities or other mineral phases, and its mineralogical composition can change based on its source.

Furthermore, smectite nanoclay manufacture and disposal may have detrimental impacts on the environment. Water contamination, habitat destruction, and soil erosion can result from the mining and processing of smectite. The ecosystem may also be negatively impacted by the disposal of waste and tainted water from the processing of smectite.

3.2.45.2 Anionic Nanoclays

Anionic nanoclays, particularly layered double hydroxides (LDHs), are characterized by a variable positive charge due to the presence of trivalent cations within their structure, which forms an octahedral sheet with divalent metal ions. In the interlayer galleries, anions like CO_3^{-2}, NO_3^-, and Cl^- balance the positive charge in these nanoclays. Anionic nanoclays can function as effective anion exchangers due to their special electrochemical structure, which makes them appropriate for use as drug delivery systems, and environmental remediation. LDHs' strong thermal stability and capacity to intercalate a variety of anions increase their usefulness in a range of industrial and biological applications.

Zeolite

Over millions of years, volcanic ash and alkaline groundwater mix to generate zeolite, a naturally occurring aluminosilicate mineral with a distinctive three-dimensional crystalline structure. Zeolites are very adaptable and helpful in various applications due to their increased surface area, porosity, and ion exchange capacity.

They serve as ion exchangers for water softening and purification, fillers in composites, adsorbents for controlling moisture and odor, molecular sieves for gas separation and purification. Zeolites are widely employed as ion exchangers, fillers in composites, adsorbents, catalysts, and molecular sieves. There are more than 40 naturally occurring zeolites and hundreds of synthesized ones.

Zeolites have been widely used as reinforcing components in composite materials to enhance the polymer matrix's thermal and mechanical characteristics. According to studies, adding zeolites can significantly improve the composite's barrier qualities, decrease gas permeability, and increase moisture resistance while also raising the composite's modulus, thermal stability and tensile strength, . Zeolites are being used as fillers in polymer composites according to recent studies. Researchers looked at different ways to create zeolite-polymer composites and how zeolite loading and type affected the composites' characteristics.

3.3 Neutral Nanoclays

Neutral nanoclays, including those in the Pyrophyllite-Talc and Kaolinite-serpentine families, have very little to no electrochemical charge and either a 1:1 or 2:1 layer type. In a 2:1 structure, 2 tetrahedral sheets encase an octahedral sheet, whereas in a 1:1 neutral nanoclay, a tetrahedral sheet faces an octahedral sheet directly. The main constituents of these nanoclays are Mg^{+2} or Al^{+3} complexes in the octahedral sheets and $[SiO_4]^{-4}$ ions in the tetrahedral sheets. They are appropriate for applications where structural stability and inertness are required, such as in ceramics and as fillers in polymer composites, because of their low charge, which also results in limited swelling properties and lower ion exchange capacities.

3.3.1 Kaolinite

Kaolinite, which has special qualities, makes it a successful candidate in various industries. Its chemical formula is $Al_2Si_2O_5(OH)_4$, and it is a member of the kaolin group of minerals. Oxygen atoms are shared between the tetrahedral and octahedral sheets that make up the mineral's layered arrangement, which eventually forms a trilayer structure. Van der Waals forces and hydrogen bonds maintain tenuous interlayer bonding.

Kaolinite is usually found in sedimentary rocks and soils, especially in tropical locations, and is typically generated by the weathering of feldspar and other minerals that are rich in aluminum. It is used in many different applications, such as a catalyst support material, a filler in composites. In addition to improving thermal stability and flame retardancy, kaolinite can increase the mechanical qualities of composites, including stiffness, strength, and toughness.

The limited flexibility of kaolinite is a significant drawback in spite of these advantageous properties. Its value in applications needing great plasticity is limited by its difficulty in molding or shaping without cracking. Furthermore, kaolinite's strength is restricted in comparison to other clay minerals, which could potentially limit its uses.

Kaolinite is a popular option for removing organic contaminants, colors, and heavy metals from water due to its high surface area, porosity, and cation exchange capacity. To enhance the adsorption capacity and selectivity of kaolinite-based adsorbents, a variety of modification approaches are investigated.

Nisar et al. provide an overview of the various industries in which kaolinite finds use. The following processing techniques are also offered as efficient ways to alter the characteristics of kaolinite for its several uses: calcination, acid treatment, and surface modification.

3.3.2 *Attapulgite*

Attapulgite is an intriguing kind of clay mineral with a needle-like structure made up of long, thin threads that are several micrometers long and have a diameter of 0.2 to 0.4 μm. Because of the random orientation of the fiber stacking, attapulgite has a non-crystalline, disordered structure.

Attapulgite's special qualities make it a popular nanoclay in many different industries, where it is employed as a thickener, adsorbent, rheological modifier, and nanofiller in polymer composites. Attapulgite nanoclays are ideal for large goods due to their remarkable surface area, increased aspect ratio, and superior heat stability.

Cavdar et al. have focused on the characteristics and uses of attapulgite-based nanocomposites, examining how attapulgite loading affects the composites' characteristics. Additionally, they emphasize the use of attapulgite-based nanocomposites in areas like drug delivery, catalysis, and polymer composites. On the other hand, Bastani et al. discuss the latest advancements in attapulgite clay-based nanocomposites, examining the different synthesis techniques and their possible uses in biomedical engineering, packaging, and building.

3.3.3 *Palygorskite*

Palygorskite, a fascinating type of clay mineral, is often used as a nanoclay due to its unique structure. Its long, thin fibers with periodic cross-bridges, having a diameter of 0.1–0.2 μm and a length of several micrometers, prove it the best choice for different applications. Palygorskite is primarily sourced from Spain, Morocco, and the USA and finds applications in various industries, including as a rheological modifier, thickener, adsorbent, and nanofiller in polymer composites.

Barbosa et al. emphasize the uses of palygorskite in areas including environmental remediation, catalysis, and biomedical engineering and go over the several modification techniques used to improve its qualities. The potential uses of palygorskite-based nanocomposites in areas like medication delivery, and polymer composites are highlighted by the authors.

3.3.4 Sepiolite

Researchers and engineers are interested in the clay mineral known as sepiolite nanoclay. The distinctive fibrous and porous structure of sepiolite is characterized by tunnels that are just 4–6 Å in diameter but can reach lengths of 100 nm. It is an effective adsorbent for tiny molecules due to its exceptional configuration, which makes it extremely helpful for a variety of applications. Sepiolite, which is frequently utilized as a nanoclay in sectors like rheology modification, thickening, and adsorption. Sepiolite nanoclays' elevated aspect ratio, surface area (specific), and thermal stability have led to many uses in different industries. Sepiolite nanoclays' remarkable surface area and porosity, together with a variety of production techniques, have made them a desirable material in applications such polymer composites, adsorption, and catalysis.

Sepiolite nanoclays have drawn more interest recently as possible reinforcing agents for polymer composites. Researchers looked into how the concentration of sepiolite affected these composites' mechanical, thermal, and barrier qualities.

Mo et al. invest the production, properties, and uses of sepiolite nanoclays. The potential uses of sepiolite-based materials in a variety of industries like medicine delivery.

3.3.5 Halloysite

The distinctive tubular structure of halloysite, a naturally occurring clay mineral with a diameter (about 50 nm) and a length (up to 1 μm), has made it a popular choice for usage as a nanoclay. Fillers, adsorbents, and medication carriers are just a few of the industries that have used halloysite nanoclays because of their elevated aspect ratio, increased surface area, and biocompatibility.

As nanofillers, halloysite nanoclays in particular have demonstrated significant promise in enhancing the barrier, mechanical, and thermal characteristics of polymer composites. Additionally, the creation of materials based on halloysite has spread to industries like medication delivery, and catalysis.

The characteristics, uses, and functions of halloysite-based materials in a variety of disciplines have been investigated in recent research by Zabihi et al., Tabarsa et al., and Guo et al. Zabihi et al. provide a thorough examination of the unique properties and potential uses of halloysite nanotubes whereas Tabarsa et al. focus on their likely use in biomedical domains such as bioimaging, tissue regeneration, and medication delivery. Guo et al. conclude by outlining the latest developments in the application of halloysite nanotubes as nanofillers in polymer composites.

3.3.5.1 Applications of Nanoclay

For many years, nanoclay material has been researched and used in various industries, like automobile engineering, aerospace and construction materials. Improved electrical, mechanical, thermal, and characteristics of the composite materials have been demonstrated by the incorporation of nanoclay.

3.3.5.2 Biomedical Application of Nanoclay

Nanoclay's remarkable biocompatibility and nontoxic qualities have generated a lot of enthusiasm in the biomedical community about its prospective applications. The use of several kinds of nanoclays in biomedical applications, such as scaffolds, soft-tissue engineering, drug delivery systems, wound dressings, and bioimaging, has increased as a result. Nanoclay's adaptability in biomedical research has created a whole new range of opportunities to improve human health and wellbeing. Because of its potential to incorporate and release therapeutic chemicals in a regulated pattern, nanoclay is currently being investigated for its utilization in drug delivery systems. Furthermore, research is looking into its potential as antibacterial agents, tissue engineering, and wound healing, highlighting nanoclay's wide range of applications and enormous potential to transform a number of medical science fields.

Wound Healing

Over time, the use of medicinal ingredients for wound healing has changed, and direct application has proven to be a successful strategy for accelerating healing. The promise of nanoclays to lessen discomfort, scarring, and infection has attracted a lot of interest. The nanoclay composite created by Mei et al. demonstrates encouraging outcomes in stopping bleeding and accelerating healing, demonstrating the ongoing efforts of researchers to enhance wound healing functionality.

Han et al. develop a tough hydrogel for wound dressing applications that incorporates polydopamine-clay-polyacrylamide (PDA-clay-PAM) nanoclay. According to the team, the PDA-clay-PAM hydrogel is a good substance for skin reformation because of its exceptional mechanical qualities, cell affinity, adhesiveness, and biocompatibility.

Scaffolds

Types of nanoclay, including Halloysite, Laponite, and Montmorillonite, have garnered a lot of interest in the creation of scaffolds for tissue engineering applications. Interestingly, researchers used scaffolds based on nanoclay to get promising results. Molla et al., for instance, create scaffolds based on Na-MMT nanoclay for in vitro bone mineralization study. Similar to this, Guo et al. use 3D printing to create a scaffold based on nanoclay that promotes bone marrow growth and aids in the repair of calvarial lesions. Another work describes a biomineralized hydroxyapatite nanoclay composite scaffold based on polycaprolactone that can be used for bone tissue engineering with stem cells.

Researchers also looked into the possibility of adding bioactive substances to scaffolds made of nanoclay for tissue engineering or bone repair. For example, to improve scaffold thermal stability, Pietraszek et al. combine halloysite nanoclay with alkaline phosphatase. Laponite nanoclay is used by Shi et al. for tissue engineering because of its special qualities, which include osteogenic bioactivity, gel-forming ability, and protein interactions. Even if the research has made great strides, there is still much room to develop new materials or polymers and enhance the effectiveness and usefulness of scaffolds based on nanoclay.

Drug Delivery

One of the most important applications for nanoclay-based nanocomposite and nanomat implementations is drug delivery. A study on hydrogel based on laponite nanoclay polymers for drug delivery was carried out by Sheikhi et al. (2018). Their research cleared the path for the creation of clay-based injectable biomaterials for drug administration and clarifies the interaction between nanoclay and polymers in physiological contexts.

Additionally, Shang Yi et al. use kaolomite, a promising substance for cancer treatment, in their research on drug delivery for cancer therapy. For the treatment of postoperative residual thyroid cancer, this kaolinite-based drug delivery system shows potential in combining targeted therapy, imaging diagnosis, and drug delivery into a single, cohesive system. Furthermore, Lobato-Aguilar reports on the antibacterial uses of chlorhexidine diacetate and montmorillonite. Within the first 5 hours, their drug delivery study shows an initial burst release, suggesting the possibility of quick antimicrobial effects.

Bioimaging

Modern life science research relies heavily on bioimaging because it allows researchers to precisely examine molecules, cells, and other biological things. Nanoclay's remarkable physical and chemical characteristics have made it a desirable material for bioimaging and optoelectronics in recent years.

In vitro cytotoxicity investigation of fluorescent gold nanoclusters stabilized on a halloysite for bioimaging is carried out by Gorbachevskii et al. (2021). According to their findings, within a specific concentration range, the AuNCs stabilized on halloysite were effectively absorbed by cells, did not cause any harmful effects, and did not harm the cell membrane. Gold nanoclusters stabilized on halloysite are suitable for bioimaging, cancer treatment, and visualizing halloysite in biological entities due to this concentration-dependent toxicity mechanism.

A simple method for creating a functional nanocarrier with laponite nanoclay is also reported by Gangadharappa et al. Their hybrids protect insoluble transition metal complexes from physical quenching by assisting in their dispersion in aqueous solutions.

Soft Tissue

The research community has been very interested in soft-tissue engineering since these tissues are essential for linking and supporting the body's many organs. By examining the function of both unmodified and modified clay nanoparticles in the creation and description of polyglycerol-sebacic acid-urethane (PGSU) nanocomposites for soft-tissue applications, Monem et al. fill this gap. They show that Cloisite Na + and Cloisite nanoclay have distinct impacts on the hydrophilicity and hydrolytic mass loss of the PGSU matrix by creating a number of elastomeric materials.

For use in soft-tissue engineering, Golbaten-Mofrad more recently reported the creation of bio-elastomer nanocomposites made of poly(glycerol sebacate-co-gelatin) that combine with graphene oxide and nanoclay. Their research demonstrates the viability of using these composites for a variety of applications, such as shielding against electromagnetic interference, drug delivery systems, and biosensing. Although additional research is required to fully explore their potential in this sector, our findings imply that nanoclay-based materials have promise for creating innovative and versatile soft-tissue engineering solutions.

3.4 Properties of Inorganic Nanofillers

Inorganic nanofillers exhibit an extraordinary spectrum of physicochemical characteristics, making them indispensable in advanced material engineering. Their intrinsic high thermal stability, superior mechanical reinforcement, and enhanced barrier functionalities derive from their structural robustness and chemically inert nature. Due to their elevated aspect ratios and extensive surface areas, these nanofillers foster substantial interfacial interactions within polymer matrices, thereby augmenting load transfer efficiency and optimizing stress distribution mechanisms (Bhushan 2017). This results in unprecedented enhancements in mechanical toughness, elongation at break, and impact resistance. In addition, their inherent electrical and thermal conductivity attributes further broaden their functional horizon, facilitating applications in conductive polymer composites and thermal interface materials. In addition, their tunable optical and catalytic behaviors, stemming from quantum confinement effects and surface plasmon resonance phenomena, provide immense potential for photonic and energy-harvesting applications. The precise control over particle morphology, crystallinity, and dispersion state enables tailored material responses, thereby optimizing their applicability across numerous industrial sectors (Pattanayak et al. 2020–2023). The synergy between nanoscale dimensions and tailored surface chemistry allows for hierarchical structuring, imparting multifunctional properties that significantly surpass conventional microfilmers. In further, their ability to modulate rheological behavior, fire retardancy, and permeability renders them pivotal in high-performance coatings, biomedical devices, and energy storage systems (Pattanayak et al. 2020–2023).

3.5 Applications of Inorganic Nanofillers

The utilization of inorganic nanofillers spans an extensive range of high-end technological domains due to their unparalleled multifunctionality and tunable physicochemical properties. In *biomedical engineering*, these nanofillers are integral to advanced drug delivery systems, bioactive scaffolds for regenerative medicine, and antimicrobial surface modifications due to their biocompatibility and controlled release mechanisms (Pattanayak et al. 2020–2023; Yadav et al. 2024). In *coatings and membranes*, they optimize corrosion resistance, UV shielding, and selective permeability control, ensuring prolonged material lifespan and environmental resilience (Jagadeesh et al. 2021). The unique physicochemical versatility of these nanofillers enables their implementation in next-generation multifunctional materials, ensuring enhanced operational stability, energy efficiency, and environmental sustainability in cutting-edge technologies (Basavegowda et al. 2021).

3.6 Comparative Analysis

3.6.1 Key Differences

The heterogeneous nature of inorganic nanofillers stems from their intrinsic physicochemical attributes, governing their applicability across advanced material systems. Metallic nanofillers, exemplified by silver and gold nanoparticles, demonstrate unparalleled electron mobility and innate antimicrobial efficacy, rendering them pivotal in bioelectronics, conductive matrices, and pathogen-resistant interfaces (Kuang et al. 2023). Their localized surface plasmon resonance (LSPR) further amplifies optoelectronic functionalities, enhancing their role in high-sensitivity biosensing platforms.

Conversely, ceramic-based nanofillers such as silica, alumina, and zirconia contribute exceptional thermal resilience, phase stability, and structural fortification, making them indispensable in thermomechanical reinforcements and oxidative-resistant coatings (Vu et al. 2023). Their high aspect ratios and surface polarities foster superior matrix adhesion, mitigating thermomechanical degradation in extreme environments.

Carbonaceous nanofillers, including graphene derivatives and carbon nanotubes (CNTs), surpass traditional fillers in charge carrier mobility and phonon transport efficiency, underpinning their dominance in next-generation nanoelectronics, thermal interface materials, and lightweight structural composites (Pattanayak et al. 2020–2023). However, their intrinsic hydrophobicity and aggregation tendencies necessitate tailored functionalization approaches, such as covalent grafting or π–π stacking modifications, to enhance dispersion kinetics and interfacial coupling (Iijima and Bethune 1993).

The anisotropic geometry, quantum confinement effects, and defect-engineered electronic states of these nanofillers critically modulate interfacial adhesion, phonon scattering, and charge delocalization within hybrid matrices (Alshehri et al. 2013–2016). Surface functionalization via heteroatom doping, ligand exchange, and core–shell architectures further diversify their applicability, enabling tunable physicochemical interactions tailored for specific industrial exigencies.

By discovering controlled defect engineering, interfacial chemistry, and hierarchical structuring, the tailored integration of these nanofillers advances material performance beyond conventional paradigms, driving innovation in multifunctional composite systems.

3.6.2 Synergistic Effects

Hybrid inorganic nanofillers exhibit unprecedented multifunctionality by synergistically integrating distinct physicochemical attributes, fostering superior material enhancements. Graphene-nanoclay hybrid architectures, for instance, manifest exceptional flame retardancy, augmented mechanical resilience, and heightened impact mitigation due to interfacial reinforcement and nanoscale confinement effects. The anisotropic morphology of graphene synergizes with the layered structure of nano-clays, optimizing load transfer dynamics and mitigating crack propagation in polymer matrices.

Metal-oxide hybrids, such as TiO_2-ZnO composites, exhibit dual photocatalytic and antimicrobial properties, driven by interfacial charge delocalization and band-gap modulation. This tailored band alignment fosters enhanced electron–hole separation, mitigating recombination losses and amplifying photocatalytic efficiencies. Such nanocomposites exhibit superior pollutant degradation and biofilm inhibition, rendering them invaluable in environmental remediation and biomedical sterilization.

The intrinsic synergy in these hybrid architectures is predominantly dictated by interfacial electronic interactions, hierarchical structuring, and controlled phase dispersion within the host matrix. These engineered nanostructures optimize mechanical compliance and uniform stress dissipation, mitigating premature material failure under dynamic loading conditions. Tailored spatial distributions further enhance matrix adhesion, ensuring structural integrity in high-performance applications.

In addition, the modulation of interfacial polarization in dielectric nanocomposites, via strategic hybridization, facilitates tunable permittivity and loss tangent minimization, enabling advanced electronic and optoelectronic applications. Functional tuning of such hybrid nanostructures is achieved through precise control over filler morphology, surface functionalization, and dispersion kinetics, ensuring optimal performance in next-generation adaptive materials (Saito and Ingham 2005–2014).

Hierarchical hybrid networks, encompassing multidimensional nanofillers, introduce synergistic reinforcement via optimized filler-matrix interactions, reducing percolation thresholds and enhancing charge mobility in conductive nanocomposites. The interplay of interfacial adhesion and charge transfer phenomena further facilitates superior mechanical robustness, making them pivotal in aerospace, automotive, and structural reinforcement domains.

The emergence of computational modeling and in-situ characterization techniques has enabled a deeper mechanistic understanding of synergistic effects, guiding the rational design of hybrid nanostructures. Advanced spectroscopic and microscopic evaluations elucidate phase interactions, defect passivation mechanisms, and localized stress distribution, contributing to predictive material engineering strategies.

Thus, the convergence of hybrid inorganic nanofillers with precision engineering methodologies heralds transformative advancements across multidisciplinary domains. These architected hybrid systems hold immense promise in revolutionizing smart coatings, high-strength composites, advanced optoelectronic platforms, and next-generation functional nanomaterials, pushing the frontier of material innovation.

3.6.3 Limitations and Challenges

Inorganic nanofillers, despite their transformative potential, pose critical challenges that necessitate strategic interventions for large-scale viability. *Dispersion stability* remains a fundamental limitation, as nanoparticle agglomeration disrupts homogeneous phase distribution, undermining interfacial adhesion and mechanical performance (Mamidi et al. 2017). Advanced surface engineering—via *polymer grafting, ligand exchange, and surfactant stabilization*—is indispensable for mitigating these stability issues, ensuring robust colloidal dispersion and compatibility.

Polymer matrix integration further complicates performance optimization, demanding *functionalization techniques* such as *silanization, plasma treatment, and covalent modification* to enhance interfacial bonding. These strategies facilitate superior mechanical reinforcement while preserving intrinsic properties. However, the *toxicological footprint* of nanoscale fillers remains a pressing concern, with potential cytotoxicity necessitating *rigorous biocompatibility profiling*. In-depth *in vitro and in vivo analyses*, supplemented by lifecycle impact assessments, are imperative to ascertain *long-term ecological and biological safety* (Kanagaraj and Liu 2011–2013).

Manufacturing scalability presents additional constraints, as *high-temperature synthesis, intricate processing protocols, and cost-prohibitive fabrication techniques* hinder widespread industrial adoption. Emerging paradigms in *solution-phase synthesis, green chemistry methodologies, and additive manufacturing* hold promise for overcoming these bottlenecks, offering sustainable and economically viable production alternatives.

The interplay between *advanced material engineering, computational simulations, and process optimization* will be instrumental in refining inorganic nanofiller applications. A holistic approach, integrating *nanostructure control, predictive modeling, and hybridization strategies*, will catalyze breakthroughs in next-generation high-performance materials.

3.7 Cutting-Edge Developments in Inorganic Nanofillers

3.7.1 Surface Functionalization and Interface Engineering

Cutting-edge developments in inorganic nanofillers have propelled sophisticated surface modification strategies to refine dispersion dynamics and interfacial cohesion within polymer matrices. Advanced functionalization modalities, including silane coupling, plasma-mediated activation, and ligand exchange, are strategically employed to modulate nanofiller compatibility with host substrates at the molecular level.

By precisely engineering surface energy parameters, these modifications suppress agglomeration tendencies and facilitate uniform nanoscale distribution. In further, tailored covalent and supramolecular grafting methodologies introduce chemically interactive moieties, reinforcing matrix adhesion and physicochemical synergy.

The integration of reactive organosilane linkers, amidation protocols, and thiol-ene click chemistry fosters robust filler-polymer interactions, augmenting mechanical resilience and thermochemical stability. In addition, zwitterionic and amphiphilic coatings have demonstrated efficacy in modulating interfacial polarity, enhancing wettability, and mitigating phase segregation (Bakshi et al. 2007).

Synergistic hybridization of functionalization techniques has also led to tunable electronic, optical, and rheological properties, vital for next-generation composite applications. Covalent immobilization of functional ligands further ensures site-specific chemical affinity, optimizing polymer network entanglement and load transfer efficiency.

Plasma-assisted methodologies, discovering high-energy species and radical-driven surface restructuring, induce selective etching and functional moiety grafting, enhancing filler-polymer compatibility. Meanwhile, electrostatic and steric stabilization mechanisms counterbalance Van der Waals forces, suppressing nanoscale particle aggregation.

Emerging research on biomimetic interface engineering explores peptide-mediated self-assembly and bioinspired adhesion motifs, fostering hierarchical structuring and enhanced interfacial robustness. Such approaches mimic natural adhesion paradigms, refining composite adaptability across diverse environmental conditions.

These advancements in nanofiller surface engineering redefine interfacial thermodynamics, imparting tunable anisotropic interactions and broadening the functional landscape of polymeric architectures.

3.7.2 Precision-Controlled Synthesis and Morphological Engineering

Advanced synthesis methodologies, encompassing sol–gel processing, hydrothermal crystallization, and atomic layer deposition (ALD), have revolutionized the fabrication of nanofillers, yielding materials with precisely engineered physicochemical properties. The sol–gel approach, a bottom-up strategy, enables the molecular-level assembly of nanostructures, ensuring unparalleled homogeneity and tunable mesoporosity through controlled hydrolysis and condensation reactions.

Hydrothermal synthesis capitalizes on thermodynamically driven nucleation and growth under elevated temperature and pressure conditions, facilitating the crystallization of nanomaterials with remarkable phase purity and anisotropic control. This method fosters the development of structurally ordered nanofillers with enhanced morphological precision, tailored electronic properties, and defect minimization.

ALD, an advanced vapor-phase deposition technique, achieves atomic-scale precision in fabricating ultra-thin, conformal coatings, thus allowing the modulation of interfacial characteristics with sub-nanometer accuracy. This methodology underpins the construction of core-shell architectures, endowing nanofillers with tunable surface reactivity, mechanical robustness, and enhanced physicochemical resilience (Drumright et al. 2000).

The synergistic integration of these synthesis routes propels the rational design of high-performance nanomaterials with application-specific functionalities. The interplay of sol–gel chemistry, hydrothermal kinetics, and ALD-mediated deposition offers an unparalleled degree of control over crystallinity, porosity, and interfacial interactions, ensuring optimal property modulation at the nanoscale.

By discovering these state-of-the-art paradigms, the next-generation nanofillers exhibit superior dispersion stability, enhanced mechanical integrity, and precise functional tunability, positioning them at the forefront of advanced materials engineering.

3.7.3 Computational and AI-Driven Material Discovery

The convergence of artificial intelligence (AI) and machine learning (ML) has revolutionized the rational design and functional optimization of next-generation nanofillers. Advanced computational paradigms, including high-throughput density functional theory (DFT) simulations and molecular dynamics (MD) modelling, are expediting the predictive engineering of nanomaterials with bespoke physicochemical attributes.

AI-enhanced predictive analytics are enabling rapid screening of novel nanofiller architectures, elucidating intricate filler-matrix interactions that dictate macroscopic performance. Deep-learning algorithms, trained on vast datasets, are now capable of establishing nontrivial correlations between nanoscale structural parameters and bulk material properties, fostering a data-driven approach to formulation science.

By integrating multiscale modelling frameworks, AI-driven methodologies facilitate accelerated materials discovery, mitigating empirical trial-and-error processes. Reinforcement learning strategies are being deployed to optimize synthesis pathways, ensuring thermodynamic stability and functional efficacy of engineered nanofillers within polymeric and composite matrices.

Inverse design principles, underpinned by generative adversarial networks (GANs) and variational autoencoders (VAEs), are redefining material informatics, enabling the predictive tailoring of nanofillers with precision-tuned functionalities. Such advancements bridge theoretical modelling and experimental realization, significantly reducing developmental timelines.

In further, quantum computing is emerging as a transformative tool in resolving electronic structure complexities, augmenting AI-guided DFT calculations for more accurate property predictions. The integration of Bayesian optimization and genetic algorithms further refines computational workflows, identifying optimal synthesis parameters with minimal experimental iterations.

These advancements are catalyzing the discovery of high-performance nanofillers with superior mechanical resilience, thermal stability, and interfacial compatibility, ensuring their seamless integration into next-generation multifunctional materials. The confluence of AI, computational chemistry, and nanotechnology is ushering in an era of accelerated innovation, wherein material performance is engineered at the atomic level with unprecedented precision.

3.7.4 Multiscale Performance Enhancement and Functional Integration

The advent of hierarchical structuring techniques has allowed for the fabrication of multiscale nanofillers exhibiting synergistic reinforcement effects across molecular, mesoscopic, and macroscopic dimensions. The incorporation of hybrid inorganic–organic motifs has been shown to impart superior mechanical robustness, thermal stability, and electrochemical functionality to composite systems. In addition, emerging self-assembly and biomimetic approaches have facilitated the development of nanofillers with dynamically tunable properties, wherein external stimuli such as pH, temperature, and light can modulate interfacial interactions and performance characteristics in real time.

3.7.5 Sustainability-Driven Innovations and Green Processing

Amidst the escalating demand for ecologically harmonious material technologies, research trajectories have increasingly converged on sustainable synthesis paradigms for inorganic nanofillers. By integrating tenets of green chemistry, methodologies such as solvent-exempt processing, biomimetic mineralization, and microwave-facilitated synthesis have been refined to attenuate environmental ramifications while augmenting the feasibility of large-scale nanofiller production.

The strategic repurposing of industrial residuals and biogenic mineral resources has been investigated to fortify circular material economies within nanotechnology, mitigating raw material depletion and waste proliferation. These alternative precursor strategies not only enhance resource efficiency but also align with global sustainability mandates.

In addition, advancements in solvent-free mechanochemical techniques have demonstrated superior efficacy in synthesizing high-purity nanofillers, circumventing hazardous reaction conditions intrinsic to conventional wet chemistry routes. This solvent-independent approach mitigates toxic effluents while preserving structural integrity and functional performance.

Microwave-assisted synthesis, discovering rapid volumetric heating and selective energy absorption, has emerged as a pivotal methodology, ensuring uniform nucleation and crystallization kinetics while substantially curbing thermal energy dissipation. This precise energy modulation facilitates controlled morphology and phase stability of the synthesized nanostructures.

Bioinspired mineralization, mimicking biologically templated hierarchical assembly, has been explored to engineer nanofillers with tunable physicochemical attributes. By employing bio-macromolecular scaffolds and peptide-mediated nucleation, researchers have successfully fabricated nanostructures with enhanced biocompatibility and mechanical resilience.

Waste valorization strategies, particularly through the utilization of metallurgical slags, silica-rich agro-waste, and marine-derived biominerals, have demonstrated remarkable potential in substituting conventional mineral sources. This not only mitigates industrial byproduct accumulation but also fosters the development of high-performance materials with a reduced carbon footprint.

Catalyst-free and template-guided synthesis modalities have been optimized to minimize auxiliary reagent dependency, reducing both synthesis complexity and ecological burden. These approaches reinforce the principles of atom economy and process intensification, streamlining nanofiller production with minimal resource expenditure.

In further, hybrid approaches integrating electrochemical deposition and ultrasonic irradiation have exhibited promising advancements in modulating particle morphology and dispersion stability, which are crucial for high-performance material applications. These methodologies contribute to enhancing interfacial interactions and reinforcing composite matrices.

Ultimately, the convergence of green synthetic innovations and sustainable precursor utilization underscores a paradigm shift in nanomaterial engineering, where environmental stewardship is seamlessly integrated with high-performance material development. The continual refinement of these eco-conscious methodologies propels the field toward a resilient and resource-efficient future.

3.7.6 Future Trajectories in Inorganic Nanofiller Engineering

The convergence of precision-engineered nanofillers, computational material design, and sustainable fabrication methodologies is anticipated to drive paradigm-shifting innovations across multifaceted technological landscapes. The strategic incorporation of multifunctional inorganic nanofillers into advanced polymer nanocomposites, biointegrative implants, and high-efficiency energy storage matrices offers unprecedented pathways for augmenting mechanical robustness, electrochemical performance, and adaptive functionality.

Synergizing nanoscale structuration with artificial intelligence-driven predictive modeling enables atomistic-level customization of nanofillers, optimizing interfacial interactions and hierarchical self-assembly. This confluence enhances charge transport kinetics, mechanical resilience, and biocompatibility, redefining material architectures for next-generation applications.

Within biomedical engineering, nanofiller-infused scaffolds and implantable substrates exhibit tailored bioactivity, controlled degradation kinetics, and enhanced cellular interfacing, fostering breakthroughs in regenerative medicine and precision therapeutics. Functionalized nanofillers further potentiate antimicrobial efficacy, osteointegration, and targeted drug delivery, revolutionizing patient-centric biomedical solutions.

In energy storage and conversion systems, nanofiller-reinforced electrode architectures and solid-state electrolytes exhibit superior ion transport pathways, thermal stability, and electrochemical durability. These advancements underpin high-capacity batteries, supercapacitors, and fuel cells, ensuring sustainability through prolonged operational lifespans and reduced environmental impact (Khater et al. 2024).

Optoelectronic applications benefit from quantum-engineered nanofillers, enabling tunable bandgaps, enhanced photonic interactions, and optimized charge carrier dynamics. These attributes drive advancements in photodetectors, light-emitting devices, and solar energy harvesting platforms, marking a shift toward next-generation semiconductor paradigms.

By integrating computational intelligence, multi-scalar characterization techniques, and bioinspired design principles, the future of nanofiller-engineered materials transcends traditional limitations. Interdisciplinary innovation will catalyze disruptive applications in aerospace composites, nanoelectronics, and theranostic platforms, bridging the gap between material science and translational technology.

With continued research into green synthesis routes and circular economy-driven material recovery, inorganic nanofillers are poised to redefine structural paradigms across multiple domains. The forthcoming era of intelligent nanomaterial engineering promises to unlock unprecedented material functionalities, driving the next wave of scientific and technological evolution.

3.7.7 Eco-Conscious Synthesis and Green Nanotechnology

The evolution of inorganic nanofiller development is progressively shifting toward environmentally responsible methodologies, integrating sustainable synthesis routes that minimize ecological footprint. Green chemistry-driven fabrication techniques, such as solvent-free processing, bioinspired mineralization, and mechanochemical synthesis, are being optimized to curtail hazardous byproducts while enhancing material efficiency. The utilization of renewable precursors, including biogenic minerals and waste-derived feedstocks, further strengthens the commitment to circular material economies, ensuring resource conservation and sustainability in large-scale applications [https://fmtmagazine.in/metallic-nanofillers-in-food-packaging/].

3.7.8 Intelligent and Stimuli-Responsive Nanofillers

The advent of smart nanofillers with dynamic responsiveness to external stimuli is poised to redefine the landscape of adaptive material systems. These advanced fillers exhibit tunable properties under varying conditions such as temperature, pH, electromagnetic fields, and mechanical stress, enabling real-time modulation of mechanical, electrical, and optical characteristics. Functionalized nanofillers with shape-memory effects, self-healing capabilities, and bio-mimetic adaptability are emerging as pivotal components in next-generation structural, biomedical, and electronic materials, fostering unprecedented material versatility [https://www.mdpi.com/journal/pharmaceutics/special_issues/metal_based_npa].

3.7.9 AI-Enabled Predictive Material Design

The synergy between nanotechnology and artificial intelligence (AI) is accelerating the rational design and optimization of inorganic nanofillers. Machine learning algorithms, coupled with high-throughput computational modelling, are facilitating predictive analytics for filler-matrix interactions, enabling the systematic tailoring of surface chemistry, dispersion kinetics, and interfacial adhesion. Advanced AI-driven material screening approaches are expediting the identification of novel nanofiller compositions, significantly reducing experimental iterations while enhancing performance predictability across diverse industrial sectors.

3.7.10 Cross-Disciplinary Integration and Technological Convergence

The intersection of inorganic nanofillers with advanced manufacturing paradigms, such as additive manufacturing and hybrid material processing, is broadening the scope of multifunctional applications. Innovations in 3D and 4D printing methodologies are enabling precise spatial distribution of nanofillers within composite frameworks, paving the way for architecturally optimized, high-performance materials. In addition, the fusion of nanofiller technology with emerging fields like quantum materials, flexible electronics, and regenerative medicine is unlocking new dimensions in functional material engineering, fostering unprecedented advancements in scientific and industrial frontiers.

3.7.11 Future Outlook and Transformative Potential

The trajectory of inorganic nanofillers is charting an ambitious course toward unparalleled material innovation, driven by sustainability, intelligent responsiveness, and AI-powered optimization. The seamless integration of these fillers into next-generation polymers, biomaterials, and electronic components is anticipated to revolutionize material performance, durability, and environmental compliance. As research continues to push the boundaries of nanoscale engineering, the transformative impact of inorganic nanofillers is poised to redefine structural, biomedical, and energy-related applications, heralding a new era of advanced material science.

3.8 Comprehensive Conclusion on Advances in Inorganic Nanofillers

3.8.1 Holistic Evolution of Inorganic Nanofiller Technologies

The field of inorganic nanofillers has undergone a profound transformation, with research efforts increasingly focusing on tailoring their structural, physicochemical, and interfacial properties. The incorporation of advanced surface modification strategies, including silanization, plasma-assisted functionalization, and ligand exchange, has significantly enhanced the dispersion and matrix compatibility of nanofillers in polymeric and composite systems. These methodologies have enabled the precise control of surface energy, charge distribution, and reactivity, ensured uniform incorporation while mitigating agglomeration challenges that traditionally hinder nanofiller efficiency. The integration of hybrid organic–inorganic fillers has further broadened the applicability of these materials, creating high-performance composites with superior mechanical, thermal, and electrical attributes.

3.8.2 State-of-the-Art Synthesis and Structural Engineering

Innovative synthesis techniques such as sol–gel processing, hydrothermal crystallization, and atomic layer deposition (ALD) have revolutionized the ability to fabricate nanofillers with defined morphologies, tunable porosity, and controlled crystallinity. These approaches facilitate the precise engineering of nanoparticle size distributions and surface functionalities, ensuring optimal interactions with host matrices. In addition, hierarchical structuring methodologies have allowed for the development of multiscale nanofillers, wherein synergistic reinforcement effects occur at molecular, mesoscopic, and macroscopic levels. The incorporation of nano-architectured fillers within polymer frameworks has paved the way for enhanced structural integrity and functional adaptability in high-performance materials.

3.8.3 AI-Driven Predictive Modelling and Computational Design

The advent of artificial intelligence (AI) and machine learning (ML) in material science has significantly expedited the rational design and optimization of inorganic nanofillers. Computational methodologies, including density functional theory (DFT) simulations and molecular dynamics (MD) modelling, have facilitated predictive insights into filler-matrix interactions, leading to enhanced performance predictability and reduced experimental iterations. AI-driven high-throughput screening has enabled the rapid identification of novel filler compositions with tailored physicochemical properties, optimizing parameters such as adhesion strength, thermal conductivity, and mechanical resilience. As a result, data-driven materials innovation is rapidly transforming the landscape of nanofiller development, bridging theoretical modelling with experimental validation.

3.8.4 Stimuli-Responsive and Smart Nanofiller Systems

The emergence of intelligent nanofillers with stimuli-responsive capabilities has opened new frontiers in the development of adaptive and self-regulating materials. These advanced fillers exhibit dynamic tunability in response to external stimuli, including temperature, pH fluctuations, electromagnetic fields, and mechanical stress. Shape-memory nanofillers, self-healing interfaces, and bioinspired hierarchical structures have been incorporated into next-generation polymeric and biomedical composites, fostering enhanced resilience and multi-functionality. In addition, the fusion of stimuli-responsive nanofillers with emerging 4D printing technologies is facilitating the fabrication of smart materials capable of autonomous structural and functional transformations.

3.8.5 Green Synthesis and Sustainable Engineering Approaches

The increasing emphasis on sustainability has propelled the development of eco-friendly nanofiller synthesis methodologies, aligning with green chemistry principles. Environmentally benign fabrication techniques, including solvent-free processing, biomimetic mineralization, and microwave-assisted synthesis, have minimized hazardous byproducts while improving energy efficiency and scalability. The utilization of renewable precursors, such as bio-derived mineral sources and industrial waste materials, has further reinforced the transition toward circular material economies. These sustainable advancements not only reduce the environmental footprint of nanofiller production but also enhance the viability of large-scale industrial applications, ensuring long-term economic and ecological benefits.

3.8.6 Multifunctional Applications and Cross-Disciplinary Integration

The integration of inorganic nanofillers into diverse application domains has underscored their transformative potential in material science, biomedical engineering, electronics, and energy storage systems. Nanofiller-reinforced composites have demonstrated exceptional mechanical reinforcement in structural applications, improved conductivity in next-generation electronic devices, and enhanced bioactivity in regenerative medicine. The incorporation of these fillers into flexible electronics, quantum materials, and advanced drug delivery systems has further expanded their technological relevance. In further, interdisciplinary collaborations between nanotechnology, biotechnology, and computational science have accelerated the development of hybrid materials with unprecedented functional adaptability (Ranjan and Dave 2024).

3.8.7 Future Horizons and Translational Potential

The trajectory of inorganic nanofiller research is poised for continued breakthroughs, driven by advances in precision synthesis, computational intelligence, and sustainability. The future of these materials lies in the seamless convergence of AI-powered predictive modelling, hierarchical material structuring, and bioinspired design principles. With the rise of autonomous manufacturing and smart material systems, nanofillers are expected to play a pivotal role in next-generation polymer engineering, biomedical implants, and energy-efficient technologies. Continued interdisciplinary research and industrial scalability efforts will be essential in unlocking the full potential of inorganic nanofillers, paving the way for their widespread adoption in high-performance and environmentally responsible applications.

References

Alghamdi HM, Rajeh A (2024) Electrical properties of $CuCoO_4$ nanoparticles. Materials

Alshehri R, Li R, Yu KN (2013–2016) Cytotoxicity and biocompatibility of carbon nanotubes. Nanotoxicology

Amstutz HC et al (1993) Long-term behavior of UHMWPE in orthopedics. Clin Orthop Relat Res

Bakshi SR et al (2007) CNT reinforced UHMWPE composites. Mater Sci Eng A

Basavegowda N et al (2021) Nanofillers in conductive composites. Polymers

Bhushan B (2017) Nanotechnology: principles and practices. Springer

Chandrakala V (2022) Metal oxide nanoparticles in drug delivery. Drug Deliv Transl Res

Dadfar SM (2019) Superparamagnetic nanoparticles in imaging. Small

Dam P (2023) Nanoparticles in wound healing. Mater Today Chem

Drumright RE et al (2000) Polylactide in drug delivery. Adv Mater

Friedrich RP (2021) Magnetic nanoparticles in tissue engineering. Adv Healthcare Mater

Gulsah G et al (2022) Drug delivery via CNT-polystyrene nanocomposites. J Mol Model

Hheidari A (2024) MONPs for targeted therapy. Pharmacogn Mag

Iijima S, Bethune DS (1993) Helical microtubules of graphitic carbon. Nature

Jagadeesh P et al (2021) Nanofiller morphology and applications. Adv Mater Sci

Jali S (2023) Barrier performance in polymer nanocomposites. Polymers

Jamroz E (2019) Zinc oxide nanoparticles in tissue regeneration and scaffolds. J Biomed Mater Res

Janska T (2024) Cerium oxide nanoparticles in CT imaging. Nanotechnology

Kanagaraj S, Liu Y (2011–2013) Wear and tribological performance of UHMWPE. Mater Design

Kuang T et al (2023) Functional CNT nanocomposites. Composite Science and Technology
Mamidi N et al (2017) UHMWPE reinforced with MWCNTs. Int J Nanomed
Mendes C (2022) ZnO nanoparticles in wound healing applications. J Tissue Eng Regen Med
Moskalyuk OA, Yudin VE (2020) Mechanical enhancement using nanofillers. Comp Sci Technol
Nayak PP (2023) Advances in nanofiller technology. Int J Mater Chem Phys
Negrescu AM, Killian M (2022) MgO nanoparticles in medicine. J Biomater Appl
Nikolova MP, Chavali MS (2020) ZnO nanoparticles for cancer therapy. Mater Sci Eng C
Pattanayak DS et al (2020–2023) Nanocomposites: Material innovation. J Mater Res Technol
Poonia N (2024) Iron oxide nanoparticles in targeted cancer therapy. Biomed Pharmacother
Ranjan R, Dave V (2024) Tissue engineering with metal fillers. Mater Sci Eng C
Saito N, Ingham E (2005–2014) UHMWPE biocompatibility in joint implants. Biomaterials
Schmidt-Mende L (2007) Dispersion of metal oxide nanofillers. Adv Mater
Sharifi R (2024) Biomedical applications of metal oxide nanoparticles. Nanomedicine
Tasnim NT (2023) Antimicrobial action of nanoparticles. Front Microbiol
Vinardell MP (2015) Cytotoxic properties of metal nanoparticles. Toxicol Vitro
Vu V et al (2023) Graphene-based polymer composites. J Appl Polym Sci
Wu Z, Huang NE (2009) Mechanical properties of tissue scaffolds. J Mech Behav Biomed Mater
Xue Y et al (2016) Osteolysis and UHMWPE. Biomater Sci
Yadav V et al (2024) Role of nanofillers in material enhancement. Mater Today: Proc
Yassin MT (2024) Safety challenges of MONPs. Curr Med Chem
Zhang J, Mou L, Jiang X (2020) Surface chemistry of nanofillers. Adv Mater Interf

Chapter 4
Synthesis Methods for Polymer Nanocomposite

Umesh S. Panchal, Ganesh R. Godge, Varsharani R. Avhad, Manish A. Raskar, Bhagyashri B. Randhawan, and Sachin N. Kothawade

Abstract Polymer nanocomposites (PNCs) are advanced materials that improve the mechanical, thermal, electrical, and barrier properties of polymer matrices by incorporating nanoparticles into them. To define the structure-property correlations and overall performance of PNCs, synthesis is essential. Depending on the kind of polymer, nanofiller, and intended use, there are several synthesis techniques that offer unique advantages. One popular method is in situ polymerization, in which nanoparticles are first distributed throughout a monomer solution. The uniform distribution of nanofiller and enhanced material properties are the outcomes of this technique, which fosters strong interfacial interactions. Another popular method is solution mixing, which involves dispersing nanoparticles in a polymer solution and then letting the solvent evaporate. Processing conditions and solvent compatibility may limit its ability to achieve good dispersion. By employing high shear mixing to incorporate nanoparticles into a polymer melt, melt blending is a solvent-free, eco-friendly method. Although it can have problems with nanoparticle aggregation, this technique offers good scalability and is compatible with thermoplastic polymers. In order to create well-distributed nanocomposites with more morphological control, emulsion polymerization involves polymerizing monomers in an aqueous solution with surfactants and scattered nanoparticles. Advanced methods include layer-by-layer (LbL) assembly, which provides exact control over

U. S. Panchal · G. R. Godge (✉)
Department of Pharmaceutics, Dr Vithalrao Vikhe Patil Foundation's College of Pharmacy, Ahilyanagar, Maharashtra, India

V. R. Avhad · B. B. Randhawan
Arihant College of Pharmacy, Ahilyanagar, Maharashtra, India

M. A. Raskar
Department of Pharmaceutical Chemistry, Dr Vithalrao Vikhe Patil Foundation's College of Pharmacy, Ahilyanagar, Maharashtra, India

S. N. Kothawade
Department of Pharmaceutics, SCSSS's Sitabai Thite College of Pharmacy, Shirur, Pune, Maharashtra, India

S. N. Kothawade et al. (eds.), *Polymer Nanocomposites*,
https://doi.org/10.1007/978-3-032-00365-2_4

nanofiller placement at the nanoscale, and electrospinning, which produces polymer nanocomposite fibers with a large surface area. Additionally, recent developments explore sol-gel processing for hybrid organic-inorganic nanocomposites, which achieve remarkable interfacial adhesion and dispersion. The characteristics, processing viability, and scalability of the nanocomposite for industrial applications are all significantly impacted by the synthesis method choice.

Keywords Polymer nanocomposites · In situ polymerization · Solution mixing · Layer-by-layer · Nanofiller · Electrospinning · Sol-gel processing

4.1 Introduction

Polymer nanocomposites are hybrid organic-inorganic materials characterized by at least one dimension of the filler phase being less than 100 nm (Mittal 2009). They are synthesized through various methods, which can be grouped into four main categories: melt intercalation, template synthesis, exfoliation adsorption, and in situ polymerization intercalation (Mittal 2009; Alexandre and Dubois 2000). Depending on the method and materials used, three microstructures can be achieved: intercalated (or micro-composite), intercalated (and/or flocculated), or exfoliated (or delaminated), as shown in Fig. 4.1. Melt intercalation is the standard technique for

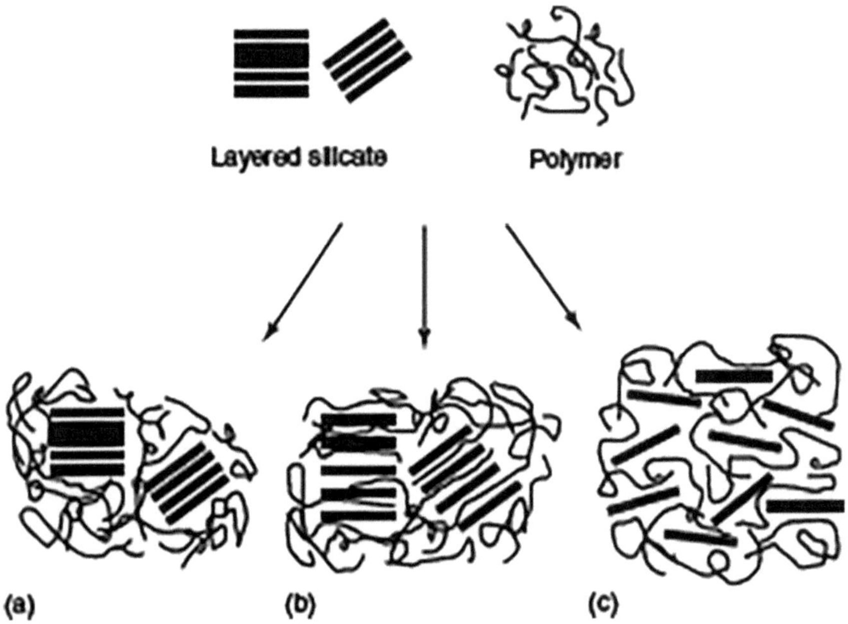

Fig. 4.1 Types of composite microstructures: (**a**) unintercalated (phase separated (microcomposite)), (**b**) intercalated (intercalated (nanocomposite)), and (**c**) exfoliated (nanocomposite) (Pomogailo et al. 2007)

producing thermoplastic polymer nanocomposites. This process involves heating the polymer matrix, adding the filler, and kneading the composite to ensure uniform distribution. This method is environmentally friendly due to the absence of solvents and is compatible with industrial processes like injection molding and extrusion, enhancing its practicality and cost-effectiveness. However, the high temperatures can damage the filler's surface modification; for instance, organoclays modified with alkyl ammonium typically decompose at temperatures exceeding 140 °C, while melt intercalation operates between 190 and 220 °C (Abedi and Abdouss 2014). Thus, optimizing processing conditions is crucial for achieving good dispersion and exfoliation. Lower temperatures or the use of more thermally stable modifications can help prevent degradation (Mittal 2009). Weak electrostatic forces between filler interlayers and compatibility with the polymer matrix enable the polymer to penetrate these interlayers, forming intercalated or exfoliated nanocomposites (Alexandre and Dubois 2000). Exfoliation adsorption, also known as polymer or prepolymer intercalation from solution, involves a solvent in which the polymer or prepolymer dissolves. Here, layered silicates are first swollen and dispersed in the solvent before being mixed with the polymer solution (Pavlidou and Papaspyrides 2008).

Polymer nanocomposites arise from the judicious combination of two or more nanoparticles or nanosized objects through appropriate techniques (Gangopadhyay and De 2000), merging polymer technology with nanotechnology. Consequently, polymer nanocomposites exhibit numerous unique properties, including catalytic (Samanta and Nandi 2009), electrochemical (Pérez-Cappe et al. 2008; Janáky and Visy 2008), mechanical (Pramanik et al. 2009; Vora and Vora 2006), magnetic (Zhou et al. 2009), optical (Mallick et al. 2007a), and biological properties (Pramanik et al. 2008). These impressive characteristics lead to a wide range of potential applications in various fields. Many studies have reported basic and novel applications of polymer nanocomposites (Pramanik et al. 2008). For instance, study in (Njagi and Andreescu 2007) documented stable enzyme biosensors based on chemically synthesized Au-PPy polymer nanocomposite. Authors in (Mallick et al. 2007b) explored their use in microelectronic devices. Ultimately, we categorize polymer nanocomposites into two types: (inorganic materials)-polymer nanocomposites and (organic materials)-polymer nanocomposites. In the case of (inorganic materials)-polymer nanocomposites, the nanometer units comprise inorganic nanoparticles and nanolayered materials. Various synthesis methods have been utilized to prepare these promising materials. From the published literature, we obtained most preparation methods from traditional techniques such as sol-gel (Prashantha et al. 2006), one-pot synthesis (Balamurugan et al. 2009), emulsion polymerization (Rana and Sahoo 2007), hot press (Mallick et al. 2007b), in situ polymerization (Segal et al. 2007), and oxidative polymerization (Cheng et al. 2006), as well as from novel methods. For instance, methods like electrochemical synthesis (Pramanik et al. 2008), self-assembly synthesis (Enlow et al. 2007), photo polymerization (Makamba et al. 2008), interfacial polymerization (Mallick et al. 2007b), and intermatrix synthesis (IMS) (Muraviev et al. 2008) have been extensively studied.

4.2 Properties and Applications of Polymer Nanocomposites

Due to the complexity and diversity of nanometer units, polymer nanocomposites are classified into two categories: (inorganic materials)-polymer nanocomposites and (organic materials)-polymer nanocomposites. Additionally, (inorganic materials)-polymer nanocomposites can be further divided based on the nature of the interaction between inorganic and polymeric components (Dovgolevsky et al. 2008): (a) The first category, referred to as (inorganic nanoparticle materials)-polymer nanocomposites in this review, consists of inorganic substances like nanoparticles, colloids, polyoxometalates, and fibers embedded within polymer matrices; (b) The second category, termed (inorganic nano-layered materials)-polymer nanocomposites in this review, involves organic polymers confined within inorganic templates. Research on polymer nanocomposites began with studies of (inorganic nanoparticle materials)-polymer nanocomposites. Authors in (Pramanik et al. 2008; Godge et al. 2024b) noted that the interfacial bonding between nanoparticles and the polymer matrix is influenced by the hydrophilicity or polarity of both fillers and polymer. Following this, pioneering work by using on clay-Nylon 6 initiated a new field of (inorganic nano-layered materials)-polymer nanocomposites (Usuki et al. 1993).

4.3 Preparation Methods of Polymer Nanocomposites

4.3.1 *Sol-Gel Synthesis Method*

The sol-gel synthesis method is a wet chemical technique utilized to prepare polymer nanocomposites on thin films, achieving a relatively homogeneous distribution of nanoparticles within the polymer matrix. In summary, the sol-gel synthesis starts with highly reactive metal alkoxide or inorganic salt and the polymer monomer, which act as precursors. These raw materials are mixed in the liquid phase, with strict control over the molar ratios. After hydrolysis, a sol is formed, which transforms into a solid gel following thermal treatment at low temperatures (Luan et al. 2010; Godge et al. 2025). The sol-gel synthesis method enables the creation of polymer nanocomposites at the molecular level, where segments, inorganic nanoparticles, and organic polymers are interconnected through covalent or hydrogen bonds (Tripathi and Shahi 2009) (Fig. 4.2).

The reparation (Tripathi and Shahi 2009) of a cross-linked poly(vinyl alcohol)-3-[[3-(triethoxysilyl)propyl]amino]propane-1-sulfonic acid (PVA-TPAPS) membrane (with TPAPS theoretically calculated to be 30 wt% ~ 70 wt%) using the sol-gel synthesis method. They dissolved a specified amount of PVA and an appropriate quantity of synthesized TPAPS in hot deionized water, mixing these solutions under

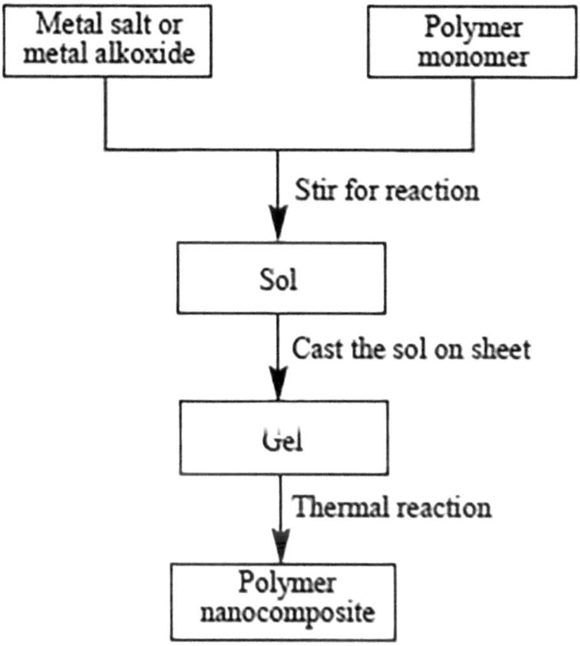

Fig. 4.2 Preparation routes of sol-gel synthesis method

stirring conditions. A known quantity of tetraethyl orthosilicate (TEOS), acting as a cross-linking agent, was then added to the zwitterionomers. The final solution was stirred for 6 h at room temperature to produce the sol. The sol-gel process was conducted through acid hydrolysis (at pH 2) of silanes. Subsequently, the hydrogen-bonded gel was cast onto a clean PVC sheet, dried under ambient conditions for 24 h, and then further dried at 70 °C for 12 h. The transparent membrane was then peeled off from the PVC sheet and cross-linked with a formal solution (HCHO + H_2SO_4) for 3 h at 60 °C.

4.3.2 One-Pot Synthesis Method

The one-pot synthesis method is widely employed for the creation of metal nanoparticle-polymer nanocomposites. This straightforward method involves just three reagents: an inorganic metal salt, a polymer monomer, and water. It can be divided into single-phase one-pot synthesis and interfacial polymerization synthesis. Although thermal treatment is not necessary, a higher temperature can speed up the reaction rate (Bai et al. 2007). Its mild experimental conditions render the one-pot synthesis method both simple and convenient (Fig. 4.3).

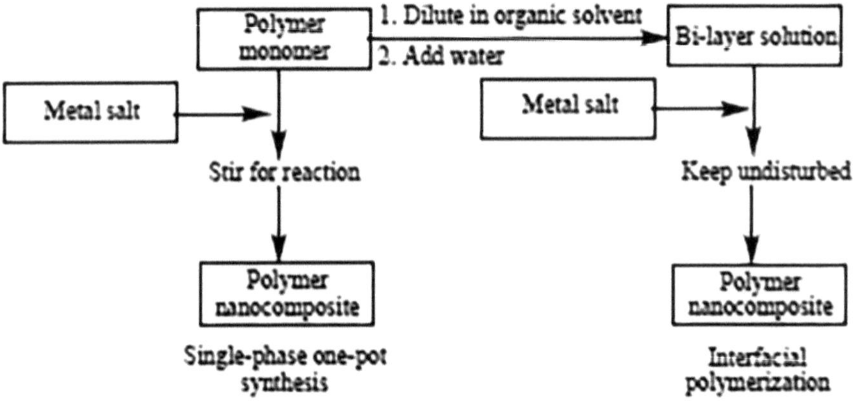

Fig. 4.3 Preparation routes of one-pot synthesis method

4.3.2.1 Single-Phase One-Pot Synthesis Method

In single-phase one-pot synthesis, the mixed solution is quickly stirred, either magnetically or vigorously, to enhance the reaction rate. The synthesis (Balamurugan et al. 2009) of Poly(3,4-ethylenedioxythiophene) (Ag-PEDOT) by adding 0.05 M AgNO3 to 10 mL of 1 M HNO_3 solution with 0.01 M 3,4-ethylenedioxythiophene. Polymerization of 3,4-ethylenedioxythiophene at neutral pH is challenging and occurs slowly. After stirring the mixture at room temperature for 15 min, the solution changed from white to a light-yellow color. Prolonged stirring led to a blackish-brown color, indicating successful polymerization and the formation of Ag-PEDOT.

4.3.2.2 Interfacial Polymerization Synthesis Method

The interfacial polymerization synthesis method, a specific type of one-pot synthesis, is conducted at the organic-aqueous interface without stirring. This process results in purely hydrophilic polymer nanocomposites that settle in the aqueous portion of the reaction medium. Importantly, the metal nanoparticles and polymers create hybrid nanocomposites with close contact, beginning their formation at the organic and aqueous boundary. The resultant materials exhibit phase selectivity, with the metal-polymer nanocomposites accumulating in the aqueous layer. This approach facilitates intimate contact between the metal and polymer, a crucial factor for synthesizing polymer nanocomposites (Mallick et al. 2007a, b). Extensively (Mallick et al. 2007a, b), investigated the interfacial polymerization synthesis method for polymer nanocomposites. For the interfacial polymerization synthesis of Au-PDA (Mallick et al. 2007a), 0.07 g of poly(1,5-diaminoanthraquinone) (PDA) was sonicated in 10 mL of toluene in a conical flask, then diluted with another 10 mL of toluene. Following this, 15 mL of ultrapure water was added, creating a bilayer of organic and aqueous components. The conical flask was left undisturbed for 10 min to maintain the bilayer. Then, 3 mL of $HAuCl_4$ ($10-^2$ mol $L-^1$) was gradually introduced to the reaction mixture, resulting in a deep reddish-brown colloidal precipitation in the aqueous layer, which settled at the bottom of the flask.

Finally, the colloidal solution was decanted from the bottom of the conical flask and dried under vacuum at 60 °C.

4.3.3 *Oxidative Polymerization Synthesis Method*

The oxidative polymerization synthesis method is commonly employed to create (inorganic nanoparticle materials)-polymer nanocomposites. This process occurs between inorganic colloids and polymer monomers in the presence of a strong oxidant. The strong oxidant facilitates the polymerization of the polymer monomers with inorganic nanoparticles. Notably, this method does not require high temperatures, as polymerization typically initiates at temperatures below 10 °C (Reddy et al. 2008; Reddy et al. 2007) (Fig. 4.4).

The synthesized (Reddy et al. 2008) the poly(o-toluidine) (Ag-PoT) polymer nanocomposite using this technique. They first prepared silver colloids stabilized by citrate to obtain Ag nanoparticles. Next, they conducted the oxidative polymerization of o-toluidine in the presence of the citrate-stabilized Ag colloids. To start the reaction, HNO_3 and 0.25 mL of o-toluidine were added to 10 mL of Ag colloid solution and stirred for 10 min at 5 °C. Polymerization progressed with the dropwise addition of 5 mL aqueous APS (0.534 g) to the solution, maintaining constant stirring for 18 h at 5 °C. This reaction yielded a green precipitate, the Ag-PoT polymer nanocomposite, which was subsequently washed with ethanol and distilled water. Finally, the nanocomposite was dried at room temperature under vacuum conditions to produce a dry Ag-PoT powder. Similarly, the investigation (Bourdo et al. 2008) the synthesis of graphite-PANI polymer nanocomposites with improved crystallinity. In contrast to their earlier study (Bourdo and Viswanathan 2005), which used aqueous dispersion graphite, this synthesis began with dry pristine graphite. The process involved combining graphite powder with distilled water to create a 25%

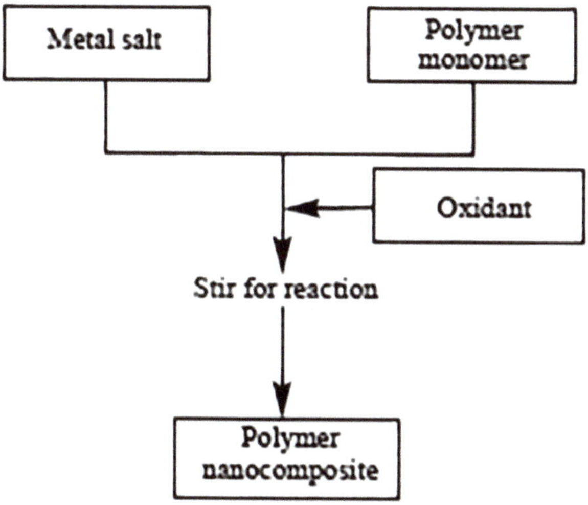

Fig. 4.4 Preparation routes of oxidative polymerization synthesis method

solid graphite suspension, which was allowed to swell overnight. The mixture was then placed in a container with grinding media and shaken overnight to achieve dispersion. Dispersed graphite was combined with 100 mL of 1 M methanesulfonic acid in a jacketed beaker. Distilled aniline, acting as a stabilizing and reducing agent, was added according to the desired graphite-to-polyaniline (PANI) ratio. The solution was cooled to 0 °C before introducing sodium persulfate at a 1.1:1 molar ratio with aniline. The reaction was stirred overnight at 0 °C, after which the mixture was filtered using vacuum suction and washed with distilled water until the filtrate was clear. The product underwent further purification by washing twice with 50 mL of 1 M methanesulfonic acid. Finally, the graphite-PANI polymer nanocomposite was dried under vacuum conditions and pressed into pellets.

4.3.4 Electrochemical Synthesis Method

Electrochemical synthesis is a simple chemical preparation technique carried out using an electrochemical workstation, often employed to create electrically conductive polymers (Bozlar et al. 2009). This workstation features three electrodes: a working electrode, a counter electrode, and a reference electrode. The method is particularly beneficial for electrocatalytic applications, enabling the direct formation of polymer nanocomposite films on the electrode surface (Janáky and Visy 2008). Bozlar et al. 2009) explored the creation of carbon nanotube (CNT)-Pmpy polymer nanocomposites via electrochemical synthesis. In this process, raw multiwalled carbon nanotubes (MWCNTs) were dispersed within a cavity microelectrode (CME) measuring approximately 15 × 30 μm, along with the monomer N-methylpyrrole. Electrochemical measurements were conducted using an EGG Princeton Applied Research PAR273 potentiostat/galvanostat in a three-electrode configuration at room temperature. The CME served as the working electrode, immersed in a 0.5 M lithium perchlorate aqueous solution, with a platinum wire as the counter electrode and an Ag|AgCl wire as the reference electrode. Subsequently, monomers (10^{-2} M) were added to the solution, and electrochemical polymerization was carried out in two steps. The first step involved classical cyclic voltammetry (one cycle from 0 to 1.2 V at a scan rate of 100 mV s^{-1}), while the second step comprised an electrochemical pulse (0 to 1.2 V for 0.5 s). Finally, the resulting hybrids were collected through sonication in tetrahydrofuran (THF) for 20 min at 45 kHz.

4.3.5 Thermally Synthesis Method

4.3.5.1 Hot Press Synthesis Method

In the current study, the hot press synthesis method is utilized to prepare polymer nanocomposite electrolytes and batteries that function at medium-to-high temperatures. This method involves high temperatures and is entirely dry and solvent-free (Appetecchi et al. 2003). Examined (Appetecchi et al. 2003), polyethylene oxide (PEO)-based polymer nanocomposite electrolyte membranes. In their research

(Appetecchi et al. 2003), they prepared the electrolyte membranes using a dry and solvent-free approach. Linear PEO (molecular mass of 1×10^5) was dried under vacuum at 60 °C for 24 h, while nanoscale fumed SiO_2 filler was dried under vacuum at 300 °C for 24 h before application. The dried materials, including PEO, SiO_2, and $LiCF_3SO_3$, were thoroughly mixed using a ball-milling technique. This mixture was then hot-pressed in an aluminum mold at approximately 80–100 °C, resulting in highly homogeneous and semi-transparent polymer nanocomposite membranes with impressive mechanical strength. Similarly, authors in (Stephan et al. 2009) reported the preparation of chitin-incorporated PEO-based nanocomposite electrolytes with an average thickness of 30–50 μm. Extensive drying was performed on all materials before synthesis. Specifically, PEO and $LiClO_4$ were dried under vacuum for 2 days at 50 °C and 100 °C, respectively, while nanochitin was dried under vacuum at 50 °C for 5 days. The polymer nanocomposite electrolytes were synthesized by dispersing the appropriate amounts of nanochitin in a PEO-$LiClO_4$ mixture, which was then hot-pressed into films. These films were dried under vacuum at 50 °C for 24 h, yielding homogeneous membranes with robust mechanical properties.

4.3.5.2 Thermal Sintering Method

Developed (Zhang et al. 2010), the highly conductive polymer nanocomposite diethyl benzyl formamide-hexamethylphosphoramide (Ag-DEBF-HMPA) using a low-temperature sintering technique. The synthesis commenced with a 4:6 molar ratio of Ag nanoparticles and flakes, which were blended with DEBF and HMPA. To ensure adequate dispersion of the conductive fillers, the mixture underwent sonication for over 1 h. After this initial step, the catalyst 1-cyanoethyl-2-ethyl-4-methyli midazole was added, and the mixture was sonicated for an additional 0.5 h. The prepared paste was then thermally cured at 150 °C and 180 °C to achieve the conductive polymer nanocomposite Ag-DEBF-HMPA.

4.3.5.3 Thermally Induced Phase Separation Synthesis Method

The HAp-PA 6,9 polymer nanocomposite, synthesized via the thermally induced phase separation method. The synthesis involved several steps (Sender et al. 2007):

- First, an nHAp suspension in N,N-dimethylacetamide was created through solvent substitution. Ethanol, used to stabilize the hydroxyapatite nanocrystals, was removed from the suspension at 75 °C while gradually adding N,N-dimethylacetamide. The mixture was then sonicated (ultrasonic bath, 30 min) to achieve optimal dispersion of nHAp nanoparticles.
- Next, the mixture was heated to 145 °C with continuous stirring, and polyamide pellets were added to the suspension to dissolve over 3 h. Finally, the viscous medium was quenched and rinsed with distilled water to precipitate the polymer nanocomposite and eliminate N,N-dimethylacetamide. The resulting solid HAp-PA 6,9 polymer nanocomposite was then dried under vacuum at 170 °C for 24 h.

4.3.6 Intermatrix Synthesis (IMS) Method

The inter-matrix synthesis procedure utilizes appropriately functionalized polymeric membranes as nonreactors. Created (Muraviev et al. 2008; Muraviev et al. 2007) polymer-stabilized magnetic nanoparticles (PSMNPs) nanocomposites, which included (stabilized monometallic Pt)-polymer nanocomposite and (stabilized bimetallic Pt-Cu core-shell nanoparticles)-polymer nanocomposite, employing this method. sulfonated polyether ether ketone (PEEK) served as the polymer matrix in their study. The core-shell (stabilized bimetallic Pt-Cu core-shell nanoparticles)-polymer nanocomposite demonstrated greater electrocatalytic activity compared to the (stabilized monometallic Pt)-polymer nanocomposite. The synthesis of (stabilized monometallic metal nanoparticles) - (sulfonated PEEK) membranes involved sequential loading of functional groups with Cu^{2+} or $[Pt(NH_3)_4]^{2+}$ ions, which were then reduced inside the polymer matrix using a 0.1 M $NaBH_4$ solution. The membranes were air-dried at room temperature. Through the intermatrix synthesis method, the (stabilized bimetallic Pt-Cu core-shell nanoparticles)-polymer nanocomposite was also produced by sequentially loading the (stabilized monometallic Cu)-polymer nanocomposite membranes with a 0.01 M $Pt(NH_3)_4^2$ solution, followed by chemical reduction of Pt with 0.1 M NaBH (Muraviev et al. 2008). Additionally, authors in (Macanás et al. 2007) synthesized metal-polymer nanocomposites using this method, emphasizing functionalized polymeric membranes for obtaining (stabilized metal nanoparticles)-polymer nanocomposites. Cu metal nanoparticles and Pt-Cu metal nanoparticles were created within the sulfonated PEEK matrix, with the polymeric membranes acting as nanoreactors for (Cu metal nanoparticles) - (sulfonated PEEK) and (Pt-Cu metal nanoparticles) - (sulfonated PEEK) composites (Macanás et al. 2007). Initially, copper ions were introduced into the matrix through a simple ion-exchange reaction, achieved by gently stirring a sulfonated PEEK membrane sample in 0.1 M $CuSO_4$ solution (step 1). Following equilibration, the mixture was rinsed with distilled water and immersed in a 0.1 M sodium borohydride solution (step 2), leading to the generation of Cu metal nanoparticles and related metal-polymer nanocomposites. During the reduction of copper, the sulfonic groups of the polymer matrix appeared to regenerate, reverting to their sodium form, thus allowing for further metal loading-reduction cycles. This facilitated the creation of core-shell metal nanoparticles, such as Pt-Cu, by altering the type of metal ions during the subsequent loading (step 3) and reduction (step 4) stages (Macanás et al. 2007).

4.3.7 Intercalation Synthesis Method

Initially, copper ions were introduced into the matrix through a simple ion-exchange reaction, achieved by gently stirring a sulfonated PEEK membrane sample in 0.1 M $CuSO_4$ solution (step 1). Following equilibration, the mixture was rinsed with distilled water and immersed in a 0.1 M sodium borohydride solution (step 2), leading to the generation of Cu metal nanoparticles and related metal-polymer nanocomposites. During the reduction of copper, the sulfonic groups of the polymer matrix appeared to regenerate, reverting to their sodium form, thus allowing for further metal loading-reduction cycles. This facilitated the creation of core-shell metal nanoparticles, such as Pt@Cu, by altering the type of metal ions during the subsequent loading (step 3) and reduction (step 4) stages (Macanás et al. 2007).

Overall, the intercalation synthesis method is simpler compared to the sol-gel synthesis method. This polymerization technique is widely applicable due to the availability of abundant and inexpensive raw materials. Polymer nanocomposites produced through intercalation polymerization can enhance mechanical properties while reducing costs. The intercalation synthesis method is a key approach for creating inorganic-organic polymer nanocomposites, allowing for the embedding of various inorganic compounds with typical layered structures, such as silicate clay, phosphate, graphite, metal oxide, disulfide, and triphosphorus sulfide complex, into organic polymers. Depending on the form of intercalation synthesis, this method can be categorized into intercalation polymerization, solution intercalation, and melt intercalation. In intercalation polymerization, monomers or intercalation agents are first inserted into the layered structure of mica-silicate, measuring 1 nm in thickness. The monomers are then polymerized between the silicate layers, leading to an expansion of the space between the layered silicates. Consequently, the layered silicate fillers are reduced to the nanometer scale and dispersed within the polymer matrix, forming polymer nanocomposites. The solution intercalation method involves directly embedding the polymer solution into the channels of the polymer. In the melt intercalation method, melt polymer is directly introduced into the inorganic channels, subsequently resulting in polymer nanocomposites. This melt intercalation method is recognized as a more versatile and environmentally friendly approach for developing layered silicate-polymer nanocomposites. Unlike in situ polymerization and solution intercalation, melt intercalation uses a mixture of layered silicate and polymer heated above the polymer's softening point. The limitations of in situ polymerization and solution intercalation stem from the lack of suitable monomers for most polymers. However, melt intercalation offers distinct advantages. Preparation (Ratna et al. 2007) of clay-PEO polymer nanocomposite using the solution intercalation method, utilizing an octadecylammonium salt of montmorillonite clay synthesized via the ion-exchange method. A clay-PEO polymer nanocomposite-based solid polymer electrolyte was created with $LiBF_4$. The organically modified clay was dispersed and swollen in acetonitrile at room temperature for 24 h and sonicated for 15 min. Following this, PEO and $LiBF_4$ ([EO]/[Li+] = 8) were mixed, adjusting the consistency by adding extra solvent. Reactions occurred in a flask placed in an oil bath, heated to 75 °C, with stirring for 10 h. The mixture was then sonicated for 30 min. Nanocomposite films were formed by pouring the solutions into horizontal glass molds and allowed to air-dry overnight. Finally, the films underwent vacuum drying at 70 °C in an oven to complete the drying process.

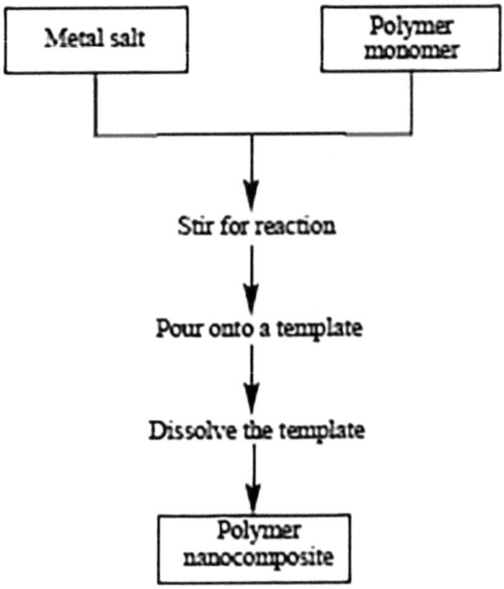

Fig. 4.5 Preparation routes of template-assisted synthesis method

4.3.8 Template-Assisted Synthesis Method

The template-assisted synthesis method provides a convenient and straightforward approach for creating monodispersed nanoparticles. Successfully (Weaver et al. 2009), synthesized hollow and solid QD-silica nanoparticles using this efficient method. The synthesis of the silica-QDs nanorods involved stirring 7.70 g of ethanol, 11.61 g of tetraethoxyorthosilicate, and 1 mL of 2.8 mM HCl vigorously in a 25 mL beaker. In a separate beaker, 1.50 g of cetyltrimethylammonium bromide was dissolved in 15 g of ethanol and 4 mL of 55 mM HCl (Fig. 4.5).

The cetyltrimethylammonium bromide solution was then added to the tetraethoxyorthosilicate mixture, which was stirred for an additional 30 min. Next, 250 mL of this mixture was combined with CdSe-ZnS QDs in a chloroform solvent inside a test tube. The resulting solution was slowly poured onto Whatman Anodisc® templates (200–300 nm pore diameter) using moderate vacuum filtration. The top and bottom surfaces of the templates were scraped and dried overnight. To release the nanoparticles, the templates were soaked in either 1 M NaOH or 10% phosphoric acid solution. Finally, the nanoparticles were collected by filtration through a nanoporous alumina filter with a 20 nm pore diameter.

4.3.9 Reversible Addition-Fragmentation Chain-Transfer (RAFT) Polymerization Synthesis Method

The RAFT polymerization mechanism (Vana et al. 2002) and consists of (I) initiation, (II) pre-equilibrium involving the initial RAFT chain-transfer agent, (III) propagation and reinitiation, (IV) chain equilibration, which is central to RAFT polymerization, and (V) termination. Synthesized (Hojjati et al. 2007) the TiO_2-PAA polymer nanocomposite through the reversible addition-fragmentation chain-transfer polymerization method, employing acrylic acid as the monomer for PAA polymerization. They then dispersed TiO_2 nanoparticles into the PAA polymer. This novel method enhanced the dispersion of nanofillers by using the bifunctional RAFT agent 2-{[(butylsulfanyl)carbonothioyl]sulfanyl}propanoic acid, which contained a carboxyl group for anchoring onto TiO_2 nanoparticles and an S=C (SC_4H_9) moiety for subsequent RAFT polymerization of acrylic acid to create the TiO_2-PAA polymer nanocomposite. The detailed synthesis process was as follows: 0.6 g of acrylic acid monomer, 0.06 g of the functionalized TiO_2-RAFT agent (10:1 wt/wt), and 0.004 g of 2,2′-azobis(2-methylpropionitrile) were dispersed in 60 mL of methanol under ultrasonication. The mixture was then transferred to a 100 mL three-neck flask equipped with a nitrogen supply, thermometer, and condenser, and stirred continuously for 24 h. After diluting with methanol, the solution was filtered through a 0.05 μm polycarbonate membrane filter to remove ungrafted polymer, and the resulting solid was washed multiple times with methanol. The final product was dried overnight in a vacuum at 60 °C. Additionally, authors in (Hojjati and Charpentier 2008) produced $nTiO_2$-PMMA polymer nanocomposite via the RAFT polymerization method. The RAFT agent 4-cyano-4-(dodecylsulfanylthiocarbonyl) sulfanyl pentanoic acid, possessing a carboxyl group for anchoring to the $nTiO_2$ surface and an S=C ($SC_{12}H_{25}$) moiety for MMA polymerization, was utilized to create the $nTiO_2$-PMMA polymer nanocomposite.

4.3.10 *Self-Assembly Synthesis Method*

Synthesized (Enlow et al. 2007), self-assembled block copolymer-templated calcium phosphate nanocomposite gels using a self-assembly method in solution. Amphiphilic block copolymer micelles in aqueous solutions served as templates for the growth of calcium phosphate nanocrystals, which self-assembled to form thermoreversible nanocomposite gels. The specific synthesis procedure for the $Ca_3(PO_4)_2$-PDEAEM polymer nanocomposite was as follows: An aqueous solution containing $NH_4H_2PO_4$, H_3PO_4, $Ca(NO_3)_2$, and PDEAEM pentablock copolymers was utilized for the precipitation of calcium phosphate. A saturated calcium phosphate solution with a Ca:P ratio of 1.67, consistent with stoichiometric hydroxyapatite, was created by combining 120 mL of 0.5 M $NH_4H_2PO_4$ solution with 200 mL of 0.5 M $Ca(NO_3)_2$ solution. This mixture was stirred for 30 min, resulting in the formation of a white precipitate. The precipitate was then separated through centrifugation. The modified $PDEAEM_{35}$-F127-$PDEAEM_{35}$ pentablock copolymer, available commercially, served as the polymer matrix phase. Gel samples were made by dissolving 3 g of the copolymer directly into the calcium phosphate solution. Control samples were also created by dissolving the copolymer in deionized water. These mixtures were stored in a refrigerator at 3 °C and stirred daily until complete dissolution of the copolymer, typically taking 3–4 days for a homogeneous solution. After the copolymer had dissolved, the samples were allowed to reach room temperature and aged for 24 h. During this time, the calcium phosphate-copolymer solution thickened into a viscous nanocomposite gel, resulting in the formation of the $Ca_3(PO_4)_2$-PDEAEM polymer nanocomposite.

4.3.11 Microwave-Induced Synthesis Method

The microwave-induced synthesis method offers a promising and innovative approach for producing polymer nanocomposites. Compared to other preparation techniques, this method yields higher production rates and results in purer products due to its short reaction time. Furthermore, it is efficient, convenient, and cost-effective, requiring only 15–20 min under controlled temperature and pressure in a microwave reactor (Chowdhury et al. 2009).

The quick formation of polymers is crucial for encapsulating nanoparticles before agglomeration occurs. With the presence of microwave radiation, inorganic nanoparticles, including nano-scale dispersions of carbon nanotubes and single-walled carbon nanotubes (SWNTs), create a reactive environment by absorbing energy, leading to rapid and localized heating that promotes the reaction.

Several methods for synthesizing SWNTs-PMMA polymer nanocomposites (Chowdhury et al. 2009). The standard procedure included the following steps: dispersing 160 mg of functionalized SWNT and 1 wt% sodium dodecyl sulfate in distilled water using sonication for 1 h. Concurrently, 4 g of the polymer monomer vinylpyrrolidone was added to this solution and sonicated for an additional hour. Subsequently, 6 mg of the free radical initiator azobisisobutyronitrile (AIBN) was dissolved in 10 mL of ethanol and incorporated into the mixture, followed by 30 min of sonication. The resulting mixture was then placed in a continuous extraction method microwave-assisted reaction system (CEM MARS) microwave reactor for 15 min at 30 psi and 90 °C, using 60% of 1600 W microwave power. Finally, the polymer nanocomposite was precipitated in acetone, filtered, and dried in a vacuum oven.

4.3.12 Melt Mixing Synthesis Method

A nanosilica-acetal copolymer (Sun et al. 2008) composed of trioxane and 1,3-dioxolane using a straightforward in situ melt mixing synthesis method. The purified trioxane was melted in a flask placed in an oil bath at 70 °C. Subsequently, 1,3-dioxolane (4.5 mol% relative to trioxane) and nanosilica (1 wt% relative to trioxane) were added to the molten trioxane. The catalyst $BF_3 \bullet O(Et)_2$ (5×10^{-4} mol% relative to trioxane) was injected into the mixture with a syringe to enhance the reaction rate. The mixture was stirred to promote polymerization. The resulting powdery reaction product was then transferred into a methanol solution (0.5 wt% triethylamine) to deactivate the catalyst, followed by multiple washes with methanol. Finally, the powdery product was dried under vacuum to yield the pure nanosilica-acetal copolymer. Property characterization revealed that the average diameter of the nanosilica-acetal copolymer was approximately 100 nm, suggesting that each polymer nanocomposite contained 2–3 nanosilica particles.

4.3.13 Catalytic Chain Transfer Polymerization (CCTP) Synthesis Method

As noted in the literature (Gridnev and Ittel 2001; Heuts et al. 2002), catalytic chain transfer polymerization (CCTP) is a highly effective synthetic technique for producing low-molecular-weight polymer chains with terminal double bonds. Polymer ligands can form robust chemical bonds with nanocrystals within polymer matrices, enhancing compatibility. Consequently, researchers are exploring the introduction of a versatile polymer ligand synthesized via CCTP, which improves interactions between polymer ligands and nanocrystals due to multiple anchor groups and double bond functionality on the exterior. However, due to their high surface area and energy, nanocrystals frequently aggregate in polymer matrices. To mitigate this issue, specially designed polymer ligands act as connectors, linking nanocrystals to polymers and preventing aggregation, resulting in well-stabilized polymer nanocomposite hybrids.

4.3.13.1 Synthesis of ZnS-PMAA-PMMA Via CCTP

Synthesis of poly(methacrylic acid)-poly(methyl methacrylate) (ZnS-PMAA-PMMA) using the CCTP method through the following steps (Chen et al. 2009):

1. **Synthesis of PMAA Macromonomer.**
   (a) A 0.3 g quantity of azo initiator VA-044 and 13 mg of cobalt catalyst bis(aqua)bis((difluoroboryl)dimethylglyoximato)cobalt (II) (CoBF) were dissolved in 2 mL acetone and combined with 150 mL distilled water.
   (b) The mixture was heated at 55 °C under a nitrogen atmosphere after six alternating cycles of evacuation and pressurization with high-purity nitrogen.
   (c) 7.5 mg CoBF (dissolved in 74.0 g MAA) was injected into the reaction mixture to initiate polymerization.
   (d) The reaction continued for 2 h under constant stirring in a nitrogen atmosphere and was then quenched with an ice-water bath.
   (e) The resulting PMAA was precipitated using diethyl ether, redissolved in distilled water, and reprecipitated multiple times to ensure purity.
   (f) The final PMAA macromonomer was isolated by centrifugation and dried under vacuum.

2. **Synthesis of ZnS Nanocrystals with PMAA Functionalized Ligand.**
   (a) A 4 mL N,N-dimethylformamide (DMF) solution containing 1.25 mM $ZnCl_2$ was dropwise added into a 40 mL DMF solution containing 12.5 mM PMAA macromonomer while stirring continuously.
   (b) Then, a 4 mL distilled water solution containing 0.9 mM $Na_2S{\cdot}9H_2O$ was added dropwise to the reaction mixture.
   (c) The solution was stirred for 2 h at room temperature to form ZnS-PMAA nanocrystals.
3. **Copolymerization of ZnS-PMAA with MMA to Form ZnS-PMAA-PMMA Nanocomposite Hybrids.**
   (a) A solution of initiator AIBN and methyl methacrylate (MMA) was added dropwise to the ZnS-PMAA nanocrystals suspension under vigorous stirring.
   (b) The reaction was carried out under a nitrogen atmosphere at 70 °C for 4.5 h with the ratio AIBN:MMA:PMAA = 0.003:1:0.2 (w/w).
   (c) The MMA monomer accounted for 10 wt% of the total solution, leading to the formation of ZnS-PMAA-PMMA polymer nanocomposite hybrids.

4.3.14 Emulsion Polymerization Synthesis Method

The emulsion polymerization synthesis method is a widely utilized technique, primarily due to water being the solvent throughout the process. This method is also highly advantageous for environmental applications because of its eco-friendly nature. With the aid of an emulsifier and mechanical stirring, the monomers are effectively dispersed in a water emulsion. Polymerization then occurs through the action of an initiator and precursor, resulting in the formation of polymer particles within a stable emulsion system (Fig. 4.6).

4.3.14.1 Inverse Emulsion Polymerization Method

Recently, the inverse emulsion polymerization technique has gained popularity in polymer synthesis. This method utilizes a nonpolar solvent as the continuous phase, with monomers dispersed in the oil phase through the use of an emulsifier. The polymerization occurs within a "water-in-oil" emulsion. This approach offers multiple benefits, such as an increased polymerization rate, higher molecular weight, narrower molecular weight distribution of the end products, and reduced polymerization temperature. Furthermore, it improves stirring and heat transfer efficiency. By incorporating additional dispersive and active monomers, the cationic properties of the polymers can be enhanced, resulting in improved flocculation characteristics.

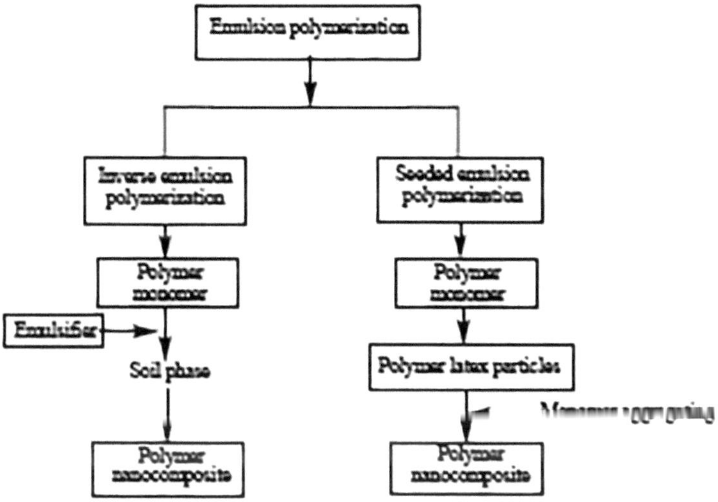

Fig. 4.6 Preparation routes of emulsion polymerization method

Developing (Owens et al. 2007), a gold-PAAM-PAA nanocomposite using an in situ inverse emulsion polymerization process. Gold nanoparticles, created through the gold chloride and citrate reduction method, were encapsulated within a PAAM-PAA interpenetrating polymer network shell during this polymerization. The polymer nanocomposite's surface was subsequently PEGylated using covalent grafting with linear methoxy-PEGN-hydroxysuccinimide, which reacted with the primary amine groups of the PAAM network. The synthesis of the gold-PAAM-PAA polymer nanocomposite was conducted through a sequential interpenetrating polymer network synthesis method. Initially, gold nanoparticles situated in the aqueous monomer droplet phase were encapsulated within polyacrylamide nanoparticles via in situ inverse emulsion polymerization. The inverse emulsion solution consisted of 81% cyclohexane (continuous phase), 13% surfactant phase (a mixture of sodium bis(2-ethylhexyl) sulfosuccinate and polyethylene glycol laurylether in a 2:1 ratio), and 6% aqueous phase. Emulsifiers were then introduced and dissolved under stirring conditions. The remainder of the aqueous phase was split into two separate premixes, each containing 11.7 wt% monomer, 4 wt% crosslinker (N,N′-methylenebisacrylamide), 5.3 wt% initiator (APS), and 79 wt% deionized distilled water. The key distinction between these premixes was that one included only acrylamide as the monomer, while the other consisted solely of acrylic acid.

The synthesis process followed these steps:

- First, the acrylamide aqueous premix was added to a three-neck round-bottom flask containing cyclohexane, gold nanoparticles, and emulsifiers. To eliminate oxygen, the mixture was purged with nitrogen gas for 30 min. Polymerization was initiated by injecting 0.3 mL of N,N,N′,N′-tetramethylethylenediamine and allowed to proceed for 2 h at room temperature. The reaction was then halted by cooling the mixture to 0 °C.

- Second, the interpenetrating polymer network synthesis was executed by introducing acrylic acid into the reaction vessel, followed by another nitrogen gas purge and the addition of N,N,N′,N′-tetramethylethylenediamine.
- Third, the gold-PAAM-PAA polymer nanocomposite was purified by removing the cyclohexane solvent phase under reduced pressure, followed by centrifugation in ethanol to precipitate the polymer nanocomposite, which was then dried and resuspended in deionized distilled water.
- Fourth, PEGylation was performed by adjusting the solution's pH to slightly above 8 and incorporating methoxy poly(ethylene glycol)-n-hydroxysuccinimide (mPEG-NHS) to reach a final concentration of 1 μM. The reaction was allowed to proceed overnight while stirring at room temperature. The final purified PEGylated polymer nanocomposite was then frozen and lyophilized into powder form.

The synthesis of a (Ai and Jiang 2009), (NiCr and ferrospinel)-conducting polymer nanocomposite using the in situ polymerization of aniline in the presence of $NiCr_{0.5}Fe_{1.5}O_4$ nanoparticles via an inverse microemulsion polymerization method. The $NiCr_{0.5}Fe_{1.5}O_4$-PANI polymer nanocomposite was produced through a microemulsion process involving a quaternary microemulsion system made up of cetyltrimethylammonium bromide (CTAB), water, cyclohexane, and n-pentanol. In this system, cyclohexane acted as the oil phase, CTAB served as the surfactant, n-pentanol worked as the cosurfactant, and an aqueous solution was the dispersed phase.

The synthesis followed these steps:

The synthesis proceeded as follows: A solution was created by dissolving 2 g of cetyltrimethylammonium bromide in 5 mL of 0.1 M HCl. Subsequently, 1 mL of n-pentanol and 50 mL of cyclohexane were added. Stirring the mixture produced a clear and transparent microemulsion solution. A specific amount of the $NiCr_{0.5}Fe_{1.5}O_4$ colloidal suspension was then added to this solution and sonicated for 1 h.

After sonication, 1 mL of aniline monomer was introduced, and the mixture was stirred for 30 min. Next, 5 mL of deionized water containing 2.49 g of ammonium persulfate was added dropwise to the stirred reaction mixture. Within minutes, the suspension turned green, signaling the polymerization of aniline.

The reaction was conducted at 0 °C under nitrogen gas while stirring for 8 h. The resulting polymer nanocomposite was collected through filtration, thoroughly washed with methanol, and vacuum dried at 60 °C for 24 h.

4.3.14.2 Seeded Emulsion Polymerization Method

The seeded emulsion polymerization technique is commonly utilized to create inorganic-organic nanostructured colloids, as it enables the synthesis of polymer nanocomposite particles with diverse morphologies and structures. Successful preparation of polymer nanocomposites requires good compatibility between hydrophilic inorganic particles and hydrophobic polymers. Interfacial promoters are necessary to foster specific interactions and enhance interfacial adhesion. A typical

method involves pretreating inorganic compounds with a coupling agent that introduces a polymerizable double bond, thereby improving compatibility. In processes involving polymer synthesis, if polymer latex particles are already present in the emulsion system and the raw materials and reaction conditions are suitably adjusted, the monomers will accumulate on the existing latex particles instead of forming new ones. This results in increased particle size without raising the particle count in the system. In seeded emulsion polymerization, the original particles serve as seeds, and the polymerization is carried out in a semibatch mode.
Synthesis of (Ohlan et al. 2008), a polymer nanocomposite featuring a PTFE copolymer core and an acrylic copolymer shell through seeded emulsion polymerization. Their procedure involved heating a mixture of water and fluoropolymer seeds (comprising 73 mol% tetrafluoroethylene, 7 mol% hexafluoropropylene, and 20 mol% vinylidene fluoride) to 78 °C while continuously bubbling nitrogen gas. Subsequently, an initiator solution of $K_2S_2O_8$ was added, followed by the addition of the second-stage monomer at a fixed rate of 4.24 mL/h. The reaction kinetics were observed gravimetrically by sampling 1–2 mL at various reaction times, with hydroquinone (~5 mg) added to halt the polymerization. Free films of the polymer were obtained by drying at room temperature for a week.

The preparation (Diaconu et al. 2008), montmorillonite (MMT)-poly(methyl methacrylate-co-butyl acrylate) polymer nanocomposite using the seeded emulsion polymerization method.

The synthesis followed this procedure:

- A seed emulsion with 20 wt% solids content (MMA/butyl acrylate comonomer) was synthesized batchwise at 75 °C for 1 h.
- During the seed preparation, all sodium montmorillonite clay (Na-MMT) was added.
- The remaining comonomer and an aqueous potassium persulfate solution were continuously fed into the reactor via two separate streams for 3 h.
- The mixture was then allowed to polymerize for another hour at 75 °C.

Furthermore, the MMT-poly(methyl methacrylate-co-butyl acrylate) polymer nanocomposite was produced by blending the blank emulsion latex with Na-MMT (3 wt% based on polymer content) under magnetic stirring (700 rpm) for 3 h at room temperature.

4.3.14.3 Other Emulsion Polymerization Methods

The synthesis of (Duan et al. 2008), calcium phosphate (Ca-P) nanoparticles and Poly(3-hydroxybutyrate-co-3-hydroxyvalerate) (Ca-P-PHBV) nanocomposite microspheres for bone tissue engineering using a solid-in-oil-in-water emulsion solvent evaporation process. The Ca-P nanoparticles were mainly encapsulated within the Ca-P-PHBV nanocomposite microspheres, with some nanoparticles embedded on the surface. The fabrication of PHBV microspheres involved an oil-in-water emulsion method, where 6.4 g of PHBV dissolved in 160 mL of chloroform at

50 °C was added to an 800 mL aqueous phase containing 1 w/v% of PVA. The emulsion was stirred at 600 rpm at room temperature until the solvent fully evaporated. The resultant PHBV microspheres were filtered, washed, and freeze-dried. For the Ca-P-PHBV nanocomposite microspheres, a solid-in-oil-in-water emulsion solvent evaporation method was also applied, with Ca-P nanoparticles produced at a pH of 10.0 to 11.0. Initially, some Ca-P nanoparticles were dispersed in the PHBV chloroform solution through ultrasonication and homogenization, forming a solid-in-oil nanosuspension (Godge et al. 2021). The subsequent procedures for creating the Ca-P-PHBV microspheres mirrored those of the PHBV microspheres. The synthesis of the copolymer (Ohlan et al. 2008), poly(AnEDOT) containing EDOT and aniline monomers using dodecylbenzene sulfonic acid and its complex barium ferrite (BaF)-poly(AnEDOT) with nanocrystalline barium ferrite (BaF) via emulsion polymerization. To address the insolubility of EDOT in water, 3,4-ethylenedioxy thiophene was homogenized in dodecylbenzene sulfonic acid before copolymerization with BaF. The interaction mechanism for poly(AnEDOT) formation is illustrated, depicting the copolymerization process between aniline and 3,4-ethylenedioxythiophene (EDOT) in the presence of dodecylbenzene sulfonic acid (DBSA) as a dopant and surfactant (Ohlan et al. 2008). This mixture was homogenized for 2 h to form a homogeneous emulsion and polymerized using 0.1 M APS solution at 0 °C while continuously stirring. The resulting blue-green precipitate was washed multiple times with isopropanol and distilled water before drying at 60 °C in a vacuum oven. The synthesis of the BaF-poly(AnEDOT) polymer nanocomposite involved homogenizing BaF particles in aqueous dodecylbenzene sulfonic acid. Then, 0.05 M of aniline and 0.05 M of 3,4-ethylenedioxy thiophene were added, using a BaF to aniline weight ratio of 2:1. The mixture was homogenized for 2 h, followed by polymerization with APS at 0 °C, resulting in a blackish-green precipitate that was washed and dried in a vacuum oven, yielding the BaF-poly(AnEDOT) polymer nanocomposite. The synthesis of (Dhawan et al. 2009), a conducting ferromagnetic-polyaniline polymer nanocomposite containing γ-Fe_2O_3 and TiO_2 nanoparticles via in situ emulsion polymerization. Two different polymer nanocomposite compositions were prepared by polymerizing aniline, TiO_2, and γ-Fe_2O_3 in weight ratios of 1:1:1 (PTF11) and 1:1:2 (PTF12) in a surfactant medium. Nanosized γ-Fe_2O_3 and TiO_2 particles were homogenized in a 0.3 M aqueous dodecylbenzene sulfonic acid solution, forming a whitish-brown emulsion. To this emulsion, 0.1 M aniline was added and homogenized for 2–3 h, leading to the formation of micelles with γ-Fe_2O_3 and TiO_2. These micelles were polymerized at temperatures below 0 °C using chemical oxidation polymerization, with $(NH_4)_2S_2O_8$ (0.1 M) acting as the oxidant. After being washed with isopropyl alcohol, the resulting products were filtered and dried in a vacuum oven at 60–65 °C. Various polymer nanocomposite formulations with aniline:TiO_2:γ-Fe_2O_3 weight ratios of 1:1:1 (PTF11) and 1:1:2 (PTF12) were synthesized in dodecylbenzene sulfonic acid medium, and the effects of ferrite constituents on the properties were analyzed (Dhawan et al. 2009; Godge et al. 2020, 2021, 2024a, b).

4.3.15 *Photopolymerization Method*

The synthesis of single-walled carbon nanotubes (SWNT) with an organic soft hydrogel polymer in glass microchannels using the photopolymerization method (Makamba et al. 2008).

In order to synthesize the SWNT-hydrogel polymer nanocomposite by the photopolymerization method, the following mixture was prepared:

- 6 w/v% acrylamide
- 20 v/v% ethylene glycol dimethacrylate
- 0.1 mg/mL SWNT-PEG-acrylate
- 1 w/v% α, α-dimethoxy-α-phenylacetophenone photoinitiator.

This mixture was dissolved in ethanol and sonicated for 15 min.

For photopatterning in the microchannels, light wavelengths of 330 to 380 nm were utilized. A mask was then placed over the microchannel to create the patterned polymer nanocomposite. The microchip was cleansed with deionized water for 15 min, followed by a 15-min ethanol cleaning. The prepared mixture was then drawn into the channel using a vacuum pump. A mask was positioned on the microchip with a 1 cm gap, about 0.5 cm from the intersection. The microscope was aligned with the mask gap, and radiation was applied for 2 min to facilitate the polymerization process. Finally, the microchannel was rinsed with ethanol, resulting in the production of the SWNT-hydrogel polymer nanocomposite.

4.4 Conclusion

This article examines polymer nanocomposites in depth, exploring their properties, preparation methods, and various applications. These nanocomposites consist of different nanometer-scale units combined with organic polymer materials. We summarize the various nanometer units of these innovative materials, including inorganic (nanoparticles or nano-layered) substances and organic nanometer units. Polymer nanocomposites exhibit numerous unique properties; the nanometer units enhance the polymers' characteristics, including catalytic, electrochemical, mechanical, magnetic, optical, and biological properties. Such distinctive attributes allow for diverse applications across different fields. Additionally, the article highlights various preparation methods for polymer nanocomposites, all aimed at altering and improving the characteristics of these new materials. By thoroughly discussing these preparation methods, researchers can create novel and ideal polymer nanocomposites. Furthermore, the review offers theoretical insights for developing new and efficient preparation strategies. Ultimately, advancements in polymer nanocomposites will lead to the design of innovative and reliable materials with enhanced properties and broader applications. Currently, polymer nanocomposites are in the early stages of development, but projections indicate significant growth as science and technology in this area progress. Consequently, polymer nanocomposites are expected to profoundly influence industrial developments, markedly enhancing properties such as mechanical strength, gas barrier capacity, flame retardancy, and electrical conductivity. The market potential for novel polymer nanocomposites is immense, and the applications are anticipated to become increasingly widespread as the technology advances.

References

Abedi S, Abdouss M (2014) A review of clay-supported Ziegler–Natta catalysts for production of polyolefin/clay nanocomposites through in situ polymerization. Appl Catal A Gen 475:386–409

Ai LH, Jiang J (2009) Facile synthesis and characterization of magnetic NiCr ferrospinel embedded in conducting polymer. J Alloys Compd 487(1–2):735–738

Alexandre M, Dubois P (2000) Polymer-layered silicate nanocomposites: preparation, properties and uses of a new class of materials. Mater Sci Eng R Rep 28(1–2):1–63

Appetecchi GB, Hassoun J, Scrosati B, Croce F, Cassel F, Salomon M (2003) Hot-pressed, solvent-free, nanocomposite, PEO-based electrolyte membranes: II. All solid-state Li/$LiFePO_4$ polymer batteries. J Power Sources 124(1):246–253

Bai J, Li Y, Du J, Wang S, Zheng J, Yang Q et al (2007) One-pot synthesis of polyacrylamide-gold nanocomposite. Mater Chem Phys 106(2–3):412–415

Balamurugan A, Ho KC, Chen SM (2009) One-pot synthesis of highly stable silver nanoparticles-conducting polymer nanocomposite and its catalytic application. Synth Met 159(23–24):2544–2549

Bourdo SE, Viswanathan T (2005) Graphite/polyaniline (GP) composites: synthesis and characterization. Carbon 43(14):2983–2988

Bourdo S, Li Z, Biris AS, Watanabe F, Viswanathan T, Pavel I (2008) Structural, electrical, and thermal behavior of graphite-polyaniline composites with increased crystallinity. Adv Funct Mater 18(3):432–440

Bozlar M, Miomandre F, Bai J (2009) Electrochemical synthesis and characterization of carbon nanotube/modified polypyrrole hybrids using a cavity microelectrode. Carbon 47(1):80–84

Chen L, Wang C, Li Q, Yang S, Hou L, Chen S (2009 Jul) In situ synthesis of transparent fluorescent ZnS–polymer nanocomposite hybrids through catalytic chain transfer polymerization technique. J Mater Sci 44(13):3413–3419

Cheng Q, Pavlinek V, Li C, Lengalova A, He Y, Saha P (2006) Synthesis and characterization of new mesoporous material with conducting polypyrrole confined in mesoporous silica. Mater Chem Phys 98(2–3):504–508

Chowdhury SR, Chen Y, Wang Y, Mitra S (2009 Mar) Microwave-induced rapid nanocomposite synthesis using dispersed single-wall carbon nanotubes as the nuclei. J Mater Sci 44(5):1245–1250

Dhawan SK, Singh K, Bakhshi AK, Ohlan A (2009) Conducting polymer embedded with nanoferrite and titanium dioxide nanoparticles for microwave absorption. Synth Met 159(21–22):2259–2262

Diaconu G, Paulis M, Leiza JR (2008) Towards the synthesis of high solids content waterborne poly (methyl methacrylate-co-butyl acrylate)/montmorillonite nanocomposites. Polymer 49(10):2444–2454

Dovgolevsky E, Kirmayer S, Lakin E, Yang Y, Brinker CJ, Frey GL (2008) Self-assembled conjugated polymer–surfactant–silica mesostructures and their integration into light-emitting diodes. J Mater Chem 18(4):423–436
Duan B, Wang M, Zhou WY, Cheung WL (2008) Synthesis of ca–P nanoparticles and fabrication of ca–P/PHBV nanocomposite microspheres for bone tissue engineering applications. Appl Surf Sci 255(2):529–533
Enlow D, Rawal A, Kanapathipillai M, Schmidt-Rohr K, Mallapragada S, Lo CT et al (2007) Synthesis and characterization of self-assembled block copolymer templated calcium phosphate nanocomposite gels. J Mater Chem 17(16):1570–1578
Gangopadhyay R, De A (2000) Conducting polymer nanocomposites: a brief overview. Chem Mater 12(3):608–622
Godge GR, Garje MA, Dode AB, Tarkase KN (2020) Nanosuspension technology for delivery of poorly soluble drugs and its applications: a review. Int J Pharm Sci Nanotech 13(4):4965–4978
Godge GR, Zarekar SB, Padale PJ (2021) Nanosuspension technologies: a innovative approach in drug delivery of sparingly soluble drugs. IJCRT 9(7):409–420
Godge GR, Hiremath SN, Panchal US, Sumbe RR, Pawar GS, Raskar MA, Bharat SC (2024a) An overview of Nanocochleates for improving Oral absorption using a Phospolipid-based drug delivery system. Bioequiv Bioavailab Int J 8(2):000237
Godge GR, Randhawan BB, Shaikh AB, Bharat SC, Raskar MA, Hiremath SN (2024b) Formulation perspective and applications of solid lipid nanoparticles for drug delivery: a review. RGUHS J Pharm Sci (RJPS) 14(1):1–10
Godge GR, Panchal US, Gade S, Sumbe RR, Pawar GS, Ranshur V (2025) Pharmaceutical Formulation Using Polymer Nanocomposites. Int Jour Res and Sci Innovations 12(4):1497–1526
Gridnev AA, Ittel SD (2001) Catalytic chain transfer in free-radical polymerizations. Chem Rev 101(12):3611–3660
Heuts JP, Roberts GE, Biasutti JD (2002) Catalytic chain transfer polymerization: an overview. Aust J Chem 55(7):381–398
Hojjati B, Charpentier PA (2008) Synthesis and kinetics of graft polymerization of methyl methacrylate from the RAFT coordinated surface of nano-TiO_2. J Polym Sci Part Polym Chem 46(12):3926–3937
Hojjati B, Sui R, Charpentier PA (2007) Synthesis of TiO2/PAA nanocomposite by RAFT polymerization. Polymer 48(20):5850–5858
Janáky C, Visy C (2008) Synthesis and characterization of poly (3-octylthiophene)/γ-Fe2O3 nanocomposite—a promising combination of superparamagnetic–thermoelectric–conducting properties. Synth Met 158(21–24):1009–1014
Luan J, Ma K, Zhang L, Li M, Li Y, Pan B (2010) Research on different preparation methods of new photocatalysts. Curr Org Chem 14(7):683–698
Macanás J, Parrondo J, Muñoz M, Alegret S, Mijangos F, Muraviev DN (2007) Preparation and characterisation of metal–polymer nanocomposite membranes for electrochemical applications. Phys Status Solidi A 204(6):1699–1705
Makamba H, Huang J, Chen HH, Chen S (2008) Photopatterning of tough single-walled carbon nanotube composites in microfluidic channels and their application in gel-free separations. Electrophoresis 29(12):2458–2465
Mallick K, Witcomb MJ, Erasmus R, Scurrell MS (2007a) Hydrophilic behaviour of gold-poly (o-phenylenediamine) hybrid nanocomposite. Mater Sci Eng B 140(3):166–171
Mallick K, Witcomb M, Scurrell M (2007b) A novel synthesis route for a gold–polymer soft composite material. Phys Status Solidi RRL—Rapid Res Lett 1(1):R1. https://doi.org/10.1002/pssr.200600028
Mittal V (2009) Optimization of polymer nanocomposite properties. John Wiley & Sons. https://books.google.com/books?hl=en&lr=&id=_m80mB1EWvkC&oi=fnd&pg=PR5&dq=Mittal,+V.+(2010)+in+Optimization+of+Polymer+Nanocomposite+Properties+(ed.+V.+Mittal),+Wiley-VCH+Verlag+GmbH+%26+Co.+KGaA,+Weinheim,+pp.+1%E2%80%9319.&ots=P_6yajGNuA&sig=fg1phxx0Aui

Muraviev DN, Macanás J, Esplandiu MJ, Farre M, Muñoz M, Alegret S (2007) Simple route for intermatrix synthesis of polymer stabilized core-shell metal nanoparticles for sensor applications. Phys Status Solidi A 204(6):1686–1692

Muraviev DN, Macanás J, Ruiz P, Muñoz M (2008 Jun) Synthesis, stability and electrocatalytic activity of polymer-stabilized monometallic Pt and bimetallic Pt/cu core–shell nanoparticles. Phys Status Solidi A. 205(6):1460–1464

Njagi J, Andreescu S (2007) Stable enzyme biosensors based on chemically synthesized au–polypyrrole nanocomposites. Biosens Bioelectron 23(2):168–175

Ohlan A, Singh K, Chandra A, Dhawan SK (2008) Conducting ferromagnetic copolymer of aniline and 3,4-ethylenedioxythiophene containing nanocrystalline barium ferrite particles. J Appl Polym Sci 108(4):2218–2225

Owens DE, Eby JK, Jian Y, Peppas NA (2007 Dec) Temperature-responsive polymer–gold nanocomposites as intelligent therapeutic systems. J Biomed Mater Res A 83A(3):692–695

Pavlidou S, Papaspyrides CD (2008) A review on polymer–layered silicate nanocomposites. Prog Polym Sci 33(12):1119–1198

Pérez-Cappe E, Mosqueda Y, Martínez R, Milian CR, Sánchez O, Varela JA et al (2008) Preparation and properties as positive electrodes of PANI–LiNi 0.8 co 0.2 O_2 nanocomposites. J Mater Chem 18(33):3965–3971

Pomogailo AD, Dzhardimalieva GI, Rozenberg AS, Kestelman VN (2007) Hafnium-containing nanocomposites. J Thermoplast Compos Mater 20(2):151–174

Pramanik N, Biswas SK, Pramanik P (2008) Synthesis and characterization of hydroxyapatite/poly(vinyl alcohol phosphate) nanocomposite biomaterials. Int J Appl Ceram Technol 5(1):20–28

Pramanik N, Mohapatra S, Bhargava P, Pramanik P (2009) Chemical synthesis and characterization of hydroxyapatite (HAp)-poly (ethylene co vinyl alcohol)(EVA) nanocomposite using a phosphonic acid coupling agent for orthopedic applications. Mater Sci Eng C 29(1):228–236

Prashantha K, Rashmi BJ, Venkatesha TV, Lee JH (2006) Spectral characterization of apatite formation on poly (2-hydroxyethylmethacrylate)–TiO2 nanocomposite film prepared by sol–gel process. Spectrochim Acta A Mol Biomol Spectrosc 65(2):340–344

Rana PK, Sahoo PK (2007) Synthesis and pressure sensitive adhesive performance of poly (EHA-*co*-AA)/silicate nanocomposite used in transdermal drug delivery. J Appl Polym Sci 106(6):3915–3921

Ratna D, Divekar S, Patchaiappan S, Samui AB, Chakraborty BC (2007) Poly(ethylene oxide)/clay nanocomposites for solid polymer electrolyte applications. Polym Int 56(7):900–904

Reddy KR, Lee K, Gopalan AI (2007) Novel electrically conductive and ferromagnetic composites of poly(aniline-*co*-aminonaphthalenesulfonic acid) with iron oxide nanoparticles: synthesis and characterization. J Appl Polym Sci 106(2):1181–1191

Reddy KR, Lee KP, Lee Y, Gopalan AI (2008) Facile synthesis of conducting polymer–metal hybrid nanocomposite by in situ chemical oxidative polymerization with negatively charged metal nanoparticles. Mater Lett 62(12–13):1815–1818

Samanta S, Nandi AK (2009) Recyclable high performance palladium heterogeneous catalyst with porous PVDF binder from the novel gel nanocomposite route. J Phys Chem C 113(11):4721–4725

Segal E, Perelman LA, Cunin F, Di Renzo F, Devoisselle J-M, Li YY et al (2007) Confinement of Thermoresponsive hydrogels in nanostructured porous silicon dioxide templates. Adv Funct Mater 17(7):1153–1162

Sender C, Dantras E, Dantras-Laffont L, Lacoste MH, Dandurand J, Mauzac M et al (2007) Dynamic mechanical properties of a biomimetic hydroxyapatite/polyamide 6,9 nanocomposite. J Biomed Mater Res B Appl Biomater 83B(2):628–635

Stephan AM, Kumar TP, Kulandainathan MA, Lakshmi NA (2009) Chitin-incorporated poly(ethylene oxide)-based nanocomposite electrolytes for lithium batteries. J Phys Chem B 113(7):1963–1971

Sun L, Yang Z, Li X (2008) Synthesis and crystallization behavior of acetal copolymer/silica nanocomposite by *in situ* cationic ring-opening copolymerization of trioxane and 1,3-dioxolane. J Appl Polym Sci 107(3):1842–1849

Tripathi BP, Shahi VK (2009) 3-[[3-(Triethoxysilyl)propyl]amino]propane-1-sulfonic acid–poly(vinyl alcohol) cross-linked Zwitterionic polymer electrolyte membranes for direct methanol fuel cell applications. ACS Appl Mater Interfaces 1(5):1002–1012

Usuki A, Kojima Y, Kawasumi M, Okada A, Fukushima Y, Kurauchi T et al (1993) Synthesis of nylon 6-clay hybrid. J Mater Res 8(5):1179–1184

Vana P, Davis TP, Barner-Kowollik C (2002) Kinetic analysis of reversible addition fragmentation chain transfer (RAFT) polymerizations: conditions for inhibition, retardation, and optimum living polymerization. Macromol Theory Simul 11(8):823–835

Vora RH, Vora M (2006) 1, 2′-Bis (4-aminophenoxy) benzene based designed fluoro-poly (ether-imide)/MMT clay nanocomposites: synthesis and properties for high performance applications. Mater Sci Eng B 132(1–2):90–102

Weaver J, Zakeri R, Aouadi S, Kohli P (2009) Synthesis and characterization of quantum dot–polymer composites. J Mater Chem 19(20):3198–3206

Zhang R, Moon KS, Lin W, Wong CP (2010) Preparation of highly conductive polymer nanocomposites by low temperature sintering of silver nanoparticles. J Mater Chem 20(10):2018–2023

Zhou J, Li R, Liu S, Li Q, Zhang L, Zhang L et al (2009) Structure and magnetic properties of regenerated cellulose/Fe_3O_4 nanocomposite films. J Appl Polym Sci 111(5):2477–2484

Chapter 5
Characterization Techniques for Polymer Nanocomposites

Vaishali P. Kudekar, Bhushan R. Rane, Ashish S. Jain, and Sachin N. Kothawade

Abstract In recent years, the development of advanced materials has been fueled by the rapidly expanding field of study known as polymer nanocomposites. The properties of the resultant nanocomposites are significantly enhanced by the addition of nanoparticles to a polymer matrix. Dental restorations, food packaging, biosensors, electronic gadgets, and other biomedical applications are just a few of the many fields in which these advancements are essential. This chapter provides a comprehensive analysis of several methods for creating polymer-based nanocomposites, including emulsion techniques, melt intercalation, sol–gel procedures, and in situ polymerization. Furthermore, characterization of these nanocomposites is necessary to comprehend their morphology, structure, and mechanical characteristics. Techniques such as transmission electron microscopy (TEM), scanning electron microscopy (SEM), X-ray diffraction (XRD), and atomic force microscopy (AFM) are frequently used to examine these characteristics. In addition to their mechanical qualities, polymer nanocomposites' thermal characteristics are very intriguing. These thermal and chemical properties are evaluated using Fourier Transform Infrared (FTIR) Spectroscopy, Thermogravimetric Analysis (TGA), and Differential Scanning Calorimetry (DSC). The review's overall goal is to provide a comprehensive understanding of how polymer nanocomposites are combined and analyzed, as well as how these procedures aid in the development and enhancement of materials for a range of industrial and medical applications.

Keywords Polymer nanocomposite · Characterization · Types of polymer nanocomposites

---

V. P. Kudekar · B. R. Rane (✉)
Department of Pharmaceutics, Shri. D. D. Vispute College of Pharmacy and Research Center, Panvel, Maharashtra, India

A. S. Jain
Shri. D. D. Vispute College of Pharmacy and Research Center, Panvel, Maharashtra, India

S. N. Kothawade
Department of Pharmaceutics, SCSSS's Sitabai Thite College of Pharmacy, Shirur, Pune, Maharashtra, India

S. N. Kothawade et al. (eds.), *Polymer Nanocomposites*,
https://doi.org/10.1007/978-3-032-00365-2_5

5.1 Introduction

Materials made of polymer matrices with trace quantities (e.g., a few weight percent of the polymer matrix) of additives with nanoscale sizes are known as polymer nanocomposites. Polymers' mechanical, thermal, and electrical characteristics are intended to be enhanced by creating polymer nanocomposites. Polymer nanocomposites differ from conventional polymer composites like carbon fiber-reinforced polymers primarily due to the significant interfacial space between polymer matrices and nanometer-sized fillers (Saito et al. 2016).

5.1.1 Types of Nanofillers Required to Manufacture Polymer Nanocomposites

One nanoscale dimension (nanoplatelet), two nanoscale dimensions (nanofiber), and three nanoscale dimensions (nanoparticulate) are the three categories into which nanofillers may be categorized. Nanofibers, for instance, are nanofillers with two nanoscale dimensions since their length is often in the micron range. Depending on the application's needs, nanofillers with the right dimensions and physical characteristics can be chosen (Saito et al. 2016).

5.1.2 Types of Polymer Nanocomposites

5.1.2.1 Polymer Nanocomposites According to Fillers

- **Carbon Nanotube**
- Carbon nanotubes (CNTs) are a class of nanomaterials that have shown great promise in improving medication delivery methods. Their distinct physicochemical and structural characteristics make them excellent choices for a range of biological uses. Because of their effectiveness as drug transporters, CNTs play a significant role in drug delivery. They can encapsulate medications on their surfaces or inside their hollow cores, preventing deterioration and facilitating regulated release. Because of their enormous surface area, therapeutic molecules can adhere to them, enabling targeted medication delivery to certain tissues or cells. Moreover, CNTs may be functionalized with biomolecules such as peptides or antibodies to increase their binding affinity and specificity to specific spots, improving therapeutic results and developing accurate drug delivery systems (Sonowal and Gautam 2024).
- **Calcium Carbonate**
- One of the most often used inorganic fillers for thermoplastics, including polypropylene (PP) and polyvinyl chloride, is calcium carbonate. The majority of

CaCO$_3$ that is sold commercially has particle sizes ranging from 1 to 50 μm. Numerous investigations' findings have shown that the mechanical characteristics of composites loaded with micron-sized CaCO$_3$ have improved the least. Poor filler–polymer interaction is thought to be one of the main causes. Surface-modified CaCO$_3$ particles have been the focus of a lot of work in an attempt to improve the interactions between the polymer and the filler (Chan et al. 2002).

- **Layered Double Hydroxide**
- To produce nanocomposites for energy storage and conversion, such as electrodes for supercapacitors, batteries, fuel cells, solar cells, photocatalysts, and electrocatalysts for water splitting, as well as for environmental remediation, such as water treatment and the elimination of harmful gases from the atmosphere, LDH can be combined with smart nanomaterials like graphene, and carbon nanotube. The third method is to create polymer nanocomposites or LDH based polymer hybrid nanocomposites for energy by introducing organically modified LDH, exfoliated LDH sheets, or LDH/other nanofillers hybrid into the polymer matrix, agricultural, biomedical, food packaging, wastewater treatment, gas barriers, flame retardants, and anti-corrosion (Thomas and Daniel 2019).
- **Cellulose Reinforcement in Polymer Nanocomposites**
- A freeze-dried cellulose suspension can be dissolved by sonication in a solution of dimethyl sulfoxide and N, N-dimethylformamide with a small quantity of water to create organic solvent-based nanocomposites. By functionalizing the surface hydroxyl group, a hydrophobic polymer matrix may be dispersed homogeneously. Techniques like electrospinning, cast drying, freeze drying, vacuum-assisted filtration, wet spinning, layer-by-layer assembly, and various micropatterning techniques are the most common among the various kinds of cellulose nanocomposite processing methods (Wang et al. 2007; Sharma et al. 2019).

5.1.2.2 Types of Polymer Nanocomposites According to Base

- **Polyolefin-based Nanocomposites**
- Nanofiller-based polyolefin nanocomposites have prospects for enhancing polyolefins (POs) with comparatively low concentrations of nanofiller. In terms of manufacturing, POs like polyethylene and polypropylene are the most prevalent synthetic polymers, with the highest tonnage worldwide. Only 23% of other thermoplastics have been utilized for compounding, whereas over 60% of created polyolefins (PE, PP) have reportedly been released onto the market. Ziegler-Natta and other catalysts are used to polymerize ethylene and propylene either by themselves or in combination with additional olefin monomers such as butene-1 and hexene-1 to create POs. The classification of polyethylenes (PEs) is based on their density and chain branching (Tripathi et al. 2017).

- **PVC (Poly-Vinyl Chloride)-Based Nanocomposites**
- One of the main thermoplastics, poly-(vinyl chloride) (PVC), is used extensively in many different industries and has a low cost of manufacture, strong insulating qualities, and resistance to chemicals and fire. Its adaptability to a range of alterations contributes to its flexibility. Because of the reduced quantity of filler required and the enhanced characteristics, the addition of mineral nanofillers to organic macromolecules, such as PVC, has produced polymer nanocomposites that are lightweight and inexpensive (Feldman 2014).
- **Nylon-based Nanocomposites**
- Synthetic aliphatic or semiaromatic polyamides make up nylon. It may be melted to create films, fibers, or other desired forms. The thermoplastics industry relies heavily on nylon because of its excellent mechanical and thermal qualities. The characteristics are highly beneficial as they dictate the uses of the generated fiber. Nylon is a material that may be used alone or in combination. The automotive sector is the primary use for nylon composites (Linganiso and Mashigo 2017).
- **Poly (Ethylene Terephthalate)-Based Nanocomposites**
- One of the most useful semicrystalline thermoplastics is polyethylene terephthalate (PET), which is used extensively in the production of soft drink bottles and food packaging material.
- Melt extrusion and in situ polycondensation are the two main methods used to manufacture PE. One of the most useful semicrystalline thermoplastics is polyethylene terephthalate (PET), which is used extensively in the production of soft drink bottles and food packaging material. Melt extrusion and in situ polycondensation are the two main methods used to manufacture PET nanocomposites; each method has pros and cons for creating polymer nanocomposites (Bunekar et al. 2019).
- **Thermoplastic Polyurethane (TPU)-Based Nanocomposites**
- Both delicate and hard components make up TPUs. While hard segments provide these polymers with rigidity, delicate segments give them flexibility. Due to the nanoparticles' specific affinity for one of the two segments, the combination of nanoparticles and TPUs has generated a lot of attention. Only when enough conductive fillers are in touch with one another can a nanocomposite be electrically conductive. Due to TPUs' well-defined phase separation, TPU-based nanocomposites may exhibit unique electrical behavior when subjected to heat processing and shear deformation (Farzaneh et al. 2021).

5.2 Preparation of Polymer Nanocomposites

Several techniques are used to prepare the polymer nanocomposites. These include the melt intercalation technique, in situ polymerization technique, hot press, dip coating, solution casting, and spin coating methods.

5.2.1 *Solution Cast Technique*

This is the simplest conventional technique for creating polymer films of varying thicknesses. First, the solution for the nanoparticles and the polymer salt must be made independently. The necessary amount of the polymer host is dissolved in the designated solvent while being continuously stirred to produce polymer salt. A uniform, clear solution is then produced by mixing the homogeneous polymer matrix with the appropriate quantity of salt and stirring. Solution nanoparticles are prepared by ultrasonically adding them to a liquid for improved dispersion. By mixing the resulting nanoparticle solution with the polymer salt solution and stirring until a homogeneous solution is achieved, the polymer nanocomposite is produced. At room temperature, the viscous solution was lastly allowed to evaporate the solvent in the petri plate. After removing any last traces of the solvent, the solvent is then maintained in a vacuum oven, and the film from the petri dish is then peeled off (Arya and Sharma 2018).

5.2.2 *Spin Coating Technique*

The spin coating process is an efficient and simple method to create consistent coatings that range in thickness from a few nanometers to a few microns. Using a spin coater set to a specific speed, a small quantity of solvent is applied to the substrate. It is essential to keep the substrate used to apply the thin layer perpendicular to the axis of rotation. The thickness of the film may be influenced by the mixture's concentration, rotating speed, spin time, and viscosity. A gel-type mixture cannot be distributed by a spin coater spinning to form a thin layer (Park et al. 2006; Krebs 2009).

5.2.3 *Hot Press Technique*

This is an inexpensive, quick technique. This method, known as solvent-free, creates superior thin films. This method includes a temperature controller and heating chamber, as well as a sample weighing cylinder that is part of the basement. This method is mostly used for polymers. The agate pestle is filled with the solvent, nanoparticle, and polymer. The mixture is heated to a temperature that is near the melting point of the polymer, and the resultant slurry after the required period. The collected sample is positioned between the blocks of stainless steel. The system's application of pressure to the slurry produces the film (Shameem et al. 2021).

5.2.4 *Dip Coating Technique*

The dip coating method uses both substrates to create a high-quality film. This method's minimal cost makes it unique. Immersion, deposition, and drainage are its three stages. The evaporation of the solvent comes next. The substrate is submerged in the solution in the first stage at a regular pace, giving enough time for coating. Drainage and deposition are the next steps. The substrate is then continuously submerged in the solvent. The interaction between the solvent and the substrate must occur. The substrate is now removed gradually and steadily to aid in the formation of a thin layer on the substrate. Evaporation, which involves the solvent, is the last stage. The surplus solvent is then eliminated by heating the substrate after the solvent has evaporated. The quality and thickness of the film are controlled by the density of the solution and the rate of removal (Shameem et al. 2021).

5.2.5 *Melt-Mixing Technique*

This method creates the appropriate polymer nanocomposites by uniformly mixing the nanomaterials in premade polymer matrices using standard melt processing equipment. The industry prefers the melt-mixing procedure because it eliminates the need for a solvent and the impact of the crosslinking reaction or polymerization process. There are no extra processes or equipment needed to use the industrial polymer processing approach. Semicrystalline polymers are melted, or amorphous polymers are softened by heating in mixing equipment (such as an extruder, Brabender plastic order, injection moulding, and kneader), and then the nanomaterials are incorporated under mechanical shear force to achieve uniform dispersion of the nanomaterial in the polymer matrix (Karak 2019).

5.2.6 *In Situ Polymerization Technique*

This method creates a polymer nanocomposite by generating the polymer in the presence of a dispersed nanomaterial. First, the low-viscosity monomer or prepolymer of the polymer is swelled and dispersed to create the well-dispersed nanomaterial. This method enables the monomer or prepolymer molecules to adsorb on the surface of the dispersed nanomaterials; the polymer chains aid in the dispersion, delamination, or unbundling of the nanomaterials during the polymerization process (Karak 2019).

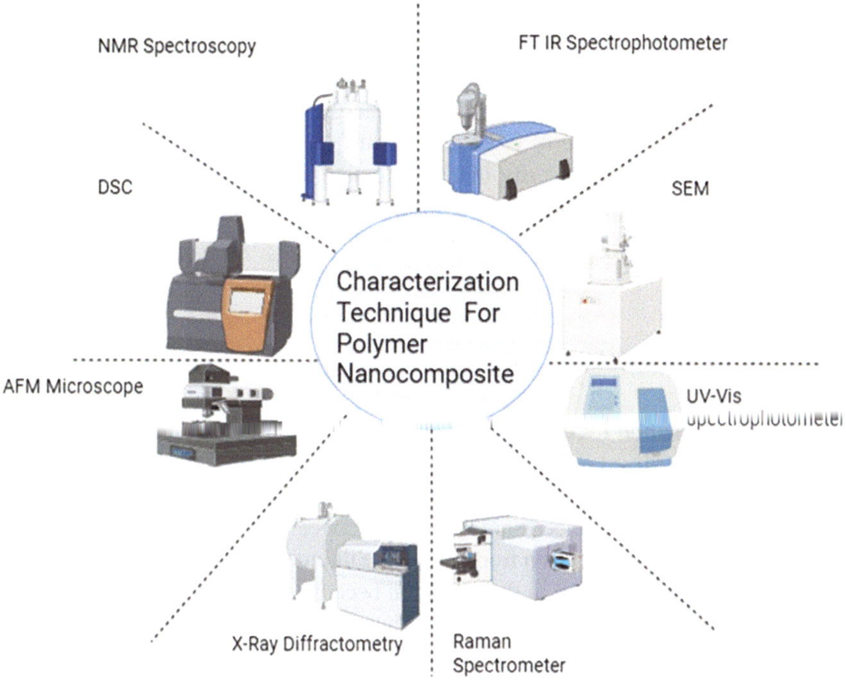

Fig. 5.1 Characterization technique of polymer nanocomposites

5.3 Characterization Technique of Polymer Nanocomposites

A variety of standard methods, as well as some unique ones, are used to characterize nanomaterials and their polymeric nanocomposites. These particular methods are crucial for comprehending the structure of such materials at the nanoscale and the interfaces between their constituent parts. The X-ray diffractometer, scanning electron microscopy, transmission electron microscopy, atomic force microscopy (AFM), FTIR spectroscopy, Raman spectroscopy, and UV-visible spectroscopy are among the frequently used diffraction, spectroscopic, and microscopic tools (Fig. 5.1).

5.3.1 UV-Visible Spectroscopy

This method, which is simple yet crucial, is used to describe polymer nanocomposites and nanomaterials. Because of their surface electrons' interaction with incoming UV light, metal or metal oxide nanoparticles exhibit characteristic surface plasmon resonance peaks in the UV–visible range. This surface plasmon resonance absorbance is the outcome of the interaction between the wave energy of the UV–visible light and the vibrational energy of the surface electrons. Since nanomaterials have very high surface electrons or surface energy, they can disperse steadily in a liquid medium. This property is known as surface plasmon resonance, and as a result, this peak is used to verify that a nanomaterial has formed. Additionally, it may be utilized to identify whether a nanomaterial is present in polymer nanocomposites (Karak 2019).

5.3.2 FTIR Spectroscopy

A widely used method for determining the existence of various chemical functional groups in functionalized nanomaterials, polymers, and their interactions is the FTIR spectroscopy. When electromagnetic radiation is absorbed, the bonds' vibration frequencies rise, causing an electronic transition between the ground state and several excited states. These absorption frequencies are unique to the kind of chemical bond and the atom group engaged in the vibration because they represent the excitations of the vibrations of the chemical bonds. Thus, FTIR spectra may also be used to analyze the physicochemical interactions between nanomaterials and the polymer matrix. Additionally, the direct usage of the nanomaterial for spectrum acquisition is made possible via attenuated total reflection (ATR)-FTIR spectroscopy. To determine the various functional groups that are present in the stabilized or functionalized molecules in nanomaterials and nanocomposites, FTIR is a useful, noninvasive, affordable, and straightforward approach (Duarah and Karak 2017).

5.3.3 NMR Spectroscopy

Similarly, the NMR spectroscopic approach is one of the more advanced and reliable methods for determining the structural properties of nanomaterials, polymers, and nanocomposites. Multinuclear NMR spectrum analysis, particularly solid-state NMR studies, reveals the conformation of various chemical environments for protons, carbons, nitrogen, silicon, and others. This is an approach based on the enticing concept of a couple of isotopic materials. The notion involves participating in solicitation and identifying the circular energy. It is bombarded with recurring electromagnetic waves at appropriate concentrations in the engaged region. Reverberation happens when the turning point and the recurring signal are the same. The quality changeover is discussed in the middle, with a survey and visual representation. Multinuclear NMR spectral analyses, especially by solid-state NMR studies, provide the conformation of different chemical environments for protons, carbons, nitrogen, silicon, etc. For example, 13C NMR spectroscopy provides information on carbon-based nanomaterials and their polymer nanocomposites, while both 1H and 13C solid-state NMR spectroscopic analyses validated the surface chemistry and interactions of polymer with nanomaterials in the nanocomposites (Qamhia et al. 2015; Ray et al. 2016).

5.3.4 *Raman Spectroscopy*

Raman spectroscopy is a novel method for characterizing and isolating common or potentially inorganic mixes in composite materials. The Raman spectrum is a representation of the Raman shift's frequency against its depth. The link between the nanoparticles and the polymer matrix may be ascertained by looking at the sharp peaks from PNCs. Plotting the Raman depth against the frequency of the Raman shift is known as the Raman spectrum. It incorporates PNCs' strong peaks, which are common for figuring out how the nanoparticles and polymer matrices relate to one another. A Raman spectrum, which may be as tiny as 1 mm, can be acquired by testing. Raman spectroscopy is an impressive method for the conceptual assessment and isolation of conventional or inorganic PNC mixtures based on carbon and cellulose (Olad et al. 2012; Chavarria et al. 2017; Lapshin et al. 2018; Wu et al. 2020).

5.3.5 *Ion Scattering Spectroscopy*

Ion scattering spectroscopy has been applied to cellulose-based nanocomposites to restrict adsorbed molecules and expose surfaces, thin film coatings, and attachment, similar to ions. The failure of the inelastic materials and the balancing rate, which depend on particle directions, determine how surface texture is measured and how low-quality particles are used (Sayer 2016).

5.3.6 *Dynamic Light Scattering*

The particle-size distribution profile of nanomaterials in solution or colloidal suspensions may be determined via dynamic light scattering (DLS). This investigation also estimates the size change of hydrated nanoparticles over time in an aqueous medium. Since the swelling of nanoparticles is typically perceived as an increase in size over time, the average hydrodynamic diameter of nanomaterials is calculated by adjusting the variation of the scattering light intensity. Given the speed of the procedure and the possibility of the scattered phase of detection, it can even track the size fluctuation in real-time. Compared to the size determined by TEM analysis, the DLS measurement is greater (Karak 2019).

5.3.7 Differential Scattering Calorimetry

A thermal assessment technique called differential scanning calorimetry (DSC) is used to measure the rate of heat transfer as temperatures vary. Among the several thermal evaluation systems available, this is one of the better techniques. These analyses provide theoretical and material modifications that evoke complex and wide-ranging structures or modifications for the heat limit. In any case, the model's temperature is adjusted by the complete lack of dissolvable stage advancements, crystallization temperature, loosening up, and contamination of the polymer. The wide limit range provided by DSC structures is 80 °C to 1600 °C (Masoodi et al. 2016; Wang 2017; Pottathara et al. 2019).

5.3.8 X-Ray Diffraction

X-ray diffraction was used to scrutinize the changes that occur to the clay due to the intercalation and/or exfoliation of the polymer into the clay galleries. XRD is a good way to investigate the spacing between the clay layers due to its ease and availability. Clays and organoclays exhibit a peak in XRD analysis because of their regular layered structures. The d-spacing or platelet separation of the clay structure is met by this peak. With the increase in d-spacing, XRD peak also shifted to lower diffraction angles (2θ). Although intercalation shows that the polymer has penetrated the gap between the clay platelets, stacking remains.

The surface chemical composition and states of oxidation of the elements are estimated by X-ray photoelectron spectroscopy (XPS) analysis or electron spectroscopy for chemical analysis (ESCA) under high vacuum conditions for both nanomaterials and polymer nanocomposites. Basically, X-ray irradiation of nanostructural materials resulted in the emission of electrons, and the XPS spectrum is obtained by measuring the kinetic energy, which provides the binding energy and the number of electrons escaping from the surface of such materials. The binding energies of specific groups are different and depend on the oxidation state of the elements; thus, such elements can be detected from the XPS spectrum (Puggal et al. 2016).

5.3.9 Gas Chromatography

Gas chromatography (GC) is used to separate a mixture into its components, which are then examined by one of the detectors. At some point, during the vaporization process in GC, the pattern is passed through a column's allotted stable bed. The evaporator heats a gas, such as helium or nitrogen, which is then regularly scattered to allow the vapor to flow through. Strong springy or nuclear sifter segments are implied to be used by GC. Chromatography separates a mixture into its constituent components, which are then decomposed. This method is employed in research to pinpoint the basic components that give intrinsic properties to carbon, cellulose, and their polymer nanocomposites. It is simple to determine the factors that influence the PNCs' qualities by recognizing these components (Li et al. 2016).

5.3.10 *Scanning Electron Microscopy*

Scanning electron microscopy (SEM) examines particle surfaces and internal structures. According to several research, if the surface of PNCs is not efficiently conductive, coating with gold nanoparticles can boost surface conductivity. This includes a finely collimated light emission that radiates throughout the examination. The pillar is inserted straight into a test that filters the exterior of a sample. The transmitted particles gather in the appropriate location, yielding surface data. Assessing nanofiller scattering on polymer surfaces is crucial for understanding the thermal and mechanical properties of materials based on their morphology (Saba et al. 2016; Wypych et al. 2016; Rane et al. 2025c; Bharath et al. 2021).

5.3.11 *Transmission Electron Microscopy*

TEM, like SEM, uses a high voltage (80–200 keV) to send an electron beam through a narrow, consistent casing (100–200 nm in thickness). Electrons are dispersed or scattered by grid patterns inside the substance's glass-like structure. Various markers are used to identify ionized X-ray photons. Most research employs TEM to analyze the interior properties of PNCs made from carbon or cellulose (Kamaraj et al. 2020; Rane et al. 2025c; Rane et al. 2025b).

5.3.12 *Secondary Ion Mass Spectroscopy*

Electron spectroscopy for chemical analysis (ESCA) spectra for identifiable compounds are difficult to repair, whereas structures strengthen polymers. Because the sampling intensity is low, the secondary ion mass spectroscopy (SIMS) approach is utilized. In a conventional assessment, the sample's surface is assailed by a strategy for a main molecule at a low thickness, to limit the change of the model surface by taking into account illumination. The surface generates nanoparticles that may be analyzed with a mass analyzer. Mass range suggests deliberate evaluations of molecular capabilities (Agubra et al. 2013; Restivo et al. 2020; Zomorodian et al. 2018).

5.3.13 Atomic Force Microscopy

Similar to SEM, AFM can also be utilized to determine the size, distribution, shape, surface area, surface roughness, and morphology of nanostructural materials. Thus, it is used to investigate the dispersion and aggregation of nanomaterials. Most interestingly, AFM can provide three-dimensional images from which particle height and volume can be calculated. In addition, oxide and electrically nonconducting surfaces can also be measured up to the subnanometer scale in aqueous fluids. However, the choice of operating mode, i.e., contact or noncontact mode, is extremely crucial for the analysis. Besides the above, the Zeta sizer is measured by the surface charge of nanomaterials, which is important as they interact with other external materials, including polymers. Generally, nanoparticles with a zeta potential greater than +30 mV or less than 30 mV are commonly made stable by electrostatic repulsion. Differential scanning calorimetric (DSC) provides the glass transition (Tg) and melting temperatures (Tm) of polymer nanocomposites, while TGA is utilized to determine the thermal degradation temperature, weight residue at different temperatures of heating, and char residue or the amount of nondegraded product present in the nanomaterials and their polymer nanocomposites (Akharame et al. 2018).

Table 5.1 summarizes several evaluation techniques that were employed to characterize the properties of the nanocomposites.

Table 5.1 Summary of characterization of polymer nanocomposites depends on recent literature

Characterization techniques	Properties analyzed	References
Tensile strength	Mechanical strength, modulus, and elongation	Stan and King (2020), Bharath and Ramajayam (2022)
XRD/WAXRD crystallinity	Interlayer spacing and dispersion degree of nanoscale fillers, kinetics of the polymer melt intercalation process	Puggal et al. (2016), Ahmad Bhat et al. (2020), Balani (2020), Bozetine et al. (2018), Ram et al. (2025), Janković et al. (2023), Ecer et al. (2022), Bhaskar et al. (2023)
DSC	Crystallization and melting properties	Masoodi et al. (2016), Wang (2017), Pottathara et al. (2019), Roshanali et al. (2020), Yerli-Soylu et al. (2022), Li et al. (2015), Nakhostin-Allaf et al. (2023)
TGA	Thermal stability (Ts)	Rashid et al. (2022), Hu et al. (2019), Khataei et al. (2020), Li et al. (2015), Zarinwall et al. (2021), Hwu et al. (2002), Janković et al. (2023), Yeşil et al. (2022)
DMTA	Thermal transition	Kazemi et al. (2021), Byberg et al. (2018)
UV-visible spectrometry	Transmittance of UV and visible light	Karak (2019), Su et al. (2021), Aparicio et al. (2012), Ram et al. (2025), Rahmatyan et al. (2022), Rane et al. (2025a, b, c), Bharath et al. (2021)

(continued)

Table 5.1 (continued)

Characterization techniques	Properties analyzed	References
Fluorescence microscopy	Enumeration and viability of bacterial cells	Zhang et al. (2020), Yao et al. (2024)
AFM	Particle size and distribution, surface morphology and microstructure, and degree of crystallization	Akharame et al. (2018), Fooladpanjeh et al. (2021), Mbodj et al. (2021), Aparicio et al. (2012)
FTIR	Chemical analysis, crystallization, and orientation of polymer, interfacial interactions	Duarah and Karak (2017), Muller et al. (2022), Zhang et al. (2014), Ismail et al. (2019), Aparicio et al. (2012), Khataei et al. (2020), Popovici et al. (2004), Jaiswal et al. (2012), Hwu et al. (2002), Ram et al. (2025), Janković et al. (2023), Shaikh et al. (2023), Yeşil et al. (2022), Rahmatyan et al. (2022), Ecer et al. (2022), Bhaskar et al. (2023)
NMR	Chemical analysis	Qamhia et al. (2015), Ray et al. (2016), Alogla et al. (2021), Shaikh et al. (2023), Rahmatyan et al. (2022)
Mass spectrometry	Chemical analysis	Zomorodian et al. (2018), Khataei et al. (2020), Rafqah et al. (2015), Puglisi et al. (2023), Zhang et al. (2021), Li et al. (2015), Letsoalo et al. (2019), Lankone et al. (2017), Pozza et al. (2015), Jaiswal et al. (2012), Hwu et al. (2002), Schmidt et al. (2009), Ram et al. (2025), Sápi et al. (2017), Janković et al. (2023), Shaikh et al. (2023), Yeşil et al. (2022), Rahmatyan et al. (2022)
GPC/SEC	Average molecular weights	Masoud et al. (2020), Puglisi et al. (2023), Alabada et al. (2023), Shahbazi (2018)
SEM	Surface morphology, microstructures like grain size and imperfections	Saba et al. (2016), Wypych et al. (2016), Budnyak et al. (2018), Rydz et al. (2019), Khataei et al. (2020), Puglisi et al. (2023), Pooja et al. (2025), Ram et al. (2025), Janković et al. (2023), Shaikh et al. (2023), Yeşil et al. (2022), Rahmatyan et al. (2022), Rane et al. (2025b, c)
TEM	Internal structure and morphology of nanoscale fillers in composites	Kamaraj et al. (2020), Zeggai et al. (2020), Nasir et al. (2020), Garcia-cardosa (2022), Jiang et al. (2015), Ram et al. (2025), Sápi et al. (2017), Janković et al. (2023)
Gas chromatography	Basic components analysis	Li et al. (2016), Wadhwani et al. (2016), Gao et al. (2025), Jaiswal et al. (2012), Pooja et al. (2025)
Raman spectroscopy	Chemical analysis	Olad et al. (2012), Chavarria et al. (2017), Lapshin et al. (2018), Wu et al. (2020), Shaikh et al. (2023)

5.4 Applications of Polymer Nanocomposites

5.4.1 *Water Treatment*

PNCs have drawn a lot of interest lately in the fields of remediation and wastewater treatment. Freshwater supplies have been decreasing in quantity and quality throughout time, owing primarily to human activities. Water is necessary for humans and plants, and it is also a fundamental resource in many industrial operations. Because of increasing industrialization, wastewater discharge has expanded dramatically, resulting in the release of a wide range of contaminants into the environment, creating negative consequences on both the ecosystem and human health. In this light, PNCs might be a useful and cost-effective material for wastewater treatment. PNCs appear to be promising in addressing the intrinsic problems of conventional particles in water treatment. Nanocomposites consisting of polymers combined with nanoparticles demonstrate superior features, such as greater resistance to fouling, thermal stability, membrane permeability, excellent mechanical behavior, higher photocatalytic activity, and higher adsorption (Wen et al. 2019).

5.4.2 *Water Disinfection*

Pathogenic bacteria in drinking water have a significant impact on human health. Water must be disinfected, either by deactivation or total elimination, or the death of harmful bacteria. In this context, PNCs with antimicrobial activity might be used to prevent the development of microbes in drinking water. PNCs' strong antibacterial action is due to the synergistic impact of their component polymer and nanomaterials. Several studies have documented the use of PNCs as antimicrobial agents in water disinfection. For example, a cellulose/silver nanocomposite with increased antibacterial activity and high water permeability, making it appropriate for use as an antimicrobial agent in the field of water treatment (Wen et al. 2019; Ahmad and Hasan 2017).

5.4.3 *Sensor Devices*

PNCs are now widely used in sensor construction. This is because polymeric materials are affordable and easy to manufacture, as well as multifunctional due to their diverse physicochemical and structural properties. Furthermore, when utilizing PNCs, it is straightforward to modify sensors, such as by adding side chains and inorganic elements to the bulk matrix. The resulting behaviors include increased conductivity, electrolytic, dielectric, and optical characteristics, higher ion selectivity, and enhanced molecule identification capabilities (Luka et al. 2015).

5.4.4 *Electromagnetic Shielding in Aerospace Applications*

The efficacy of shielding is determined by conductivity; more conductivity leads to greater shielding efficiency. The development of lightweight materials with good electromagnetic radiation shielding capabilities to prevent interference is very desirable. In this approach, PNCs might serve a variety of functions, including aeronautical and environmental solutions. Various PNCs, including conductive nanofillers, such as metal nanoparticles, magnetic nanoparticles, or carbon nanomaterials, might be used to protect against electromagnetic interference (Nasouri and Shoushtari 2017).

5.4.5 *Food Packaging*

PNCs have recently gained significant interest in the food packaging business. The addition of exfoliated graphene oxide to polycaprolactone resulted in a nanocomposite that might be used as a packaging material. The cellulose/clay nanocomposite has also exhibited improved mechanical capabilities, making it appropriate for food packaging. PNCs have been utilized in food packaging for a variety of reasons, including their excellent mechanical and thermal capabilities, antibacterial activity, barrier properties, and optical qualities. Certain PNCs, on the other hand, may be made up of synthetic polymers or biopolymers that have been modified with nanomaterials. Because of their negative effects on human health and the environment, the use of such PNCs in food packaging, especially those based on synthetic petroleum-based polymeric films, must be limited (Mujtaba et al. 2019).

5.5 Conclusion

It is well established that adding small amounts of nanoscale fillers to polymers improves their mechanical and physical characteristics. It offers a fundamental comprehension of the many techniques used to produce polymer nanocomposites. Additionally, a variety of methodologies have been used to provide the characterization of polymer nanocomposites. A nanocomposite's suitability for a given application is influenced by its preparation technique and the kinds of nanofillers used. Food packaging, water treatment, sensors, and electromagnetic shielding are some applications for polymer nanocomposites. This is because, in comparison to the original polymer, the polymer nanocomposites showed better characteristics.

5.6 Future Perspective

Polymer nanocomposites in pharmaceuticals appear to have a bright future because of recent developments in material science and technology. PNCs are positioned to play a significant role in the development of pharmaceutical applications because of trends that emphasize improved drug delivery systems, multifunctional capabilities, smart materials, sustainability, integration with manufacturing processes, and strong research activities. In the upcoming years, no doubt, continued investigation of these tendencies will result in better patient outcomes and more potent treatments.

References

Agubra VA, Owuor PS, Hosur MV (2013) Influence of Nanoclay dispersion methods on the mechanical behaviour of E-glass/epoxy nanocomposites. Nano 3:550–563. https://doi.org/10.3390/nano3030550

Ahmad Bhat S et al (2020) NiO nanoparticle doped-PVA-MF polymer nanocomposites: Preparation, Congo red dye adsorption and antibacterialactivity. Arab J Chem 13(6):5724–5739. https://doi.org/10.1016/j.arabjc.2020.04.011

Ahmad R, Hasan I (2017) Efficient remediation of an aquatic environment contaminated by Cr (VI) and 2, 4-dinitrophenol by XG-g-polyaniline@ ZnO nanocomposite. J Chem Eng Data 62:1594–1607. https://doi.org/10.1021/acs.jced.7b00099

Akharame MO, Fatoki OS, Opeolu BO, Olorunfemi DI, Oputu OU (2018) Polymeric nanocomposites (PNCs) for wastewater remediation: an overview. Polym Technol Eng 57:1801–1827. https://doi.org/10.1080/03602559.2017.1394355

Alabada R et al (2023) Investigation of effective parameters in the production of alumina gel through the sol-gel method. Case Stud Chem Environ Eng 8:100405

Alogla AA, Baumers M, Tuck C, Elmadih W (2021) The impact of additive manufacturing on the flexibility of a manufacturing supply chain. Appl Sci 11:3707. https://doi.org/10.3390/app11083707

Aparicio FJ, Blaszczyk-Lezak I, Sanchez-Valencia JR, Alcaire M, Gonzalez JC, Serra C, Barranco A (2012) Plasma deposition of perylene–adamantane nanocomposite thin films for NO2 room-temperature optical sensing. J Phys Chem C 116(15):8731–8740

Arya A, Sharma AL (2018) J Solid State Electrochem 2018:1–21

Balani SB (2020) Additive manufacturing of the high-performance thermoplastics: experimental study and numerical simulation of the Fused Filament fabrication. Institut National Polytechnique de Toulouse – INPT

Bharath KK, Ramajayam S (2022) Peek in dental implant—a review of literature. World J Adv Res Rev 15(02):139–143

Bharath M, Rane B, Jain A (2021) Formulation development and evaluation of transfersome nanogel loaded with levocetrizine. Int J Pharm Sci Nanotechnol 14(2):5383–5395

Bhaskar B, Patra R, Soma Raju KRC et al (2023) Biofilm inhibiting nanocomposite coatings on stainless steel surgical instruments: a possible strategy to prevent TASS. J Coat Technol Res 20:559–572. https://doi.org/10.1007/s11998-022-00689-9

Bozetine I et al (2018) Structural and optical properties of SiC-SiO_2 nanocomposite thin films. J Physics Conf Ser 987(1):1

Budnyak TM et al (2018) Peculiarities of synthesis and properties of Lignin-Silica nanocomposites prepared by Sol-Gel method. Nanomaterials 8:950. https://doi.org/10.3390/nano8110950

Bunekar N, Tsai TY, Lee SH, Chang YL (2019) In situ synthesis of polyethylene terephthalate nanocomposites using Bis(2-hydroxyethyl) phthalate monomer. J Chin Chem Soc 66(7):698–704. https://doi.org/10.1002/jccs.201800510

Byberg KI, Gebisa AW, Lemu HG (2018) AC SC. Polym Test. https://doi.org/10.1016/j.polymertesting.2018.10.040

Chan CM, Wu J, Li JX, Cheung YK (2002) Polypropylene/calcium carbonate nanocomposites. Polymer 43(10):2981–2992. https://doi.org/10.1016/S0032-3861(02)00103-6

Chavarria F, Shah R, Hunter D, Paul D (2017) Effect of melt processing conditions on the morphology and properties of nylon 6 nanocomposites. Polym Eng Sci 47(12):1847–1864. https://doi.org/10.1002/pen.20959

Duarah R, Karak N (2017) Facile and ultrafast green approach to synthesize bio-based luminescent reduced carbon Nanodot: an efficient photocatalyst. ACS Sustain Chem Eng 5(11):9454–9466. https://doi.org/10.1021/acssuschemeng.7b02548

Ecer Ü, Şahan T, Zengin A, Gubbuk İH (2022) Decolorization of rhodamine B by silver nanoparticle-loaded magnetic sporopollenin: characterization and process optimization. Environ Sci Pollut Res Int 29(52):79375–79387. https://doi.org/10.1007/s11356-022-21416-2. Epub 2022 Jun 10

Farzaneh A, Rostami M, Nazockdast H (2021) Thermoplastic polyurethane/multiwalled carbon nanotubes nanocomposites: effect of nanoparticle content, shear, and thermal processing. Polym Compos 42(9):4804–4813. https://doi.org/10.1002/pc.25904

Feldman D (2014) Poly(vinyl chloride) nanocomposites. J Macromol Sci, Part A 51(8):659–667. https://doi.org/10.1080/10601325.2014.925265

Fooladpanjeh S, Yousefi F, Molaei F, Zarghami M (2021) Journal of molecular graphics and modelling thermal conductivity of random polycrystallineBC 3 nanosheets: a step towards realistic simulation of 2D structures. J Mol Graph Model. 107(June):107977. https://doi.org/10.1016/j.jmgm.2021.107977

Gao X, Gao W, Lu D, Xu R (2025) Di-cyclohexene oxide bridged by DDSQ: preparation, characterization, and application as fillers for cyanate Ester resin. Molecules 30(5):1113. https://doi.org/10.3390/molecules30051113. PMID: 40076336; PMCID: PMC11902247

Garcia-cardosa M (2022) A review on additive manufacturing of micromixing devices. Micromachines 13:73. https://doi.org/10.3390/mi13010073

Hu C, Sun J, Long C, Wu L, Zhou C, Zhang X (2019) Synthesis of nano zirconium oxide and its application in dentistry. Nanotechnol Rev. 8(1):396–404. https://doi.org/10.1515/ntrev-2019-0035

Hwu JM et al (2002) The characterization of organic modified clay and clay-filled PMMA nanocomposite. J Appl Polym Sci 83(8):1702–1710

Ismail YA, Luqman M, Mane RS, Greish YE, Pathan HM (2019) Advances in applications of polymer nanocomposites. Adv Mater Sci Eng 2019:7136981. https://doi.org/10.1155/2019/7136981

Jaiswal M, Chauhan D, Sankararamakrishnan N (2012) Copper chitosan nanocomposite: synthesis, characterization, and application in removal of organophosphorous pesticide from agricultural runoff. Environ Sci Pollut Res Int 19(6):2055–2062. https://doi.org/10.1007/s11356-011-0699-6. Epub 2012 Jan 4

Janković B et al (2023) Biomineral nanocomposite scaffold (CaCO3/PVA based) carrier for improved stability of vitamin D3: characterization analysis and material properties. J Mater Sci 58(15):6580–6601

Jiang Z, Yang T, Zhang Y, Wang J (2015) Characterization and evaluation of the efficiency of SiO2/tetra-α-(2,4-di-tert-butylphenoxy)-phthalocyaninato zinc nanocomposite as photosensitizers for oxidation of 2,4,6-trichlorophenol. Environ Technol 36(13–16):1643–1650. https://doi.org/10.1080/09593330.2014.1002864. Epub 2015 Feb 6

Kamaraj M, Dodson EA, Datta S (2020) Effect of graphene on the properties of flax fabric reinforced epoxy composites. Adv Compos Mater 29:443–458. https://doi.org/10.1080/09243046.2019.1676601

Karak N (2019) Fundamentals of nanomaterials and polymer nanocomposites. In: Nanomaterials and polymer nanocomposites. Elsevier, pp 1–45

Kazemi F, Naghib SM, Zare Y, Rhee KY (2021) Biosensing applications of polyaniline (PANI)-based nanocomposites: a review. Polym Rev 61(3):553–597. https://doi.org/10.1080/15583724.2020.1858871

Khataei MM, Yamini Y, Ghaemmaghami M (2020) Reduced graphene-decorated covalent organic framework as a novel coating for solid-phase microextraction of phthalate esters coupled to gas chromatography-mass spectrometry. Microchim Acta 187:1–9

Krebs FC (2009) Sol Energy Mater Sol Cells 93(4):394–412. https://doi.org/10.1016/j.solmat.2008.12.010

Lankone RS et al (2017) Photodegradation of polymer-CNT nanocomposites: effect of CNT loading and CNT release characteristics. Environ Sci Nano 4(4):967–982

Lapshin S, Swain SK, Isayev AI (2018) Ultrasound aided extrusion process for preparation of polyolefin-clay nanocomposites. Polym Eng Sci 48(8):1584–1591. https://doi.org/10.1002/pen.21056

Letsoalo MR, Godeto TW, Magadzu T, Ambushe AA (2019) Selective speciation of inorganic arsenic in water using nanocomposite based solid-phase extraction followed by inductively coupled plasma-mass spectrometry detection. J Environ Sci Health A Tox Hazard Subst Environ Eng 54(9):924–932. https://doi.org/10.1080/10934529.2019.1609321. Epub 2019 May 3

Li X, Yang H, Li Y-c (2015) Characterization of thermal reaction of aluminum/copper (II) oxide/poly(tetrafluoroethene) nanocomposite by thermogravimetric analysis, differential scanning calorimetry, mass spectrometry and X-ray diffraction. Thermochim Acta 621:68–73. ISSN 0040-6031. https://doi.org/10.1016/j.tca.2015.10.012

Li M-C, Zhang Y, Cho UR (2016) Mechanical, thermal, and friction properties of Rice bran carbon/nitrile rubber composites: influence of particle size and loading. Mater Des 63:565–574. https://doi.org/10.1016/j.matdes.2014.07.019

Linganiso ZL, Mashigo AP (2017) In situ polymerization of nylon-cellulose nano composite. Arch Med 3(1):2

Luka G, Ahmadi A, Najjaran H, Alocilja E, DeRosa M, Wolthers K, Malki A, Aziz H, Althani A, Hoorfar M (2015) Microfluidics integrated biosensors: a leading technology towards lab-on-a-chip and sensing applications. Sensors 15:30011–30031. https://doi.org/10.3390/s151229783

Masoodi R, El-Hajjar R, Pillai K, Sabo R (2016) Mechanical characterization of cellulose nanofiber and bio-based epoxy composite. Mater Des 36:570–576

Masoud EM, Liu L, Peng B (2020) Synthesis, characterization, and applications of polymer nanocomposites. J Nanomater 2020:10–11. https://doi.org/10.1155/2020/5439136

Mbodj NG, Abuabiah M, Plapper P, El Kandaoui M, Yaacoubi S (2021) Bead geometry prediction in laser-wire additive manufacturing processusing machine learning: case of study. Appl Sci 11:11949. https://doi.org/10.3390/app112411949

Mujtaba M, Morsi RE, Kerch G, Elsabee MZ, Kaya M, Labidi J, Khawar KM (2019) Current advancements in chitosan-based film production for food technology: a review. Int J Biol Macromol 121:889–904. https://doi.org/10.1016/j.ijbiomac.2018.10.109

Muller O, Hege C, Guerchoux M, Merlat L (2022) Synthesis, characterization and nonlinear optical properties of polylactide and PMMA basedazophloxine nanocomposites for optical limiting applications. Mater Sci Eng B Solid-State Mater Adv Technol. 276(September2021):115524. https://doi.org/10.1016/j.mseb.2021.115524

Nakhostin-Allaf F, Eskandari H, Shamkhali AN, Heydari A (2023) Determination of Orange II and sulfasalazine in food, tablets, urine, soil, and water by in situ preparation of nickel hydroxide Nanoflakes/magnetite nanoparticles for magnetic solid-phase extraction followed by electrochemical detection. Anal Lett 56(16):2554–2571

Nasir R, Hasan R, Chowdhury S (2020) Handbook of polymer and ceramic nanotechnology. Springer, Cham. https://doi.org/10.1007/978-3-030-10614-0

Nasouri K, Shoushtari AM (2017) Fabrication of magnetite nanoparticles/Polyvinylpyrrolidone composite nanofibers and their application as electromagnetic interference shielding material. J Thermoplast Compos Mater 31:431–446. https://doi.org/10.1177/0892705717699682

Olad A, Azar RH, Babaluo AA (2012) Investigation on the mechanical and thermal properties of intercalated epoxy/layered silicate nanocomposites. Int J Polym Mater Polym Biomater 61(12):1035–1049. https://doi.org/10.1080/00914037.2012.680435

Park CH, Park M, Yoo SI, Joo SK (2006) J Power Sources 158(2):1442–1446. https://doi.org/10.1016/j.jpowsour.2005.10.020

Pooja G, Senthil Kumar P, Boobalan C, Rangasamy G (2025) Enhanced flotation for the removal of pharmaceutical contaminants from water systems using graphene oxide-CTAB nanocomposites. Langmuir 41(8):5032–5047. https://doi.org/10.1021/acs.langmuir.4c04066. Epub 2025 Feb 21

Popovici M et al (2004) Preparation and characterization of magnetite-based silica nanocomposite. Acta Period Tech 35:121–129. https://doi.org/10.2298/APT0435121P

Pottathara YB, Grohens Y, Kokol V, Kalarikkal N, Thomas S (2019) Synthesis and processing of emerging two-dimensional nanomaterials. In: Pottathara YB, Thomas S, Kalarikkal N, Grohens Y, Kokol V (eds) Nanomaterials synthesis. Micro and nano technologies, 1st edn. Elsevier, New York, pp 1–25. https://doi.org/10.1016/B978-0-12-815751-9.00001-8

Pozza GM, Crotty S, Rawiso M, Schubert US, Lutz PJ (2015) Molecular and structural characterization of hybrid poly(ethylene oxide)-polyhedral oligomeric silesquioxanes star-shaped macromolecules. J Phys Chem B 119(4):1669–1680. https://doi.org/10.1021/jp505191d. Epub 2015 Jan 9

Puggal S, Dhall N, Singh N, Litt MS (2016) A review on polymer nanocomposites: synthesis, characterization, and mechanical properties. Indian J Sci Technol 9(4):1–6. https://doi.org/10.17485/ijst/2016/v9i4/91212

Puglisi R, Scamporrino AA, Dintcheva NT, Filippone G, Bruno E, Scarfato P, Cerruti P, Carroccio SC (2023) Photo- and water-degradation phenomena of ZnO bio-blend based on poly(lactic acid) and polyamide 11. Polymers (Basel) 15(6):1434. https://doi.org/10.3390/polym15061434. PMID: 36987214; PMCID: PMC10058673

Qamhia II, Sham SS, El-Hajjar RF (2015) Quasi-isotropic Triaxially braided cellulose-reinforced composites. Mech Adv Mater Struct 22(12):988–995. https://doi.org/10.1080/15376494.2014.945075

Rafqah S, Seddigi ZS, Ahmed SA, Danish E, Sarakha M (2015) Use of quadrupole time of flight mass spectrometry for the characterization of transformation products of the antibiotic sulfamethazine upon photocatalysis with Pd-doped ceria-ZnO nanocomposite. J Mass Spectrom 50(2):298–307. https://doi.org/10.1002/jms.3521

Rahmatyan MS et al (2022) Synthesis and characterization of Tetrachloromenthoxyphosphor@TiO2 nanocomposite as a high-performance photocatalyst for catalytic degradation of methyl violet. Int J Nano Dimension 13(1):105–116

Ram R et al (2025) Synergistic heterojunction effects in Ag_3PO_4/SnO_2 nanocomposites: a photocatalytic study on isoproturon degradation. Front Bioeng Biotechnol 13:1458965

Rane B, Patil V, Jain A (2025a) Nanocarrier breakthroughs: revolutionizing cancer treatment with nanotechnology. Curr Nanomed 10:1. https://doi.org/10.2174/0124054615335110250115053941

Rane B, Mhatre N, Jain A (2025b) Next-generation nanocarriers for skin cancer therapy via transdermal delivery. Nano Biomed Eng 232:1–16. https://doi.org/10.26599/NBE.2025.9290116

Rane B., Padave A., Jain A. Nano-encapsulated ebastine niosomal transdermal nanogel: Qbd model for allergy treatment and evaluation. Biology and Life Sciences Forum, MDPI, 2025c 38(1), 9. https://doi.org/10.3390/blsf2024038009

Rashid IA et al (2022) Electrically conductive epoxy/polyaniline composite fabrication and characterization for electronic applications. J Reinf Plast Compos 41(1–2):34–45. https://doi.org/10.1177/07316844211023991

Ray D, Bhattacharya D, Mohanty AK, Drzal LT, Misra M (2016) Static and dynamic mechanical properties of Vinylester resin matrix composites filled with Fly ash. Macromol Mater Eng 291(7):784–792. https://doi.org/10.1002/mame.201500400

Restivo J, Gonçalves Pinto Soares OS, Ribeiro Pereira MF (2020) Processing methods used in the fabrication of macrostructures containing 1D carbon nanomaterials for catalysis. PRO 8:1329. https://doi.org/10.3390/pr8101329

Roshanali M, Nodehi A, Atai M (2020) Synthesis and characterization of core-shell nanoparticles and their application in dental resins. J Mech Behav Biomed Mater 110(July):103926. https://doi.org/10.1016/j.jmbbm.2020.103926

Rydz J, Šiškova A, Andicsova EA (2019) Scanning electron microscopy and atomic force microscopy: topographic and dynamical surface studies of blends, composites, and hybrid functional materials for sustainable future. Adv Mater Sci Eng 2019:6871785. https://doi.org/10.1155/2019/6871785

Saba N, Paridah M, Abdan K, Ibrahim N (2016) Effect of oil palm Nano filler on mechanical and morphological properties of Kenaf reinforced epoxy composites. Constr Build Mater 123:15–26. https://doi.org/10.1016/j.conbuildmat.2016.06.026

Saito T, Hirota M, Tamura N, Kimura S, Fukuzumi H, Heux L, Isogai A (2016) Individualization of Nano-sized plant cellulose fibrils by direct surface carboxylation using TEMPO catalyst under neutral conditions. Biomacromolecules 10:1992–1996. https://doi.org/10.1021/bm100631j

Sápi A et al (2017) Determination of the platinum concentration of a Pt/silica nanocomposite decorated with ultra small Pt nanoparticles using single particle inductively coupled plasma mass spectrometry. J Anal At Spectrom 32(5):996–1003

Sayer M (2016) Elastic properties and buckling load evaluation of ceramic particles filled glass/epoxy composites. Compos Part B 59:296–305. https://doi.org/10.1016/j.compositesb.2013.12.002

Schmidt B, Petersen JH, Bender Koch C, Plackett D, Johansen NR, Katiyar V, Larsen EH (2009) Combining asymmetrical flow field-flow fractionation with light-scattering and inductively coupled plasma mass spectrometric detection for characterization of nanoclay used in biopolymer nanocomposites. Food Addit Contam Part A Chem Anal Control Expo Risk Assess 26(12):1619–1627. https://doi.org/10.1080/02652030903225740

Shahbazi Y (2018) Characterization of nanocomposite films based on chitosan and carboxymethylcellulose containing Ziziphora clinopodioides essential oil and methanolic Ficus carica extract. J Food Process Preserv 42(2):e13444

Shaikh S et al (2023) A novel ultrasound-assisted approach for the synthesis of Biscoumarin and Bislawsone derivatives using rGO/TiO2 nanocomposite as a heterogeneous catalyst. Chem Africa 6(3):1479–1494

Shameem MM, Sasikanth SM, Annamalai R, Raman RG (2021) A brief review on polymer nanocomposites and its applications. Mater Today Proc 45:2536–2539. https://doi.org/10.1016/j.matpr.2020.11.639

Sharma A, Thakur M, Bhattacharya M, Mandal T, Goswami S (2019) Commercial application of cellulose Nano-composites–a review. Biotechnol Rep 21:e00316. https://doi.org/10.1016/j.btre.2019.e00316

Sonowal L, Gautam S (2024) Advancements and challenges in carbon nanotube-based drug delivery systems. Nano-Struct Nano-Objects 38:101117. https://doi.org/10.1016/j.nanoso.2023.101117

Stan G, King SW (2020) Atomic force microscopy for nanoscale mechanical property characterization. J Vac Sci Technol B 38(6):060801. https://doi.org/10.1116/6.0000544

Su X et al (2021) A comparative study of polymer nanocomposites containing multi-walled carbon nanotubes and graphene nanoplatelets. Nano Mater Sci. 4:185. https://doi.org/10.1016/j.nanoms.2021.08.003

Thomas S, Daniel S (eds) (2019) Layered double hydroxide polymer nanocomposites. Woodhead Publishing

Tripathi SN, Rao GS, Mathur AB, Jasra R (2017) Polyolefin/graphene nanocomposites: a review. RSC Adv 7(38):23615–23632. https://doi.org/10.1039/C7RA01656A

Wadhwani PM, Panchal VK, Shah NK (2016) Investigation of anti-corrosive properties of o-anisidine-N-salicylidene and its nanocomposite o-anisidine-N-salicylidene/NiONPs on mild steel in 2 N HCl. RSC Adv 6(93):90897–90915. https://doi.org/10.1039/C6RA13630C

Wang X (2017) Effects of nanoparticles on the properties of cement-based materials. Doctoral Dissertation, Iowa State University, Ames, IA, USA
Wang N, Ding E, Cheng R (2007) Surface modification of cellulose nanocrystals. Front Chem Eng China 1(3):228–232. https://doi.org/10.1007/s11705-007-0051-x
Wen Y, Yuan J, Ma X, Wang S, Liu Y (2019) Polymeric nanocomposite membranes for water treatment: a review. Environ Chem Lett 17:1539–1551. https://doi.org/10.1007/s10311-019-00881-4
Wu Q, Miao W-S, Zhang Y-D, Gao H-J, Hui D (2020) Mechanical properties of nanomaterials: a review. Nanotechnol Rev 9(1):259–273. https://doi.org/10.1515/ntrev-2020-0023
Wypych F, Seefeld N, Denicoló I (2016) Preparation of nanocomposites based on the encapsulation of conducting polymers into 2H-MoS2 and 1T-TiS2. Appl Organomet Chem 20:356–360. https://doi.org/10.1002/aoc.3444
Yao B, Gu L, Huang L, Li R, Fan Z, Chen Z, Qin D, Gao L (2024) Using magnetic molecularly imprinted polymer Technology for Determination of fish serum glucose levels. Polymers (Basel) 16(11):1538. https://doi.org/10.3390/polym16111538. PMID: 38891484; PMCID: PMC11174484
Yerli-Soylu N, Akturk A, Kabak O, Erol-Taygun M, Karbancioglu-Guler F, Kucukbayrak S (2022) TiO_2 nanocomposite ceramics doped with silver nanoparticles for the photocatalytic degradation of methylene blue and antibacterial activity against Escherichia coli. Eng Sci Technol An Int J. 35:101175. https://doi.org/10.1016/j.jestch.2022.101175
Yeşil TA, Dilek Ö, Tilki T, Yavuz M (2022) Synthesis, characterization and investigation of electrical conductivity of poly(2-Oxo-2-phenyl ethyl –2-amino benzoate)/bentonite nanocomposite. Süleyman Demirel Univ Facul Arts Sci J Sci 17(2):568–576. https://doi.org/10.29233/sdufeffd.1148894
Zarinwall A, Waniek T, Saadat R, Braun U, Sturm H, Garnweitner G (2021) Comprehensive characterization of APTES surface modifications of hydrous Boehmite nanoparticles. Langmuir 37(1):171–179. https://doi.org/10.1021/acs.langmuir.0c02682. Epub 2020 Dec 23
Zeggai FZ, Belbachir M, Carbonnier B, Rahmouni A, Kaldoun B (2020) Dataset for synthesis of conducting polymers nanocomposites basedon aniline and 4-amino-benzylamine catalyzed by chromium (III) exchanged maghnite (Algerian MMT) via in situ polymerization. DataBr. 29:105161. https://doi.org/10.1016/j.dib.2020.105161
Zhang Y, Du W, Zhou X, Yu H (2014) Review of research on the mechanical properties of the human tooth. Int J Oral Sci 6:61. https://doi.org/10.1038/ijos.2014.21
Zhang Z, Ouyang Y, Cheng Y, Chen J, Li N, Zhang G (2020) Size-dependent phononic thermal transport in low-dimensional nanomaterials. Phys Rep 860:1–26. https://doi.org/10.1016/j.physrep.2020.03.001
Zhang Y, Liu W, Huang L, Qian K (2021) Plasmonic ZnO nanocomposite - assisted mass spectrometry for detection of low molecular compounds. In: 2021 IEEE international conference on manipulation, manufacturing and measurement on the nanoscale (3M-NANO), Xi'an, China, pp 426–430. https://doi.org/10.1109/3M-NANO49087.2021.9599786
Zomorodian K, Veisi H, Mousavi SM, Ataabadi MS, Yazdanpanah S, Bagheri J, Mehr AP, Hemmati S, Veisi H (2018) Modified magnetic nanoparticles by PEG-400-immobilized ag nanoparticles (Fe3O4@PEG-Ag) as a core/shell nanocomposite and evaluation of its antimicrobial activity. Int J Nanomedicine 13:3965–3973. https://doi.org/10.2147/IJN.S161002. PMID: 30022820; PMCID: PMC6042532

Chapter 6
Polymer Nanocomposite Reinforcement Strategies

Dattatraya Manohar Shinkar, Unmesh Gulabrao Bhamare, Prakash Pannalal Muleva, Deepali Dattatray Bhandari, Ramanlal Narayan Kachave, and Sunil Vishvnath Amrutkar

Abstract Polymer nanocomposites are advanced materials that integrate nanoscale particles into polymer matrices, offering significant performance improvements across various technological domains. These composites show improved mechanical, thermal, and electrical properties by adding nanomaterials such as metal oxides, graphene, and carbon nanotubes (CNTs). Their promise in vital applications, such as energy systems, environmental solutions, and biomedical technology, has been highlighted by a recent study. Researchers are investigating novel synthesis methods, including in situ polymerization and electrospinning, which allow for significant enhancements in material functioning through the strategic integration of nanomaterials. Improved fracture toughness, increased thermal stability, and special self-healing properties are the most encouraging advancements. To achieve previously unheard-of material performance, new research is concentrating on improving nanoparticle dispersion and interfacial interactions, putting polymer nanocomposites at the forefront of modern materials science. Achieving uniform nanoparticle dispersion and comprehending intricate nanoscale interactions continue to be major obstacles, but the possibility of creating advanced composites continues to spark intense scientific curiosity and technical advancement.

Keywords Nanocomposites · Polymer · Reinforcement · Carbon nanotubes · Nanoparticle

D. M. Shinkar (✉) · U. G. Bhamare · P. P. Muleva
Department of Pharmaceutics, GES's Sir Dr. M. S. Gosavi College of Pharmaceutical Education and Research, Nashik, Maharashtra, India

D. D. Bhandari · R. N. Kachave · S. V. Amrutkar
Department of Pharmaceutical Chemistry, GES's Sir Dr. M. S. Gosavi College of Pharmaceutical Education and Research, Nashik, Maharashtra, India

S. N. Kothawade et al. (eds.), *Polymer Nanocomposites*,
https://doi.org/10.1007/978-3-032-00365-2_6

6.1 Introduction

Polymer nanocomposites have emerged as transformative materials in modern science, characterized by the integration of nanostructured fillers—such as carbon nanotubes, metal oxides, graphene, and nanoclays—into polymer matrices to enhance their structural, thermal, and electrical properties (Hussain et al. 2006; Paul and Robeson 2008). These materials exhibit a remarkable balance between weight and performance, making them ideal for applications in aerospace, electronics, biomedical engineering, and energy systems (Hussain et al. 2006; Paul and Robeson 2008; Ray and Okamoto 2003).

The enhancement in properties is largely attributed to the large surface area and high aspect ratio of the nanoparticles, which facilitate improved load transfer and interfacial bonding within the polymer matrix (Wang et al. 2004). Moreover, these nanocomposites offer unique characteristics such as flame retardancy, self-healing capability, and improved barrier properties against gases and solvents (Alexandre and Dubois 2000; Utracki et al. 2007).

Despite their promise, the development of polymer nanocomposites faces critical challenges, including nanoparticle agglomeration, poor dispersion, and weak interfacial adhesion (Hussain et al. 2006; Sinha Ray et al. 2003). These issues can significantly compromise the expected performance gains and remain a central focus in nanocomposite research. As a result, advanced fabrication strategies such as in situ polymerization, solution casting, electrospinning, and sol-gel processing are being explored to address these hurdles (Hussain et al. 2006; Thostenson et al. 2001).

Current research is also investigating hybrid nanofillers and surface modification techniques to improve filler–matrix compatibility and achieve more uniform dispersion (Koo 2006; Nogueira et al. 2014). The potential of these nanocomposites to revolutionize material performance continues to drive innovation across interdisciplinary domains, positioning them at the forefront of next-generation materials technology (Hussain et al. 2006; Ray and Okamoto 2003).

6.1.1 Overview

Polymer nanocomposites are a class of advanced materials formed by embedding nanostructured fillers into a polymer matrix, resulting in significantly improved properties over conventional composites (Paul and Robeson 2008; Ray and Okamoto 2003). These enhancements arise from the nanoscale dimensions of the fillers, which provide a high surface area for interaction, leading to improved mechanical strength, thermal stability, and electrical conductivity (Wang et al. 2004; Utracki et al. 2007).

The concept originated in the 1990s with the incorporation of layered silicates into polymers, which demonstrated unexpected performance gains even at low filler loadings (Alexandre and Dubois 2000). Since then, the field has rapidly expanded

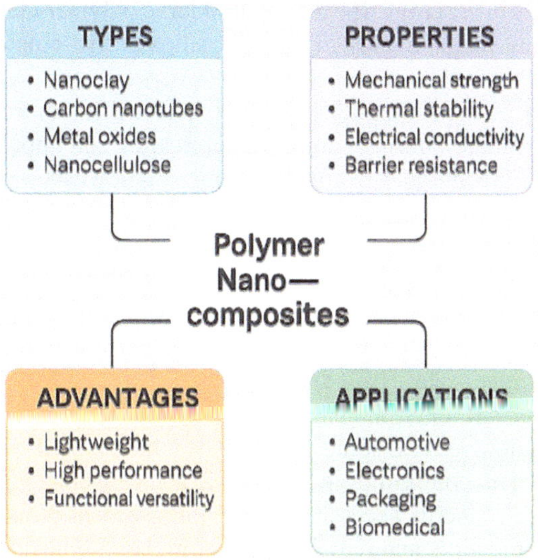

Fig. 6.1 Overview of polymer nanocomposites

to include a variety of nanofillers such as carbon nanotubes (CNTs), graphene, metal oxides, and nanocellulose (Paul and Robeson 2008; Koo 2006; Nogueira et al. 2014).

The versatility of polymer nanocomposites allows their application in high-performance sectors such as aerospace, automotive, packaging, and biomedicine. Their customizable nature—through control of the filler type, loading, dispersion method, and interfacial chemistry—offers a platform for designing materials with tailored functionality (Thostenson et al. 2001; Koo 2006). Figure 6.1 gives you an overview of this chapter.

6.1.2 Challenges

Despite their promising attributes, polymer nanocomposites face several technical and processing-related challenges that hinder their widespread commercial adoption. One of the foremost issues is the uniform dispersion of nanoparticles within the polymer matrix (Hussain et al. 2006; Wang et al. 2004). Nanoparticles tend to agglomerate due to van der Waals forces and high surface energies, leading to non-homogeneous structures that negatively impact the composite's mechanical and thermal properties (Utracki et al. 2007).

Another major challenge is the weak interfacial adhesion between the polymer matrix and the nanofillers. Poor interface compatibility leads to inefficient stress transfer, limiting the reinforcement effect of the nanomaterials (Ray and Okamoto 2003). This issue is especially critical in hydrophobic polymer matrices where polar nanofillers such as metal oxides or layered silicates are employed (Sinha Ray et al. 2003).

Moreover, processing limitations such as inadequate control over nanofiller alignment and orientation during composite fabrication can impair material performance (Koo 2006). Techniques like melt mixing, solution casting, or in situ polymerization must be carefully optimized to prevent degradation of either the polymer or the nanomaterial (Nogueira et al. 2014).

Scalability is another concern, especially when translating lab-scale processes to industrial production. Many effective methods for nanoparticle dispersion—such as high-shear mixing, sonication, or the use of surfactants—are not always viable or cost-effective at larger scales (Paul and Robeson 2008).

Finally, toxicity and environmental concerns related to certain nanomaterials, such as carbon nanotubes and some metal oxides, have raised questions about the safety of polymer nanocomposites in biomedical and consumer applications (Zhang et al. 2019). These concerns necessitate careful assessment of nanocomposite biocompatibility, degradation, and recyclability before large-scale utilization.

6.1.3 Key Features

Polymer nanocomposites distinguish themselves from traditional composites through a set of unique features driven by the nanoscale nature of the reinforcing agents. One of the most defining attributes is the high surface area-to-volume ratio of nanofillers, which leads to more effective stress transfer and improved mechanical strength even at low filler loadings (Ray and Okamoto 2003; Wang et al. 2004).

A significant feature is their multifunctionality—they can simultaneously enhance mechanical, thermal, electrical, and barrier properties of polymers, making them suitable for diverse applications such as smart packaging, flexible electronics, and drug delivery systems (Paul and Robeson 2008; Thostenson et al. 2001). For example, the incorporation of graphene or carbon nanotubes enables superior electrical conductivity, while metal oxide nanoparticles like TiO_2 or ZnO can impart UV shielding and antimicrobial properties (Wang et al. 2004; Nogueira et al. 2014).

Another important characteristic is tailorability. By adjusting filler type, concentration, aspect ratio, surface modification, and orientation, nanocomposites can be customized for specific end-use requirements (Koo 2006). This makes them highly adaptable to niche applications in biotechnology, aerospace, and renewable energy sectors.

Furthermore, some polymer nanocomposites exhibit self-healing and stimuli-responsive behavior, where external stimuli such as heat, pH, or electric fields trigger a functional response like healing of cracks or controlled drug release (Kashiwagi et al. 2002). Such smart functionalities are increasingly being explored in biomedical devices and wearable technologies.

In terms of thermal and flame resistance, the inclusion of nanofillers often reduces flammability and enhances thermal stability, which is critical for applications in construction, electronics, and automotive interiors (Utracki et al. 2007; Visakh and Thomas 2013).

6.2 Nanocomposites

Nanocomposites are multiphase materials composed of a continuous polymer matrix and a dispersed nanoscale filler phase, where at least one dimension of the filler is less than 100 nm (Utracki et al. 2007). These materials leverage the unique size-dependent properties of nanomaterials to enhance the overall performance of the composite, even at relatively low filler concentrations (Ray and Okamoto 2003).

The key feature distinguishing nanocomposites from traditional microcomposites is the significant increase in surface area-to-volume ratio, which improves interfacial interactions between the filler and the matrix. This leads to improved mechanical strength, modulus, thermal stability, gas barrier performance, flame retardancy, and in some cases, electrical conductivity (Paul and Robeson 2008; Wang et al. 2004).

Common types of nanofillers include layered silicates (e.g., montmorillonite), carbon-based nanomaterials (e.g., carbon nanotubes, graphene), metal and metal oxide nanoparticles (e.g., ZnO, TiO_2, Ag), and nanocellulose (Nogueira et al. 2014; Pinnavaia and Beall 2000). Each nanofiller contributes differently to the composite system. For instance, carbon nanotubes (CNTs) and graphene offer exceptional tensile strength and electrical conductivity, while metal oxides provide antimicrobial and UV-protective properties (Alexandre and Dubois 2000; Koo 2006).

Nanocomposites are classified based on the nature of interaction and dispersion of the nanofillers in the matrix. These include intercalated, exfoliated, or phase-separated systems, especially prominent in clay-based nanocomposites (Lee et al. 2020). The degree of exfoliation significantly affects final properties and is a focus area in material design (Fig. 6.2).

Applications of nanocomposites are expanding rapidly across industries. In packaging, they offer improved barrier properties; in biomedicine, they are being explored for tissue scaffolds and drug delivery systems; and in automotive and aerospace, they provide lightweight alternatives with enhanced strength and durability.

6.2.1 *Formulations of Nanocomposites*

The formulation of polymer nanocomposites involves the careful selection and integration of nanofillers into a polymer matrix to achieve desired performance characteristics. Successful formulation depends on several critical factors: filler type,

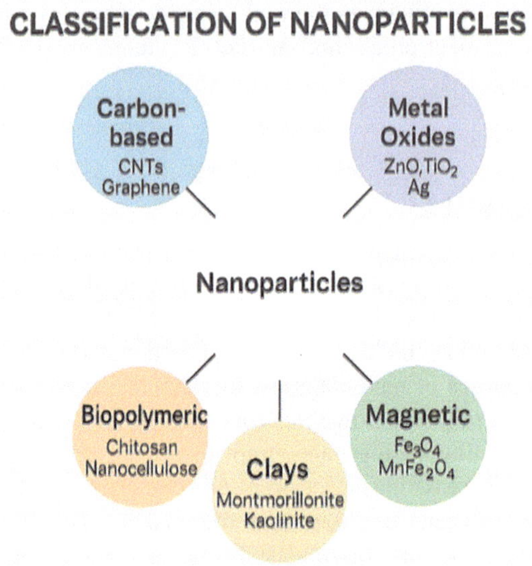

Fig. 6.2 Classification of nanoparticles

polymer compatibility, filler loading, dispersion method, and processing conditions (Wang et al. 2004).

Nanofillers can be categorized broadly into inorganic (e.g., silica, metal oxides, layered silicates), carbon-based (e.g., carbon nanotubes, graphene, fullerenes), and biopolymeric (e.g., nanocellulose, chitosan nanoparticles) (Nogueira et al. 2014). These fillers are incorporated into thermoplastic, thermosetting, or elastomeric matrices, depending on the application and property requirements (Sinha Ray et al. 2003).

Formulation methods are typically grouped into three major approaches.

1. Solution Intercalation—Both polymer and nanofiller are dispersed in a common solvent. Upon solvent evaporation, nanofillers can be intercalated or exfoliated within the matrix (Utracki et al. 2007). This method offers good dispersion but is limited by solvent toxicity and environmental concerns.
2. Melt Blending—The polymer and nanofillers are mixed in the melt state using extruders or mixers. This solvent-free method is industrially scalable but often struggles with achieving complete exfoliation or dispersion, especially for hydrophilic fillers in nonpolar matrices (Paul and Robeson 2008).
3. In Situ Polymerization—Monomers are polymerized in the presence of nanofillers, often leading to better matrix–filler interaction and dispersion due to chemical bonding or encapsulation during polymer growth (Thostenson et al. 2001).

Surface functionalization of nanofillers—such as through silane coupling agents, acid treatments, or polymer grafting—is often employed to improve compatibility and dispersion within the polymer matrix (Vaia and Giannelis 1997; Landfester

2009). For example, oxidized CNTs or silanized clays exhibit enhanced interactions with polar polymers like polyvinyl alcohol (PVA) or epoxy resins.

Filler loading typically ranges from 1 to 10 wt%. Beyond a certain threshold, agglomeration becomes prominent, reducing mechanical performance and introducing stress concentrators (Ray and Okamoto 2003). Therefore, achieving an optimal balance between dispersion, interfacial bonding, and loading is critical in formulation design.

Formulations are tailored based on end-use applications. For example, high-barrier food packaging uses nanoclay–polyolefin systems; high-strength structural materials use CNT–epoxy composites; and biomedical applications prefer biodegradable matrices like PLA or chitosan with bioactive nanoparticles (Pinnavaia and Beall 2000; Gojny et al. 2005).

6.3 Polymerization

Polymerization is a crucial step in the formation of polymer nanocomposites, especially when using in situ methods, where the polymer is synthesized in the presence of dispersed nanofillers. This approach enables improved interfacial bonding between the nanomaterial and polymer matrix, often leading to superior dispersion and enhanced performance characteristics (Thostenson et al. 2001).

The presence of nanofillers during polymerization can influence the reaction kinetics, molecular weight distribution, and morphology of the resulting polymer. Proper control over these factors is essential to achieve desirable composite properties. For instance, nanoclays or carbon nanotubes can act as nucleating agents, affecting crystallinity and improving thermal stability (Utracki et al. 2007; Gojny et al. 2005).

Polymerization methods are chosen based on compatibility with the nanofiller, desired matrix structure (thermoplastic vs thermoset), and end-use application. This leads to the exploration of several polymerization strategies as discussed in the next section.

6.3.1 Various Techniques of Polymerization

There are several polymerization techniques employed in the synthesis of polymer nanocomposites. Each method presents distinct advantages and limitations in terms of filler dispersion, scale-up feasibility, and material properties.

1. In Situ Polymerization

   In this method, nanofillers are dispersed in a monomer solution, followed by the initiation of polymerization. This approach enhances interfacial interaction and often improves the dispersion of nanofillers due to the encapsulation of nanoparticles during polymer growth (Thostenson et al. 2001; Landfester 2009).

It is widely used for clay-polyamide, CNT-epoxy, and graphene-polystyrene composites.

2. Solution Polymerization

   Monomers and fillers are dissolved or dispersed in a common solvent. The polymerization occurs in solution, often producing well-dispersed nanocomposites. However, this method has environmental limitations due to solvent use and recovery challenges (Ray and Okamoto 2003).

3. Bulk Polymerization

   This solvent-free method allows for high molecular weight polymer formation but often leads to issues with heat dissipation and viscosity control. The nanofiller must be well dispersed beforehand to avoid aggregation (Sinha Ray et al. 2003)

4. Emulsion and Miniemulsion Polymerization

   These methods involve polymerizing monomers in micellar structures stabilized in an aqueous phase. They are particularly suitable for latex nanocomposites, offering excellent control over particle size and dispersion (Antonietti and Landfester 2002; Chiefari et al. 1998).

5. Melt Polymerization

   Carried out at high temperatures, this process eliminates the need for solvents and can be integrated with melt compounding. It is favorable for thermoplastics but may not achieve ideal dispersion of all nanofillers (Paul and Robeson 2008).

Recent developments include controlled/living polymerization techniques such as ATRP (Atom Transfer Radical Polymerization) and RAFT (Reversible Addition-Fragmentation Chain Transfer) polymerization, which offer precise control over molecular architecture and are promising for creating functional nanocomposites (Wu et al. 2008). Figure 6.3 gives the technique of polymerization.

6.3.2 *Challenges*

Although polymerization is central to the synthesis of nanocomposites, it is accompanied by multiple technical challenges that affect the final structure and performance of the composite material. These challenges arise from both chemical incompatibilities and processing limitations.

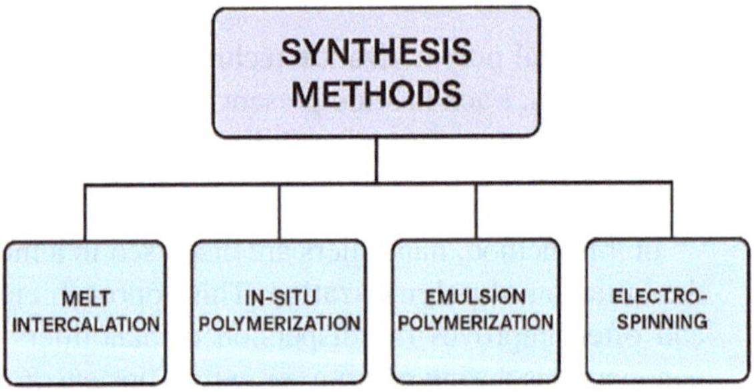

Fig. 6.3 Technique of polymerization

One of the most persistent issues is nanofiller agglomeration during polymerization. Even when dispersion is initially achieved, nanofillers such as carbon nanotubes or layered silicates can re-agglomerate due to thermodynamic incompatibility or increasing viscosity during polymer growth (Wang et al. 2004). This results in poor distribution and reduced reinforcing effects in the final composite.

Another challenge is the interference of nanofillers with polymerization kinetics. Nanoparticles may adsorb initiators or catalysts on their surface, altering reaction rates and leading to heterogeneous polymer structures (Vaia and Giannelis 1997). This is particularly evident in free-radical polymerization, where radical trapping or inhibition can occur due to the high surface activity of nanomaterials (Sui et al. 2011).

Thermal and mechanical stability of fillers during high-temperature processes like melt or bulk polymerization can also be problematic. For instance, graphene oxide and certain metal oxides may undergo structural transformations or degradation under high heat, compromising the intended properties of the composite (Zhang et al. 2013).

Additionally, achieving uniform filler orientation and alignment is especially difficult in bulk systems. In situ polymerization may offer better control, but alignment is often random unless specific techniques (e.g., magnetic, electric fields, or shear-induced alignment) are used (Koo 2006).

Scale-up issues represent another critical limitation. Laboratory-scale polymerization methods like solution and emulsion techniques often face difficulties in translating to industrial-scale production due to factors like solvent recovery, heat control, and reproducibility of filler dispersion (Ray and Okamoto 2003).

Overcoming these challenges requires a multipronged strategy, involving surface modification of nanofillers, optimization of polymerization parameters, and advanced processing methods like reactive extrusion or supercritical fluid-assisted polymerization (Antonietti and Landfester 2002; Gahleitner et al. 2010).

6.4 Rheological Properties

Rheological properties are critical in understanding the processability, dispersion state, and microstructural behavior of polymer nanocomposites. The introduction of nanofillers into polymer matrices often results in significant changes to the viscoelastic behavior, flow characteristics, and processing response of the composite system (Paul and Robeson 2008).

One of the most observed effects is the increase in melt viscosity and storage modulus (G′), especially at low frequencies. This behavior is attributed to the formation of a filler network or percolation structure, where nanofillers interact and restrict polymer chain mobility (Thostenson et al. 2001; Krishnamoorti and Vaia 2007). For instance, even at low loadings (1–5 wt%), carbon nanotubes or nanoclays can cause a dramatic increase in complex viscosity due to their high aspect ratio and surface interactions (Kim and Macosko 2008).

The rheological percolation threshold—the minimum filler concentration required to form a continuous network—depends on filler type, shape, surface chemistry, and dispersion quality. Proper dispersion results in a lower threshold and better reinforcement, whereas agglomerated systems show poor rheological reinforcement and inferior mechanical properties (Zhang et al. 2007).

Nanocomposites often exhibit non-Newtonian behavior, especially shear thinning, which is beneficial for processes like extrusion and injection molding. The degree of shear thinning increases with nanofiller content and network formation, aiding flow during processing but potentially causing orientation-related anisotropy in the final product (Krupa and Miková 2014).

Dynamic mechanical analysis (DMA) provides insights into temperature-dependent viscoelastic properties, including storage modulus (G'), loss modulus (G''), and damping factor ($\tan \delta$). The addition of nanofillers typically raises the glass transition temperature (T_g) and enhances modulus, reflecting restricted chain motion due to polymer–filler interactions (Gojny et al. 2005).

Moreover, rheological behavior can be tailored by modifying filler surfaces (e.g., with surfactants or coupling agents) or by using polymer blends to modulate matrix viscosity and compatibility. These approaches help fine tune the flow behavior, which is crucial for scalable processing (Landfester 2009).

Understanding rheological behavior is not only important for processing but also serves as an indirect method to evaluate filler dispersion, network formation, and nanostructure evolution, making it a powerful tool in nanocomposite formulation and quality control (Goh et al. 2016).

6.5 Design, Synthesis, and Application of Nanocomposites

The design and synthesis of polymer nanocomposites require a multidisciplinary approach combining materials science, polymer chemistry, and nanotechnology to tailor their structure–property relationships for specific applications. The primary objective is to achieve a homogeneous dispersion of nanofillers and a strong interfacial interaction with the polymer matrix to maximize reinforcement effects (Wang et al. 2004).

6.5.1 Design Considerations

Key parameters in nanocomposite design include the following.

Selection of Nanofillers: Based on the required function—mechanical strength (CNTs, graphene), barrier properties (nanoclays), thermal stability (metal oxides), or biocompatibility (nanocellulose) (Koo 2006).

Surface Modification: Functionalization of nanofillers improves matrix compatibility, promotes dispersion, and enables chemical bonding (e.g., amine- or silane-functionalized CNTs) (Vaia and Giannelis 1997).

Polymer Type: Thermoplastics (e.g., PP, PET), thermosets (e.g., epoxy, polyester), and biodegradable polymers (e.g., PLA, PCL) are selected according to application (Nogueira et al. 2014).

Advanced simulation tools and computational modeling now aid in predicting percolation thresholds, mechanical performance, and nanostructure evolution, reducing trial-and-error experimentation in nanocomposite development (Ramakrishna et al. 2005).

6.5.2 Synthesis Techniques

The major synthesis methods include the following.

In Situ Polymerization: Offers strong filler–matrix interaction and better filler dispersion, ideal for high-performance composites (Landfester 2009).

Solution Casting: Effective for lab-scale synthesis, especially with water-soluble or solvent-compatible systems like PVA–clay composites (Zhang et al. 2007).

Melt Blending: Industrially scalable, used for thermoplastics; often requires compatibilizers or shear-enhanced dispersion (Sinha Ray et al. 2003).

Electrospinning: Produces nanofiber mats with well-dispersed nanoparticles, useful in biomedical scaffolds and filtration membranes (Al-Saleh and Sundararaj 2009).

6.5.3 Applications

Polymer nanocomposites are now integral to various engineering and functional applications.

Aerospace and Automotive: Lightweight yet mechanically robust parts, reduced fuel consumption, and enhanced thermal resistance (Krupa and Mikovά 2014).

Electronics: Flexible circuits, electromagnetic interference (EMI) shielding, and conductive films using CNT- or graphene-based nanocomposites (Rhim et al. 2013).

Biomedical Devices: Drug delivery, tissue engineering scaffolds, wound dressings, and biosensors using biocompatible matrices like chitosan, PLA, and PEG with nano-Ag or hydroxyapatite (Gojny et al. 2005).

Packaging: Food packaging with nanoclay or TiO_2 improves gas and UV barrier properties and provides antimicrobial functionality (Arico et al. 2005).

Energy Storage: Electrolyte membranes, separators, and electrodes in lithium-ion batteries and fuel cells using polymer-graphene or polymer-SiO_2 nanocomposites (Rai et al. 2009).

The adaptability of polymer nanocomposites to a wide range of sectors is driven by their tunable properties and processability, making them critical to future materials development.

6.6 Application of Nanoparticle Components

Nanoparticles play a pivotal role in the performance of polymer nanocomposites, with each type offering distinct functional, mechanical, thermal, or biological advantages. The choice and incorporation of these nanoparticles are guided by the specific application for which the composite is being designed.

1. Carbon-Based Nanoparticles

   Carbon nanotubes (CNTs) and graphene are widely used due to their extraordinary tensile strength, electrical conductivity, and thermal stability. CNT–polymer composites are employed in electromagnetic shielding, antistatic materials, and flexible electronics, while graphene-based composites enhance mechanical toughness and conductivity in structural and electronic devices (Koo 2006; Rhim et al. 2013).
2. Metal and Metal Oxide Nanoparticles

Nanoparticles such as ZnO, TiO_2, and Ag impart multifunctional properties, including:

Antimicrobial activity: Ag and ZnO disrupt microbial membranes, used in food packaging, wound dressings, and medical coatings (Emamifar et al. 2011; Saha et al. 2012).

UV protection: TiO_2 and ZnO are effective UV absorbers, suitable for sunscreens and protective films (Hossain et al. 2014).

Catalytic and sensing applications: Metal oxide nanoparticles act as photocatalysts and electrochemical sensors in environmental and biomedical fields (Habibi et al. 2010).

3. Clay Nanoparticles

   Layered silicates (e.g., montmorillonite) improve barrier properties, flame retardancy, and dimensional stability. These are widely applied in food packaging, automotive interiors, and construction materials. When exfoliated and well dispersed, they create a tortuous path for gas or solvent permeation, significantly reducing permeability (Utracki et al. 2007).
4. Biopolymeric Nanoparticles

   Nanoparticles derived from natural sources—such as nanocellulose, chitosan, and starch nanoparticles—are gaining traction due to their biocompatibility, biodegradability, and renewability. Applications include:

   Tissue engineering scaffolds
   Controlled drug delivery
   Sustainable packaging (Dash et al. 2011; Pankhurst et al. 2003)

5. Magnetic Nanoparticles
   Fe_3O_4 (magnetite) and $MnFe_2O_4$ nanoparticles are incorporated into polymers for magnetically responsive systems, including MRI contrast agents, magnetic targeting for drug delivery, and actuation in soft robotics (Mody et al. 2010).
6. Hybrid Nanoparticles.
   Recent advances focus on hybrid nanoparticles that combine multiple functions—for instance, core–shell structures (e.g., silica-coated silver or gold nanoparticles), which integrate optical, antimicrobial, and mechanical features in a single system. These are useful in biosensors, smart packaging, and theranostic devices (Siró and Plackett 2010).

The versatility and tunability of nanoparticle components allow engineers and researchers to design application-specific nanocomposites, unlocking innovations across healthcare, energy, agriculture, consumer goods, and environmental sustainability.

6.7 Future Perspective

The future of polymer nanocomposites lies in developing smart, sustainable, and multifunctional materials that meet the evolving needs of advanced technologies and environmental stewardship. Ongoing research is focused on overcoming current limitations while unlocking new frontiers in material performance.

1. Precision Design and Tailoring
   Emerging tools such as machine learning, multiscale modeling, and AI-guided materials design are transforming how polymer nanocomposites are developed. These computational approaches allow for predictive control over filler dispersion, interfacial bonding, and macroscopic properties, significantly reducing trial-and-error experimentation (Ramakrishna et al. 2005).
2. Bio-Based and Green Nanocomposites
   There is a growing push toward using biodegradable polymers (e.g., PLA, PCL, chitosan) and renewable nanofillers (e.g., nanocellulose, lignin nanoparticles) to create eco-friendly composites. These materials align with circular economy principles and offer solutions for green packaging, agriculture, and biomedical applications (Dash et al. 2011; Ligon et al. 2017).
3. Multifunctional and Responsive Systems
   The integration of stimuli-responsive nanoparticles into polymer matrices is opening the door to "smart" nanocomposites that react to heat, pH, light, or magnetic fields. These are particularly promising in wearable electronics, biosensing, drug delivery, and self-healing systems (Kashiwagi et al. 2002).
4. 3D Printing and Additive Manufacturing
   Advancements in additive manufacturing technologies now enable the direct printing of polymer nanocomposites with controlled microstructures and complex geometries. Incorporating nanomaterials like CNTs, graphene, or magnetic

nanoparticles into printable filaments allows for the development of customized functional components with embedded sensing, actuation, or thermal regulation (Leigh et al. 2012; Fadeel et al. 2018).

5. Scalable and Sustainable Fabrication

   Future focus will also include developing low-energy, solvent-free, and scalable synthesis methods, such as reactive extrusion and supercritical fluid-assisted processing, which minimize environmental impact while maintaining performance (Gahleitner et al. 2010). Green solvents, microwave-assisted synthesis, and in situ compatibilization will be key to achieving industrial feasibility.

6. Regulatory and Safety Considerations

   As nanocomposites find increasing use in healthcare and consumer products, standardized toxicity evaluation, long-term biocompatibility studies, and regulatory frameworks will be essential to ensure safety and compliance with global norms (Dhand et al. 2013). Sustainable disposal and recycling protocols must also be developed.

   In conclusion, the evolution of polymer nanocomposites is poised to converge with advances in sustainability, intelligent systems, and digital manufacturing, offering a transformative pathway toward next-generation materials.

6.8 Conclusion

Polymer nanocomposites have emerged as a transformative class of materials, offering enhanced mechanical, thermal, electrical, and functional properties through the incorporation of nanoscale fillers. Their versatility stems from the ability to tailor matrix–filler interactions, filler dispersion, and structural architecture to meet specific performance requirements across a wide range of industries.

This chapter has highlighted the fundamental concepts, design strategies, synthesis techniques, and applications of polymer nanocomposites. Emphasis was placed on the importance of nanoparticle selection, interfacial engineering, and rheological behavior, which collectively determine the final properties of the composite. Despite remarkable progress, challenges such as nanofiller agglomeration, limited compatibility, and processing scalability persist and must be addressed to unlock the full commercial potential of these materials.

Looking ahead, the integration of smart functionality, biodegradability, and sustainable processing will define the next generation of polymer nanocomposites. Emerging innovations in machine learning, 3D printing, and green chemistry offer exciting opportunities to design high-performance, environmentally friendly materials with multifunctional capabilities.

Continued interdisciplinary research, supported by robust safety evaluations and regulatory standards, will ensure the safe and effective deployment of nanocomposites in critical sectors such as healthcare, energy, electronics, and environmental engineering. Thus, polymer nanocomposites stand at the forefront of modern material innovation, poised to meet the demands of a rapidly advancing technological world.

References

Alexandre M, Dubois P (2000) Polymer-layered silicate nanocomposites: preparation, properties and uses of a new class of materials. Mater Sci Eng R Rep 28(1–2):1–63

Al-Saleh MH, Sundararaj U (2009) EMI shielding with CNT composites. Carbon 47(7):1738–1746

Antonietti M, Landfester K (2002) Polymerization in dispersed systems: emulsion and miniemulsion techniques. Adv Mater 14(16):1197–1211

Arico AS, Bruce P, Scrosati B, Tarascon JM, Van Schalkwijk W (2005) Nanostructured materials for advanced energy conversion and storage devices. Nat Mater 4(5):366–377

Chiefari J, Chong YK, Ercole F, Krstina J, Jeffery J, Le TPL et al (1998) Living free-radical polymerization by reversible addition–fragmentation chain transfer: the RAFT process. Macromolecules 31(16):5559–5562

Dash M, Chiellini F, Ottenbrite RM, Chiellini E (2011) Chitosan-based nanocomposites for biomedical applications. Biotechnol Adv 29(4):427–435

Dhand V, Rhee KY, Park SJ (2013) A comprehensive review on graphene nanocomposites: research status and trends. J Nanomater 2013:763953

Emamifar A, Kadivar M, Shahedi M, Soleimanian-Zad S (2011) ZnO nanoparticles in antimicrobial packaging. Food Control 22(6):408–413

Fadeel B, Bussy C, Chen C, Duke A, Gazzi A, Guo C et al (2018) Safety assessment of graphene-based materials: focus on human health and the environment. Adv Mater 30(36):1703459

Gahleitner M, Paulik C, Wolfahrt M, Fiebig J (2010) Influence of nanoparticles on polymer rheology. Rheol Acta 49(9):925–934

Goh M, Tan Y, Lee J (2016) Predictive modeling in polymer nanocomposites. Compos Part B 91:385–394

Gojny FH, Wichmann MH, Köpke U, Fiedler B, Schulte K (2005) Functionalized carbon nanotubes in epoxy composites. Carbon 43(7):1537–1545

Habibi Y, Lucia LA, Rojas OJ (2010) Cellulose nanocrystals: chemistry, self-assembly, and applications. Chem Soc Rev 39(4):1519–1542

Hossain MZ, Kumar GS, Raja IA (2014) Metal oxide nanoparticles in sensors. Sens Actuators B Chem 192:65–81

Hussain F, Hojjati M, Okamoto M, Gorga RE (2006) Review article: polymer-matrix nanocomposites, processing, manufacturing, and application: an overview. J Compos Mater 40(17):1511–1575

Kashiwagi T, Du F, Douglas JF, Winey KI, Harris RH, Shirai Y (2002) Thermal and flammability properties of polymer nanocomposites with nanoclays. Chem Mater 14(3):1361–1367

Kim H, Macosko CW (2008) Morphology and properties of clay/polyamide nanocomposites. Polymer 49(14):3981–3992

Koo JH (2006) Polymer nanocomposites: processing, characterization, and applications. McGraw-Hill, New York

Krishnamoorti R, Vaia RA (2007) Polymer nanocomposites: structure, rheology and processing. J Polym Sci B Polym Phys 45(24):3252–3266

Krupa I, Mikovа́ G (2014) Rheology of polymer composites and nanocomposites. In: Nalwa HS (ed) Handbook of polymernanocomposites. Springer

Landfester K (2009) Miniemulsion polymerization and the structure of polymer and hybrid nanoparticles. Angew Chem Int Ed 48(25):4488–4507

Lee JY, Tan LP, Mehta M, Yang Y (2020) Biomedical applications of polymer-based nanocomposites. Prog Polym Sci 107:101231

Leigh SJ, Bradley RJ, Purssell CP, Billson DR, Hutchins DA (2012) 3D printing of conductive polymer composites. PLoS One 7(11):e49365

Ligon SC, Liska R, Stampfl J, Gurr M, Mülhaupt R (2017) Polymers for 3D printing and additive manufacturing. Chem Rev 117(15):10212–10290

Mody VV, Siwale R, Singh A, Mody HR (2010) Introduction to theranostics. J Nanomed Nanotechnol 1(5):1000104

Nogueira DR, Tavano L, Silva AM, Veiga F, Garcia ML, Lucangioli S et al (2014) Nanotoxicology of polymeric nanomaterials. J Biomed Nanotechnol 10(10):2891–2911
Pankhurst QA, Connolly J, Jones SK, Dobson J (2003) Applications of magnetic nanoparticles in biomedicine. J Phys D Appl Phys 36(13):R167–R181
Paul DR, Robeson LM (2008) Polymer nanotechnology: nanocomposites. Polymer 49(15):3187–3204
Pinnavaia TJ, Beall GW (2000) Polymer–clay nanocomposites. Wiley, Hoboken
Rai M, Yadav A, Gade A (2009) Silver nanoparticles as a new generation of antimicrobials. Biotechnol Adv 27(1):76–83
Ramakrishna S, Fujihara K, Teo WE, Yong T, Ma Z, Ramaseshan R (2005) An introduction to electrospinning and nanofibers. World Scientific, Singapore
Ray SS, Okamoto M (2003) Polymer/layered silicate nanocomposites: a review from preparation to processing. Prog Polym Sci 28(11):1539–1641
Rhim JW, Park HM, Ha CS (2013) Bio-nanocomposites for food packaging applications. Food Hydrocoll 33(2):327–335
Saha A, Dhanalakshmi D, Rajendiran N (2012) Nanostructured ZnO/TiO_2 composites for UV protection. J Photochem Photobiol B 113:55–63
Sinha Ray S, Yamada K, Okamoto M, Ueda K (2003) Control of biodegradability of poly(L-lactide) via nanocomposite technology. Macromol Mater Eng 288(3):203–208
Siró I, Plackett D (2010) Microfibrillated cellulose and composites: a review. Cellulose 17(3):459–494
Sui Z, Zhang X, Lei Y, Luo Y (2011) Thermal stability of graphene oxide. Carbon 49(13):4314–4322
Thostenson ET, Ren Z, Chou TW (2001) Advances in the science and technology of carbon nanotubes and their composites: a review. Compos Sci Technol 61(13):1899–1912
Utracki LA, Sepehr M, Boccaleri E (2007) Synthetic, layered nanoparticles for polymeric nanocomposites (PNCs). Polym Adv Technol 18(1):1–37
Vaia RA, Giannelis EP (1997) Polymer melt intercalation in organically-modified layered silicates. Macromolecules 30(25):8000–8009
Visakh PM, Thomas S (2013) Advances in elastomers I: blends and interpenetrating networks. Springer, Berlin
Wang Z, Liang Z, Wang B, Zhang C, Kramer L (2004) Processing and property investigations of single-walled carbon nanotube (SWNT)/epoxy resin nanocomposites. Compos Part A Appl Sci Manuf 35(10):1225–1232
Wu C, Liu Y, Li Y, Zhang H (2008) Influence of CNTs on radical polymerization. J Polym Sci A Polym Chem 46(3):984–992
Zhang Q, Zhang K, Gu Y, Guo J (2007) Rheological and electrical behavior of CNT-filled composites. Macromolecules 40(24):8270–8276
Zhang Y, Shen L, Phang IY, Liu T (2013) Supercritical CO_2-assisted dispersion of nanofillers. J Supercrit Fluids 76:138–145
Zhang M, Yu L, Ding B (2019) Self-healing polymer nanocomposites: a review. Polym Chem 10(36):5028–5040

Chapter 7
Effect of Original Reinforcement and Surface Functionalization on Properties

Kazi Tasnuva Alam, Zarin Tasnim Tisha, and Tanvir Ahmed

Abstract Polymer nanocomposites, consisting of a polymer matrix reinforced with nanofillers, significantly improve the properties of the material. This chapter explores the critical relationship between the original reinforcement's characteristics and how surface functionalization affects the properties of the final or resultant nanocomposite. The intrinsic properties of different nanofillers, including carbon nanotubes (CNTs), graphene, and clay minerals, and their influence on the initial characteristics of the nanocomposite have been discussed in this chapter. It further covers the impact of several surface modification techniques, including chemical grafting, polymer coating, and electrostatic interactions, on interfacial adhesion, dispersion, and the resulting electrical, mechanical, thermal, and barrier properties of nanocomposites. To attain the optimum performance improvements, the chapter highlights the significance of a tailored strategy for surface functionalization that considers the unique characteristics of both the nanofiller and polymer matrix in a polymer nanocomposite system.

Keywords Nanocomposites · Polymer matrix · Surface functionalization · Nanofiller

K. T. Alam · T. Ahmed (✉)
Department of Pharmaceutical Sciences, School of Health and Life Sciences, North South University, Dhaka, Bangladesh
e-mail: tasnuva.alam.241@northsouth.edu; tanvir.ahmed111@northsouth.edu

Z. T. Tisha
School of Chemistry and Chemical Sciences, University of Surrey, Guildford, UK
e-mail: zt00362@surrey.ac.uk

S. N. Kothawade et al. (eds.), *Polymer Nanocomposites*,
https://doi.org/10.1007/978-3-032-00365-2_7

7.1 Introduction

Polymer nanocomposites or PNCs have emerged as a fundamental component of nanotechnology, attracting considerable scientific attention stemming from the necessity for developing materials that demonstrate distinctive mechanical, electrical, chemical, and optical properties. This is accomplished through the incorporation of nanoscale materials that substantially enhance the properties of conventionally used materials. Polymer nanocomposites are composite or hybrid materials made of polymers that have been strengthened or combined with nanoparticles that serve as nanofillers. Nanofillers are primarily organic and inorganic materials with excellent thermal and mechanical properties. They are utilized for enhancing the existing properties of polymers in nanocomposite formulations. Polymer nanocomposites display unique properties in contrast to conventional polymers, mainly due to their extensive internal interfacial area and the nanoscale size of the materials integrated within. The distinctive multifunctional characteristics arising from the nanostructure of nanocomposites present a chance to exceed the usual features linked to regular composites. Applications across various fields have demonstrated that PNCs enhance mechanical properties, gas permeability, biodegradability, biocompatibility, heat resistance, dimensional stability, thermal expansion, electrical conductivity, optical properties, rheological properties, crystallization control, processability, and more when incorporated into nanoparticle compounds or formulations. In the recent decade, polymer nanocomposites have been extensively used across various fields, including aerospace, electronics, biomedicine, construction, packaging, and energy storage. The introduction of scanning tunneling microscopy and scanning probe microscopy in the early 1980s has significantly contributed to the growth and expansion of the field of polymer composites. For instance, polymer nanocomposites are widely utilized in tissue engineering to develop biocompatible scaffolds that facilitate tissue regeneration. They offer a structural framework for cell growth and tissue repair. Hydroxyapatite (HAP)-polymer nanocomposites serve as a biocompatible and osteoconductive alternative for bone repair and implantation purposes. Additionally, PNCs can be engineered to target specific diseases, such as cancer, by integrating nanoparticles that can identify and attach to cancer cells through the modification of their surface characteristics. They are also used in biosensors, imaging agents, and wound-healing applications. Beyond biomedical uses, NASA Langley Center has created transparent organically modified layered silicate nanocomposites. They are lightweight and robust, making them suitable for aircraft applications. Improved mechanical and thermal expansion qualities reduced microcrack formations and thermal cycling, enabling long-term stability under harsh conditions (Manero and Sanchez-Solis 2013; Urio and Murphy 2024; Tofighy and Mohammadi 2020).

When designing polymer nanocomposites, a crucial factor to consider is how well the nanofillers are dispersed or distributed throughout the bulk polymer matrix. Hence, these nanofillers are essential components of a composite formulation. These components are characterized as fillers with particle sizes ranging from 1 to

100 nm. Their principal role is to enhance the polymer matrix at a minimal concentration (<10 wt%) while ultimately augmenting the polymer's capabilities, as previously discussed, without modifying its density. Polymer nanocomposites incorporate nanofillers characterized by a significantly high interfacial area relative to their volume. The separation between these fillers is approximately equivalent to the dimensions of polymer chains, measuring around 20 nm in diameter, which results in a high level of interaction. With the increase of filler content, the interparticle spacing diminishes to a level below the average particle radius, leading to the development of a spatially interconnected structure. At low concentrations, nanofillers exert a substantial impact on the polymer matrix, resulting in distinctive material properties that are not present in conventional macrocomposites. Furthermore, the multifunctionality of these nanomaterials broadens their applicability to novel domains. Typical nanofillers used and under investigation are carbon nanotubes (CNTs) (single-walled nanotube (SWNT) and multiwalled nanotube (MWNT)), nanoballs, nanoneedles, silicon carbide, polyhedral oligomeric silsesquioxanes (POSS), nanofibers, and layered materials (layered silicate, layered titanate). Researchers at Toyota Central Research & Development (TCRD) have created Nylon-6-clay nanocomposites. Compared to pure Nylon-6, the composite showed a 50% increase in strength, a twofold enhancement in modulus, and an 80 °C increase in thermal distortion temperature, although only having 4.2% organically modified layered silicate, thereby gaining widespread popularity in the automobile industry. In recent years, organoclay has found extensive application in various polymer systems, including epoxy and polyurethanes, for the design and development of nanocomposites. The use of these fillers results in excellent mechanical properties at minimal loadings, along with enhanced scratch resistance, superior barrier characteristics, improved fire resistance, and better heat distortion performance relative to clean polymers. Nanomaterials (nanofillers) enhance reinforcement efficacy due to their elevated aspect ratio. The characteristics of nanocomposite materials are significantly influenced by the dimensional measure of their constituent phases and the extent of their intermixing. The characteristics of the used components and the synthesis method, such as polymer intercalation, in situ polymerization, or melt intercalation, can lead to the production of various types of polymers with distinct properties, including flocculated, intercalated, and exfoliated nanocomposites (Manero and Sanchez-Solis 2013; Bernhardt et al. 2014; Sinha Ray and Okamoto 2003). Nanofillers can improve composite formulations; however, nanomaterials exhibit a propensity to coalesce as a result of strong van der Waals forces between particles, resulting in a reduction in the characteristics of polymer nanocomposites, necessitating the surface functionalization of the nanomaterials. The surface of PNCs can be modified through various methods, including grafting, covalent bonding, polymer wrapping, electrostatic interactions, surface adsorption (utilizing surfactants and other substances), and in situ polymerization. The improvement in dispersion is achieved through the surface functionalization of nanomaterials, which enhances the compatibility between the polymer matrix and the filler. Ultimately, this procedure facilitates the production of nanocomposites with exceptional performance for cutting-edge applications (Tofighy and Mohammadi 2020).

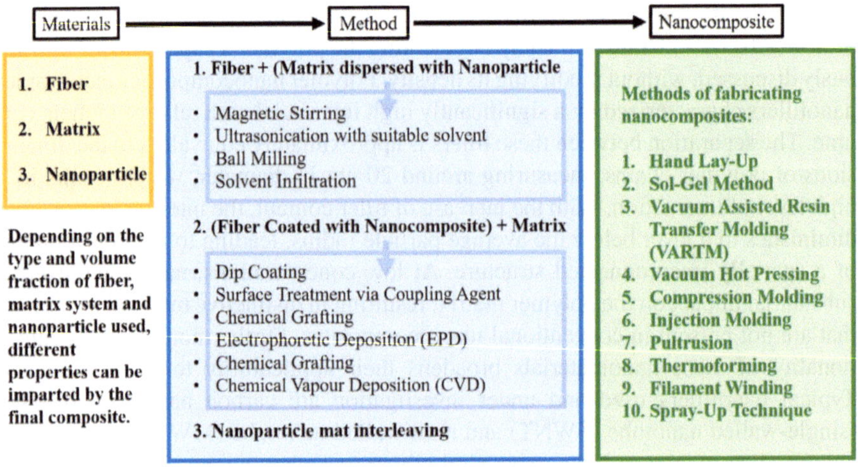

Fig. 7.1 Processing techniques for nanocomposites with various methods. (Reprinted from Bin et al. (2024), Copyright, © 2024 The Author(s). Published by Elsevier Ltd. This is an open-access article distributed under the terms of the Creative Commons CC-BY license)

This chapter explores the impact of original reinforcement materials and surface functionalization on the characteristics of polymer nanocomposites. It covers many categories of reinforcing nanoparticles (nanofillers), their function in enhancing mechanical, thermal, electrical, and barrier properties, and the difficulties related to their dispersion and compatibility. This literature thoroughly examines several surface functionalization strategies that enhance filler–polymer interactions and investigates these parameters' impact on composite performance, offering insights for optimizing nanocomposite design across various industrial applications.

Figure 7.1 shows various synthesis techniques of polymer nanocomposites that are currently employed.

7.2 Original Reinforcement: Influence of Nanofiller Properties

The concept of reinforcement is fundamental in the study of composites within material science. Here, a matrix material—be it a polymer, metal, or ceramic—is integrated with a reinforcing phase, which may include fibers, nanoparticles, or nanotubes, to improve mechanical, thermal, or electrical properties. Among all, nanofiller reinforcement offers several advantages over other particulate reinforcements owing to their higher surface area, better interaction ability, multifunctionality, improved processability, and improved thermal, mechanical, and electrical properties.

The following sections and Table 7.1 look into various nanoparticles utilized as nanofillers in the production of polymer nanocomposites, which have applications across multiple industries today.

Table 7.1 Types of original reinforcement in polymer nanocomposites

S no	Reinforcement type	Composition	Size (nm)	Morphology	Common applications	References
01	Carbon nanotubes (CNTs)	Rolled-up graphene sheets consisting of a single layer of carbon	1–100 nm	Cylindrical hollow tubes can be grouped into two types: single-walled, which consist of a single graphene layer, and multiwalled, which consist of multiple concentric graphene layers	Automotive, aerospace, biomedicine, electronics	Moses et al. (2019)
02	Graphene and its derivatives	Single or multilayer graphene sheets	1–100 nm	Honeycomb lattice structural configuration with a single, 2D layer of graphene sheets	Electronics, biosensors, biomedicine, energy storage	Lee et al. (2022)
03	Nanoclays	Layered silicate materials (magnesium and aluminum silicate)	1–100 nm	Platelet-like structures with an octahedral alumina layer between stacked sheets of tetrahedral silica layers	Packaging, barrier films, coatings, biomedicine, waste water treatment	Patel et al. (2006)
04	Metallic nanoparticles	Gold, silver, copper, zinc	1–100 nm	Spherical, rod-like, cube, and other complex structures	Biosensors, biomedicine, antimicrobial coatings, printed electronics, catalysis	Tamayo et al. (2018)
05	Ceramic nanoparticles	Inorganic compounds (silica, alumina, calcium carbonate, titania, zirconia)	1–100 nm	Spherical, rod-like, irregular structures	Biomedical implants, tissue engineering, flame retardant, barrier films, antistatic coatings	Carvalho et al. (2018)

7.2.1 Carbon Nanotubes (CNTs)

7.2.1.1 Structure and Properties

Recent decades have witnessed a shift toward the development of light-weight smart nanocomposite materials characterized by self-healing and self-sustaining properties. These materials present numerous advantages over conventional materials, including polymers, ceramics, and metals. Carbon nanotubes (CNTs) precisely address this specific niche, meeting the requirements of contemporary technological advancements. CNTs represent one-dimensional cylindrical nanostructures that are developed from the rolling of graphene sheets, characterized by sp^2 hybridization of carbon atoms. CNTs have sizes ranging from 1 nm to 100 nm and lengths reaching several millimeters, characterized by remarkably high aspect ratios and specific surface areas. These structures are categorized into single-walled carbon nanotubes (SWCNTs) or multiwalled carbon nanotubes (MWCNTs) depending on the number of concentric graphene layers they contain. The distinct chirality of CNTs determines whether they display metallic or semiconducting characteristics. According to their nanometer-scale dimensions and atomic structure, carbon nanotubes exhibit exceptional and superior mechanical (high tensile strength, elasticity, flexibility), electrical, and thermal capabilities, rendering them particularly advantageous in creating CNT-reinforced polymer nanocomposites for applications in nanoelectronics, biomedicine, aerospace, automobile, energy harvesting, and many more (Radhamani et al. 2018; Imani Yengejeh et al. 2017; Yim et al. 2025).

7.2.1.2 Influence on Nanocomposite Properties

Due to their extraordinary combination of properties, carbon nanotube-reinforced polymer composites have revolutionized material science. The integration of carbon nanotubes into a polymer matrix produces a composite material that merges the mechanical strength, electrical conductivity, and thermal stability of CNTs with the intrinsic flexibility of the polymer.

Carbon nanotubes (CNTs) have been extensively utilized as reinforcement agents to optimize the mechanical characteristics of polymer composites. Numerous studies show that functionalization and dispersion methods substantially affect the efficacy of CNT reinforcement. Functionalization methods, including grafting with dendrimers, fluorinated surfactants, and modifications using carboxylic acids or silanes, have enhanced dispersion, interfacial adhesion, and load transfer in polymer matrices. Moreover, processing methods such as ultrasonic treatment, extrusion, and three-roll milling have enhanced CNT distribution. Research on several polymer matrices, such as epoxy, PLA, PVA, PDMS, UHMWPE, PEEK, and PPS, has indicated significant enhancements in Young's modulus, tensile strength, and impact resistance. Results reveal that, whereas CNTs markedly boost stiffness and modulus, advancements in tensile strength are occasionally constrained by aggregation or inadequate dispersion. Therefore, improving the surface chemistry and processing conditions of CNTs is essential for enhancing their reinforcing capabilities in polymer nanocomposites. Nevertheless, the inclusion of CNTs in polymer nanocomposites significantly improves the tensile strength, rigidity, fracture resistance, durability, and load-bearing capacity of the nanoformulation, attributed to their high surface area and aspect ratio, and remarkable intrinsic mechanical properties (Sun et al. 2008; Hayasaki et al. 2021; Hur et al. 2023).

The high conductivity, low percolation threshold, and strong charge transport capabilities of CNTs enhance the electrical performance of polymer nanocomposites. These attributes establish effective conductive networks within the polymer matrix, enhancing charge mobility, electrical resistance, and electromagnetic shielding. Carbon nanotubes (CNTs) are optimal for flexible electronics, sensors, and electromagnetic interference (EMI) shielding materials owing to their capacity to facilitate electron transmission and alter the dielectric properties of polymers. The efficacy of electrical performance is contingent upon the dispersion, alignment, and matrix compatibility of CNTs, which influence composite characteristics. Several investigations have explored the electrical conductivity and electromagnetic interference shielding qualities of CNT-based polymer composites, emphasizing that elements like CNT dispersion, filler type, processing techniques, and surface treatments play a crucial role in improving the electrical and EMI shielding capabilities of the formed composites. For instance, researchers like Hur et al. optimized SWCNT/PDMS and MWCNT/PDMS composites via the three-roll milling technique, attaining maximum electrical conductivities of 85 S/m for SWCNT and 65 S/m for MWCNT at designated milling durations. Moreover, Krause et al. noted a reduced electrical percolation threshold in SWCNT/polypropylene composites, along with conductivity enhancement as the SWCNT content increased (Hur et al. 2023; Krause et al. 2016).

Additionally, numerous investigations have explored methods to improve the thermal characteristics of CNT/polymer composites through functionalization, dispersion techniques, and surface modifications. CNTs demonstrate outstanding thermal conductivity and stability, which markedly enhances the thermal efficacy of polymer nanocomposites. Researchers like Yanmaz et al. altered single-walled carbon nanotubes (SWCNTs) via hydroxylation and silane processes, thereafter dispersing the materials in polyvinyl alcohol (PVA) resin with sodium dodecyl sulfate (SDS) to attain uniform distribution. The findings demonstrated that the temperature for glass transition rose with increased carbon nanotube (CNT) concentration and surface changes. Hussain et al. conducted an additional analysis of multiwalled carbon nanotubes/polyaniline (MWCNT/PANI) and Ag-MWCNT/PANI composites, illustrating that CNTs improve thermal stability. The Ag-MWCNT/PANI composites exhibited a 12–15% decrease in weight loss relative to MWCNT/PANI composites, suggesting enhanced thermal resistance. The effectiveness of carbon nanotubes (CNTs) in improving thermal properties is dependent upon the dispersion quality and the interaction between carbon nanotubes and the matrix, and the methods used for functionalization. Consistent dispersion and robust interfacial bonds are critical for optimizing heat transfer, minimizing thermal degradation, and increasing the composite's resistance to elevated temperatures. CNT-based composites exhibit significant enhancements that render them highly appropriate for applications necessitating superior thermal stability, particularly in the electronics, aerospace, and automotive sectors, where effective heat dissipation and thermal resistance are essential (Yim et al. 2025; Yanmaz et al. 2021).

7.2.2 *Graphene and Its Derivatives*

7.2.2.1 Structure and Properties

Graphene-based PNCs represent a rapidly advancing field within the realm of nanoengineered materials. Graphene, distinguished by its elevated aspect ratio and minimal density, has garnered significant interest because of its extraordinary mechanical, electrical, and electronic attributes. This positions it as a prime candidate for the fabrication of functional and structural graphene-reinforced composites, especially when compared to carbon nanotubes. Considering the basic building block of all other graphitic carbon allotropes of different dimensions, such as graphite, carbon nanotubes (CNTs), and fullerenes, graphene is an atomically thin, two-dimensional (2D) sheet structure composed of sp2 carbon atoms arranged in a honeycomb lattice. Graphene sheets have improved surface-to-volume ratios relative to carbon nanotubes because the inside surface of the nanotubes is inaccessible to polymer materials. They could offer benefits in altering various matrix properties, including mechanical strength, permeability, rheological behavior, and thermal stability. Consequently, graphene-based polymer nanocomposites are currently the subject of extensive research. Moreover, custom-designed graphene-based nanocomposites that leverage the exceptional characteristics associated with both graphene fillers and polymer matrices can exhibit improved performance across various applications, including packaging, electronics, and biomedicine, among others (Dubey et al. 2024).

7.2.2.2 Influence on Nanocomposite Properties

Graphene and its derivative compounds, such as graphene oxide (GO) and reduced graphene oxide (rGO), as well as functional graphene, significantly enhance polymer nanocomposites by improving their mechanical, electrical, thermal, barrier, and flame-retardant properties. Their effective performance is closely linked to factors such as dispersion, alignment, and interfacial bonds with the polymer matrix. Graphene's high aspect ratio, sp^2-hybridized carbon bonds, and superior load transfer contribute to increased tensile strength, rigidity, and impact resistance. Their intrinsic electrical conductivity facilitates applications in electromagnetic interference (EMI) shielding, flexible electronics, and conductive coatings. Additionally, its superior thermal conductivity improves heat dissipation and thermal stability. Graphene's impermeable structure enhances gas and moisture barrier properties, which are advantageous for packaging and anti-corrosion applications. Furthermore, its capacity to form a charred protective layer improves flame retardancy. Strong π–π interactions result in agglomeration, necessitating surface functionalization to achieve uniform dispersion, good filler integration, and robust interfacial bonding. Moreover, the morphology of graphene fillers—exfoliated, intercalated, or stacked—significantly influences composite performance, as exfoliated structures improve mechanical strength, conductivity, and barrier properties. In addition, fabrication methods such as melt mixing, solution blending, and in situ polymerization influence the dispersion state. Also, advanced techniques including atomic force microscopy (AFM) and nanoindentation are utilized to enhance filler dispersion and composite performance, positioning graphene-based nanocomposites as highly promising for applications in aerospace, automotive, electronics, coatings, bioimaging, biosensing, and tissue engineering (Toto et al. 2022; Dubey et al. 2024).

7.2.3 *Clay Minerals*

7.2.3.1 Structure and Properties

Clay minerals are fine-grained, layered silicate minerals composed mostly of alumina (Al_2O_3) and silica (SiO_2), with various proportions of other elements. These layered structures are made up of tetrahedral and octahedral sheets that stack using van der Waals forces. More specifically, clay minerals consist of layers formed by tetrahedral and octahedral arrangements, containing OH (hydroxyl) ions at the peak and Al (aluminum) ions centrally located. An infinite, hexagonal, two-dimensional network configuration is produced in the crystal by a tetrahedral structure, which is made up of a central cation encircled by four coordinating oxygen atoms and joined to the neighboring tetrahedra by a three-cornered joint. In octahedral sheets, the edges between each octahedral and its adjoining octahedral arrangement are shared, resulting in the formation of hexagonal sheets or quasi-hexagonal symmetry. They possess distinctive characteristics attributable to their diminutive size (<2 μm), categorizing them as one-dimensional nanomaterials. Their crystal structure features high cation exchange capacity, plasticity, swelling behavior, and diminished permeability, rendering them advantageous for diverse applications, including the development of polymer nanocomposites. The uniform distribution of clay particles within the polymer matrix functions as a reinforcing agent in the formulation of polymer nanocomposites, notably improving the creation of a significant interfacial area per unit volume, which serves as the principal reinforcement technique of clay-based polymer nanocomposites. Furthermore, due to their elevated aspect ratio and remarkable mechanical, thermal, and barrier properties, when effectively incorporated, these layered silicates can enhance stiffness, strength, thermal stability, and gas barrier attributes in polymer matrices. Common clay minerals utilized in nanocomposites comprise montmorillonite, hectorite, saponite, laponite, bentonite, kaolinite, rectorite, and chlorite (Aboudi Mana et al. 2017; Bee et al. 2018; Yang et al. 2010).

7.2.3.2 Influence on Nanocomposite Properties

Clay minerals, particularly phyllosilicates such as montmorillonite, are crucial in improving the characteristics of polymer nanocomposites. Their elevated aspect ratio (length-to-thickness ratio) substantially influences the thermal, barrier, and mechanical characteristics of the polymer matrix. Incorporating these clays at low concentrations (often below 5 wt%) results in significant enhancements without markedly elevating the composite material's cost. The most significant improvement is evident in mechanical parameters, including tensile strength, and the polymer chains. These interactions limit the flexibility of polymeric chains, leading to a material that is more rigid and durable. Additionally, the presence of clay minerals significantly enhances thermal properties, contributing to improved heat distortion temperature and overall thermal stability. The outstanding heat transfer properties of clay particles are the main reason for this, as they also facilitate a more efficient dispersion of heat throughout the material. Furthermore, the nanocomposites demonstrate enhanced flame retardancy, as the clay particles create a protective charred layer upon exposure to heat, inhibiting further combustion. Furthermore, the gas barrier properties of the polymer are enhanced, rendering these materials suitable for use in packaging, especially within the food and medical sectors, where moisture and gas permeability are essential considerations. The morphological structure of clay within the polymer matrix, including exfoliated, intercalated, or mixed, significantly influences the degree of property enhancement. The distribution of clay particles, particularly when they are exfoliated, leads to increased tensile, flexural, and impact strengths, as well as enhanced durability, flame retardancy, and stability. Moreover, clay minerals improve mechanical, thermal, solvent, UV, and environmental stress cracking resistance in polymers. This makes clay-based polymer nanocomposites useful in automotive, construction, electronics, and coating applications. Clay minerals are abundant, cheap, and environmentally friendly, making these nanocomposites economically and ecologically appealing. The uses and benefits of polymer-clay nanocomposites are growing as research explores the interaction between clay-based materials, their modifications, and polymer matrices, promising creative solutions across numerous industries (Theng 2012).

7.2.4 *Other Nanofillers*

7.2.4.1 Metal Nanoparticles

In addition to other materials, metallic nanoparticles have achieved significant prominence in developing polymer nanocomposites. Metal-based nanoparticles, including gold, silver, copper, silica, zinc, and platinum, when integrated with polymers, form nanocomposites that demonstrate superior features, such as antibacterial activity, catalytic efficiency, and higher conductivity. These nanocomposites preserve the low density and processability of the polymer matrix, facilitating the creation of specialized, functional products at low nanoparticle concentrations (1–10 wt%). Polymers additionally aid in stabilizing and functionalizing metal nanoparticles. Composite materials are widely employed in applications such as water treatment, textiles, food packaging, sensors, and medical devices owing to their distinctive and enhanced features. Metal nanoparticles improve mechanical strength and stiffness, boost electrical conductivity for applications such as sensors and electronics, and facilitate thermal conductivity transfer to polymers for heat dissipation. Furthermore, they demonstrate distinctive optical characteristics and antibacterial capabilities, especially silver nanoparticles, which are advantageous in medicinal, packaging, and textile applications. Metal nanoparticles enhance flame-retardant characteristics and facilitate the development of sustainable materials across several industries. Moreover, these metal-based polymeric nanocomposites are utilized in water treatment management. However, efficient dispersion and functionalization are essential for optimizing these advantages (Tamayo et al. 2018).

7.2.4.2 Ceramic Nanoparticles

Ceramic nanoparticles are inorganic, metalloid solids made of oxides such as alumina (Al_2O_3), silica (SiO_2), titanium dioxide (TiO_2), zirconia (ZrO_2), nitrides, or carbides, possessing at least one dimension within the nanoscale (below 100 nm). Different compositions of these nanomaterials with varying morphologies and shapes have been produced through sequential heating and cooling processes. The incorporation of ceramic-based nanomaterials into a polymer matrix markedly improves the effectiveness and functionality of the resulting polymer nanocomposite. Consequently, ceramic polymer nanocomposites have garnered considerable academic and industrial interest due to their ability to markedly improve polymer properties even at lower filler loadings, leading to extensive research and investigation in this area. They enhance mechanical strength, rigidity, and impact resistance while also increasing thermal stability, rendering them suitable for high-temperature applications. These nanoparticles improve barrier properties, enhancing gas and moisture resistance, which is advantageous for packaging. Moreover, the incorporation of conductive fillers enhances chemical resistance, durability, and electrical properties. Due to these notable features, they are utilized across several industries, including automobile, aerospace, and biomedical applications, such as biomedical implants, tissue engineering, flame retardants, barrier films, and antistatic coatings. Furthermore, they have been extensively studied in recent times for their roles in drug delivery systems, wound healing, bioimaging, and photocatalysis. As with metal-based nanocomposites, effective dispersion and functionalization are essential for optimizing their performance in polymer matrices (Hadi et al. 2022).

Figure 7.2 presents a detailed schematic of the fabrication, testing, and tribological behavior of polymer nanocomposites.

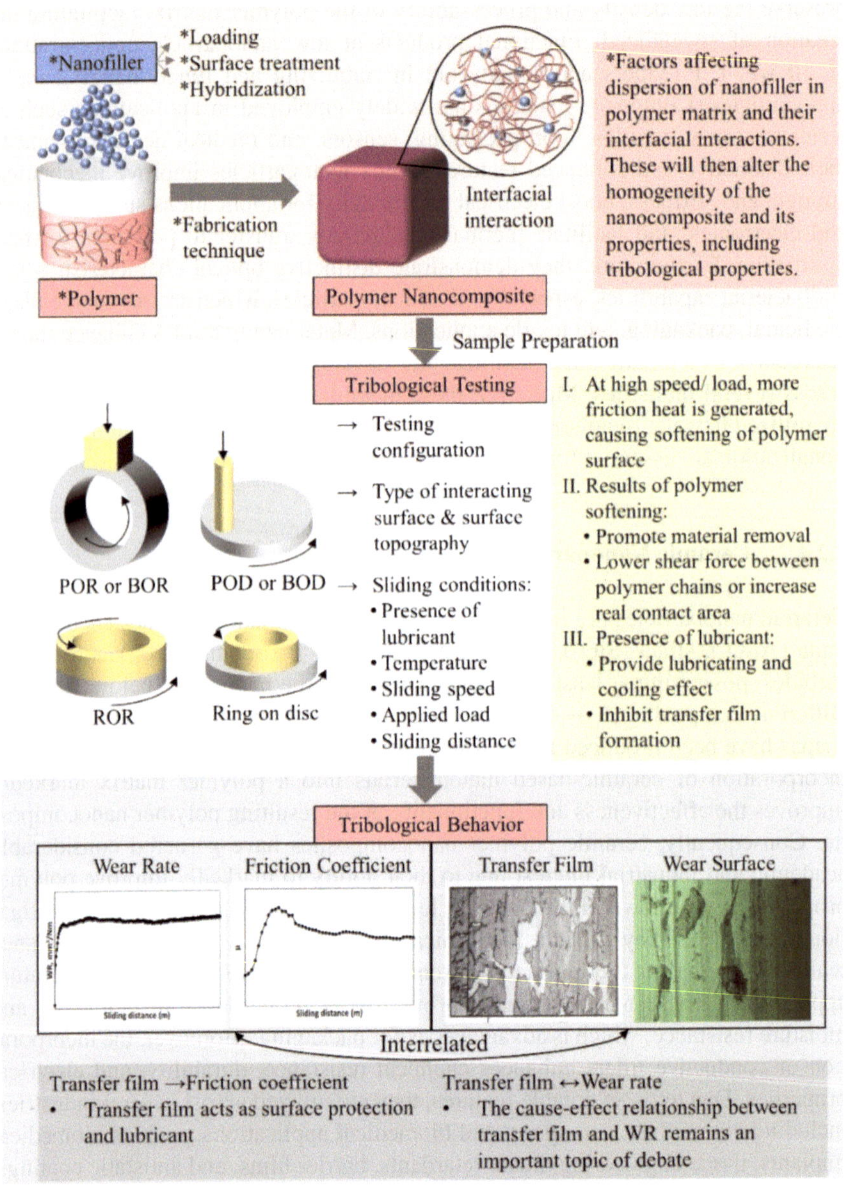

Fig. 7.2 A fundamental tribosystem for nanocomposites of polymers. (Reprinted from Chan et al. (2021), Copyright MDPI 2021. This open-access article is licensed under the terms of the CC BY 4.0 license)

7.3 Covalent Modification

7.3.1 Grafting of Polymers

Previous researches have proved that polymer chain grafting into the surface of nanoparticles increases the dispersibility and controls nanoparticle spacing. The interaction between the graft-matrix polymer should be favorable for enhanced dispersibility (Mitra et al. 2021). The grafting of polymer chains in nanocomposites can be achieved in four ways such as grafting to, grafting from, templated, and in situ polymerization or encapsulation. The covalent bond formation results from the polymer chain grafting. This bond makes up the entropy loss due to the stretching of the polymer chain (Tawade et al. 2021). In the grafting-to method, the end-functionalized polymer chain is attached onto the polymer surface through chemical reactions resulting in surface modification through silylation. This is followed by reactions including thiolene, epoxy−amine, etc. The grafting of polystyrene (PS), poly(methyl methacrylate) (PMMA), and poly(vinylidene fluoride) (PVDF) to BaTiO3 NPs. But due to high graft density, this process entails steric hindrance (Tawade et al. 2023; Eskandari et al. 2019). The grafting-from method, which is also known as surface-initiated control radical polymerization (SI-CRP). It involves growing polymer chains directly from the surface of nanoparticles (Eskandari et al. 2019; Apata et al. 2023). For the encapsulation technique, NPs are encapsulated within polymer materials which was formed beforehand, or encapsulation materials are grown over the NPs through in situ techniques. In this technique, the polymer has to be hyperdispersant, having the functional groups of hydroxyl, carboxyl, carbonyl, and nitro to ensure the proper bonding between the NP and the polymeric materials (Hiremath et al. 2021). Lastly, template-assisted grafting crystallizes BaTiO3 NPs within the specific area of the template, and this grafting efficiently fabricates the core of the shells of NPs with precise control of structures and morphologies (Tawade et al. 2023).

7.3.2 Chemical Reactions

The interaction between polymer matrix and nanofiller, as well as the relationship among individual components, often requires chemical modification of the components (Arslan and Tasdelen 2017). Because most of the prominent polymers lack stiffness, thermal and electrical conductivity, resulting in unsatisfactory mechanical properties. Therefore, the utilization of nano fillers in a polymeric matrix is being studied to increase the efficiency. For example, partially oxidized graphene is chemically modified, where surface modification is achieved. This increases solubility and compatibility with other polymeric components in nanocomposites (Su et al. 2022). To improve the physicochemical and mechanical properties of polymers, synthetic chemical functionalization can play a vital role in polymeric science, especially during commercialization. The poly(vinyl chloride) (PVC), which is a common commercial polymer, has been chemically modified through the Barbier reaction. This modification is based on the addition of magnesium metal to generate alkyl-magnesium chloride groups, which are nucleophilic on the polymer backbone, and these nucleophilic centers are added to electrophilic carbonyl groups. And, this modification facilitated the importance of PVC (Arslan et al. 2022).

7.3.3 Noncovalent Modification

7.3.3.1 Polymer Coating

Nanofillers can be noncovalently modified, and the modifying agents are linked through van der Waals interactions, ionic bonds, π–π interactions, hydrogen bonding, or hydrophobic interactions. Mostly, the interaction between the nanofillers and surfactants occurs via ionic (Chandrashekar et al. 2023). Utilizing noncovalent modification of polymer nanocomposites has reduced the interfacial thermal resistance (Wang et al. 2016). Between noncovalent and covalent methods, the noncovalent modifications do not hamper the intrinsic structure of graphene, so the properties are preserved. Therefore, noncovalent interactions have been utilized extensively for polymeric nanocomposites (Ji et al. 2015). As the polymer has a microporous structure, corrosive ions can easily penetrate, leading to their poor serviceability. To overcome these drawbacks, advanced structural modifications such as using polymer coating have enormously attracted the attention (Ahmadi and Ahmad 2020). Polymer-based nanocomposite coatings (PBNCs) are a type of hybrid coating system where two distinct phases, the matrix and reinforcing agents, at the nanoscale, are utilized. The matrix is a polymeric resin, and reinforcing agents are inorganic nanoparticles. This coating system is an excellent choice for ensuring material stability and corrosion protection, while also providing functional properties such as self-cleaning capability, self-healing, hydrophobicity, transparency, superior adhesion, ultraviolet (UV) stability, and low infrared emissivity (Ammar et al. 2019). Coating nanofiller by using low molecular weight polymers, such as poly(catechol amine) wrapped alumina– graphene oxide hybrids, polydopamine-coated copper nanowire, polyvinylpyrrolidone-wrapped carbon nanotubes, etc., on their surface via noncovalent interactions has been proven beneficial to increase the interfacial adhesion between the nanofiller and polymer matrix. This resulted in improved thermal activity of the polymer nanocomposites (Chandrashekar et al. 2023). Then, carbon nanotubes (CNTs) have wide applications in the biomedical field. But due to the poor solubility, the application of CNTs has been limited. To overcome this limitation, polymer wrapping by noncovalent interaction has been investigated in the case of stability and homogeneity (Fujigaya and Nakashima 2015).

7.3.3.2 Electrostatic Interactions

The internalization of nanoparticles during interaction with cells is a crucial factor in biomedical applications. Generally, the positive nanoparticles are more taken up by cells due to the electrostatic interaction with the cells, as the cell membrane is negatively charged (Forest and Pourchez 2017). Incorporating both acid and base in NPs is a common way for designing NPs that can cause electrostatic interaction to target biomolecules efficiently (Honda et al. 2020). The positive NPs establish the electrostatic interactions with lipid bilayers, and lipid headgroups are responsible for that due to the straightforwardness of anionic phosphatidylglycerol (PG) lipids and cationic trimethylammonium-propane (TAP) lipids (Wang and Malmstadt 2017). The electrostatic interaction between polymeric nanocomposites is caused by overlapping double electrical layers, and London–van der Waals interactions due to quantum and thermodynamic fluctuations of electromagnetic fields (Filippov and Starov 2023).

7.3.3.3 Surfactant Adsorption

Polymer adsorption can be achieved through noncovalent modification, and this process has been proven to be an effective method for introducing unique functionalities to cellulose materials. In surface adsorption, nano fillers or nano polymers are adhered on the surface of the polymer matrix. This technology is very effective for improving the characteristics of polymers, including thermal stability, mechanical strength, electrical conductivity, and barrier properties, by the integration of nanoscale elements such as graphene, carbon nanotubes, clay, or metal oxides (Nelson 2013).

7.3.4 *In Situ Polymerization*

In situ polymerization is a chemical encapsulation technique, and it is similar to interfacial polymerization. In this technique, the unique characteristic is that no reactants are used while synthesizing the core material. It is a continuous phase, and it does not happen between the continuous phase and the core material, as in interfacial polymerization (Nelson 2013). In this polymerization, the nano filler is dispersed more, and the interaction with the polymer interface is strong. This does not require a large amount of organic solvents like the intercalation method, and is environmentally friendly (Mao and Wang 2020). In situ generally occurs in the polymerization mixture, where the process is carried out in different polymerization steps. This results in a hybrid relationship among nanoparticles and polymer molecules where the monomers have low molecular weights. And these are distributed within the aqueous solution. After that, the polymerization reaction is triggered, as a suitable oxidizing agent is added, and this agent is then exposed to heat or radiation. Subsequently, the solution is filtered out, and the product is the nanocomposite having polymer molecules bound to nano particles. Through this method, several nanostructures can be achieved, and these structures will have higher quality and better chemical properties, like nanorods, nano particles, and nanofibers (Manopriya and Hareesh 2021). In-situ polymerization provides an appealing and promising platform to improve the pharmacological characteristics of biopharmaceuticals (Jia et al. 2018), and robust fabrication tool for surpassing the intricacies related to the integration of polymer electrolytes in lithium polymer batteries (Vijayakumar et al. 2021).

7.4 Effect of Surface Functionalization on Properties

7.4.1 Mechanical Properties

Surface functionalization involves modifying the surface of a material to activate its unique functions and improve its properties, as well as improving the existing properties of polymers significantly. It can enhance the compatibility among the existing nanoparticles within a polymer matrix and change their properties, for example, increased solubility, compatibility, dispersion, assembly, and so on. Functionalization can increase the surface area and mechanical strength, thus increasing durability (Wieszczycka et al. 2021). Integration of nanoparticles in a polymer matrix increases physical activities (Kwon et al. 2018). Carbon nanotubes based on nanocomposites have been an integral area of research due to nanotubes' high aspect ratios, low mass density, and intrinsically superior electrical, mechanical, and thermal properties. When multi walled carbon nanotubes are surface modified with acid and then modified with plasma treatment under nitrogen atmosphere, the mechanical properties of nanocomposites have been increased (Pitchan et al. 2016).

7.4.1.1 Tensile Strength and Modulus

Surface modification of nanofibers with acrylate-derived polymers integrated with aromatic side chains can increase tensile strength and modulus (Athulya Wickramasingha et al. 2022). Polymer nano composites combined with the carbon nano materials are being proven to show outstanding mechanical properties such as higher tensile, impact, and bending strength, low weight, and less cost due to incorporation of carbon nano particles (Hegde et al. 2024). Nanofiber-reinforced composites can demonstrate a higher tensile strength under static, impact, and fatigue loads (Lasenko et al. 2023).

7.4.1.2 Impact Strength

Impact strength is a major parameter in choosing a polymer in the plastic industry. In polymer nanocomposites, it has already been established that combining clay filler often increases the stiffness of the polymer, but it is ambiguous whether it retains or enhances toughness, resulting in uncertainty to assess impact strength (Chen and Evans 2009). The impact strength of polypropylene (PP) can be enhanced by blending ethylene-propylene-diene monomer (EPDM), which is an elastomer. However, the addition of EPDM compromises the mechanical strength of PP/EPDM polymer nanocomposites. Some nanoparticles, like silicon dioxide (SiO_2), titanium dioxide (TiO_2), carbon nanotubes (CNT), carbon black (CB), and zinc oxide (ZnO), can be added to the PP/EPDM composites, and that can restore the required stiffness and impact strength. So, to maintain stiffness and overall balance between the impact strength and toughness, reinforcing other nanoparticles is important (Taher et al. 2024).

7.4.1.3 Toughness

Nanocomposites exhibit excellent fracture toughness even with a low fraction of NPs. This unique characteristic has gained widespread attention. Fracture-type toughness of polymer nanocomposites loses energy through damage mechanisms. Interfacial debonding can be the cause of the damage mechanisms leading to subsequent plastic yielding of nanovoids. As they have excellent toughness, polymer nanocomposites have been employed in various sectors under diverse strain rates as structural materials (Lee et al. 2024). Besides, fracture toughness shows remarkable behavior. Because the energy absorbed within the crack propagation, specifically the energy that is necessary to create new surface area, is denoted by fracture toughness. This energy is then controlled by the particular kind of energy-dissipating interaction between the crack front and nanoparticles. In order to toughen fiber-reinforced composites and virgin polymer matrices, nanofillers are utilized in comparatively modest amounts (Stern and Marom 2024).

7.4.2 Thermal Properties

7.4.2.1 Thermal Stability

The thermal stability of a polymer indicates its ability to endure heat exposure while maintaining its properties, such as strength, toughness, and elasticity, at designated temperatures (Jones et al. 2009; Charles et al. 2009). The evaluation of the thermal stability of polymers is commonly conducted using thermogravimetric analysis (TGA). The relationship between weight loss and temperature is observed, with specific temperatures indicated as $Tx\%$ (where x signifies mass loss, such as $x = 1\%$, 3%, 5%, 10%, 20%, 50%) being calculated from the thermogravimetric profile. The thermal stability of a polymer is significantly influenced by its chemical structure, crystallinity, and molecular weight. The incorporation of aromatic structures in the polymer backbone, coupled with cross-linking activities, is recognized to enhance the thermal stability of polymers. The presence of double bonds or oxygen-containing structures in the main chain reduces the heat resistance of polymers (Hurley et al. 2015; Król-Morkisz and Pielichowska 2018).

7.4.2.2 Thermal Degradation

Three main processes are involved in the nonoxidative thermal breakdown of polymers: random chain scission, side-group elimination, and depolymerization. In the course of degradation, a recombination process can take place, leading to cross-linking or cyclization (Kaczmarek et al. 2012). Typically, multiple degradation mechanisms occur concurrently. Depolymerization, commonly known as unzipping, is a free radical process that serves as the reverse of polymerization. During this process, polymer chains undergo degradation into monomers, dimers, and/or oligomers when subjected to elevated temperatures. This reaction usually starts at the end of the chain when the free radical is created. After that, during the depropagation phase, the monomers are systematically removed from the main chain. Depolymerization represents a one-way process that persists until the complete breakdown of the polymer chain occurs, leading to its termination. Thermal depolymerization serves as a key characteristic of specific polymers generated via chain polymerization, including poly (methyl methacrylate) (PMMA), polystyrene, and polyoxymethylene (POM). This is in contrast to polycondensation polymers, which do not experience this depolymerization process (Crompton 2012). The second process for the heat breakdown of polymers is random chain scission. Free radicals are produced at random along the core polymer chains in this multistep process, which causes the polymer to break apart into smaller molecules. The chain lengths of these molecules vary, and some of them are sufficiently small to display adequate volatility. In addition to random scission, end-chain scission represents another mechanism for the thermal degradation of polymers, characterized by the effective removal of monomers from the chain ends. The processes of chain scission led to a notable decrease in both the molecular mass and mechanical strength of polymers. Random chain scission is a common degradation mechanism observed in polyolefins (Ueno et al. 2010). The elimination of side groups constitutes the third mechanism of degradation. This reaction involves the cleavage of bonds between the side groups and the principal chain. The sequestered side groups interact with one another. In particular instances, the free-side groups participate and transform into cyclic structures during the cyclization process. The compounds generated during the elimination process are often volatile breakdown products (Król-Morkisz and Pielichowska 2018; Visakh and Nazarenko 2015).

7.4.2.3 Thermal Conductivity

The incorporation of highly thermally conductive fillers, combined with the engineering of polymer chains and morphology at the atomic and molecular levels, can enhance the thermal conductivity of polymers. Composites incorporating macro- or micro-fillers have been the subject of extensive research for nearly a century, particularly through the lens of effective medium theory (Nan et al. 1997; Wang and Pan 2008; Pietrak and Wiśniewski 2015; Wang et al. 2014). The thermal properties of nanocomposites with nanofillers are still not fully understood and can show considerable variations. The significant specific surface area of nanofillers can enhance interfacial thermal resistance in a nanocomposite, contrary to composites at the macro- or micro-scales. Moreover, the mathematical characterization of heat conduction becomes increasingly intricate when nanofillers create a network within a nanocomposite structure. Apart from the interactions among the fillers themselves, which produce a filler network, the polymer matrix, the fillers, and their interactions with each other define the thermal conductivity. This section first presents the impact of fillers on thermal conductivity, followed by an elaborate discussion of the thermal transport mechanisms associated with various filler networks.

Fillers commonly utilized in nanocomposites are grouped into metallic, ceramic, and carbon-based categories. Metallic fillers, including copper nanoparticles (Luyt et al. 2006), copper nanowires (Chen et al. 2016), aluminum fibers (Danes et al. 2003), silver particles (Jouni et al. 2013), and gold and palladium powders, can be employed to enhance the thermal conductivity of polymer nanocomposites. Nonetheless, these metallic nanofillers may result in enhanced electrical conductivity, thereby limiting their applicability in contexts requiring electrical insulation. The electrical conductivity of metallic fillers can be modified through oxidation or surface treatment; however, high thermally conductive ceramic fillers are preferred due to their superior electrical insulation properties and thermal stability (Boudenne et al. 2005; Krupa et al. 2013). Common ceramic nanofillers possessing high thermal conductivity include magnesium oxide (MgO) (Ge et al. 2014; Lee and Choi 2013), aluminum oxide (Al_2O_3) (Yu et al. 2016a; Sun et al. 2016; Sim et al. 2005), silicon nitride (Si_3N_4) (Kusunose et al. 2013; Zhou et al. 2009), silicon carbide (SiC) (Zhou et al. 2015; Mun et al. 2014), zinc oxide (ZnO) (Liu and Yan 2011), aluminum nitride (AlN) (Zhang and Qi 2014), and boron nitride (BN) (Kim et al. 2014; Jiang et al. 2017). In contrast to metallic and ceramic fillers, nanostructured carbon fillers have garnered considerable attention due to their superior thermal conductivity. The thermal conductivity of expanded graphite (EG) is approximately 300 Wm^{-1} K^{-1} (Kim et al. 2017). Additionally, the thermal conductivity of graphene nanoplates (GNP) ranges from 1000 to 5000 Wm^{-1} K^{-1} (Song et al. 2013; Liu and Yan 2011). The remarkable mechanical strength, chemical stability, and thermal conductivity of carbon nanotubes (CNTs), which vary from 1000 to 3000 Wm^{-1} K^{-1}, make them extremely promising (Balandin et al. 2008; Shahil and Balandin 2012).

Nonetheless, the extensively recorded superior thermal conductivity of individual CNTs and graphene nanoplatelets does not readily convert to elevated thermal conductivity in nanocomposite materials (Winey et al. 2007). Among conventional metallic or ceramic fillers, such as aluminum or silicon carbide, CNT-polymer nanocomposites have clearly lower thermal conductivity than others. Two probable explanations could be: (1) The effectiveness of individual carbon nanotubes (CNT) and graphene nanoparticles (GNP) applied in nanocomposites can differ significantly from those generated for standalone thermal property assessment (Feng et al. 2013). (2) Carbon nanotubes (CNTs) typically exhibit various defects that can significantly diminish their effective thermal conductivity, including Stone-Wales defects, vacancy defects, and intertube junctions. The intrinsic flaws of carbon nanotubes (CNTs), along with the presence of kinks, twists, and waviness within the CNTs, may diminish their aspect ratio, resulting in a lower-than-anticipated thermal conductivity of CNT-in-epoxy composites (Kumari et al. 2008; Yu et al. 2016b).

The reduction in thermal conductivity associated with waviness is similarly noted in GNP (Kim et al. 2015). Methods for surface modification have been established to tackle defects in CNTs and GNPs by changing their morphology or structure and defect density, including acid treatment and plasma treatment. The heat resistance at the interface between the nanofiller and polymer can significantly diminish the advantages provided by nanofillers with elevated thermal conductivity. At low filler concentrations, the absence of a filler network typically causes significant interfacial thermal resistance between the fillers and the matrix, thus resulting in reduced thermal conductivity. The polymer molecules surrounding the CNT or GNP can lead to strain and shape changes, which may result in a reduction of the thermal conductivity of both the CNT and GNP (Zhang et al. 2014).

Enhancing the number of fillers to establish a network can markedly improve heat conductivity (Shtein et al. 2015; Shin et al. 2013; Huang et al. 2015). Nonetheless, an elevated concentration of fillers may adversely affect other critical properties of a polymer, such as its mechanical, electrical, and optical features. The development of a 3D filler network via the uniform distribution of fillers is an effective method for improving thermal conductivity while preserving a low filler concentration. Nonetheless, there is still a significant gap in the theoretical comprehension of the heat transfer mechanisms present in nanocomposites that incorporate 3D filler networks. Theoretical models primarily depend on effective medium theory (EMT) (Ngo et al. 2016a; TabkhPaz et al. 2016; Suplicz et al. 2016) or percolation-based theory (Gao et al. 2007; Xie et al. 2008), which might not be well-suited for nanocomposites that include a three-dimensional filler network. A variety of numerical techniques, including the Monte Carlo method (Wang et al. 2009; Péraud and Hadjiconstantinou 2011; Yang et al. 2005), the finite element method (Ngo and Byon 2015; Ngo et al. 2016b), and the lattice Boltzmann method (Zhou and Cheng 2014; Wang et al. 2007), has been recently advanced by the nanoscale heat transfer community. These methods may improve physical comprehension, yet their implementation in design continues to pose difficulties. This discussion delves into polymer nanocomposites, analyzing both types that do not possess an interfiller network and those that do, to clarify the existing knowledge regarding their thermal conductivity (Huang et al. 2018).

7.4.3 Electrical Properties

7.4.3.1 Electrical Percolation

Percolation theory investigates the connectivity of entities within a network and its influence on the macroscopic characteristics of the system. Percolation models efficiently characterize a polymer composite system's electrical conductivity, showing a clear insulator-to-conductor transition as filler percentage increases. The findings hold significance for assemblies of thin-film metal nanowires and carbon nanotubes without polymer, which are gaining recognition as promising alternatives to indium tin oxide (ITO) for flexible and transparent conductors. The electrical percolation in conducting polymer composites with elongated fillers is significantly influenced by the properties of the fillers, including the aspect ratio (L/D), electrical characteristics, and variability in filler attributes. The nanocomposite's microstructure, encompassing the filler's dispersion and orientation, plays a vital role in this phenomenon. This is especially relevant for composites that utilize nanotubes, given the considerable variability in their properties, which can vary from semiconducting to metallic, in addition to the obstacles encountered in attaining adequate dispersion of nanotubes within a polymer matrix.

Since the 1950s, percolation events in disordered media have been extensively studied using various analytical and simulation methods. Initial percolation models utilized lattice networks where particles were restricted to specific periodic locations. Analytical solutions for one-dimensional and two-dimensional lattice systems have been found (Kesten 1980; Wierman 1981; Fisher and Essam 1961), and numerically addressed via Monte Carlo simulations in three-dimensional contexts (Dean and Bird 1967; Kirkpatrick 1973). Although numerous precise analytical and numerical solutions exist for lattice percolation systems, these models do not effectively capture the complexities of actual composite systems. As a result, continuum (or off-lattice) models were developed to improve the flexibility in the spatial arrangement and orientation of objects. Investigating the continuum percolation of rods is essential for predicting insulator–conductor transitions in real composite systems containing rod-like fillers. This collection includes early fiber-reinforced polymer composites and modern polymer nanocomposites that integrate carbon nanotubes and metal nanowires. Two-dimensional sticks (Balberg and Binenbaum 1983; Behnam et al. 2007; Du et al. 2005) and three-dimensional rods have been the subject of countless analytical and computational studies since the early 1980s (Foygel et al. 2005; Berhan and Sastry 2007; White et al. 2009).

The conditions for classical percolation theory relate to the onset of geometrical connectedness, leading to a substantial increase in electrical conductivity. In polymer nanocomposites, particles may exhibit segregation because of the presence of a polymeric layer, even when subjected to high loadings (Balberg 2009; Ambrosetti et al. 2010). Under these conditions, geometric and electrical connections are incompatible; electric transmission among conducting particles is mostly accomplished via the hopping mechanism of the polymer matrix and quantum mechanical tunneling. Electrons are effectively transmitted through the intervening polymer (Balberg 2009; Balberg et al. 2004). In order to address this paradox, Balberg et al. (Vionnet-Menot et al. 2005; Johner et al. 2008) proposed a tunneling-percolation model that combines tunneling probability's exponential decay with the distribution function's unique characteristics for interparticle distances. Generally, polymer composite systems follow classical percolation theory, displaying a power law relationship between conductivity and concentration, defined by universal and nonuniversal critical exponent values. The compliance of experimental systems with percolation theory within this framework is ascribed to a tunneling network of particles analogous to percolation, wherein neighbors are distanced by intervals comparable to or less than the tunneling decay distance, while other adjacent particles are effectively excluded from the tunneling network. The closest particles establish a primary conduction pathway or an efficient percolation cluster inside the system. Recent verification by experiment of the tunneling percolation principle was accomplished through capacitance probe microscopy studies of carbon black/polymer composites (Balberg et al. 2004; Toker et al. 2003), revealing that the percolating clusters within the material comprised nearest-neighbor particles. Ambrosetti et al. (2009) successfully replicated the essential characteristics of percolation-like volume fraction dependence of conductivity in nanocomposites, doing this without a priori establishing a microscopic threshold, thereby neglecting the conductance contribution from greater interparticle distances. This methodology, similar to Balberg's model of tunneling percolation, relies on the diminishing tunneling conductivity as the distances between particles grow, resulting in discrepancies in the recorded percolation thresholds and conductivities for analogous composite systems. Significant interparticle attraction forces lead to the aggregation of nanoscopic particles that possess large surface-to-volume ratios. Carbon nanotubes demonstrate significant van der Waals interparticle interactions, around 0.5 eV for each nanotube-to-nanotube contact per nanometer (Dyke and Tour 2004). Because of their high aspect ratios and inherent characteristics, carbon nanotubes tend to aggregate significantly. The clustering of nanoparticles at the nanoscale, along with the external forces exerted during composite manufacturing and the internal interactions between particles and the medium, is essential in influencing the distribution at micro- and macro-scales. The electrical properties of the completed composite are significantly affected by the details of the nanoparticle microstructure. As a result, various percolation thresholds and conductivities have been recorded in the literature for analogous nanoparticle/matrix systems.

Accurate control and a thorough comprehension of filler dispersion across various scales within the polymer matrix are crucial for reliably forecasting composite properties and crafting composites with uniform characteristics. Carbon nanotubes (CNTs) are predominantly used in research concerning elongated nanoparticles within polymer nanocomposites. The concepts presented herein can be easily generalized to systems involving other rod-shaped particulates, despite this section reflecting this specific context (Mutiso and Winey 2015).

7.4.3.2 Dielectric Properties

Nanocomposites show a greater interfacial volume than microcomposites (Lewis 2005). The interactions between the polymer and nanometric particulates are emphasized as crucial elements that enhance the properties of nanodielectric systems. Nanometric interfaces are believed to affect composites' dielectric characteristics both passively and actively (Lewis 2005; Smith et al. 2008). However, it is noted that the surfaces of bare nanoparticles exhibit significant conductivity. The improvement of dielectric properties in nanodielectrics can be attributed to several factors. These include changes in polymer structure at the interface and local distribution of charges affected by nanoparticle surfaces, modifications in the density and energy depth of trap areas because of alterations in local structure at the interface of polymers, which influence charge mobility and the long-term stability of trapped states, as well as an increased probability of scattering mechanisms, among others. As a result, one can create enhanced dielectric substances can be developed by altering the interface between the nanofiller and polymer matrix while preserving the dielectric properties of the bulk material (Tuncer et al. 2007).

The benefits offered by extensive interfacial sites in nanodielectrics largely depend on tackling the issues related to attaining a uniform distribution of nanoparticles and avoiding particle agglomeration (Smith et al. 2008; Yuan et al. 2011). Integrating high surface-energy ceramic particles into low surface-energy polymer results in significantly inhomogeneous electric fields at the interfaces. This phenomenon can enhance charge conduction due to insufficient and uneven distribution, ultimately causing weakened dielectric breakdown strength of the polymer nanocomposites (Yuan et al. 2011; Schuman et al. 2010). The incorporation of surfactants or surface coupling agents, such as organophosphates, enhances surface tension, resulting in improved dispersion and film quality relative to ceramic-particle composites that lack dispersing aids (Yuan et al. 2011; Schuman et al. 2010; Siddabattuni et al. 2013).

7.4.4 Barrier Properties

The barrier properties of a polymer improve when its macromolecules are oriented, as opposed to when they are unoriented. The degree of crystallinity in polymers plays a key role in permeability since it depends on the speed at which molecules of gas, liquid, or vapor can infiltrate the polymer matrix, a factor dictated by the polymer's physical structure. Comparatively, to the amorphous domain, crystalline regions show more resistance to penetrating molecules. With an increase in polymer crystallinity, there is a corresponding decrease in permeability.

In the glassy or crystalline state, a polymer exhibits rigidity and fragility due to the constrained mobility of macromolecular chains. The diffusivity of the penetrant is therefore low. In contrast, polymers in the rubbery or elastic state exhibit both resilience and flexibility due to the unimpeded movement of their chains. The diffusivity of penetrants in rubbers could be affected due to the existence of a glassy, immobilized rubber layer encasing the clay particles (Zadrapa et al. 2011). The chemical degradation of the polymer is influenced by its permeability and crystallinity. Cellulose undergoes acid hydrolysis more efficiently in amorphous regions compared to crystalline domains. In polymers such as polyethylene (PE) and polypropylene (PP), the amorphous regions exhibit greater susceptibility to O_2 attack than the crystalline regions. Polymer processing techniques, including extrusion and compression molding, modify and eventually define the structural properties of the polymer. The interplay between molecular orientation, temperature, applied stress, and duration significantly affects the crystallinity of polymers.

The segmental motion of macromolecular chains is predominantly affected by temperature and the concentration of the absorbed penetrant. Increased temperature enhances segmental motion, enabling structural transitions in the polymer, including glass and melting transitions, which subsequently influence solution and diffusion processes. The temperature and the presence of the absorbed penetrant enhance the system's free volume.

Barrier properties considered satisfactory overall are achieved through polymers that incorporate polar groups, including esters, nitriles, chlorine, and fluorine. Poly (vinyl alcohol) (PVA) and uncoated cellophane serve as effective gas barriers; however, these compounds exhibit significant deficiencies as moisture barriers. Contributing factors include high chain stiffness, elevated glass transition temperature (Tg), and high specific gravity, all of which indicate close packing between chains.

A significant portion of polymer production is associated with polyethylene (PE), poly (vinyl chloride) (PVC), and polypropylene (PP), which are predominant in the film market. Additional significant polymers comprise polyethylene terephthalate (PET) and polystyrene (PS), alongside engineering polymers like polyamide (PA) and polycarbonate (PC). Polyolefins, including high-density polyethylene (HDPE), display superior moisture barrier characteristics alongside significant permeability to hydrocarbons. Aliphatic polyamide (PA 6, PA 66) demonstrates superior resistance to hydrocarbons; however, it demonstrates significant permeability to water vapor and experiences considerable degradation in gas barrier properties when exposed to moisture or swelling from water (Feldman 2001).

The primary challenge in polymer nanocomposites involves achieving efficient distribution and exfoliation of platelets within the polymer matrix. This process is essential for producing properly aligned, high aspect ratio particles that enhance mechanical reinforcement or create a tortuous diffusion pathway to improve barrier properties. According to the simple composite hypothesis, the tortuosity factor is independent of the type of penetrant or the size of its particles and is instead influenced by the particle content, shape (such as aspect ratio), position, and orientation (Takahashi et al. 2006). The intercalated portion of the filler is reported (Osman et al. 2003) to have minimal impact on permeability reduction, while the exfoliated layers are identified as the primary contributors to the creation of the permeation barrier. Consequently, better gas transport properties are associated with a lower degree of exfoliation. The enhancement of the aspect ratio of nanofiller particles in composites suggests the presence of exfoliated platelets rather than intercalated ones, which exert the most profound influence on the properties of the nanocomposite. Numerous research examining barrier properties used minimal quantities of exfoliated nanofillers. In contrast, a novel approach involved the creation of a self-supporting fabric film made of clay, which was subsequently infiltrated with a mixture of epoxy resin and amine hardener, followed by polymerization (Paul and Robeson 2008; Feldman 2013).

Table 7.2 outlines an evaluation signifying how functionalization enhances the material properties and efficacy of the polymer nanocomposites.

Figure 7.3 illustrates nanocomposite membranes that are conventionally fabricated for improving the material performance of the composite structure.

Table 7.2 Comparative analysis of functionalized vs. nonfunctionalized nanocomposites

S no	Reinforcement type	Functionalization	Tensile strength (GPa)	Young's modulus (GPa)	Electrical conductivity (S/m)	Thermal stability (°C)	References
01	Graphene	Dispersion	130	953 ± 21.4	1.09×10^7	800	Chen et al. (2023) and Liu et al. (2019)
02	Carbon nanotubes	Addition of surface functional group	100	270–950	10–106	1600–2800	Dubey et al. (2021), Bailey et al. (2013) and Sivasubramaniyam et al. (2023)
03	Layered silicates	Adsorbents, ion exchangers and catalysts	n/a	100–300	10^{-5} to 10^{-6}	300	Sirinakorn et al. (2018), Yadav et al. (2018), Yun and Maier (1999) and Davis et al. (2004)
04	Carbon nanofiber	Coating	100–200	90	1.2×10^{-3}	500	Takakura et al. (2019), Krasnikova et al. (2019), Zhang et al. (2006) and Sádovská et al. (2023)

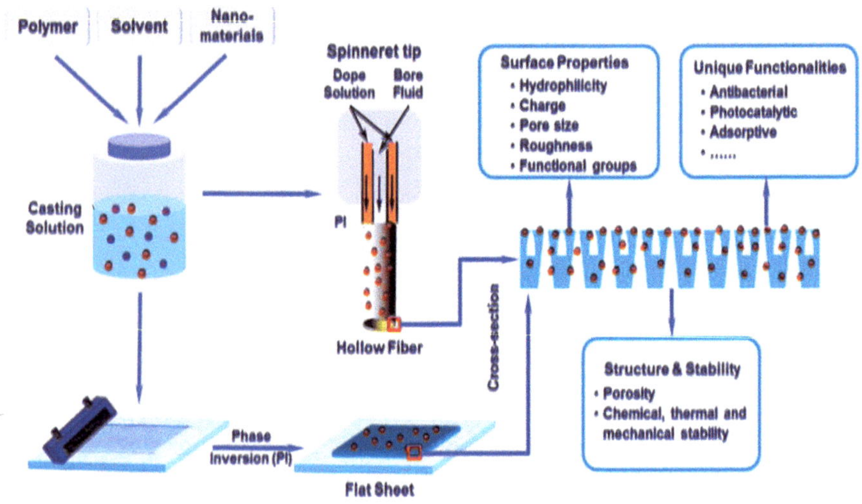

Fig. 7.3 Nanocomposite membranes that are conventionally fabricated. (Reprinted from Kausar and Bocchetta (2022), Copyright MDPI 2022. This open-access article is licensed under the terms of the CC BY 4.0 license)

7.5 Challenges and Potential

Nanocomposites have the potential to attain significant commercial prominence in the field of advanced composites. Carbon fiber-reinforced composites display constraints in their attainable characteristics, specifically in cross-ply configurations, owing to the low strength and modulus of the matrix phase. Combining carbon nanotubes at the nanoscale with carbon nanofibers at the larger scale can greatly increase the matrix's strength and modulus, improving the composite's overall characteristics. This method holds promise for enhancing unidirectional composites; however, it could also have a considerable impact on cross-ply composites, which are the most widely used type of composite structure in advanced applications. This concept is presently being evaluated and has the potential to drive a notable progression in the domain of advanced composites. The core idea has been applied in the commercial sector for specialized equipment for sports, including hockey sticks and tennis rackets. Future use of airplanes and wind energy turbines depends on these improvements. This methodology uses a hierarchical manufacturing technique, including several dimensional scales beginning at the nanoscale, therefore matching naturally occurring composite structures. A similar idea applies to composites reinforced with commodities, in which exfoliated clay can be included in the matrix of unsaturated polyester–fiberglass composites or another polymer strengthened with fiberglass.

Grasping the principles of nature's techniques for enhancing material characteristics via nanoscale fabrication, especially in biomimetics, is anticipated to drive progress in polymer nanocomposites. Nanostructured surfaces display superhydrophobic characteristics reminiscent of the lotus leaf and showcase impressive adhesion comparable to that of a gecko's foot. The integration of biological and polymer material science frequently entails the development of polymeric blends and composite substances at the nanoscale dimension that emulates biological systems.

Carbon nanotubes and exfoliated graphite or graphene present significant possibilities in the fields of electrical, electronics, and optoelectronics, along with potential applications in a range of emerging technologies. A significant focus is the replacement of ITO as a transparent conductor to enable the advancement of more cost-effective and adaptable devices. Proposals have been made regarding carbon nanotube sheets, and the exploration of CNT-conjugated polymer composites is essential, depending on the achievement of sufficient electrical conductivity. Nonetheless, the potential for cost-effective graphene production remains unfulfilled, as the realization of broad applications hinges on advancements in synthesis techniques (Schaefer and Justice 2007; Vaia and Maguire 2007).

Achieving optimal properties in nanocomposites typically necessitates superior dispersion of the nanostructures. The natural tendency of nanoparticles, such as platelets and fibers at the nanoscale, to aggregate into larger agglomerates can significantly affect the attainable qualities. In certain cases, the optimal isotropic dispersion may not be the preferred configuration; instead, a hierarchical morphology can be sought for its distinctive attributes, such as percolation pathways that enhance electrical conductivity or patterned structures that yield innovative optical or electronic characteristics (Paul and Robeson 2008; Vaia and Maguire 2007).

Recent advancements in the field of electrochemical applications have been observed in conducting polymers and rare earth oxides, facilitated by modifications to the surface of the working electrode. Nanocomposites and nanohybrid materials derived from conducting polymers enhance the beneficial properties of polymers across various domains. The composites demonstrated improved thermal and cyclic stability, as well as decreased internal resistance when applied in dielectric applications, antistatic functions, semiconductor roles, and energy storage systems. This chapter discusses biopolymeric nanocomposites utilized in various domains such as energy storage, energy conservation, energy production, power generation, and corrosion inhibition. ICP nanocomposites have garnered considerable attention due to their exceptional electrical properties, affordability, straightforward production methods, and environmentally friendly nature. Potential possibilities for nanocomposites are becoming more widely recognized, including graphene and specific nanofillers. Numerous opportunities for customizing the engineering features of PNC frameworks have been made possible by recent advancements in the creation, assessment, testing, characterization, and computational modeling of nanostructures.

Corrosive inhibitors and anodic shielding are the two principal techniques for reducing metal corrosion rates. Composite-conducting polymers (C.C.P.s) exhibit enhanced corrosion resistance relative to conductive polymeric coating materials. This seems to result from the augmented surface area of the nanoadditive facilitating dopant release and the creation of barriers that influence diffusion transport.

Future studies will concentrate on a range of reinforcing particulates, especially concerning the use of C.C.P.s on particular metal surfaces across different fields. In addition, considering that the corrosion resistance of cathodic protection systems largely depends on anodic shielding, it is crucial to perform an in-depth investigation into the stabilization of the passive oxide layer under the polymeric coating and the efficacy of inhibitors in obstructing aggressive anions from infiltrating the polymeric membrane.

The toxicological implications of electroactive poly(aniline) oligomers have been extensively studied, highlighting significant concerns regarding their application in biomedical fields. In this field, proposals for innovative electroactive oligomers based on enhanced stabilization and biocompatible electroactive monomeric subunits and/or oligomers may warrant attention. Another approach involves employing a relatively low concentration of aniline-based oligomers or investigating novel options to enhance biocompatibility. Electroactive macromonomers offer a viable method for synthesizing grafting copolymers characterized by conductivity and biodegradability. Significant research is necessary before these biomaterials can be utilized in cellular studies, including tissue engineering, regenerative medicine, bioengineering, biomedical devices, and scaffolding. The advancement of this domain could be significantly beneficial. A significant trend is the importance of not only synthesizing novel materials but also thoroughly characterizing them. Recent investigations have focused on the bio-interface of materials with cells, examining nanosized scaling characteristics and interactions with biomolecules to enhance the understanding of optimal cellular interface interactions. The findings may promote the advancement of intelligent biomaterials soon. C.P.s typically exhibit infusibility and insolubility, which complicates the treatment of these materials, especially when they are combined with other polymeric matrices for anti-static applications. Advancements in manufacturing methods are effectively addressing previous concerns, while conductive polymers are showing promising potential in anti-static applications. The simplicity of their synthesis, reduced density, light coloration, and controllable electrical properties render water-based conductive coatings especially appealing, justifying additional investigation for efficient anti-static devices. Nonetheless, several obstacles remain to be addressed. The initial issue is the scarcity of advanced nanofillers, alongside the high operational costs and complexities associated with scaling up. The subsequent challenge pertains to the management and treatment of PNCs. Effectively controlling the distribution, dispersion, and alignment of nanofillers is essential for improving the functional performance of polymer nanocomposites (PNCs). The ultimate challenge lies in the insufficient comprehensive understanding and predictive abilities concerning the consistent, viable interactions between core processing and structural traits, which are crucial for fully harnessing the commercialization potential of these groundbreaking biomaterials. The produced models often did not predict scientific outcomes accurately. Additionally, extensive research aims to develop innovative biopolymer nanocomposites utilizing cleaner, sustainable, economical, and environmentally sound renewable energy sources, which is essential for achieving sustainable development. Conducting polymers with multifunctional applications face limitations due to their toxic and carcinogenic properties, as well as inconsistencies between in vitro and in vivo studies. Physical characteristics, cytocompatibility, bioactivity, and biocompatibility are the main obstacles that C.P.s must overcome. While C.P.s present significant potential for multifunctional applications, they also exhibit considerable drawbacks, including an initial rapid release rate of drugs and the hydrophobicity of the polymer, which impede their utilization. Drug-releasing mechanisms are of significant interest to researchers, offering potential applications in cancer treatment and minimally invasive methods for various neurological and cardiovascular therapies. Another notable limitation of current C.P.s is that their cyclic stabilization is not as effective as that of carbon materials. Furthermore, the energy capacity, power density, specific capacitance, and specific power are significantly reduced in comparison to transition metal oxides (Sharma et al. 2021).

7.6 Conclusion

Incorporating original reinforcement materials and surface functionalization techniques has significantly enhanced the advancement of polymer nanocomposites, allowing for precise regulation of their mechanical, thermal, electrical, and chemical characteristics. This chapter emphasizes the role of tailored surface modifications in enhancing interfacial interactions between the polymer matrix and nanofiller materials, leading to improved dispersion, decreased agglomeration, and enhanced overall performance. The relationship between reinforcement and functionalization shapes the ultimate characteristics of polymer nanocomposites, expanding their potential uses in diverse fields such as biomedical engineering, electronics, automotive, and aerospace industries. Future research should prioritize optimizing functionalization techniques for sustainable and scalable production alongside exploring novel reinforcement materials to enhance nanocomposite performance. Advancements in nanotechnology and material science will drive the evolution of polymer nanocomposites, providing innovative solutions for next-generation functional materials.

Acknowledgments Not applicable.

Data Availability Statement Data sharing is not applicable to this chapter as no datasets were generated or analyzed during the current study period.

Ethics Statement Not applicable.

Funding This research did not receive any specific grant from funding agencies in the public, commercial, or not-for-profit centers.

Conflict of Interest The authors declare no conflict of interest.

Author Contributions **CRediT author statement**: Conceptualization: Tanvir Ahmed; Writing—Original Draft: Kazi Tasnuva Alam, Zarin Tasnim Tisha, Tanvir Ahmed; Writing—Review & Editing: Kazi Tasnuva Alam, Zarin Tasnim Tisha, Tanvir Ahmed; Visualization: Tanvir Ahmed.

References

Aboudi Mana SC, Hanafiah MM, Chowdhury AJK (2017) Environmental characteristics of clay and clay-based minerals. Geol Ecol Landsc 1:155–161

Ahmadi Y, Ahmad S (2020) Polymeric nanocomposite coatings. In: Corrosion protection at the nanoscale, pp 363–378

Ambrosetti G, Johner N, Grimaldi C, Maeder T, Ryser P, Danani A (2009) Electron tunneling in conductor-insulator composites with spherical fillers. J Appl Phys 106:16103. https://doi.org/10.1063/1.3159040

Ambrosetti G, Grimaldi C, Balberg I, Maeder T, Danani A, Ryser P (2010) Solution of the tunneling-percolation problem in the nanocomposite regime. Phys Rev B 81:155434. https://doi.org/10.1103/PhysRevB.81.155434

Ammar S, Ma IAW, Ramesh K, Ramesh S (2019) Polymers-based nanocomposite coatings. In: Nanomaterials-based coatings: fundamentals and applications. Elsevier, Amsterdam

Apata IE, Tawade BV, Cummings SP, Pradhan N, Karim A, Raghavan D (2023) Comparative study of polymer-grafted BaTiO3 nanoparticles synthesized using normal ATRP as well as ATRP and ARGET-ATRP with sacrificial initiator with a focus on controlling the polymer graft density and molecular weight. Molecules 28:4444

Arslan M, Tasdelen MA (2017) Polymer nanocomposites via click chemistry reactions. Polymers 9:499

Arslan R, Atilla Tasdelen M, Arslan M (2022) Covalent modification of poly(vinyl chloride) via organometallic Barbier reaction. Eur Polym J 177:111475

Athulya Wickramasingha Y, Dharmasiri B, Randall JD, Yin Y, Andersson GG, Nepal D et al (2022) Surface modification of carbon fiber as a protective strategy against thermal degradation. Compos Part A Appl Sci Manuf 153:106740

Bailey J, Oliveri A, Levin E (2013) Thermal stability of functionalized carbon nanotubes studied by in-situ transmission electron microscopy. Bone 23:1–7

Balandin AA, Ghosh S, Bao W, Calizo I, Teweldebrhan D, Miao F et al (2008) Superior thermal conductivity of single-layer graphene. Nano Lett 8:902–907. https://doi.org/10.1021/nl0731872

Balberg I (2009) Tunnelling and percolation in lattices and the continuum. J Phys D Appl Phys 42:64003. https://doi.org/10.1088/0022-3727/42/6/064003

Balberg I, Binenbaum N (1983) Computer study of the percolation threshold in a two-dimensional anisotropic system of conducting sticks. Phys Rev B 28:3799–3812. https://doi.org/10.1103/PhysRevB.28.3799

Balberg I, Azulay D, Toker D, Millo O (2004) Percolation and tunneling in composite materials. Int J Mod Phys B 18:2091–2121. https://doi.org/10.1142/S0217979204025336

Bee SL, Abdullah MAA, Bee ST, Sin LT, Rahmat AR (2018) Polymer nanocomposites based on silylated-montmorillonite: a review. Prog Polym Sci 85:57–82

Behnam A, Guo J, Ural A (2007) Effects of nanotube alignment and measurement direction on percolation resistivity in single-walled carbon nanotube films. J Appl Phys 102:44313. https://doi.org/10.1063/1.2769953

Berhan L, Sastry AM (2007) Modeling percolation in high-aspect-ratio fiber systems. II. The effect of waviness on the percolation onset. Phys Rev E 75:41121. https://doi.org/10.1103/PhysRevE.75.041121

Bernhardt M, Justnes H, Tellesbø H, Wiik K (2014) The effect of additives on the properties of lightweight aggregates produced from clay. Cement and concrete composites 53:233–8

Bin RA, Haque M, Islam SMM, Uddin Labib KMR (2024) Nanotechnology-enhanced fiber-reinforced polymer composites: recent advancements on processing techniques and applications. Heliyon 10:e24692. https://doi.org/10.1016/j.heliyon.2024.e24692

Boudenne A, Ibos L, Géhin E, Fois M, Majesté JC (2005) Anomalous behavior of thermal conductivity and diffusivity in polymeric materials filled with metallic particles. J Mater Sci 40:4163–4167

Carvalho A, Fernandes AR, Baptista PV (2018) Nanoparticles as delivery systems in cancer therapy: focus on gold nanoparticles and drugs. In: Applications of targeted nano drugs and delivery systems. Nanoscience and nanotechnology in drug delivery. Elsevier, Amsterdam

Chan JX, Wong JF, Petrů M, Hassan A, Nirmal U, Othman N et al (2021) Effect of nanofillers on tribological properties of polymer nanocomposites: a review on recent development. Polymers 13:2867

Chandrashekar A, Hegde M, Krishna S, Ayippadath Gopi J, Kotresh TM, Prabhu TN (2023) Non-covalent surface functionalization of nanofillers towards the enhancement of thermal conductivity of polymer nanocomposites: a mini review. Eur Polym J 198:112379

Charles J, Ramkumaar GR, Azhagiri S, Gunasekaran S (2009) FTIR and thermal studies on Nylon-66 and 30% glass fibre reinforced Nylon-66. J Chem 6:23–33

Chen B, Evans JRG (2009) Impact strength of polymer-clay nanocomposites. Soft Matter 5:3572–3584

Chen W, Wang Z, Zhi C, Zhang W (2016) High thermal conductivity and temperature probing of copper nanowire/upconversion nanoparticles/epoxy composite. Compos Sci Technol 130:63–69

Chen L, Li N, Yu X, Zhang S, Liu C, Song Y et al (2023) A general way to manipulate electrical conductivity of graphene. Chem Eng J 462:142139

Crompton TR (2012) Thermal stability of polymers. Smithers Rapra, Shrewsbury

Danes F, Garnier B, Dupuis T (2003) Predicting, measuring, and tailoring the transverse thermal conductivity of composites from polymer matrix and metal filler. Int J Thermophys 24:771–784

Davis RD, Gilman JW, Sutto TE, Callahan JH, Trulove PC, De Long HC (2004) Improved thermal stability of organically modified layered silicates. Clays Clay Miner 52:171–179

Dean P, Bird NF (1967) Monte Carlo estimates of critical percolation probabilities. Math Proc Camb Phil Soc 63:477–479. https://www.cambridge.org/core/product/848067B321E3E218806A8C34D5A6D207

Du F, Fischer JE, Winey KI (2005) Effect of nanotube alignment on percolation conductivity in carbon nanotube/polymer composites. Phys Rev B 72:121404. https://doi.org/10.1103/PhysRevB.72.121404

Dubey R, Dutta D, Sarkar A, Chattopadhyay P (2021) Functionalized carbon nanotubes: synthesis, properties and applications in water purification, drug delivery, and material and biomedical sciences. Nanoscale Adv 3:5722–5744

Dubey A, Ovais M, Maurya S, Bisen AC, Bhatta RS (2024) Graphene-based polymer nanocomposites. In: Properties and applications of graphene nanosheets, pp 65–87

Dyke CA, Tour JM (2004) Covalent functionalization of single-walled carbon nanotubes for materials applications. J Phys Chem A 108:11151–11159. https://doi.org/10.1021/jp046274g

Eskandari P, Abousalman-Rezvani Z, Roghani-Mamaqani H, Salami-Kalajahi M, Mardani H (2019) Polymer grafting on graphene layers by controlled radical polymerization. Adv Colloid Interface Sci 273:102021

Feldman D (2001) Polymer barrier films. J Polym Environ 9:49–55. https://doi.org/10.1023/A:1020231821526

Feldman D (2013) Polymer nanocomposite barriers. J Macromol Sci Part A Pure Appl Chem 50:441–448

Feng DL et al (2013) Effects of doping, Stone-Wales and vacancy defects on thermal conductivity of single-wall carbon nanotubes. Chin Phys B 22:16501. https://doi.org/10.1088/1674-1056/22/1/016501

Filippov AV, Starov V (2023) Interaction of nanoparticles in electrolyte solutions. J Phys Chem B 127:6562–6572

Fisher ME, Essam JW (1961) Some cluster size and percolation problems. J Math Phys 2:609–619. https://doi.org/10.1063/1.1703745

Forest V, Pourchez J (2017) Preferential binding of positive nanoparticles on cell membranes is due to electrostatic interactions: a too simplistic explanation that does not take into account the nanoparticle protein corona. Mater Sci Eng C 70:889–896

Foygel M, Morris RD, Anez D, French S, Sobolev VL (2005) Theoretical and computational studies of carbon nanotube composites and suspensions: electrical and thermal conductivity. Phys Rev B 71:104201. https://doi.org/10.1103/PhysRevB.71.104201

Fujigaya T, Nakashima N (2015) Non-covalent polymer wrapping of carbon nanotubes and the role of wrapped polymers as functional dispersants. Sci Technol Adv Mater 16:24802

Gao L, Zhou X, Ding Y (2007) Effective thermal and electrical conductivity of carbon nanotube composites. Chem Phys Lett 434:297–300. https://www.sciencedirect.com/science/article/pii/S0009261406018446

Ge Z, Ye F, Ding Y (2014) Composite materials for thermal energy storage: enhancing performance through microstructures. ChemSusChem 7:1318–1325

Hadi AG, Al-Ramadhan Z, Hashim A (2022) Studies on characteristics of polymer/ceramics nanocomposites for biomedical and industrial applications. Nanosistemi Nanomater Nanotechnol 20:535–548

Hayasaki T, Yamada Y, Kai X, Almarasy AA, Akasaka S, Fujimori A (2021) Study on the improvement of dispersibility and orientation control of fluorocarbon-modified single-walled carbon nanotubes in a fluorinated polymer matrix. Polym Compos 42:4845–4859

Hegde G, Bhavana HT, Maligi AS, Vandana M, Hegde G, Hegde G (2024) Enhancement of tensile strength and elastic modulus using bio-waste based carbon nanospheres doped polymer nanocomposites. J Polym Res 31:1–7

Hiremath A, Murthy AA, Thipperudrappa S, Bharath KN (2021) Nanoparticles filled polymer nanocomposites: a technological review. Cogent Eng 8:1991229

Honda R, Gyobu T, Shimahara H, Miura Y, Hoshino Y (2020) Electrostatic interactions between acid-/base-containing polymer nanoparticles and proteins: impact of polymerization pH. ACS Appl Bio Mater 3:3827–3834

Huang X, Wang S, Zhu M, Yang K, Jiang P, Bando Y et al (2015) Thermally conductive, electrically insulating and melt-processable polystyrene/boron nitride nanocomposites prepared by in situ reversible addition fragmentation chain transfer polymerization. Nanotechnology 26:15705. https://doi.org/10.1088/0957-4484/26/1/015705

Huang C, Qian X, Yang R (2018) Thermal conductivity of polymers and polymer nanocomposites. Mater Sci Eng R Rep 132:1–22. https://doi.org/10.1016/j.mser.2018.06.002

Hur O, Kang BH, Park SH (2023) Optimization of electrical and mechanical properties of a single-walled carbon nanotube composite using a three-roll milling method. Mater Chem Phys 309:128354

Hurley MJ, Gottuk DT, Hall JR Jr, Harada K, Kuligowski ED, Puchovsky M et al (2015) SFPE handbook of fire protection engineering. Springer, Berlin

Imani Yengejeh S, Kazemi SA, Öchsner A (2017) Carbon nanotubes as reinforcement in composites: a review of the analytical, numerical and experimental approaches. Comput Mater Sci 136:85–101

Ji X, Cui L, Xu Y, Liu J (2015) Non-covalent interactions for synthesis of new graphene based composites. Compos Sci Technol 106:25–31

Jia X, Wang L, Du J (2018) In situ polymerization on biomacromolecules for nanomedicines. Nano Res 11:5028–5048

Jiang Y, Liu Y, Min P, Sui G (2017) BN@PPS core-shell structure particles and their 3D segregated architecture composites with high thermal conductivities. Compos Sci Technol 144:63–69. https://www.sciencedirect.com/science/article/pii/S0266353816315548

Johner N, Grimaldi C, Balberg I, Ryser P (2008) Transport exponent in a three-dimensional continuum tunneling-percolation model. Phys Rev B 77:174204. https://doi.org/10.1103/PhysRevB.77.174204

Jones RG, Wilks ES, Metanomski WV, Kahovec J, Hess M, Stepto R et al (2009) Compendium of polymer terminology and nomenclature: IUPAC recommendations, 2008. Royal Society of Chemistry, Cambridge

Jouni M, Boudenne A, Boiteux G, Massardier V, Garnier B, Serghei A (2013) Electrical and thermal properties of polyethylene/silver nanoparticle composites. Polym Compos 34:778–786

Kaczmarek H, Chylińska M, Ziegler-Borowska M (2012) Thermal properties of novel polymers based on poly (hydantoin-methyl-p-styrene) and their substrates. J Therm Anal Calorim 110:1315–1326

Kausar A, Bocchetta P (2022) Polymer/graphene nanocomposite membranes: status and emerging prospects. J Compos Sci 6:76

Kesten H (1980) The critical probability of bond percolation on the square lattice equals 1/2. Commun Math Phys 74:41–59

Kim K, Kim M, Hwang Y, Kim J (2014) Chemically modified boron nitride-epoxy terminated dimethylsiloxane composite for improving the thermal conductivity. Ceram Int 40:2047–2056. https://www.sciencedirect.com/science/article/pii/S0272884213009115

Kim SY, Noh YJ, Yu J (2015) Thermal conductivity of graphene nanoplatelets filled composites fabricated by solvent-free processing for the excellent filler dispersion and a theoretical approach for the composites containing the geometrized fillers. Compos Part A Appl Sci Manuf 69:219–225. https://www.sciencedirect.com/science/article/pii/S1359835X14003698

Kim HS, Kim JH, Kim WY, Lee HS, Kim SY, Khil M-S (2017) Volume control of expanded graphite based on inductively coupled plasma and enhanced thermal conductivity of epoxy composite by formation of the filler network. Carbon N Y 119:40–46. https://www.sciencedirect.com/science/article/pii/S0008622317303706

Kirkpatrick S (1973) Percolation and conduction. Rev Mod Phys 45:574–588. https://doi.org/10.1103/RevModPhys.45.574

Krasnikova IV, Mishakov IV, Vedyagin AA (2019) Functionalization, modification, and characterization of carbon nanofibers. In: Carbon-based nanofillers and their rubber nanocomposites: fundamentals and applications. Elsevier, Amsterdam

Krause B, Pötschke P, Ilin E, Predtechenskiy M (2016) Melt mixed SWCNT-polypropylene composites with very low electrical percolation. Polymer 98:45–50

Król-Morkisz K, Pielichowska K (2018) Thermal decomposition of polymer nanocomposites with functionalized nanoparticles. In: Polymer composites with functional nanoparticles: synthesis, properties, and applications. Elsevier, Amsterdam. https://doi.org/10.1016/B978-0-12-814064-2.00013-5

Krupa I, Cecen V, Boudenne A, Prokeš J, Novák I (2013) The mechanical and adhesive properties of electrically and thermally conductive polymeric composites based on high density polyethylene filled with nickel powder. Mater Des 51:620–628

Kumari L, Zhang T, Du GH, Li WZ, Wang QW, Datye A et al (2008) Thermal properties of CNT-alumina nanocomposites. Compos Sci Technol 68:2178–2183. https://www.sciencedirect.com/science/article/pii/S0266353808001218

Kusunose T, Yagi T, Firoz SH, Sekino T (2013) Fabrication of epoxy/silicon nitride nanowire composites and evaluation of their thermal conductivity. J Mater Chem 1:3440–3445

Kwon NK, Kim H, Han IK, Shin TJ, Lee HW, Park J et al (2018) Enhanced mechanical properties of polymer nanocomposites using dopamine-modified polymers at nanoparticle surfaces in very low molecular weight polymers. ACS Macro Lett 7:962–967

Lasenko I, Sanchaniya JV, Kanukuntla SP, Ladani Y, Viluma-Gudmona A, Kononova O et al (2023) The mechanical properties of nanocomposites reinforced with PA6 electrospun nanofibers. Polymers 15:673
Lee SH, Choi Y (2013) Electro-physical properties of composites with nano-sized oxides. J Nanosci Nanotechnol 13:7610–7614
Lee SJ, Yoon SJ, Jeon I (2022) Graphene/polymer nanocomposites: preparation, mechanical. Polymers 1:1–24
Lee W, Yoo T, Baek K, Cho M, Chung H, Shin H et al (2024) Strain rate effects on fracture toughness of polymer nanocomposites: a multiscale study. Eng Fract Mech 298:109924
Lewis TJ (2005) Interfaces: nanometric dielectrics. J Phys D Appl Phys 38:202. https://doi.org/10.1088/0022-3727/38/2/004
Liu L-H, Yan M (2011) Functionalization of pristine graphene with perfluorophenyl azides. J Mater Chem 21:3273–3276. https://doi.org/10.1039/C0JM02765K
Liu F, Wang M, Chen Y, Gao J (2019) Thermal stability of graphene in inert atmosphere at high temperature. J Solid State Chem 276:100–103
Luyt AS, Molefi JA, Krump H (2006) Thermal, mechanical and electrical properties of copper powder filled low-density and linear low-density polyethylene composites. Polym Degrad Stab 91:1629–1636
Manero O, Sanchez-Solis A (2013) Polymer nanocomposites. In: Handbook of polymer synthesis, characterization, and processing, pp 585–604
Manopriya S, Hareesh K (2021) The prospects and challenges of solar electrochemical capacitors. J Energy Storage 35:102294
Mao HN, Wang XG (2020) Use of in-situ polymerization in the preparation of graphene/polymer nanocomposites. Xinxing Tan Cailiao/New Carbon Mater 35:336–343
Mitra A, Trifkovic M, Ponnurangam S (2021) Surface functionalization-induced effects on nanoparticle dispersion and associated changes in the thermophysical properties of polymer nanocomposites. Macromolecules 54:3962–3971
Moses JC, Gangrade A, Mandal BB (2019) Carbon nanotubes and their polymer nanocomposites. In: Nanomaterials and polymer nanocomposites: raw materials to applications. Elsevier, Amsterdam, pp 145–175
Mun SY, Lim HM, Ahn H, Lee DJ (2014) Thermal conductivities of epoxy composites comprising fibrous carbon and particulate silicon carbide fillers. Macromol Res 22:613–617
Mutiso RM, Winey KI (2015) Electrical properties of polymer nanocomposites containing rod-like nanofillers. Prog Polym Sci 40:63–84. https://doi.org/10.1016/j.progpolymsci.2014.06.002
Nan C-W, Birringer R, Clarke DR, Gleiter H (1997) Effective thermal conductivity of particulate composites with interfacial thermal resistance. J Appl Phys 81:6692–6699
Nelson G (2013) Microencapsulated colourants for technical textile application. In: Advances in the dyeing and finishing of technical textiles. Woodhead Publishing, Sawston
Ngo I-L, Byon C (2015) A generalized correlation for predicting the thermal conductivity of composite materials. Int J Heat Mass Transf 83:408–415. https://www.sciencedirect.com/science/article/pii/S0017931014010874
Ngo I-L, Jeon S, Byon C (2016a) Thermal conductivity of transparent and flexible polymers containing fillers: a literature review. Int J Heat Mass Transf 98:219–226. https://www.sciencedirect.com/science/article/pii/S0017931015304932
Ngo I-L, Prabhakar Vattikuti SV, Byon C (2016b) Effects of thermal contact resistance on the thermal conductivity of core–shell nanoparticle polymer composites. Int J Heat Mass Transf 102:713–722. https://www.sciencedirect.com/science/article/pii/S0017931016303301
Osman MA, Mittal V, Morbidelli M, Suter UW (2003) Polyurethane adhesive nanocomposites as gas permeation barrier. Macromolecules 36:9851–9858. https://doi.org/10.1021/ma035077x
Patel HA, Somani RS, Bajaj HC, Jasra RV (2006) Nanoclays for polymer nanocomposites, paints, inks, greases and cosmetics formulations, drug delivery vehicle and waste water treatment. Bull Mater Sci 29:133–145

Paul DR, Robeson LM (2008) Polymer nanotechnology: nanocomposites. Polymer 49:3187–3204. https://www.sciencedirect.com/science/article/pii/S0032386108003157

Péraud J-PM, Hadjiconstantinou NG (2011) Efficient simulation of multidimensional phonon transport using energy-based variance-reduced Monte Carlo formulations. Phys Rev B 84:205331. https://doi.org/10.1103/PhysRevB.84.205331

Pietrak K, Wiśniewski TS (2015) A review of models for effective thermal conductivity of composite materials. J Power Technol 95:14

Pitchan MK, Bhowmik S, Balachandran M, Abraham M (2016) Effect of surface functionalization on mechanical properties and decomposition kinetics of high performance polyetherimide/MWCNT nano composites. Compos Part A Appl Sci Manuf 90:147–160

Radhamani AV, Lau HC, Ramakrishna S (2018) CNT-reinforced metal and steel nanocomposites: a comprehensive assessment of progress and future directions. Compos Part A Appl Sci Manuf 114:170–187

Sádovská G, Honcová P, Morávková J, Jirka I, Vorokhta M, Pilař R et al (2023) The thermal stability of carbon materials in the air: quantitative structural investigation of thermal stability of carbon materials in air. Carbon N Y 206:211–225

Schaefer DW, Justice RS (2007) How nano are nanocomposites? Macromolecules 40:8501–8517. https://doi.org/10.1021/ma070356w

Schuman TP, Siddabattuni S, Cox O, Dogan F (2010) Improved dielectric breakdown strength of covalently-bonded interface polymer–particle nanocomposites. Compos Interfaces 17:719–731. https://doi.org/10.1163/092764410X495315

Shahil KMF, Balandin AA (2012) Graphene–multilayer graphene nanocomposites as highly efficient thermal interface materials. Nano Lett 12:861–867. https://doi.org/10.1021/nl203906r

Sharma S, Sudhakara P, Omran AAB, Singh J, Ilyas RA (2021) Recent trends and developments in conducting polymer nanocomposites for multifunctional applications. Polymers 13:1–31

Shin H, Yang S, Chang S, Yu S, Cho M (2013) Multiscale homogenization modeling for thermal transport properties of polymer nanocomposites with Kapitza thermal resistance. Polymer 54:1543–1554. https://www.sciencedirect.com/science/article/pii/S0032386113000566

Shtein M, Nadiv R, Buzaglo M, Kahil K, Regev O (2015) Thermally conductive graphene-polymer composites: size, percolation, and synergy effects. Chem Mater 27:2100–2106. https://doi.org/10.1021/cm504550e

Siddabattuni S, Schuman TP, Dogan F (2013) Dielectric properties of polymer-particle nanocomposites influenced by electronic nature of filler surfaces. ACS Appl Mater Interfaces 5:1917–1927

Sim L, Ramanan SR, Ismail H, Seetharamu KN, Goh TJ (2005) Thermal characterization of Al2O3 and ZnO reinforced silicone rubber as thermal pads for heat dissipation purposes. Thermochim Acta 430:155–165

Sinha Ray S, Okamoto M (2003) Polymer/layered silicate nanocomposites: a review from preparation to processing. Prog Polym Sci 28:1539–1641

Sirinakorn T, Imwiset K, Bureekaew S, Ogawa M (2018) Inorganic modification of layered silicates toward functional inorganic-inorganic hybrids. Appl Clay Sci 153:187–197

Sivasubramaniyam V, Ramasamy S, Venkatraman M, Gatto G, Kumar A (2023) Carbon nanotubes as an alternative to copper wires in electrical machines: a review. Energies 16:1–17

Smith RC, Liang C, Landry M, Nelson JK, Schadler LS (2008) The mechanisms leading to the useful electrical properties of polymer nanodielectrics. IEEE Trans Dielectr Electr Insul 15:187–196

Song SH, Park KH, Kim BH, Choi YW, Jun GH, Lee DJ et al (2013) Enhanced thermal conductivity of epoxy-graphene composites by using non-oxidized graphene flakes with non-covalent functionalization. Adv Mater 25:732–737

Stern T, Marom G (2024) Fracture mechanisms and toughness in polymer nanocomposites: a brief review. J Compos Sci 8:395

Su X, Wang R, Li X, Araby S, Kuan HC, Naeem M et al (2022) A comparative study of polymer nanocomposites containing multi-walled carbon nanotubes and graphene nanoplatelets. Nano Mater Sci 4:185–204

Sun L, Warren GL, O'Reilly JY, Everett WN, Lee SM, Davis D et al (2008) Mechanical properties of surface-functionalized SWCNT/epoxy composites. Carbon N Y 46:320–328

Sun R, Yao H, Zhang H-B, Li Y, Mai Y-W, Yu Z-Z (2016) Decoration of defect-free graphene nanoplatelets with alumina for thermally conductive and electrically insulating epoxy composites. Compos Sci Technol 137:16–23

Suplicz A, Hargitai H, Kovacs JG (2016) Methodology development for through-plane thermal conductivity prediction of composites. Int J Therm Sci 100:54–59. https://www.sciencedirect.com/science/article/pii/S1290072915002732

TabkhPaz M, Shajari S, Mahmoodi M, Park D-Y, Suresh H, Park SS (2016) Thermal conductivity of carbon nanotube and hexagonal boron nitride polymer composites. Compos Part B Eng 100:19–30. https://www.sciencedirect.com/science/article/pii/S1359836816309349

Taher F, Afshari M, Houmani A, Samadi MR, Bakhshi S, Afshari H (2024) Simultaneous enhancement of the impact strength and tensile modulus of PP/EPDM/TiO2 nanocomposite fabricated by fused filament fabrication. Colloid Polym Sci 302:393–407

Takahashi S, Goldberg HA, Feeney CA, Karim DP, Farrell M, O'Leary K et al (2006) Gas barrier properties of butyl rubber/vermiculite nanocomposite coatings. Polymer 47:3083–3093. https://www.sciencedirect.com/science/article/pii/S0032386106002527

Takakura A, Beppu K, Nishihara T, Fukui A, Kozeki T, Namazu T et al (2019) Strength of carbon nanotubes depends on their chemical structures. Nat Commun 10:1–7

Tamayo L, Palza H, Bejarano J, Zapata PA (2018) Polymer composites with metal nanoparticles. In: Polymer composites with functionalized nanoparticles. Elsevier, Amsterdam

Tawade BV, Apata IE, Pradhan N, Karim A, Raghavan D (2021) Recent advances in the synthesis of polymer-grafted low-L and high-K nanoparticles for dielectric and electronic applications. Molecules 26:2942

Tawade BV, Singh M, Apata IE, Veerasamy J, Pradhan N, Karim A et al (2023) Polymer-grafted nanoparticles with variable grafting densities for high energy density polymeric nanocomposite dielectric capacitors. JACS Au 3:1365–1375

Theng BKG (2012) Polymer-clay nanocomposites. In: Developments in clay science, 2nd edn. Elsevier B.V, Amsterdam

Tofighy MA, Mohammadi T (2020) Carbon nanotubes-polymer nanocomposite membranes for pervaporation. In: Polymer nanocomposite membranes for pervaporation, pp 105–133

Toker D, Azulay D, Shimoni N, Balberg I, Millo O (2003) Tunneling and percolation in metal-insulator composite materials. Phys Rev B 68:41403. https://doi.org/10.1103/PhysRevB.68.041403

Toto E, Laurenzi S, Santonicola MG (2022) Recent trends in graphene/polymer nanocomposites for sensing devices: synthesis and applications in environmental and human health monitoring. Polymers 14:1030

Tuncer E, Sauers I, James DR, Ellis AR, Paranthaman MP, Aytuğ T et al (2007) Electrical properties of epoxy resin based nano-composites*. Nanotechnology 18:25703. https://doi.org/10.1088/0957-4484/18/2/025703

Ueno T, Nakashima E, Takeda K (2010) Quantitative analysis of random scission and chain-end scission in the thermal degradation of polyethylene. Polym Degrad Stab 95:1862–1869. https://www.sciencedirect.com/science/article/pii/S0141391010001849

Urio PJ, Murphy SP (2024) Introduction and challenge. In: Gender-inclusive Higher Education in Tanzania: Transforming Academia, vol 32, pp 1–12

Vaia RA, Maguire JF (2007) Polymer nanocomposites with prescribed morphology: going beyond nanoparticle-filled polymers. Chem Mater 19:2736–2751. https://doi.org/10.1021/cm062693+

Vijayakumar V, Anothumakkool B, Kurungot S, Winter M, Nair JR (2021) In situ polymerization process: an essential design tool for lithium polymer batteries. Energ Environ Sci 14:2708–2788

Vionnet-Menot S, Grimaldi C, Maeder T, Strässler S, Ryser P (2005) Tunneling-percolation origin of nonuniversality: theory and experiments. Phys Rev B 71:64201. https://doi.org/10.1103/PhysRevB.71.064201

Visakh PM, Nazarenko OB (2015) Thermal degradation of polymer blends, composites and nanocomposites. Springer, Berlin, pp 1–16

Wang L, Malmstadt N (2017) Interactions between charged nanoparticles and giant vesicles fabricated from inverted-headgroup lipids. J Phys D Appl Phys 50:415402

Wang M, Pan N (2008) Predictions of effective physical properties of complex multiphase materials. Mater Sci Eng R Rep 63:1–30

Wang M, He J, Yu J, Pan N (2007) Lattice Boltzmann modeling of the effective thermal conductivity for fibrous materials. Int J Therm Sci 46:848–855. https://www.sciencedirect.com/science/article/pii/S1290072906002122

Wang M, Kang Q, Pan N (2009) Thermal conductivity enhancement of carbon fiber composites. Appl Therm Eng 29:418–421. https://www.sciencedirect.com/science/article/pii/S1359431108001312

Wang X, Zhang L, Pei L (2014) Thermal conductivity augmentation of composite polymer materials with artificially controlled filler shapes. J Appl Polym Sci 131

Wang Y, Yang C, Mai YW, Zhang Y (2016) Effect of non-covalent functionalisation on thermal and mechanical properties of graphene-polymer nanocomposites. Carbon N Y 102:311–318

White SI, DiDonna BA, Mu M, Lubensky TC, Winey KI (2009) Simulations and electrical conductivity of percolated networks of finite rods with various degrees of axial alignment. Phys Rev B 79:24301. https://doi.org/10.1103/PhysRevB.79.024301

Wierman JC (1981) Bond percolation on honeycomb and triangular lattices. Adv Appl Probab 13:298–313. https://www.cambridge.org/core/product/D2F18A5392DEFC352B9C89CAC621FDA0

Wieszczycka K, Staszak K, Woźniak-Budych MJ, Litowczenko J, Maciejewska BM, Jurga S (2021) Surface functionalization—the way for advanced applications of smart materials. Coord Chem Rev 436:213846

Winey KI, Kashiwagi T, Mu M (2007) Improving electrical conductivity and thermal properties of polymers by the addition of carbon nanotubes as fillers. MRS Bull 32:348–353. https://www.cambridge.org/core/product/35BFC064B23B8F761BF1F0B3FF1E73ED

Xie SH, Liu YY, Li JY (2008) Comparison of the effective conductivity between composites reinforced by graphene nanosheets and carbon nanotubes. Appl Phys Lett 92:243121. https://doi.org/10.1063/1.2949074

Yadav HKS, Almokdad AA, Shaluf SIM, Debe MS (2018) Polymer-based nanomaterials for drug-delivery carriers. In: Nano-carriers for drug delivery: nanoscience and nanotechnology in drug delivery. Elsevier, Amsterdam

Yang R, Chen G, Dresselhaus MS (2005) Thermal conductivity of simple and tubular nanowire composites in the longitudinal direction. Phys Rev B 72:125418. https://doi.org/10.1103/PhysRevB.72.125418

Yang Z, Peng H, Wang W, Liu T (2010) Crystallization behavior of poly(ε-caprolactone)/layered double hydroxide nanocomposites. J Appl Polym Sci 116:2658–2667

Yanmaz E, Doğan M, Turhan Y (2021) Effect of sodium dodecyl sulfate on thermal properties of polyvinyl alcohol (PVA)/modified single-walled carbon nanotube (SWCNT) nanocomposites. Diamond Relat Mater 115:108359

Yim YJ, Yoon YH, Kim SH, Lee JH, Chung DC, Kim BJ (2025) Carbon nanotube/polymer composites for functional applications. Polymers 17:119

Yu W, Qi Y, Zhou Y, Chen L, Du H, Xie H (2016a) Synergistic improvement of thermal transport properties for thermoplastic composites containing mixed alumina and graphene fillers. J Appl Polym Sci 133:43242

Yu J, Choi HK, Kim HS, Kim SY (2016b) Synergistic effect of hybrid graphene nanoplatelet and multi-walled carbon nanotube fillers on the thermal conductivity of polymer composites and theoretical modeling of the synergistic effect. Compos Part A Appl Sci Manuf 88:79–85. https://www.sciencedirect.com/science/article/pii/S1359835X16301555

Yuan J-K, Yao S-H, Dang Z-M, Sylvestre A, Genestoux M, Bail J (2011) Giant dielectric permittivity nanocomposites: realizing true potential of pristine carbon nanotubes in polyvinylidene fluoride matrix through an enhanced interfacial interaction. J Phys Chem C 115:5515–5521. https://doi.org/10.1021/jp1117163

Yun SK, Maier J (1999) Synthesis and ionic conductivity of supramolecular layered silicate hybrids of phosphotungstates and poly(ethylene glycol) dicarboxylates. Chem Mater 11:1644–1647

Zadrapa P, Malac J, Konecny P (2011) Filler and mobility of rubber matrix molecules. Polym Bull 67:927–936. https://doi.org/10.1007/s00289-011-0533-9

Zhang J, Qi S (2014) Mechanical, thermal, and dielectric properties of aluminum nitride/glass fiber/epoxy resin composites. Polym Compos 35:381–385

Zhang Y, Suhir E, Xu Y (2006) Effective young's modulus of carbon nanofiber array. J Mater Res 21:2948–2954

Zhang J, Jiang C, Jiang D, Peng H-X (2014) Nano-engineering thermal transport performance of carbon nanotube networks with polymer intercalation: a molecular dynamics study. Phys Chem Chem Phys 16:4378–4385. https://doi.org/10.1039/C3CP53714E

Zhou F, Cheng G (2014) Lattice Boltzmann model for predicting effective thermal conductivity of composite with randomly distributed particles: considering effect of interactions between particles and matrix. Comput Mater Sci 92:157–165. https://www.sciencedirect.com/science/article/pii/S0927025614003577

Zhou W, Wang C, Ai T, Wu K, Zhao F, Gu H (2009) A novel fiber-reinforced polyethylene composite with added silicon nitride particles for enhanced thermal conductivity. Compos Part A Appl Sci Manuf 40:830–836

Zhou W, Chen Q, Sui X, Dong L, Wang Z (2015) Enhanced thermal conductivity and dielectric properties of Al/β-SiCw/PVDF composites. Compos Part A Appl Sci Manuf 71:184–191

Chapter 8
Functional Polymer Nanocomposite

Ashwini R. Patil, Bhushan R. Rane, Ashish S. Jain, and Sachin N. Kothawade

Abstract Nanocomposites have three types polymer, ceramic, and metal matrix nanocomposites. Nowadays polymer nanocomposites have received tremendous attention in academics and industries with several applications due to their improved properties like high strength-to-weight ratio, thermal stability, electrical conductivity, chemical stability, easily customizable product properties, flexible manufacturing processes, high resistance to corrosion or erosion and lower cost. Such nanocomposites have a wide range of drug delivery and tissue engineering applications. They can be used to create scaffolds that support the growth and regeneration of tissues, gene therapy, food and beverage preservation, biosensing, and bioimaging, providing enhanced sensitivity and specificity for detecting biological molecules and imaging tissues. Nanocomposites are alternatives to micro-composites and monolithic materials due to their outstanding properties. These nanocomposites were fabricated using carbon nanotubes in 1991 and composite materials with a nanometer range. Recently, nanocomposites have exhibited great potential for early detection and accurate imaging in cancer diagnosis. This chapter discusses the various fabrication techniques and additives required for the fabrication of polymer nanocomposites, their manufacturing method of preparation and characterization in detail. Recent challenges for polymer nanocomposites include biocompatibility, toxicity, controlled release, scalability, and stability in biological environments. Polymer nanocomposites are expected to play a crucial role in developing advanced drug delivery systems that can target specific cells or tissues, improving the efficacy and reducing side effects of treatments.

A. R. Patil · B. R. Rane (✉)
Department of Pharmaceutics, Shri. D. D. Vispute College of Pharmacy and Research Center, Panvel, Maharashtra, India

A. S. Jain
Shri. D. D. Vispute College of Pharmacy and Research Center, Panvel, Maharashtra, India

S. N. Kothawade
Department of Pharmaceutics, SCSSS's Sitabai Thite College of Pharmacy, Shirur, Pune, Maharashtra, India

S. N. Kothawade et al. (eds.), *Polymer Nanocomposites*,
https://doi.org/10.1007/978-3-032-00365-2_8

Keywords Nanocomposites · Polymer · Functional · Targeting · Diagnosis

8.1 Introduction

Materials that combine a polymer matrix with inorganic nanoparticles (NPs) to form a material with improved properties, like high strength, thermal stability, electrical conductivity, and chemical stability, are called nanomaterials (Kumari et al. 2019). Nanomaterials incorporate nanosized particles (size one, two, three, or less than 100 nm) into a matrix of standard materials having unique characteristics and having biocompatible, stimuli-responsive responsiveness, and functionality properties. These properties are ideal for diagnosing and treating chronic diseases, including cancer, diabetes, and cardiovascular diseases (Gutierrez et al. 2022).

Nanocomposites have emerged as a significant focus in modern research and development, particularly within the expansive field of nanotechnology. Among them, polymer nanocomposites have gained prominence due to their ability to enhance mechanical properties, making them highly valuable for various applications. One of their most promising uses is in drug delivery systems, where their unique surface characteristics and rheological properties enable efficient drug encapsulation and controlled release. Structurally, a nanocomposite consists of a matrix-filler system, in which fillers such as nanoparticles, fibers, or fragments are uniformly dispersed within the polymer matrix, enhancing its overall performance. Ongoing research aims to precisely control nanostructures to optimize these materials for specific biomedical and industrial applications. The functional properties of polymer nanocomposites extend beyond mere mechanical improvements; they also exhibit remarkable flexibility, tunable morphology, and enhanced interfacial characteristics. These advancements position polymer nanocomposites as a transformative class of materials with vast potential in biomedicine, engineering, and beyond (Kumari et al. 2019).

Functional polymer nanocomposites are advanced materials that integrate polymers with nanoscale fillers, such as nanoparticles, nanotubes, or graphene, to achieve enhanced properties (Andoh et al. 2024). Polymer nanocomposites are classified as inorganic nanofillers (e.g., silica, titanium dioxide, clay), carbon-based nanofillers (e.g., graphene, carbon nanotubes), metallic nanofillers (e.g., silver, gold, copper nanoparticles), and organic nanofillers (e.g., dendrimers, polymeric nanoparticles) (Kango et al. 2013). These are prepared by different methods like in situ polymerization, solution blending, melt blending, and electro-spinning (Hiremath et al. 2021), and manufactured by extrusion and injection molding, layer-by-layer assembly, and sol-gel processing (Armstrong 2015).

Physicochemical and biocompatible properties have been introduced to maximize therapeutic efficiency with minimum side effects. The fundamental understanding relating to properties, structure, polymer matrix material interactions, and further surface modifications facilitates the assessment of desired biological interaction for potential healthcare applications (Kumar et al. 2018; Mezzenga and Ruokolainen 2009).

Polymer nanocomposites are reported to reform the diagnosis and treatment strategies of diverse disease and treatment strategies of diverse diseases through targeted and controlled release (Bose et al. 2018). Coherent design of drug loaded polymer nanocomposites could assist in overcoming several physicochemical limitations that are associated with free drugs (without enwrapping by any carrier) (Wang et al. 2016; Sarode et al. 2014; Mura et al. 2013).

Nanocomposites designed for drug delivery must exhibit multiple functionalities to ensure optimal therapeutic efficacy. These include: (1) highly efficient and fluent drug cargo encapsulation to maximize drug loading and stability, (2) protection against premature drug release to prevent degradation or loss of therapeutic effect before reaching the target site, and (3) the ability to overcome various intra- and extracellular barriers within the human body. These attributes are crucial for enhancing bioavailability, improving targeted delivery, and minimizing potential side effects, thereby making nanocomposites a promising platform for advanced drug delivery systems (Shi et al. 2013; Kumar et al. 2016).

Polymer nanocomposites hold immense potential for advancing healthcare quality due to their multifunctional capabilities. These materials offer (1) high accuracy and reliability in the diagnostic process, enabling early and precise disease detection; (2) efficient targeting of therapeutic agents, ensuring controlled and site-specific drug delivery with minimal side effects; and (3) effective healing of injured tissues, making them highly suitable for tissue engineering applications. By integrating these attributes, polymer nanocomposites contribute to the development of innovative medical solutions, improving patient outcomes, and revolutionizing modern healthcare practices (Yukird et al. 2017; Prabhu et al. 2017).

Polymer nanocomposites have emerged as highly promising materials for nanocarrier applications due to their exceptional advantages. These include excellent bioavailability, ensuring efficient absorption and therapeutic efficacy; high water solubility, which enhances drug dispersion and stability in biological environments; and a nontoxic nature, making them biocompatible and safe for medical use. Additionally, polymer nanocomposites exhibit minimal deactivation potential, preserving the integrity and activity of the encapsulated drugs, along with good acceleration ability, facilitating rapid and targeted drug delivery. Furthermore, their minimal antigenic activity reduces the risk of immune system activation, making them ideal candidates for advanced drug delivery systems. These unique properties position polymer nanocomposites as a versatile and efficient platform for nanomedicine applications (Mura et al. 2013).

The incorporation of polymer nanocomposites in drug delivery applications would illustrate their role in enhancing targeted drug delivery. This technique efficiently transports active therapeutic agents to specific body sites, ensuring prolonged retention and improved therapeutic outcomes. Targeted drug delivery primarily operates through the following mechanisms (Fig. 8.1).

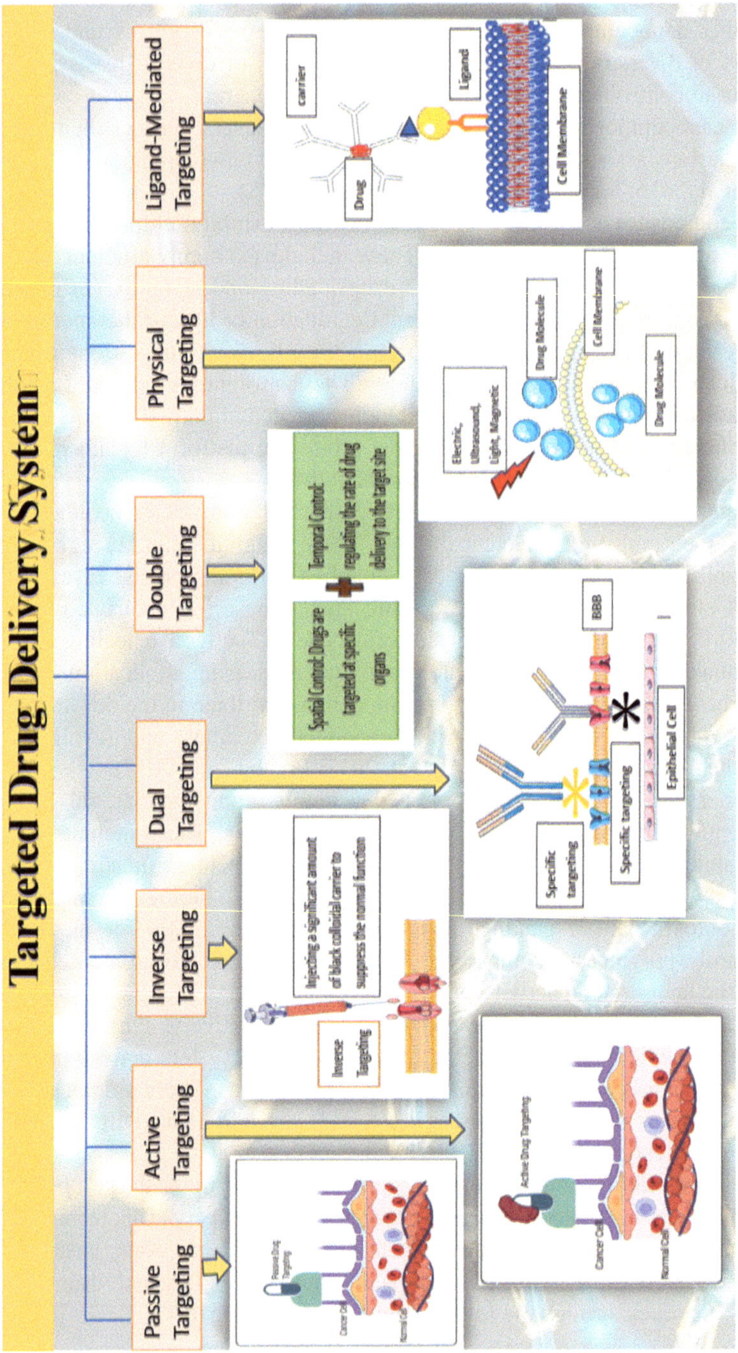

Fig. 8.1 Targeted drug delivery system

1. *Passive Targeting*: This approach relies on the enhanced permeability and retention (EPR) effect, where nanocarriers accumulate in tumor tissues or inflamed sites due to leaky vasculature and impaired lymphatic drainage. Drug delivery systems target the drug to the systemic circulation. Drug delivery systems entrap the drug till reaching the target site, for example, salinomycin delivered by passive targeting micelles used for the suppression of breast cancer and stem cell cancer (Jawahar and Meyyanathan 2012).
2. *Active Targeting*: In this method, nanocarriers are functionalized with specific ligands (such as antibodies, peptides, or aptamers) that bind to receptors on target cells, ensuring precise drug delivery with minimal off-target effects (Merino et al. 2015). Drug delivery system identification of the target group includes bioadhesive non-ionic surfactant, antibodies, or albumin protein, which attach to the surface of the drug delivery system to the receptor (Bhargav et al. 2013). There are three types of active drug targeting systems: (1) first-order targeting, which that targets organ; (2) second-order targeting, which targets cell; and (3) third-order targeting, which is intracellular targeting (Minko et al. 2004).
3. *Inverse Targeting*: Inverse drug targeting is useful for non-reticulum-endothelial system (RES) organs (Lee et al. 1995). This system suppresses the normal uptake function of RES by injection of a large amount of blank drug delivery system or large molecule of dextran sulfate to make a saturation of RES (Minko et al. 2004).
4. *Ligand-Mediated Targeting*: Ligand-mediated drug targeting depends on the receptor uptake of natural low-density lipoprotein (LDL) particles and synthetic micro-emulsions of LDL particles covered with apoproteins (Vyas and Sihorkar 2000).
5. *Physical Targeting*: Physical targeting is achieved by controlling the external physical changes including temperature change, change in pH, and applying an electric field in the drug delivery systems to allow targeting them to the specific site (Kumari et al. 2016).
6. *Dual Targeting*: In this system the carrier has a synergistic effect on the entrapped drug and hence increases the therapeutic effect (Minko et al. 2004).
7. *Double Targeting*: This system is a combination of both temporal and spatial delivery. Spatial delivery involves the targeting of the drug to the target site, while temporal delivery involves the control of drug release at the target site (Minko et al. 2004).

In targeted drug delivery, circulation time plays a crucial role in determining the success of a particular drug. To enhance circulation time, the hydrophilic nature of nanocomposites facilitates the formation of hydration films, rendering them antiphagocytic. This antiphagocytic property arises from the hydrophobic interactions inherent to the reticuloendothelial system, which prevents rapid clearance and allows drug-loaded nanocomposites to remain in circulation for an extended period. Beyond circulation time, several key factors significantly influence the effectiveness of targeted drug delivery (Raemdonck et al. 2014; Chomoucka et al. 2010). These include the size and geometry of the nanocomposites, their physicochemical properties, blood flow rate, the depth of the targeted tissue, field strength, and vascular supply. Optimizing these parameters ensures precise and efficient drug delivery, enhancing therapeutic efficacy while minimizing off-target effects (Ben-Nissan et al. 2016; Manuja et al. 2016).

Active targeting offers a significant advantage in drug delivery by enhancing the specificity and efficiency of nanocomposite-based therapies. This approach utilizes cell-specific ligands, such as carrier proteins or antibodies, to direct drug-loaded nanocomposites to a particular site, thereby minimizing off-target effects and improving therapeutic outcomes. Several strategies can be employed to achieve active targeting, one of the most effective being the identification of receptor characteristics on the targeted cells. By utilizing cell-specific ligands that bind to these receptors, nanocarriers can selectively accumulate at the desired location, ensuring precise drug delivery. This targeted approach not only increases drug efficacy but also reduces systemic toxicity, making it a highly promising method for advanced drug delivery applications (Anselmo and Mitragotri 2014).

The use of contrast agents plays a crucial role in monitoring and ensuring the precise delivery of drug-loaded nanocomposites to targeted cells. These agents enable real-time imaging and tracking, allowing researchers and clinicians to confirm the magnetic positioning of nanocomposites at the intended site. By incorporating contrast agents, such as magnetic nanoparticles or fluorescent markers, drug delivery systems can be visualized through techniques like magnetic resonance imaging (MRI) or fluorescence imaging. This not only enhances the accuracy of targeted therapy but also facilitates adjustments to optimize drug release. The integration of contrast agents with nanocomposites thus represents a significant advancement in precision medicine, improving therapeutic outcomes while minimizing off-target effects (Mahdavinia et al. 2015).

The use of stimuli-responsive materials in nanocomposites has emerged as a powerful strategy to achieve target-specific drug delivery. These materials are designed to respond to specific environmental triggers such as pH, temperature, or an external electric field, ensuring controlled and precise drug release at the targeted site. pH-responsive nanocomposites take advantage of the acidic tumor microenvironment or intracellular compartments like endosomes and lysosomes to trigger drug release. Similarly, temperature-sensitive materials enable drug delivery in response to localized hyperthermia, while electrically responsive nanocomposites can be controlled through external electric fields to enhance targeting efficiency. By integrating these stimuli-responsive properties, nanocomposites can improve therapeutic outcomes, minimize systemic toxicity, and enhance the efficacy of drug delivery systems (Ulbrich et al. 2016). By integrating both passive and active targeting strategies, drug-loaded nanocomposites can effectively circulate throughout the body until they reach and accumulate at the desired tissue site. Passive targeting relies on the enhanced permeability and retention (EPR) effect, allowing nanocarriers to accumulate in diseased tissues due to leaky vasculature, while active targeting employs cell-specific ligands to enhance precision in drug delivery. Compared to conventional drugs, nanocomposites offer significant advantages by ensuring higher drug concentration at the diseased site while minimizing adverse effects on surrounding healthy tissues and organs. This dual-targeting approach enhances therapeutic efficacy, reduces systemic toxicity, and improves overall treatment outcomes, making nanocomposites a promising platform for advanced drug delivery systems (Habibi et al. 2016).

This chapter explores the diverse biomedical applications of polymer nanocomposites, highlighting their potential in advancing modern healthcare and related fields. We delve into their role in cancer and diabetes diagnosis and treatment, where polymer nanocomposites enhance drug delivery, imaging, and targeted therapies. Additionally, we examine their significance in ophthalmic drug delivery systems, offering controlled and sustained drug release for improved therapeutic outcomes. Beyond biomedical applications, we discuss their utility in smart pesticide delivery, ensuring precision and efficiency in agricultural practices. The chapter also covers their contributions to tissue engineering, where polymer nanocomposites serve as scaffolds that promote cell growth, regeneration, and biocompatibility. Their antimicrobial and antibacterial properties are explored in the context of healthcare and food packaging, helping to prevent infections and extend shelf life. Furthermore, we analyze the responsiveness of polymer nanocomposites to pH and redox stimuli, enabling smart and targeted drug delivery. Lastly, we highlight the biomedical applications of advanced nanocomposites, including gold-polymer and graphene-polymer nanocomposites, emphasizing their role in diagnostics, imaging, and therapeutic interventions. Through this discussion, we aim to provide a comprehensive understanding of the evolving landscape of polymer nanocomposites and their transformative impact across multiple disciplines.

8.2 Biomedical Applications of Polymeric Nanocomposites

Nanotechnology has paved the way for groundbreaking advancements in medicine, giving rise to the field of nanomedicine, which harnesses the potential of nanocomposites to revolutionize healthcare. Nanomedicine involves the use of nanoscale materials to diagnose, treat, and prevent diseases, as well as to manage traumatic injuries, alleviate pain, and enhance overall human health. Among the primary classes of nanocomposite materials, polymer nanocomposites, metal nanocomposites, and ceramic nanocomposites play crucial roles in various biomedical applications. While polymer nanocomposites have already found widespread use in packaging, optoelectronic devices, construction, and aerospace, their potential in medicine is becoming increasingly evident. Recent research has demonstrated the efficacy of nanocomposites in drug delivery, bioimaging, regenerative medicine, and antimicrobial applications. The integration of molecular tools and an in-depth understanding of human biology further enhances the precision and effectiveness of nanomedicine, promising innovative solutions for some of the most challenging medical conditions. As research continues to expand, nanomedicine is set to transform conventional healthcare, offering personalized, targeted, and minimally invasive treatment strategies (Feldman 2016) (Fig. 8.2).

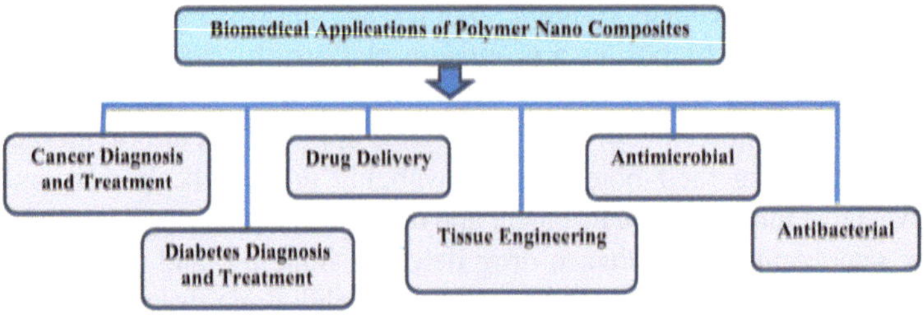

Fig. 8.2 Biomedical applications of polymer nanocomposites

Polymer nanocomposites have gained significant attention in the medical field due to their versatile properties and wide-ranging applications. Recent studies have highlighted their potential in developing advanced biomedical products for antibacterial treatments, tissue engineering (TE), cancer therapy, drug delivery, medical imaging, and dental applications. The incorporation of nanoscale fillers into polymer matrices enhances their mechanical strength, biocompatibility, and functional properties, making them ideal for various medical uses. In antibacterial applications, polymer nanocomposites exhibit strong antimicrobial activity, reducing the risk of infections in medical devices and implants. In tissue engineering, they serve as scaffolds that promote cell adhesion, proliferation, and regeneration. Additionally, their ability to encapsulate and release drugs in a controlled manner makes them highly effective in targeted drug delivery and cancer therapy, improving treatment outcomes while minimizing side effects. Polymer nanocomposites are also used in medical imaging for enhanced contrast in diagnostic procedures and dental applications for stronger and more durable restorative materials. As research continues, polymer nanocomposites are expected to drive innovations in nanomedicine, offering new opportunities for improved patient care and advanced therapeutic solutions.

8.2.1 Cancer Diagnosis and Treatment

Global Cancer Observatory (GLOBOCAN) predicted that cancer cases in India would increase to 2.08 million, accounting for a rise of 57.5% in 2040 from 2020 (Piñeros et al. 2021). Diagnosing a tumor at its initial proliferation phase is crucial for effective treatment and improved patient outcomes. However, in cases where the tumor remains asymptomatic, early detection becomes highly challenging, often delaying intervention until the disease has progressed to a more critical stage. Various imaging techniques, such as X-rays, magnetic resonance imaging (MRI), positron emission tomography (PET), and computed tomography (CT) scans, play a vital role in identifying abnormal tissue growth. Additionally, Pap tests are commonly used to detect morphological changes associated with malignant and preneoplastic conditions. To enhance diagnostic accuracy, these conventional methods are often combined with advanced techniques such as autofluorescence bronchoscopy, single photon emission computed tomography (SPECT), biomolecular markers, confocal micro-endoscopy, endobronchial ultrasonography, and optical coherence tomography. These cutting-edge technologies provide higher resolution imaging and molecular insights, facilitating the early detection and characterization of tumors even at a microscopic level, thus improving the chances of timely and targeted therapeutic intervention (Kumar et al. 2017).

These techniques perform exceptionally well in detecting diverse fluctuations in tissue and organ biochemistry, as well as metabolic changes, particularly at the molecular level. However, increased metabolism alone is not always a definitive indicator of tumor presence, making early-stage detection a persistent challenge (Liao et al. 2017). As a result, the need for novel diagnostic techniques has become crucial to improving accuracy and reducing false positives. In this context, the application of polymer nanocomposites has gained significant attention in cancer detection and treatment (Liu et al. 2011). These advanced materials offer promising capabilities for identifying tumor cells with high specificity and enabling selective irradiation of targeted tumor areas. By leveraging their unique physicochemical properties, polymer nanocomposites enhance imaging sensitivity, facilitate real-time monitoring, and improve therapeutic precision, ultimately contributing to more effective and less invasive cancer management strategies. Therefore, the use of magnetic nanoparticles reinforced within a polymer matrix nanocomposite has emerged as an attractive platform for diverse cancer therapies. These hybrid materials combine the advantages of magnetic nanoparticles, such as targeted drug delivery and hyperthermia treatment, with the versatility of polymer matrices, which enhance biocompatibility and controlled release properties (Lim et al. 2011).

8.2.1.1 Properties and Mechanisms of Nanocomposites in Cancer Diagnosis

High Surface Area

Nanoparticles have high surface area to volume ratio that allows them to interact with many molecules; therefore it is ideal for carrying and delivering diagnostic agents. High surface area enhances the following properties and functionalities in nanocomposites:

1. Enhances strength and stiffness.
2. Enhances thermal conductivity.
3. Improves electrical conductivity.
4. Enhances gas barrier properties.
5. Increases catalytic activity.
6. Improves the drug delivery system.

Example 8.1 Titanium dioxide (TiO_2) nanocomposites widely used photocatalyst incorporated into nanocomposites to enhance their photocatalytic activity.

Example 8.2 TiO_2 and zeolite polymer nanocomposites have large surface area that provide more active sites for photocatalysts and improved separation and catalytic properties (Abla et al. 2023).

Biocompatibility

Nanocomposites are designed to be biocompatible, which ensures their safe use in the body, and it is important for cancer diagnosis.

E.g., Metal oxide nanocomposites zinc oxide (ZnO) and titanium dioxide (TiO_2), and carbon-based nanocomposites, e.g., carbon nanotubes and graphene are shown to be biocompatible and are used in biosensor for detecting various analytes including glucose, DNA, and protein (Andoh et al. 2024).

Targeted Delivery

Nanocomposites enhance the precision and effectiveness of cancer detection, leading to earlier diagnosis and improved treatment outcomes.

E.g., Molybdenum disulfide based nanocomposites employed for cancer diagnosis and imaging these nanomaterial integrated with targeting ligands, enable selective binding to cancer cells that improves the accuracy of diagnostic imaging (Wang et al. 2021).

Enhanced and Targeted Imaging

It includes enhanced permeability and retention effect, active targeting, and targeted delivery. Nanocomposites in targeted imaging can incorporate various imaging agents such as fluorescent dyes, radioisotopes, or magnetic nanoparticles used to visualize tumor using different imaging agents such as fluorescence imaging, positron emission tomography (PET) or magnetic resonance imaging (MRI) (Qamar 2021).

Bio-Responsive Imaging

Nanocomposites can be designed with an aim, to respond to specific changes in tumor environment such as pH and temperature/hypoxia, and release imaging agent in response to the changes; bioresponsive nanocomposites will respond to specific biological in the tumor environment, altering their properties and enhancing their imaging capabilities and hold great promise for cancer diagnosis (Li et al. 2021).

Electrochemical Sensing

Nanocomposites are used to create electrochemical biosensors that detect cancer biomarkers in blood or other body fluids. Graphene electrochemical biosensors for cancer diagnosis demonstrate the potential for rapid, sensitive, and cost-effective detection of cancer biomarkers.

E.g., Graphene assisted sensors explored for detection of pancreatic cancer molecules such as K-Ras gene, CEA, and micro-RNA.

Electrochemical biosensors also utilize polymer nanocomposites containing metallic nanoparticles and carbon nanomaterials to enhance sensor performance, improving diagnostic sensitivity and selectivity (Sadeghi et al. 2023).

Table 8.1 presents the different nanocomposites targeted to diagnose/treat different cancers and their key outcomes.

Table 8.2 presents different nanocomposites used to treat different types of cancer.

Table 8.1 Application of nanocomposites in cancer diagnosis

S. no	Polymer nanocomposites	Oncology type/ study model	Diagnostic technique/ treatment method	Conclusion	References
1	89Zr-Df-IAB22M2C	Metastatic melanoma, non-small cell lung cancer, and hepatocellular carcinoma/ clinical trial	Theranostics	Radiotracer uptake in tumors was noted in 10 of 15 subjects; CD8 PET imaging is safe and visualizes the whole-body biodistribution of CD8+ leukocytes in tumors and normal tissues	Zeng et al. (2020)
2	AS1411-Au NPs/Au@PC NP	Breast cancer/ clinical breast tissue samples	Targeted mass spectrometry imaging	Labeling agent for tumor tissue imaging by LDI-MS; binds tumor cell membranes	Zhu et al. (2017)
3	AuNP@ (AS1411-AgNCs)n	HeLa and NIH-3T3 cells/in vitro	Targeted imaging and selective killing—Theranostics	Specific cancer cell-targeted imaging and selective killing in a low AuNP@ (AS1411-AgNCs) concentration	Farwell et al. (2022)
4	AuNPs/ PB-PEDOT	CEA/clinical serum samples	Biosensor	Serves as an immunosensor for detecting CEA	Hasanzadeh et al. (2019)

(continued)

Table 8.1 (continued)

S. no	Polymer nanocomposites	Oncology type/ study model	Diagnostic technique/ treatment method	Conclusion	References
5	AuPtPd/ rGO-modified GCE	Breast cancer (MDA-MB-231 and T47D cells)/in situ	Biosensor	Monitors the release of H_2O_2 from cancer cells; excellent electrocatalytic activity towards the reduction of H_2O_2	Doan et al. (2021)
6	BH-NO2@BSA and BH-NH2 @ BSA	Orthotopic hepatic tumor/BALB/c nude mice, HCC-LM3-fLuc tumor cells	An orthotopic hepatic tumor/ BALB/c nude mice, HCC-LM3-fLuc tumor cells, optoacoustic tomography imaging, and NIR-I/NIR-II fluorescence imaging	Capable of detecting early-stage orthotopic liver tumors	Qi et al. (2020)
7	Fe_2O_3@ PPy-DOX-PEG	Hepatocellular carcinoma	Imaging and cancer cell killing theranostics	Kills cancer cells and realizes an MRI in vivo	Dong et al. (2018a)
8	GCE/Ti3 C2 TX/Pt-Pd/SOx	Prostate cancer/ clinical (human serum)	Biosensor	Shows potential in the early diagnosis of prostate cancer	Han et al. (2017)
9	IR-CS-PPy NCs	Breast cancer/in vitro and in vivo	Imaging and tumor killing—Theranostics	Dual-modal imaging; destruction of the tumor tissues with IR-CS-PPy NCs without further recurrence	Yang et al. (2017)
10	Magnetic PPy/ Fe_3O_4-core and a gold nanoshell (MGNSHs)	Hela cells in vitro; Kunming mice in vivo	Multimodal imaging and photothermal cancer therapy	Enhances the contrast for both MRI and X-ray CT imaging, promising for multimodal imaging-guided cancer treatment	Zhou et al. (2015)

Table 8.1 (continued)

S. no	Polymer nanocomposites	Oncology type/ study model	Diagnostic technique/ treatment method	Conclusion	References
11	MFG-LPM NPs	Nasopharyngeal epidermal carcinoma cell line (KB cell), normal human embryonic lung fibroblasts (MRC-5 cell), and murine breast cancer cell line (4T1 cell); BALB/c mice	MRI-guided cancer diagnosis and therapy	Impressive MRI performance as a T2 contrast agent; high cancer cell killing efficacy	Ran et al. (2022)
12	MFR-AS1411	C6 rat glioma cells/ BALB/c nude mice	Targeted imaging	Specific targeting of cancer cells; enhanced fluorescence activity	Tseng et al. (2017)
13	P[DA-β-CD / CTAB])-AgNPs-GCE	BRCA1/clinical plasma samples and MCF-7 cancer cell	Biosensor	Satisfactory results in BRCA1 detection; potential utility in clinical analysis	Wang et al. (2015)
14	PSA-Mn-Msn-Cy7	Prostate cancer/ LNCap and HSF cells, nude mice	Fluorescence and MR dual-imaging	PSA targeted-optical and MR imaging of cell and tumor-bearing mice	Li et al. (2019)
15	S1/PEG/AuNPs/ GCEs	BRCA1/clinical serum samples	Biosensor	Biosensor ultrahigh sensitivity and selectivity for BRCA1 detection	Du et al. (2020)
16	SarOx/CHIT/ CuNPs/ c-MWCNT/Au	Prostate cancer/ clinical (sera of prostate cancer patients)	Biosensor	Potential prostate cancer diagnosis via sarcosine detection	Ko et al. (2010)
17	SiO2@MPGs	Pancancer/3T3 cells, tumor-bearing mice	Biosensor	Excellent enzyme-mimic catalytic efficiency for ROS-responsive fluorescence imaging, both in vitro and in vivo	Narwal et al. (2018)

(continued)

Table 8.1 (continued)

S. no	Polymer nanocomposites	Oncology type/ study model	Diagnostic technique/ treatment method	Conclusion	References
18	SiO2@ T1-RGDk NPs	Prostate cancer/ PC-3 cell, nude mice	Fluorescent imaging	Excellent targeting ability for visualizing cancer cells	Uhlirova et al. (2018)
19	SPIONs/Au/CS/ SOX/NPs	Prostate cancer/ clinical (urine specimens)	Biosensor	Potential prostate cancer diagnosis via sarcosine detection	Nan et al. (2017)

AuNP@(AS1411-AgNCs)n gold nanoparticle@(AS1411 silver nanoclusters) nanocomposite, *HeLa* human epithelial carcinoma, *NIH-3T3* mouse embryo fibroblasts, *AS1411-Au NPs/Au@PC NP* poly(catechin) capped-gold nanoparticles (Au@PC NPs) and smaller nucleolin-binding aptamer (AS1411) conjugated gold NPs (AS1411-Au NPs), *IR-CS-PPy NCs* IR783 conjugated chitosan-polypyrrole nanocomposites, *Fe_2O_3@PPy-DOX-PEG* iron(III) oxide or ferric oxide (Fe_2O_3)@ polypyrrole (PPy)-doxorubicin hydrochloride(DOX)-polyethylene glycol (PEG), *HepG2* human hepatocellular carcinoma cell line, *MFG-LPM NPs* inner core of an anti-cancer drug MTX surrounding by a nanometer-thin layer of gold as the shell with Fe_3O_4 magnetic nanoparticles (NPs) evenly distributed in the gold layer, and the outermost hybrid LA-PEG-MTX molecules, *AuPtPd/rGO-modified GCE* trimetallic AuPtPd nanocomposites platform on reduced graphene oxide (rGO) nanosheets with modified rGO and trimetallic AuPtPd nanoparticles on a glassy carbon electrode (GCE), *GCE/Ti3 C2 TX/Pt-Pd/Sox* Ti3 C2 TX/Pt-Pd nanocomposite synthesized by a facile one-step alcohol reduction and used on a glassy carbon electrode (GCE) with sarcosine oxidase (SOx), *SPIONs/Au/CS/SOX/NPs* superparamagnetic iron oxide nanoparticles (SPIONs)/gold (Au)/chitosan (CS)/sarcosine oxidase (SOX)/nanoparticles (NPs), SarOx/CHIT/CuNPs/c-MWCNT/Au electrode, sarcosine oxidase (SarOx)/chitosan (CHIT)/copper nanoparticles (CuNPs)/carboxylated multi-walled carbon nanotubes (cMWCNT)/gold (Au) electrode, *S1/PEG/AuNPs/GCEs* modified polyethylene glycol (PEG)/gold nanoparticle (AuNPs)/glassy carbon electrode (GCE), *P[DA-β-CD/CTAB])-AgNPs-GCE* polymer nanocomposite-modified GCE poly(dopamine beta cyclodextrine-cetyltrimethylammonium bromide) fixed with silver NPs, *SiO_2@MPGs* silicon dioxide (SiO_2)@metal crosslinked polymeric nanogels (MPGs)

8.2.1.2 Recent Advance in PH and Redox Responsive Polymer Nanocomposites for Cancer Therapy

Over the past half-decade, the cancer mortality rate has significantly decreased due to advancements in early cancer diagnosis, improved diagnostic techniques, better treatment strategies, and a deeper understanding of tumor biology. The development of polymer nanocomposites has emerged as an effective therapeutic approach, offering several advantages such as reduced side effects, increased biocompatibility, and enhanced therapeutic efficacy. Nanoparticles incorporated within these nanocomposites interact with the surface of cancer cells, leading to intracellular changes such as a drop in cytosolic pH and an increase in redox potential. These biochemical alterations create a tumor-specific microenvironment that facilitates targeted drug release, enhances cellular uptake, and triggers selective cancer cell apoptosis while sparing healthy tissues. This cutting-edge approach not only improves treatment

Table 8.2 Types of nanocomposites used in cancer treatment

S. no	Polymer nanocomposites	Oncology type/study model	Action of nanocomposites	Conclusion	References
1	A photothermal agent (indocyanine green) complexed with PDMAEMA/rGO nanocomposites		Targeted release	Exhibited pH stimuli behavior and quenching effects to target abnormal cancer cells to ensure their detection and treatment	Sherwani et al. (2015)
2	CuInS2/ZnS/liposome-rGO (carbon-based nanocomposite)	Human esophagus carcinoma cells. In vitro: Eca-109 cells; In vivo: Mice	PTT and PDT	Reduced the toxicity caused by CuInS2/ZnS	Thorat et al. (2017)
3	DOX@MoS2-PEI-HA (hybrid nanocomposite)	Metastatic breast cancer	PTT	Significantly enhanced breast tumor growth inhibition	Dong et al. (2018b)
4	DOX-GO-PEG-FA (carbon-based nanocomposite)	Melanoma. In vitro: B16F0 cells; In vivo: male C57BL/6J mice	PTT and PDT	GO-PEG-FA initiates the generation of singlet oxygen required to demonstrate dual-mode photodynamic and photothermal treatment in vivo for the treatment of tumors upon NIR light activation	Wu et al. (2016)
5	Fe_3O_4@Bio-MOF-FC (metal-based nanocomposite)	Lung carcinoma (M109)-bearing BALB/c mice	DDS	Tumor shrinking	Chan et al. (2019)
6	FePt@Kao-Dox (ceramic nanocomposite)	Hepatocellular carcinoma	CT	Enhances the magnetic signal and killing of tumor cells	He et al. (2019)
7	La0.7Sr0.3MnO3/PEG nanocomposites	Doxorubicin	Targeted release	Good biocompatibility, colloidal stability, and high cancer cell death rate (89%) for 24 h relative to free doxorubicin (75%)	Shakoori et al. (2017)
8	MoS2-PEI-HA (lipid-based nanocomposite)	Mice bearing the MCF7-ADR tumor	DDS	Completely eradicate the tumor	Sharker et al. (2015)
9	MPEG–PCL (polymeric nanocomposite)	CT26 mouse colon carcinoma cell line	CT	Sustained the release of the drug	Kalluru et al. (2016)

(continued)

Table 8.2 (continued)

S. no	Polymer nanocomposites	Oncology type/study model	Action of nanocomposites	Conclusion	References
10	$mTiO_2$@PPY (metal-based nanocomposite)	Human-hepatoma HepG2/4T1 cell lines	Synergistic antitumor therapy	Synergistic antitumor therapy	Nejadshafiee et al. (2019)
11	PLGA–TPGS (polymeric nanocomposite)	Human breast adenocarcinoma/MCF-7 cell line	CT	Induced apoptosis/overcome multi-drug resistance	Wang et al. (2014)
12	Poly(N-isopropylacrylamide) (PNIPAM)-based nanocomposites using the free radical copoly merization cross-linking technique	Cisplatin		The PNIPAM-based magnetic nanocomposites exhibited stimuli (e.g., pH, magnetic field, and temperature)-responsive behavior while possessing fluorescent properties to monitor drug release	El-Ghannam et al. (2010)
13	Polyfunctional Ab@polyamidoamine (PAMAM) dendrimer-siRNA@ Cis-poly(d,l lactide-co-glycolide) (PLGA) nanocomposites	Combination of siRNA based therapy and chemotherapy		Treat liver cancer cells	Loiseau et al. (2019)
14	Polylactide-cisplatin loaded-cationic lipid assisted nanocomposites		Targeted release	Treat drug-resistant lung cancer cells	Cao et al. (2016)
15	SCPC (ceramic nanocomposite)	Breast cancer/4T1 mammary tumor cells	DDS	Stopped tumor growth	Tang et al. (2015)
16	TiONts-AuNPs-PEG3000-DTX (hybrid nanocomposite)	Human PC-3 prostate adenocarcinoma cells	IT injection/ radiotherapy	Delayed tumor growth	Xu et al. (2015)

CuInS2/ZnS/liposome-rGO copper indium sulfur zinc sulfide/liposome on reduced graphene oxide, *PTT* photothermal therapy, *PDT* photodynamic therapy, *DOX-GO-PEG-FA* doxorubicin-graphene oxide-polyethylene glycol-folic acid, *B16F0 cells* murine melanoma cell line, *PLGA–TPGS* poly(lactide-co-glycolide)-d-a-tocopheryl polyethylene glycol 1000 succinate, *MCF-7 cell line* Michigan cancer foundation-7 cell line, *MPEG–PCL* methoxy poly(ethylene glycol)-poly(ε-caprolactone), *SDT* sonodynamic therapy, *CT* chemotherapy, *$mTiO_2$@PPY* mesoporous TiO_2 coated with polypyrrole, *Fe_3O_4@Bio-MOF-FC* bio-metal-organic framework coated with folic acid-chitosan conjugate, *MoS2-PEI-HA* molybdenum disulfide (MoS2)–polyethyleneimine (PEI)– hyaluronic acid, *MCF7-ADR* multidrug-resistant MCF-7 breast adenocarcinoma cells, *DDS* drug delivery systems, *DOX@MoS2-PEI-HA* doxorubicin molybdenum sulfide-polyethyleneimine-hyaluronidase, *TiONts-AuNPs-PEG3000-DTX* titanate nanotubes-gold nanoparticles-heterobifunctional polymer-docetaxel, *FePt@Kao-Dox* iron/platinum-based kaolinite-doxorubicin, *SCPC* silica–calcium phosphate nanocomposite, *PEG* polyethylene glycol, *P(NIPAM-co-AAD)* poly(N-isopropylacrylamide-co-acrylic acid), *PVP* polyvinylpyrrolidone, *PNIPAM* poly(N-isopropylacrylamide), *PLGA* poly(D,L-lactic-co-glycolic acid), *PAMAM* polyamidoamine, *RNA* ribonucleic acid

precision but also paves the way for more efficient and personalized cancer therapies (Gaddimath et al. 2024).

Polymer nanocomposites (PNCs) play a significant role in cancer therapy by safeguarding delicate drugs until they reach their intended delivery sites, ensuring stability and controlled release. However, the targeted delivery of therapeutic agents faces numerous physicochemical and biological challenges that hinder optimal treatment efficacy. Key obstacles include poor drug solubilization, limited biocompatibility, and the complexity of achieving precise site-specific delivery to cells and tissues. The tumor microenvironment, with its unique physiological conditions such as abnormal pH, hypoxia, and dense extracellular matrix, further complicates effective drug transport and uptake. Additionally, systemic clearance mechanisms, immune responses, and off-target effects pose additional barriers to efficient drug delivery. Overcoming these challenges requires innovative strategies, such as surface functionalization of PNCs with targeting ligands, stimuli-responsive drug release mechanisms, and advanced nanoengineering techniques. By addressing these limitations, PNC-based drug delivery systems can significantly enhance therapeutic precision, reduce toxicity, and improve overall treatment outcomes in cancer therapy (Demirci et al. 2014).

Two types of polymers are commonly used for specific drug delivery in cancer therapy: cationic and anionic polymers. Anionic polymers, such as poly(glutamic acid) (PGA), poly(ethylacrylic acid) (PEAA), poly(methacrylic acid) (PMAA), and poly(acrylic acid) (PAA), are particularly useful for pH-responsive drug delivery (Ratemi 2018). These polymers undergo a transition from hydrophilic to hydrophobic as the pH decreases within the tumor microenvironment, leading to polymer destabilization, swelling, or changes in solubility. This response facilitates drug release through polymer disassembly (Deirram et al. 2019).

In contrast, cationic polymers, such as poly(β-amino ester) and poly(L-histidine) (PHis), exhibit a shift from hydrophobic to hydrophilic behavior at lower pH levels. Their positive charge in acidic cellular compartments enhances cellular uptake, making them ideal carriers for intracellular drug delivery. A notable example is the pH-sensitive release of the anticancer drug doxorubicin (DOX), where protonation of carboxylic groups in nanocarriers accelerate drug release due to reduced electrostatic attraction between protonated acid groups and DOX molecules. The tunable properties of these polymers make them highly effective in targeted and controlled drug release strategies, improving therapeutic precision while minimizing systemic toxicity (Kim et al. 2009).

Table 8.3 presents the different nanocomposites that are used to treat different cancers by using different study models.

8.2.2 Diabetic Diagnosis and Treatment

Table 8.4 presents the different polymer nanocomposites used for detection of diabetes.

Table 8.3 In vitro studies of the different pH/redox-responsive polymeric nanoparticles drug delivery

S. no	pH/redox-responsive polymer nanocomposites	Cancer type/ study model	Treatment method	Conclusion	References
1	(PAE-ss-mPEG)	Doxorubicin	Controlled drug release	Anticancer treatment	Dai et al. (2019)
2	(PDPA)capsules Cargo release	Rhodamine B isothiocyanate labeled OVA	Cargo release		Chen et al. (2017)
3	Acrylic acid (AAc) and 2-methacryloylethyl acrylate (MEA) Doxorubicin Conjugate	Doxorubicin	Rapid drug release	Anticancer treatment	Du et al. (2011)
4	Conjugated (C-dots-HBA-dox) and (C-dots-S–S-dox)	Doxorubicin	Controlled drug release	Cancer chemotherapy	Cheng et al. (2013)
5	Cross-linked polyphosphazene	Curcumin and Ce6	Controlled drug release	Cancer therapy	Shi et al. (2020)
6	DEXssPEGCOOH	Doxorubicin	Targeted release	Cancer therapy	Saito et al. (2003)
7	DOX@Dex-SSnanogel	Doxorubicin	Cumulative amount of drug release	Cancer chemotherapy	Hettiarachchi et al. (2021)
8	MSNs(DOX@PRMSNs)	Doxorubicin	Targeting ligands	Cancer therapy	Shi et al. (2021)
9	PCL-b-P(OEGMA-co MAEBA)	Camptothecinor Doxorubicin	Accelerated drug release	Cancer chemotherapy	Guo et al. (2015)
10	PEG–PAA(SH)–PDEA	FITC-BSA/CC	Controlled drug release	Cancer therapy	Fu et al. (2018)
11	Poly(2-(pyridin-2-yldisulfanyl)ethyl acrylate)	RPDSG/DOX	Controlled drug release	Cancer therapy	Xiong et al. (2018)
12	Poly(β-aminoesters)	CD44	Controlled drug release	Breast cancer; lung metastasis	Yu et al. (2020)
13	RPAE-PEG	Doxorubicin	Controlled drug release	Cancer therapy	John et al. (2017)

Table 8.4 Polymer used in nanocomposite-based sensors for diabetes management

S. no	Polymer nanocomposites	Detection	References
1	Au/MXene nanocomposites	Gox	Rakhi et al. (2016)
2	Gold nanoparticles-PVP-PANI nanocomposites	Gox	Gopalan et al. (2009)
3	MWCNTs-COOH-poly(2 aminothiophenol)-gold nanoparticles	Amperometric glucose	Miao et al. (2015)
4	Nafion-silica/MWCNTs-g-PANI nanocomposites	Gox	Sedghi and Pezeshkian (2015)
5	PANI nanotubes	Gox	Wang et al. (2009)
6	Polymeric ionic liquid/graphene nanocomposites	Gox	Zhang et al. (2011a)
7	Polypyrrole-chitosan-TiO_2 nanocomposites	Non-enzymatic glucose	Al-Mokaram et al. (2017)

8.2.3 Polymer Nanocomposites in Antibacterial Medicine

Table 8.5 presents different polymer nanocomposites used in antibacterial medicine.

Table 8.5 Polymer nanocomposites in antibacterial medicine

S. no	Polymer nanocomposites	Antibacterial medicine	References
1	Ag and Catechol ratio 2:1 the poly (ethylene glycol) (PEG)/TA (tannin)/nano Ag coating	Resist Gram −ve and Gram +ve bacteria achieved 100% at 37 °C after incubating for 12 h	Chen et al. (2015)
2	$AgNO_3$ added to CS/poly (ethyleneoxide) (PEO) aqueous acetic solution	Antimicrobial activity against *E. coli* and *Staphylococcus aureus* with bigger inhibition zone	Arvand et al. (2015)
3	Biopolymer polyhydroxyalkanoate was used as a capping and stabilizing agent to minimize the agglomeration of Ag nanoparticles (22–43 nm)	Antimicrobial activity against both *E. coli* XLIB strain (Gram-negative) and *Lysinibacillus fusiformis* strain (Gram positive)	Pramanik et al. (2015)
4	CS/Phosphoramide/Ag prepared with 1–5% Ag nanoparticles	In vitro antibacterial activity against two Gram +ve *Staphylococcus aureus*, *Bacillus cereus* and two Gram −ve *E. coli* and *Pseudomonas aeruginosa*	Shariatinia et al. (2015)
5	Graphene nanoplatelets in PLA containing PEG and epoxidized palm oil	Antimicrobial activity toward *E. coli*, *Staphylococcus aureus*, *Salmonella typhimurium*, and *Listeria monocytogenes*	Chieng et al. (2015)
6	Loading Ag nanoparticles onto cotton fiber preliminary coated with maleic anhydride (MA) plasma polymer	Antibacterial activity	Piñeros et al. (2021)
7	Low density PE (PELD)/oMMT/carvacrol oil polymer nanocomposites	Excellent activity against *E. coli* bacteria	Shemesh et al. (2015)
8	Medical grade PVC reinforced with Ag nanoparticles and semiconducting ZnO micro particles	Antibacterial system with excellent activity against *E. coli* and sufficient against *Staphylococcus aureus*	Bazant et al. (2014)
9	Nanocomposites derived from biodegradable cellulose with Cu, Ag, Fe, In, and noble metals, as nanofillers	Antibacterial, antifungal	Nadagouda and Varma (2007)
10	Natural rubber (NR)/TiO_2 polymer nanocomposite	Antibacterial activity	Seentrakoon et al. (2013)
11	PLA/nano ZnO/Cu chlorophyll acid (CCA) polymer nanocomposite	Antibacterial rate (R) against *E. coli*	Sun et al. (2015)
12	Poly(4-allyloxy-2H-chromen 2-one)/Ag nanocomposite	Active selectively against bacterial pathogen *Enterococcus fecalis*	Srivastava et al. (2012)

(continued)

Table 8.5 (continued)

S. no	Polymer nanocomposites	Antibacterial medicine	References
13	Poly(lactic acid) (PLA)/ZnO (3%) nanoparticles	Antibacterial properties	Doumbia et al. (2015)
14	Polyethylene (PE) or polypropylene (PP) PP/PVP/Ag nanoparticles film	No cytotoxicity for mammalian cells, but present antimicrobial effect against Gram-negative *E. coli* and Gram-positive *Staphylococcus aureus*	Piñeros et al. (2021)
15	Polyimidosulfide/Ag nanocomposites	Antibacterial activity	Piñeros et al. (2021)
16	Silver release from PVP/Ag nanocomposite	Antimicrobial activity against *Staphylococcus aureus*	Jovanović et al. (2014)
17	Sodium alginate and poly (acrylamide-co-N-vinyl caprolac tone-co-acrylamidoglycolic acid) based dual responsive semi interpenetrating polymer network hydrogels/Ag nanocomposites	Antibacterial activities toward *E. coli* and Bacilli	Reddy et al. (2014)

8.2.4 Polymer Nanocomposites Application in Ophthalmic Drug Delivery Systems

Table 8.6 presents nanocomposite fabrication techniques using different material used to study the ophthalmic drug delivery system.

8.2.5 Polymer Nanocomposites and Tissue Engineering (TE)

Polymer nanocomposites play an important role in TE and regenerative medicine. They can be used for producing scaffolds providing a 3D framework on which cells can proliferate and differentiate, secrete extracellular matrix (ECM), and form functional tissue. The knowledge of cell biology, materials science, and bioengineering provide the basics for TE that aims to regenerate different body tissues with specific properties and versatility. Synthetic scaffolds must possess mechanical, chemical, surface, and electrical characteristics that match the extracellular matrix or basement membrane of a specific desired tissue.

Table 8.7 presents different nanocomposites and their uses in tissue engineering.

Table 8.6 Polymer nanocomposites application in ophthalmic drug delivery systems

S. no	Materials used for fabrication techniques	Fabrication techniques	Drug	System	Outcomes	References
1	Carboxymethyl tamarind kernel polysaccharide nanocomposites	Ionotropic gelation method	Tropicamide	Ex vivo corneal permeation studies	Non irritant and mucoadhesive nature of the nanocomposites	Ogunjimi et al. (2017)
2	CG-VV-PRN-LDH nanocomposites	Co-precipitation hydrothermal method	Trichlorobutanol as bacteriostatic agent and glucose as osmotic pressure regulator	In vivo precorneal retention study on rabbits	Excellent increase in area under the drug concentration in ocular tissues vs time curve (14.7-fold) and mean residence time (2.2-fold) than those of commercial products	Chou et al. (2017)
3	Delonix polysaccharides	Quality-by-design modified nanoprecipitation method	BODIPY dyes as test molecules	In vitro release, cell toxicity, and cellular uptake studies	Good stability of Delonix nanoparticles in the presence of lachrymal fluid and lysozyme along with sustained release profile of BODIPY	Xu et al. (2016)
4	GA-functionalized gelatin-g-PNIPAM	Redox technique	Pilocarpine	In vivo and in vitro biocompatibility and drug release studies	Strong effect of redox radical initiation reaction temperature over physicochemical properties of nanocarriers	Al Khateb et al. (2016)
5	Gelation-g-PNIPAM	Carbodiimide coupling chemistry	Epigallocatechin gallate	In vivo and in vitro degradation and drug release studies	Biodegradable in situ gelling system with enhanced pharmacological efficacy of topically instilled drug	Prosperi-Porta et al. (2016)
6	HM-HPMC/-CD hydrogel temperature	Reversible sol-gel phase transition method	Diclofenac sodium	In vivo and in vitro studies	Good ocular bioavailability of the drug due to rapid gel formation on the eye surface at physiological	Zhang et al. (2016)

(continued)

Table 8.6 (continued)

S. no	Materials used for fabrication techniques	Fabrication techniques	Drug	System	Outcomes	References
7	Micellar supramolecular hydrogel of MPEG-PCL block copolymer/-CD	Host-guest inclusion method	Diclofenac	In vivo and in vitro release and pharmacokinetics studies	Sustained drug release profile over 216 h, low cytotoxicity towards L-929 and HCEC cells, and good ocular bioavailability	Luo and Lai (2017)
8	PLA-b-p(MAA-PBA) copolymers	RAFT polymerization	Cyclosporine A	In vivo and in vitro studies	Good biocompatibility with the corneal surface and undetectable acute in vivo irritation in Sprague-Dawley rats	Iohara et al. (2017)
9	Pluronic F127	–	Ofloxacin	In vivo and in vitro studies	Good in situ transparent gel formation under physiological conditions with negligible irritation to mucosal tissues	Yuan et al. (2015)
10	PVA nanowafers	Hydrogel template method	Axitinib as antibiotic drug	In vivo study in murine ocular burn model	Effective treatment of corneal neovascularization with negligible adverse immune response and twice therapeutic efficiency than a topical eye drop system for axitinib delivery	Kaur et al. (2012)

Table 8.7 Polymer nanocomposites used in tissue engineering

S. no	Polymer nanocomposites	Tissue engineering	References
1	45S5 BG is a promising material for obtaining bone TE scaffolds	Bone tissue engineering	Li et al. (2014)
2	Au nanoparticles can help proteins to retain their biological activity upon adsorption. Because of their large surface area and good electronic properties	Used in the construction of electrochemical sensors and biosensors	Piñeros et al. (2021)
3	Camforsulfone acid (CPSA)/PANi/PLCL nanofibers with electrically conductive properties	Platform substrate to study the effect of electrical signals on cell activities and to direct desirable cell function	Piñeros et al. (2021)
4	Composite electrodes made of the polysaccharide agarose (A) and carbon nanotubes (CNTs)	Using neuro-relevant proteins (laminin and a-melanocyte stimulating hormone) and implanted in brain tissue, ability to modify brain tissue response through surface modification	Piñeros et al. (2021)
5	Conductive polymers (like PANi) serving to effectively deliver electrical signals	Cardiomyocytes, neurons, and osteoblasts respond to electrical signals by improving their functional outcome	Qazi et al. (2014)
6	Poly(L-lactic acid) (PLLA) and random PLLA/activated platelet rich plasma (PRGF) mats, aligned PLLA/PRGF scaffolds	In vitro promoted cell alignment and elongation while enhancing cell attachment and proliferation	Díaz-Gómez et al. (2015)
7	Poly L-lactide reinforced with nanofillers such as TCP and silica	Regenerative medicine used to fill bone losses and stimulate accelerated bone regeneration	Stodolak-Zych et al. (2012)
8	Poly(3-hydroxy butyrate)/oMMT nanocomposite surfaces	Supported cell attachment and proliferation for tissue engineering	Panayotidou et al. (2015)

8.2.6 Polymer Nanocomposites in Drug Delivery

Table 8.8 presents different nanocomposites and their uses in drug delivery.

8.2.7 Gold Polymer Nanocomposites Biomedical Application

Gold polymer nanocomposites possess various biomedical applications (Fig. 8.3).

Table 8.9 presents biomedical applications of gold polymer nanocomposites.

Table 8.8 Polymer nanocomposites in drug delivery

S. no	Polymer nanocomposites	Drug delivery	References
1	(Lidocaine, Caprofen) release from poly(lactic-co-glycol acid)/TiO_2 nanotubes	Drug carrier was controlled by the solubility of the drug, the degree of polymer swelling/degradation, and the electrostatic force polymer-drug	Cheng et al. (2015)
2	Alginate-CS (Chitisan)-pluronic nanoparticles	For intraarticular nonsteroidal anti-inflammatory (osteoarthritis) meloxican delivery	Fattahpour et al. (2015)
3	Doxorubicin incorporated poly(lactic acid (PLA)/poly (ethylene glycol) (PEG)/multi-walled-carbon nanotube (MWCNT) nanofibrous scaffold	Superior candidate for antitumor drug delivery	Irani et al. (2018)
4	Irbesartan/polymer nanofibers	It showed an increasing relative bio-availability than the plain irbesartan suspension	Adeli (2015)
5	Poly(acrylic acid-co-acrylamide-co-2-acrylamido-2-methyl-1-propansulfonic acid)-grafted nanocellulose/PVA composite	Fascinating vehicle for the in vitro administration of amoxicillin into the gastrointestinal tract	Adeli (2015)
6	Poly(vinyl acetate) coated indomethacin loaded poly(lactide-co-glicolide) nanoparticles	Able to penetrate through the transfollicular pathway and to migrate the nanoparticles to follicles	Tomoda et al. (2014)
7	Polyblend CS (Chitisan)/poly (ethylene oxide) (PEO) of the ciprofloxacin hydroxide or moxifloxacin hydrochloride drug	Good antimicrobial activity against both the Gram negative bacteria *E. coli* and Gram-positive bacteria *Staphylococcus aureus* by controlled drug release	Jia and Kerr (2015)
8	Polyblend PVA/poly (vinyl acetate) (PVAc) exhibited a useful and convenient system for electrospinning	Control the rate and period of drug release in wound healing applications	Jia and Kerr (2015)

8.2.8 Graphene Polymer Nanocomposites Biomedical Application of Polymer Nanocomposites

Graphene polymer nanocomposites enhance mechanical, thermal, and electrical properties of biomaterial and increase cellular attachment and growth in biomaterial surface.

Table 8.10 presents biomedical applications of polymer nanocomposites and their uses.

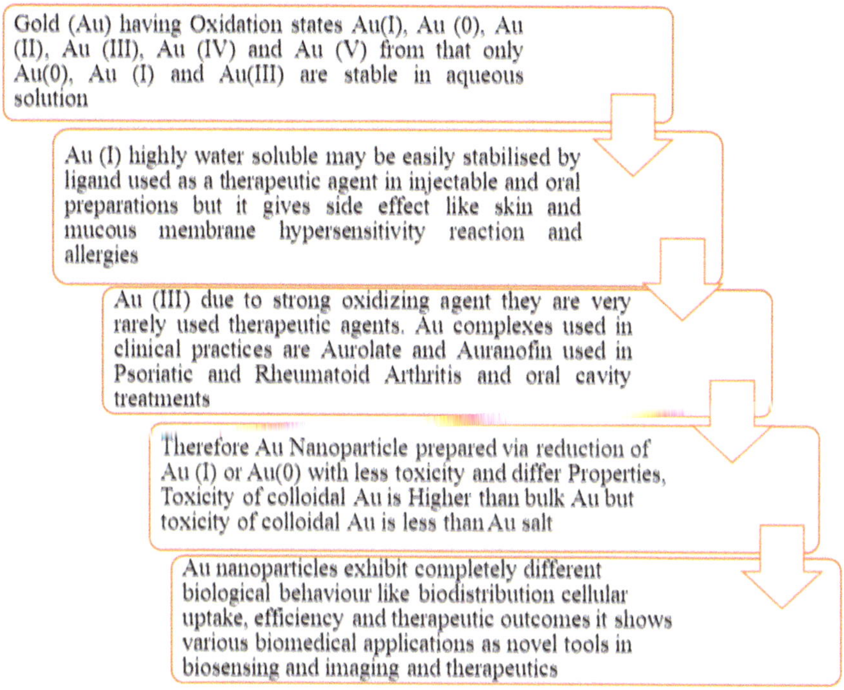

Fig. 8.3 Gold polymer nanocomposite

Table 8.9 Gold polymer nanocomposites biomedical application

S. no	Polymer nanocomposites	Therapeutic applications	Conclusion	References
1	PEG nanoparticle carrier	Drug delivery	Slow release of methotrexate (MTX) for anticancer activity in HeLa cells	Yoon et al. (2016)
2	Loaded simultaneously onto PEGylated AuNPs, Doxorubicin (DOX), and MTX were incorporated into AuNPs via ionic interactions, whereas 6-mercaptopurine was incorporated into AuNPs via disulfide bonds with the polymeric shell		Anticancer activity	Ghorbani and Hamishehkar (2017)
3	PEG-coated AuNPs that could simultaneously deliver bleomycin and DOX		HeLa cancer cell line	Farooq et al. (2018)
4	PEG-coated AuNPs to simultaneously deliver DOX and kaempferol		Efficient colon cancer therapy	Meena et al. (2021)
5	Chitosan (CS)-oligosaccharide-stabilized AuNPs were used in the controlled release of paclitaxel		MDA-MB-231 breast cancer cell lines	Manivasagan et al. (2016)
6	Gelatin-coated AuNPs as a controlled-release system for DOX		MCF-7 breast cancer cells	Suarasan et al. (2016)
7	Poly(acrylamide-acrylonitrile)-PEG-lipoic acid encapsulating AuNRs		Penetrate the tumor cells deeply to achieve the in situ controlled release of the loaded drug (DOX)	Wu et al. (2021)
8	7-(diethylamino)-4-(hydroxymethyl)-2H-chromen-2-one (DEACM); which then immobilized on cyclodextrin (-CDs) bearing AuNRs		Inhibition effect on the proliferation of 4T1 breast cancer cells	Liang et al. (2016)

Table 8.9 (continued)

S. no	Polymer nanocomposites	Therapeutic applications	Conclusion	References
9	AuNPs modified with oligonucleotides	Gene therapy	Gene regulation agents to control protein expression in cells	Rosi et al. (2006)
10	Layer-by-layer biodegradable polymer coatings containing two different polymers, 1-(3-aminopropyl)-4-methylpiperazine end-modified poly(N,N-bis(acryloyl) cystamine-co-3-amino-1-propanol) containing disulfide bonds and 1-(3-aminopropyl)-4 methylpiperazine end-modified poly(1,4-butanediol diacrylate-co-4-amino-1-butanol)		Used in RNA delivery	Bishop et al. (2015)
11	Folic acid (FA)-modified AuNPs mounted on poly(amidoamine) dendrimers		Polymer coating protected mRNA in the cellular environment against RNases	Mbatha et al. (2021)
12	CS-coated AuNPs		Used as DNA carriers	Abrica-González et al. (2019)
13	AuST-polymer hybrid structure was demonstrated as a controllable siRNA delivery system		Used in cancer therapy	Sardo et al. (2017)

(continued)

Table 8.9 (continued)

S. no	Polymer nanocomposites	Therapeutic applications	Conclusion	References
14	Au nanorod	Photothermal (PTT) and photodynamic therapies (PDT)	Used in photothermal (PTT) utilize light energy to destroy abnormal cells	Zeng et al. (2013)
15	Au Coarse Grain		Used in photothermal (PTT) and photodynamic therapies (PDT) utilize light energy to destroy abnormal cells	Chen et al. (2010)
16	Au nanoparticle		Used in photothermal (PTT) utilize light energy to destroy abnormal cells	Richardson et al. (2009)
17	Au nanostars		Used in photothermal (PTT) and photodynamic therapies (PDT) utilize light energy to destroy abnormal cells	Ayala-Orozco et al. (2014)
18	Au nanoflower		Used in photothermal (PTT) utilize light energy to destroy abnormal cells	Huang et al. (2015)

Table 8.9 (continued)

S. no	Polymer nanocomposites	Therapeutic applications	Conclusion	References
19	Synthesized bacteria inspired nanofibers that reduce metal ions. Using these nanofibers, AuNPs were prepared	Tissue engineering	Are on a stable, conductive scaffold for use in tissue engineering	Navaei et al. (2016)
20	AuNP-containing hydrogel could be injected into swine ears, with the presence of AuNPs		Improved longevity of the material; in addition, the irritation level of the material was retained, suggesting that it could be used as a soft tissue filler	Grant et al. (2018)
21	The peptide sequence Arg-Glu-Asp-Val was incorporated in AuNPs, and the AuNPs were then used to produce hydrogel from alginate		Selective adhesion and proliferation of human umbilical vein endothelial cells	Wang et al. (2017)
22	Chitosan/pectin thermosensitive hydrogels containing AuNPs		Enhanced proliferation and growth rates were observed in preosteoblastic mouse cells cultured	Tentor et al. (2017)
23	Photocurable gelatin hydrogel bearing AuNPs		Used as alternatives to BMP-2 in bone tissue engineering	Heo et al. (2014)
24	AuNPs modified with N-acetyl cysteine were loaded into a gelatin hydrogel		Influenced new bone formation	Lee et al. (2018)
25	Synthesized carbon nanofiber/AuNP-conductive scaffolds via 21 of 30 electrospinning and electrospraying method		For bone tissue engineering applications	Nekounam et al. (2020)
26	Prepared a hybrid nanocomposite using Au and polyaniline		Studied as a nerve regeneration model	Saderi et al. (2018)

Table 8.10 Graphene polymer nanocomposites biomedical applications of polymer nanocomposites

S. no	Biomedical application	Polymer nanocomposites	Uses	References
1	Graphene for biosensing	GO-based electrochemical sensor	Label-free detection of cancer cells	Feng et al. (2011)
2		Graphene/PET electrodes using non-contact electric field stimulation	Enhance human neuroblastoma cell-to-cell interactions	Heo et al. (2011)
3		Graphene/PVDF/ZnO	Temperature sensitive	Lee et al. (2015)
4		HPCD-GO/TPP	Hemoglobin sensor	Xu et al. (2010)
5		Graphene/PEI/GOD graphene/GOD/CHI GOD/Pt/graphene/CHI	Glucose	Kang et al. (2009)
6		Graphene/Nafion /GC	Guanine and adenine	Yin et al. (2010)
7		Graphene/CHI	Uric acid	Lian et al. (2012)
8		Graphene/Pt-Ptoxide/ Nafion	Thrombonodulin (associated to progression/metastasis of tumors)	Yang et al. (2012)
9		Graphene/PTCA graphene/PET	Cancer cells	Lee et al. (2015)
10	Tissue engineering with graphene	Synthesized a GO/ $CaCO_3$ construct	Good compatibility with osteoblast cells and increased bone bioactivity in vitro	Kim et al. (2011a)
11		CHI-PVA membrane containing grapheme	Completes skin wound healing after 10 days of application, which was not observed with the control groups without graphene	Lu et al. (2012)
12	Molecular imaging with graphene-based nanomaterials	Graphene and GO	Many molecular imaging techniques, including magnetic resonance imaging, optical, photoacoustic, and radionuclide-based (e.g., positron emission tomography) imaging	Kim et al. (2011b)
13		Nanoscale GO was covalently functionalized with PEG	Conjugating the intrinsic photoluminescence of GO with a B-cell specific anti-CD20 antibody, it enables near-infrared imaging in live lymphoma cells	
14		GO/PEI composite	As a bioimaging tool	
15	Drug/gene delivery with graphene-based nanomaterials	CHI-grafted GO	Delivery of an anti-inflammatory drug, ibuprofen	Rana et al. (2011)
16		GO-PEI nanocarrier	Sequential delivery of therapy siRNA and DOX led to the significant enhanced anticancer efficacy	Zhang et al. (2011b)

GOD glucose oxidase, *HPCD* hydroxypropyl-b-cyclodextrin, *PTCA* perylenetetracarboxylicacid, *PVDF* poly(vinylidenefluoride), *TPP* tetraphenylporphyrin, *ZnO* zincoxide

8.3 Challenges and Future Prospects of Polymer Nanocomposites

Functional polymer nanocomposites (PNCs) have significant potential across various fields, including electronics, biotechnology, and materials science. However, they also face several challenges that need to be addressed for their practical application and commercialization. Some of the challenges and future prospects for functional polymer nanocomposites are synthesis challenges, i.e., uniform dispersion of nanoparticles, surface interaction, scalability of synthesis method, characterization challenges like accuracy in characterizing size, shape, and distribution, interfacial characterization, property evaluation, cancer diagnosis challenges like sensitivity, specificity, imaging resolution, biocompatibility, and long-term safety.

Polymeric nanocomposites have been extensively explored over the past decade due to their growing significance in the healthcare sector. However, large-scale production remains a critical challenge due to the complex fabrication processes involved. Beyond fabrication complexity, selecting the appropriate manufacturing techniques and optimizing operating conditions are essential for achieving high-quality nanocomposites. One of the major hurdles in large-scale production is maintaining particle size at the nanoscale, as aggregation tends to occur over time. Dispersion instability of nanofillers in the polymer matrix is another issue, often influenced by gravity, which is difficult to counteract. While surfactants and stabilizers can help maintain dispersion stability, their addition may compromise the native properties of the nanomaterial. Similarly, coagulation or precipitation methods can alter the structural integrity of nanocomposites, potentially diminishing their exceptional properties. Despite their immense potential in healthcare applications, further research is necessary to assess their toxicity profile before clinical trials can commence. Addressing these challenges will be crucial for the successful commercialization and widespread adoption of polymer nanocomposites in medical and pharmaceutical fields (Fujisawa et al. 2017).

Reliability remains a fundamental challenge in ensuring the dependability of nanotechnology-based systems for cancer diagnosis. Achieving consistent and precise results is essential for their practical application in clinical settings. However, overcoming challenges related to reproducibility, sensitivity, and stability is crucial for advancing their use in real-world applications. Addressing these limitations will pave the way for more effective and reliable cancer diagnosis and therapy, ultimately improving patient outcomes (Wang et al. 2021). Maintaining high sensitivity in the design of nanocomposites for cancer detection remains a significant challenge, as ensuring reliable differentiation between cancerous and healthy cells is crucial. The risk of false-positive and false-negative results can lead to misdiagnoses and inappropriate treatment decisions, potentially compromising patient outcomes. Given the extensive diversity among cancer types, it is essential to develop adaptable nanocomposites capable of detecting distinct tumor attributes and genetic mutations. Tailoring these nanomaterials to recognize specific biomarkers will enhance diagnostic precision, enabling early detection and personalized treatment strategies. Advancements in nanotechnology, combined with interdisciplinary research, are key to overcoming these challenges and improving the reliability of cancer diagnostics (Naeimi et al. 2022). One major drawback is the release of acidic byproducts during degradation, which can lead to localized inflammation and hinder tissue regeneration. Additionally, these materials often exhibit poor cell affinity, limiting their ability to support cell adhesion and proliferation. To overcome these challenges, future research should focus on integrating nano-bioceramics with faster degradation rates to neutralize acidic byproducts and create a more favorable microenvironment. Furthermore, the incorporation of bioactive molecules, such as growth factors or peptides, can enhance cellular interactions and promote tissue integration, ultimately improving the overall performance of polymer composite-based biomaterials.

One of the major challenges in cancer treatment and diagnosis is the development of engineered gene and drug delivery systems that can precisely target diseased cells without affecting healthy tissues, especially within the complex tumor microenvironment. Traditional therapies, such as chemotherapy and radiation, often cause significant damage to normal cells, leading to severe side effects and reduced patient quality of life. To overcome these limitations, researchers are focusing on advanced nanotechnology-based systems, including polymer nanocomposites and functionalized nanoparticles, which offer enhanced selectivity and controlled drug release. These engineered systems are designed to exploit the unique biochemical and physiological characteristics of tumor tissues, such as abnormal pH, hypoxia, and enhanced permeability, ensuring that therapeutic agents are delivered specifically to cancerous cells. By integrating responsive nanomaterials, targeted ligands, and genetic engineering techniques, these novel platforms hold immense potential for improving treatment precision, minimizing side effects, and ultimately enhancing patient outcomes in cancer therapy (Rane et al. 2025).

8.4 Conclusions

Functional polymer nanocomposites (PNCs) have emerged as a transformative class of materials in healthcare, offering diverse applications due to their tunable properties, biocompatibility, and enhanced performance. The nature of polymers and the various factors influencing polymer nanoparticles allow for the development of innovative technical applications. By incorporating multiple nano-sized polymers in a single nanocomposite, unique combinations of chemical compositions, physical properties, and morphological characteristics can be achieved, enabling a wide range of biomedical applications.

In this discussion, we explored the significant role of PNCs in cancer diagnosis and treatment, diabetic diagnosis and therapy, ophthalmic drug delivery systems, tissue engineering, and antimicrobial applications, including antibacterial coatings for food packaging and healthcare industries. Additionally, the potential of gold and graphene-based polymer nanocomposites was highlighted, showcasing their advanced functionalities in targeted drug delivery, imaging, and biosensing.

Despite these promising applications, the advantages and disadvantages of polymer nanoparticles must be carefully evaluated in future studies. Standardized protocols for assessing the toxicity of polymer nanocomposites are essential to ensure their safety and minimize potential health risks. Establishing regulatory guidelines will facilitate their responsible use in biomedical applications while mitigating unforeseen challenges. As research progresses, polymer nanocomposites are expected to play a crucial role in advancing personalized medicine, improving treatment efficacy, and revolutionizing various aspects of healthcare.

References

Abla F, Elsayed Y, Abu Farha N, Obaideen K, Mohamed AA, Lee H, Han C, Egilmez M, Kanan S (2023) Fabrication of high surface area TiO2-MoO3 nanocomposite as a photocatalyst for organic pollutants removal from water bodies. Catalysts 13(2):362

Abrica-González P, Zamora-Justo JA, Sotelo-López A, Vázquez-Martínez GR, Balderas-López JA, Muñoz-Diosdado A, Ibáñez-Hernández M (2019) Gold nanoparticles with chitosan, N-acylated chitosan, and chitosan oligosaccharide as DNA carriers. Nanoscale Res Lett 14(1):258

Adeli E (2015) Irbesartan-loaded electrospun nanofibers-based PVP K90 for the drug dissolution improvement: fabrication, in vitro performance assessment, and in vivo evaluation. J Appl Polym Sci 132(27)

Al Khateb K, Ozhmukhametova EK, Mussin MN, Seilkhanov SK, Rakhypbekov TK, Lau WM, Khutoryanskiy VV (2016) In situ gelling systems based on Pluronic F127/Pluronic F68 formulations for ocular drug delivery. Int J Pharm 502(1–2):70–79

Al-Mokaram AM, Yahya R, Abdi MM, Mahmud HN (2017) The development of non-enzymatic glucose biosensors based on electrochemically prepared polypyrrole–chitosan–titanium dioxide nanocomposite films. Nanomaterials 7(6):129

Andoh V, Ocansey DK, Naveed H, Wang N, Chen L, Chen K, Mao F (2024) The advancing role of nanocomposites in cancer diagnosis and treatment. Int J Nanomedicine 31:6099–6126

Anselmo AC, Mitragotri S (2014) Cell-mediated delivery of nanoparticles: taking advantage of circulatory cells to target nanoparticles. J Control Release 190:531–541

Armstrong G (2015) An introduction to polymer nanocomposites. Eur J Phys 36(6):063001

Arvand M, Mirzaei E, Derakhshan MA, Kharrazi S, Sadroddiny E, Babapour M, Faridi-Majidi R (2015) Fabrication of antibacterial silver nanoparticle-modified chitosan fibers using eucalyptus extract as a reducing agent. J Appl Polym Sci 132(25)

Ayala-Orozco C, Urban C, Knight MW, Urban AS, Neumann O, Bishnoi SW, Mukherjee S, Goodman AM, Charron H, Mitchell T, Shea M (2014) Au nanomatryoshkas as efficient near-infrared photothermal transducers for cancer treatment: benchmarking against nanoshells. ACS Nano 8(6):6372–6381

Bazant P, Kuritka I, Hudecek O, Machovsky M, Mrlik M, Sedlacek T (2014) Microwave-assisted synthesis of ag/ZnO hybrid filler, preparation, and characterization of antibacterial poly (vinyl chloride) composites made from the same. Polym Compos 35(1):19–26

Ben-Nissan B, Macha I, Cazalbou S, Choi AH (2016) Calcium phosphate nanocoatings and nanocomposites, part 2: thin films for slow drug delivery and osteomyelitis. Nanomedicine 11(5):531–544

Bhargav E, Madhuri N, Ramesh K, Manne A, Ravi V (2013) Targeted drug delivery—a review. World J Pharm Pharm Sci 3(1):150–169

Bishop CJ, Tzeng SY, Green JJ (2015) Degradable polymer-coated gold nanoparticles for co-delivery of DNA and siRNA. Acta Biomater 11:393–403
Bose S, Das A, Basu S, Drzal LT (2018) Covalent functionalization of graphene using polyacryloyl chloride and performance of functionalized graphene–epoxy nanocomposite. Polym Compos 39(9):3119–3128
Cao ZT, Chen ZY, Sun CY, Li HJ, Wang HX, Cheng QQ, Zuo ZQ, Wang JL, Liu YZ, Wang YC, Wang J (2016) Overcoming tumor resistance to cisplatin by cationic lipid-assisted prodrug nanoparticles. Biomaterials 94:9–19
Chan MH, Hsieh MR, Liu RS, Wei DH, Hsiao M (2019) Magnetically guided theranostics: optimizing magnetic resonance imaging with sandwich-like kaolinite-based iron/platinum nanoparticles for magnetic fluid hyperthermia and chemotherapy. Chem Mater 32(2):697–708
Chen J, Glaus C, Laforest R, Zhang Q, Yang M, Gidding M, Welch MJ, Xia Y (2010) Gold nanocages as photothermal transducers for cancer treatment. Small 6(7):811–817
Chen J, Peng Y, Zheng Z, Sun P, Wang X (2015) Silver-releasing and antibacterial activities of polyphenol-based polyurethanes. J Appl Polym Sci 132(4)
Chen J, Ding J, Wang Y, Cheng J, Ji S, Zhuang X, Chen X (2017) Sequentially responsive shell-stacked nanoparticles for deep penetration into solid tumors. Adv Mater 29(32):1701170
Cheng R, Meng F, Deng C, Klok HA, Zhong Z (2013) Dual and multi-stimuli responsive polymeric nanoparticles for programmed site-specific drug delivery. Biomaterials 34(14):3647–3657
Cheng F, Gao J, Wang L, Hu X (2015) Composite chitosan/poly (ethylene oxide) electrospun nanofibrous mats as novel wound dressing matrixes for the controlled release of drugs. J Appl Polym Sci 132(24)
Chieng BW, Ibrahim NA, Yunus WM, Hussein MZ, Then YY, Loo YY (2015) Reinforcement of graphene nanoplatelets on plasticized poly (lactic acid) nanocomposites: mechanical, thermal, morphology, and antibacterial properties. J Appl Polym Sci 132(11)
Chomoucka J, Drbohlavova J, Huska D, Adam V, Kizek R, Hubalek J (2010) Magnetic nanoparticles and targeted drug delivering. Pharmacol Res 62(2):144–149
Chou SF, Luo LJ, Lai JY (2017) In vivo pharmacological evaluations of pilocarpine-loaded antioxidant-functionalized biodegradable thermogels in glaucomatous rabbits. Sci Rep 7(1):42344
Dai L, Li X, Duan X, Li M, Niu P, Xu H, Cai K, Yang H (2019) A pH/ROS cascade-responsive charge-reversal nanosystem with self-amplified drug release for synergistic oxidation-chemotherapy. Adv Sci 6(4):1801807
Deirram N, Zhang C, Kermaniyan SS, Johnston AP, Such GK (2019) pH-responsive polymer nanoparticles for drug delivery. Macromol Rapid Commun 40(10):1800917
Demirci S, Celebioglu A, Aytac Z, Uyar T (2014) pH-responsive nanofibers with controlled drug release properties. Polym Chem 5(6):2050–2056
Díaz-Gómez L, Ballarin FM, Abraham GA, Concheiro A, Alvarez-Lorenzo C (2015) Random and aligned PLLA: PRGF electrospun scaffolds for regenerative medicine. J Appl Polym Sci 132(5)
Doan VH, Nguyen VT, Mondal S, Vo TM, Ly CD, Vu DD, Ataklti GY, Park S, Choi J, Oh J (2021) Fluorescence/photoacoustic imaging-guided nanomaterials for highly efficient cancer theragnostic agent. Sci Rep 11(1):15943
Dong W, Ren Y, Bai Z, Yang Y, Wang Z, Zhang C, Chen Q (2018a) Trimetallic AuPtPd nanocomposites platform on graphene: applied to electrochemical detection and breast cancer diagnosis. Talanta 189:79–85
Dong X, Yin W, Zhang X, Zhu S, He X, Yu J, Xie J, Guo Z, Yan L, Liu X, Wang Q (2018b) Intelligent MoS2 nanotheranostic for targeted and enzyme-/pH-/NIR-responsive drug delivery to overcome cancer chemotherapy resistance guided by PET imaging. ACS Appl Mater Interfaces 10(4):4271–4284
Doumbia AS, Vezin H, Ferreira M, Campagne C, Devaux E (2015) Studies of polylactide/zinc oxide nanocomposites: influence of surface treatment on zinc oxide antibacterial activities in textile nanocomposites. J Appl Polym Sci 132(17):41776

Du JZ, Du XJ, Mao CQ, Wang J (2011) Tailor-made dual pH-sensitive polymer–doxorubicin nanoparticles for efficient anticancer drug delivery. J Am Chem Soc 133(44):17560–17563

Du D, Fu HJ, Ren WW, Li XL, Guo LH (2020) PSA targeted dual-modality manganese oxide–mesoporous silica nanoparticles for prostate cancer imaging. Biomed Pharmacother 121:109614

El-Ghannam A, Ricci K, Malkawi A, Jahed K, Vedantham K, Wyan H, Allen LD, Dréau D (2010) A ceramic-based anticancer drug delivery system to treat breast cancer. J Mater Sci Mater Med 21:2701–2710

Farooq MU, Novosad V, Rozhkova EA, Wali H, Ali A, Fateh AA, Neogi PB, Neogi A, Wang Z (2018) Retracted article: gold nanoparticles-enabled efficient dual delivery of anticancer therapeutics to HeLa cells. Sci Rep 8(1):2907

Farwell MD, Gamache RF, Babazada H, Hellmann MD, Harding JJ, Korn R, Mascioni A, Le W, Wilson I, Gordon MS, Wu AM (2022) CD8-targeted PET imaging of tumor-infiltrating T cells in patients with cancer: a phase I first-in-humans study of 89Zr-Df-IAB22M2C, a radiolabeled anti-CD8 minibody. J Nucl Med 63(5):720–726

Fattahpour S, Shamanian M, Tavakoli N, Fathi M, Sheykhi SR, Fattahpour S (2015) Design and optimization of alginate–chitosan–pluronic nanoparticles as a novel meloxicam drug delivery system. J Appl Polym Sci 132(28)

Feldman D (2016) Polymer nanocomposites in medicine. J Macromol Sci A 53(1):55–62

Feng L, Chen Y, Ren J, Qu X (2011) A graphene functionalized electrochemical aptasensor for selective label-free detection of cancer cells. Biomaterials 32(11):2930–2937

Fu X, Hosta-Rigau L, Chandrawati R, Cui J (2018) Multi-stimuli-responsive polymer particles, films, and hydrogels for drug delivery. Chem 4(9):2084–2107

Fujisawa S, Togawa E, Kuroda K (2017) Facile route to transparent, strong, and thermally stable nanocellulose/polymer nanocomposites from an aqueous Pickering emulsion. Biomacromolecules 18(1):266–271

Gaddimath S, Payamalle S, Channabasavana Hundi Puttaningaiah KP, Hur J (2024) Recent advances in pH and redox responsive polymer nanocomposites for cancer therapy. J Compos Sci 8(1):28

Ghorbani M, Hamishehkar H (2017) Redox and pH-responsive gold nanoparticles as a new platform for simultaneous triple anti-cancer drugs targeting. Int J Pharm 520(1–2):126–138

Gopalan AI, Lee KP, Ragupathy D, Lee SH, Lee JW (2009) An electrochemical glucose biosensor exploiting a polyaniline grafted multiwalled carbon nanotube/perfluorosulfonate ionomer–silica nanocomposite. Biomaterials 30(30):5999–6005

Grant SA, Zhu J, Gootee J, Snider CL, Bellrichard M, Grant DA (2018) Gold nanoparticle-collagen gels for soft tissue augmentation. Tissue Eng Part A 24(13–14):1091–1098

Guo X, Wei X, Jing Y, Zhou S (2015) Size changeable nanocarriers with nuclear targeting for effectively overcoming multidrug resistance in cancer therapy. Adv Mater 27(41):6450–6456

Gutierrez AM, Frazar EM, Klaus MV, Paul P, Hilt JZ (2022) Hydrogels and hydrogel nanocomposites: enhancing healthcare through human and environmental treatment. Adv Healthc Mater 11(7):2101820

Habibi N, Kamaly N, Memic A, Shafiee H (2016) Self-assembled peptide-based nanostructures: smart nanomaterials toward targeted drug delivery. Nano Today 11(1):41–60

Han L, Zhang Y, Zhang Y, Shu Y, Chen XW, Wang JH (2017) A magnetic polypyrrole/iron oxide core/gold shell nanocomposite for multimodal imaging and photothermal cancer therapy. Talanta 171:32–38

Hasanzadeh M, Feyziazar M, Solhi E, Mokhtarzadeh A, Soleymani J, Shadjou N, Jouyban A, Mahboob S (2019) Ultrasensitive immunoassay of breast cancer type 1 susceptibility protein (BRCA1) using poly (dopamine-beta cyclodextrine-Cetyl trimethylammonium bromide) doped with silver nanoparticles: a new platform in early stage diagnosis of breast cancer and efficient management. Microchem J 145:778–783

He Y, Wan J, Yang Y, Yuan P, Yang C, Wang Z, Zhang L (2019) Multifunctional polypyrrole-coated mesoporous TiO2 nanocomposites for photothermal, sonodynamic, and chemotherapeutic

treatments and dual-modal ultrasound/photoacoustic imaging of tumors. Adv Healthc Mater 8(9):1801254
Heo C, Yoo J, Lee S, Jo A, Jung S, Yoo H, Lee YH, Suh M (2011) The control of neural cell-to-cell interactions through non-contact electrical field stimulation using graphene electrodes. Biomaterials 32(1):19–27
Heo DN, Ko WK, Bae MS, Lee JB, Lee DW, Byun W, Lee CH, Kim EC, Jung BY, Kwon IK (2014) Enhanced bone regeneration with a gold nanoparticle–hydrogel complex. J Mater Chem B 2(11):1584–1593
Hettiarachchi SD, Cilingir EK, Maklouf H, Seven ES, Paudyal S, Vanni S, Graham RM, Leblanc RM (2021) pH and redox triggered doxorubicin release from covalently linked carbon dots conjugates. Nanoscale 13(10):5507–5518
Hiremath A, Murthy AA, Thipperudrappa S, Bharath KN (2021) Nanoparticles filled polymer nanocomposites: a technological review. Cogent Eng 8(1):1991229
Huang J, Guo M, Ke H, Zong C, Ren B, Liu G, Shen H, Ma Y, Wang X, Zhang H, Deng Z (2015) Rational design and synthesis of γFe2 O3@ au magnetic gold nanoflowers for efficient cancer theranostics. Adv Mater 27(34):5049–5056
Iohara D, Okubo M, Anraku M, Uramatsu S, Shimamoto T, Uekama K, Hirayama F (2017) Hydrophobically modified polymer/α-cyclodextrin thermoresponsive hydrogels for use in ocular drug delivery. Mol Pharm 14(8):2740–2748
Irani M, Mir Mohamad Sadeghi G, Haririan I (2018) Electrospun biocompatible poly (ε-caprolactonediol)-based polyurethane core/shell nanofibrous scaffold for controlled release of temozolomide. Int J Polym Mater Polym Biomater 67(6):361–366
Jawahar N, Meyyanathan S (2012) Polymeric nanoparticles for drug delivery and targeting: a comprehensive review. Int J Health Allied Sci 1(4):217
Jia H, Kerr LL (2015) Kinetics of drug release from drug carrier of polymer/TiO2 nanotubes composite—pH dependent study. J Appl Polym Sci 132(7)
John JV, Uthaman S, Augustine R, Chen H, Park IK, Kim I (2017) pH/redox dual stimuli-responsive sheddable nanodaisies for efficient intracellular tumor-triggered drug delivery. J Mater Chem B 5(25):5027–5036
Jovanović Ž, Radosavljević A, Stojkovska J, Nikolić B, Obradovic B, Kačarević-Popović Z, Mišković-Stanković V (2014) Silver/poly (N-vinyl-2-pyrrolidone) hydrogel nanocomposites obtained by electrochemical synthesis of silver nanoparticles inside the polymer hydrogel aimed for biomedical applications. Polym Compos 35(2):217–226
Kalluru P, Vankayala R, Chiang CS, Hwang KC (2016) Nano-graphene oxide-mediated in vivo fluorescence imaging and bimodal photodynamic and photothermal destruction of tumors. Biomaterials 95:1
Kang X, Wang J, Wu H, Aksay IA, Liu J, Lin Y (2009) Glucose oxidase–graphene–chitosan modified electrode for direct electrochemistry and glucose sensing. Biosens Bioelectron 25(4):901–905
Kango S, Kalia S, Celli A, Njuguna J, Habibi Y, Kumar R (2013) Surface modification of inorganic nanoparticles for development of organic–inorganic nanocomposites—a review. Prog Polym Sci 38(8):1232–1261
Kaur H, Ahuja M, Kumar S, Dilbaghi N (2012) Carboxymethyl tamarind kernel polysaccharide nanoparticles for ophthalmic drug delivery. Int J Biol Macromol 50(3):833–839
Kim JO, Kabanov AV, Bronich TK (2009) Polymer micelles with cross-linked polyanion core for delivery of a cationic drug doxorubicin. J Control Release 138(3):197–204
Kim S, Ku SH, Lim SY, Kim JH, Park CB (2011a) Graphene–biomineral hybrid materials. Adv Mater 23(17):2009–2014
Kim H, Namgung R, Singha K, Oh IK, Kim WJ (2011b) Graphene oxide–polyethylenimine nanoconstruct as a gene delivery vector and bioimaging tool. Bioconjug Chem 22(12):2558–2567
Ko HY, Lee JH, Kang H, Ryu SH, Song IC, Lee DS, Kim S (2010) A nucleolin-targeted multimodal nanoparticle imaging probe for tracking cancer cells using an aptamer. J Nucl Med 51(1):98–105

Kumar S, Bhanjana G, Kumar A, Taneja K, Dilbaghi N, Kim KH (2016) Synthesis and optimization of ceftriaxone-loaded solid lipid nanocarriers. Chem Phys Lipids 200:126–132

Kumar S, Rani R, Dilbaghi N, Tankeshwar K, Kim KH (2017) Carbon nanotubes: a novel material for multifaceted applications in human healthcare. Chem Soc Rev 46(1):158–196

Kumar S, Nehra M, Dilbaghi N, Tankeshwar K, Kim KH (2018) Recent advances and remaining challenges for polymeric nanocomposites in healthcare applications. Prog Polym Sci 80:1–38

Kumari P, Ghosh B, Biswas S (2016) Nanocarriers for cancer-targeted drug delivery. J Drug Target 24(3):179–191

Kumari PV, Rao YS, Akhila S (2019) Role of nanocomposites in drug delivery. GSC Biol Pharm Sci 8(3):94–103

Lee MJ, Lee MH, Shim CK (1995) Inverse targeting of drugs to reticuloendothelial system-rich organs by lipid microemulsion emulsified with poloxamer 338. Int J Pharm 113(2):175–187

Lee JS, Shin KY, Cheong OJ, Kim JH, Jang J (2015) Highly sensitive and multifunctional tactile sensor using free-standing ZnO/PVDF thin film with graphene electrodes for pressure and temperature monitoring. Sci Rep 5(1):7887

Lee D, Heo DN, Nah HR, Lee SJ, Ko WK, Lee JS, Moon HJ, Bang JB, Hwang YS, Reis RL, Kwon IK (2018) Injectable hydrogel composite containing modified gold nanoparticles: implication in bone tissue regeneration. Int J Nanomedicine 13:7019–7031

Li W, Garmendia N, de Larraya UP, Ding Y, Detsch R, Grünewald A, Roether JA, Schubert DW, Boccaccini AR (2014) 45S5 bioactive glass-based scaffolds coated with cellulose nanowhiskers for bone tissue engineering. RSC Adv 4(99):56156–56164

Li H, Li K, Zeng Q, Zeng Y, Chen D, Pang L, Chen X, Zhan Y (2019) Novel vinyl-modified RGD conjugated silica nanoparticles based on photo click chemistry for in vivo prostate cancer targeted fluorescence imaging. RSC Adv 9(44):25318–25325

Li Y, Zhang C, Li G, Deng G, Zhang H, Sun Y, An F (2021) Protease-triggered bioresponsive drug delivery for the targeted theranostics of malignancy. Acta Pharm Sin B 11(8):2220–2242

Lian H, Sun Z, Sun X, Liu B (2012) Graphene doped molecularly imprinted electrochemical sensor for uric acid. Anal Lett 45(18):2717–2727

Liang Y, Gao W, Peng X, Deng X, Sun C, Wu H, He B (2016) Near infrared light responsive hybrid nanoparticles for synergistic therapy. Biomaterials 100:76–90

Liao J, Wei X, Ran B, Peng J, Qu Y, Qian Z (2017) Polymer hybrid magnetic nanocapsules encapsulating IR820 and PTX for external magnetic field-guided tumor targeting and multifunctional theranostics. Nanoscale 9(7):2479–2491

Lim EK, Huh YM, Yang J, Lee K, Suh JS, Haam S (2011) pH-triggered drug-releasing magnetic nanoparticles for cancer therapy guided by molecular imaging by MRI. Adv Mater 23(21):2436

Liu M, Jia Z, Jia D, Zhou C (2014) Recent advance in research on halloysite nanotubes-polymer nanocomposite. Prog Polym Sci 39(8):1498–1525

Loiseau A, Boudon J, Oudot A, Moreau M, Boidot R, Chassagnon R, Mohamed Saïd N, Roux S, Mirjolet C, Millot N (2019) Titanate nanotubes engineered with gold nanoparticles and docetaxel to enhance radiotherapy on xenografted prostate tumors. Cancer 11(12):1962

Lu B, Li T, Zhao H, Li X, Gao C, Zhang S, Xie E (2012) Graphene-based composite materials beneficial to wound healing. Nanoscale 4(9):2978–2982

Luo LJ, Lai JY (2017) Epigallocatechin gallate-loaded gelatin-g-poly (N-isopropylacrylamide) as a new ophthalmic pharmaceutical formulation for topical use in the treatment of dry eye syndrome. Sci Rep 7(1):9380

Mahdavinia GR, Etemadi H, Soleymani F (2015) Magnetic/pH-responsive beads based on caboxymethyl chitosan and κ-carrageenan and controlled drug release. Carbohydr Polym 128:112–121

Manivasagan P, Bharathiraja S, Bui NQ, Lim IG, Oh J (2016) Paclitaxel-loaded chitosan oligosaccharide-stabilized gold nanoparticles as novel agents for drug delivery and photoacoustic imaging of cancer cells. Int J Pharm 511(1):367–379

Manuja A, Kumar B, Chopra M, Bajaj A, Kumar R, Dilbaghi N, Kumar S, Singh S, Riyesh T, Yadav SC (2016) Cytotoxicity and genotoxicity of a trypanocidal drug quinapyramine sulfate loaded-sodium alginate nanoparticles in mammalian cells. Int J Biol Macromol 88:146–155

Mbatha LS, Maiyo F, Daniels A, Singh M (2021) Dendrimer-coated gold nanoparticles for efficient folate-targeted mRNA delivery in vitro. Pharmaceutics 13(6):900

Meena D, Vimala K, Kannan S (2021) Combined delivery of DOX and Kaempferol using PEGylated gold nanoparticles to target colon cancer. J Clust Sci 1-5:173

Merino S, Martín C, Kostarelos K, Prato M, Vázquez E (2015) Nanocomposite hydrogels: 3D polymer–nanoparticle synergies for on-demand drug delivery. ACS Nano 9(5):4686–4697

Mezzenga R, Ruokolainen J (2009) Nanoparticles in the right place. Nat Mater 8(12):926–928

Miao Z, Wang P, Zhong A, Yang M, Xu Q, Hao S, Hu X (2015) Development of a glucose biosensor based on electrodeposited gold nanoparticles–polyvinylpyrrolidone–polyaniline nanocomposites. J Electroanal Chem 756:153–160

Minko T, Dharap SS, Pakunlu RI, Wang Y (2004) Molecular targeting of drug delivery systems to cancer. Curr Drug Targets 5(4):389–406

Mura S, Nicolas J, Couvreur P (2013) Stimuli-responsive nanocarriers for drug delivery. Nat Mater 12(11):991–1003

Nadagouda MN, Varma RS (2007) Biomacromolecules 8:2762–2767

Naeimi R, Najafi R, Molaei P, Amini R, Pecic S (2022) Nanoparticles: the future of effective diagnosis and treatment of colorectal cancer? Eur J Pharmacol 936:175350

Nan X, Zhang X, Liu Y, Zhou M, Chen X, Zhang X (2017) Dual-targeted multifunctional nanoparticles for magnetic resonance imaging guided cancer diagnosis and therapy. ACS Appl Mater Interfaces 9(11):9986–9995

Narwal V, Kumar P, Joon P, Pundir CS (2018) Fabrication of an amperometric sarcosine biosensor based on sarcosine oxidase/chitosan/CuNPs/c-MWCNT/au electrode for detection of prostate cancer. Enzyme Microb Technol 113:44–51

Navaei A, Saini H, Christenson W, Sullivan RT, Ros R, Nikkhah M (2016) Gold nanorod-incorporated gelatin-based conductive hydrogels for engineering cardiac tissue constructs. Acta Biomater 41:133–146

Nejadshafiee V, Naeimi H, Goliaei B, Bigdeli B, Sadighi A, Dehghani S, Lotfabadi A, Hosseini M, Nezamtaheri MS, Amanlou M, Sharifzadeh M (2019) Magnetic bio-metal–organic framework nanocomposites decorated with folic acid conjugated chitosan as a promising biocompatible targeted theranostic system for cancer treatment. Mater Sci Eng C 99:805–815

Nekounam H, Allahyari Z, Gholizadeh S, Mirzaei E, Shokrgozar MA, Faridi-Majidi R (2020) Simple and robust fabrication and characterization of conductive carbonized nanofibers loaded with gold nanoparticles for bone tissue engineering applications. Mater Sci Eng C 117:111226

Ogunjimi AT, Melo SM, Vargas-Rechia CG, Emery FS, Lopez RF (2017) Hydrophilic polymeric nanoparticles prepared from Delonix galactomannan with low cytotoxicity for ocular drug delivery. Carbohydr Polym 157:1065–1075

Panayotidou E, Kroustalli A, Baklavaridis A, Zuburtikudis I, Achilias DS, Deligianni D (2015) Biopolyester-based nanocomposites: structural, thermo-mechanical and biocompatibility characteristics of poly (3-hydroxybutyrate)/montmorillonite clay nanohybrids. J Appl Polym Sci 132(11)

Piñeros M, Mery L, Soerjomataram I, Bray F, Steliarova-Foucher E (2021) Scaling up the surveillance of childhood cancer: a global roadmap. JNCI J Natl Cancer Inst 113(1):9–15

Prabhu S, Ananthanarayanan P, Aziz SK, Rai S, Mutalik S, Sadashiva SR (2017) Enhanced effect of geldanamycin nanocomposite against breast cancer cells growing in vitro and as xenograft with vanquished normal cell toxicity. Toxicol Appl Pharmacol 320:60–72

Pramanik N, Bhattacharyya A, Kundu PP (2015) Spectroscopic analysis and catalytic application of biopolymer capped silver nanoparticle, an effective antimicrobial agent. J Appl Polym Sci 132(8)

Prosperi-Porta G, Kedzior S, Muirhead B, Sheardown H (2016) Phenylboronic-acid-based polymeric micelles for mucoadhesive anterior segment ocular drug delivery. Biomacromolecules 17(4):1449–1457

Qamar SU (2021) Nanocomposites: potential therapeutic agents for the diagnosis and treatment of infectious diseases and cancer. Colloid Interface Sci Commun 43:100463

Qazi TH, Rai R, Boccaccini AR (2014) Tissue engineering of electrically responsive tissues using polyaniline based polymers: a review. Biomaterials 35(33):9068–9086

Qi M, Pan H, Shen H, Xia X, Wu C, Han X, He X, Tong W, Wang X, Wang Q (2020) Nanogel multienzyme mimics synthesized by biocatalytic ATRP and metal coordination for bioresponsive fluorescence imaging. Angew Chem 132(29):11846–11851

Raemdonck K, Braeckmans K, Demeester J, De Smedt SC (2014) Merging the best of both worlds: hybrid lipid-enveloped matrix nanocomposites in drug delivery. Chem Soc Rev 43(1):444–472

Rakhi RB, Nayak P, Xia C, Alshareef HN (2016) Novel amperometric glucose biosensor based on MXene nanocomposite. Sci Rep 6(1):36422

Ran B, Chen C, Liu B, Lan M, Chen H, Zhu Y (2022) A Ti3C2TX/Pt–Pd based amperometric biosensor for sensitive cancer biomarker detection. Electrophoresis 43(20):2033–2043

Rana VK, Choi MC, Kong JY, Kim GY, Kim MJ, Kim SH, Mishra S, Singh RP, Ha CS (2011) Synthesis and drug-delivery behavior of chitosan-functionalized graphene oxide hybrid nanosheets. Macromol Mater Eng 296(2):131–140

Rane B, Patil V, Jain A (2025) Nanocarrier breakthroughs: revolutionizing cancer treatment with nanotechnology. Curr Nanomed 10

Ratemi E (2018) pH-responsive polymers for drug delivery applications. In: Stimuli responsive polymeric nanocarriers for drug delivery applications, vol 1, pp 121–141

Reddy PR, Rao KM, Rao KK, Shchipunov Y, Ha CS (2014) Synthesis of alginate based silver nanocomposite hydrogels for biomedical applications. Macromol Res 22:832–842

Richardson HH, Carlson MT, Tandler PJ, Hernandez P, Govorov AO (2009) Experimental and theoretical studies of light-to-heat conversion and collective heating effects in metal nanoparticle solutions. Nano Lett 9(3):1139–1146

Rosi NL, Giljohann DA, Thaxton CS, Lytton-Jean AKR, Han MS, Mirkin CA (2006) Oligonucleotide-modified gold nanoparticles for intracellular gene regulation. Science 312:1027–1030

Sadeghi M, Sadeghi S, Naghib SM, Garshasbi HR (2023) A comprehensive review on electrochemical nano biosensors for precise detection of blood-based on comarkers in breast cancer. Biosensors 13(4):481

Saderi N, Rajabi M, Akbari B, Firouzi M, Hassannejad Z (2018) Fabrication and characterization of gold nanoparticle-doped electrospun PCL/chitosan nanofibrous scaffolds for nerve tissue engineering. J Mater Sci Mater Med 29:1

Saito G, Swanson JA, Lee KD (2003) Drug delivery strategy utilizing conjugation via reversible disulfide linkages: role and site of cellular reducing activities. Adv Drug Deliv Rev 55(2):199–215

Sardo C, Bassi B, Craparo EF, Scialabba C, Cabrini E, Dacarro G, D'Agostino A, Taglietti A, Giammona G, Pallavicini P et al (2017) Gold nanostar-polymer hybrids for siRNA delivery: polymer design towards colloidal stability and in vitro studies on breast cancer cells. Int J Pharm 519:113–124

Sarode AL, Wang P, Obara S, Worthen DR (2014) Supersaturation, nucleation, and crystal growth during single-and biphasic dissolution of amorphous solid dispersions: polymer effects and implications for oral bioavailability enhancement of poorly water soluble drugs. Eur J Pharm Biopharm 86(3):351–360

Sedghi R, Pezeshkian Z (2015) Fabrication of non-enzymatic glucose sensor based on nanocomposite of MWCNTs-COOH-poly (2-aminothiophenol)-au NPs. Sens Actuators B 219:119–124

Seentrakoon B, Junhasavasdikul B, Chavasiri W (2013) Enhanced UV-protection and antibacterial properties of natural rubber/rutile-TiO2 nanocomposites. Polym Degrad Stab 98(2):566–578

Shakoori Z, Ghanbari H, Omidi Y, Pashaiasl M, Akbarzadeh A, Jomeh Farsangi Z, Rezayat SM, Davaran S (2017) Fluorescent multi-responsive cross-linked P (N-isopropylacrylamide)-based nanocomposites for cisplatin delivery. Drug Dev Ind Pharm 43(8):1283–1291

Shariatinia Z, Nikfar Z, Gholivand K, Abolghasemi TS (2015) Antibacterial activities of novel nanocomposite biofilms of chitosan/phosphoramide/ag NPs. Polym Compos 36(3):454–466

Sharker SM, Lee JE, Kim SH, Jeong JH, In I, Lee H, Park SY (2015) pH triggered in vivo photothermal therapy and fluorescence nanoplatform of cancer based on responsive polymer-indocyanine green integrated reduced graphene oxide. Biomaterials 61:229–238

Shemesh R, Goldman D, Krepker M, Danin-Poleg Y, Kashi Y, Vaxman A, Segal E (2015) LDPE/clay/carvacrol nanocomposites with prolonged antimicrobial activity. J Appl Polym Sci 132(2)

Sherwani MA, Tufail S, Khan AA, Owais M (2015) Dendrimer-PLGA based multifunctional immuno-nanocomposite mediated synchronous and tumor selective delivery of siRNA and cisplatin: potential in treatment of hepatocellular carcinoma. RSC Adv 5(49):39512–39531

Shi J, Yu X, Wang L, Liu Y, Gao J, Zhang J, Ma R, Liu R, Zhang Z (2013) PEGylated fullerene/iron oxide nanocomposites for photodynamic therapy, targeted drug delivery and MR imaging. Biomaterials 34(37):9666–9677

Shi H, Xu M, Zhu J, Li Y, He Z, Zhang Y, Xu Q, Niu Y, Liu Y (2020) Programmed co-delivery of platinum nanodrugs and gemcitabine by a clustered nanocarrier for precision chemotherapy for NSCLC tumors. J Mater Chem B 8(2):332–342

Shi J, Ren Y, Ma J, Luo X, Li J, Wu Y, Gu H, Fu C, Cao Z, Zhang J (2021) Novel CD44-targeting and pH/redox-dual-stimuli responsive core–shell nanoparticles loading triptolide combats breast cancer growth and lung metastasis. J Nanobiotechnol 19:188

Srivastava A, Mishra V, Singh P, Kumar R (2012) Coumarin-based polymer and its silver nanocomposite as advanced antibacterial agents: synthetic path, kinetics of polymerization, and applications. J Appl Polym Sci 126(2):395–407

Stodolak-Zych E, Góra A, Zych Ł, Szumera M (2012) Bioactivity of fibrous polymer based nanocomposites for application in regenerative medicine. In: Materials science forum, vol 714. Trans Tech Publications Ltd, p 229

Suarasan S, Focsan M, Potara M, Soritau O, Florea A, Maniu D, Astilean S (2016) Doxorubicin-incorporated nanotherapeutic delivery system based on gelatin-coated gold nanoparticles: formulation, drug release, and multimodal imaging of cellular internalization. ACS Appl Mater Interfaces 8(35):22900–22913

Sun Y, Zhang Y, Xu K, Xu W, Yu D, Zhu L, Xie H, Cheng R (2015) Thermal, mechanical properties, and low-temperature performance of fibrous nanoclay-reinforced epoxy asphalt composites and their concretes. J Appl Polym Sci 132(12)

Tang X, Liang Y, Feng X, Zhang R, Jin X, Sun L (2015) Co-delivery of docetaxel and Poloxamer 235 by PLGA–TPGS nanoparticles for breast cancer treatment. Mater Sci Eng C 49:348–355

Tentor FR, de Oliveira JH, Scariot DB, Lazarin-Bidoia D, Bonafe EG, Nakamura CV, Venter SA, Monteiro JP, Muniz EC, Martins AF (2017) Scaffolds based on chitosan/pectin thermosensitive hydrogels containing gold nanoparticles. Int J Biol Macromol 102:1186–1194

Thorat ND, Bohara RA, Noor MR, Dhamecha D, Soulimane T, Tofail SA (2017) Effective cancer theranostics with polymer encapsulated superparamagnetic nanoparticles: combined effects of magnetic hyperthermia and controlled drug release. ACS Biomater Sci Eng 3(7):1332–1340

Tomoda K, Yabuki N, Terada H, Makino K (2014) Application of polymeric nanoparticles prepared by an antisolvent diffusion with preferential solvation for iontophoretic transdermal drug delivery. Colloid Polym Sci 292:3195–3203

Tseng YT, Harroun SG, Wu CW, Mao JY, Chang HT, Huang CC (2017) Satellite-like gold nanocomposites for targeted mass spectrometry imaging of tumor tissues. Nanotheranostics 1(2):141

Uhlirova D, Stankova M, Docekalova M, Hosnedlova B, Kepinska M, Ruttkay-Nedecky B, Ruzicka J, Fernandez C, Milnerowicz H, Kizek R (2018) A rapid method for the detection of sarcosine using SPIONs/au/CS/SOX/NPs for prostate cancer sensing. Int J Mol Sci 19(12):3722

Ulbrich K, Hola K, Subr V, Bakandritsos A, Tucek J, Zboril R (2016) Targeted drug delivery with polymers and magnetic nanoparticles: covalent and noncovalent approaches, release control, and clinical studies. Chem Rev 116(9):5338–5431

Vyas SP, Sihorkar V (2000) Endogenous carriers and ligands in non-immunogenic site-specific drug delivery. Adv Drug Deliv Rev 43(2–3):101–164

Wang Z, Liu S, Wu P, Cai C (2009) Detection of glucose based on direct electron transfer reaction of glucose oxidase immobilized on highly ordered polyaniline nanotubes. Anal Chem 81(4):1638–1645

Wang D, Xu Z, Yu H, Chen X, Feng B, Cui Z, Lin B, Yin Q, Zhang Z, Chen C, Wang J (2014) Treatment of metastatic breast cancer by combination of chemotherapy and photothermal ablation using doxorubicin-loaded DNA wrapped gold nanorods. Biomaterials 35(29):8374–8384

Wang W, Fan X, Xu S, Davis JJ, Luo X (2015) Low fouling label-free DNA sensor based on polyethylene glycols decorated with gold nanoparticles for the detection of breast cancer biomarkers. Biosens Bioelectron 71:51–56

Wang YJ, Larsson M, Huang WT, Chiou SH, Nicholls SJ, Chao JI, Liu DM (2016) The use of polymer-based nanoparticles and nanostructured materials in treatment and diagnosis of cardiovascular diseases: recent advances and emerging designs. Prog Polym Sci 57:153–178

Wang B, Wang W, Yu Y, Zhang Y, Zhang J, Yuan Z (2017) The study of angiogenesis stimulated by multivalent peptide ligand-modified alginate. Colloids Surf B Biointerfaces 154:383–390

Wang J, Sui L, Huang J, Miao L, Nie Y, Wang K, Yang Z, Huang Q, Gong X, Nan Y, Ai K (2021) MoS2-based nanocomposites for cancer diagnosis and therapy. Bioact Mater 6(11):4209–4242

Wu Q, Chu M, Shao Y, Wo F, Shi D (2016) Reduced graphene oxide conjugated with CuInS2/ZnS nanocrystals with low toxicity for enhanced photothermal and photodynamic cancer therapies. Carbon 108:21–37

Wu D, Xu S, Zhang X, Li Y, Zhang W, Yan Q, Yang Q, Guo F, Yang G (2021) A near-infrared laser-triggered size-shrinkable nanosystem with in situ drug release for deep tumor penetration. ACS Appl Mater Interfaces 13(14):16036–16047

Xiong D, Zhang X, Peng S, Gu H, Zhang L (2018) Smart pH-sensitive micelles based on redox degradable polymers as DOX/GNPs carriers for controlled drug release and CT imaging. Colloids Surf B Biointerfaces 163:29–40

Xu C, Wang X, Wang J, Hu H, Wan L (2010) Synthesis and photoelectrical properties of β-cyclodextrin functionalized graphene materials with high bio-recognition capability. Chem Phys Lett 498(1–3):162–167

Xu G, Shi H, Ren L, Gou H, Gong D, Gao X, Huang N (2015) Enhancing the anti-colon cancer activity of quercetin by self-assembled micelles. Int J Nanomedicine 16:2051–2063

Xu T, Zhang J, Chi H, Cao F (2016) Multifunctional properties of organic-inorganic hybrid nanocomposites based on chitosan derivatives and layered double hydroxides for ocular drug delivery. Acta Biomater 36:152–163

Yang YC, Dong SW, Shen T, Jian CX, Chang HJ, Li Y, He FT, Zhou JX (2012) A label-free amperometric immunoassay for thrombomodulin using graphene/silver-silver oxide nanoparticles as a immobilization matrix. Anal Lett 45(7):724–734

Yang T, Gao Y, Liu Z, Xu J, Lu L, Yu Y (2017) Three-dimensional gold nanoparticles/prussian blue-poly (3, 4-ethylenedioxythiophene) nanocomposite as novel redox matrix for label-free electrochemical immunoassay of carcinoembryonic antigen. Sens Actuators B 239:76–84

Yin H, Zhou Y, Ma Q, Ai S, Ju P, Zhu L, Lu L (2010) Electrochemical oxidation behavior of guanine and adenine on graphene–Nafion composite film modified glassy carbon electrode and the simultaneous determination. Process Biochem 45(10):1707–1712

Yoon JH, Ganbold EO, Joo SW (2016) PEGylation density-modulated anticancer drug release on gold nanoparticles in live cells. J Ind Eng Chem 33:345–354

Yu K, Yang X, He L, Zheng R, Min J, Su H, Shan S, Jia Q (2020) Facile preparation of pH/reduction dual-stimuli responsive dextran nanogel as environment-sensitive carrier of doxorubicin. Polymer 200:122585

Yuan X, Marcano DC, Shin CS, Hua X, Isenhart LC, Pflugfelder SC, Acharya G (2015) Ocular drug delivery nanowafer with enhanced therapeutic efficacy. ACS Nano 9(2):1749–1758

Yukird J, Wongtangprasert T, Rangkupan R, Chailapakul O, Pisitkun T, Rodthongkum N (2017) Label-free immunosensor based on graphene/polyaniline nanocomposite for neutrophil gelatinase-associated lipocalin detection. Biosens Bioelectron 87:249–255

Zeng J, Goldfeld D, Xia Y (2013) A plasmon-assisted optofluidic system for measuring the photothermal conversion efficiencies of gold nanostructures and controlling an electrical switch. Angew Chem 52(15):4169

Zeng Z, Ouyang J, Sun L, Zeng C, Zeng F, Wu S (2020) Activatable nanocomposite probe for preoperative location and intraoperative navigation for orthotopic hepatic tumor resection via MSOT and aggregation-induced near-IR-I/II fluorescence imaging. Anal Chem 92(13):9257–9264

Zhang Q, Wu S, Zhang L, Lu J, Verproot F, Liu Y, Xing Z, Li J, Song XM (2011a) Fabrication of polymeric ionic liquid/graphene nanocomposite for glucose oxidase immobilization and direct electrochemistry. Biosens Bioelectron 26(5):2632–2637

Zhang L, Lu Z, Zhao Q, Huang J, Shen H, Zhang Z (2011b) Enhanced chemotherapy efficacy by sequential delivery of siRNA and anticancer drugs using PEI-grafted graphene oxide. Small 7(4):460–464

Zhang Z, He Z, Liang R, Ma Y, Huang W, Jiang R, Shi S, Chen H, Li X (2016) Fabrication of a micellar supramolecular hydrogel for ocular drug delivery. Biomacromolecules 17(3):798–807

Zhou J, Li J, Ding X, Liu J, Luo Z, Liu Y, Ran Q, Cai K (2015) Multifunctional Fe2O3@ PPy-PEG nanocomposite for combination cancer therapy with MR imaging. Nanotechnology 26(42):425101

Zhu YJ, Li WJ, Hong ZY, Tang AN, Kong DM (2017) Stable, polyvalent aptamer-conjugated near-infrared fluorescent nanocomposite for high-performance cancer cell-targeted imaging and therapy. J Mater Chem B 5(46):9229–9237

Chapter 9
Green Synthesis Approaches for Polymer Nanocomposites

Swathi Yempallya, Kailas Karandeb, Ajit Kulkarnic, and Veeranna Yempallyd

Abstract The sustainable synthesis of polymer nanocomposites (PNCs) has gained considerable recognition for its emphasis on environmentally friendly materials, energy-efficient methods, and the use of renewable resources. Innovations in green chemistry have paved the way for creating PNCs using natural polymers such as cellulose, chitosan, starch, alginate, and collagen, alongside biobased nanofillers like montmorillonite, silica nanoparticles, cellulose nanofibers, and lignin nanoparticles. These renewable materials not only minimize environmental impact but also improve the biodegradability of the resulting composites.

Recent developments have highlighted the adoption of green methodologies that enhance the performance and efficiency of PNCs. Techniques such as microwave-assisted synthesis and ultrasonic processing have been optimized to reduce energy consumption and production time while ensuring better dispersion of natural nanofillers within polymer matrices. Mechanochemical approaches, which avoid the use of solvents, have seen notable advancements, and ionic liquid-based synthesis is now more adept at controlling nanoparticle size and uniformity. Furthermore, supercritical fluid processing has emerged as an effective way to improve material properties without relying on harmful solvents.

The use of plant extracts, microbial processes, and enzymatic catalysts in the synthesis of nanofillers is gaining momentum due to their renewable and non-toxic attributes. In situ polymerization techniques, which enable direct formation around

S. Yempallya
Center for Advanced Materials, Qatar University, Doha, Qatar

K. Karandeb (✉)
Department of Pharmaceutics, Arvind Gavali College of Pharmacy, Satara, Maharashtra, India

A. Kulkarnic
School of Pharmacy, MIT Vishwaprayag University, Solapur, Maharashtra, India
e-mail: ajit.kulkarni@mitvpu.ac.in

V. Yempallyd
School of Computing, MIT Vishwaprayag University, Solapur, Maharashtra, India
e-mail: veeranna.yempally@mitvpu.ac.in

S. N. Kothawade et al. (eds.), *Polymer Nanocomposites*,
https://doi.org/10.1007/978-3-032-00365-2_9

nanofillers, ensure even distribution and eliminate the need for additional processing steps. Moreover, integrating renewable energy sources, such as solar energy, to drive synthesis reactions is becoming a promising strategy to reduce overall energy demands. This chapter delves into these advanced green techniques, offering a comprehensive overview of their applications and benefits.

Keywords Green synthesis · Polymer nanocomposites (PNCs) · Renewable nanofillers · Biodegradable polymers · Microwave-assisted synthesis · Ionic liquids

9.1 Introduction

Green synthesis represents a forward-thinking, environmentally friendly approach to material development, emphasizing sustainability and minimizing ecological impact (Anastas and Eghbali 2010; Horvath and Anastas 2007). As a cornerstone of modern material science, it advocates the use of non-toxic methods and renewable resources, aligning with global efforts to reduce pollution, conserve energy, and foster sustainable development (Singh et al. 2018).

In polymer nanocomposites (PNCs), green synthesis has become pivotal, combining environmental sustainability with advanced material innovation. PNCs, which consist of a polymer matrix and nanoscale fillers working together to enhance material properties, benefit significantly from green synthesis principles (Caseri 2000; Guo et al. 2018a; Kango et al. 2013). By integrating these principles, the industry not only reduces environmental risks but also innovatively addresses complex material requirements (Hussain 2020; Mittal 2011). The alignment of green chemistry with the United Nations Sustainable Development Goals (SDGs) has fueled the demand for sustainable PNCs in areas like environmental remediation, climate regulation, and biomedicine. However, the mere use of biopolymers or biodegradable polymers does not automatically equate to a truly green synthesis process. It is essential to critically assess reagents used in processes like crosslinking and polymerization to ensure their environmental compatibility. Merely categorizing a synthesis as "green" based on organically sourced polymers is insufficient, highlighting a significant knowledge gap in the field.

To close this gap, it is crucial to educate the next generation of entrepreneurs in green chemistry principles, enabling them to design sustainable PNCs using advanced tools like artificial intelligence (AI) and machine learning. Techniques such as regression analysis, neural networks, and predictive modeling can accelerate the discovery of sustainable PNCs, unlocking new possibilities for biopolymers and driving data-driven innovations. By leveraging AI, the industry can enhance recovery, recycling, and waste reduction, promoting greener synthesis methods and contributing to a circular economy.

Achieving a circular economy necessitates innovative green synthesis methods. One emerging strategy involves utilizing a mix of biomass and e-waste as feedstocks for sustainable PNC production. For instance, tea leaf extracts, rich in organic functional groups, can act as effective metal-chelating agents. These extracts facilitate the recovery of noble metals from e-waste by forming complexes with organic compounds like carboxylates and phenolics, contributing to the sustainable production of PNCs. Another promising approach is the use of bio-enzymes and biomass-derived materials as alternatives to hazardous initiators and nanofillers in PNC synthesis. This method replaces harmful chemicals with readily available bioresources, which often possess intrinsic therapeutic properties. Moreover, incorporating biobased raw materials for medical applications in PNC manufacturing allows for recycling with minimal environmental and economic consequences.

To advance green chemistry in PNC synthesis, several key actions are essential. These include embracing sustainable feedstocks, replacing harmful reagents with biobased alternatives, integrating AI for material innovation, and prioritizing recycling and waste reduction. Such efforts not only ensure adherence to green principles but also pave the way for a more sustainable and environmentally conscious future in material science. Figure 9.1 summarizes the key principles of green chemistry and forms the basis for recognition of the "Greenness" of PNC synthesis. The establishment of strict regulations and enforcement measures is crucial to hold industries accountable for adhering to green chemistry principles during the production of polymer nanocomposites (PNCs). Violations should be met with appropriate legal consequences to ensure compliance. Recognizing and rewarding companies that adopt green manufacturing techniques for sustainable PNCs can stimulate

Fig. 9.1 Twelve principles of Green Chemistry as coined by Anastas, P. T. and Warner, J. C (Anastas and Eghbali 2010)

innovation. Such incentives encourage industries to prioritize environmentally friendly practices.

The integration of renewable nanofillers into polymer matrices has gained significant attention due to their ability to create sustainable materials. These nanofillers, derived from natural resources, not only improve the mechanical, thermal, and functional properties of polymers but also reduce environmental impact. By promoting biodegradability and resource efficiency, renewable polymers align with green chemistry principles. This chapter delves into the advancements in renewable nanofillers and their applications in industries like packaging, automotive, biomedical, and environmental remediation (Mohan and Kanny 2020). The pursuit of sustainable materials has driven the advancement of green synthesis techniques for PNCs, emphasizing renewable resources and eco-friendly methods. These techniques not only minimize environmental harm but also enhance the properties of the resulting materials, making them suitable for applications in packaging, biomedical devices, and automotive components (Idumah 2021). Figure 9.2 summarizes the key applications of biorenewable resource-derived polymer nanocomposites.

Plant extract-mediated synthesis is an eco-friendly and sustainable method for producing nanomaterials. Utilizing natural reducing agents found in plant extracts, this approach eliminates harmful chemicals while adhering to green chemistry principles. Recent innovations focus on leveraging plant extracts rich in bioactive compounds to develop nanomaterials for applications in biomedicine, catalysis, and

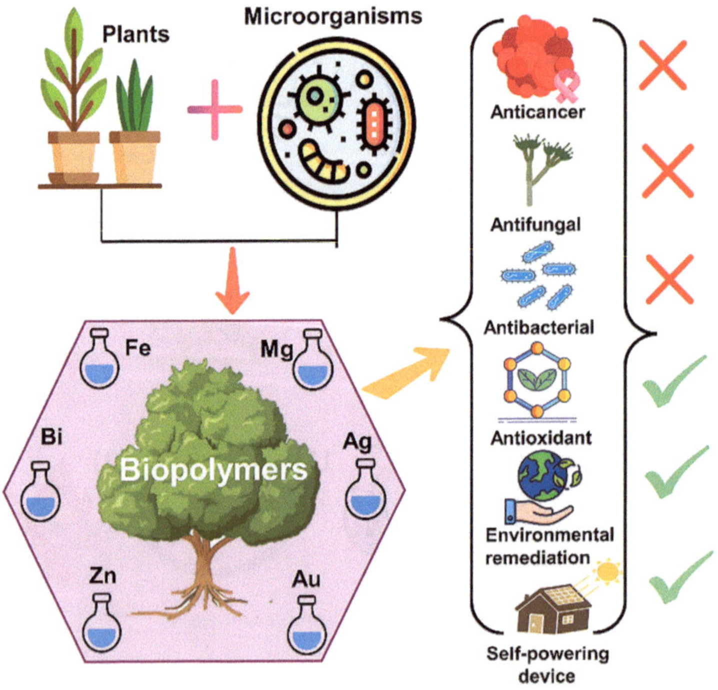

Fig. 9.2 Applications of biorenewable resource-derived polymer nanocomposites

environmental protection (Jha and Jha 2021; Khan et al. 2022). Microbial and enzymatic synthesis represents a sustainable alternative to traditional chemical methods, relying on microorganisms and enzymes as natural catalysts. This method minimizes energy consumption and eliminates hazardous reagents, providing precise control over nanoparticle size, shape, and distribution. Advances in this field demonstrate scalable, environmentally friendly production of metal and metal oxide nanoparticles, with applications spanning biomedicine, catalysis, and environmental remediation (Verma et al. 2019; Edison and Pradeep 2020; Shukla and Iravani 2018).

In situ polymerization is a powerful technique for creating PNCs by polymerizing monomers in the presence of nanofillers, resulting in enhanced interactions between the filler and the matrix. This method ensures optimal nanofiller dispersion, improved interface adhesion, and precise control over the material's properties. Recent studies have explored optimizing temperature and solvent parameters to enhance mechanical and thermal performance. Industries such as packaging, automotive, and biomedicine widely employ in situ polymerization to develop tailored, high-performance materials (Kumar and Kumar 2022; Fu et al. 2019; Ali et al. 2024; Zhao et al. 2021; Lv et al. 2019). The reaping benefits of implementing green synthesis methods for the fabrication of polymer nanocomposites and challenges associated with the commercial-scale production of PNCs will be discussed briefly in this chapter. Moreover, the latest trends in the production of sustainable polymer nanocomposites adhering to principles of green chemistry from the past 5 years will also be discussed. This chapter gives a general overview of the latest advancements in the field of green synthesis of PNCs and specific examples of green synthesis methods wherever applicable. However, given the plethora of research articles in green synthesis of PNCs, it is impossible to summarize all the relevant information in one book chapter. Therefore, this chapter focuses on key techniques utilized in the fabrication of PNCs following green chemistry principles and socioeconomic implications of green synthesis of PNCs.

9.2 Renewable Nanofillers and Polymer Matrices: A Discussion on Biobased Nanofillers and Natural Polymers

9.2.1 Introduction

The increasing need for sustainable materials has generated considerable interest in renewable nanofillers and polymer matrices (Saba et al. 2019). Traditional materials, often sourced from non-renewable resources, pose considerable environmental issues, such as trash generation and carbon emissions. As industry and academics shift toward sustainable alternatives, renewable nanofillers and natural polymer matrices have surfaced as viable options (Khan et al. 2024; Naeem et al. 2024). These materials fulfill the need for sustainability while providing enhanced performance for mechanical strength, thermal stability, and biocompatibility (Arif et al. 2022). Biobased nanofillers, including montmorillonite and cellulose nanofibers, signify an innovative advancement in materials research. These nanofillers, sourced from natural materials, have remarkable capabilities attributed to their nanoscale size and extensive surface area (Ramakrishnan and Sumitha 2024). Montmorillonite, a stratified silicate, has been extensively investigated for its mechanical and barrier characteristics, while cellulose nanofibers, derived from plant cell walls, provide exceptional strength and flexibility (Baruah and Mahanta 2022). Investigation of polymer-clay nanocomposites has persisted for many decades. This unique combination imparts enhanced qualities to the resulting material compared to the individual components alone. The incorporation of nanoparticles into a polymer matrix creates a significant interfacial area between the two phases. This interface enables strong interactions between nanoparticles and polymer chains, leading to substantial improvements in mechanical, thermal, barrier, and flammability properties (Xie et al. 2024). The effectiveness of a polymer nanocomposite depends on several critical factors, including the type of nanoparticle (e.g., clay, carbon nanotubes, and graphene), the size and morphology of the nanoparticles, surface chemistry and functionalization, the type of polymer and its molecular weight/structure, and compatibility with the nanoparticles. The distribution of nanofiller inside the polymer may be intercalated. Nanocomposites consist of polymer chains intercalated between the layers of layered nanoparticles, such as clay, leading to a distinct multilayered architecture. Exfoliated composites include nanoparticles that are entirely and evenly distributed inside the polymer matrix, creating a singular, continuous phase. Micro composites arise when the polymer fails to intercalate between the silicate layers, leading to a composite with discrete phases. The dispersion type shown in Fig. 9.3 has the greatest potential for property enhancement due to the ideal interfacial area between the nanoparticles and the polymer (de Oliveira and Beatrice 2018). This section focuses on nanofillers, the essential component of polymer nanocomposites.

This sub-chapter examines the production, characteristics, and uses of renewable nanofillers and polymer matrices. It underscores their capacity to replace traditional materials, accentuating their contribution to the establishment of a circular economy. The conversation also examines the problems related to their adoption, such as scalability and cost, while investigating novel solutions and future possibilities. By enhancing the comprehension and utilization of these materials, researchers may aid in the development of sustainable technologies that reconcile environmental stewardship with superior performance.

9.2.2 Biobased Nanofillers

9.2.2.1 Clay-Based Nanofillers

Common varieties include montmorillonite, bentonite, kaolinite, hectorite, and halloysite (Table 9.1). These materials are categorized as "layered silicates" and are often used in polymer nanocomposites to improve material characteristics owing to

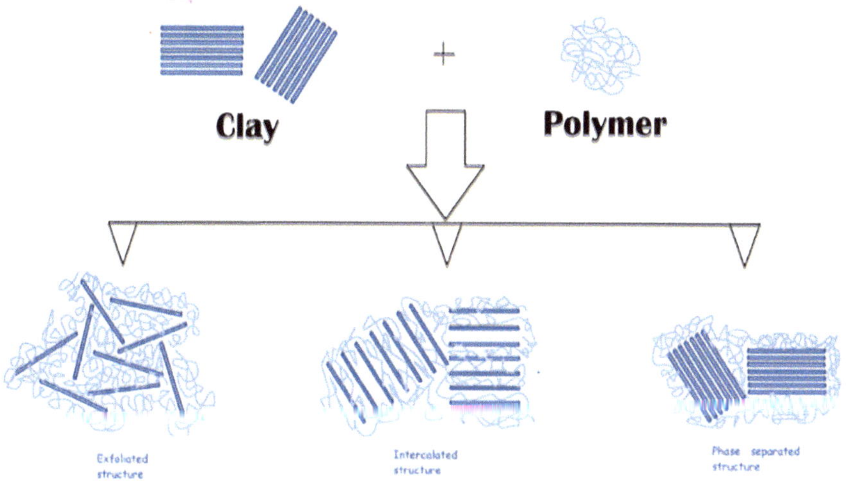

Fig. 9.3 Probable structures of polymer nanocomposites from layered nanoclays (de Oliveira and Beatrice 2018)

Table 9.1 Classification of nanoclay (Idumah et al. 2022)

Species of clay	Nanoclay architecture	Group of nanoclay
Fraipontite etc.	1:1	Rectorite, halloysite, kaolinite
Vermiculite etc.	2:1	Smectite, chrysotile, vermiculite, pyrophyllite, brittle mica, talc, mica
Cookeite, Amesite	2:1:1	Chlorite

their layered structure, which may be intercalated or exfoliated inside the polymer matrix. This subsection discusses Montmorillonite, one of the oldest and most extensively studied clay-based nanofibers in wastewater treatment and medicine (Al Kausor et al. 2022).

Clays are naturally occurring, ecologically sustainable materials distinguished by a high specific surface area and exceptional cation exchange capacity. Montmorillonite is an easily accessible clay. Montmorillonite is a clay mineral classified under the smectite group. It is characterized by its distinctive layered architecture, including two silica tetrahedral sheets around a center alumina octahedral sheet, resulting in a 2:1 layer structure (Wypych and de Freitas 2022). This configuration allows the penetration of water and other polar molecules into the layers, facilitating their expansion. This characteristic is advantageous in several applications since it allows for the modification of the material's qualities (Fig. 9.4).

Montmorillonite has a substantial cation exchange capacity (CEC), making it particularly efficient at adsorbing various ions and compounds. This feature is used in pollution control operations, whereby montmorillonite may adsorb and retain contaminants, including heavy metals and organic chemicals (Bhattacharyya and Gupta 2008). The adsorptive properties are affected by its extensive surface area,

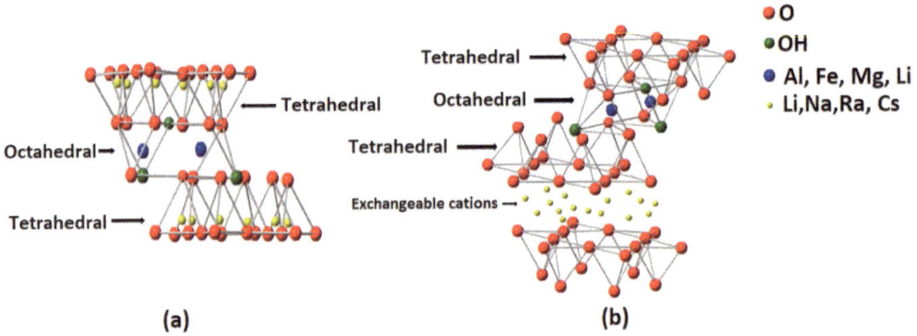

Fig. 9.4 Structures of mineral clay: type 1:1 (**a**) and type 2:1 (**b**) (Perera et al. 2024)

porous structure, and the existence of certain acidic spots. These characteristics make montmorillonite very successful in wastewater or effluent treatment and soil remediation applications as well (Sarkar et al. 2024). The inherent characteristics of the clay may be augmented by several alteration processes. Intercalation entails the insertion of molecules between layers to augment space, hence enhancing adsorptive capacity (Barakan and Aghazadeh 2021). Pillaring establishes sturdy columns between layers, resulting in a porous structure that withstands collapse during heating or hydration. This technique generates materials exhibiting elevated thermal and chemical stability, which are advantageous in catalysis and filtration (Cool et al. 2004). Montmorillonite may be altered using organic chemicals, including surfactants, to render its surface hydrophobic (Shen and Gao 2019). This enhances the adsorption of non-polar molecules, hence broadening their use in the elimination of organic contaminants from water. Furthermore, acid or heat treatments may modify its structure, augmenting its surface area and improving its capacity to engage with contaminants. Si_4^+4+ in the silica tetrahedron may be substituted by Fe^{3+} and Al^{3+}, whereas Al^{3+} in the alumina octahedron may be replaced with low-valent cation such as Mg^{2+} and Fe^{2+}, resulting in a negatively charged montmorillonite structure. Montmorillonite cells absorb exchangeable cations to attain electrical charge balance. They are situated between the unit layers, forming interlayer domains with beneficial ion-exchange capabilities and adsorption characteristics. This distinctive characteristic was analyzed to extract the antibiotic tetracycline present in wastewater treatment, therefore addressing environmental concerns (Shi et al. 2024). Figure 9.5 illustrates a schematic depiction of charge distribution.

Multiple clay nanofillers, such as paragonite, clintonite, hectorite, halloysite, pyrophyllite, bentonite, and their organically modified forms, may be used in the production of polymer nanocomposites, with potential applications such as rheological modifiers, gas absorbents, and drug delivery systems (Kataki et al. 2022). The attributes of alternative fillers, including high specific surface area, low density, and barrier properties, are considered for reinforcement with the polymer.

Applications in Drug Delivery and Wastewater Treatment Montmorillonite, characterized by its adsorption capabilities, moisture retention, swelling, and rheological characteristics, has been thoroughly investigated in the medical domain

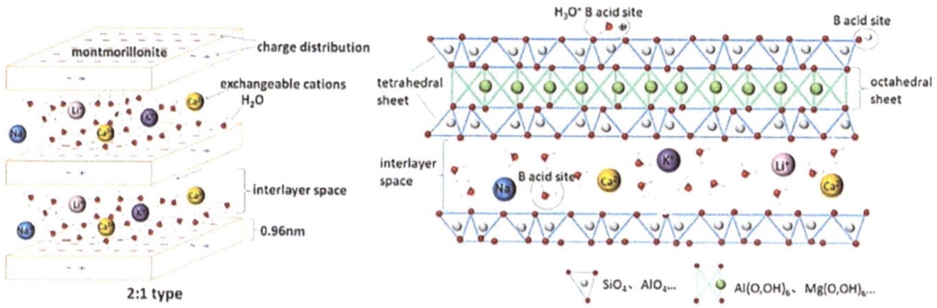

Fig. 9.5 Representation diagram of montmorillonite layered structure (Shi et al. 2024)

Table 9.2 Summary of application of clay nanofillers in drug delivery

Type of clay	Application	Description	Reference
Magnesium-incorporated bentonite	Adsorb fluoride	Fluoride greater than 1.5 mg L^{-1} in potable water results in health problems, like dental and skeletal fluorosis	Thakre et al. (2010)
Montmorillonite	Heavy metal removal. Can be used as a healing agent	Adsorbs toxic heavy metals like Pb, As, Mn, Cd, Cr and Ni from aqueous solutions	Bhattacharyya and Gupta (2008)
	Reduce toxicity	Adsorption of aflatoxin	Shi et al. (2005)
Sodium montmorillonite	Anti-viral	Adsorbs bovine rotaviruses and coronaviruses and reduces their toxicity	Clark et al. (1998)
Dioctahedral Diosmectite (SMECTA®), a commercially available clay mineral group	Treatment of acute infectious diarrhea	Adsorb light	Khediri et al. (2011)
MMT loaded with capecitabine	Anticancer	The nanocomposite hydrogel for drug administration demonstrated sufficient effectiveness against 4 T1 cancer cell line in both in vivo and in vitro settings	Taleblou et al. (2020)
Intercalation of timolol maleate into the MMT, MMT-alginate nanocomposite, MMT-enveloped zeolitic imidazolate framework	Active ingredients: Timolol maleate, irinotecan, brimonidine, imidazolate	Sustained/controlled drug delivery/enhanced mucosal adherence systems	Hassen et al. (2023)

(Park et al. 2016). The ion-exchange characteristics of MMT and its derivatives are essential for wastewater management. The principal applications of MMT in medical administration and water purification are enumerated in Tables 9.2 and 9.3, respectively.

Table 9.3 Summary of applications wastewater treatment (Elella et al. 2021)

Aspect	Details
Nanomaterials for water purification	Nanomaterials have been extensively used for water filtration.
Sorbents used for dye adsorption	Clays, zeolites, metal oxides, and activated carbons are used as sorbents for dye adsorption in water.
Clay minerals (e.g., MMT)	Clay minerals such as montmorillonite (MMT) are used for the elimination of organic colors and colors from water owing to their substantial surface area and cation exchange capacity.
Antibacterial properties of clays	Clays such as MMT show antibacterial action against pathogenic bacteria like *S. Aureus* and *E. Coli.*
Montmorillonite nanoclay	MMT has a distinctive surface architecture characterized by charged layers, making it an efficient adsorbent for the elimination of contaminants in aqueous solution.
Three-dimensional cross-linked composites	These composites exhibit efficacy in dye adsorption owing to their stability, substantial adsorption capacity, and robust mechanical strength.
XG and antibacterial polymers (PVI)	The hybridization of thermally stable synthetic polymers, like poly(vinylimidazole) (PVI), with XG improves the adsorption effectiveness of the composite. XG has hydrophilic groups that enhance its efficacy in dye removal.

9.2.2.2 Organic Nanofillers

Organic nanofillers, sourced from renewable resources such as flora, fauna, or microbes, enhance the performance of biopolymers owing to their biodegradability, biocompatibility, availability, and mechanical attributes. Examples include cellulose, chitin, lignin, starch, protein-based nanomaterials (silk fibroin and collagen), bacterial nanocellulose, clay minerals (montmorillonite), and lipid-based nanomaterials (liposomes) (Shen et al. 2020) (Fig. 9.6).

Montmorillonite, characterized by its adsorption capabilities, moisture retention, swelling, and rheological characteristics, has been thoroughly investigated in the medical domain (Alavi 2019). The ion-exchange characteristics of MMT and its derivatives are essential for wastewater management. The principal applications of MMT in medical administration and water purification are enumerated in Tables 9.4 and 9.5, respectively.

9.2.2.3 Chitosan and Chitin

These originate from crustacean exoskeletons and fungal byproducts. Its characteristics include antibacterial activity, biofilm formation capability, and superior biodegradability (Ikram et al. 2021). Nanochitin and nanochitosan, originating from the biopolymer chitin, have exceptional physicochemical and biological characteristics. These nanoparticles are produced at various sizes, leading to diverse applications from biomedical to environmental solutions. Chitin is transformed into nanochitin and nanochitosan by controlled top-down and bottom-up methodologies (Lee et al.

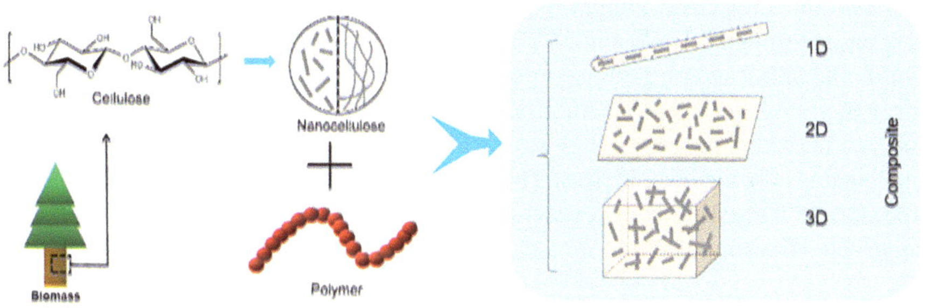

Fig. 9.6 Polymer composites from nanocellulose and polymers

Table 9.4 Practical application of nanocellulose (e.g., polymer composites incorporating nanocellulose, along with the observed improvements) (Eid 2023)

Polymer matrix	Effect of nanocellulose
Polyurethane	Enhanced tensile strength by 53%, improved shape recovery
Poly(lactic acid) (PLA)	Improved nanofiller distribution with the use of surfactants
Polyurethane-based SMP	Increased material stiffness with temperature, enhanced strength
Polyvinyl alcohol	Creation of non-toxic electroactive hydrogels with superior displacement in electric fields
Water-triggered SMP	Improved shape recovery, young's modulus, and elongation at break
Core–shell nanocellulose/ polyaniline	Improved film-forming characteristics, exceptional electrochromic properties
CNC in UV-curable acrylic coatings	Enhanced mechanical properties

Table 9.5 Methods of nanochitin production (Lee et al. 2023)

Method	Reactants	Average size (nm)		Crystallinity
		Width	Length	
Inorganic acid hydrolysis (HCl)	Alfa chitin, HCL crab shells, shrimp shells	6–50	100–500	Highly crystalline
TEMPO-mediated oxidation	α-Chitin, (2,2,6,6-tetramethylpiperidin-1-yl) oxyl radical (TEMPO)/NaClO/NaBr	7–20	5.-500	Crystalline
Deep eutectic solvent (DES)	α-Chitin, choline chloride + acids	29–83	165–844	Moderately crystalline

2023). The top-down approach preserves native crystal structures, while the bottom-up technique involves disintegration and reassembly.

Applications in Medicine Nanochitosan functions as an efficient drug carrier owing to its bioadhesive and controlled-release characteristics (Yee Kuen and Masarudin 2022). Chitosan nanoparticles improve oral bioavailability (Guadarrama-Escobar et al. 2023) and facilitate targeted drug delivery (Zheng et al.

2016). Low molecular weight (LMW) chitosans (3–50 kDa) activate immune responses through NF-κB and MAPK signaling pathways (Yan et al. 2021). Both chitin and chitosan compromise microbial membranes, inhibit DNA/RNA synthesis, and exhibit broad-spectrum efficacy (Olza et al. 2023). Additionally, they are utilized in wound dressings (Lee et al. 2023), surgical sutures (Basem et al. 2024), and coatings for medical devices. The mechanical properties of nanochitin render it appropriate for scaffolds in regenerative medicine. Chitosan-based hydrogels promote cell adhesion, proliferation, and differentiation.

Applications in Wastewater Treatment The extensive surface area and functional groups of chitosan facilitate the efficient adsorption of heavy metals (e.g., Pb and Cu) and dyes. Chitosan-coated membranes inhibit biofouling and microbiological contamination in water treatment systems (Basem et al. 2024). Chitosan-derived products serve as environmentally sustainable flocculants for the elimination of suspended particles and organic contaminants (Zhu 2023).

9.2.2.4 Carbon-Based Nanofillers

Graphene oxide (GO), graphene nanoplates (GNPs), carbon nanofillers (CNFs), carbon nanotubes (CNTs), and carbon black (CB) exemplify carbon-based nanofillers extensively used in the production of multifunctional polymer composites (Zheng et al. 2023). Graphene and carbon nanotubes (CNTs) serve as two prime examples of carbon-based nanofillers. Graphene Oxide (GO) enhances tensile strength and Young's modulus while maintaining flexibility. Carbon nanotubes improve stiffness, electrical conductivity, and thermal resistance. An example instance of CBNs is shown below. Carbon nanotubes (CNTs), being one-dimensional cylindrical structures derived from the rolling of single-layered carbon atom sheets, are an ideal selection for polymer composites. Studies demonstrate the advantages of carbon nanotubes (CNTs) as hybrid fillers in polymer composites, leading to highly linked structures. A study produced MWCNT/epoxy composites with very high filler aspect ratios (10–68 wt%) using hot-press infiltration. This method involves injecting epoxy resin into multiwalled carbon nanotubes (MWCNTs) contained inside a semipermeable membrane, resulting in composites with both aligned and randomly oriented carbon nanotubes (CNTs). Aligned CNT composites at 68 wt% exhibited a notable Young's modulus of 36 GPa and enhanced electrical conductivity (Ali et al. 2024). Table 9.6 delineates the methodologies for the synthesis of carbon nanotubes (CNTs), including environmentally sustainable approaches, alongside their associated advantages and drawbacks.

9.2.2.5 Inorganic Nanofillers

Inorganic nanofillers are materials with at least one dimension in the 1–100 nm range that consist of inorganic constituents, including metal oxides, metals, or ceramics. They are included in polymer matrices to improve the diverse

Table 9.6 Preparation methods of CNTs (Ali et al. 2024)

Method	Description	Green aspects	Limitations
Arc discharge in liquid media	Variant of arc discharge where water or ionic liquids replace inert gases to lower energy consumption and reduce carbon emissions	Reduces inert gas usage; environmentally friendlier compared to traditional arc discharge	Still energy-intensive; requires purification of byproducts
Solar thermal synthesis	Use concentrated solar energy to replace conventional heat sources, reducing reliance on fossil fuels	Eliminates fossil fuel use; reduces carbon emissions	Limited scalability; requires consistent solar energy availability
Hydrothermal synthesis	Involves the use of water-based reactions under high pressure and temperature for CNT growth	Using water as a solvent minimizes hazardous chemicals	Requires high energy for pressurization; scalability issues
Bacterial synthesis	Employ microorganisms to produce CNTs through biological processes, reducing energy and chemical requirements	Low-temperature process; uses renewable biological systems; minimal chemical waste	Still in experimental stages, low yield, limited scalability
Green chemical vapor deposition (CVD)	Uses renewable hydrocarbons like methane from biogas or captured CO_2 as carbon sources for CNT growth, alongside efficient catalyst recycling methods	Use renewable feedstocks; reduces greenhouse gas emissions; improves catalyst lifecycle	Requires advanced reactor designs; it still involves high temperatures
Plasma-enhanced CVD with green precursors	Utilizes plasma to enhance CNT growth, combined with eco-friendly precursor gases like CO_2 or bioderived methane	Lower environmental impact with green precursors; reduced process emissions	Expensive equipment; scalability challenges
Layer-by-layer (LBL) assembly with green solvents	Use biodegradable or low-toxicity solvents in the LBL process to reduce environmental hazards	Reduces solvent toxicity; suitable for thin-film applications	Limited scalability; requires precise solvent control
Swelling method with green solvents	Employs non-toxic or biodegradable solvents (e.g., ethanol, water-based systems) for swelling polymers and dispersing CNTs	Environmentally safer solvents, suitable for sensitive polymers	Limited to specific solvent-polymer systems; slower evaporation rates
Green functionalization techniques	Functionalizes CNTs with biomolecules or non-toxic reagents to improve dispersion and bonding without using hazardous chemicals	Improves compatibility with polymer matrices; avoids hazardous chemical usage	Limited to specific applications; functionalization may alter CNT properties

characteristics of the resultant nanocomposites. These advantages may include improved mechanical strength, thermal stability, electrical conductivity, optical characteristics, and barrier qualities. Typical instances include metal oxides such as ZnO, TiO_2, $Al2O_3$, and SiO_2, in addition to carbon nanotubes and clays. Inorganic fillers like zinc oxide and titanium dioxide improve mechanical qualities and exhibit antibacterial efficacy. ZnO functions as a fortifying and antimicrobial agent. Titanium dioxide (TiO_2) enhances ultraviolet resistance and thermal stability. This discussion highlights the use of PVA, a water-soluble and biodegradable polymer, as a standard preparation technique, indicating a more environmentally sustainable alternative to several other polymer matrices. A succinct case study (Rajesh et al. 2023) on the synthesis of PVA/TiO2/ZnO polymer nanocomposites is shown in Table 9.7.

Conclusion

This section has examined several biobased and inorganic nanofillers and their uses in polymer nanocomposites, focusing on sustainable and eco-friendly production techniques. Clay nanofillers, particularly montmorillonite, improve polymer characteristics due to their layered architecture and adsorptive qualities, and they have

Table 9.7 Summary of preparation of PVA/TiO2/ZnO polymer nanocomposites (Rajesh et al. 2023)

Step	Description	Details from article	Potential green aspects
Solution preparation	Dissolve PVA in a solvent. Disperse TiO2 and ZnO nanoparticles in a solvent, using ultrasonication	No specific details on solvents are explicitly given, but given PVA'S properties, it is likely water	Using water as a solvent is preferable to organic solvents, reducing environmental impact and health hazards. Ultrasonication is a relatively energy-efficient method for nanoparticle dispersion
Mixing	Combine the PVA solution and nanoparticle dispersion	Not specified in detail	Simple mixing processes are generally preferred in green chemistry.
Casting	Deposit the mixture onto a substrate to create a thin layer	A solution casting method is employed	Solution casting is a relatively simple and low-energy method for film preparation
Drying	Permit the solvent to evaporate, resulting in the residual nanocomposite film	No specific drying conditions mentioned	Natural air drying or low-temperature drying methods are more energy-efficient
Concentrations	Vary the weight percentages of TiO2 and ZnO	PVA/(x)TiO2(15-x) ZnO nanocomposites were prepared for x = 0, 1, 5, 7.5, 10, 14, and 15 wt% filling concentrations	Minimizing the use of materials is a general principle of green chemistry

applications in wastewater treatment and drug administration. Organic nanofillers such as chitosan and chitin, sourced from waste materials, provide antibacterial and biodegradable characteristics, with nanoforms generated using both top-down and bottom-up methodologies for various applications. Carbon-based nanofillers, including graphene and carbon nanotubes, markedly enhance the mechanical and electrical characteristics of polymers. The synthesis of these CNTs is increasingly adopting environmentally friendly approaches, such as bacterial synthesis, green chemical vapor deposition using renewable carbon sources, and processes that use less harmful solvents like water or ethanol. Inorganic nanofillers, such as ZnO and TiO_2, improve polymer characteristics, including antibacterial efficacy and UV resistance. Preparation techniques, such as solution casting, provide possible environmental advantages by using water as solvent and enabling energy-efficient processing. The transition to biobased and sustainably manufactured nanofillers, together with environmentally friendly preparation procedures, signifies an essential advancement in the creation of eco-friendly, high-performance materials for many applications. Despite constraints like scalability and cost, current research aims to develop novel ways to promote broader usage and unlock the full potential of these sustainable, high-performance materials.

9.3 Recent Trends in Green Synthesis of PNCs

9.3.1 General Overview of Green Synthesis

Green synthesis is an environmentally sustainable method for producing polymer nanocomposites by integrating biodegradable polymers such as polylactic acid (PLA) and polyhydroxyalkanoates (PHA) with inorganic nanoparticles (NPs) like silica and silver, alongside plant-derived antioxidants, and microorganism-derived nanofillers (Armentano et al. 2018; Jafarzadeh et al. 2024). This eco-friendly approach eliminates harmful chemicals, reduces environmental impact, and creates high-performance materials suitable for applications such as food packaging, environmental remediation, and biomedical use (Vieira et al. 2022). Plant-based antioxidants further enhance the stability, longevity, and shelf life of these materials, making them ideal for applications like prolonged food preservation. Biodegradable, non-toxic, and cost-effective, these nanocomposites offer a sustainable alternative to traditional materials. While challenges like scalability and property optimization persist, advancements in green chemistry and nanotechnology are driving their wider adoption (Rawat and Singh 2022; Jafarzadeh et al. 2024). Figure 9.7 depicts the common processing methods utilized in the green synthesis of PNCs.

Biodegradable polymers, including PLA, poly(vinyl alcohol) (PVA), and PHA, are pivotal in creating environmentally sustainable polymer nanocomposites. These polymers offer an eco-friendly substitute for conventional plastics by being sourced from renewable resources, thereby reducing reliance on fossil fuels and lowering the carbon footprint of production. PLA is a widely used biodegradable polymer

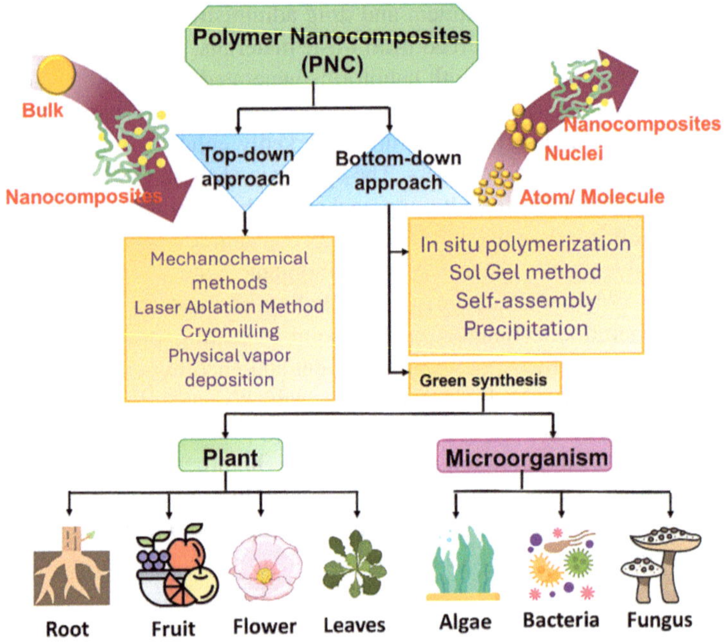

Fig. 9.7 Different synthetic methods utilized in the production of PNCs using green synthesis

derived from lactic acid, which is produced by fermenting carbohydrates, typically from plant-based sources like corn starch. Its renewable nature and biodegradability make it suitable for diverse applications (Taib et al. 2023). PVA is synthesized from vinyl alcohol, which can be sourced from renewable materials such as starch or glucose, offering another sustainable option (Chiellini et al. 2003). PHA is produced by bacteria that metabolize organic substances such as glucose or lipids. This results in a polymer that can biodegrade naturally in environmental conditions (Kumar et al. 2021; Samrot et al. 2021). The primary advantage of biodegradable polymers lies in their dual ability to enhance material properties while benefiting the environment. When combined with nanoparticles like silica, silver, and carbon nanotubes, these polymers exhibit improved mechanical strength, thermal stability, and biodegradability. Nanoparticles reinforce the polymer matrix, enhance attributes such as tensile strength and flexibility, and impart additional functionalities like antibacterial properties (Omerović et al. 2021), UV resistance, and electrical conductivity. Another key benefit is the ability of biodegradable polymers to naturally degrade at the end of their lifecycle, producing non-toxic byproducts. This helps reduce plastic waste and minimizes the long-term environmental impact associated with conventional plastics, which take centuries to decompose. The natural degradation of these materials in landfills or natural environments makes them a sustainable solution to global pollution concerns.

Despite their benefits, significant challenges exist in scaling up the production of biodegradable polymer nanocomposites for industrial use. Manufacturing processes are often expensive and technically complex, which restricts their widespread application in industries such as packaging (Ahmed et al. 2023), healthcare, and electronics (Muncke et al. 2020; Groh et al. 2021). Additionally, achieving desirable properties in nanocomposites, such as uniform nanoparticle dispersion, long-term stability, and consistent mechanical performance, remains a persistent challenge.

To address these issues, researchers are actively improving synthesis techniques to reduce costs and enhance material properties. Overcoming these hurdles is essential for making biodegradable polymer nanocomposites more accessible and practical for large-scale industrial applications. Advancements in this area are expected to bridge the gap between laboratory research and commercial adoption, ensuring that sustainable materials play a greater role in future industrial processes.

9.3.2 *Importance of Plant Extracts as Green Reducing and Stabilizing Agents*

The green synthesis of nanoparticles (NPs) is a crucial step toward creating eco-friendly nanocomposites, offering a safer and more sustainable alternative to conventional methods that often rely on hazardous chemicals detrimental to human health and the environment. Plant extracts, rich in bioactive compounds such as polyphenols, carboxylic acids, and hydroxyl groups, act as natural reducing and stabilizing agents in the formation of nanoparticles. These compounds replace toxic chemicals with natural alternatives, ensuring a greener synthesis process.

For instance, polyphenols possess strong antioxidant properties, enabling the reduction of metal salts like silver nitrate into nanoparticles such as silver nanoparticles. Green tea extract, which is abundant in catechins and flavonoids, is commonly used for synthesizing metal nanoparticles. These nanoparticles can then be incorporated into polymer matrices for applications in photovoltaic devices, sensors, and antimicrobial coatings (Vieira et al. 2022; Alharbi et al. 2022). Other plant extracts, such as those derived from citrus fruits, rosemary, and green coffee, also contain polyphenolic compounds that facilitate the reduction of metal ions. The resulting nanoparticles enhance the properties of biodegradable polymer nanocomposites, providing features like antimicrobial activity, UV protection, and antibacterial effects (Vieira et al. 2022; El-Sayed and Youssef 2023).

This eco-friendly approach eliminates the need for harmful chemicals, ensuring that the resulting nanocomposites are biodegradable and capable of supporting a sustainable lifecycle from production to disposal. However, scaling up this process for industrial applications presents challenges, including achieving uniform nanoparticle distribution and maintaining stability in the final materials. Current research efforts are focused on refining synthesis techniques and exploring novel plant extracts to overcome these challenges. These advancements aim to promote the adoption of biodegradable, low-toxicity nanocomposites in critical fields such as environmental restoration and healthcare (Huang et al. 2014; Muncke et al. 2020).

9.3.3 Latest Trends in Green Synthesis

Bacterial factories represent a promising recent trend in the sustainable synthesis of biodegradable polymer nanocomposites (PNCs). This approach leverages bacteria to facilitate the assembly of building blocks (polymers and nanofillers), mimicking natural biomineralization processes where organisms create intricate inorganic-organic hybrid structures. This concept is embodied in bacterial factories, where bacteria consume nanofillers and nutrients derived from bioresources containing monomers, producing desired biopolymers incorporating the nanofillers in a controlled manner (Guan et al. 2022). Automated programming allows for precise control of the nanofiller ratio and other critical factors. This combination of machine learning and automated programming enables the low-cost production of bioresource-derived PNCs with desirable physicochemical properties using a novel technique termed aerosol-assisted biosynthesis (AABS) (Guan et al. 2022). The authors Qing et al. demonstrated a compelling case for establishing bacterial factories for the rapid, cost-effective, and mass-scale production of PNCs with the assistance of machine learning, opening new avenues for sustainable materials manufacturing. The bacterial cellulose (BC) nanocomposite synthesis from microbial organisms and PNCs production is strategically combined in one step to reduce operating costs, and attempts were made to gain insights into spatiotemporal control over the fabrication of PNCs. Similarly, Domeneguetti and coworkers employed in situ green synthesis method using synthetic clay material, laponite, in a culture media containing *Komagataeibacter xylinus* microorganism to fabricate BC/laponite PNCs with control over the morphology (Domeneguetti et al. 2023).

In a novel approach, Xia Liao et al. combined two environmentally friendly techniques: supercritical carbon dioxide ($scCO_2$) and high-speed mechanical mixing. This strategic integration leveraged the strengths of both methods to create high-performance polymer nanocomposites (PNCs). Their approach yielded MWCNT/polystyrene nanocomposites with significantly improved electrical conductivity and unique segregated network structures. These characteristics make them ideal candidates for applications in electromagnetic interference (EMI) shielding (Chen et al. 2019).

3D printing and additive manufacturing represent a cutting-edge approach to sustainable polymer nanocomposite (PNC) fabrication. These techniques have recently garnered significant attention as alternatives for PNC synthesis due to key advantages, including reduced production costs and time, and the ability to precisely produce PNCs with tailored physicochemical properties. This precision is achieved through the careful selection of biorenewable resources, such as polylactic acid (PLA), for the 3D printing matrix, combined with the strategic addition of suitable additives like nanofillers and metallic nanoparticles. This approach allows for the fabrication of PNCs with desired characteristics, such as conductive accessible networks, antimicrobial properties, and optoelectronic properties. For example, thermosetting poly urethane (TPU)/PLA/r-GO blend has been fabricated using an additive manufacturing technique by Chen and coworkers (Chen et al. 2017). Blending of PLA with thermosetting polyurethane foam (TPU) and inorganic nanofiller reduced graphene oxide (r-GO) using a fused deposition model of additive manufacturing technique. The TPU/PLA/r-GO PNCs were produced by sequential addition of TPU and R-GO by a solvent-based mixing process and extruded into PLA matrix filaments for FDM printing. The resulting PNCs exhibited excellent biocompatibility with NIH3T3 cells and enhanced mechanical properties.

Recent interest in CO_2 sequestration and utilization has spurred research into its conversion into useful chemicals. Alroiathi et al. (Alroaithi and Xu 2023) employed a green synthesis method to produce polypropylene carbonate (PPC) from CO_2, subsequently using it as a polymer matrix for polyurethane foam. They further fabricated PPC/nanoclay polymer nanocomposites (PNCs) via a solution intercalation process. The lower viscosity of the CO_2-derived PPC resulted in PNCs with enhanced anti-corrosion properties. These studies reveal new approaches to tackle the ill effects of waste material by converting it into useful products. The Soy protein isolates (SPI) were used in the fabrication of SPI-oxidized sugar (OS) nanocomposite films, using a green synthesis approach (Bukartyk et al. 2022). The soy-based sugars were oxidized (glucose and sucrose) in aqueous solutions for the purpose of crosslinking soy protein isolates (SPIs), which served as the model material for polymer matrix and clay material as nanofiller to produce PNCs without the use of any organic solvents. This study illustrates the potential application of green synthesis methods for producing PNCs at low cost using bioresources as feedstock, therefore avoiding negative impact on the environment. The advancement in the research on SPI-based nanocomposites will pay for the way for innovative solutions for the development of PNCs without dependence on fossil fuels at an affordable cost.

A synergistic approach combining two distinct green synthesis methods was recently demonstrated by Livi and coworkers (2016). They employed tetraalkylphosphonium and dialkylimidazolium ionic liquids as surfactants to facilitate the reaction between montmorillonite (MMT) clay and poly(butylene adipate-co-terephthalate) (PBAT). The Ionic liquid also functions as a nucleating site. Supercritical CO_2 was then used to further improve the homogeneous distribution of the clay within the polymer matrix. This combined approach resulted in PBAT-clay nanocomposites with enhanced thermal stability, highlighting the benefits of integrating these methods to address challenges encountered when using either technique independently. This synergistic approach opens promising avenues for the development of high-performance, sustainable polymer nanocomposites for various applications.

9.4 Methods for Synthesizing Polymer Nanocomposites from Natural Polymers

Recently, natural polymers such as celluloses, chitosan, starch, alginate, and collagen have surfaced as sustainable matrices for PNCs, providing benefits such as biodegradability, biocompatibility, and environmental sustainability (Meher et al. 2024a; Rahmani et al. 2022). These natural materials mitigate increasing apprehensions about the environmental repercussions of synthetic polymers while facilitating uses in several domains, including biomedical engineering, energy storage, and packaging (Baranwal et al. 2022). Synthesis techniques are essential in defining the structural, mechanical, and functional characteristics of PNCs. Conventional methods, such as solvent evaporation and nanoprecipitation, have been modified to suit natural polymers. The integration of green synthesis methods has revolutionized the area by diminishing dependence on hazardous solvents, decreasing energy usage, and using biobased nanofillers and renewable energy sources (Pulingam et al. 2022).

This section explores nanofillers and sophisticated ways for synthesizing polymer nanocomposites (PNCs) from natural polymers, emphasizing current innovations like microwave-assisted synthesis, ultrasonic processing, mechanochemical processes, ionic liquid systems, and supercritical fluid technologies. This section emphasizes creative methods to elucidate the successful use of natural polymers in the development of high-performance, eco-friendly PNCs (Fig. 9.8 and Table 9.8).

9.4.1 Microwave-Assisted Synthesis: A Green Approach for Polymer Nanocomposites from Natural Polymers

Microwave-assisted synthesis has arisen as a novel and sustainable technique in materials research. It enables efficient energy transfer, decreases reaction durations, and lowers solvent use, making it a viable green method for the synthesis of polymer nanocomposites (Sridhar et al. 2024). When combined with natural polymers, Microwave-Assisted Synthesis (MAS) serves as an effective instrument for producing environmentally sustainable nanocomposite materials with enhanced mechanical, thermal, and functional characteristics. This approach adheres to the tenets of green chemistry, emphasizing reduced environmental impact and resource conservation (Verma et al. 2024a).

Principles of Microwave-Assisted Synthesis (MAS)

Microwaves function throughout a frequency spectrum of 300 MHz to 300 GHz, with 2.45 GHz often used for synthesis. The technique utilizes dielectric heating, whereby polar molecules and ions engage with microwave radiation, producing localized heat immediately inside the reaction medium (Zhu and Chen 2014). This targeted energy absorption facilitates fast and consistent heating, providing several benefits such as decreased energy use, time efficiency, and less undesirable side effects (Kumar et al. 2020).

Natural Polymers in Nanocomposite Synthesis

Natural polymers, including cellulose, chitosan, starch, alginate, and proteins, serve as a sustainable matrix to produce nanocomposites. The researchers effectively examined these natural resources for the benefit of the scientific and social

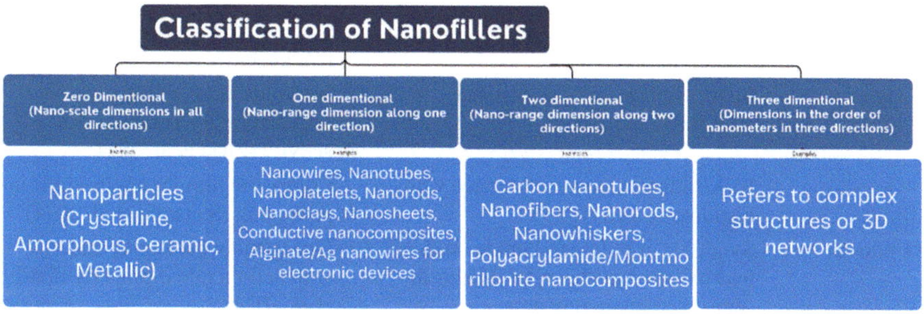

Fig. 9.8 Classification of nanofillers

Table 9.8 Classification based on type of nanofillers

Type	Description	Examples
Metal sulfide	Improve the stability of nanocomposites, often embedded in polymer matrices	CdS, ZnS, zinc sulfide, sodium carboxymethyl cellulose nanocomposites
Metal hydroxide	Known for flame-retardant properties, used in various polymer matrices	Magnesium hydroxide, aluminum hydroxide
Metal oxide	Synthesized via hydrothermal, sol–gel, sonochemical methods, etc.	Silica
Silicate	Often prepared with alcohols and used in various applications like oil recovery	Silica powder

communities. A selection of natural polymers used in composite synthesis is shown in Table 9.9, highlighting their distinctiveness and applications.

9.4.2 Ultrasonic Processing in the Synthesis of Polymer Nanocomposites

Ultrasonic processing has been shown to be an efficient method for boosting nanofiller dispersion, accelerating reaction rates, and minimizing total processing duration in the manufacture of polymer nanocomposites (Soman et al. 2023). This method employs high-frequency sound waves (ultrasound) to generate mechanical vibrations that induce cavitation in a liquid. The cavitation effect results in the creation and implosion of small bubbles, producing localized high shear forces that may disintegrate aggregates and improve dispersion at the nanoscale (Pawlaczyk-Łuszczyńska and Dudarewicz 2020). Ultrasonic processing's capacity to regulate nanofiller dispersion has facilitated the development of more efficient nanocomposite materials in biomedical and wastewater treatment applications.

Table 9.9 Natural polymers for stable and uniform dispersion used in synthesis of composites

Polymer	Synthesis uniqueness	Applications
Alginate	Silver alginate composite using microwave irradiation	Scaffolds, drug delivery, tissue engineering for antimicrobial activity (Zhao et al. 2014)
	PVA/sodium alginate integrated with histidine-coated AgNPs	Composite film for applications in packaging (Suganthi et al. 2024)
Chitosan	Synthesis of poly(acrylamide-co-2-hydroxyethyl methacrylate)/chitosan semi-interpenetrating network ZnO nanocomposite by microwave technology	Eco-friendly packaging material with antimicrobial property (Patil et al. 2022)
	Chitosan-based polymer nanocomposites	Waste water treatment, mercury detoxification. (Goci et al. 2023)
	Chitin nanowhiskers deacetylation to a chitosan nanoscaffold using microwave method	Fibrous network with a nanoporous structure in environmental remediation (Lertwattanaseri et al. 2009)
	Chitosan nanoforms/chitosan nanocomposites	Drug delivery systems (Sivanesan et al. 2021)
Gelatin	LED-curable methacrylated gelatin	Bio applications (Baghban et al. 2021)
	Hybrid hydrogels comprising carboxymethyl cellulose and gelatin grafted with acrylic acid by microwave irradiation	Catalyst for dye degradation (Sethi et al. 2019)
	Elemental silver stabilized by gelatin	
Hyaluronic acid	Chitosan/hyaluronic acid (HA) hybrid	Skin shield, healing, and regeneration (Alkabli 2024)
	Thermosensitive and injectable hydrogel synthesized using microwave-assisted green process utilizing HA, extract of corn silk, and AgNPs	Bioengineering material for skin regeneration and wound healing (Makvandi et al. 2019)
	Silver nanoparticles were biosynthesized using a microwave irradiation approach for an injectable nanocomposite based on hyaluronic acid and the extract of corn silk	Biomimetic antibacterial scaffold in bone regeneration (Makvandi et al. 2020)
Collagen	Biocompatible collagen-based bioactive composite synthesis using microwave irradiation	Collagen-based bioactive composite for tooth regeneration (Ilyas et al. 2018)
Fibrin	Silk fibroin nanocomposite enhanced with gold nanorods	Peripheral nerve tissue engineering applications (Afjeh-Dana et al. 2019)

Table 9.9 (continued)

Polymer	Synthesis uniqueness	Applications
Dextran	α-Glucan (type of dextran); first commercialized bacterial exopolysaccharide	Wide-ranging applications as biomaterials (Ukkund et al. 2022)
	Carboxymethylation and Fe3O4 MNPs synthesis using microwave	Application in drug delivery system (Do et al. 2019)
Cellulose	Mucilage-loaded silver nanocomposite/ nanosilver	Drug delivery system (Karande and Gawade 2020)
	Microwave-facilitated hydrothermal synthesis of cellulose/hydroxyapatite nanocomposites	Biomedicine (Fu et al. 2016)
Polylactic Acid (PLA)	Synthesis of L-poly(lactic acid) using microwave irradiation of lactic acid derived from lignocellulosic waste	Used to produce degradable packaging materials, and its porous scaffolds used in orthopedics, medicine carriers, nanofibers, foams, etc. (Senila et al. 2023)
	Microwave-assisted functionalization (MAF) method is used in the production of MAFBs characterized by elevated surface oxygenation, dimensional consistency, and structural integrity	Boplastic, an alternative to plastic (Ke et al. 2022)
	Ring-opening polymerization of ε-caprolactone	Biomaterial in tissue engineering. Packaging systems (Yang et al. 2022)
	Microwave heating and actuation of shape-memory polycaprolactone foam	Shape-memory polymers (SMPs); smart materials: SMP foams explored in various biomedical applications, especially in aneurysm treatment (Zhang et al. 2015)
Polyethylene Glycol (PEG)	Microwave-assisted NP synthesis using PEG and β-cyclodextrin	Hydrogel scaffolds, drug delivery (Aivazoglou et al. 2018)
Heparin	Heparin nanocomposites	Drug delivery and medicine (Meher et al. 2024b)
	Bifunctional low molecular weight heparin-coated nanosilver	Catheter lock solution, systemic antibiotics (Meher and Poluri 2022)
Agarose	Core–shell decoration Fe_3O_4 NP (Au/ CS-agar@Fe_3O_4) in chitosan-agarose composite	Drug targeting (Sheng et al. 2024)
	Nanosized hydroxyapatite/agarose powder, microwave sintered composites	Bone filler, drug delivery (Kolanthai et al. 2016)

Ultrasonic waves induce the development of bubbles (cavitation) in the solvent or polymer solution. The rapid disintegration of these bubbles produces significant shear forces that fragment the agglomerates of nanofillers and distribute them evenly throughout the polymer matrix, perhaps accounting for the observed dispersion and homogeneity (Soman et al. 2023). In an ultrasonic apparatus, a transducer transforms electrical energy into ultrasonic vibrations, a probe/horn/sonotrode conveys those vibrations to the target material by amplifying the amplitude, while a booster further enhances the vibration amplitude prior to reaching the horn, and an ultrasonicate constitutes the comprehensive system that produces and transmits ultrasonic waves, typically comprising a power supply, generator, transducer, and horn/probe. Figure 9.9 illustrates a schematic arrangement and its operation.

In drug delivery systems, ultrasonic dispersion facilitates the homogeneous distribution of nanocarriers, including liposomes or polymeric nanoparticles, hence assuring consistent drug release patterns. In wastewater applications, ultrasonic dispersion may promote the integration of photocatalytic nanoparticles (e.g., TiO_2) with polymers, resulting in nanocomposite membranes that boost pollutant removal.

Methods of Ultrasonic Processing in Polymer Nanocomposite Synthesis

1. Ultrasonic processing may be used via several approaches depending on the desired outcome and the properties of the polymer and nanofiller involved. The primary ultrasonic techniques include:
2. **Ultrasonic Dispersion**: Frequently used to distribute nanoparticles inside a polymer matrix. Dispersion is accomplished by applying ultrasonic waves directly to the polymer solution and nanoparticle combination. This method is particularly efficacious for spreading two-dimensional fillers, including graphene oxide (GO), nanoplatelets of graphene (GNPs), and nanotubes of carbon (CNTs) (Korayem et al. 2017). The stability of the nanocomposite is contingent

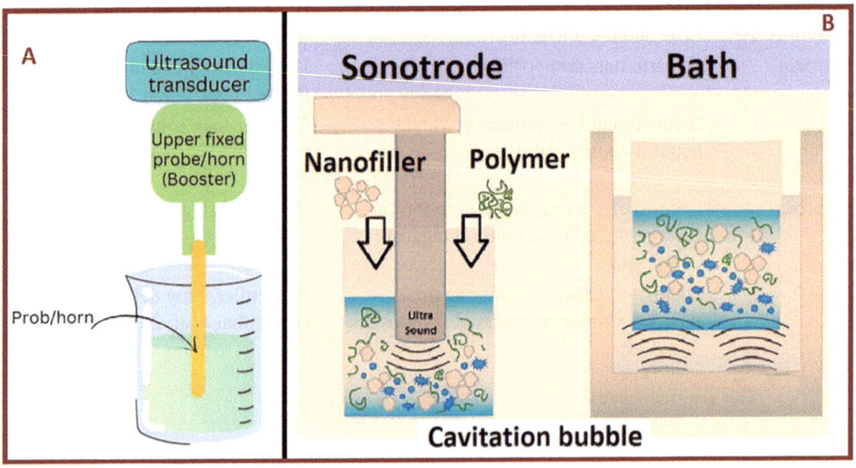

Fig. 9.9 The main components, A and functioning B, of a sonochemical process: an ultrasound transducer, a booster, a horn/probe, and beakers containing the liquid sample

upon the duration of sonication. High consistency of DTG and TGA curves over three sets of 1-h ultrasonically treated samples indicates analogous thermal degradation behavior, attributable to the even dispersion and distribution of GNPs (Guo et al. 2024).

3. **Ultrasonic-Assisted Melt Mixing**: This method employs ultrasonic waves during the melt mixing of polymer matrices with nanofillers (Zhang et al. 2020). The intense shear pressures generated by ultrasound facilitate the disaggregation of nanoparticle agglomerates and promote uniform dispersion inside the polymer matrix. This method is extensively used for thermoplastic polymers like polyethylene (PE) and polypropylene (PP).
4. **Ultrasonic-Assisted Polymerization**: Ultrasonic waves may be used during the polymerization process to enhance the synthesis of nanocomposites (Kumar et al. 2023; Yuan et al. 2022). This method significantly improves the interaction between nanoparticles and the expanding polymer chains, leading to superior dispersion and increased material characteristics. Instances include the manufacture of polylactic acid (PLA) (Li et al. 2017) and polyvinyl alcohol (PVA) (Soman et al. 2023) composites.
5. **Ultrasonic-Assisted Solvent Evaporation**: This technique utilizes ultrasonic waves to scatter nanoparticles inside a polymer solution, thus leading to the evaporation of the solvent. This method is optimal for the fabrication of membranes and thin films (Mohamat et al. 2024). The dispersion of titanium dioxide (TiO2) in polyvinylidene fluoride (PVDF) was accomplished using ultrasonic-assisted solvent evaporation (Ding et al. 2023), leading to improved photocatalytic capabilities.
6. **Ultrasonic-Assisted Coagulation/Flocculation**: Ultrasound technology has garnered significant interest in water treatment in recent years, owing to its efficiency and ease, with coal slime water undergoing ultrasonic prior treatment to enhance flocculation (Shahdan et al. 2021).

Conclusion

This section examines the synthesis of polymer nanocomposites (PNCs) derived from sustainable natural polymers, including cellulose, chitosan, starch, alginate, and collagen, while addressing environmental issues associated with synthetic polymers. Although conventional techniques such as solvent evaporation and nanoprecipitation are used, emphasis is placed on green synthesis methods, namely microwave-assisted synthesis (MAS), which provides benefits like efficient energy transfer, shortened reaction durations, and decreased solvent consumption. A range of nanofillers, including metal sulfides, hydroxides, oxides, and silicates, is examined, with a thorough analysis of the application of MAS to several natural polymers such as alginate, chitosan, gelatin, hyaluronic acid, collagen, fibrin, dextran, cellulose, PLA, PEG, heparin, and agarose. These PNCs have several uses, including biomedical engineering, packaging, environmental cleanup, and medication delivery. Moreover, the efficacy of ultrasonication in augmenting nanofiller dispersion within polymer matrices is emphasized (refer to Table 9.10), resulting in improved mechanical, thermal, and functional attributes. Direct ultrasonication

Table 9.10 Approaches in polymer nanocomposites and case studies

Case study	Method, nanofiller, and polymer matrix	Ultrasonic approach	Applications	Reference
Graphene nanoplatelets (GNPs) in poly(lactic acid) (PLA) (Elella et al. 2021)	Ultrasonic dispersion; polymer matrix: PLA	Ultrasonic homogenizer, 40 kHz, 100 rpm, 1 h. even GNP dispersion in the PLA matrix at 0.4 wt.%	Drug delivery systems, tissue engineering	Guo et al. (2024)
Ultrasonic-assisted compounding of CNTs with polypropylene (PPs) of different molecular weights and viscosities using twin-screw extrusion	Of three methods for CNT dispersion (ultrasonic treatment during masterbatch dilution, during masterbatch preparation, and direct compounding), masterbatch preparation with ultrasonication (method 2) was most effective. While low-viscosity PPs aid dispersion, ultrasonication improves it even in high-viscosity PPs, significantly reducing CNT agglomerates	Carbon nanotubes (NC7000) with 9.5 nm diameter and 1.5 μm length used as the nanofiller. PP with varying molecular weights and viscosities was used, including Ziegler-Natta and metallocene-catalyzed PPs. Viscosity impacts CNT dispersion and the effectiveness of ultrasonic treatment	Wastewater filtration, pollutant removal, automotive and aerospace applications	Zhong and Isayev (2016)
Porous graphene oxide framework (GOF), synthesized via crosslinking with triazine-based linkers	Ultrasonic bath at 60 Hz to enhance dispersion and adsorption efficiency. High surface area (14 m^2/g), mesoporosity, thermal stability, and functional groups facilitating chelation	Ultrasonic waves ensured uniform dispersion of GOF and maximized the exposure of active sites	Removal of Cd^{2+}, Pb^{2+} from polluted water. Environmental remediation	Santos et al. (2024)

Table 9.10 (continued)

Case study	Method, nanofiller, and polymer matrix	Ultrasonic approach	Applications	Reference
Development of self-protected photocatalytic PVDF membranes by incorporating polyaniline (PANI) and polydopamine (PDA) layers between the membrane and TiO2 coating. Focus on dye wastewater treatment and antifouling properties	PDA or PANI self-polymerized on PVDF hollow fiber membrane surfaces, and subsequently, TiO2 nanoparticles were coated onto layers through ultrasonic dispersion. Ultrasonic agitation (40 kHz, 40 °C, 1 h) for dispersion of TiO2 nanoparticles. Improved uniform distribution of TiO2 nanoparticles on the PDA-modified membranes, whereas PANI-modified membranes showed a more clustered distribution	Ultrasonic agitation (40 kHz, 40 °C, 1 hour) for dispersion of TiO2 nanoparticles. Improved uniform distribution of TiO2 nanoparticles on the PDA-modified membranes, whereas PANI-modified membranes showed a more clustered distribution	Dye removal: 95.7% efficiency (PANI-TiO2) and 88% (PDA-TiO2). Antifouling: Flux recovery rates (PDA: 56.3%, PANI: 37.1%). - water purification: Photocatalytic degradation	Le et al. (2024)
Priorly dispersed organo-montmorillonite (organo-MMT) nanofiller used to improve flexibility, toughness, thermal stability, and biostability of biomedical ethyl vinyl acetate (EVA) nanocomposites	Pre-dispersion of organo-MMT nanofiller via ultrasonication and magnetic stirring in water and toluene, followed by melt compounding with EVA matrix at 160 °C using an internal mixer and compression molding into sheets for testing. Nanofiller: Organically modified montmorillonite (organo-MMT) with dimethyl dialkyl ammonium chloride as the surfactant. Polymer matrix: Ethyl vinyl acetate (EVA) copolymer with 18 wt% vinyl acetate and 82 wt% ethylene content	1 h, 25 kHz ultrasonic bath disperses MMT in PS, enhancing thermal and mechanical strength. Water-based ultrasonication optimizes MMT dispersion (highly disordered nanoplatelets, minimum agglomeration, improved EVA interaction), leading to increased flexibility (~14% elongation increase), thermal stability (higher degradation temperature), and biostability (reduced PBS degradation)	Biomedical applications such as insulation material for implantable devices, due to enhanced mechanical properties and reduced cytotoxicity, making it suitable for close contact with human tissue	Osman et al. (2017)

(continued)

Table 9.10 (continued)

Case study	Method, nanofiller, and polymer matrix	Ultrasonic approach	Applications	Reference
Investigation of ultrasonication modes (direct and indirect) to improve the dispersion of nano-TiO2 in PVDF membranes, enhancing hydrophilicity, permeability, and antifouling properties	PVDF membranes modified with nano-TiO2 using dispersion methods, mechanical stirring, and ultrasonication. Membranes analyzed for their structure, hydrophilicity, and fouling resistance. Nanofiller: Titanium dioxide (TiO2) nanoparticles (size range, 5–10 nm, anatase form). Polymer matrix: Polyvinylidene fluoride (PVDF) membrane (by phase inversion)	Indirect ultrasonication: water bath for transmitting ultrasound. Direct ultrasonication: Use energy directly by a titanium-alloy ultrasonic probe. Duration: 1–5 hr. direct ultrasonication showed better dispersion and fewer agglomerates compared to indirect ultrasonication or mechanical stirring alone	Enhanced PVDF-TiO2 membranes suitable for water and wastewater treatment, with applications in ultrafiltration and protein rejection due to improved hydrophilicity, permeability, and fouling resistance	Xie et al. (2020)
Preparation and characterization of PVA-PVP blend nanocomposite films filled with thiamine-modified zirconia (ZrO2) nanoparticles for enhanced thermal stability and hydrophilicity	Modification of ZrO2 nanoparticles with thiamine (Th). Preparation of PVA/PVP blend solution (50/50, w/w). Casting/solvent evaporation of the mixture to form films. Nanofiller: Zirconia (ZrO2) nanoparticles modified with thiamine (Th). Polymer matrix: Blend of poly(vinyl alcohol) (PVA) and poly(N-vinyl-2-pyrrolidone) (PVP) in a 50/50 weight ratio	Ultrasonic irradiation (20 kHz, 100 W) effectively dispersed modified ZrO2-Th nanoparticles in PVA/PVP, enhancing thermal stability, hydrophilicity (reduced contact angle), and UV absorption. Nanoparticles with showed compatibility with polymer matrix. Resulting composite retained biocompatibility and biodegradability of PVA/PVP	Enhanced properties suggest potential applications in areas requiring biocompatible, thermally stable, and hydrophilic materials. For optical applications	Mallakpour and Shafiee (2018)

Table 9.10 (continued)

Case study	Method, nanofiller, and polymer matrix	Ultrasonic approach	Applications	Reference
Study on ultrasonically induced polymerization and grafting methods for fabricating polymer nanocomposites with carbonaceous nanoparticles, such as CNTs and graphene	Sonochemical methods employing cavitation to assist polymerization and functionalization of nanofillers. Techniques include grafting-from approaches and in situ polymerization. Nanofiller: Carbon nanotubes (CNTs), including single-walled (SWCNT) and multiwalled (MWCNT), and graphene derivatives like GO and reduced graphene oxide, rGO. Polymer matrix: Polystyrene (PS), polyaniline (PANI), poly(3-hexylthiophene) (P3HT). Functional polymers like poly(2,2,6,6-tetramethylpiperidin-4-yl methacrylate) (PTMA) and dendritic poly(amidoamine) (PAMAM)	Ultrasonic waves (20 kHz to 500 MHz) for promoting dispersion, functionalization, and polymerization. No re-stacking in graphene sheets, ensures uniform coating of CNTs within polymer matrices. Improved electrical, thermal, and mechanical properties, temperature sensitivity or electrochemical activity for sensors	Electronics: Optoelectronic devices, supercapacitors. Sensors: Electrochemical strain sensors, biochemical sensing. Coatings: Anti-static and transparent films Biomedical: Drug delivery, antimicrobial surfaces. Energy: Electrode materials for batteries and capacitors	Cohen et al. (2020)
Coal slime water flocculation	Ultrasonic-assisted flocculation; Nanofiller: N/A; polymer matrix: Coal slime water	Ultrasound (192 W, 8 min) with flocculant; dispersion: Improved flocculation efficiency, reduced clay on coal slime particles; properties: Floc size 80.45 μm, fractal dimension 2.01, flocculation efficiency 92.64%	Water treatment, coal mine wastewater	Ding et al. (2023)

often yields enhanced dispersion relative to indirect techniques or mechanical agitation. The growing significance of synthesizing PNCs by green technologies, particularly MAS and ultrasonication, signifies a prospective avenue for the development of high-performance, eco-friendly materials in several domains.

9.4.3 Mechanochemical Techniques

Various techniques have been employed for the successful incorporation of inorganic nanofillers/or additives into a natural polymer matrix to synthesize sustainable PNCs for suitable applications. Among these methods, the mechanochemical synthesis method is highly favorable because of green chemistry principles as it involves a solvent-free approach. Moreover, it offers additional advantages such as enhancement in properties such as mechanical strength and stability due to efficient mixing, and can be applied to a variety of polymers irrespective of their solubility. The field of mechanochemical synthesis is still in the infant stage, and there are immense opportunities to develop innovative techniques to pave the way for the design of PNCs with enhanced physicochemical properties, which widen their application in various sectors that are inaccessible otherwise.

The mechanical force is supplied through ball milling or kneading to break down the filler particles to a nano size (see Fig. 9.10). The smaller size particles provide a high surface area and facilitate the surface adhesion of nanoparticles in the polymer matrix, therefore enhancing the probability of chemical reaction for the mechanochemical synthesis of PNCs. These chemical reactions further enhance the homogenous dispersion of nanofillers inside the polymer matrix. Since solvent usage is minimal in this approach, the chemical hazards associated with organic solvents are completely eradicated. Moreover, the ease of scalability to larger batches and the low energy involved in mechanochemical synthesis promote its usage in commercial applications at the industrial scale.

Among the various techniques reported in the literature, ball milling is extensively studied due to its straightforward operating principle. This method utilizes mechanical forces generated by the rapid movement of milling media (balls) within a rotating mill to grind inorganic fillers, thereby promoting their homogeneous dispersion within the polymer matrix (Gao et al. 2023). The physicochemical properties of polymer nanocomposites (PNCs) can be tailored by adjusting parameters such as ball size, milling speed, and the filler-to-polymer ratio. For instance, cellulose nanofiber (CNF) nanofillers were successfully dispersed in polyamide-6 (PA6) to create biobased PA6-CNF PNCs using planetary ball milling (Jalaee et al. 2023). This direct milling method proved superior to solution casting and melt mixing for fabricating these biobased CNF-polymer nanocomposites, yielding PNCs with enhanced thermal stability and superior mechanical performance compared to those produced by conventional methods (Jalaee et al. 2023). Similarly, direct milling effectively incorporated carbon nanotubes (CNTs) into an epoxy resin matrix to create an epoxy/CNT film sensor with excellent strain sensitivity (Chow and Tan

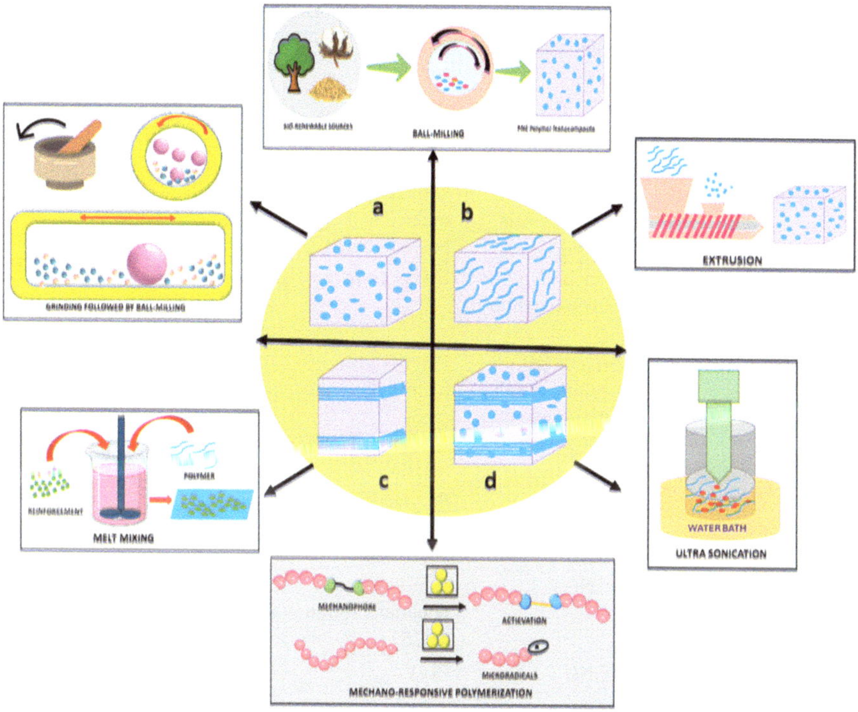

Fig. 9.10 Various processes involved in the mechanochemical synthesis of PNCs

2010). Lagron and coworkers used high-energy ball milling for the selective incorporation of bacterial cellulose nanowhiskers (BCNW) into poly(3-hydroxybutyrate-co-3-hydroxyvalerate) (PHBV) (Ambrosio-Martín et al. 2015). The BCNW acted as a nucleating agent, improving PHBV crystallinity and enhancing the thermal stability of the resulting PHBV/BCNW PNC. Furthermore, ball milling also improved the oxygen barrier properties and reduced water vapor diffusion in the PNC films. Table 9.11 summarizes recent applications of mechanochemical synthesis for developing polymer nanocomposites and highlights the key improvements in their physicochemical properties.

Another key process used in the mechanochemical synthesis of PNCs is the extrusion method. This method involves introducing shear stress on the material to create composite materials of fixed cross-sections by subjecting the material to pass through the die of a desirable cross-section. This method allows the selective distribution of nanofillers into the polymer matrix and offers several advantages over conventional methods. As solvent-free techniques, extrusion and injection molding offer industrially viable, economical, and high-throughput processing solutions. The use of renewable resources like nanocellulose fibers derived from different bioresources like plant extracts and bacteria affords environment-friendly reinforcement fillers that can be mixed with polymer matrix using an extrusion process to design and develop PNCs of desired shape and size for selective applications.

Table 9.11 Fabrication of PNCs through mechanochemical synthesis

S.No	Mechanochemical method	Biobased fillers used	Polymer composite	Significance	Ref.
1	Direct ball milling	Cellulose nanofibers	PA6-CNF	Enhanced thermal stability	Jalaee et al. (2023)
2	Planetary ball milling	Cellulose nanocrystals (CNC)	Ultrahigh molecular weight polyethylene	Improved biocompatibility	Wang et al. (2016)
3	High-speed ball milling	Nanocellulose from bamboo	Epoxy resin	Enhanced mechanical properties	Verma et al. (2024b)
4	High-energy ball milling	Bacterial cellulose nano whiskers (BCNW)	Poly(3-hydroxybutyrate-co-3-hydroxyvalerate) (PHBV)	Improved thermal stability	Ambrosio-Martín et al. (2015)
5	Centrifugal ball milling	Cellulose nanocrystals (CNC)	Polylactic acid (PLA)		
6	Extrusion	CNC	Acrylonitrile butadiene styrene (ABS)	Increased storage modulus and viscosity	Leao et al. (2020)
7	Twin-screw extrusion followed by injection molding	CNC	PLA	Higher values of storage modulus and dispersion of nanoparticles	Mariano et al. (2017)
8	Melt compounding	PN-FR@CNF	PLA	Exceptional flame-retardant property	Feng et al. (2017)
9	Ball milling	Almond shell nanoparticles	PLA	Increased water solubility and biodegradability	Edward et al. (2022)
10	Wet ball milling	Polymeric bis(adamantyl)-1,2-dioxetane	Poly(methyl acrylate)	Enhanced mechanochemical strength and mechanochemical luminescence	Yuan et al. (2019)
11	Melt-extrusion	Multiwalled carbon nanotubes (MWCNTs)	Polypropylene	Enhanced rheology, tensile strength, and young's modulus	Pérez-Medina et al. (2016)
12	Ultrasound-assisted extrusion	ZnO	Nylon-6	Enhanced mechanical and antibacterial properties	Andrade-Guel et al. (2021)

However, there are some limitations in the extrusion process that need to be addressed for further advancement in this field. Specifically, with CNC as a nanofiller, the thermal degradation of CNC and heterogeneous dispersion of CNC nanofillers in polymer matrix during the extrusion are often encountered, which needs the attention of researchers in this field. Moreover, there is also a possibility of elongation of nanoparticles in the direction of extrusion, which inhibits the formation of desirable cross-sections. Better control over the processing parameters using automation through machine learning will help in overcoming the above-mentioned problem encountered during the extrusion process. Another challenge in implementing the extrusion/injection molding process for the fabrication of PNCs is the reversible agglomeration of nanoparticles produced during the extrusion process (Dufresne 2018). These limitations need to be addressed before extending these processing methods for large-scale applications in the polymer industry.

Melt compounding is another method employed in polymer nanocomposite (PNC) synthesis, sharing a similar principle with ball milling. Agglomerated nanofiller particles are blended into the polymer matrix, and the kneading action within the melt breaks down these agglomerates, achieving uniform dispersion. This method addresses limitations such as nanofiller agglomeration and non-homogeneous distribution within the polymer matrix. A key advantage is the preservation of nanofiller surface activity during dispersion, enabling the use of these PNCs as catalysts in various applications. Melt compounding has been successfully used to fabricate graphene-polymer and silica/polymer nanocomposites (Tanahashi 2010).

Mechanochemical synthesis encompasses a range of techniques beyond traditional milling, including ultrasonication and the triggering of in situ polymerization through mechanical bond disruption. One strategy involves selectively functionalizing a monomer with a mechanoresponsive tag, which can be activated through son chemical processing or grinding to produce polymer nanocomposites (Soman et al. 2023). For instance, during extrusion, the shear stress induced in the presence of clay nanofillers can rupture polypropylene chains, thereby facilitating clay particle dispersion within the polypropylene matrix. Yuan and coworkers further exemplified the use of mechanical force for bond disruption in fabricating polymer nanocomposites for optoelectronic applications (Yuan et al. 2019). They tagged a conjugated polymer matrix (CPM) with 1,2-dioxyketone and used ball milling to generate nanosheets, which were subsequently incorporated into poly (methyl acrylate) (PMA) to investigate the optoelectronic properties of the resulting polymer nanocomposites. Currently, the ultrasonication method is also the most prevalent process utilized in the dispersion of inorganic/organic nanofiller into polymer matrix. The vibrational energy generated in ultrasonication is effective in the homogenous distribution of nanofillers in the polymer matrix. For example, Perez Medina et al. demonstrated that the PP/MWCNTs electrical conductivity is enhanced due to suppression of agglomeration of MWCNTs employing acoustic cavitation resulting from ultrasonic waves, and electrical properties were dependent on the conditions employed during the extrusion process (Pérez-Medina et al. 2016). The controlled cavitation and dispersion of nanofillers offer unique advantages to the ultrasonic-assisted extrusion techniques, and further research is in progress to gain control over the spatiotemporal characteristics of PNCs using additive manufacturing techniques.

9.4.4 *Ionic Liquid-Based Synthesis*

Green chemistry, which aims to develop chemical processes that are sustainable for the environment, and efficient in terms of energy and resource-conserving, recently found a possible adversary in ionic liquids (ILs) (Plechkova and Seddon 2008). These salts, which remain liquid at or near room temperature, are made up of a large organic cation and a small inorganic or organic anion. Their distinctive features, such as minimal vapor pressure, great thermal stability, and variable solubility, make them ideal candidates for substituting standard organic solvents in several chemical processes (Plechkova and Seddon 2008; Rogers and Seddon 2003). The use of ILs in nanocomposite production represents a significant step toward sustainable material development (Montolio et al. 2019). Traditional fabrication methods frequently use toxic organic solvents and energy-intensive procedures, which produce hazardous waste and contribute to environmental damage (Qamar et al. 2020). In contrast, ILs offer a cleaner, more environmentally friendly option due to their low vapor pressure, which eliminates volatile organic compound (VOC) emissions, and their superior thermal stability (Xu and Cheng 2021), which allows for efficient and regulated synthesis (Mallakpour and Dinari 2012). Their versatility stems from their tunable qualities, which allow the selection of cations and anions to impact solubility, dispersion, and interactions between the nanofiller and the matrix (Mallakpour and Dinari 2012). This versatility enables researchers to fine-tune nanocomposites' physicochemical properties, ensuring that they satisfy the unique needs of applications like mechanical reinforcement, thermal stability, and increased conductivity (Veeman et al. 2021; Nguyen Thi et al. 2020) (Fig. 9.11).

In addition, ILs promote greener synthesis pathways by enabling solvent-free or one-pot reactions, thereby lowering the number of steps, energy consumption, and waste creation during manufacturing (Mallakpour and Dinari 2012; Zhang et al. 2014). This technique is consistent with green chemistry principles, such as waste reduction, energy efficiency, and the use of safer solvents (Anastas and Eghbali 2010). Beyond environmental benefits, incorporating ILs into nanocomposite synthesis fosters material science innovation, allowing for the development of new materials with specific capabilities. These materials have a wide range of uses, including energy storage (Armand et al. 2009), environmental remediation (Ghandi 2014), and biomedical devices (Petkovic et al. 2011), demonstrating the scope of this sustainable approach. By incorporating ILs, researchers and companies may

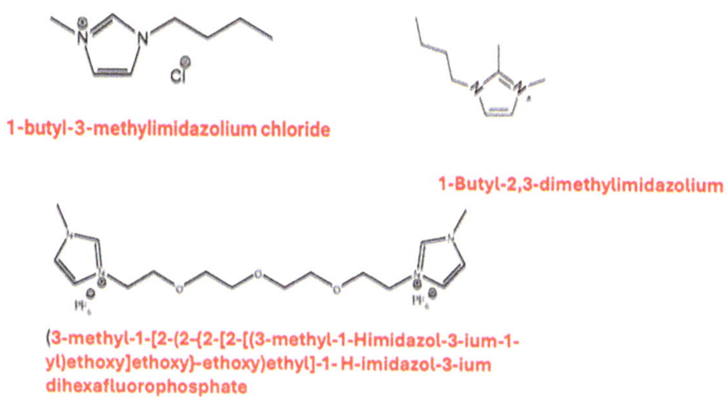

Fig. 9.11 Chemical structures of commonly used ionic liquids employed in the synthesis of PNCs

solve global environmental issues while pushing the limits of material performance and design.

Ionic Liquid Properties for Nanocomposites and Fabrication Mechanisms in Ionic Liquids

ILs have unique qualities that make them excellent for nanocomposite synthesis, providing more control over manufacturing processes and material properties. One of their distinguishing features is their broad solubility range, which allows them to dissolve a wide range of organic, inorganic, and polymeric compounds (Marciniak 2010). This capability provides homogeneous reactions and uniform nanocomposite structures, which are required to achieve desired material properties (Dupont et al. 2002). Furthermore, ILs have outstanding thermal and chemical stability, allowing them to survive high temperatures and resist reactive species, making them ideal for synthesis processes requiring strict conditions (Wasserscheid and Keim 2000). Another distinguishing feature is their adjustable polarity, which may be altered by varying the cation-anion composition.

ILs play an important part in the various mechanisms of nanocomposite production, providing distinct advantages over traditional approaches. One such method is the *ionic liquid-assisted sol–gel process,* a flexible technology that uses ILs as both solvents and stabilizing agents. Their capacity to control the gelation process enables homogeneous nanoparticle dispersion, producing nanocomposites with improved structural and functional properties (Huddleston et al. 2001). Another notable approach is *electrodeposition in IL* media, which benefits from ILs' high ionic conductivity and electrochemical stability. These qualities allow for precise and efficient nanomaterial deposition, making this approach ideal for producing metal-matrix nanocomposites with better properties (Katayama 2005). Furthermore, *microwave-assisted synthesis in ILs* produces quick and uniform heating, which speeds up reaction kinetics and promotes the creation of highly crystalline nanocomposites. This technique is not only energy-efficient but also environmentally benign, following green chemistry principles by lowering energy consumption and waste (Zhu and Chen 2014). Together, these methods demonstrate ILs' transformative potential for enhancing sustainable and efficient nanocomposite production.

Case Studies and Applications

To effectively optimize and implement ionic liquid (IL) processes, a more thorough mechanistic understanding is crucial. Widespread application hinges on the benefits of ILs outweighing their drawbacks, particularly their cost compared to the value of the processed material. Process modeling is vital for estimating energy requirements and costs, allowing for comparison with alternative pretreatment methods. While ILs offer numerous advantages and potential applications, their high cost—typically 5–20 times greater than conventional molecular solvents—currently presents a significant barrier to commercial viability. The Ionic liquids formulated from green chemical sources are the need of the hour to reduce costs. However, the advantages of IL usage in polymer dissolution outweigh its operating costs. The ILS has been used by several researchers to dissolve and disperse inorganic nanofillers into the polymer matrix. The chemical reactions along with the schematic representation of key steps involved in the fabrication of CNC-LA/IL nanocomposite are shown in Fig. 9.12 (Wang et al. 2020a). The IL, [VBIm][BF_4], is carefully grafted on the surface of the CNC nanocrystal, and it was further decorated by lactic acid to increase surface activity. The CNC-LA/IL PNCs exhibit better suspension ability compared to bare CNC. Simple physical mixing of the imidazolium-based ionic liquid [BMIm][BF4] with a CNC suspension enhanced the thermal stability and dispersion of the CNC within a poly(lactic acid) matrix.
Ionic liquids also act as a reaction medium to precipitate inorganic nanaofillers in the biobased polymers to design novel PNCs which are inaccessible otherwise through conventional preparation methods. For example, 1-butyl-3-methylimidazoliumchloride [C4 mim] Cl ionic liquid, has been employed in the mineralization of calcium phosphate in the cellulose polymer matrix to prepare a cellulose/chloropatite PNCs (Salama et al. 2014). Ma et al. (2013) employed a microwave-assisted ionic liquid method to prepare cellulose/$CaCO_3$

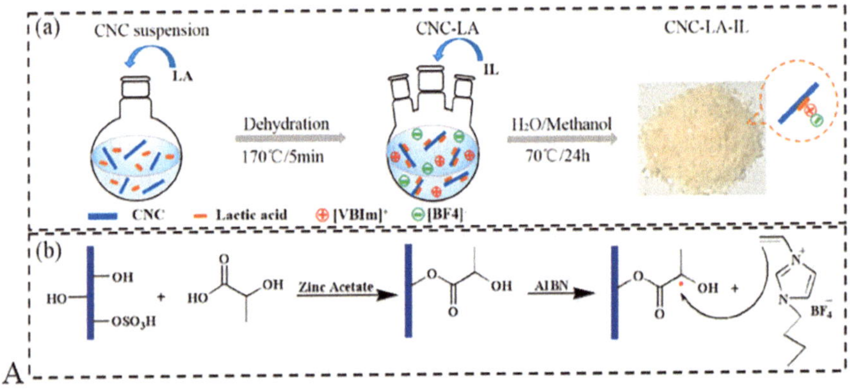

Fig. 9.12 Fabrication of ionic liquid grafted PLA nanocomposites

nanocomposites using a green synthesis approach. The authors investigated the impact of cellulose concentration on $CaCO_3$ morphology. Their results revealed a morphological shift in $CaCO_3$ from polyhedral to cubic, and finally to particulate, with increasing cellulose content. The resulting nanocomposites exhibited biocompatibility and good cytotoxic activity, suggesting potential for biomedical uses. Ionic liquid-mediated synthesis of biobased polymers is gaining considerable attention, providing opportunities to create innovative nanomaterial composites. Precipitating inorganic or organic nanofillers enables new applications for existing biopolymers.

9.4.5 *Supercritical Fluid Processing*

Supercritical fluid processing techniques complement the green chemistry principles for the fabrication of PNCs as they minimize the usage of organic solvents and are environmentally friendly. Supercritical fluids offer unique advantages over other conventional solvents, such as low viscosity, high diffusivity, and negligible surface tension (Haldorai and Shim 2014). Moreover, they are non-toxic, and the majority of the time, they can be recycled. Supercritical carbon dioxide (scCO2) is widely employed as a solvent successfully for the fabrication of PNCs, and it also supplements the sequestration of carbon dioxide (Yeo et al. 1993). The growing demand for the reuse of carbon dioxide to produce useful chemicals also catalyzed this field of green synthesis of PNCs using supercritical CO_2 either as a feedstock or as a solvent for promoting the uniform distribution of nanofillers within the polymer matrix. The high diffusivity of $ScCO_2$ aids in promoting rapid mass transfer and efficient transport of reactants and products, which in turn promotes uniform distribution of nanofillers such as cellulose nanocrystals and inorganic nanofillers in the polymer matrix (Yuvaraj et al. 2012). Another major advantage of using SCF (supercritical fluids) is that they offer low processing temperatures, which inhibit polymer degradation. Baron Charles Cagniard de Latour experimentally verified the critical temperature and pressure of carbon dioxide at 31.1 °C, and 73.8 bar, respectively. Specifically, this is very beneficial in the case of cellulose-based polymer nanocomposite fabrication. CO_2, present in the atmosphere at 3.5×10^5 parts per billion by volume, is readily recyclable as a solvent, mitigating its environmental impact. Its non-flammability, non-toxicity, and low cost further enhance its appeal for large-scale synthetic applications (Yuvaraj et al. 2012).

In the past few years, several researchers investigated the deployment of CO_2 as SCF for polymer nanocomposite synthesis. Major processing methods like supercritical impregnation, supersupercritical solution processing, supercritical polymerization, and supercritical drying have been reported so far in the literature for the synthesis of PNCs by using CO_2 as SCF. Figure 9.13 summarizes the key principles involved in each of these techniques and their advantages. Recently, Baldino et al. demonstrated the fabrication of polymer nanocomposite membranes consisting of homogeneous cellulose acetate (CA)-graphene oxide (GO) with controlled GO

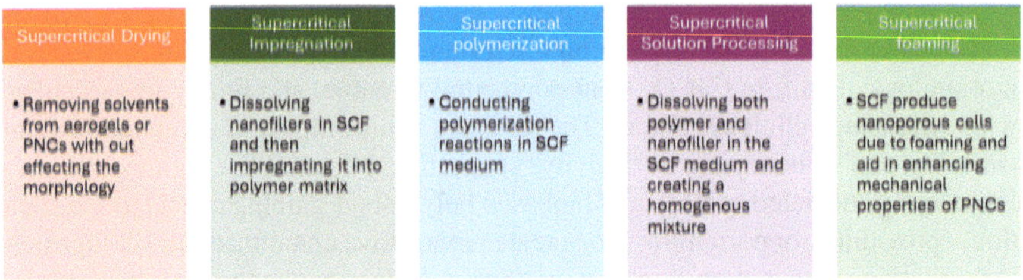

Fig. 9.13 Different processing techniques used in SCF-assisted polymer nanocomposite synthesis and their advantages

layer dispersion and interfacial interactions within the polymer matrix. For processing CA/GO PNC nanocomposite, supercritical CO_2 (SC-CO_2)-assisted phase inversion was employed at an operating condition of 200 bar pressure and 40 ° C temperature, with the 9% w/v GO as inorganic nanofiller (Baldino et al. 2015). The GO loading was maximum in the CA polymer matrix, and CA morphology was unaffected with the SCF used as solvent, and sulfur and Cl impurities which were present in the GO were removed completely after SCF treatment. These results exhibit additional advantages of employing SCF for PNCs fabrication for removing unwanted impurities in the nanofillers, which affect the performance of PNCs (Baldino et al. 2015).

Ma and coworkers also recently reported the fabrication of CNF/PVDF polymer nanocomposite material (CNF = cellulose nanofiber, PVDF = poly(vinyl difluoride)) by using scCO2 as a defoaming medium to enhance nanofiller dispersion in PVDF matrix (Ma et al. 2022). They have utilized the supercritical fluid-assisted thin cell wall stretching-defoaming method (**SFS-D method**) to enhance the dielectric properties of PNCs, and they attributed this phenomenon to an increase in the nanomaterial interface due to SCF, and uniform distribution of CNF nanofiller in PVDF, as shown in Fig. 9.14. Moreover, it was observed that electronic properties like electric breakdown strength and EMI (Electromagnetic interference) shielding property, as well as thermal stability, were enhanced compared to the agglomerated nanocomposite structures, which were untreated with SCF, as shown in Fig. 9.14 (Ma et al. 2022).

To improve photoluminescence and stability for application in solar cells, Mumin and coworkers fabricated core/shell CdS/ZnS quantum dots (QDs) encapsulated in poly(ethylene-co-vinyl acetate) using $scCO_2$ (Mumin et al. 2018). The QDs were first coated with silica. These silica-coated QDs were then vinyl-functionalized with vinyltrimethoxysilane (either before or during polymerization). Characterization using SEM, TEM, and CLSM revealed that supercritical CO_2 (scCO2) significantly enhanced QD dispersion within the EVA polymer network. The two-step process (functionalization before polymerization) yielded the most uniform QD distribution. Furthermore, better encapsulation was observed in the case of the two-step approach in comparison to the in situ one-step process, and enhanced optoelectrical properties and thermal insulation were reported for the polymer nanocomposites synthesized using the above method.

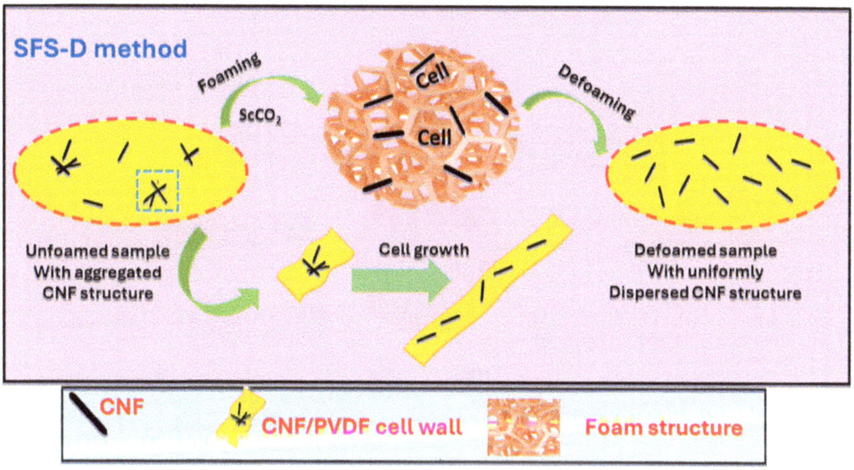

Fig. 9.14 Fabrication of CNF/PVDF nanocomposite using SCF

A novel magnetic polymer nanocomposite consisting of Fe_3O_4 nanoparticles and natural rubber was synthesized using a rapid, versatile, reliable, safe, environmentally friendly, and cost-effective in situ latex method. The scCO2 was utilized to improve the dispersion of magnetic nanofillers inside the natural rubber, resulting in the nanocomposite's enhanced magnetorheological behavior and mechanical properties. Notably, the tensile strength of the nanocomposite, without additional reinforcing fillers, reached approximately 14.6 MPa (Wang et al. 2019). The magnetic PNCs exhibited high sensitivity and magnetic properties suitable for application in the fabrication of biocompatible sensors.

9.5 Plant Extract-Mediated Synthesis of Polymer Nanocomposites: A Focus on Green Methods

9.5.1 Introduction

The amalgamation of nanotechnology and polymer science has resulted in the creation of sophisticated materials with customized attributes. Nevertheless, conventional synthetic methods often use aggressive chemicals and energy-demanding procedures, prompting environmental apprehension. The concepts of green chemistry prioritize the creation of sustainable and eco-friendly methodologies. In this context, the synthesis mediated by plant extracts has garnered considerable interest as a viable green option for the production of polymer nanocomposites (Ameta et al. 2021). Green synthesis, using plant extracts as reducing and stabilizing agents, has emerged as a viable alternative to conventional procedures. Plant extracts are abundant in phytochemicals, including flavonoids, terpenoids, and alkaloids, which

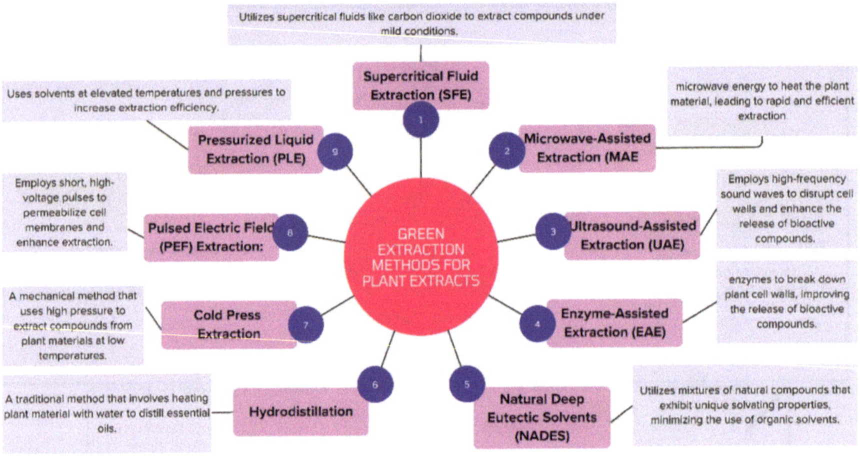

Fig. 9.15 Extraction of bioactive compounds from plant materials, a crucially important and most fundamental step in green synthesis

exhibit intrinsic reducing and capping characteristics. These phytochemicals may efficiently decrease metal ions to nanoparticles while concurrently stabilizing them to avert agglomeration. This bioinspired methodology has numerous significant benefits, including environmental sustainability, biocompatibility, ecological friendliness, and economic efficiency (Singh et al. 2023).

9.5.2 *Green Extraction Methods for Plant Extracts*

Efficient and sustainable extraction of bioactive compounds from plant materials is crucial for green synthesis. Several green extraction techniques have emerged and are depicted in Fig. 9.15.

These approaches have several benefits compared to traditional procedures, including reduced solvent use, abbreviated extraction durations, and enhanced extraction yields.

9.5.3 *Novel Case Studies of Plant Extract-Mediated Synthesis*

Numerous recent research have shown the capability of plant extract-mediated synthesis for generating functional polymer nanocomposites:

9.5.3.1 Phyto-Extract-Mediated Synthesis, Electrochemical and Optical Properties

The research conducted by Stozhko et al. (2024) underscores the green production of silver nanoparticles (AgNPs) using plant extracts, accentuating the significance of antioxidants in modulating nanoparticle characteristics.

Methodology

Four plant species—gooseberry (Ribes uva-crispa), blue honeysuckle (Lonicera caerulea), strawberry (Fragaria vesca), and sea buckthorn (Hippophae rhamnoides)—were chosen for their antioxidant activity (AOA). Extracts were obtained by boiling plant powders in water at 80 °C, then followed by centrifugation. These extracts diminished and stabilized AgNPs when combined with silver nitrate at 70 °C.

Key Findings

Morphology: TEM (Fig. 9.16) revealed spherical AgNPs ranging from 7 nm (sea buckthorn) to 13 nm (gooseberry). Higher AOA resulted in smaller nanoparticles.

Optical Properties: UV-visible spectroscopy showed a "blue shift" in SPR wavelengths with increasing AOA, reflecting reduced particle size. Figure 9.17 presents the optical properties of AgNPs, with a "blue shift" indicating smaller nanoparticle sizes for higher AOA.

Electrochemical Properties: EIS and CV indicated enhanced conductivity and lower charge transfer resistance for AgNPs synthesized with higher AOA extracts. Figure 9.18 shows Nyquist plots comparing charge transfer resistance, demonstrating improved conductivity with higher AOA, while Fig. 9.19 illustrates CV curves correlating higher AOA with increased electroactivity and redox behavior.

Applications

This sustainable method allows accurate regulation of nanoparticle characteristics, which is essential for medicinal, catalytic, and environmental uses. Green-synthesized AgNPs may augment antibacterial and optical properties when included in polymer nanocomposites, providing sustainable material alternatives.

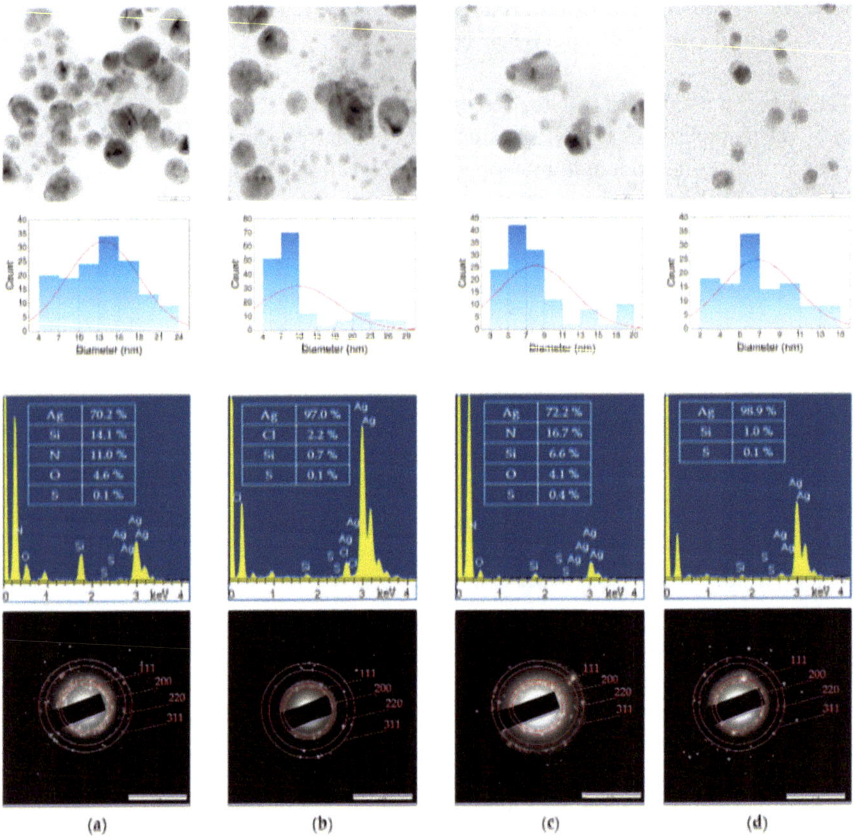

Fig. 9.16 TEM pictures, histogram (particle size distribution), EDS spectra, and SAED patterns of AgNPs synthesized from (**a**) gooseberry, (**b**) blue honeysuckle, (**c**) strawberry, and (**d**) sea buckthorn

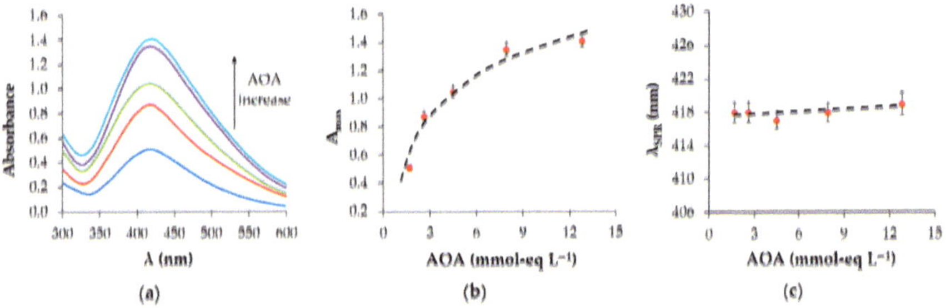

Fig. 9.17 Absorption spectra of phyto-AgNPs(b) sols obtained as a result of synthesis using sea buckthorn extracts with different AOA (**a**). Dependence of A_{max} (**b**) and λ_{SPR} (**c**) of phyto-AgNPs(b) sols on AOA of sea buckthorn extracts

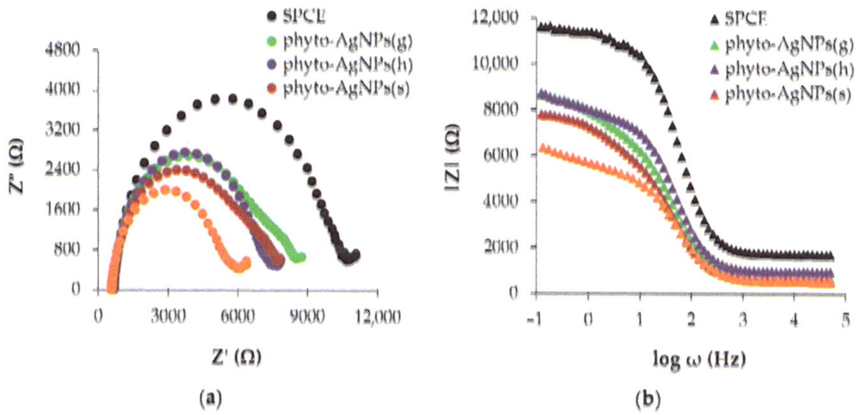

Fig. 9.18 (**a**) Nyquist plot and (**b**) Bode plot for unmodified and AgNP-modified SPCE in 0.1 mol L−1 KCl comprising 5 mmol L−1 K3[Fe (CN)6]

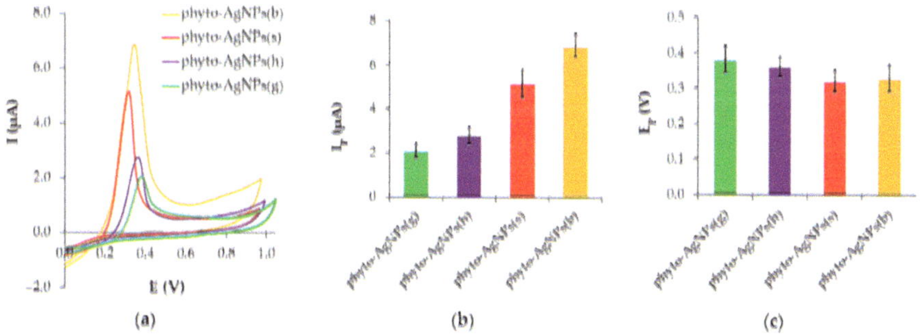

Fig. 9.19 Cyclic voltammograms of AgNPs using plant extracts at pH 5 (**a**). Dependencies of phyto-AgNP oxidation on plant extract AOA for Ip (**b**) and Ep (**c**)

9.5.3.2 Case Study: Green Synthesis of Magnetite Iron Oxide/Reduced Graphene Oxide Nanocomposites for Wastewater Treatment (Akhtar et al. 2024)

Introduction

Industrial effluents substantially contribute to water pollution, posing a critical environmental challenge. The study by Akhtar et al. (2024) demonstrates the effectiveness of green-synthesized nanocomposites—magnetite iron oxide/reduced graphene oxide (FeNPs/rGO)—in degrading methylene blue dye from wastewater. This eco-friendly synthesis employs *Azadirachta indica* (Neem) leaf extract for the reduction and stabilization of nanomaterials, aligning with sustainable development goals.

Methodology

Synthesis of FeNPs/rGO Nanocomposites
FeNPs and rGO were produced from iron chloride hexahydrate and graphene oxide, respectively, via a green deposition approach. The extract of *A. indica* functioned as a reducing, capping, and stabilizing agent. The resultant FeNPs/rGO nanocomposites were studied using methods such as UV-vis spectroscopy, XRD, FTIR, and EDX-SEM.

Photocatalytic Activity
Photocatalytic efficacy was evaluated using dye, methylene blue, under visible light. Nanocomposites were incorporated into MB solutions, and degradation efficiency was assessed by monitoring absorbance variations at 665 nm over time.

Crucial Findings

Efficient Dye Degradation The FeNPs/rGO nanocomposites accomplished 75% degradation of methylene blue in 1 h, markedly surpassing the performance of FeNPs alone, which reached just 16% degradation. This higher activity is due to synergistic interaction between magnetite iron oxide and reduced graphene oxide, leading to improved photocatalytic capabilities and a decreased band gap.

Eco-Friendly and Scalable The use of *Azadirachta indica* ensured an environmentally benign synthesis without toxic reducing agents. Magnetic properties of the FeNPs allowed for easy recovery and reuse, reducing secondary waste.

Mechanistic Insights

FTIR analysis (Fig. 9.20) identified the existence of functional groups such as –OH, –COOH, and C=O, which are essential for pollutant adsorption and degradation. SEM-EDX (Fig. 9.21) demonstrated a consistent dispersion of particles and their elemental content. The surface morphology (Fig. 9.22) exhibited a rough texture across all materials, with particles displaying cohesive, clustered, and spherical forms. SEM XRD (Fig. 9.23) validated the crystalline structure of the nanocomposites.

Implications

This study highlights the effectiveness of FeNPs/rGO nanocomposites as practical and sustainable photocatalysts for wastewater treatment. Their ability to efficiently decompose colors in visible light, along with an eco-friendly production method, makes them beneficial for widespread use in industrial wastewater treatment.

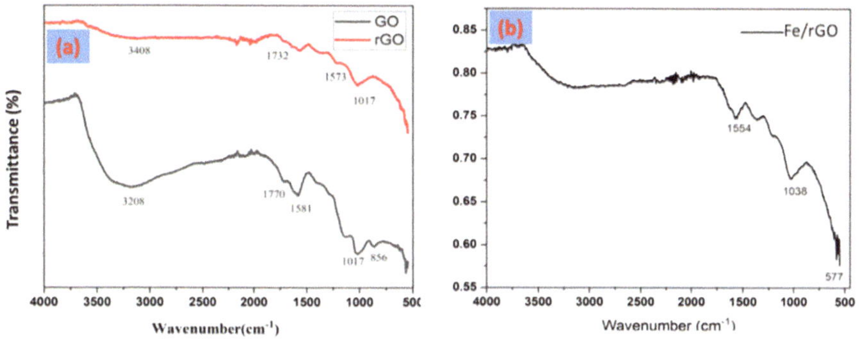

Fig. 9.20 FTIR spectrum of graphene oxide produced by chemical approach, and reduced graphene oxide (rGO) manufactured via green method (**a**), together with the IR spectrum of magnetite FeO/reduced graphene oxide (FeNPs/rGO) nanocomposite generated with green deposition (**b**)

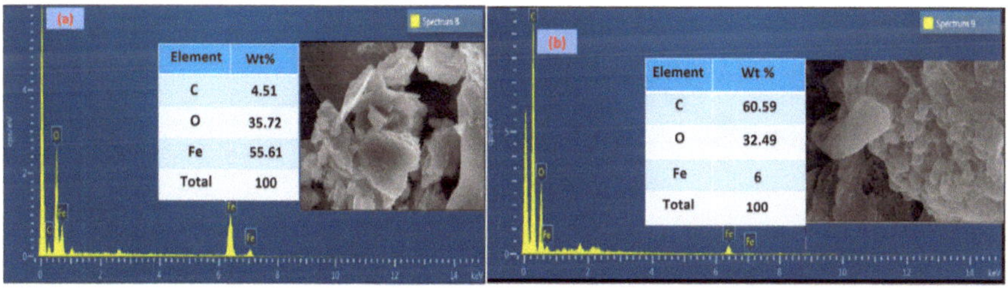

Fig. 9.21 The pattern of EDX of FeNPs produced using a green technique (**a**) and the EDX diagram of FeNPs/rGO nanocomposites generated by a green deposition process (**b**)

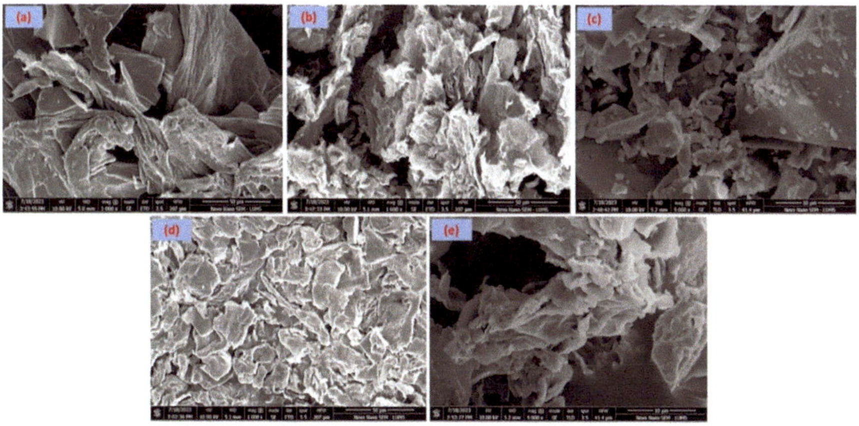

Fig. 9.22 Surface morphology of Graphene Oxide nanoparticles (**a**), rGO nanoparticles (**b**), Fe nanoparticles (**c**), Fe/GO nanocomposites (**d**), and Fe/rGO nanocomposites (**e**), respectively

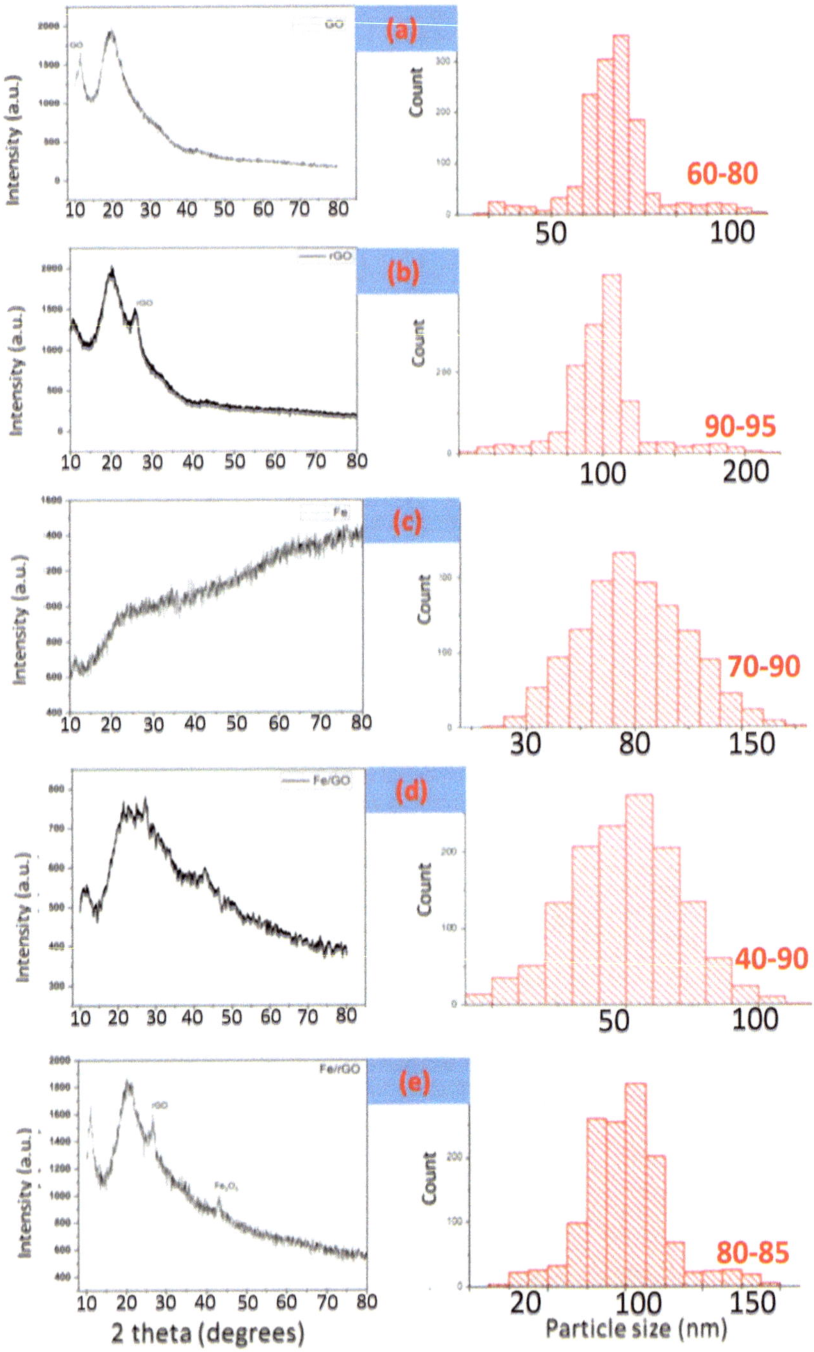

Fig. 9.23 XRD analysis and average particle size for (**a**) GO nanosheets, (**b**) rGO nanoparticles, (**c**) Fe nanoparticles, (**d**) FeGO, and (**e**) the FerGO nanocomposite synthesized through a green deposition method

Conclusion of the Case Study The ecologically sustainable synthesis of FeNPs/rGO nanocomposites is a viable solution for wastewater cleanup. The study demonstrates the integration of nanotechnology and green chemistry in reducing environmental pollutants and promoting sustainability.

9.5.3.3 Case Study

Eco-friendly production of AgNPs from *M. oleifera* and their effectiveness against gram-negative bacteria by targeting quorum sensing and biofilm formation.

The case study is derived from the scholarly publication titled "Green Synthesis of Silver Nanoparticles Using *Moringa Oleifera* and Its Efficacy Against Gram-Negative Bacteria Targeting Quorum Sensing and Biofilms," authored by Zarrin Haris and Iqbal Ahmad.

The rise of multidrug-resistant (MDR) microorganisms poses a considerable worldwide health threat. Resolving this problem necessitates novel solutions, one of which involves the use of green-synthesized nanoparticles. This case study examines the synthesis of silver nanoparticles (AgNPs) using the aqueous leaf extract of *Moringa oleifera* (MO) and their effectiveness against Gram-negative bacterial infections. MO-AgNPs have been promised as alternative therapeutic agents in addressing antimicrobial resistance (AMR) by interfering with quorum sensing (QS) and biofilm formation.

Methodology

Synthesis of Silver Nanoparticles The leaves of *Moringa oleifera* were harvested, shade-dried, and ground into powder. An aqueous extract was obtained by boiling 5 g of leaf powder in 100 mL of double-distilled water at 90 °C for 1 hour. The extract was filtered and combined with a 1 mM $AgNO_3$ solution. The reaction displayed a color transition from colorless to dark brown, indicating the synthesis of AgNP. The nanoparticles underwent centrifugation, followed by washing and drying at 55 °C overnight (Fig. 9.24).

Prodigiosin Production In *S. marcescens*, pigment synthesis was inhibited by 77.4% at 16 μg/mL (Fig. 9.25).

Exoprotease Activity MO-AgNPs reduced exoprotease activity by 76.02% at the highest tested concentration.

Swarming Motility Bacterial motility was inhibited by 86.5% at sub-MIC levels.

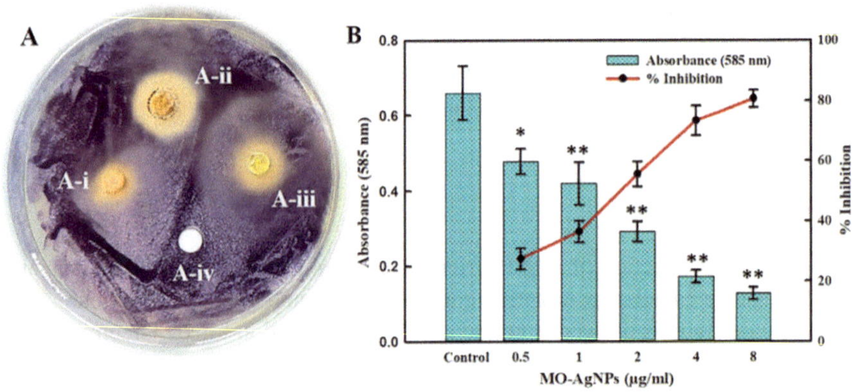

Fig. 9.24 The impact of MO-AgNPs on violacein production (Haris and Ahmad 2024)

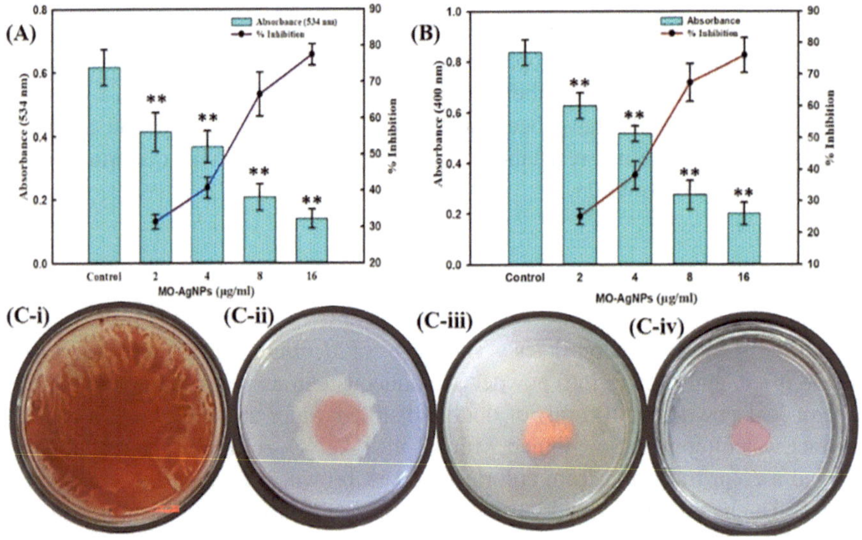

Fig. 9.25 Impact of MO-AgNPs on prodigiosin synthesis in *Serratia marcescens*. The impact of MO-AgNPs on exoprotease synthesis in *S. marcescens* (Haris and Ahmad 2024)

Biofilm Inhibition

- **Crystal Violet Assay**: Biofilm formation in *S. marcescens*, *E. fergusonii*, and *C. violaceum* was reduced by 81.5%, 77.9%, and 82.4%, respectively.
- **Microscopic Analysis**: SEM and CLSM revealed significant disruption of biofilm structure (Fig. 9.26).

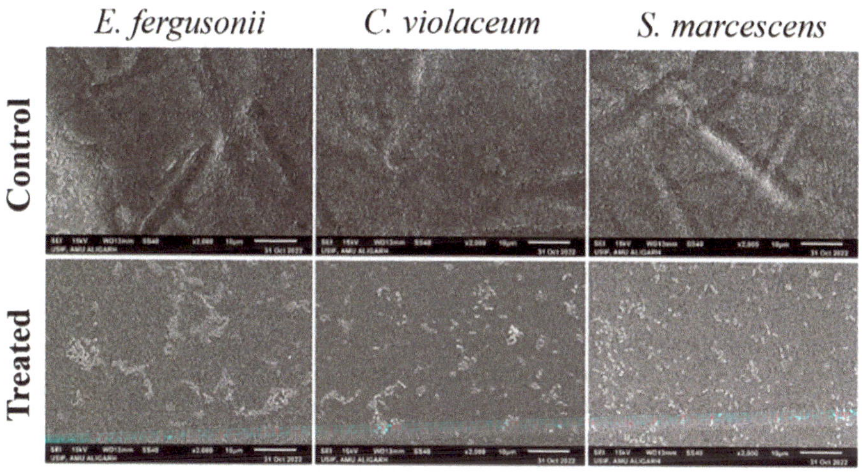

Fig. 9.26 SEM images of *E. fergusonii, C. violaceum, and S. marcescens* biofilm with and without sub-MICs of MO-AgNP

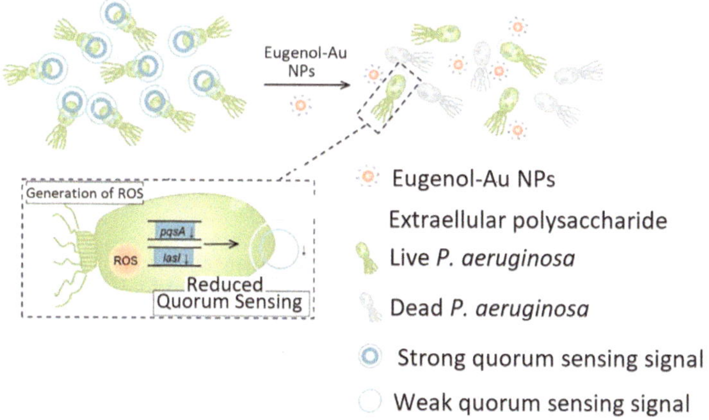

Fig. 9.27 Au nanoparticles impair the bacterial quorum sensing mechanism and increase intracellular reactive oxygen species production (Chen et al. 2024)

Biomedical Significance of the Silver/Gold Nanoparticles

Quorum Sensing (QS) Inhibition Gold nanoparticles (AuNPs) may interfere with microbial intercellular communication routes (Sousa et al. 2023), thereby reducing biofilm development and the generation of virulence factors (Harriott 2019). The advanced quorum-sensing (QS) mechanism present in microorganisms confers resistance to various drugs, hampering infection control efforts (Chen et al. 2024) (Fig. 9.27).

Reactive Oxygen Species (ROS) Generation Reactive oxygen species generated by gold nanoparticles promote oxidative stress, resulting in damage to microbial membranes and DNA. This stress may significantly impair microbial membranes and DNA, leading to cell death and further cytotoxic consequences. The toxicity of AuNPs is partially attributable to their capacity to infiltrate cell membranes, inducing apoptotic-like mechanisms and considerable DNA damage (Canaparo et al. 2020) (Figs. 9.28 and 9.29).

Anticancer Activity Functionalized AuNPs use the increased permeability and retention (EPR) effect in tumor tissues, facilitating selective accumulation. Tumors often exhibit inadequately structured blood vessels characterized by substantial intercellular spaces, facilitating the leakage and accumulation of nanoparticles into tumor tissue more efficiently than in normal tissues with densely arranged blood vessels (Lee et al. 2014). Gold nanoparticles (AuNPs) elevate reactive oxygen species (ROS) levels, resulting in mitochondrial malfunction and subsequent cancer cell death by apoptotic mechanisms (Sun et al. 2018; Karande and Gawade 2020).

Drug Delivery AuNPs' surface can be functionalized with ligands to deliver chemotherapeutics directly to cancer cells, reducing off-target toxicity (Huang et al. 2023).

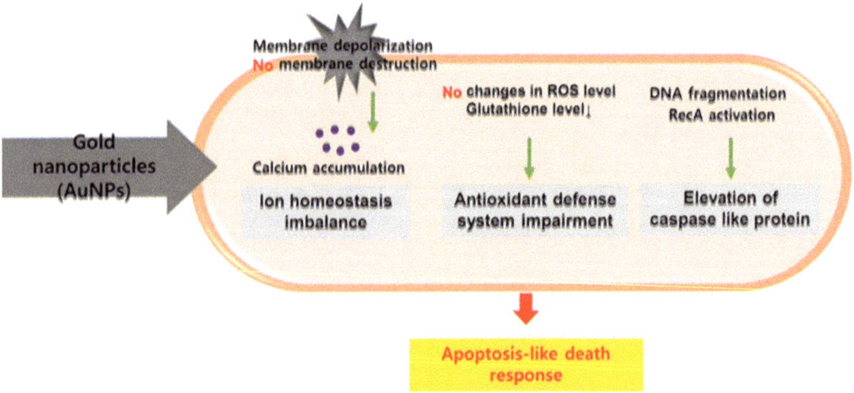

Fig. 9.28 Role gold nanoparticles in ROS (Lee and Lee 2018)

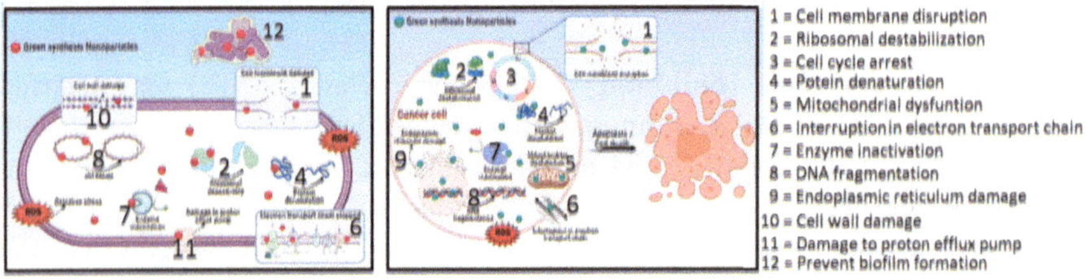

Fig. 9.29 Antimicrobial and anticancer potential of metal nanoparticles (Hosseingholian et al. 2023)

9.5.3.4 Conclusion and Future Directions

This case study posits that the production of AuNPs mediated by *Azadirachta indica* may provide a dual-action medicinal platform. Although preliminary results indicate this potential, more in vitro and in vivo investigations are necessary to confirm effectiveness and safety. The integration of AuNPs with traditional pharmaceuticals may provide synergistic benefits, fulfilling the pressing need for multifunctional agents in antimicrobial resistance and cancer treatment.

9.6 Microbial and Enzymatic Synthesis of Polymer Nanocomposites

9.6.1 Introduction

The amalgamation of nanoscale metals and polymers presents a persuasive substitute for conventional composite materials in attaining innovative functionality. Metal-based polymer nanocomposites have become essential instruments for material scientists, facilitating progress in several domains like electronics, biomedicine, catalysis, and energy storage (Pramanik and Das 2019). Microorganisms are essential for the synthesis of metal nanostructures and function as bionanofactories. Microbial production transpires via extracellular and intracellular routes, resulting from reduction, enzymatic oxidation, and chelation. Reductases are essential for extracellular production, enabling the reduction of ions into their respective metal nanoparticles or composites (Khalil et al. 2022). These metal-based nanoparticles or composites have several uses across various sectors, including wastewater treatment, medicine, drug delivery, and textiles. One of the notable uses in medicine is the antibacterial mechanism of metal nanoparticles. Authors elucidated the many applications of these composters, as seen in Fig. 9.30 (Suvetha and Mani 2024).

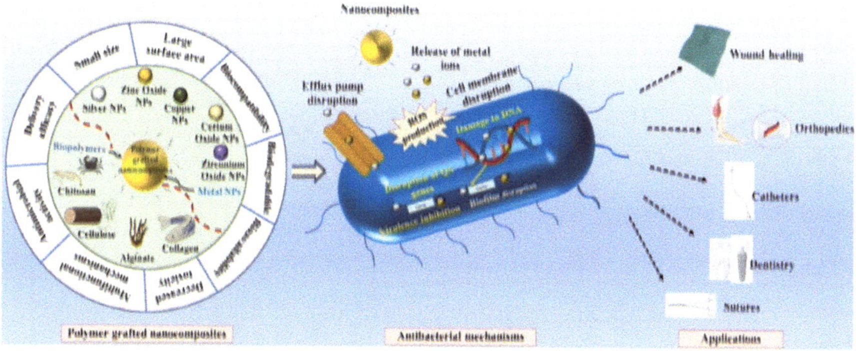

Fig. 9.30 Polymer grafted metal nanocomposites with antibacterial mechanism and diverse applications in medicine (Suvetha and Mani 2024)

Traditional chemical synthesis techniques often use hazardous chemicals, extreme temperatures, and energy-demanding procedures that pose environmental and health hazards (Ahmed et al. 2022). Microorganisms (bacteria, fungi, and algae) and enzymes are being investigated as natural agents for the production of polymer nanocomposites to address these issues (Singh et al. 2023).

The creation of microbes and enzymes relies on the intrinsic capacity of biological systems to generate reduced agents and enzymes that function as catalysts. These technologies allow the creation of biocompatible and non-toxic nanostructures in mild, aqueous environments, making them environmentally friendly and cost-effective. This part provides insights into the mechanism of metallic nanocomposite synthesis from microbial sources and the many bacteria engaged in this process (Fig. 9.31).

9.6.2 *Mechanisms of Microbial and Enzymatic Synthesis*

9.6.2.1 Bacterial Synthesis

Bacteria excrete enzymes and metabolites that degrade polymeric precursors, facilitating the synthesis of nanocomposites (Alsaiari et al. 2023). These may serve as secure and economical instruments for the synthesis of metallic nanoparticles such as gold, silver, copper, zinc, titanium, palladium, and nickel. The extracellular reduction of metal ions for nanoparticle synthesis utilizes microbial enzymes and proteins, as well as components from bacterial or fungal cell walls (Ying et al. 2022). In contrast, the intracellular process begins with the electrostatic attraction of metal ions by carboxyl groups in the microbial cell wall, followed by the translocation of metal ions through the cell wall and their reduction by intracellular proteins, enzymes, or cofactors to form nanoparticles. The culture filtrate is collected and combined with an aqueous salt solution for extracellular synthesis. The synthesis is observed through the color change of the combined solution (Siddiqi et al. 2018).

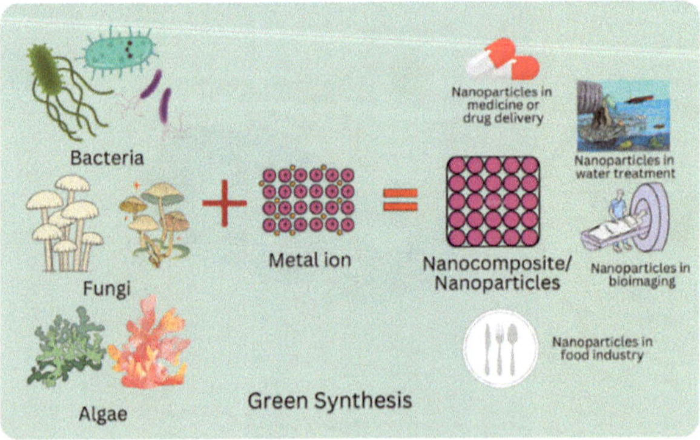

Fig. 9.31 Components of metal nanoparticles from microns

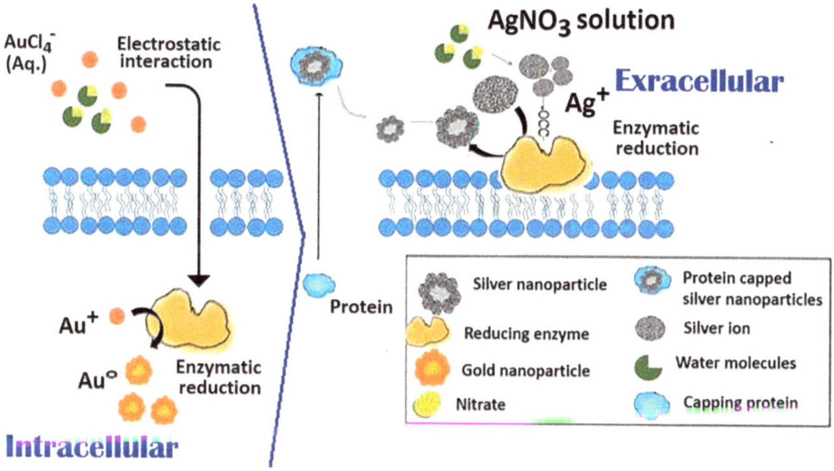

Fig. 9.32 Microbial enzymes in bio reduction of metal, metalloid, and nonmetal ions to nanoparticles (Lahiri et al. 2021)

The transition from bright yellow to dark brown signifies the creation of silver nanoparticles (AgNPs). Targeting intracellular synthesis, biomass is meticulously washed with sterile water following the incubation of microorganisms under optimal growth conditions, and subsequently incubated with a metal ion solution. The color change serves as an indicator of synthesis. Nanoparticles are subsequently gathered using suitable techniques such as ultrasonication, centrifugation, and washing. Synthesized particles exhibit spherical, disk, cuboidal, hexagonal, and triangular geometries (Siddiqi et al. 2018; Ali et al. 2020). Biomolecules in the microbial culture medium or cell wall constituents can reduce metal ions (Fig. 9.32) through an extracellular mechanism. Intracellularly, electrostatic interactions of the functional groups on the cell wall attract metals, metalloids, and metal ions, which then interact with intracellular proteins to facilitate nanoparticle synthesis (Fig. 9.32). The simplicity of extraction and high efficacy characterize extracellular reduction, making it more advantageous than intracellular reduction. Deceased bacteria may also be used for the creation of nanoparticles, similar to active bacteria. Substantial amounts of nanoparticles (100–200 nm in size) may be synthesized in a pure manner by the extracellular secretion of bacterial enzymes. The bacterial cell wall plays a critical role, allowing metals to permeate into the cytoplasm and then be transported back to the wall for extracellular secretion. The metal-binding ability of the bacterial cell and S-layer renders them advantageous for applications in bioremediation. The cell wall with metal-binding sites can be changed by chemical reactions for specific groups, such as amines and carboxyl groups, which converts positive charge into negative charge (Kapoor et al. 2021) (Fig. 9.33).

Few examples of the synthesis of silver nanoparticles using bacterial extracts are listed in Table 9.12.

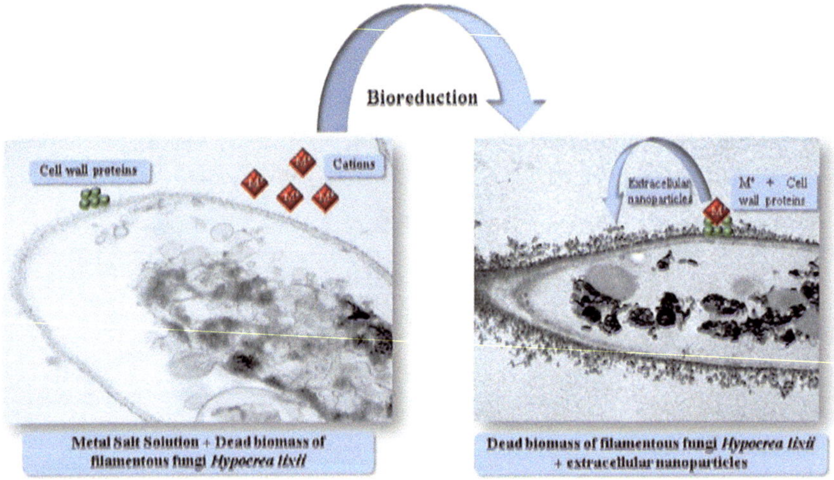

Fig. 9.33 Illustration of the mechanism of extracellular synthesis of metal nanoparticles by dead biomass of *Hypocrea lixii*

9.6.2.2 Extracellular Enzymes

Extracellular enzymes function as reducing agents and are crucial in the synthesis of metallic nanoparticles. Enzymes such as acetyl xylan esterase, cellobiohydrolase D, and glucosidase found in fungi contribute to the creation of metallic nanoparticles. *Rhodopseudomonas capsulata* synthesizes AuNP extracellularly via NADH-dependent reductase enzymes for electron transfer from NADH. Upon receiving electrons, gold ions were reduced to become AuNPs. *Fusarium oxysporum* is used as a reducing agent for the synthesis of AgNPs and AuNPs. Shuttle quinone and nitrate-dependent reductase derived from *Fusarium* species were used for the extracellular synthesis of nanoparticles. Enzymes from *Fusarium semitectum* and *Fusarium solani* were used for the extracellular synthesis of AgNPs.

Intracellular Enzymes Alkalotolerant Actinomycetes, namely *Rhodococcus* species, and alkalo-thermophilic *Thermomonospora* species were used for the intracellular synthesis of gold nanoparticles. Exposure of *Verticillium* species to a silver ion solution resulted in intracellular reduction and the synthesis of silver nanoparticles. A similar method was used for the creation of gold nanoparticles with *Verticillium* as a prolific supply of reducing enzymes, indicating a potential natural approach for the intracellular synthesis of metal nanoparticles with yeasts. A potential mechanism for intracellular nanoparticle creation involves the electrostatic interaction between metal cations and amid groups in yeast cell wall enzymes, leading to the reduction of ions by these enzymes, resulting in the buildup of metal ions and the formation of nanoparticles (Fig. 9.34). The diagram illustrates a potential process for the intracellular creation of nanoparticles by yeasts (Kapoor et al. 2021).

Table 9.12 Bacteria-mediated synthesis of silver nanoparticles (Siddiqi et al. 2018)

Location	Size (nm) and shape	Bacteria
Extracellular	8–12; spherical	*Acinetobacter calcoaceticus*
	4–40	*A. haemolyticus* MMC8
Extracellular and intracellular	6.4	*Aeromonas* sp. SH10
Extracellular	63–90	*Bordetella* sp.
	25–35; spherical	*Enterobacter aerogenes*
	42.2–89.6; spherical	*Escherichia coli*
		Geobacter sulfurreducens
	10	*Gluconobacter roseus*
Intracellular	25	*Idiomarina* sp.
Extracellular	15–37; spherical	*Klebsiella pneumoniae*
	5–32	
	10–40; quasispherical	*Morganella* sp.
Extracellular and intracellular	10–20; spherical	*Proteus mirabilis*
Extracellular	6.3 ± 4.9; spherical, disk-shaped	*Pseudomonas aeruginosa* SM1
	8–24; spherical	
Intracellular	5–25; quasispherical	
Extracellular	Spherical 3–15	*Rhodobacter sphaeroides*
	Spherical 5–20	*Rhodopseudomonas palustris*
	2–16; spherical (Ag2S)	*Shewanella oneidensis* MR-1
	93; cuboidal	*Stenotrophomonas maltophilia*
Extracellular and intracellular	50–100; Spherical	*Vibrio alginolyticus*
Extracellular	14.86; spherical, triangular, rod-shaped	*Xanthomonas oryzae*
	10–80	*Yersinia enterocolitica*
Extracellular and periplasmic space	5–15	*Bacillus* sp.
Intracellular	4–5; spherical	*B. cereus*
Extracellular	12 and 65; spherical and triangular	*B. flexus*
Cell-free extract	18.69–63.42; spherical	*B. licheniformis* Dahb1
Extracellular	5–30; spherical	*B. safensis* LAU 13
–	10–30; spherical	*B. methylotrophicus* DC3
Extracellular	Triangular, hexagonal	*B. subtilis*
–	20–60; polydispersed (AgCl)	*B. subtilis* MTCC 3053
Extracellular	44–143	*B. thuringiensis*

(continued)

Table 9.12 (continued)

Location	Size (nm) and shape	Bacteria
Intracellular	10–50; spherical	*Brevibacterium casei*
Extracellular	10–15	*Corynebacterium* SH09
	10–80	*Enterococcus faecalis*
	5–50; spherical	*Exiguobacterium* sp.
	5–35; spherical	*Geobacillus stearothermophilus*
	2–20; spherical (Ag_2O)	*Lactobacillus mindensis*
	10–15; spherical	*Rhodococcus* sp.
	10–80	*Staphylococcus epidermidis*
	20–40; spherical	*Thermoactinomyces* sp.
	10–100; spherical	*Ureibacillus thermosphaericus*

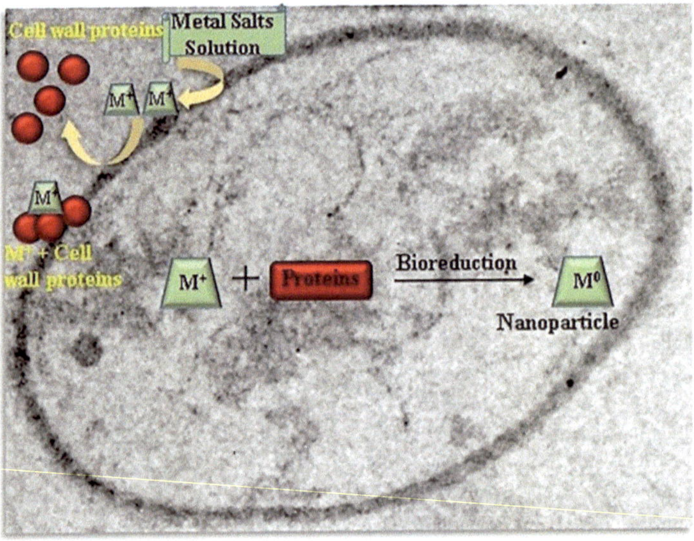

Fig. 9.34 Schematic illustration of the mechanism of intracellular nanoparticle formation by yeasts (Kapoor et al. 2021)

9.6.2.3 Fungal Synthesis

Fungi are eukaryotic microorganisms favored for the large-scale synthesis of nanoparticles. Fungal biomass typically exhibits a growth rate surpassing that of bacteria, and fungi continue to proliferate even subsequent to the production of nanoparticles (Tsivileva et al. 2021). The growth of mycelium, characterized by its extensive surface area for interactions, facilitates the accumulation of diverse fungal biopolymers that provide many reactions for the synthesis of nanoparticles and their encapsulation inside the shell structure. Live fungi and isolated fungal-cell fractions often exhibit comparable efficacy in nanoparticle formation and characteristics. The accumulated nanoparticles induce various physiological and biochemical alterations in bacteria; yet, fungi may withstand harmful nanoparticle materials, such as heavy metals, in the reaction media at concentrations far beyond the established allowed limits in polluted environments. In the presence of fungus, several kinds of nanoparticles may grow on the surface of mycelia, with a visibly detectable transformation indicated by a color shift in the fungal biomass, such as the transition from pink to red due to the deposition of elemental selenium. The mechanism of nanoparticle production from fungus is intricate. This process consists of two primary steps: first, the concentration of metal ions on the fungal mycelia, since positively charged metal ions are attracted to the slightly negatively charged cell wall; second, the depletion of metal ions by the fungi, resulting in nanoparticle creation (Adeleke et al. 2024). Employing endophytic fungi for the remediation of waste effluents and environmental contaminants is a potential approach for pollution management and the rehabilitation of degraded ecosystems. Table 9.13 (Adeleke et al. 2024) is an overview of the nanoparticles generated from the fungus.

Mechanism of Biosynthesis Using Fungi

Fungi, often classified under *Ascomycota* and *Basidiomycota*, react to various metalloid and metal precursors, including their salts and organic compounds. Figure 9.35 illustrates the absorption and transport of these precursors (metal ions) into fungal cells via various channels and proteins, facilitating subsequent metabolic reactions.

Table 9.13 Nanoparticle synthesis using fungi isolated from respective plant sources (Adeleke et al. 2024)

NPs	Plant source	Fungi	Plant source
TiO_2	*Sorghum bicolor*	*Trichoderma citrinoviride*	*Sorghum bicolor*
Au	*Cleistes fragrans*	*Fusarium solani*	*Cleistes fragrans*
Au	*Commiphora wightii*	*Cladosporium* sp.	*Commiphora wightii*
Au	*Rauvolfia tetraphylla*	*Alternaria* sp.	*Rauvolfia tetraphylla*
Fe	*Tecomella undulata*	Penicillium oxalicum	*Tecomella undulata*
CoO	*Neoponera foetida*	*A. nidulans*	*Neoponera foetida*
ZnO	*Millingtonia hortensis*	*Xylaria acuta*	*Millingtonia hortensis*
Zn	*Nothapodytes foetida*	*A. flavus*	*Nothapodytes foetida*
Cu	*Calendula arvensis*	*Streptomyces capillispiralis*	*Calendula arvensis*
CuO	*Aegle marmelos*	*A. terreus*	*Aegle marmelos*
ZnO	*Balanites aegyptiaca*	*Periconium* sp.	*Balanites aegyptiaca*
Ag	*Avicennia marina, Suaeda monica* and *Rhizophora mucronata*	*A. conicus, Penicillium janthinellum,* and *Phomosis* sp.	*Avicennia marina, Suaeda monica* and *Rhizophora mucronata*
Ag	*Centella asiatica*	*A. versicolor*	*Centella asiatica*
Ag	*Stypandra glauca*	*A. niger*	*Stypandra glauca*
Au	*Chonemorpha fragrans*	F. solani	*Chonemorpha fragrans*
Fe_3O_4	*Origanum majorana*	*A. terreus*	*Origanum majorana*
Au	*Sargassum wightii*	*C. cladosporioides*	*Sargassum wightii*

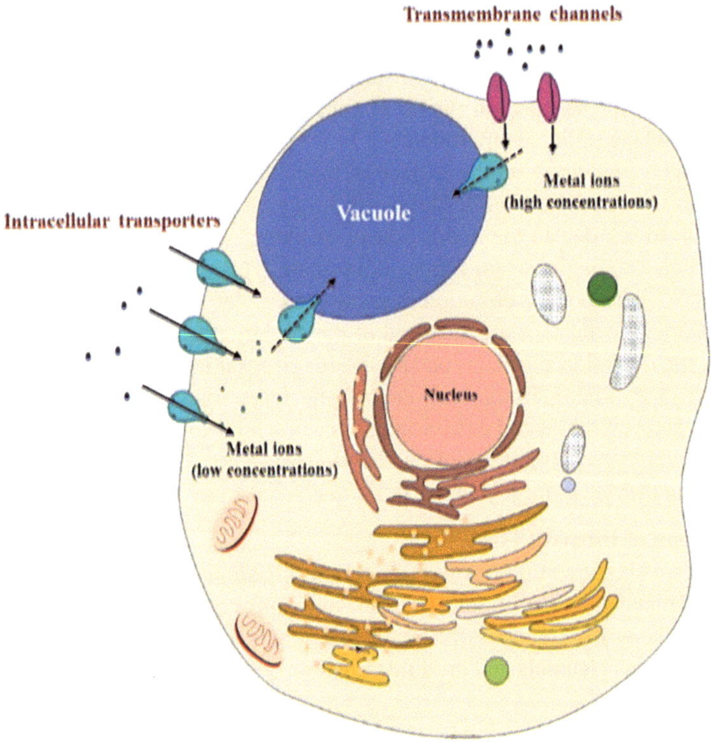

Fig. 9.35 Transport of precursors across the membrane (Rami et al. 2024)

In the creation of Ag and Au nanoparticles, fungi often use precursor chemicals such as silver nitrate or gold chloride. These ions are incorporated into the fungal biomass, initiating a series of metabolic processes. Likewise, for metalloids like selenium and tellurium, fungi interact with selenites, selenates, tellurites, or tellurates. The complex mechanisms that began with this recognition phase provide the groundwork for the next stages in NP synthesis (Rami et al. 2024).

9.6.2.4 Algal Synthesis

Algae have historically been used commercially and industrially for food, animal feed, additives, cosmetics, medicines, fertilizers, and bioremediation agents. The justification for using algae in green synthesis includes the absence of exogenous reducing or capping agents, little energy input, cost-effective synthesis, and straightforward reaction processes. Nevertheless, the use of algae for nanoparticle manufacturing, a field termed phyco-nanotechnology, remains nascent (Chaudhary et al. 2020). Algae are an excellent source of phytonutrients, proteins, pigments, and other nutrients; the green synthesis process has advanced throughout time, making it a viable biofabricator for nanoparticle manufacturing (Sampath et al. 2022). There are several applications of nanoparticles in various domains, including medication delivery, textiles, and imaging. A notable use of nanoparticles generated from the brown algae Cystoseira baccata. These brown algae-derived AuNPs have significant potential for colorectal cancer therapy (Michael et al. 2022).

Mechanism of Biosynthesis

The production of nanoparticles from algae is researched via both intracellular and extracellular methods, depending on the site of nanoparticle formation. The intracellular production pathway is dose-dependent, whereby NP biosynthesis transpires inside the algal cell (Fig. 9.35). NADPH, or NADPH-dependent reductase, generated during metabolic activities such as nitrogen fixation, photosynthesis, and respiration, functions as a reducing agent (Chaudhary et al. 2020) (Fig. 9.36).

Algal-synthesized nanoparticles are regarded as safe, biocompatible, and environmentally friendly; hence, they have been explored and examined for their uses in medicine, including antibacterial, antifungal, anticancer, antioxidative, and antidiabetic properties (Nawaz et al. 2022). A summary of many algal-mediated biosynthetic processes is shown in Table 9.14.

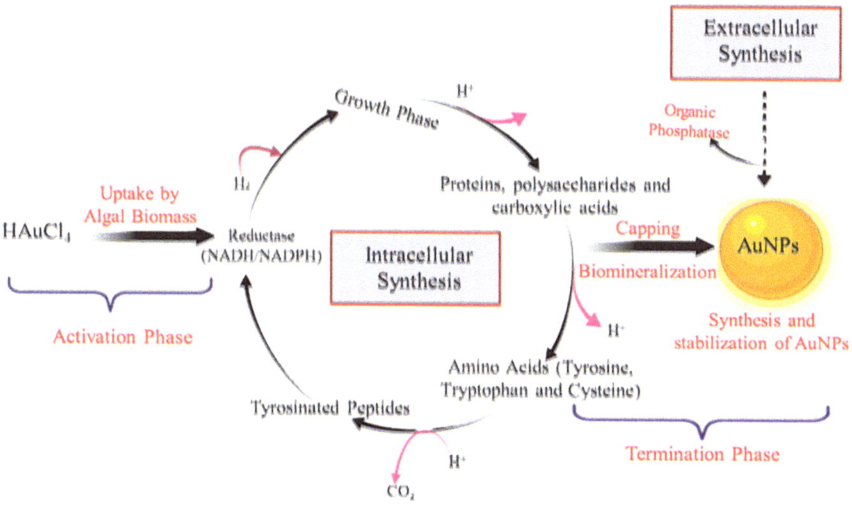

Fig. 9.36 Illustration of mechanism: extracellular and intracellular synthesis of algae-mediated nanoparticles (Chaudhary et al. 2020)

Table 9.14 Algae-mediated, extracellular biosynthesis of nanoparticles (Chaudhary et al. 2020)

Algae	Reducing agent	NPs	Shape and Size
Turbinaria conoides	Carbonyl groups, polyamines	Ag	Spherical, 96 nm
Sargassum spp.	Hydroxyl and amine functional groups	Au	Hexagonal, 300–400 nm
Fucus vesiculosus	Hydroxyl groups	Au	Spherical, 20–50 nm
Sargassum polycystum	Extracellular terpenoids	Ag	Spherical, 7 nm
Turbinaria ornata	Organic moieties	Ag	Spherical, polydispersed, 22 nm
Sargassum muticum	Sulfate and hydroxyl groups	Ag	Spherical, crystalline, 43–79 nm
Sargassum muticum	Sulfate and hydroxyl groups	ZnO	Hexagonal, 30–57 nm
Laminaria japonica	Polysaccharides	Au	Spherical, 15–20 nm
Turbinaria conoides	Carboxylic, fucoidan, polyphenolic groups	Au	Polydispersed, rectangular, triangular, 6–10 nm
Ecklonia cava	Hydroxyl and phenolic groups	Au	Spherical, 20–50 nm
Gilidiella acerosa	Metabolites	Ag	Spherical, 18–46 nm
Sargassum wightii	–	Au	Thin planner, 8–12 nm
Sargassum muticum	Polysaccharides	Ag/Au	41.0 ± 5.7 nm (Ag), 10.4 ± 1.2 nm (Au)
Sargassum myriocystum	1-cyclopentyl group	Au	Polydispersed, spherical and triangular, 15 nm
Turbinaria conoides	Biochemical groups	Au	Rectangular, square, triangular, 15 nm
Stoechospermum marginatum	Hydroxyl groups	Au	Hexagonal, triangular, spherical, 18.7–93.7 nm
Padina gymnospora	Fucoxanthin and flavonoids	Au	Spherical, 8–21 nm
Dictyota bartayresiana	Phenolics, carboxylic, and amine groups	Au	Polydispersed and spherical nanoparticles
Ecklonia cava	Amine and hydroxyl groups	Au	Spherical, FCC, and triangular, 20–50 nm
Padina gymnospora	Fucoxanthin and flavonoids	Au	Spherical, 8–21 nm
Sargassum muticum	Proteins and bio-organic compounds	Au	FCC and spherical, 5.42 nm
Sargassum tenerrimum	Secondary metabolites	Au	Anisotropic and polydispersed, 4–45 nm
Cystoseira baccata	Polyphenols, polysaccharides	Au	Poly-crystalline and spherical, 8.4 ± 2.2 nm
Sargassum wightii grevilli	Oxidation of the alcoholic group	Ag	Spherical, 8–27 nm
Sargassum ilicifolium	Bioactive compounds	Ag	Spherical, 33–40 nm
Sargassum polycystum	Hexadecane and octadecanol	Ag	FCC and spherical, 5–7 nm

(continued)

Table 9.14 (continued)

Algae	Reducing agent	NPs	Shape and Size
Sargassum plagiophyllum	Carboxylic and benzene rings	Ag	Spherical, 20–50 nm
Padina pavonica	Terpenoids	Ag	Polydispersed and spherical, 10–72 nm
Padina tetrastromatica	Alkanes	Ag	Spherical, 4 nm
Padina gymnospora	Aqueous extract	Ag	Spherical, 25–40 nm
Colpmenia sinusa	Polysaccharides	Ag	Spherical, 20 nm
Cystophora moniliformis	Metabolites, phenolic compounds	Ag	FCC, 75 nm
Sargassum cinereum	–	Ag	Triangular and spherical, 45–76 nm
Sargassum longifolium	Ketone and carboxylic groups	Ag	Cubical, 30 nm
Scaberia agardhii	Proteins	Ag	Polydispersed, 40–50 nm
Sargassum polycystum C. agardh	Metabolites and proteins	Ag	Spherical
Sargassum vulgare	Secondary OH groups	Ag	Spherical, 10 nm

9.6.2.5 Methods and Applications of Microbial Synthesized Metal Composites

Wastewater Treatment

Polymer nanocomposites produced by microbiological and enzymatic techniques have exceptional efficacy in wastewater treatment owing to their extensive surface area, adsorptive characteristics, and photocatalytic capabilities (Table 9.15).

Biomedicine

Microbially produced metal nanocomposites are appropriate for several biological applications, including antibacterial and anticancer uses (Fig. 9.37). Other medical conditions may also be addressed utilizing biosynthesized metal composites. Table 9.16 highlights current metal nanocomposites produced from microbial sources exhibiting antibacterial properties.

Applications in Cancer Treatment and Waste Water Management
Cancer is a major cause of mortality worldwide, with millions of new diagnoses occurring each year. Numerous traditional anticancer agents have limited effectiveness and may damage healthy cells. Metal nanoparticles have significant potential in oncological therapy. Figure 9.37 (Roshani et al. 2023) illustrates metal nanoparticles and their potential use in oncology. These metal composites may be produced via microbiological sources (Fig. 9.38).

Table 9.15 Bionanoparticles for removal of wastewater pollutants (Malik et al. 2022)

Biosynthesized NPs	Microorganism used in synthesis	Pollutant(s)	Mechanism of pollutant adsorption
Iron Oxide NPs	Manglicolous fungi	Cr(VI)	Chemisorption
Manganese Oxide NPs	*Marinobacter* sp. MnI-79	Ag+	Electrostatic attraction, redox
Iron Oxide NPs	*Streptomyces thermolineatus*	Cu	Electrostatic attraction, surface complexation
Cadmium Sulfide NPs	*Pseudomonas aeruginosa* JP-11	Cd(II)	Adsorption
Palladium NPs	*Spirulina plantesis*	Pb	Adsorption
Iron Oxide NPs	*Aspergillus tubingenesis* STSP 25	Pb(II), Ni(II), Cu(II), Zn(II)	Adsorption
Magnetite/rGO nanocomposite	*Shewanella oneidensis*	Methylene blue	Electrostatic attraction
Palladium NPs	*Caldicellulosiruptor saccharolyticus*	Methyl orange, Diatrizoate	Reduction
Silver NPs	*Bacillus marisflavi* TEZ7	Direct Blue-1, Methyl Red, Reactive Black 5	Photocatalytic degradation
Silver NPs	*Bacillus* spp., *Sphingomonas paucimobilis*	Malachite green	Adsorption
Silver NPs	*Spirulina plantensis*, *Anabaena variabilis*	Malachite green	Biosorption
Zinc Oxide NPs	*Pseudochrobactrum* sp. C5	Methanol blue, reactive black 5	Catalytic degradation
Bio-Pd/Au NPs	*Shewanella oneidensis* MR-1	Diclofenac	Catalytic degradation
Manganese Oxide NPs	*Pseudomonas putida*	Estrone, 7α-ethinylestradiol	Absorption, oxidation
Platinum NPs	*Desulfovibrio vulgaris*	17-β estradiol, sulfamethoxazole	Catalytic reduction
Biogenic Palladium NPs	*Escherichia coli*	Ciprofloxacin	Reductive degradation

Microbial-assisted nanoparticles precisely target cancer and may serve as selective carriers for various anticancer medications, therefore minimizing damage to healthy cells. Silver nanoparticles (AgNPs) derived from the endophytic fungus *Botryosphaeria rhodina*, isolated from *Catharanthus roseus*, were investigated for their effects on lung cancer using the A549 cell line. The proposed method of scavenging free radicals and inducing apoptosis may account for the cancer activity (Hosseingholian et al. 2023). A native strain, GFCr-4, was used in another investigation to manufacture metallic and bimetallic nanoparticles with putative anticancer properties (Fahim et al. 2024).

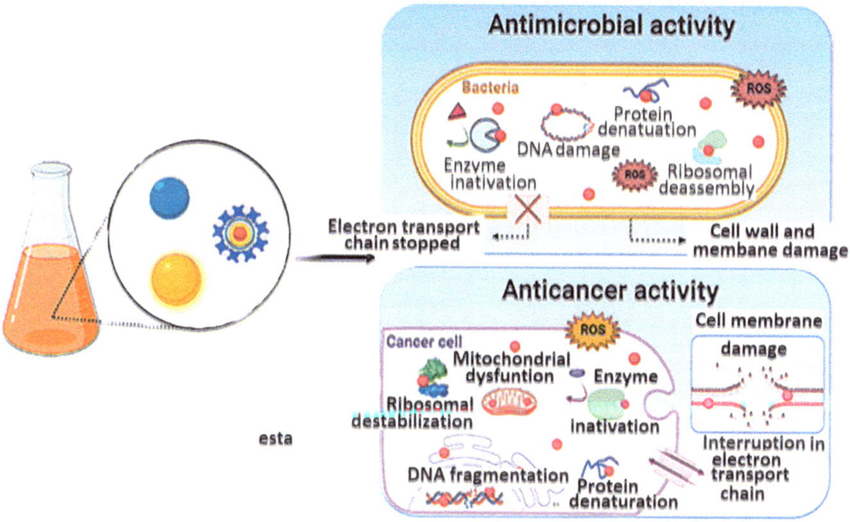

Fig. 9.37 Biosynthesized metal composites in medicine (Hosseingholian et al. 2023)

Table 9.16 Research studies on green synthesis of polymer nanocomposites

Microorganism/ enzyme	Polymer composite	Green synthesis mechanism	Application	Reference
Talaromyces sp. 7S1 (Accession No. OP295112)	Chitosan-	AgNPs conjugate (Ch-AgNPs) Biosynthesis using Talaromyces sp. 7S1.	Antimicrobial, Antibiofilm, Antioxidant, Anticancer	Shawky and Hamed (2024)
Fungi	Chitin and chitosan	Nanocomposites	Treatment and early diagnosis of fungal infections	Gómez-Gaviria and Mora-Montes (2024)
endophytic fungus, *A. atrobrunnea* (ZS06)	Chitosan capped AgNPs	AgNPs from A. *atrobrunnea* ZS06 cell-free filtrate mediated reduction	Antifungal	Al-Zahrani and Al-Garni (2023)
P. aeruginosa ATCC 27853	AgNPs	Silver-containing nanoparticles (fluorinated phthalocyanine–silver)	Antioxidant effects	Albayrak et al. (2023)
S. aureus ATCC 29213	*Zn-doped CuO NPs*	Microbial reduction	Inhibition of biofilm formation	Shalom et al. (2017)

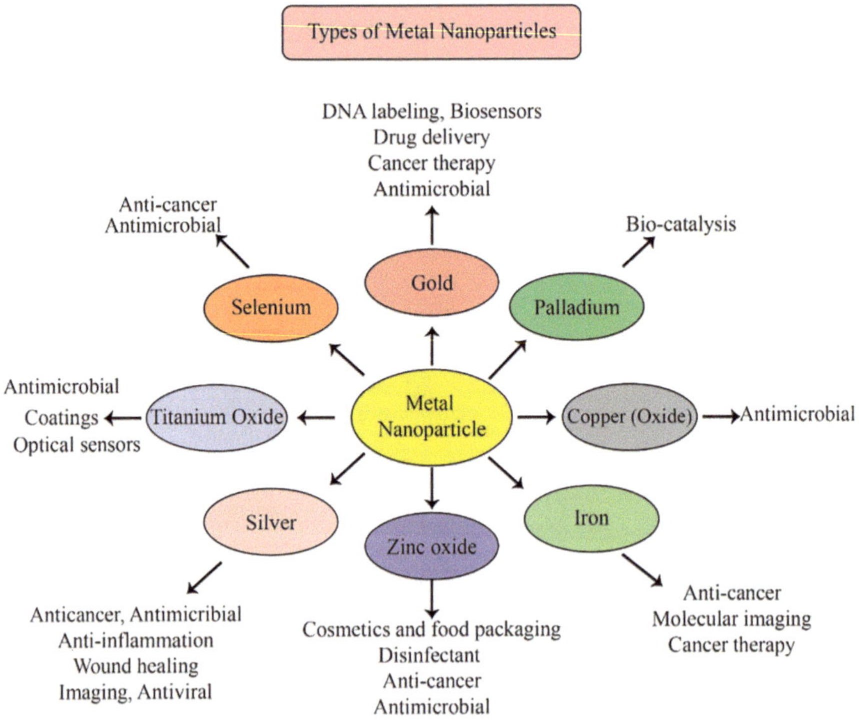

Fig. 9.38 Metal nanoparticles in cancer therapy (Roshani et al. 2023)

Conclusion

Microbial synthesis is a potential eco-friendly method for the development of polymer nanocomposites with substantial applications in wastewater treatment and biomedicine. The use of natural reducing agents and biocatalysts adheres to the tenets of green chemistry, promoting sustainability and environmental safety. The microbial production of polymer nanocomposites has several benefits compared to traditional chemical techniques. These materials are environmentally sustainable, nontoxic, biocompatible, safe, cost-effective, and possess biomedical and environmental uses. Nonetheless, issues like yield optimization, large-scale manufacturing, and the preservation of microbial culture stability must be resolved. Future research should concentrate on improving synthesis efficiency and broadening applications in practical wastewater treatment and improved biomedical technology.

9.7 Sol–Gel Process with Green Precursors

The sol–gel process involves the transition from a liquid "sol" (a colloidal suspension) to a solid "gel" phase via hydrolysis and condensation processes, which commonly use metal alkoxide precursors (Brinker and Scherer 2013; Dimitriev et al. 2008). For environmentally friendly applications, this technique is modified to reduce the usage of harmful solvents and chemicals (Tundo et al. 2000). For example, water-based sol–gel processes use water as a solvent rather than toxic organic solvents, considerably lowering the environmental effect. Furthermore, plant extracts can be natural reducing agents in sol–gel synthesis, providing a greener alternative to manufactured chemicals (Bao et al. 2019). These plant-derived chemicals, rich in polyphenols and other active biomolecules, accelerate the gelation process and improve the biocompatibility of the resultant composites (Zhao et al. 2023).

This eco-friendly modification of the sol–gel technique adheres to green chemistry principles and allows for the creation of high-purity materials with precise qualities (Brinker and Scherer 2013; Bao et al. 2019). By fine-tuning reaction parameters such as pH, temperature, and precursor concentrations, researchers can attain exact control over the material's shape, porosity, and surface properties. Such customized features are critical for a wide range of applications, including biomedical devices and environmental cleanup (Gnanasekaran 2019). The use of renewable and non-toxic precursors assures that the process is both sustainable and ecologically friendly, making it a key technology in the development of biodegradable nanocomposite materials (Sharma et al. 2025, 2023a).

Applications of the Sol-Gel Method for Green Synthesis of Polymer Nanocomposites

Polymer nanocomposites created via the sol–gel technique are an expert class of materials that combine the distinct features of natural polymers and inorganic nanoparticles (Pomogailo 2005; Matei et al. 2022; Hashim et al. 2020). This fusion produces composites with remarkable mechanical strength, thermal stability, and functional adaptability (Picken et al. 2010), addressing different applications such as environmental cleanup, biomedical devices, energy storage, and improved coatings (Idris et al. 2023). The addition of nanoparticles such as silica, titanium dioxide, and zinc oxide to the polymer matrix greatly improves the material's characteristics. For example, silica nanoparticles improve thermal and dimensional stability, whereas titanium dioxide has photocatalytic activity, making it excellent for environmental applications. Zinc oxide's antibacterial characteristics make these composites useful for biomedical applications. These advancements provide resistance to UV radiation, chemical degradation, and microbiological attacks, matching the needs of modern technology (Mahy et al. 2021). Furthermore, the sol–gel technique provides fine control over the material's microstructure, resulting in tunable porosity, high surface area, and improved homogeneity (Brinker and Scherer 2013). The scientific community is continuously investigating novel combinations of biopolymers and eco-friendly precursors to broaden the applications of these nanocomposites (Althomali et al. 2022). Sol–gel-derived polymer nanocomposites are evolving to provide creative solutions for sustainable development and technological advancement while adhering to green chemistry principles.

9.7.1 Biodegradable Packaging

The sol–gel process for synthesizing biodegradable polymer nanocomposites offers an environmentally benign and sustainable approach to generating innovative packaging materials. Incorporating inorganic nanoparticles such as silica, titanium dioxide, and zinc oxide with biodegradable polymers such as starch, cellulose, and chitosan results in nanocomposites with improved mechanical characteristics, water resistance, and biodegradability. These materials show considerable promise in the packaging business, providing a viable alternative to standard plastics while also contributing to environmental sustainability. With further modification and investigation, sol–gel-derived biodegradable nanocomposites have the potential to play an important role in the future of sustainable packaging solutions.

Food Packaging

The development of biodegradable polymer nanocomposites using the sol–gel technique has transformed food packaging by improving performance, sustainability, and biodegradability food packaging materials must meet several important requirements to ensure food safety, quality, and shelf life, including barrier properties to prevent entry of moisture, oxygen, and light, mechanical strength for durability, moisture resistance to maintain food integrity, UV protection to guard against harmful radiation, and freshness preservation while being environmentally sustainable (Mukherjee et al. 2023). Sol–gel-derived biodegradable polymer nanocomposites are increasingly being used in food packaging due to their ability to meet these requirements. According to Huang et al. the sol–gel process is a potential strategy for developing biodegradable polymer nanocomposites that may efficiently perform these tasks in food packaging. By integrating nanoparticles such as silica, clay, and metal oxides into biodegradable polymers such as starch, cellulose, and chitosan, the sol–gel technique improves barrier characteristics, mechanical strength, and moisture resistance (Huang et al. 2018). According to Derungs et al. including montmorillonite clay into starch-based matrices enhances both mechanical characteristics and moisture barrier performance (Derungs et al. 2021).

In addition to increased mechanical and barrier qualities, sol–gel produced biodegradable polymer nanocomposites have antibacterial benefits. Metal oxide nanoparticles, such as TiO_2 and ZnO, have antibacterial characteristics that prevent contamination and extend shelf life. Zhang et al. showed that adding zinc oxide nanoparticles into chitosan films improves antibacterial action, hence preventing the growth of foodborne microorganisms. Adding UV-blocking nanoparticles, such as TiO_2, enhances the stability and protects light-sensitive food products (Zhang et al. 2021). As highlighted by Gore et al. UV protection in packaging materials contributes to the preservation of food quality, particularly for light-sensitive products such as dairy and vegetables. The increased demand for environmentally friendly packaging solutions has accelerated the application of sol–gel generated nanocomposites in the food industry (Gore and Prajapat 2022). These materials are biodegradable, organically degrade without creating long-term environmental damage, and are made from renewable resources, offering them a viable alternative to typical plastic packaging (Mukherjee et al. 2023).

Medical Packaging

Biodegradable polymer nanocomposites are rapidly being employed in medical packaging due to their improved safety, usefulness, and environmental friendliness (Kudumula 2016). These materials, which incorporate nanoparticles such as silver, zinc oxide, and silica into biodegradable polymers such as polylactic acid (PLA), chitosan, and starch, provide superior antibacterial and antimicrobial properties, which are critical for medical product sterility and integrity (Cruz-Delgado et al. 2023). In the medical industry, packaging is critical for extending the shelf life and safety of fragile items such as wound dressings, implants, and pharmaceutical packaging (Cruz-Delgado et al. 2023; Kaczmarek-Szczepańska and Wekwejt 2024). Biodegradable nanocomposites are very useful for reducing contamination by microbes, a common source of product degradation. de Souza et al. and Bari et al. emphasized the expanding use of materials such as these nanocomposites citing their potential to replace traditional plastic packaging with more sustainable and effective alternatives (de Souza et al. 2022; Bari et al. 2016). One notable application is drug packaging, in which biodegradable nanocomposites that contain ZnO and Ag nanoparticles deliver not only mechanical strength and flexibility but also antimicrobial activity, ensuring pharmaceutical companies product safety during shipment and storage. Badawy et al. showed that introducing Ag nanoparticles into chitosan films significantly enhanced the packaging's antimicrobial characteristics, protecting against microbial deterioration. These substances are also useful in wound therapy, with biodegradable dressing incorporating Ag nanoparticles efficiently avoiding infections while stimulating healing (Badawy et al. 2019; Zhao et al. 2024). Furthermore, the introduction of silica nanoparticles in PLA improves oxygen barrier qualities, which are critical when packaging sensitive medical products. These features, together with the biodegradability of the materials, make them a viable and sustainable alternative to standard non-biodegradable packaging. The sol–gel technique is critical in the manufacturing of these nanocomposites, providing fine control over the integration of nanoparticles, which improves features such as antibacterial activity and biodegradability (Zhao et al. 2024; Brandelli 2024; Kamarudin et al. 2022; Marano et al. 2022).

Several factors such as pH, solution concentration, surface interaction approaches, and nature of nanofillers used have a tremendous impact on the physiochemical properties of PNCs. For example, cellulose/silica nanocomposites, several parameters that influence the properties of PNCs were elaborately reviewed by Shange and Coworkers (Shange et al. 2024). The authors found that temperature control, surface interaction, and careful selection of solvents are the crucial factors that need to be taken care of for the production of cellulose/Silica nanocomposites with control over the desired size of nanocomposites and uniform dispersion of inorganic filler. The future studies should aim at further exploring these parameter optimizations with the latest green synthesis methods to come up with novel materials for biosensors and wearable electronic materials, which is going to have a positive impact on the socioeconomic front.

9.8 In Situ Polymerization for PNC Manufacturing

In situ polymerization is a reliable approach for synthesizing polymer nanocomposites, in which monomers polymerize in the presence of distributed nanofillers, assuring their confinement inside the polymer matrix. This method efficiently inhibits nanofiller aggregation, enabling uniform dispersion and increasing interfacial adhesion between the polymer and fillers, which is critical for improving the composite's mechanical, thermal, and electrical properties (de Oliveira and Beatrice 2018). For example, polymer-clay nanocomposites generated using in situ polymerization have improved heat stability and barrier properties, making them appropriate for advanced applications (Gobena and Woldeyonnes 2024) such as packaging and automotive sectors (Ajayan et al. 2006; Tjong 2006). As shown in Fig. 9.39, the monomer aniline polymerization was initiated in situ in the presence of montmorillonite (MMT) nanoclay, using ammonium persulfate (APS) as an oxidizing agent. The PANI-MMT polymer nanocomposite was formed in higher yield at below ambient temperature (0 °C), and the polymerization process was completed in less time (Rallini and Kenny 2017). Furthermore, this method is highly adjustable, allowing the use of a variety of polymers and fillers, including carbon nanotubes and graphene, to create tailored characteristics for specific applications (Rallini and Kenny 2017). Polyaniline-based nanocomposites, for example, created utilizing in

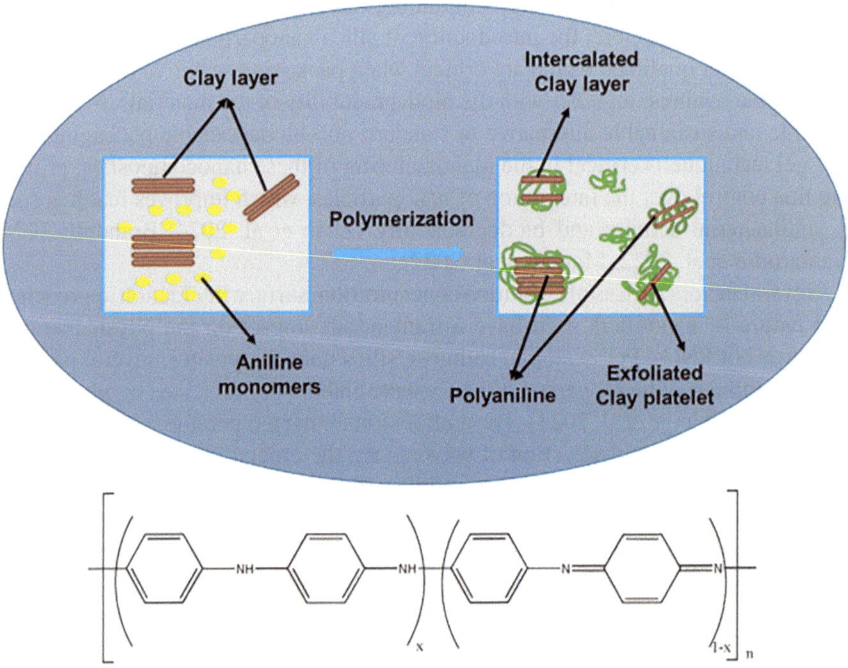

Fig. 9.39 Schematic illustration of in situ polymerization of aniline in the presence of clay

situ polymerization and carbon fillers, have high electrical conductivity, making them potential candidates for flexible electronics and sensors (Singh and Shukla 2020; Kalotra and Mehta 2021).

9.8.1 *Surface Preparation*

The surface preparation of natural nanofillers is an important stage in the in situ polymerization process because it ensures compatibility with the monomers. Nanofillers are functionalized by adding certain chemical groups to their surfaces, such as hydroxyl (-OH), carboxyl (-COOH), or amine (-NH_2) groups, to increase their propensity for monomers. This procedure is required to overcome the inherent hydrophobic-hydrophilic mismatch between nanofillers and various polymer matrices, resulting in homogeneous dispersion during polymerization (Kazemi et al. 2021; Paszkiewicz et al. 2024). As an illustration, functionalizing clay nanofillers with organosilanes enhances their dispersion in hydrophobic polymer matrices by enhancing interface flexibility (de Oliveira and Beatrice 2018; Zapata et al. 2008). Similarly, carbon-based nanofillers such as graphene or carbon nanotubes are frequently oxidized to introduce oxygen-containing groups, allowing for stronger interactions with monomers and improved polymer matrix integration (Kim et al. 2010; Park and Ruoff 2009). These functionalization procedures ensure that the nanofillers are effectively absorbed into the polymer framework, eliminating agglomeration and maximizing their performance-enhancing potential in the finished composite material.

9.8.2 *Polymerization Process*

In situ polymerization consists of three important stages: initiation, propagation, and termination, each of which contributes to the synthesis of polymer nanocomposites. The initiation stage of the reaction begins with the activation of monomers, which can be accomplished using thermal (Gong et al. 2022), photochemical (Chatani et al. 2014), or catalytic methods (Fouassier and Lalevée 2014). Thermal initiators, such as benzoyl peroxide or azobisisobutyronitrile (AIBN), are often employed to create free radicals, which activate monomers and start the chain reaction (Ajayan et al. 2006; Gong et al. 2022). Exposure to ultraviolet (UV) light causes photoinitiators such as camphor quinone to break down and release reactive radicals, which is very useful for precision polymerization in optoelectronic applications (Chatani et al. 2014). Catalytic systems, such as Ziegler-Natta catalysts or metallocenes, are commonly used in coordination polymerization to improve control over the polymer's molecular weight and tacticity, notably in polyolefin synthesis (Shamiri et al. 2014). These approaches enable the development of reactive intermediates, such as free radicals, cations, or anions, which are critical to the polymerization process.

In the propagation phase of in situ polymerization, monomers are sequentially added to activated species, resulting in polymer chain elongation and the encapsulation of functionalized nanofillers inside the expanding polymer matrix (Fan et al. 2020). Functional compounds, such as hydroxyl or amine groups, placed onto nanofillers during surface preparation, improve encapsulation by promoting strong interfacial interaction with polymer chains, increasing compatibility and dispersion (Jamróz et al. 2019). The uniform incorporation of nanofillers avoids agglomeration and enhances their strengthening potential, resulting in substantial developments in mechanical characteristics such as tensile force and modulus, also seen in polymer-clay and carbon nanotube composites (Mapossa et al. 2023; Dai and Sun 2016). The process closes with the termination phase, in which reacting intermediates become neutralized, usually through combination or disproportionation reactions, stabilizing the polymer and completing the nanocomposite structure. This methodical procedure assures that nanocomposites have improved mechanical, thermal, and functional properties, making them appropriate for use in energy storage (Rajpoot et al. 2023), flexible electronics (Carrascosa et al. 2024), and high-performance coatings (Alnassar et al. 2015). Figure 9.40 demonstrates the key steps involved in the synthesis of NFC/PANI nanocomposite using the in situ polymerization technique.(Luong et al. 2013) The composite material possesses enhanced mechanical strength and high conductivity and can be fabricated as paper also.

Nanofillers that have inherent chemical reactivity can also act as a catalyst to induce in situ polymerization. The biorenewable nanofillers provide unique features to PNCs owing to their biocompatibility and biodegradability (Ates et al. 2020).

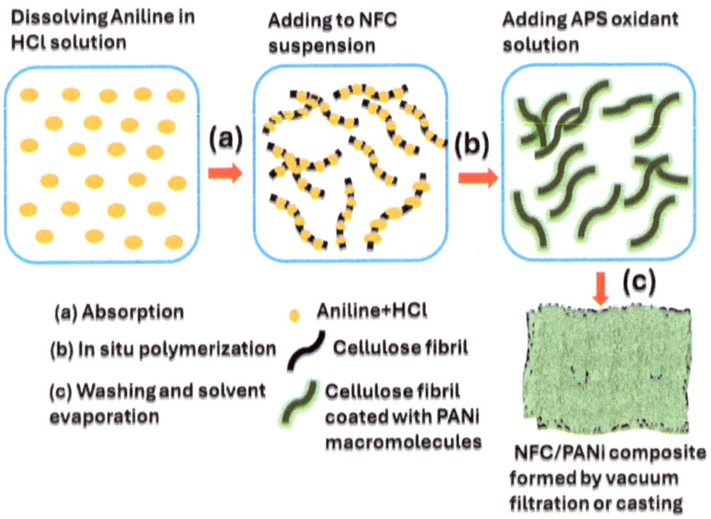

Fig. 9.40 The key stages of in situ polymerization reaction involved in the synthesis of NFC/PANI nanocomposites

Furthermore, monomers derived from renewable sources can themselves be polymerized in situ to create biobased nanocomposites. For example, terpenes and terpenoids can be polymerized with the addition of nanostructures such as clay, carbon nanotubes (CNTs), and graphene. Similarly, vegetable oil-derived monomers (containing double bonds or epoxide groups) can be converted into nanocomposites in the presence of bioresource-derived nanofillers such as bacterial cellulose (BC), cellulose nanofibers (CNF) as additives· A significant advantage of in situ polymerization is the minimization of additive agglomeration, resulting in highly homogeneous materials.

9.8.3 Different Types of Polymerizations

The preparation of PMMA/cellulose nanocomposites was achieved through in situ free radical polymerization, employing the biobased solvent cyrene (Sušac et al. 2024). In this process, cellulose nanofibers were dispersed within the cyrene solvent, and the polymerization of methyl methacrylate was initiated using free radicals. The resulting nanocomposites demonstrated improvements in Young's modulus and thermal stability, with the extent of enhancement directly related to the concentration of incorporated cellulose nanocrystals. In situ emulsion polymerization represents another widely employed technique for nanocomposite synthesis, providing a convenient pathway to multifunctional materials characterized by homogeneous nanofiller distribution (Xu et al. 2023). This method offers several advantages, including cost-effectiveness, enhanced dispersion of nanofillers within the polymer matrix, and adaptability to various applications. Figure 9.41 provides a schematic representation of the key steps involved in PNC fabrication via in situ emulsion polymerization. The utilization of water as the polymerization medium makes it environment-friendly, and its compatibility with a range of monomers enables its widespread application in different areas. As illustrated in Fig. 9.41, the uniform

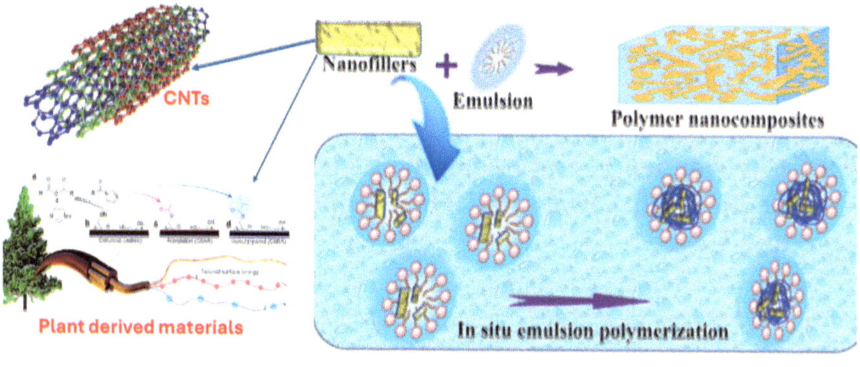

Fig. 9.41 Schematic representation of in situ emulsion polymerization

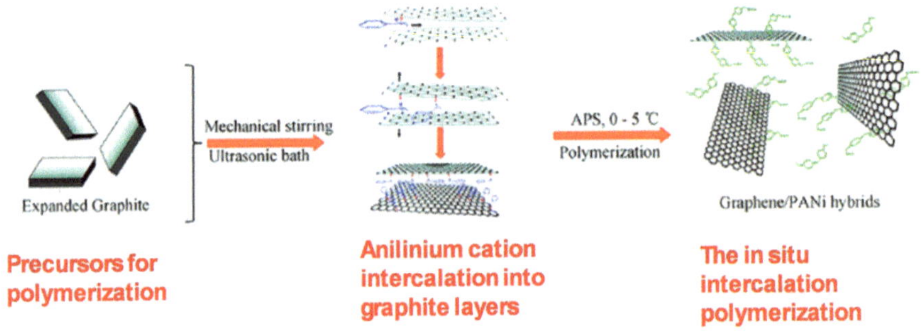

Fig. 9.42 Fabrication of graphene/polymer nanocomposites through in situ intercalation polymerization (Guo et al. 2018b)

dispersion of nanofillers achieved through this technique results in enhanced mechanical properties of the resulting nanocomposites.

Intercalative polymerization is also a viable method for the green synthesis of PNCs (Qi et al. 2005). The in situ intercalation method is employed in the production of thin-film nanocomposite materials, and this method is associated with fast reaction kinetics. Figure 9.42 clearly explains the various steps involved in the intercalation-assisted polymer nanocomposite fabrication (Guo et al. 2018b). The aniline is converted into anilinium cation and inserted between the layers of graphene nanolayers. The in situ intercalation polymerization of aniline results in the formation of an inorganic-organic hybrid polymer nanocomposite (Graphite/PANI). The in situ polymerization techniques have some disadvantages such as limited control over the polymerization kinetics and the possibility of incorporation of impurities in the polymer matrix. More research is necessary to understand the mechanism of polymerization and control the shape and size of PNCs produced through the polymerization.

In situ polymerization processing methods where the organic solvents have been omitted are strategically to produce novel biopolymeric nanocomposites are necessary to promote the green synthesis of polymeric nanocomposites. This approach offers advantages over traditional blending methods by promoting better control over nanofiller dispersion and minimizing agglomeration, ultimately tailoring the nanocomposite properties for specific applications.

9.9 Environmental and Economic Benefits: Green Synthesis Strategies for PNCs and Their Impact on Energy, Waste, and Cost

Polymer nanocomposites (PNCs) incorporate nanoscale fillers, including nanoparticles, nanofibers, and nanotubes, into polymer matrices, forming a class of sophisticated materials with outstanding mechanical, thermal, and electrical characteristics.

These improved qualities facilitate many applications across several domains, including aerospace, electronics, and biology. Nevertheless, conventional synthesis techniques for PNCs—such as chemical vapor deposition, sol–gel procedures, and high-energy mechanical milling—are often marked by significant energy consumption, reliance on toxic substances, and production of hazardous waste. With the increasing need for sustainable manufacturing practices, green synthesis methods provide a potential alternative by delivering environmentally friendly, economically feasible, and energy-efficient techniques for producing high-performance materials. Multiple sustainable methodologies have arisen for PNC manufacturing. Research designs in these methodologies may be meticulously crafted by taking into account the following considerations and benefits.

Solution Processing Using Green Solvents This involves dispersing nanoparticles in a green solvent (e.g., water, ethanol, supercritical CO_2) and mixing with a polymer solution (Chakravarty et al. 2019). Subsequent solvent removal yields the nanocomposite. Examples include using aqueous dispersions of cellulose nanocrystals (CNCs) or graphene oxide (GO) with water-soluble polymers (Pal et al. 2021; Montes et al. 2015; Yang et al. 2019). This approach is especially appealing for polymers that are soluble or dispersible in water or other biobased solvents, therefore reducing the need for hazardous organic solvents. Nevertheless, meticulous regulation of processing conditions is crucial to avert nanoparticle agglomeration and guarantee uniform dispersion inside the polymer matrix.

Melt Processing with Biobased Polymers Melt processing, a common industrial technique, becomes greener by utilizing biobased polymers like polylactic acid (PLA), polyhydroxyalkanoates (PHAs), and starch (Wang et al. 2020b; Castro-Aguirre et al. 2016; Wu et al. 2023). These polymers originate from renewable resources and are often biodegradable or compostable, providing an end-of-life option that mitigates plastic waste buildup. This approach is very scalable and completely eliminates the need for solvents. Nonetheless, attaining uniform dispersion of nanoparticles in the melt may be difficult owing to variations in viscosity and surface tension between the polymer and the nanoparticles. Modifying the surface of nanoparticles or using compatibilizers might enhance dispersion.

In Situ Polymerization with Green Initiators/Catalysts This involves polymerizing monomers in the presence of nanoparticles, employing green initiators (e.g., enzymes and UV irradiation) or catalysts (Gazzotti et al. 2017; McHale et al. 2017). For instance, ring-opening polymerization of lactide initiated by enzymes (Engel et al. 2019) can produce PLA-based nanocomposites. This method allows for strong interfacial interactions between the polymer matrix and the nanoparticles, leading to enhanced mechanical properties. The use of biocatalysts like enzymes offers high selectivity and specificity, reducing unwanted side reactions (Kobayashi 1999; Li et al. 2010).

Biotemplating This method employs biological entities (viruses, bacteria, and DNA) as templates for the synthesis of nanoparticles, which are then integrated into a polymer matrix. This provides exact control over nanoparticle organization and may produce intricate hierarchical structures. This technology is currently under development for large-scale manufacturing but has considerable promise for producing sophisticated nanocomposites with customized features (Sharma et al. 2023b; Herrera-Beurnio et al. 2021).

9.9.1 Environmental and Economic Benefits of Green Synthesis

Reduced Energy Usage Eco-friendly procedures often function at ambient temperatures or need less energy input relative to conventional high-temperature processes. Water-based solution processing necessitates less energy for solvent elimination compared to the use of high-boiling organic solvents. Melt processing necessitates heat but may be enhanced for energy economy, particularly when integrated with biobased polymers that exhibit lower processing temperatures. In situ polymerization with UV irradiation may be more energy-efficient than thermal start. This decreased energy use immediately results in reduced operational expenses and a smaller carbon imprint (Rodríguez-Rojas et al. 2024; Anastas and Lankey 2000).

Minimized Waste Generation and Reduced Toxicity The use of non-toxic solvents (water, ethanol) and biobased products markedly reduces hazardous waste. This reduces disposal expenses and mitigates environmental degradation, including air and water contamination. Substituting petroleum-derived polymers with biodegradable biopolymers reduces reliance on fossil resources and mitigates plastic waste buildup, thereby addressing the escalating issue of plastic pollution. Green synthesis further eliminates the need for hazardous catalysts and initiators, hence significantly reducing environmental and health hazards (Mehrabani et al. 2022; Prasad and Sharma 2019).

Cost Efficiency Although an initial investment in green technology may be required, the long-term advantages include diminished material expenses (utilizing renewable resources that are often more economical and accessible than petroleum-based feedstocks), reduced energy use, and cheaper waste disposal costs. Eliminating costly and dangerous chemicals results in immediate financial savings in acquisition and management. Moreover, reduced regulatory constraints related to the management of hazardous materials might result in financial savings (Wu et al. 2021; Dadkhah and Tulliani 2022; Collias et al. 2021).

Enhanced Sustainability and Circularity Green synthesis adheres to the tenets of a circular economy by advocating for renewable resource use, waste reduction, and the design for biodegradability or recyclability at the end of life. This fosters a more sustainable materials economy and reduces dependence on restricted resources. By completing the cycle of material lifecycles, green nanocomposites may reduce environmental impact and promote a more sustainable future.

Green synthesis methods for polymer nanocomposites have significant advantages from both environmental and economic standpoints. These technologies enhance the sustainable manufacture of high-performance materials applicable in many sectors by decreasing energy usage, lowering waste formation, and optimizing cost efficiency. The transition to green synthesis signifies both a scientific and technical progression, as well as a crucial development toward realizing overarching objectives of sustainability and environmental stewardship. As research and development in this domain progress, green synthesis is expected to be essential in influencing the future of nanocomposite manufacturing, fostering both innovation and sustainability within the materials science sector.

9.10 Challenges and Opportunities: Addressing the Technical and Economic Barriers to Industrial Implementation of Green Synthesis

Green synthesis, which emphasizes sustainability and low environmental effect, presents a tempting alternative to traditional chemical techniques, particularly in the fabrication of polymer nanocomposites. Translating laboratory-scale green synthesis methodologies to viable commercial processes poses considerable technological and economic obstacles. Surmounting these obstacles is essential for the widespread adoption and actualization of green chemistry's full potential. The most critical problems include,

- ***Scalability and Reproducibility:*** Many green synthesis methods, particularly those using biobased materials or complex chemical pathways, are developed at a restricted scale. Expanding these processes often introduces unforeseen challenges related to mass transfer, stoichiometric ratios, mixing efficiency, temperature control, and maintaining consistent product quality. Ensuring repeatability across several batches and production locations may be difficult (Afonso et al. 2024; Ying et al. 2022).
- ***Process Optimization and Control:*** Green synthesis pathways often include intricate reaction processes and several interrelated factors. Enhancing these processes for yield, dimensional regulation, purity, and specific nanocomposite characteristics necessitate advanced process control methodologies (Osman et al. 2024). Real-time monitoring and feedback systems may be essential to

guarantee consistent product quality and reduce waste creation (Varadharajan et al. 2024).

- ***Material Characterization and Standardization:*** Characterizing nanocomposites produced by eco-friendly methods poses challenges owing to the intrinsic complexity of biobased materials and the likelihood of compositional variances. Implementing standardized characterization protocols and quality control measures is crucial for guaranteeing product consistency and compliance with regulatory standards (Dubey et al. 2023).
- ***Stability and Shelf Life:*** Biobased materials utilized in green synthesis may be prone to degradation over time, which can impact the stability and shelf life of the resultant nanocomposites. Formulating efficient preservation methods and comprehending the long-term properties of these materials is essential for industrial applications (Osman et al. 2024). The detrimental effects of chemical corrosion and pathogenic bacterial aggregation on material surfaces are of paramount concern. These difficulties lead to significant economic losses and adversely affect human health. Corrosion yearly accounts for around 4% of the global GDP, with over 50% attributed to microbially driven corrosion (Idumah 2021).
- ***Solvent Selection and Recovery:*** Although green chemistry prioritizes the use of environmentally friendly solvents such as water or supercritical CO_2, some applications may nevertheless need the use of organic solvents. Formulating effective solvent recovery and recycling methodologies is crucial for mitigating environmental effects and lowering production expenses (Idoko et al. 2024).
- ***Initial Investment Expenditures:*** The adoption of green synthesis technologies often necessitates substantial preliminary capital for new apparatus, infrastructure, and process innovation. This may provide a hindrance for smaller enterprises and might deter the implementation of sustainable practices in traditional sectors.
- ***Availability and Cost of Raw Materials:*** The supply and pricing of biobased raw materials may vary according to seasonal influences, farming methods, and market demand. Securing a reliable and economical supply chain is essential for the economic sustainability of green synthesis (Burgess et al. 2023).
- ***Process Efficiency and Yield:*** Green synthesis pathways may sometimes demonstrate diminished yields or reduced reaction speeds relative to traditional approaches. Enhancing process efficiency and optimizing product yield is crucial for attaining economic competitiveness (Al Harby et al. 2024).
- ***Market acceptability and Consumer perspective:*** The perspective of consumers and the acceptability of items produced by green synthesis technologies might impact their economic viability. Informing customers on the advantages of green chemistry and fostering trust in these goods is crucial for stimulating market demand (Savale et al. 2023).
- ***Absence of Established Supply Chains and Infrastructure:*** In several instances, the requisite supply chains and infrastructure for procuring biobased components and executing green synthesis methods remain inadequately developed. This may result in logistical difficulties and elevate production expenses (Makepa and Chihobo 2024).

Notwithstanding these limitations, considerable prospects exist for enhancing the practical use of green synthesis in the creation of polymer nanocomposites. Current research and development in biocatalysis, nanotechnology, and process engineering are yielding more efficient, scalable, and cost-effective green synthesis approaches (Rathod et al. 2024). Heightened environmental restrictions and governmental incentives are prompting firms to embrace more sustainable operations. This fosters a conducive atmosphere for the development of green chemical technology. Rising consumer awareness of environmental concerns and heightened demand for sustainable goods are propelling market demand for eco-friendly alternatives (Achmad et al. 2023). Cooperation among academia, business, and governmental entities is crucial for expediting the development and use of green synthesis technology. Utilizing LCA methodology to illustrate the environmental and economic advantages of green synthesis over traditional methods might serve as a compelling motivation for industrial implementation.

Confronting these difficulties requires a multidisciplinary strategy that includes chemists, engineers, materials scientists, economics, and policymakers. Emphasizing the creation of resilient, scalable, and economical green synthesis platforms, with favorable legislation and market incentives, is essential for unlocking the whole potential of green chemistry in the polymer nanocomposite sector. A comprehensive strategy that encompasses the whole life cycle of materials, from raw material procurement to end-of-life management, is crucial for attaining genuine sustainability. The use of digital technologies, including artificial intelligence and machine learning, is pivotal in enhancing green synthesis processes and expediting their industrial application.

9.11 Future Directions in Green Synthesis

While significant advancements have been made in the green synthesis of polymer nanocomposites (PNCs), current literature often overlooks crucial sustainability factors. These include raw material processing costs, the environmental impact of bioresource exploitation, and comprehensive toxicity assessments. Defining a synthesis as 'green' solely based on one bioderived component is misleading. Future research must prioritize holistic sustainability assessments, including mandatory carbon footprint reporting and standardized guidelines for classifying green synthesis processes. Despite advancements in bioassisted polymer synthesis, poor nanoparticle dispersion and interfacial incompatibility continue to hinder material performance. Enhancing spatiotemporal matching during synthesis is therefore crucial for realizing the full potential of these nanocomposites. One must explore and exploit the lustrous advantages offered by 3D bioprinting techniques. "3D printing represents a significant advancement in PNC fabrication, allowing for controlled nanofiller distribution and the creation of designed functionalities, often with the aid of machine learning. Future research should focus on two key areas: achieving precise nanofiller positioning within the polymer matrix at specific concentrations and enabling efficient nanofiller recovery for reuse or recycling at the end of the PNC's lifespan."

To advance polymer nanocomposite technology, precise control over nanofiller size, shape, and distribution is essential. Biomimicry offers a powerful strategy but requires further research to effectively translate natural phenomena into practical applications. Studying the proliferation and immune evasion tactics of cancer cells provides a compelling example of the potential for bioinspired materials design. A key challenge in biobased PNC production is scaling up laboratory-developed processes to commercial levels. This is frequently hindered by the absence of readily available industrial-scale technologies. To overcome this hurdle, future research should prioritize developing simulated large-scale production environments using advanced computational modeling and machine learning. Combining multiple processing methods synergistically offers another promising avenue, allowing each method to compensate for the limitations of others.

The bacterial factory concept is recently gaining attention for large-scale production of PNCs. In this method bacteria are employed as machinery to produce PNCs of desired shape and size with spatiotemporal control. The primary challenge in realizing the full potential of "bacteria factories" for nanocomposite synthesis lies in the limited automation of their mechanical components. Future advancements will likely involve integrating sophisticated detection and identification tools, coupled with machine learning and artificial intelligence algorithms. This would essentially provide the system with an "artificial brain," enabling autonomous data processing and precise control over nanocomposite growth parameters. Such an intelligent and highly integrated hybrid system promises to revolutionize the large-scale, convenient, and efficient production of sustainable functional materials. Sustainable PNC production requires addressing the significant energy demands associated with both mass production and the computational infrastructure for automation. Nature offers a valuable lesson: microorganisms efficiently conduct metabolic processes and cell production with minimal energy and resource input. Future research should strive to emulate this efficiency by developing material design strategies that minimize both building blocks and energy consumption.

Achieving the full potential of PNCs hinges not only on controlled synthesis but also on advanced characterization. Precise determination of physicochemical properties, including viscosity, rheology, and mechanical strength, along with accurate impurity analysis, is paramount. However, the structural complexity of PNCs often limits the effectiveness of conventional nanomaterial characterization techniques. This necessitates the development of novel, more sophisticated analytical methods capable of elucidating the formation mechanisms of these complex nanostructures. In situ and operando techniques like wide-angle X-ray scattering (WAXS), resonant inelastic X-ray scattering (RIXS), and 3D Tomography have to be explored further to understand the possible mechanism of formation of polymer nanocomposites and make necessary changes for future design of innovative PNCs for a wide range of applications.

9.12 Conclusion

This chapter has outlined the core principles of green chemistry as they apply to the sustainable development of polymer nanocomposites. It aims to introduce these concepts to advance research in this field. Specifically, the chapter details key strategies such as utilizing biorenewable resources as nanofillers and employing ionic liquids as greener alternatives to traditional organic solvents in nanocomposite fabrication. Various biobased feedstocks, including cellulose, algae, microorganisms, bacterial cellulose, and agricultural waste, are explored as sources for nanofillers, emphasizing their inherent sustainability and their contribution to creating eco-friendly polymeric nanomaterials when combined with biopolymers. Furthermore, the chapter provides an overview of recent trends in nanofiller selection and the synthesis of polymer nanocomposites, focusing on achieving precise control over nanofiller size, shape, and dispersion within the polymer matrix. It also highlights cutting-edge advancements, such as synergistic approaches that combine different processing methods to overcome individual limitations, illustrated with relevant examples. Finally, the chapter briefly addresses the influence of operating conditions, such as temperature, pH, and the use of catalysts, on obtaining polymer nanocomposites with desired porosity and shape while minimizing impurities.

This chapter also acknowledges the limitations of green synthesis methods, including challenges in precisely controlling nanofiller loading and difficulties in scaling up production. However, it highlights the significant potential of integrating automation-assisted printing techniques, such as additive manufacturing and 3D bioprinting, with machine learning models. This combined approach offers a promising pathway to address these challenges, enabling mass production and achieving greater spatiotemporal control over the size and shape of polymer nanocomposite materials. Supercritical fluid (SCF)-assisted polymer nanocomposite synthesis has recently emerged as a highly promising technique for addressing challenges related to nanofiller solubility and achieving homogeneous dispersion. Its ability to operate at lower temperatures and utilize greenhouse gases like carbon dioxide as foaming agents for creating porous structures makes this method particularly efficient and attractive for the green synthesis of polymer nanocomposites. There is a need to extend the application of SCF method for design of multifunctional materials for application in the fabrication of wearable biomedical devices.

This chapter also covers the principles and fundamentals of in situ polymerization, detailing various techniques currently employed to fabricate nanocomposites with enhanced mechanochemical, magnetic, and optoelectronic properties. Mechanochemical synthesis processes, such as ultrasonication, extrusion, and melt mixing, are increasingly competitive with other green synthesis methods, opening new research avenues in the mechanochemical synthesis of polymer composites using waste plastic and biowaste. The future trends are directed at mimicking microorganisms to maximize the output with minimal input in terms of feed stock, energy supplied, and time consumed. The future advancement in this field is directed by advanced characterization techniques for understanding the minute details about the

This chapter has outlined the core principles of green chemistry as they apply to the sustainable development of polymer nanocomposites. It aims to introduce these concepts to advance research in this field. Specifically, the chapter details key strategies such as utilizing biorenewable resources as nanofillers and employing ionic liquids as greener alternatives to traditional organic solvents in nanocomposite fabrication. Various biobased feedstocks, including cellulose, algae, microorganisms, bacterial cellulose, and agricultural waste, are explored as sources for nanofillers, emphasizing their inherent sustainability and their contribution to creating eco-friendly polymeric nanomaterials when combined with biopolymers. Furthermore, the chapter provides an overview of recent trends in nanofiller selection and the synthesis of polymer nanocomposites, focusing on achieving precise control over nanofiller size, shape, and dispersion within the polymer matrix. It also highlights cutting-edge advancements, such as synergistic approaches that combine different processing methods to overcome individual limitations, illustrated with relevant examples. Finally, the chapter briefly addresses the influence of operating conditions, such as temperature, pH, and the use of catalysts, on obtaining polymer nanocomposites with desired porosity and shape while minimizing impurities.

This chapter also acknowledges the limitations of green synthesis methods, including challenges in precisely controlling nanofiller loading and difficulties in scaling up production. However, it highlights the significant potential of integrating automation-assisted printing techniques, such as additive manufacturing and 3D bioprinting, with machine learning models. This combined approach offers a promising pathway to address these challenges, enabling mass production and achieving greater spatiotemporal control over the size and shape of polymer nanocomposite materials. Supercritical fluid (SCF)-assisted polymer nanocomposite synthesis has recently emerged as a highly promising technique for addressing challenges related to nanofiller solubility and achieving homogeneous dispersion. Its ability to operate at lower temperatures and utilize greenhouse gases like carbon dioxide as foaming agents for creating porous structures makes this method particularly efficient and attractive for the green synthesis of polymer nanocomposites. There is a need to extend the application of the SCF method for the design of multifunctional materials for the fabrication of wearable biomedical devices.

This chapter also covers the principles and fundamentals of in situ polymerization, detailing various techniques currently employed to fabricate nanocomposites with enhanced mechanochemical, magnetic, and optoelectronic properties. Mechanochemical synthesis processes, such as ultrasonication, extrusion, and melt mixing, are increasingly competitive with other green synthesis methods, opening new research avenues in the mechanochemical synthesis of polymer composites using waste plastic and biowaste. The future trends are directed at mimicking microorganisms to maximize the output with minimal input in terms of feedstock, energy supplied, and time consumed. The future advancement in this field is directed by advanced characterization techniques for understanding the minute details about the nanofiller-polymer interactions at the nanosecond time scale level, and judicial selection of materials to aid in the development of multifunctional materials for diversified applications.

Looking ahead, the confluence of green chemistry with advanced computational and AI-driven techniques is likely to speed up innovation in the polymer nanocomposite area. These technologies allow us to optimize material qualities, expedite processing procedures, and forecast results with great precision, lowering environmental and economic costs. Furthermore, promoting interdisciplinary cooperation will be key to unlocking new opportunities in this field, such as the implementation of closed-loop systems for material recycling and the incorporation of sustainable practices into large-scale production. These advances have the potential to reshape industries ranging from biomedical engineering to environmental remediation, proving the transformative power of sustainable materials in addressing global concerns.

References

Achmad GN, Yudaruddin R, Nugroho BA, Fitrian Z, Suharsono S, Adi AS, Hafsari P, Fitriansyah F (2023) Government support, eco-regulation and eco-innovation adoption in SMEs: the mediating role of eco-environmental. J Open Innov: Technol Mark Complex 9(4):100158

Adeleke BS, Olowe OM, Ayilara MS, Fasusi OA, Omotayo OP, Fadiji AE, Onwudiwe DC, Babalola OO (2024) Biosynthesis of nanoparticles using microorganisms: a focus on endophytic fungi. Heliyon 10(21):e39636. https://doi.org/10.1016/j.heliyon.2024.e39636

Afjeh-Dana E, Naserzadeh P, Nazari H, Mottaghitalab F, Shabani R, Aminii N, Mehravi B, Rostami FT, Joghataei MT, Mousavizadeh K, Ashtari K (2019) Gold nanorods reinforced silk fibroin nanocomposite for peripheral nerve tissue engineering applications. Int J Biol Macromol 129:1034–1039. https://doi.org/10.1016/j.engreg.2024.07.001

Afonso IS, Cardoso B, Nobrega G, Minas G, Ribeiro JE, Lima RA (2024) Green synthesis of nanoparticles from olive oil waste for environmental and health applications: a review. J Environ Chem Eng 12:114022

Ahmed SF, Mofijur M, Rafa N, Chowdhury AT, Chowdhury S, Nahrin M, Islam AS, Ong HC (2022) Green approaches in synthesising nanomaterials for environmental nanobioremediation: technological advancements, applications, benefits and challenges. Environ Res 204:111967

Ahmed AS, Negm AN, Mohammed M, Abd El-Majeed M, Ali AK, Abdelmotalleib M (2023) Biodegradable polymers for industrial applications. In: Handbook of biodegradable materials. Springer, Cham, pp 451–476

Aivazoglou E, Metaxa E, Hristoforou E (2018) Microwave-assisted synthesis of iron oxide nanoparticles in biocompatible organic environment. AIP Adv 8(4). https://doi.org/10.1063/1.4994057

Ajayan PM, Schadler LS, Braun PV (2006) Nanocomposite science and technology. Wiley

Akhtar MS, Fiaz S, Aslam S, Chung S, Ditta A, Irshad MA, Al-Mohaimeed AM, Iqbal R, Al-Onazi WA, Rizwan M, Nakashima Y (2024) Green synthesis of magnetite iron oxide nanoparticles using Azadirachta indica leaf extract loaded on reduced graphene oxide and degradation of methylene blue. Sci Rep 14(1):18172. https://doi.org/10.1038/s41598-024-69184-y

Al Harby NF, Fetouh HA, El-Batouti M (2024) Facile green synthesis route for new ecofriendly photo catalyst for degradation acid red 8 dye and nitrogen recovery. Sci Rep 14(1):1091

Al Kausor M, Gupta SS, Bhattacharyya KG, Chakrabortty D (2022) Montmorillonite and modified montmorillonite as adsorbents for removal of water soluble organic dyes: a review on current status of the art. Inorg Chem Commun 143:109686. https://doi.org/10.1016/j.inoche.2022.109686

Alavi M (2019) Modifications of microcrystalline cellulose (MCC), nanofibrillated cellulose (NFC), and nanocrystalline cellulose (NCC) for antimicrobial and wound healing applications. e-Polymers 19(1):103–119

Albayrak S, Farajzadeh N, Yasemin Yenilmez H, Özdemir S, Gonca S, Altuntaş Bayır Z (2023) Fluorinated phthalocyanine/silver nanoconjugates for multifunctional biological applications. Chem Biodivers 20(7):e202300389. https://doi.org/10.1002/cbdv.202300389

Alharbi NS, Alsubhi NS, Felimban AI (2022) Green synthesis of silver nanoparticles using medicinal plants: characterization and application. J Radiat Res Appl Sci 15(3):109–124

Ali MA, Ahmed T, Wu W, Hossain A, Hafeez R, Islam Masum MM, Wang Y, An Q, Sun G, Li B (2020) Advancements in plant and microbe-based synthesis of metallic nanoparticles and their antimicrobial activity against plant pathogens. Nano 10(6):1146. https://doi.org/10.3390/nano10061146

Ali Z, Yaqoob S, Yu J, D'Amore A (2024) Critical review on the characterization, preparation, and enhanced mechanical, thermal, and electrical properties of carbon nanotubes and their hybrid filler polymer composites for various applications. Comp C Open Access:100434

Alkabli J (2024) Recent advances in the development of chitosan/hyaluronic acid-based hybrid materials for skin protection, regeneration, and healing: a review. Int J Biol Macromol:135357. https://doi.org/10.1016/j.ijbiomac.2024.135357

Alnassar M, Alfadhel A, Ivanov YP, Kosel J (2015) Magnetoelectric polymer nanocomposite for flexible electronics. J Appl Phys 117(17)

Alroaithi M, Xu W (2023) Nanocomposite of CO2-based polycarbonate polyol with highly exfoliated Nanoclay. ACS Omega 8(6):5247–5256

Alsaiari NS, Alzahrani FM, Amari A, Osman H, Harharah HN, Elboughdiri N, Tahoon MA (2023) Plant and microbial approaches as green methods for the synthesis of nanomaterials: synthesis, applications, and future perspectives. Molecules 28(1):463

Althomali RH, Alamry KA, Hussein MA, Tay GS (2022) Versatile applications of biopolymer nanocomposites: a review. Chem Select 7(31):e202200843

Al-Zahrani SS, Al-Garni SM (2023) Antifungal potentiality of mycogenic silver nanoparticles capped with chitosan produced by endophytic Amesia atrobrunnea. Saudi J Biol Sci 30(9):103746. https://doi.org/10.1016/j.sjbs.2023.103746

Ambrosio-Martín J, Lopez-Rubio A, Fabra MJ, Gorrasi G, Pantani R, Lagaron JM (2015) Assessment of ball milling methodology to develop polylactide-bacterial cellulose nanocrystals nanocomposites. J Appl Polym Sci 132(10)

Ameta R, Rai AK, Vyas S, Bhatt JP, Ameta SC (2021) Green synthesis of nanocomposites using plant extracts and their applications. In: Handbook of greener synthesis of nanomaterials and compounds. Elsevier, pp 663–682

Anastas P, Eghbali N (2010) Green chemistry: principles and practice. Chem Soc Rev 39(1):301–312

Anastas PT, Lankey RL (2000) Life cycle assessment and green chemistry: the yin and yang of industrial ecology. Green Chem 2(6):289–295

Andrade-Guel M, Ávila-Orta CA, Cabello-Alvarado C, Cadenas-Pliego G, Esparza-González SC, Pérez-Alvarez M, Quiñones-Jurado ZV (2021) Non-woven fabrics based on nanocomposite nylon 6/ZnO obtained by ultrasound-assisted extrusion for improved antimicrobial and adsorption methylene blue dye properties. Polymers 13(11):1888

Arif ZU, Khalid MY, Sheikh MF, Zolfagharian A, Bodaghi M (2022) Biopolymeric sustainable materials and their emerging applications. J Environ Chem Eng 10(4):108159

Armand M, Endres F, MacFarlane DR, Ohno H, Scrosati B (2009) Ionic-liquid materials for the electrochemical challenges of the future. Nat Mater 8(8):621–629

Armentano I, Puglia D, Luzi F, Arciola CR, Morena F, Martino S, Torre L (2018) Nanocomposites based on biodegradable polymers. Materials 11(5):795

Ates B, Koytepe S, Ulu A, Gurses C, Thakur VK (2020) Chemistry, structures, and advanced applications of nanocomposites from biorenewable resources. Chem Rev 120(17):9304–9362

Badawy ME, Lotfy TM, Shawir S (2019) Preparation and antibacterial activity of chitosan-silver nanoparticles for application in preservation of minced meat. Bull Natl Res Cent 43(1):1–14

Baghban SA, Ebrahimi M, Bagheri-Khoulenjani S, Khorasani M (2021) A highly efficient microwave-assisted synthesis of an LED-curable methacrylated gelatin for bio applications. RSC Adv 11(25):14996–15009. https://doi.org/10.1039/d1ra01269j

Baldino L, Sarno M, Cardea S, Irusta S, Ciambelli P, Santamaria J, Reverchon E (2015) Formation of cellulose acetate–graphene oxide nanocomposites by supercritical CO2 assisted phase inversion. Ind Eng Chem Res 54(33):8147–8156

Bao Z, Xian C, Yuan Q, Liu G, Wu J (2019) Natural polymer-based hydrogels with enhanced mechanical performances: preparation, structure, and property. Adv Healthc Mater 8(17):1900670

Barakan S, Aghazadeh V (2021) The advantages of clay mineral modification methods for enhancing adsorption efficiency in wastewater treatment: a review. Environ Sci Pollut Res 28(3):2572–2599

Baranwal J, Barse B, Fais A, Delogu GL, Kumar A (2022) Biopolymer: a sustainable material for food and medical applications. Polymers 14(5):983

Bari SS, Chatterjee A, Mishra S (2016) Biodegradable polymer nanocomposites: an overview. Polym Rev 56(2):287–328

Baruah P, Mahanta D (2022) Montmorillonite-cellulose based nano-composites and applications. Adv Appl Micro Nano Clay Biopolym Based Comp 125:302–322

Basem A, Jasim DJ, Majdi HS, Mohammed RM, Ahmed M, Al-Rubaye AH (2024) Adsorption of heavy metals from wastewater by chitosan: a review. Resul. Eng. 14:102404

Bhattacharyya KG, Gupta SS (2008) Adsorption of a few heavy metals on natural and modified kaolinite and montmorillonite: a review. Adv Colloid Interf Sci 140(2):114–131

Brandelli A (2024) Nanocomposites and their application in antimicrobial packaging. Front Chem 12:1356304

Brinker CJ, Scherer GW (2013) Sol-gel science: the physics and chemistry of sol-gel processing. Academic Press

Bukartyk M, Zholobko O, Wu XF (2022) Green synthesis of soy protein nanocomposites: effects of cross-linking and clay nanoparticles on the mechanical performance. ACS Omega 7(7):5883–5893

Burgess PR, Sunmola FT, Wertheim-Heck S (2023) A review of supply chain quality management practices in sustainable food networks. Heliyon 9(11):e21179

Canaparo R, Foglietta F, Limongi T, Serpe L (2020) Biomedical applications of reactive oxygen species generation by metal nanoparticles. Materials 14(1):53

Carrascosa A, Sánchez JS, Morán-Aguilar MG, Gabriel G, Vilaseca F (2024) Advanced flexible wearable electronics from hybrid nanocomposites based on cellulose nanofibers, PEDOT: PSS and reduced graphene oxide. Polymers 16(21):3035

Caseri W (2000) Nanocomposites of polymers and metals or semiconductors: historical background and optical properties. Macromol Rapid Commun 21(11):705–722

Castro-Aguirre E, Iniguez-Franco F, Samsudin H, Fang X, Auras R (2016) Poly (lactic acid)—mass production, processing, industrial applications, and end of life. Adv Drug Deliv Rev 107:333–366

Chakravarty P, Famili A, Nagapudi K, Al-Sayah MA (2019) Using supercritical fluid technology as a green alternative during the preparation of drug delivery systems. Pharmaceutics 11(12):629

Chatani S, Kloxin CJ, Bowman CN (2014) The power of light in polymer science: photochemical processes to manipulate polymer formation, structure, and properties. Polym Chem 5(7):2187–2201

Chaudhary R, Nawaz K, Khan AK, Hano C, Abbasi BH, Anjum S (2020) An overview of the algae-mediated biosynthesis of nanoparticles and their biomedical applications. Biomolecules 10(11):1498

Chen Q, Mangadlao JD, Wallat J, De Leon A, Pokorski JK, Advincula RC (2017) 3D printing biocompatible polyurethane/poly(lactic acid)/graphene oxide nanocomposites: anisotropic properties. ACS Appl Mater Interfaces 9(4):4015–4023. https://doi.org/10.1021/acsami.6b11793

Chen J, Liao X, Xiao W, Yang J, Jiang Q, Li G (2019) Facile and green method to structure ultralow-threshold and lightweight polystyrene/MWCNT composites with segregated conductive networks for efficient electromagnetic interference shielding. ACS Sustain Chem Eng 7(11):9904–9915

Chen H, Hu P, Wang Y, Liu H, Zheng J, Huang Z, Zhang X, Liu Y, Zhou T (2024) From quorum sensing inhibition to antimicrobial defense: the dual role of eugenol-gold nanoparticles against carbapenem-resistant pseudomonas aeruginosa. Colloids Surf B: Biointerfaces 247:114415. https://doi.org/10.1016/j.colsurfb.2024.114415

Chiellini E, Corti A, D'Antone S, Solaro R (2003) Biodegradation of poly (vinyl alcohol) based materials. Prog Polym Sci 28(6):963–1014

Chow WS, Tan PL (2010) Epoxy/multiwall carbon nanotube nanocomposites prepared by sonication and planetary mixing technique. J Reinf Plast Compos 29(15):2331–2342

Clark KJ, Sarr AB, Grant PG, Phillips TD, Woode GN (1998) In vitro studies on the use of clay, clay minerals and charcoal to adsorb bovine rotavirus and bovine coronavirus. Vet Microbiol 63(2–4):137–146

Cohen S, Zelikman E, Suckeveriene RY (2020) Ultrasonically induced polymerization and polymer grafting in the presence of carbonaceous nanoparticles. PRO 8(12):1680. https://doi.org/10.3390/pr8121680

Collias DI, James MI, Layman JM (2021) Introduction—circular economy of polymers and recycling technologies. In: Circular economy of polymers: topics in recycling technologies. American Chemical Society, pp 1–21

Cool P, Vansant EF, Auerbach SM, Carrado KA, Dutta PK (2004) Pillared clays and porous clay heterostructures. CRC Press, Boca Raton, FL

Cruz-Delgado VJ, Rodríguez-Fuentes N, Pat-Cetina VC, Cervantes-Uc JM (2023) Nanocomposites based on biodegradable polymers for biomedical applications. In: Green-based nanocomposite materials and applications. Springer, Cham, pp 317–337

Dadkhah M, Tulliani JM (2022) Green synthesis of metal oxides semiconductors for gas sensing applications. Sensors 22(13):4669

Dai L, Sun J (2016) Mechanical properties of carbon nanotubes-polymer composites. In: Carbon nanotubes–current progress of their polymer composites. IntechOpen, pp 155–194

de Oliveira AD, Beatrice CA (2018) Polymer nanocomposites with different types of nanofiller. In: Nanocomposites-recent evolutions. IntechOpen, pp 103–104. https://doi.org/10.5772/intechopen.81329

de Souza AL et al (2022) Antimicrobial biodegradable packaging with nanotechnology application. Res Soc Dev 11(8):e3511830406–e3511830406

Derungs I, Rico M, López J, Barral L, Montero B, Bouza R (2021) Influence of the hydrophilicity of montmorillonite on structure and properties of thermoplastic wheat starch/montmorillonite bionanocomposites. Polym Adv Technol 32(11):4479–4489

Dimitriev Y, Ivanova Y, Iordanova R (2008) History of sol-gel science and technology. J Univ Chem Technol Metallurgy 43(2):181–192

Ding S et al (2023) Ultrasonic-assisted flocculation and sedimentation of coal slime water using the Taguchi method. Ener Sourc A Rec Util Environ Eff 45(4):10523–10536. https://doi.org/10.1080/15567036.2023.2247363

Do HD, Le TT, Pham HN, Ha PT (2019) Microwave-assisted dextran modification and nanoparticle synthesis for application in drug delivery system. Adv Nat Sci Nanosci Nanotechnol 10(1):015008. https://doi.org/10.1088/2043-6254/ab0282

Domeneguetti RR, Sakai VY, Perotti GF, Silva IC, Tercjak A, Barud HS, Pavan F, Constantino VR, Ribeiro SJ (2023) Structural and morphological properties of in-situ biosynthesis of biocompatible bacterial cellulose/Laponite nanocomposites. Appl Clay Sci 234:106851

Dubey RK, Shukla S, Hussain Z (2023) Green synthesis of silver nanoparticles; A sustainable approach with diverse applications. Chinese. J Appl Physiol:e20230007

Dufresne A (2018) Cellulose nanomaterials as green nanoreinforcements for polymer nanocomposites. Philos Trans R Soc A Math Phys Eng Sci 376(2112):20170040

Dupont J, de Souza RF, Suarez PA (2002) Ionic liquid (molten salt) phase organometallic catalysis. Chem Rev 102(10):3667–3692
Edison LK, Pradeep NS (2020) Actinobacterial nanoparticles: green synthesis, evaluation and applications. In: Green nanoparticles: synthesis and biomedical applications. Springer, Cham, pp 371–384
Edward MSG, Louis ACF, Srinivasan H, Venkatachalam S (2022) A mechanochemical approach for synthesizing almond shell nanoparticles and their potential application on the enhancement of polylactic acid film properties. Iran Polym J 31(12):1523–1535
Eid BM (2023) Biomedical applications of organic Nanofillers in polymers. In: Nanofillers. CRC Press, pp 297–322
Elella MH, Goda ES, Abdallah HM, Shalan AE, Gamal H, Yoon KR (2021) Innovative bactericidal adsorbents containing modified xanthan gum/montmorillonite nanocomposites for wastewater treatment. Int J Biol Macromol 167:1113–1125. https://doi.org/10.1016/j.ijbiomac.2020.11.065
El-Sayed SM, Youssef AM (2023) Eco-friendly biodegradable nanocomposite materials and their recent use in food packaging applications: a review. Sustain Food Technol 1(2):215–227
Engel J, Cordellier A, Huang L, Kara S (2019) Enzymatic ring-opening polymerization of lactones: traditional approaches and alternative strategies. ChemCatChem 11(20):4983–4997
Fahim M, Shahzaib A, Nishat N, Jahan A, Bhat TA, Inam A (2024) Green synthesis of silver nanoparticles: a comprehensive review of methods, influencing factors, and applications. JCIS Open 16:100125
Fan C, Tang H, Wang L, Li Y, Wang X, Wang S, Liang X (2020) The preparation of a core–shell stationary phase by the in situ polymerization of a hydrophilic polymer on the surface of silica and its chromatographic performance. New J Chem 44(27):11704–11709
Feng J, Sun Y, Song P, Lei W, Wu Q, Liu L, Yu Y, Wang H (2017) Fire-resistant, strong, and green polymer nanocomposites based on poly (lactic acid) and core–shell nanofibrous flame retardants. ACS Sustain Chem Eng 5(9):7894–7904
Fouassier JP, Lalevée J (2014) Photochemical production of interpenetrating polymer networks; simultaneous initiation of radical and cationic polymerization reactions. Polymers 6(10):2588–2610
Fu LH, Liu YJ, Ma MG, Zhang XM, Xue ZM, Zhu JF (2016) Microwave-assisted hydrothermal synthesis of cellulose/hydroxyapatite nanocomposites. Polymers 8(9):316
Fu S, Sun Z, Huang P, Li Y, Hu N (2019) Some basic aspects of polymer nanocomposites: a critical review. Nano Mater Sci 1(1):2–30
Gao Z, Han Q, Liu J, Zhao K, Yu Y, Feng Y, Han S (2023) Dispersion of carbon nanotubes improved by ball milling to prepare functional epoxy nanocomposites. Coatings 13(3):649
Gazzotti S, Farina H, Lesma G, Rampazzo R, Piergiovanni L, Ortenzi MA, Silvani A (2017) Polylactide/cellulose nanocrystals: the in situ polymerization approach to improved nanocomposites. Eur Polym J 94:173–184
Ghandi K (2014) A review of ionic liquids, their limits and applications. Green Sustain Chem 2014
Gnanasekaran D (2019) Green biopolymers and its nanocomposites in various applications: state of the art. In: Green biopolymers and their nanocomposites. Springer, Singapore, pp 1–27
Gobena ST, Woldeyonnes AD (2024) A review of synthesis methods, and characterization techniques of polymer nanocomposites for diverse applications. Discov Mater 4(1):52
Goci MC, Leudjo Taka A, Martin L, Klink MJ (2023) Chitosan-based polymer nanocomposites for environmental remediation of mercury pollution. Polymers 15(3):482. https://doi.org/10.3390/polym15030482
Gómez-Gaviria M, Mora-Montes HM (2024) Exploring the potential of chitin and chitosan in nanobiocomposites for fungal immunological detection and antifungal action. Carbohydr Res 19:109220. https://doi.org/10.1016/j.carres.2024.109220
Gong M, Zhang D, Zhang J, Li J, Chen X (2022) Optimization of initiator contents in room temperature polymerization of methyl methacrylate. Polym Polym Compos 30:09673911221143201
Gore AH, Prajapat AL (2022) Biopolymer nanocomposites for sustainable UV protective packaging. Front Mater 9:855727

Groh KJ, Geueke B, Martin O, Maffini M, Muncke J (2021) Overview of intentionally used food contact chemicals and their hazards. Environ Int 150:106225

Guadarrama-Escobar OR, Serrano-Castañeda P, Anguiano-Almazán E, Vázquez-Durán A, Peña-Juárez MC, Vera-Graziano R, Morales-Florido MI, Rodriguez-Perez B, Rodriguez-Cruz IM, Miranda-Calderón JE, Escobar-Chávez JJ (2023) Chitosan nanoparticles as oral drug carriers. Int J Mol Sci 24(5):4289

Guan Q-F, Han Z-M, Ling Z-C, Yang H-B, Shu-Hong Y (2022) Growing bacterial cellulose-based sustainable functional bulk nanocomposites by biosynthesis: recent advances and perspectives. Acc Mater Res 3(6):608–619. https://doi.org/10.1021/accountsmr.2c00028

Guo JZ, Song K, Liu C (eds) (2018a) Polymer-based multifunctional nanocomposites and their applications. Elsevier

Guo Y, Peng F, Wang H, Huang F, Meng F, Hui D, Zhou Z (2018b) Intercalation polymerization approach for preparing graphene/polymer composites. Polymers 10(1):61. https://doi.org/10.3390/polym10010061

Guo J, Li Z, Xie Y, Gao J, Yang X, Li MJ (2024) Improved phase change performance of ultrasonic-assisted melting: a visualized experimental study. Energy 306:132500

Haldorai Y, Shim JJ (2014) Synthesis of polymer nanocomposites in supercritical CO2. In: Synthesis techniques for polymer nanocomposites. Wiley, pp 267–290

Haris Z, Ahmad I (2024) Green synthesis of silver nanoparticles using Moringa oleifera and its efficacy against gram-negative bacteria targeting quorum sensing and biofilms. J Umm Al-Qura Univ Appl Sci 10(1):156–167. https://doi.org/10.1007/s43994-023-00089-8

Harriott MM (2019., ISBN 9780128012383) Biofilms and antibiotics. In: Reference module in biomedical sciences. Elsevier. https://doi.org/10.1016/B978-0-12-801238-3.62124-4

Hashim AF, Youssef K, Roberto SR, Abd-Elsalam KA (2020) Hybrid inorganic-polymer nanocomposites: synthesis, characterization, and plant-protection applications. In: Multifunctional hybrid nanomaterials for sustainable Agri-food and ecosystems. Elsevier, pp 33–49

Hassen JH, Abdalkadir HK, Abed SF (2023) An overview of medical applications of montmorillonite clay. J Med Sci 92(2):e826. https://doi.org/10.20883/medical.e826

Herrera-Beurnio MC, Hidalgo-Carrillo J, López-Tenllado FJ, Martin-Gómez J, Estévez RC, Urbano FJ, Marinas A (2021) Bio-templating: an emerging synthetic technique for catalysts. A review. Catalysts 11(11):1364

Horvath IT, Anastas PT (2007) Innovations and green chemistry. Chem Rev 107(6):2169–2173

Hosseingholian A, Gohari SD, Feirahi F, Moammeri F, Mesbahian G, Moghaddam ZS, Ren Q (2023) Recent advances in green synthesized nanoparticles: from production to application. Mater Today Sustain 24:100500. https://doi.org/10.1016/j.mtsust.2023.100500

Huang L, Weng X, Chen Z, Megharaj M, Naidu R (2014) Green synthesis of iron nanoparticles by various tea extracts: comparative study of the reactivity. Spectrochim Acta A Mol Biomol Spectrosc 130:295–301

Huang Y, Mei L, Chen X, Wang Q (2018) Recent developments in food packaging based on nanomaterials. Nano 8(10):830

Huang H, Liu R, Yang J, Dai J, Fan S, Pi J, Wei Y, Guo X (2023) Gold nanoparticles: construction for drug delivery and application in cancer immunotherapy. Pharmaceutics 15(7):1868. https://doi.org/10.3390/pharmaceutics15071868

Huddleston JG, Visser AE, Reichert WM, Willauer HD, Broker GA, Rogers RD (2001) Characterization and comparison of hydrophilic and hydrophobic room temperature ionic liquids incorporating the imidazolium cation. Green Chem 3(4):156–164

Hussain CM (ed) (2020) Handbook of polymer nanocomposites for industrial applications. Elsevier

Idoko FA, Ezeamii GC, Ojochogwu OJ (2024) Green chemistry in manufacturing: innovations in reducing environmental impact. World J Adv Res Rev 23(3):2826–2841

Idris DS, Roy A, Pandit S, Alghamdi S, Almehmadi M, Alsaiari AA, Abdulaziz O, Alsharif A, Khandaker MU, Faruque MRI (2023) Polymer-based nanocarriers for biomedical and environmental applications. E-Polymers 23(1):20230049

Idumah CI (2021) Novel advancements in green and sustainable polymeric nanocomposites coatings. Curr Res Green Sustain Chem 4:100173
Idumah CI, Okonkwo UC, Obele CM (2022) Recently emerging advancements in montmorillonite polymeric nanoarchitectures and applications. Clean Mater 4:100071. https://doi.org/10.1016/j.clema.2022.100071
Ikram R, Mohamed Jan B, Abdul Qadir M, Sidek A, Stylianakis MM, Kenanakis G (2021) Recent advances in chitin and chitosan/graphene-based bio-nanocomposites for energetic applications. Polymers 13(19):3266
Ilyas K, Qureshi SW, Afzal S, Gul R, Yar M, Kaleem M, Khan AS (2018) Microwave-assisted synthesis and evaluation of type 1 collagen–apatite composites for dental tissue regeneration. J Biomater Appl 33(1):103–115. https://doi.org/10.1177/0885328218773220
Jafarzadeh S, Nooshkam M, Zargar M, Garavand F, Ghosh S, Hadidi M, Forough M (2024) Green synthesis of nanomaterials for smart biopolymer packaging: challenges and outlooks. J Nanostructure Chem 14(2):113–136
Jalaee A, Kamkar M, French V, Rojas OJ, Foster EJ (2023) Direct milling: efficient, facile, and green method for processing fibrillated cellulose/polymeric nanocomposites with boosted thermomechanical and rheological performance. Carbohydr Polym 314:120932
Jamróz E, Kulawik P, Kopel P (2019) The effect of nanofillers on the functional properties of biopolymer-based films: a review. Polymers 11(4):675
Jha SK, Jha A (2021) Plant extract mediated synthesis of metal nanoparticles, their characterization and applications: a green approach. Curr Green Chem 8(3):185–202
Kaczmarek-Szczepańska B, Wekwejt M (2024) Biodegradable polymers to biomedical and packaging applications. MDPI
Kalotra S, Mehta R (2021) Synthesis of polyaniline/clay nanocomposites by in situ polymerization and its application for the removal of Acid Green 25 dye from wastewater. Polym Bull 78:2439–2463. https://doi.org/10.1007/s00289-020-03222-3
Kamarudin SH, Rayung M, Abu F, Ahmad SB, Fadil F, Karim AA, Norizan MN, Sarifuddin N, Mat Desa MSZ, Mohd Basri MS, Samsudin H (2022) A review on antimicrobial packaging from biodegradable polymer composites. Polymers 14(1):174
Kango S, Kalia S, Celli A, Njuguna J, Habibi Y, Kumar R (2013) Surface modification of inorganic nanoparticles for development of organic–inorganic nanocomposites—A review. Prog Polym Sci 38(8):1232–1261
Kapoor RT, Salvadori MR, Rafatullah M, Siddiqui MR, Khan MA, Alshareef SA (2021) Exploration of microbial factories for synthesis of nanoparticles–a sustainable approach for bioremediation of environmental contaminants. Front Microbiol 12:658294. https://doi.org/10.3389/fmicb.2021.658294
Karande KM, Gawade SP (2020) Synthesis of nanosilver and its comparative evaluation of cytotoxic activity. Res J Pharm Technol 13(2):659–663. https://doi.org/10.5958/0974-360X.2020.00126.2
Kataki MS, Kakoti BB, Deka K, Rajkumari A (2022) Nanotechnology applications in natural Nanoclays production and application for better sustainability. In: Sustainable nanotechnology: strategies, products, and applications. Wiley, pp 159–171
Katayama Y (2005) Electrodeposition of metals in ionic liquids. In: Electrochemical aspects of ionic liquids. Wiley, p 111
Kazemi F, Naghib SM, Zare Y, Rhee KY (2021) Biosensing applications of polyaniline (PANI)-based nanocomposites: a review. Polym Rev 61(3):553–597
Ke L, Shang H, Tang M, Li X, Jiang L, Lu S, Tang D, Huang D, Zhu J, Liu C, Xu H (2022) High-heat and UV-barrier poly (lactic acid) by microwave-assisted functionalization of waste natural fibers. Int J Biol Macromol 220:827–836. https://doi.org/10.1016/j.ijbiomac.2022.08.114
Khalil AT, Ovais M, Iqbal J, Ali A, Ayaz M, Abbas M, Ahmad I, Devkota HP (2022) Microbes-mediated synthesis strategies of metal nanoparticles and their potential role in cancer therapeutics. In: Seminars in cancer biology, vol 86. Academic Press, pp 693–705. https://doi.org/10.1016/j.semcancer.2021.06.006

Khan F, Shariq M, Asif M, Siddiqui MA, Malan P, Ahmad F (2022) Green nanotechnology: plant-mediated nanoparticle synthesis and application. Nanomaterials 12:673

Khan UA, Srivastava A, Asthana N (2024) Harnessing green composites for energy efficiency in the era of industry 4.0. In: Engineering materials for efficient energy storage and conversion. IGI Global, pp 195–222

Khediri F, Mrad AI, Azzouz M, Doughi H, Najjar T, Mathiex-Fortunet H, Garnier P, Cortot A (2011) Efficacy of diosmectite (Smecta)® in the treatment of acute watery diarrhoea in adults: a multicentre, randomized, double-blind, placebo-controlled, parallel group study. Gastroenterol Res Pract 2011(1):783196

Kim F, Cote LJ, Huang J (2010) Graphene oxide: surface activity and two-dimensional assembly. Adv Mater 22(17):1954–1958

Kobayashi S (1999) Enzymatic polymerization: a new method of polymer synthesis. J Polym Sci A Polym Chem 37(16):3041–3056

Kolanthai E, Ganesan K, Epple M, Kalkura SN (2016) Synthesis of nanosized hydroxyapatite/agarose powders for bone filler and drug delivery application. Mater Today Commun 8:31–40

Korayem AH, Tourani N, Zakertabrizi M, Sabziparvar AM, Duan WH (2017) A review of dispersion of nanoparticles in cementitious matrices: nanoparticle geometry perspective. Constr Build Mater 153:346–357

Kudumula K (2016) Scope of polymer nano-composite in bio-medical applications. J Mech Civil Eng 13:18–21

Kumar A, Kumar N (2022) Advances in transparent polymer nanocomposites and their applications: a comprehensive review. Polym Plast Technol Mater 61(9):937–974

Kumar A, Kuang Y, Liang Z, Sun X (2020) Microwave chemistry, recent advancements, and eco-friendly microwave-assisted synthesis of nanoarchitectures and their applications: a review. Mater Today Nano 11:100076

Kumar V, Sehgal R, Gupta R (2021) Blends and composites of polyhydroxyalkanoates (PHAs) and their applications. Eur Polym J 161:110824

Kumar AR, Padmakumar A, Kalita U, Samanta S, Baral A, Singha NK, Ashokkumar M, Qiao GG (2023) Ultrasonics in polymer science: applications and challenges. Prog Mater Sci 136:101113

Lahiri D, Nag M, Sheikh HI, Sarkar T, Edinur HA, Pati S, Ray RR (2021) Microbiologically-synthesized nanoparticles and their role in silencing the biofilm signaling cascade. Front Microbiol 12:636588

Le TM, Chuchak R, Sairiam S (2024) Empowering TiO2–coated PVDF membranes stability with polyaniline and polydopamine for synergistic separation and photocatalytic enhancement in dye wastewater purification. Sci Rep 14(1):15969

Leao RM, Jesus LC, Bertuoli PT, Zattera AJ, Maia JM, del Menezzi CHS, Amico SC, da Luz SM (2020) Production and characterization of cellulose nanocrystals/acrylonitrile butadiene styrene nanocomposites. J Compos Mater 54(27):4207–4214

Lee H, Lee DG (2018) Gold nanoparticles induce a reactive oxygen species-independent apoptotic pathway in Escherichia coli. Colloids Surf B: Biointerfaces 167:1–7. https://doi.org/10.1016/j.colsurfb.2018.03.049

Lee J, Chatterjee DK, Lee MH, Krishnan S (2014) Gold nanoparticles in breast cancer treatment: promise and potential pitfalls. Cancer Lett 347(1):46–53. https://doi.org/10.1016/j.canlet.2014.02.006

Lee S, Hao LT, Park J, Oh DX, Hwang DS (2023) Nanochitin and nanochitosan: chitin nanostructure engineering with multiscale properties for biomedical and environmental applications. Adv Mater 35(4):2203325

Lertwattanaseri T, Ichikawa N, Mizoguchi T, Tanaka Y, Chirachanchai S (2009) Microwave technique for efficient deacetylation of chitin nanowhiskers to a chitosan nanoscaffold. Carbohydr Res 344(3):331–335

Li Z, Renneckar S, Barone JR (2010) Nanocomposites prepared by in situ enzymatic polymerization of phenol with TEMPO-oxidized nanocellulose. Cellulose 17:57–68

Li J, Li Y, Song Y, Niu S, Li N (2017) Ultrasonic-assisted synthesis of polyvinyl alcohol/phytic acid polymer film and its thermal stability, mechanical properties and surface resistivity. Ultrason Sonochem 39:853–862

Livi S et al (2016) Supercritical CO2–ionic liquids: a successful wedding to prepare biopolymer foams. ACS Sustain Chem Eng 4(2):461–470

Luong ND, Korhonen JT, Soininen AJ, Ruokolainen J, Johansson LS, Seppälä J (2013) Processable polyaniline suspensions through in situ polymerization onto nanocellulose. Eur Polym J 49(2):335–344

Lv J, Zhou P, Zhang L, Zhong Y, Sui X, Wang B, Chen Z, Xu H, Mao Z (2019) High-performance textile electrodes for wearable electronics obtained by an improved in situ polymerization method. Chem Eng J 361:897–907

Ma MG, Dong YY, Fu LH, Li SM, Sun RC (2013) Cellulose/CaCO3 nanocomposites: microwave ionic liquid synthesis, characterization, and biological activity. Carbohydr Polym 92(2):1669–1676

Ma H, Qin C, Jin B, Gong P, Lan B, Huang Y, Park CB, Li G (2022) Using a supercritical fluid-assisted thin cell wall stretching–defoaming method to enhance the nanofiller dispersion, emi shielding, and thermal conduction property of CNF/PVDF nanocomposites. Ind Eng Chem Res 61(10):3647–3659

Mahy JG, Lejeune L, Haynes T, Lambert SD, Marcilli RHM, Fustin CA, Hermans S (2021) Eco-friendly colloidal aqueous sol-gel process for TiO2 synthesis: the peptization method to obtain crystalline and photoactive materials at low temperature. Catalysts 11(7):768

Makepa DC, Chihobo CH (2024) Barriers to commercial deployment of biorefineries: a multi-faceted review of obstacles across the innovation chain. Heliyon 10(12):e32649

Makvandi P, Ali GW, Della Sala F, Abdel-Fattah WI, Borzacchiello A (2019) Biosynthesis and characterization of antibacterial thermosensitive hydrogels based on corn silk extract, hyaluronic acid and nanosilver for potential wound healing. Carbohydr Polym 223:115023

Makvandi P, Ali GW, Della Sala F, Abdel-Fattah WI, Borzacchiello A (2020) Hyaluronic acid/corn silk extract based injectable nanocomposite: a biomimetic antibacterial scaffold for bone tissue regeneration. Mater Sci Eng C 107:110195. https://doi.org/10.1016/j.msec.2019.110195

Malik S, Dhasmana A, Preetam S, Mishra YK, Chaudhary V, Bera SP, Ranjan A, Bora J, Kaushik A, Minkina T, Jatav HS (2022) Exploring microbial-based green nanobiotechnology for wastewater remediation: a sustainable strategy. Nano 12(23):4187. https://doi.org/10.3390/nano12234187

Mallakpour S, Dinari M (2012) Ionic liquids as green solvents: progress and prospects. In: Green solvents II: properties and applications of ionic liquids. Springer, Dordrecht, pp 1–32

Mallakpour S, Shafiee E (2018) A simple method for the sonochemical synthesis of PVA/ZrO2-vitamin B1 nanocomposites: morphology, mechanical, thermal and wettability investigations. Ultrason Sonochem 40:881–889. https://doi.org/10.1016/j.porgcoat.2018.04.023

Mapossa AB, da Silva Júnior AH, de Oliveira CRS, Mhike W (2023) Thermal, morphological and mechanical properties of multifunctional composites based on biodegradable polymers/bentonite clay: a review. Polymers 15(16):3443

Marano S, Laudadio E, Minnelli C, Stipa P (2022) Tailoring the barrier properties of PLA: a state-of-the-art review for food packaging applications. Polymers 14(8):1626

Marciniak A (2010) The solubility parameters of ionic liquids. Int J Mol Sci 11(5):1973–1990

Mariano M, Pilate F, de Oliveira FB, Khelifa F, Dubois P, Raquez JM, Dufresne A (2017) Preparation of cellulose nanocrystal-reinforced poly (lactic acid) nanocomposites through non-covalent modification with PLLA-based surfactants. ACS Omega 2(6):2678–2688

Matei E, Predescu AM, Râpă M, Ţurcanu AA, Mateş I, Constantin N, Predescu C (2022) Natural polymers and their nanocomposites used for environmental applications. Nano 12(10):1707

McHale E, Steindal CC, Kutzke H, Benneche T, Harding SE (2017) In situ polymerisation of isoeugenol as a green consolidation method for waterlogged archaeological wood. Sci Rep 7(1):46481

Meher MK, Poluri KM (2022) Bifunctional Dalteparin/enoxaparin coated nanosilver formulation to prevent bloodstream infections during hemodialysis. Carbohydr Polym 291:119546. https://doi.org/10.1016/j.carbpol.2022.119546

Meher A, Anwesha N, Sagadevan S, Moharana S (2024a) Polymer blend nanocomposites are analogs to natural polymers. In: Novel bio-nanocomposites for biomedical applications. Springer, Cham, pp 37–60

Meher MK, Naidu G, Mishra A, Poluri KM (2024b) A review on multifaceted biomedical applications of heparin nanocomposites: progress and prospects. Int J Biol Macromol 17:129379. https://doi.org/10.1016/j.ijbiomac.2024.129379

Mehrabani SAN, Vatanpour V, Koyuncu I (2022) Green solvents in polymeric membrane fabrication: a review. Sep Purif Technol 298:121691

Michael A, Singh A, Roy A, Islam MR (2022) [retracted] fungal-and algal-derived synthesis of various nanoparticles and their applications. Bioinorg Chem Appl 2022(1):3142674. https://doi.org/10.1155/2022/3142674

Mittal V (2011) In-situ synthesis of polymer nanocomposites. In-Situ Synth Poly Nanocomp:1–25

Mohamat R, Bakar SA, Mohamed A, Muqoyyanah M, Othman MH, Rohani R, Mamat MH, Ahmad MK, Malek MF, Khalil HA, Ali K (2024) Recent advances for wastewater treatment on polyvinylidene fluoride-based membrane: a review. Int J Integ Eng 16(5):517–537

Mohan TP, Kanny K (2020) Green nanofillers for polymeric materials. In: Green nanomaterials. Advanced structured materials. Springer, Singapore, pp 99–138

Montes S, Carrasco PM, Ruiz V, Cabañero G, Grande HJ, Labidi J, Odriozola I (2015) Synergistic reinforcement of poly (vinyl alcohol) nanocomposites with cellulose nanocrystal-stabilized graphene. Compos Sci Technol 117:26–31

Montolio S, Altava B, García-Verdugo E, Luis SV (2019) Supported ILs and materials based on ILs for the development of green synthetic processes and procedures. In: Green synthetic processes and procedures. Royal Society of Chemistry

Mukherjee A, Panda S, Jeelani PG, Mossa AT, Chidambaram R (2023) Biodegradable polymers/silica nanocomposites: applications in food packaging. In: Nanotechnology applications for food safety and quality monitoring. Academic Press, pp 395–414

Mumin MA, Akhter KF, Oyeneye OO, Xu WZ, Charpentier PA (2018) Supercritical fluid assisted dispersion of nano-silica encapsulated CdS/ZnS quantum dots in poly (ethylene-co-vinyl acetate) for solar harvesting films. ACS Appl Nano Mater 1(7):3186–3195

Muncke J, Andersson AM, Backhaus T, Boucher JM, Carney Almroth B, Castillo A, Chevrier J, Demeneix BA, Emmanuel JA, Fini JB, Gee D (2020) Impacts of food contact chemicals on human health: a consensus statement. Environ Health 19:1–12

Naeem A, Saeed B, AlMohamadi H, Lee M, Gilani MA, Nawaz R, Khan AL, Yasin M (2024) Sustainable and green membranes for chemical separations: a review. Sep Purif Technol 3:126271

Nawaz K, Chaudhary R, Khan AK, Hano C, Abbasi BH, Anjum S (2022) Algal-mediated biosynthesis of nanoparticles and their potential therapeutic applications. In: Bioprospecting algae for nanosized materials. Springer, Cham, pp 171–210. https://doi.org/10.1007/978-3-030-81557-8_8

Nguyen Thi HY, Nguyen BTD, Kim JF (2020) Sustainable fabrication of organic solvent nanofiltration membranes. Membranes 11(1):19

Olza S, Salaberria AM, Alonso-Varona A, Samanta A, Fernandes SC (2023) The role of nanochitin in biologically-active matrices for tissue engineering-where do we stand? J Mater Chem B 11(25):5630–5649

Omerović N, Djisalov M, Živojević K, Mladenović M, Vunduk J, Milenković I, Knežević NŽ, Gadjanski I, Vidić J (2021) Antimicrobial nanoparticles and biodegradable polymer composites for active food packaging applications. Compr Rev Food Sci Food Saf 20(3):2428–2454

Osman AF, Fitri TF, Rakibuddin M, Hashim F, Johari SA, Ananthakrishnan R, Ramli R (2017) Pre-dispersed organo-montmorillonite (organo-MMT) nanofiller: morphology, cytocompat-

ibility and impact on flexibility, toughness and biostability of biomedical ethyl vinyl acetate (EVA) copolymer. Mater Sci Eng C 74:194–206
Osman AI, Zhang Y, Farghali M, Rashwan AK, Eltaweil AS, Abd El-Monaem EM, Mohamed IM, Badr MM, Ihara I, Rooney DW, Yap PS (2024) Synthesis of green nanoparticles for energy, biomedical, environmental, agricultural, and food applications: a review. Environ Chem Lett 22(2):841–887
Pal N, Banerjee S, Roy P, Pal K (2021) Cellulose nanocrystals-silver nanoparticles-reduced graphene oxide based hybrid PVA nanocomposites and its antimicrobial properties. Int J Biol Macromol 191:445–456
Park S, Ruoff RS (2009) Chemical methods for the production of graphenes. Nat Nanotechnol 4(4):217–224
Park JH, Shin HJ, Kim MH, Kim JS, Kang N, Lee JY, Kim KT, Lee JI, Kim DD (2016) Application of montmorillonite in bentonite as a pharmaceutical excipient in drug delivery systems. J Pharm Investig 46:363–375
Paszkiewicz S, Walkowiak K, Barczewski M (2024) Biobased polymer nanocomposites prepared by in situ polymerization: comparison between carbon and mineral nanofillers. J Mater Sci 59(30):13805–13823
Patil MB, Rajamani SB, Mathad SN, Patil AY, Hussain MA, Alorfii HS, Khan A, Asiri AM, Khan I, Puttegowda M (2022) Microwave-assisted synthesis of poly (acrylamide-co-2-hydroxyethyl methacrylate)/chitosan semi-IPN ZnO nanocomposite membranes for food packaging applications. J Mater Res Technol 20:3537–3548. https://doi.org/10.1016/j.jmrt.2022.08.079
Pawlaczyk-Łuszczyńska M, Dudarewicz A (2020) Impact of very high-frequency sound and low-frequency ultrasound on people–the current state of the art. Int J Occup Med Environ Health 33(4):389
Perera KY, Hopkins M, Jaiswal AK, Jaiswal S (2024) Nanoclays-containing bio-based packaging materials: properties, applications, safety, and regulatory issues. J Nanostructure Chem 14(1):71–93. https://doi.org/10.1007/s40097-023-00525-5
Pérez-Medina JC, Waldo-Mendoza MA, Cruz-Delgado VJ, Quiñones-Jurado ZV, González-Morones P, Ziolo RF, Martínez-Colunga JG, Soriano-Corral F, Avila-Orta CA (2016) Metamaterial behavior of polymer nanocomposites based on polypropylene/multi-walled carbon nanotubes fabricated by means of ultrasound-assisted extrusion. Materials 9(11):923
Petkovic M, Seddon KR, Rebelo LPN, Pereira CS (2011) Ionic liquids: a pathway to environmental acceptability. Chem Soc Rev 40(3):1383–1403
Picken SJ, Korobko A, Mendes E, Norder B, Makarova V, Vasilyev GB, Karbushev V, Tolstykh M (2010) Mechanical and thermal properties of polymer nanocomposites. In: 43rd IUPAC world polymer congress
Plechkova NV, Seddon KR (2008) Applications of ionic liquids in the chemical industry. Chem Soc Rev 37(1):123–150
Pomogailo AD (2005) Polymer sol-gel synthesis of hybrid nanocomposites. Colloid J 67:658–677
Pramanik S, Das P (2019) Metal-based nanomaterials and their polymer nanocomposites. In: Nanomaterials and polymer nanocomposites. Elsevier, pp 91–121. https://doi.org/10.1016/B978-0-12-814615-6.00003-5
Prasad K, Sharma M (2019) Green solvents for the dissolution and processing of biopolymers. Curr Opin Green Sustain Chem 18:72–78
Pulingam T, Foroozandeh P, Chuah JA, Sudesh K (2022) Exploring various techniques for the chemical and biological synthesis of polymeric nanoparticles. Nano 12(3):576
Qamar SA, Asgher M, Khalid N (2020) Bioinspired nanocomposites: functional materials for sustainable greener technologies. In: Renewable energy-resources, challenges and applications. IntechOpen
Qi R, Jin X, Nie J, Yu W, Zhou C (2005) Synthesis and properties of polystyrene–clay nanocomposites via in situ intercalative polymerization. J Appl Polym Sci 97(1):201–207

Rahmani S, Maroufkhani M, Mohammadzadeh-Komuleh S, Khoubi-Arani Z (2022) Polymer nanocomposites for biomedical applications. In: Fundamentals of bionanomaterials. Elsevier, pp 175–215
Rajesh K, Crasta V, Bairy R, Patil PS (2023) Effect of TiO2/ZnO nanofillers on structural, optical and nonlinear optical properties of PVA. Opt Mater 145:114481
Rajpoot A, Banik N, Ezeani OE (2023) A review on the recent development on polymer nanocomposite for energy storage application. Mater Sci Res India 20(Special Issue 1):13–26
Rallini M, Kenny JM (2017) Nanofillers in polymers. In: Modification of polymer properties. William Andrew Publishing, pp 47–86
Ramakrishnan RK, Sumitha NS (2024) Nanoclay-reinforced polymers. In: Nanoclay-based sustainable materials. Elsevier, pp 91–114
Rami MR, Meskini M, Sharafabad BE (2024) Fungal-mediated nanoparticles for industrial applications: synthesis and mechanism of action. J Infect Public Health 17(10):102536
Rathod S, Preetam S, Pandey C, Bera SP (2024) Exploring synthesis and applications of green nanoparticles and the role of nanotechnology in wastewater treatment. Biotechnol Rep 41:e00830
Rawat S, Singh J (2022) Green biocomposite materials for sustainable remediation application. In: Sustainable nanotechnology for environmental remediation. Elsevier, pp 273–297
Rodríguez-Rojas MDP, Bustos-Terrones V, Díaz-Cárdenas MY, Vázquez-Vélez E, Martínez H (2024) Life cycle assessment of green synthesis of TiO2 nanoparticles vs. Chem Synth Sustain 16(17):7751
Rogers RD, Seddon KR (2003) Ionic liquids--solvents of the future? Science 302(5646):792–793
Roshani M, Rezaian-Isfahni A, Lotfalizadeh MH, Khassafi N, Abadi MH, Nejati M (2023) Metal nanoparticles as a potential technique for the diagnosis and treatment of gastrointestinal cancer: a comprehensive review. Cancer Cell Int 23(1):280. https://doi.org/10.1186/s12935-023-03115-1
Saba N, Jawaid M, Asim M (2019) Nanocomposites with nanofibers and fillers from renewable resources. In: Green composites for automotive applications. Woodhead Publishing, Cambridge, MA, pp 145–170
Salama A, Neumann M, Günter C, Taubert A (2014) Ionic liquid assisted formation of cellulose/calcium phosphate hybrid materials. Beilstein J Nanotechnol 5:1553–1568
Sampath S, Madhavan Y, Muralidharan M, Sunderam V, Lawrance AV, Muthupandian S (2022) A review on algal mediated synthesis of metal and metal oxide nanoparticles and their emerging biomedical potential. J Biotechnol 360:92–109. https://doi.org/10.1016/j.jbiotec.2022.10.009
Samrot AV, Samanvitha SK, Shobana N, Renitta ER, Senthilkumar P, Kumar SS, Abirami S, Dhiva S, Bavanilatha M, Prakash P, Saigeetha S (2021) The synthesis, characterization and applications of polyhydroxyalkanoates (PHAs) and PHA-based nanoparticles. Polymers 13(19):3302
Santos BA, França AM, Silva LT, Gouveia AG, Vidal CB, Loiola AR, Nascimento RF (2024) Magnetic zeolite from fly ash as a cost-effective adsorbent for Cu^{2+}, Pb^{2+}, Zn^{2+}, and Cd^{2+} removal from aqueous media: a comprehensive study. Int J Environ Analy Chem 104(16):3913–3935. https://doi.org/10.24200/sci.2023.60682.6933
Sarkar A, Mushahary N, Basumatary F, Das B, Basumatary SF, Venkatesan K, Selvaraj M, Rokhum SL, Basumatary S (2024) Efficiency of montmorillonite-based materials as adsorbents in dye removal for wastewater treatment. J Environ Chem Eng 12(3):112519
Savale TK, Byram Anand P, Perumalla Varalaxmi AB, Quaye JA (2023) Green marketing strategies: assessing consumer perception and adoption of eco-friendly products. Remittances Rev 8(4)
Senila L, Cadar O, Kovacs E, Gal E, Dan M, Stupar Z, Simedru D, Senila M, Roman C (2023) L-poly (lactic acid) production by microwave irradiation of lactic acid obtained from lignocellulosic wastes. Int J Mol Sci 24(12):9817. https://doi.org/10.3390/ijms24129817
Sethi S, Kaith BS, Kumar V (2019) Fabrication and characterization of microwave assisted carboxymethyl cellulose-gelatin silver nanoparticles imbibed hydrogel: its evaluation as dye degradation. React Funct Polym 142:134–146. https://doi.org/10.1016/j.reactfunctpolym.2019.06.014

Shahdan D, Chen RS, Ahmad S (2021) Optimization of graphene nanoplatelets dispersion and nano-filler loading in bio-based polymer nanocomposites based on tensile and thermogravimetry analysis. J Mater Res Technol 15:1284–1299. https://doi.org/10.1016/j.jmrt.2021.08.130

Shalom Y, Perelshtein I, Perkas N, Gedanken A, Banin E (2017) Catheters coated with Zn-doped CuO nanoparticles delay the onset of catheter-associated urinary tract infections. Nano Res 10:520–533. https://doi.org/10.1007/s12274-016-1310-8

Shamiri A, Chakrabarti MH, Jahan S, Hussain MA, Kaminsky W, Aravind PV, Yehye WA (2014) The influence of Ziegler-Natta and metallocene catalysts on polyolefin structure, properties, and processing ability. Materials 7(7):5069–5108

Shange MG, Khumalo NL, Mohomane SM, Motaung TE (2024) Factors affecting silica/cellulose nanocomposite prepared via the sol–gel technique: a review. Materials 9:1937. https://doi.org/10.3390/ma17091937

Sharma B, Thomas S, Bajpai PK, Ghosal K, Shekhar S (eds) (2023a) Advances in bionanocomposites: materials, applications, and life cycle. Elsevier

Sharma VK, Sharma DK, Sharma S (2023b) Bio-mimetic and bio-templating of natural materials: an approach to sustainable development using bionanotechnology. In: Bionanotechnology towards green energy. CRC Press, pp 187–206

Sharma D, Bansal S, Jain P (2025) Green nanocomposites: incorporating biopolymers into plastics for enhanced environmental sustainability. In: Sustainability in polymer technology and plastic engineering. Apple Academic Press, pp 95–116

Shawky RM, Hamed AA (2024) Chitosan-silver nanoparticles conjugate mediated by endophytic fungus Talaromyces sp. 7S1: antimicrobial, antibiofilm and anticancer. Egypt J Chem 67(3):413–423. https://doi.org/10.21608/ejchem.2023.227121.8361

Shen T, Gao M (2019) Gemini surfactant modified organo-clays for removal of organic pollutants from water: a review. Chem Eng J 375:121910

Shen R, Xue S, Xu Y, Liu Q, Feng Z, Ren H, Zhai H, Kong F (2020) Research progress and development demand of nanocellulose reinforced polymer composites. Polymers 12(9):2113

Sheng Y, Wu B, Zhang C (2024) Gold nanoparticles decorated chitosan-agarose modified magnetic nanocomposite: catalytic application for the preparation of Tetrazoles and biological evaluation for the treatment of lung carcinoma. J Polym Environ 32:1–14. https://doi.org/10.1007/s10924-024-03252-2

Shi YH, Xu ZR, Feng JL, Xia MS, Hu CH (2005) Effects of modified montmorillonite nanocomposite on growing/finishing pigs during aflatoxicosis. Asian Australas J Anim Sci 18(9):1305–1309

Shi Y, Wang X, Feng C, Yang S (2024) Nano-clay montmorillonite removes tetracycline in water: factors and adsorption mechanism in aquatic environments. IScience 27(2). https://doi.org/10.1016/j.isci.2024.108952

Shukla AK, Iravani S (eds) (2018) Green synthesis, characterization and applications of nanoparticles. Elsevier

Siddiqi KS, Husen A, Rao RAK (2018) A review on biosynthesis of silver nanoparticles and their biocidal properties. J Nanobiotechnol 16:14. https://doi.org/10.1186/s12951-018-0334-5

Singh P, Shukla SK (2020) Advances in polyaniline-based nanocomposites. J Mater Sci 55(4):1331–1365

Singh J, Dutta T, Kim KH, Rawat M, Samddar P, Kumar P (2018) 'Green' synthesis of metals and their oxide nanoparticles: applications for environmental remediation. J Nanobiotechnol 16:1–24

Singh H, Desimone MF, Pandya S, Jasani S, George N, Adnan M, Aldarhami A, Bazaid AS, Alderhami SA (2023) Revisiting the green synthesis of nanoparticles: uncovering influences of plant extracts as reducing agents for enhanced synthesis efficiency and its biomedical applications. Int J Nanomedicine:4727–4750. https://doi.org/10.2147/IJN.S419369

Sivanesan I, Gopal J, Muthu M, Shin J, Mari S, Oh J (2021) Green synthesized chitosan/chitosan nanoforms/nanocomposites for drug delivery applications. Polymers 13(14):2256

Soman V, Vishwakarma K, Poddar MK (2023) Ultrasound assisted synthesis of polymer nanocomposites: a review. J Polym Res 30(11):406

Sousa M, Afonso AC, Saavedra MJ, Simões LC, Simões M (2023) Hydrocinnamic acid and perillyl alcohol are effective against Escherichia coli biofilms when used alone and combined with antibiotics. J Appl Microbiol 134(10):lxad234

Sridhar SK, Bhavani PD, Noothi S, Gajula LR, Goudanavar P, Gowthami B, Naveen NR (2024) Microwave revolution: transforming biomedical synthesis for tissue engineering advancements. Curr Microw Chem 11(2):95–115

Stozhko N, Tarasov A, Tamoshenko V, Bukharinova M, Khamzina E, Kolotygina V (2024) Green silver nanoparticles: plant-extract-mediated synthesis. Optic Electrochem Prop Physchem 4(4):402–419. https://doi.org/10.3390/physchem4040028

Suganthi S, Vignesh S, Al-Ansari MM, Al-Humaid LA, Oh TH, Raj V (2024) Development of PVA/sodium alginate incorporated with histidine capped silver nanoparticles for food packaging application. Polym Adv Technol 35(5):e6413

Sun H, Jia J, Jiang C, Zhai S (2018) Gold nanoparticle-induced cell death and potential applications in nanomedicine. Int J Mol Sci 19(3):754. https://doi.org/10.3390/ijms19030754

Sušac K, Vidović E, Vrsaljko D, Jukić A (2024) Synthesis and properties of transparent PMMA/cellulose nanocomposites prepared by in situ polymerization in green solvent. Poly Comp 45:11434

Suvetha S, Mani AK (2024) Metallic nanocomposites– a prevailing tool to decipher the bacterial crosstalk. J Ind Eng Chem. https://doi.org/10.1016/j.jiec.2024.09.044

Taib NAAB, Rahman MR, Huda D, Kuok KK, Hamdan S, Bakri MKB, Julaihi MRMB, Khan A (2023) A review on poly lactic acid (PLA) as a biodegradable polymer. Polym Bull 80(2):1179–1213

Taleblou N, Sirousazar M, Hassan ZM, Khaligh SG (2020) Capecitabine-loaded anti-cancer nanocomposite hydrogel drug delivery systems: in vitro and in vivo effi cacy against the 4T1 murine breast cancer cells. J Biomater Sci Polym Ed 31(1):72–92. https://doi.org/10.1080/09205063.2019.1675225

Tanahashi M (2010) Development of fabrication methods of filler/polymer nanocomposites: with focus on simple melt-compounding-based approach without surface modification of nanofillers. Materials 3(3):1593–1619

Thakre D, Rayalu S, Kawade R, Meshram S, Subrt J, Labhsetwar N (2010) Magnesium incorporated bentonite clay for defluoridation of drinking water. J Hazard Mater 180(1–3):122–130

Tjong SC (2006) Structural and mechanical properties of polymer nanocomposites. Mater Sci Eng R Rep 53(3–4):73–197

Tsivileva O, Pozdnyakov A, Ivanova A (2021) Polymer nanocomposites of selenium biofabricated using fungi. Molecules 26(12):3657. https://doi.org/10.3390/molecules26123657

Tundo P, Anastas P, Black DS, Breen J, Collins TJ, Memoli S, Miyamoto J, Polyakoff M, Tumas W (2000) Synthetic pathways and processes in green chemistry. Introductory overview. Pure Appl Chem 72(7):1207–1228

Ukkund SJ, Alke B, Taqui SN, Syed UT (2022) Dextran nanoparticles: preparation and applications. In: Polysaccharide nanoparticles. Elsevier, pp 1–31. https://doi.org/10.1016/B978-0-12-822351-2.00019-X

Varadharajan S, Vasanthan KS, Mathur V, Hariperumal N, Mazumder N (2024) Green synthesis and multifaceted applications: challenges and innovations in carbon dot nanocomposites. Disc Nano 19(1):205

Veeman D, Shree MV, Sureshkumar P, Jagadeesha T, Natrayan L, Ravichandran M, Paramasivam P (2021) Sustainable development of carbon nanocomposites: synthesis and classification for environmental remediation. J Nanomater 2021(1):5840645

Verma ML, Sharma S, Dhiman K, Jana AK (2019) Microbial production of nanoparticles: mechanisms and applications. Microb Nanobio 1:159–176

Verma DK, Verma C, Fuertes PO (eds) (2024a) Green chemical synthesis with microwaves and ultrasound. Wiley

Verma YK, Singh AK, Paswan MK, Gurmaita PK (2024b) Preparation and characterization of bamboo based nanocellulose by ball milling and used as a filler for preparation of nanocomposite. Polymer 308:127396

Vieira IRS, de Carvalho APAD, Conte-Junior CA (2022) Recent advances in biobased and biodegradable polymer nanocomposites, nanoparticles, and natural antioxidants for antibacterial and antioxidant food packaging applications. Compr Rev Food Sci Food Saf 21(4):3673–3716

Wang S, Feng Q, Sun J, Gao F, Fan W, Zhang Z, Li X, Jiang X (2016) Nanocrystalline cellulose improves the biocompatibility and reduces the wear debris of ultrahigh molecular weight polyethylene via weak binding. ACS Nano 10(1):298–306

Wang Y, Ma Y, Sun Y, Guo G, Liu Y (2019) In situ latex synthesis of magnetic polymer nanocomposites for application in magnetorheological materials. Polym Adv Technol 30(8):2005–2016

Wang L, Jie H, Liu Y, Shu J, Hao W, Wang Z, Pan X, Zhang N, Zhou L, Zhang J (2020a) Ionic liquids grafted cellulose nanocrystals for high-strength and toughness PVA nanocomposite. ACS Appl Mater Interfaces 12(34):38796–38804. https://doi.org/10.1021/acsami.0c11217

Wang L, Gardner DJ, Wang J, Yang Y, Tekinalp HL, Tajvidi M, Li K, Zhao X, Neivandt DJ, Han Y, Ozcan S (2020b) Towards industrial-scale production of cellulose nanocomposites using melt processing: a critical review on structure-processing-property relationships. Compos Part B 201:108297

Wasserscheid P, Keim W (2000) Ionic liquids—new "solutions" for transition metal catalysis. Angew Chem Int Ed 39(21):3772–3789

Wu N, Demchuk Z, Voronov A, Pourhashem G (2021) Sustainable manufacturing of polymeric materials: a techno-economic analysis of soybean oil-based acrylic monomers production. J Clean Prod 286:124939

Wu Y, Gao X, Wu J, Zhou T, Nguyen TT, Wang Y (2023) Biodegradable polylactic acid and its composites: characteristics, processing, and sustainable applications in sports. Polymers 15(14):3096

Wypych F, de Freitas RA (2022) Clay minerals: classification, structure, and properties. In: Developments in clay science, vol 10. Elsevier, pp 3–35

Xie W, Li J, Sun F, Dong W (2020) Ultrasonication favors TiO 2 nano-particles dispersion in PVDF ultrafiltration membrane to effectively enhance membrane hydrophilicity and anti-fouling capability. Environ Sci Pollut Res 27:9503–9519. https://doi.org/10.1007/s11356-019-06862-9

Xie F, Gao C, Avérous L (2024) Alginate-based materials: enhancing properties through multiphase formulation design and processing innovation. Mater Sci Eng R Rep 159:100799

Xu C, Cheng Z (2021) Thermal stability of ionic liquids: current status and prospects for future development. PRO 9(2):337

Xu H, Hao Z, Wang C, Deng J, Wang T, Zhang J (2023) In situ emulsion polymerization to multifunctional polymer nanocomposites: a review. Macromol Chem Phys 224(20):2300185

Yan D, Li Y, Liu Y, Li N, Zhang X, Yan C (2021) Antimicrobial properties of chitosan and chitosan derivatives in the treatment of enteric infections. Molecules 26(23):7136

Yang MC, Tseng YQ, Liu KH, Cheng YW, Chen WT, Chen WT, Hsiao CW, Yung MC, Hsu CC, Liu TY (2019) Preparation of amphiphilic chitosan–graphene oxide–cellulose nanocrystalline composite hydrogels and their biocompatibility and antibacterial properties. Appl Sci 9(15):3051

Yang G, Ma R, Zhang S, Liu Z, Pei D, Jin H, Liu J, Du W (2022) Microwave-assisted in situ ring-opening polymerization of ε-caprolactone in the presence of modified halloysite nanotubes loaded with stannous chloride. RSC Adv 12(3):1628–1637. https://doi.org/10.1039/D1RA07469E

Yee Kuen C, Masarudin MJ (2022) Chitosan nanoparticle-based system: a new insight into the promising controlled release system for lung cancer treatment. Molecules 27(2):473

Yeo SD, Lim PG, Debenedetti PG, Bernstein H (1993) Formation of microparticulate protein powders using a supercritical fluid anti-solvent. Biotechnol Bioeng 41:341–346

Ying S, Guan Z, Ofoegbu PC, Clubb P, Rico C, He F, Hong J (2022) Green synthesis of nanoparticles: current developments and limitations. Environ Technol Innov 26:102336. https://doi.org/10.1016/j.eti.2022.102336

Yuan Y, Chen W, Ma Z, Deng Y, Chen Y, Chen Y, Hu W (2019) Enhanced optomechanical properties of mechanochemiluminescent poly (methyl acrylate) composites with granulated fluorescent conjugated microporous polymer fillers. Chem Sci 10(7):2206–2211

Yuan X, Lin S, Zhao K, Han Y (2022) Emulsion-ultrasonic spray method to prepare polylactic acid microspheres. Mater Lett 309:131461

Yuvaraj H, Shim JJ, Lim KT (2012) Synthesis of polymer–inorganic filler nanocomposites in supercritical CO2. J Supercrit Fluids 71:45–63

Zapata P, Quijada R, Retuer J, Moncada E (2008) Preparation of nanocomposites by in situ polimerization. J Chil Chem Soc 53(1):1359–1360

Zhang S, Sun J, Zhang X, Xin J, Miao Q, Wang J (2014) Ionic liquid-based green processes for energy production. Chem Soc Rev 43(22):7838–7869

Zhang F, Zhou T, Liu Y, Leng J (2015) Microwave synthesis and actuation of shape memory polycaprolactone foams with high speed. Sci Rep 5(1):11152. https://doi.org/10.1038/srep11152

Zhang D, Sun J, Lee LJ, Castro JM (2020) Overview of ultrasonic assisted manufacturing multifunctional carbon nanotube nanopaper based polymer nanocomposites. Eng Sci 10(29):35–50

Zhang X, Li Y, Guo M, Jin TZ, Arabi SA, He Q, Ismail BB, Hu Y, Liu D (2021) Antimicrobial and UV blocking properties of composite chitosan films with curcumin grafted cellulose nanofiber. Food Hydrocoll 112:106337

Zhao X, Xia Y, Li Q, Ma X, Quan F, Geng C, Han Z (2014) Microwave-assisted synthesis of silver nanoparticles using sodium alginate and their antibacterial activity. Colloids Surf A Physicochem Eng Asp 444:180–188. https://doi.org/10.1016/j.colsurfa.2013.12.008

Zhao Z, Xia K, Hou Y, Zhang Q, Ye Z, Lu J (2021) Designing flexible, smart and self-sustainable supercapacitors for portable/wearable electronics: from conductive polymers. Chem Soc Rev 50(22):12702–12743

Zhao L, Zhou Y, Zhang J, Liang H, Chen X, Tan H (2023) Natural polymer-based hydrogels: from polymer to biomedical applications. Pharmaceutics 15(10):2514

Zhao Y, Dai Z, Huang H, Tian J, Cai H (2024) Injectable silver nanoparticle-based hydrogel dressings with rapid shape adaptability and antimicrobial activity. Appl Biochem Biotechnol:1–16

Zheng B, Wen ZS, Huang YJ, Xia MS, Xiang XW, Qu YL (2016) Molecular weight-dependent immunostimulative activity of low molecular weight chitosan via regulating NF-κB and AP-1 signaling pathways in RAW264. 7 macrophages. Mar Drugs 14(9):169

Zheng S, Wang Y, Zhu Y, Zheng C (2023) Recent advances in the construction and properties of carbon-based nanofillers/polymer composites with segregated network structures. Mater Today Commun 29:106773

Zhong J, Isayev AI (2016) Ultrasonically assisted compounding of CNT with polypropylenes of different molecular weights. Polymer 107:130–146. https://doi.org/10.1016/j.polymer.2016.11.006

Zhu Y (2023) Application of chitosan-based materials in wastewater treatment. Highlight Sci Eng Technol 69:489–497

Zhu YJ, Chen F (2014) Microwave-assisted preparation of inorganic nanostructures in liquid phase. Chem Rev 114(12):6462–6555

Chapter 10
Biodegradable Polymer Nanocomposites

Kuldeep Vinchurkar, Hardik Rana, Dhrusha Patel, Samya Sheth, and Sachin N. Kothawade

Abstract The demand for sustainable and environmentally friendly materials has led to significant advancements in biodegradable polymer nanocomposites. These innovative materials combine the biodegradability of natural and synthetic polymers with the enhanced mechanical, thermal, and barrier properties provided by nanoscale fillers. This chapter explores the development and applications of biodegradable polymer nanocomposites, providing a comprehensive understanding of the types of polymers used, nanofillers, and fabrication techniques. Key topics include the natural and synthetic biodegradable polymers, the roles of organic and inorganic nanofillers, and methods for synthesizing nanocomposites with desirable properties. Furthermore, we examine the mechanical and thermal performance, biodegradation profiles, and environmental impact of these materials, showcasing their potential in diverse fields. By analyzing current challenges and future research directions, this chapter aims to highlight the pivotal role of biodegradable polymer nanocomposites in advancing sustainable material science.

Keywords Biodegradable polymer · Nanocomposites · Nanofillers · Applications of biodegradable polymer

K. Vinchurkar (✉)
Department of Pharmaceutics, Sandip Institute of Pharmaceutical Sciences (SIPS), Nashik, Maharashtra, India

H. Rana · D. Patel · S. Sheth
Department of Pharmaceutics, Anand Pharmacy College, Anand, Gujarat, India

S. N. Kothawade
Department of Pharmaceutics, SCSSS's Sitabai Thite College of Pharmacy, Shirur, Pune, Maharashtra, India

S. N. Kothawade et al. (eds.), *Polymer Nanocomposites*,
https://doi.org/10.1007/978-3-032-00365-2_10

10.1 Introduction to Biodegradable Polymer Nanocomposites

10.1.1 Overview of Biodegradable Polymers

Once heralded as the future, synthetic polymers today represent a serious environmental risk because of their sluggish decomposition rate and dependence on non-renewable resources. Biodegradable polymers from renewable resources such as maize, soy, and sugarcane provide a more sustainable option. These polymers have less impact on the environment since they are easier to biodegrade. Nevertheless, they frequently break down rapidly and might not be as mechanically strong as synthetic polymers, particularly in damp environments. Notwithstanding these drawbacks, biodegradable polymers have several benefits, such as lower energy costs and the removal of hazardous chemical fillers. We can lessen our reliance on fossil fuels and help create a cleaner world by encouraging the creation of new biodegradable polymers with better qualities (Leja and Lewandowicz 2010).

Biodegradable polymers undergo microbial degradation into biomass and gases under aerobic and anaerobic conditions. Their degradation rate is influenced by molecular weight, hydrophobicity, and crystallinity. Polymers with both hydrophobic and hydrophilic domains tend to degrade faster than those with only one type of domain. To be considered biodegradable biomaterials, these polymers must possess specific properties, including a controlled degradation rate, biocompatibility, appropriate mechanical properties, and non-toxic degradation products. These properties ensure the polymers can effectively serve their intended purpose while minimizing environmental impact and potential health risks.

Recent research has explored ways to enhance the properties of biodegradable polymers, such as thermoplastic starch (TPS), polylactic acid (PLA), and polyhydroxyalkanoates (PHA), through the addition of copolymers, additives, or fibers. While TPS was the first biopolymer on the market, PLA and PHA have emerged as the most promising biodegradable polymers due to their versatility and wide applications. Biodegradable polymers can be classified based on their origin: natural or synthetic. Natural biodegradable polymers are derived from natural sources, while synthetic biodegradable polymers are produced from petroleum-based sources (Chandra and Rustgi 1998).

10.1.2 Importance of Nanocomposites in Sustainability

Composites, created naturally or artificially, are mixtures of two or more materials (heterogeneous systems) with a multiphase matrix structure. Each composite matrix component influences the aggregate material's bulk characteristics. One of the primary characteristics of composites is the physical distinguishability of their constituents and interfaces between various components; in other words, composites are made up of combinations of elements distributed throughout a continuous matrix.

Microcomposites (size range: 1–5 μm, particleboard), nanocomposites (size range: ≤1 μm, natural bone tissues), and macrocomposites (size range: ≥5 μm, concrete) are the three main categories into which composites may be divided based on particle size. The ideal nanocomposite is a solid structure with nanometer-scale dimensions and distinguishable distances between its comparable or different phases or material with at least one component phase in the nanoscale range (up to 100 nm). These structural phases might be one-dimensional (nanowires), two-dimensional (layered graphene), three-dimensional (fullerenes), or zero-dimensional (graphene quantum dots) (Nandhini et al. 2024).

Adding nanoparticles to a host polymer can result in performance improvements that are twice as good for the environment: first, sustainable nanocomposites made of recycled or bioplastics could take the place of common petroleum-based polymers; second, significant plastic savings could be realized by taking advantage of the nanocomposites' superior specific properties. However, nanoparticles' intrinsic environmental load may jeopardize the anticipated advantages. Here, we discuss the contentious topic of "green" polymer nanocomposites made from recycled bioplastics and their environmental sustainability. The life-cycle evaluation research on nanocomposites and their constituent parts is critically reviewed. Despite their often modest composition, nanoparticles significantly influence the environment. The benefits of employing green polymer matrices are negated by the high-energy requirements and significant greenhouse gas emissions associated with the manufacture of typical nanofillers, such as nanocellulose, titanium dioxide, silver, and, most importantly, carbon nanotubes, except organo-clays and graphene. Therefore, achieving good performance is essential to creating material-saving polymer nanocomposites that are genuinely sustainable. For this reason, adding more nanoparticles or functionalizing them to improve their dispersion in the host polymer may have unanticipated positive effects on the environment (Solomon et al. 2024).

10.1.3 Advantages Over Conventional Materials

Biodegradable nanocomposite materials present a persuasive case for improving material performance and addressing environmental issues. Biodegradable polymers and nanofillers are combined to create materials with improved mechanical qualities, such as excellent impact resistance and tensile strength. This makes them appropriate for various uses, such as structural elements and food packaging. Additionally, by encouraging recyclability and decreasing waste, their biodegradability significantly lessens the environmental impact. These nanocomposites, which come from renewable resources like proteins and polysaccharides, support a circular economy.

It enhances food safety and shelf life and has positive environmental effects. Foodborne disease risk is decreased by the efficient inhibition of microbial growth caused by integrating antimicrobial compounds into nanocoatings. This creative solution tackles the health issues related to conventional plastic packaging, which frequently contains dangerous microorganisms. We can promote sustainable habits and improve product functioning by incorporating nanotechnology into biodegradable systems, opening the door to a healthier and more sustainable future (Armentano et al. 2018).

In addition to transforming material science, they are making significant advancements in medication administration and healthcare. Researchers are creating novel drug delivery systems that may precisely target particular cells or tissues by modifying the characteristics of these materials, increasing therapeutic efficacy, and minimizing adverse effects. These devices may be made to deliver medications at regulated rates, guaranteeing continuous drug release and enhancing treatment plans. Additionally, there are several benefits to biodegradable implants and medical devices composed of these materials, such as decreased infection risk, greater tissue integration, and improved biocompatibility. As nanotechnology develops, we anticipate many more innovative uses of biodegradable nanocomposites in the healthcare industry, eventually enhancing patient outcomes and quality of life (Patwary et al. 2020).

10.2 Types of Biodegradable Polymers

10.2.1 Natural Polymers (e.g., Chitosan, Cellulose, and Starch)

Natural biodegradable polymers are directly extracted from natural sources. Biopolymers are created in nature due to the growth cycles of all organisms. Its formation involves polymerization events with a chain of enzyme-catalyzed growth from active monomers produced within cells through intricate metabolic processes. They have strong biocompatibility and are biodegradable by nature. Some of the most common natural biodegradable polymers and their properties are discussed in Table 10.1.

10.2.2 Synthetic Biodegradable Polymers (e.g., Polylactic Acid and Polycaprolactone)

Synthetic biodegradable polymers are derived from petroleum-based sources and are not directly extracted from nature. Conventional polymerization techniques, such as aliphatic copolymer, polylactide, and aliphatic polyesters, provide synthetic biodegradable polymers. Compared to traditional plastics, degradable polyesters are considered the most promising materials for use because of their suitable

Table 10.1 Natural polymers and their properties

Polymer	Source	Properties
Collagen	Mammalian tissue	Non-toxic, biocompatible, versatile processing, suitable for cell attachment
Chitosan	Shellfish exoskeletons	Easily controllable properties, chemical modification potential, low foreign body reaction, easy processing
Gelatin	Animal collagen (pigskin, bovine hides, bones)	Tasteless, odorless, soluble in aqueous solutions, gel strength, viscosity, emulsifying, foaming, wetting agent
Alginate	Brown seaweed	Quickly processed, non-toxic, non-inflammatory, biodegradable, controllable porosity, cell encapsulation potential
Lactose	Milk	Water soluble, low glycemic index, energy source for newborns
Starch	Plants (e.g., corn, potato, and wheat)	Biodegradable, versatile, forms gels, film-forming properties
Cellulose	Plants	Abundant, renewable, strong, rigid
Pectin	Plant cell walls (citrus peel and apple pomace)	Biodegradable, forms gels, water-soluble, food-grade

breakdown time and industrial manufacturing. They are made in various forms, including polybutylene succinate, polycaprolactone, and polylactic acid. Their ecological pollution is minimal. Because polyester contains readily hydrolyzable esters makes for over 90% of biopolymers. Oil-based monomers like PCL and bio-based polymers like PLA are two types of synthetic biodegradable polymers. Some of the most common synthetic biodegradable polymers and their properties are discussed in Table 10.2.

10.2.3 Hybrid Systems

Due to the exceptional qualities of the materials produced, the immobilization of inorganic nanoparticles on polymeric substrates has garnered much interest recently. The hybrid materials, derived from their inorganic equivalents, combine the advantages of plastic matter, including flexibility, low cost, mechanical stability, and excellent durability. Biocompatibility and biodegradability, two characteristics of biopolymers, create new opportunities for these hybrid materials, particularly in regenerative medicine and environmentally friendly materials (green nanocomposites). Research on nanocomposites is a new multidisciplinary field closely connected to important subjects like biomimetic systems, biomineralization processes, and bioinspired materials. An interesting area of study that capitalizes on the synergistic assembly of biopolymers with inorganic nanometer-sized solids is the impending creation of innovative bionanocomposites that introduce multifunctionality (Cantarella et al. 2021). Table 10.3 provides an overview of various hybrid nanocomposite systems, their key components, improved properties, and potential applications.

Table 10.2 Synthetic biodegradable polymers

Polymer	Properties	Applications
Polylactic acid (PLA)	Renewable resource-based, moderate mechanical strength, thermal stability	Packaging, 3D printing, medical devices
Polycaprolactone (PCL)	Flexible, low melting point, good processability	Tissue engineering, drug delivery, medical implants, food packaging, nanofiber production
Polyglycolic acid (PGA)	High crystallinity, rigid, high tensile strength	Medical sutures, drug delivery, tissue engineering
Poly(butylene) succinate (PBS)	Good thermal stability, good mechanical properties, biodegradable to non-toxic products, good processability	Packaging, agriculture, fishing, construction, medicine, 3D printing, biocomposites
Poly(butylene-adipate-co-terephthalate) (PBAT)	Good mechanical properties (e.g., toughness and tear resistance), flexibility, high elongation at break	Packaging, films, fibers, antimicrobial films
Polyanhydrides	Controlled release, bioerodible, low mechanical strength	Drug delivery, tissue engineering
Polyorthoesters	Acid-labile, tunable degradation rate, controlled release	Drug delivery

Table 10.3 Biodegradable polymer hybrid system

Hybrid system	Key components	Improved properties	Potential application
Natural fiber polymer-nanocomposite	Wood, Kenaf, Coir, polypropylene, montmorillonite nanoclay	Improved structural and mechanical properties, increased crystallinity	Construction, automotive, packaging
Chitosan-gelatin hybrid nanocomposite	Chitosan, gelatin, ZnO nanoparticles	Improved heat stability, elasticity, structural integrity, antimicrobial properties	Food packaging
Chitosan-gelatin-PEG-AgNP hybrid nanocomposite	Chitosan, gelatin, polyethylene glycol, silver nanoparticles	Improved mechanical properties, reduced light transmittance, antimicrobial properties	Food packaging
Polyaniline/porous silicon hybrid nanocomposite	Polyaniline, porous silicon nanoparticles	Photothermal effect, biodegradability, pH-responsive drug delivery	Cancer therapy
PCL-graphene-silver hybrid nanocomposite	Polycaprolactone, graphene, silver nanoparticles	Improved mechanical properties, electrical conductivity, and antibacterial properties	Tissue engineering, fracture repair
PLA-based nanocomposite	Polylactides, functional nanoparticles	Improved mechanical, thermal, and barrier properties	Packaging, medical devices, and automotive components

10.3 Nanofillers Used in Biodegradable Nanocomposites

Nanocomposites are emerging materials gaining attention because of their superior mechanical, thermal, and functional properties. Nanocomposites are materials that incorporate nanosized particles into a matrix of standard material. This significantly improves mechanical strength, toughness, and electrical or thermal conductivity properties. Incorporating nanofillers into biodegradable polymers greatly improves the performance of these materials, which can be used in many applications. Nanofillers are nanoscale particles added to a material matrix to enhance physical, mechanical, thermal, or electrical properties (Dantas de Oliveira and Augusto Gonçalves Beatrice 2019). These fillers can be made from carbon, clay, or metal oxides and are usually measured in the nanometer range. Nanofillers are of numerous types but can be broadly categorized into organic and inorganic types based on their composition. Each type has different properties and applications, making them suitable for various functions in nanocomposites (Saharudin et al. 2023; Jamróz et al. 2019).

10.3.1 Organic Nanofillers (e.g., Nanocellulose and Chitin Nanowhiskers)

Organic nanofillers have been applied to enhance the properties of the biopolymers. Natural biopolymer nanofibrils of molecules like collagen, cellulose, and chitin are continuously used as organic nanofillers due to their biodegradability, biocompatibility, availability, strength, and mechanical properties. Organic nanofillers comprise many of these. These nanofillers provide excellent mechanical, thermal, and barrier properties with biodegradability and eco-friendliness. By incorporating organic nanofillers into biodegradable matrices, stronger, more durable, and more functional composites can be developed without losing the product's sustainability. These organic materials, including cellulose derivatives, chitin, starch, and protein-based fillers, are used in food packaging, biomedical applications, and eco-friendly coatings. Table 10.4 summarizes various organic nanofillers used in biodegradable nanocomposites and their sources, properties, and applications (Jagadeesh et al. 2021; Dai et al. 2023; Stanisz et al. 2022; Eichhorn et al. 2022).

10.3.2 Inorganic Nanofillers (e.g., Clay, Silica, and Graphene)

Integrating inorganic nanofillers can result in the functionalization of polymers with great success. In general, the type of nanostructure filler can be classified into metal oxide and clay-based. These inorganic fillers and additives generally form composite biopolymer electrolytes by binding with the polymer matrix. Various forms of

Table 10.4 Organic nanofillers

Nanofiller type	Source	Key properties	Role in nanocomposites	Applications
Nanocellulose (CNC, CNF, and BC)	Wood, cotton, bacteria (e.g., *Gluconacetobacter*)	High mechanical strength, biodegradability, nanoscale size, renewable	Reinforcement material improves mechanical and barrier properties	Food packaging, biodegradable plastics, biomedical devices, paper products
Chitin/ chitosan	Exoskeletons of crustaceans, insects, fungi	Biodegradable, antimicrobial, biocompatible, natural polymer	Antimicrobial agent, mechanical reinforcement, biodegradable additive	Wound dressings, food packaging, drug delivery, water filtration
Nanowhiskers	Wood pulp, cotton, agricultural waste	High aspect ratio, mechanical strength, biodegradability	Reinforcement in biodegradable matrices for improved mechanical properties	Sustainable packaging, coatings, medical applications
Hemp nanofibers	Hemp plant	High tensile strength, renewable, biodegradable	Reinforcement in biodegradable composites for mechanical properties	Biodegradable plastics, construction materials, packaging
Sisal nanofibers	Sisal plant	Strength, biodegradable, renewable	Reinforcement in biocomposites improves mechanical and thermal properties	Eco-friendly packaging, automotive parts, construction materials
Jute nanofibers	Jute plant	High-strength, lightweight, biodegradable	Reinforcement in biodegradable composites for improved strength	Biodegradable composites in textiles, packaging, and construction
Whey protein nanoparticles	Dairy by-product (whey)	Biocompatible, biodegradable, renewable	Nanofiller in biodegradable matrices, enhancing mechanical and barrier properties	Biodegradable food packaging, drug delivery, coatings

(continued)

Table 10.4 (continued)

Nanofiller type	Source	Key properties	Role in nanocomposites	Applications
Starch nanocrystals/fibers	Corn, potato, rice, other grains	Biodegradable, renewable, biodegradable composite material	Reinforcement material increases mechanical properties and good barrier performance	Eco-friendly packaging, biomedical applications, films and coatings
Gelatin nanoparticles	Animal collagen	Biodegradable, biocompatible, renewable	Reinforcement in biodegradable films, emulsifying agent in composites	Wound healing, food packaging, drug delivery systems
Protein-based nanofillers	Soy protein, corn protein, casein	Biodegradable, biocompatible, water-soluble, renewable	Reinforcement, improving film-forming ability, enhancing mechanical properties	Biodegradable films, food packaging, biodegradable plastics
Alginate nanofillers	Brown algae	Biodegradable, biocompatible, gelling agent	Provides viscosity and strength to biodegradable composites, improves mechanical properties	Medical applications, wound healing, drug delivery, food packaging
Bacterial nanocellulose (BC)	Bacteria (e.g., *Gluconacetobacter xylinus*)	High purity, strength, biocompatible, water-retaining	Reinforcement in biodegradable composites, films, and coatings	Biomedical applications, food packaging, textiles, tissue engineering
Cellulose acetate nanofibers	Wood pulp, cotton	Biodegradable, thermoplastic, high tensile strength	Reinforcement in composite matrices for biodegradable films and coatings	Biodegradable packaging, coatings, automotive applications
Nano lignins	Lignocellulosic biomass (e.g., wood)	Antioxidant properties, biodegradability, renewable	Nanofiller in biopolymer matrices improves mechanical and barrier properties, antioxidant and antibacterial activity	Food packaging, biodegradable plastics, medical applications, coatings

fillers, such as spheroidal, fibrous, or porous, have been studied for composite enhancements. They are often used because of their superior thermal, mechanical, and barrier properties. Table 10.5 elucidates the properties of inorganic nanofillers (Kausar 2020; Piekarska et al. 2016).

10.3.3 Functionalized Nanomaterials

With their enhanced properties that enhance the performance of biodegradable polymers, functionalized nanomaterials and nanofillers have become critical components in the development of biodegradable nanocomposites. To enhance their dispersion, compatibility, and interaction with the polymer matrix, functionalization involves changing the surface properties of nanomaterials, such as nanoparticles, nanoclays, and nanofibers, by different functional groups, including hydroxyl, carboxyl, and amino. Customization of nanofillers through surface functionalization opens up the possibility of creating sophisticated, high-performance materials with multifarious applications in electronics, coatings, food packaging, and biomedical devices (Basavegowda and Baek 2021). Table 10.6 collects all the functionalized nanofillers and the enhanced performance upon addition to the biodegradable matrices.

10.4 Synthesis Techniques of Biodegradable Nanocomposites

10.4.1 Solution Casting Method

The solvent casting approach involves gradually adding nanofillers to the polymer solution while continuously swirling a specific solvent to dissolve the polymer. In order to evaporate the solvent, the prepared homogeneous solution is also cast in a mold. Thermosets and thermoplastics can be made with water and an organic solvent. This method is widely utilized and has long been used to create polymer thin films of various thicknesses (Nguyen et al. 2023).

10.4.2 Melt Compounding

Because it does not use solvents, this method is safe for the environment. It allows the addition of additives like silicates, clay, or nanofillers to the polymer by heating it to a high temperature and stirring it continuously until a homogeneous dispersion

Table 10.5 Inorganic nanofillers

Nanofiller type	Source	Key properties	Role in nanocomposites	Applications
Clay (montmorillonite, MMT)	Natural clay minerals (e.g., bentonite and montmorillonite)	High aspect ratio, layered structure, high surface area, and ability to form strong interactions with polymers	Enhances mechanical strength, thermal stability, and barrier properties	Food packaging, coatings, and automotive parts
Silica (SiO_2) (Mukherjee et al. 2022)	Natural minerals, sand, synthetic methods	High surface area, hydrophilic, high thermal stability, UV resistance	Improves thermal stability, mechanical properties, and transparency	Electronics, biodegradable packaging, coatings
Graphene oxide (GO) (Saberi et al. 2024)	Graphite (exfoliation)	High surface area, excellent mechanical strength, conductivity	Increases mechanical strength, electrical properties, and barrier performance	Biomedical devices, energy storage, packaging
Carbon nanotubes (CNTs)	Carbon sources (e.g., graphite and hydrocarbons)	High aspect ratio, high tensile strength, excellent electrical conductivity	Reinforces polymer matrix for increased strength, electrical conductivity, and barrier properties	Packaging, electronics, biomedical devices
Nano zinc oxide (ZnO) (Aremu et al. 2021)	Zinc ore	High UV blocking ability, antimicrobial properties, high surface area	Provides UV protection antibacterial properties, enhances thermal stability	Sunscreens, biodegradable plastics, medical applications
Titanium dioxide (TiO_2)	Titanium ore	Photocatalytic properties, UV resistance, white colorant	Enhances UV stability and mechanical strength, provides color stability	Food packaging, cosmetics, medical applications
Calcium carbonate ($CaCO_3$)	Limestone, marble	High surface area, low cost, good dispersion in polymer matrix	Improves rigidity, reduces cost, enhances mechanical properties	Packaging, construction materials, automotive parts
Magnesium oxide (MgO) (Saberi et al. 2024)	Magnesite	High thermal stability, flame retardant properties, biocompatibility	Increases thermal stability and flame resistance of nanocomposites	Fire-resistant packaging, biodegradable coatings

Table 10.6 Functionalized materials

Functionalized nanofiller type	Surface modification	Key properties	Role in nanocomposites	Applications	References
Functionalized nanoclay (MMT)	Surface modification with silane, amino, or epoxy groups	Enhanced dispersion, strong interfacial interaction, increased compatibility with polymer matrix	Improves mechanical, thermal, and barrier properties	Food packaging, coatings, biodegradable plastics, automotive parts	Basavegowda and Baek (2021)
Functionalized graphene oxide (GO)	Surface functionalization with carboxyl, hydroxyl, or amine groups	High mechanical strength, excellent electrical conductivity, enhanced polymer interaction	Enhances mechanical strength, electrical properties, and dispersion	Biomedical devices, food packaging, energy storage	Jing et al. (2015)
Functionalized silica (SiO_2)	Functionalization with organic groups (e.g., silane, alkyl groups)	High surface area, thermal stability, hydrophilic behavior	It improves thermal stability, mechanical properties, and dispersibility	Eco-friendly packaging, automotive components, biomedical devices	Liberman et al. (2014)
Functionalized carbon nanotubes (CNTs)	Functionalization with carboxyl, amino, or epoxy groups	High tensile strength, electrical conductivity, and high surface area	Reinforcement material improves mechanical, electrical, and barrier properties	Food packaging, biodegradable electronics, medical applications	Ibrahim et al. (2023)
Functionalized chitosan nanofillers	Modification with thiol, carboxyl, or amino groups	Biocompatible, biodegradable, antimicrobial properties	Antimicrobial agent, reinforcement material, and biodegradable additive	Food packaging, biomedical applications, drug delivery systems	Brun et al. (2021)
Functionalized starch nanofillers	Surface modification with functional groups like ester, acetyl, or glycerol	Biodegradable, renewable, low cost, enhances composite properties	Enhances matrix compatibility, improves mechanical and barrier properties	Biodegradable packaging, coatings, medical films	Ibrahim et al. (2023)
Functionalized calcium carbonate ($CaCO_3$)	Surface treatment with stearic acid or silane	Improved dispersion, increased mechanical properties, low cost	Reinforcement material reduces production cost, improves rigidity	Construction materials, automotive parts, biodegradable plastics	Ippolito et al. (2020)

forms. Because no solvent is involved, this procedure is environmentally friendly and has a high degree of durability. This technique can be applied to unsuitable polymers for in-situ polymerization and solution interaction (Gobena and Woldeyonnes 2024).

10.4.3 In Situ Polymerization

A reagent that aids in polymerization and creates well-defined multidimensional structure nanocomposites is added after the polymer has been mixed in an appropriate solvent. This technique can create nanocomposites of various sizes, shapes, and morphologies. Additionally, it facilitates the formation of uniform dispersion inside a polymer matrix (Zapata et al. 2008).

10.4.4 Advanced Methods (e.g., Electrospinning and 3D Printing)

There are certain types of new methods that have been developed to synthesize nanocomposites, which are depicted in Table 10.7.

Table 10.7 Advanced methods

Synthesis technique	Description	Key materials	Applications	References
Electrospinning	A high-voltage process that creates nanofibers by stretching a polymer solution into thin fibers.	Biodegradable polymers (PLA, PCL, and PHA), Nanoparticles (cellulose nanocrystals, clay, carbon nanotubes)	Tissue engineering, drug delivery, filtration	Cao et al. (2019)
3D Printing (additive manufacturing)	A layer-by-layer printing technique to create complex structures from biodegradable polymers mixed with nanoparticles.	PLA, PCL, PHA, nanofillers (graphene oxide, cellulose, chitin)	Medical devices, biodegradable packaging, scaffolds	Hou et al. (2024)
Electrospraying	Similar to electrospinning, instead of fibers, it produces droplets that can form thin films or coatings.	Biodegradable polymers (PCL, PLA), nanoparticles (metallic nanoparticles, chitosan)	Coatings, drug delivery, antimicrobial films	Nguyen and Yang (2021)

(continued)

Table 10.7 (continued)

Synthesis technique	Description	Key materials	Applications	References
Freeze drying (lyophilization)	A process where a biodegradable polymer-nanoparticle mixture is frozen and the solvent is removed through sublimation to create a porous material.	Biodegradable polymers (PVA, alginate), nanoclays, silica	Controlled drug release, tissue engineering scaffolds	Yu et al. (2021)
Laser-assisted bioprinting (LAB)	A method of printing complex 3D structures using laser light to process biodegradable materials and create scaffolds with controlled microstructures.	Collagen, gelatin, PLA, PCL, nanoparticles (hydroxyapatite, silver nanoparticles)	Bone tissue engineering, wound healing, drug delivery	Yu et al. (2020)
Direct ink writing (DIW)	A 3D printing technique where bioinks, including biodegradable polymers with nanoparticles, are extruded through a nozzle to form solid structures.	PLA, PCL, gelatin, hydroxyapatite nanoparticles	Bone regeneration, tissue engineering, prosthetics	Hou et al. (2024)
Spin coating	A technique where a thin layer of biodegradable polymer-nanocomposite is spread over a substrate by spinning at high speeds.	PLA, PCL, chitosan, carbon nanotubes	Thin films for sensors, coatings, biosensors	Khosravi et al. (2020)
Bioprinting	A form of 3D printing using living cells and biomaterials, including biodegradable composites, to create tissue-like structures.	Hydrogel-based polymers, gelatin, collagen, nanoparticles (hyaluronic acid)	Tissue engineering, regenerative medicine	Loukelis et al. (2023)

10.5 Characterization and Properties

The microstructure of biodegradable nanocomposites significantly impacts their characteristics, such as the degree of crosslinking, the amorphous phase, the presence of polar or non-polar groups, and the degree of crystallinity. These properties are strongly related to the chemistry of the polymer matrix, the particular production techniques employed, and the high aspect ratios of the nanofillers. Nanofillers must be distributed homogeneously in the polymer matrix for better mechanical and physical properties, with strength enhancement, flexibility, and barrier properties. Moreover, the important barriers, mechanical, thermal, optical, and functional qualities of these nanocomposites characterize them for usage, specifically in food packaging systems where they add shelf life and give way to environmental sustainability. Techniques of characterization such as scanning electron microscopy (SEM), X-ray diffraction (XRD), and thermogravimetric analysis (TGA) are commonly employed to assess the distribution, morphology, and performance of the nanocomposites, providing valuable insights into their behavior and suitability for specific applications (Mukherjee et al. 2023).

10.5.1 Mechanical Properties

Packaging protects food items from external flaws and impacts, such as cracks and occasionally isolated breaks in the packaging material. Strain at break, tenacity, and ultimate stress can all improve a polymer's mechanical qualities for food packaging systems. Therefore, nanofillers are distributed or reinforced in polymer matrices via a reinforcing mechanism to enhance characteristics like strength and stiffness. The size, shape, concentration, orientation, surface area, dispersion state, and polydispersity of the nanofillers are the main factors that determine this reinforcement and may cause them to graft with the matrix polymers. Nanocomposites made with minor fillers have better mechanical properties; the more filler added, the worse the overall mechanical properties are (Le et al. 2020).

Through the process of relaxation change, and molecular mobility, these nanofillers significantly improve the mechanical property hardness of the nanocomposite. In order to significantly improve the dispersion of such constituents in the matrix material, a particle is blended and mixed with polymer matrices for interfacial adhesion. The hybrid material should be the same type as both components to improve adhesion at the interface and create a strong bond. Examples of this include hydrophilic-hydrophilic materials, hydrophobic fillers, and hydrophobic matrices. Moreover, loading and particle size significantly impact the particle-polymer matrix composites' mechanical characteristics. The strength of the finished composite depends on the connection between the fillers and the matrix.

10.5.2 Optical Properties

The optical properties of polymeric nanocomposites include fluorescence and transparency. A UV-vis spectrophotometer usually determines the transparency and percentage of light transmittance through the 200–800 nm sample. Optical transparency for CNFs and hyperbranched bio-based epoxy thermosetting nanocomposites. The polymer matrix can be considerably radiated by adding the appropriate nanomaterials such as circular dichroism (CD), sodium zincate (Na-ZnOz), TiO, or CD oxides. Polymer nanocomposites with orange-red fluorescence by incorporating Na-ZnO in PVA using an aqueous solution casting technique. Besides fluorescence, the polymer nanocomposites exhibited excellent thermal stability, self-cleaning properties, and strong UVA screening ability (Kumbhakar et al. 2024).

10.5.3 Thermal Stability and Barrier Properties

Thermal properties are crucial for using polymeric materials, from significant industrial buildings to consumer goods packaging, in various applications. One significant technological issue is the low heat conductivity of polymers or the difference in thermal expansion properties between fillers and polymeric components. The thermal conductivity of most nanoparticles and nanofillers ranges from 100 to 400 W/m•K. Nanofillers stop Heat-induced polymer breakdown, which functions as a barrier to mass and heat transport. They significantly improve the chemical and physical mechanisms for polymeric nanocomposites by reducing the diffusion of gaseous products and the molecular mobility of polymers (Yesaswi et al. 2022).

The structure, solubility, and diffusivity of polymers are among the variables that affect how easily gases and tiny molecules pass through the matrix. Therefore, the inherent capacity of polymers to promote the exchange of low-molecular-weight molecules is intimately associated with their barrier characteristics. The size and kind of nanofillers and the topologies of the nanocomposites affect the change in barrier characteristics. Because of their high aspect ratio, well-dispersed nanofillers in the polymer matrix will likely lengthen the convoluted course of the penetration molecules. This could affect the solubility and diffusivity of molecules in the interfacial domains and produce an impermeable structure in the polymer matrix (Abdollahi et al. 2013; Ma et al. 2022).

10.5.4 Electromagnetic Interference Shielding Property

Because of the explosive growth of electronic industries over the past three decades, there is now greater reliance on electrical equipment. Electrical devices require higher frequencies for operation, but electromagnetic interference signals prevent them from doing so. Therefore, it has been reported that a variety of polymer nanocomposites with conductive nanofillers, such as carbon black, graphene nanoplatelets, graphene oxide (GO), nanoclay, or metal oxides like CO, NiO, WOz, TiOz, and others, offer efficient EMI shielding.

Conductive fillers are essential for effective EMI shielding because they reflect and absorb the polymer matrix. Moreover, the shielding effectiveness of nanocomposites of poly(lactic acid) (PLA)/ester linkage-based thermoplastic polyurethane (TPU) at different filler loadings with conductive carbon black Vulcan XC-72 (VCB). The introduction of the conductive carbon black filler into the PLA/TPU matrix is increasing its EMI shielding properties in the X band more than that of the pristine matrix, and the biodegradable EMI shielding material in use absorbed or reflected the maximum number of electromagnetic waves and transmitted very minimum of them (Kausar and Ahmad 2023).

10.5.5 Biodegradability Assessments

The degradation occurs when polymers degrade under light, heat, and mechanical stress, producing smaller pieces that impair the material's mechanical qualities, make it brittle, and shorten its lifespan. Polymers or their composites must resist environmental factors to ensure long-term durability in outdoor applications. Studying polymer stabilization and degradation is critical to increasing the product's longevity. The biodegradability of nanocomposites can be measured by assessing the breakdown capability of nanocomposites in several environments, such as aquatic settings, composting, and soil burial. Weight loss, mechanical and physical properties changes, and chemical analysis are applied to monitor the degradation process. These environmental elements include temperature, humidity, soil type, and microbial activity. These factors can significantly influence the rate at which biodegradation can occur and slow it down or speed it up (Pandey et al. 2005).

The biodegradation process can be followed using molecular weight analysis, weight loss measurements, and surface shape changes. However, microbes play an important role because they determine how well the process functions. Considering the nature of the base polymer, the type of nanoparticle, and the environment in which it is applied, it is critical to evaluate biodegradability to understand biodegradable nanocomposites' environmental impact and degradation behavior. Testing techniques involving soil burial, water immersion, and enzymatic degradation are often used to measure chemical changes, weight loss, and gas production. It is essential to monitor the effects of nanoparticles since, based on size, surface area, and chemical composition, they may either enhance or inhibit biodegradation. Extensive testing for biodegradability ensures that these materials break down safely and efficiently, decreasing their long-term ecological footprint (Machado et al. 2015). Different examples of nanocomposites are shown in Table 10.8.

10.5.6 Surface Morphology and Structural Analysis

Understanding biodegradable nanocomposites' surface morphology and structural analysis is crucial for understanding their performance and characteristics in various applications, especially in the environmental and biomedical domains. Techniques such as atomic force microscopy (AFM), optical microscopy (OM), and electron microscopy (SEM and TEM) provide fine-grained images of surface characteristics such as filler alignment, crystallization behavior, and particle dispersion. Advanced techniques like Raman spectroscopy and X-ray photoelectron spectroscopy (XPS) provide details about the molecular structure and elemental

Table 10.8 Different nanocomposite degradation

Polymer-nanocomposite	Degradation mechanism	Degradation time	Monomers	Detection technique	Reference
PLA/clay nanocomposite	Hydrolytic degradation, microbial degradation	6 months to 2 years	Lactic acid (PLA)	SEM, FTIR, TGA, XRD, mass loss measurements	Iñiguez-Franco et al. (2017)
Starch/clay nanocomposite	Enzymatic and hydrolytic degradation	3 to 12 months	Starch (glucose units)	SEM, FTIR, XRD, Water absorption test	Chung et al. (2010)
Polycaprolactone (PCL)/ graphene oxide nanocomposite	Hydrolytic degradation, enzymatic degradation	6 months to 1 year	Caprolactone (PCL), graphene oxide	TGA, SEM, XRD, biodegradation assays, FTIR	Correa et al. (2019)
Chitosan/nano-TiO2 nanocomposite	Enzymatic, UV, and microbial degradation	3 to 6 months	Chitosan (N-acetylglucosamine), TiO2	TGA, FTIR, SEM, UV-Vis	Castillo et al. (2022)
Poly(lactic acid) (PLA)/ ZnO nanocomposite	UV and hydrolytic degradation, oxidation	6 months to 1 year	Lactic acid (PLA), Zinc oxide (ZnO)	SEM, FTIR, XRD, TGA, UV-Vis spectroscopy	Huang et al. (2015)

composition, ensuring that nanofillers are successfully integrated into the polymer matrix. When these techniques are combined, a comprehensive evaluation of nanocomposites can be made, which is the basis for their development for better mechanical, biological, and environmental properties (Hiremath et al. 2022; Khichar et al. 2020; Benetti et al. 2005; Deshmukh and Pandey 2023). Few test are elucidated in Table 10.9.

Table 10.9 Types of test

Test name	How to perform	How to indicate the results
Atomic absorption spectroscopy (AAS)	A system using a cathode light, pneumatic nebulizer with a low flame, grating monochromator, and photomultiplier tube detector is used. Atomization occurs in a flame or graphite furnace, and the characteristic radiation absorbed by atoms is measured	Results are indicated by the amount of absorption at specific wavelengths, following Beer–Lambert's law. The concentration of the absorbing species is proportional to the light absorption
Circular dichroism (CD) spectroscopy	The sample absorbs right and left-polarized light differently, leading to circularly polarized light. Differential absorption can be measured to study the structure of biomolecules.	Results are presented as CD spectra with peaks at certain wavelengths. The peaks are characteristic of the secondary structure of biomolecules, such as α-helix or β-sheet.
CD spectroscopy for nanocomposites and biosensors	Measure CD spectra of nanocomposites, such as chitosan/graphene oxide, in a solution at different pH levels and analyze changes in molecular structure and interactions	CD spectra exhibit intensity changes at key wavelengths, such as 210 nm for chitosan. Changes in such bands can indicate the influence of pH or matrix interactions on molecular conformation
Electron paramagnetic resonance spectroscopy (EPR)	Expose the sample to microwave radiation in a magnetic field. Measure the absorption or emission of microwave radiation by unpaired electrons changing between energy states	Complex EPR spectral patterns reflect the results, pointing to unpaired electrons and yielding information on hyperfine interactions, the molecular environment, and metal coordination
Atomic emission spectroscopy (AES)	Excitation sources such as flame, spark, or plasma can break samples into free atoms. Measure the photons these atoms emit after they decay to a lower energy state	Results are shown as emission spectra with characteristic wavelengths. The intensity of specific peaks correlates to the concentration of the elements in the sample
Energy dispersive X-ray spectroscopy (EDXS)	Expose the sample to a primary beam of X-rays or electrons. Measure the energy released as electrons fall into lower energy levels and emit X-rays.	The results will be presented as an energy spectrum showing peaks at characteristic energies for each element. Peak intensities indicate the relative abundance of elements in the sample

(continued)

Table 10.9 (continued)

Test name	How to perform	How to indicate the results
Fluorescence spectroscopy	Use a spectrofluorometer, where the sample is excited with a broad wavelength from a light source in the UV-vis region. After the sample is emitted, emission photons are detected	Emission spectra are recorded and analyzed for shifts in wavelengths or intensity changes, indicating the nature and characteristics of the polymer-nanocomposite
Fourier transform infrared spectroscopy (FTIR)	An FTIR spectrometer and a Michelson interferometer measure infrared absorption spectra. This decodes the sample's vibrational modes	Peaks in the spectra at given wave numbers indicate functional groups or molecular structure changes, thus confirming successful synthesis or modifications.
Mass spectroscopy	Mass spectrometry is used to ionize and fragment the analyte	Samples are analyzed in terms of molecular mass and fragmentation patterns
Mossbauer spectroscopy	Gamma-ray photon absorption was used to study the nanocomposites' iron oxide and magnetic properties. Take the spectrum at room temperature, per temperature, and low temperature. Look for characteristic doublets and sextets	A doublet indicates super-magnetic behavior, whereas a sextet indicates antiferromagnetic properties. Isomer shifts and hyperfine fields help identify phases
NMR spectroscopy	Subject atom nuclei to absorption of radiofrequency radiation. Record proton or carbon NMR spectra to identify chemical groups	Peaks in the NMR spectrum correspond to specific protons or carbons. The chemical shifts (ppm) help identify the functional groups and the degree of grafting or polymerization
X-ray photoelectron spectroscopy (XPS)	Expose the sample to an X-ray photon beam and detect ejected electrons. Analyze the sample's surface composition	We provide quantitative data on elemental composition and oxidation states. Peaks in the XPS spectrum correspond to specific elements and oxidation states
Raman spectroscopy	Utilize inelastically scattered radiation to examine the vibrational modes of molecules	Peaks in the Raman spectrum indicate particular molecular vibrations. A shift or change in peak intensities indicates the presence of specific nanofillers
UV-Vis spectroscopy	Expose the sample to UV-vis radiation and measure the absorption spectrum	Peaks in the absorption spectrum indicate electronic transitions. A blue shift in the peaks indicates changes in conjugation length and interactions between nanoparticles and the polymer matrix

Table 10.9 (continued)

Test name	How to perform	How to indicate the results
Electron microscopy (SEM)	A high-energy electron beam scans the sample's surface and records the secondary electrons, generating micrographs	SEM micrographs provide information on the morphology of the surface, dispersal of particles, and agglomeration
Transmission electron microscopy (TEM)	Pass a high-energy electron beam through an ultra-thin sample and analyze diffraction and scattering patterns	The results are displayed in detailed images of the sample's atomic structure, crystallographic patterns, and elemental composition
Scanning probe microscopy/atomic force microscopy (SPM/AFM)	A probe scans the surface to detect atomic interactions and measure mechanical properties, adhesion, and surface roughness	The results are represented as topographical images, showing surface features and providing roughness data and force interactions
Optical microscopy (OM)	Use light to examine samples' surface morphology and microstructure in transmitted or reflected modes	The results include high-resolution images of surface features such as crystal structures, spherulitic growth, and filler dispersion
Confocal laser scanning microscopy (CLSM)	The capture of high-resolution, three-dimensional images of surfaces and cell interactions using laser scanning	Results provide three-dimensional images that expose the detailed surface morphology, absorption of proteins, and cell attachment
Photoacoustic microscopy (PAM)	Irrigate the sample with light and measure the ultrasound waves generated to create high-resolution images with tissue penetration	Results include high-resolution images detailing the tissue's cellular structures and molecular properties

10.6 Applications of Biodegradable Polymer Nanocomposites

10.6.1 Biomedical Applications

10.6.1.1 Drug Delivery Systems

Biodegradable polymer nanocomposites have become increasingly important in drug delivery systems due to their ability to improve therapeutic outcomes through controlled release, targeted delivery, and enhanced bioavailability. These systems, which include beads, films, implants, microparticles, and nanostructured systems, are designed for various routes of administration, such as oral, nasal, ocular, and transdermal, ensuring precise medication delivery with minimal side effects. Polymers like chitosan, PLA, PCL, and dextran are commonly used for their biodegradability, biocompatibility, and capacity to encapsulate various drugs, including hydrophilic and lipophilic substances (Ulery et al. 2011). Advances in nanotechnology have led to the development of nanoparticles capable of targeting specific organs or cells, such as the lung, kidney, or pancreas, enabling controlled drug release over time. Functionalizing these polymers with targeting ligands enhances drug delivery accuracy, allowing higher drug concentrations at the intended site and reducing systemic toxicity. Furthermore, these biodegradable polymers are being explored for gene and protein delivery, offering promising solutions for treating genetic disorders, vaccines, and other chronic conditions (Dananjaya et al. 2024; Alaswad et al. 2022).

The use of biodegradable nanocomposites in drug delivery systems is increasing because of their ability to target certain body parts, improve bioavailability, and offer a prolonged release. This is especially the case for those containing PLA and PLGA. These nanocomposites minimize the disadvantages of conventional medication formulations, which often require frequent dosing and lead to irregular drug levels, side effects, and poor patient compliance. PLA and PLGA nanoparticles ensure controlled and sustained drug release by maintaining the desired medication concentrations in the body over time. They are very effective for encapsulating various pharmaceuticals, including proteins, hormones, and psychiatric drugs. They are handy for treating cancer, neurological diseases, and chronic illnesses. Because PLA nanoparticles degrade slowly, they provide prolonged drug retention and lower organ toxicity, while PLGA copolymers can be tailored to achieve sustained release and optimize drug incorporation efficiency.

Block copolymers such as PLGA-PEG, PLA, and PLGA have emerged as promising candidates for thermosensitive drug delivery as they allow drug-controlled release in response to environmental stimuli such as temperature changes. Since drugs such as naltrexone hydrochloride can be delivered in a controlled manner, enhancing efficacy while minimizing side effects, this method is beneficial for conditions that require prolonged treatment, such as opioid addiction or chronic pain management. Biocompatible materials, such as collagen, fibrin, and chitosan, are being added to nanocomposites for drug delivery applications because of their biodegradability and compatibility with the human body. Montmorillonite-PLGA nanocomposites, for example, have been developed to enhance the bioavailability and residence time of drugs like atenolol in the stomach environment. Despite these advantages, there are still problems, especially concerning the degradation of PLGA, which can create an acidic microenvironment that might destabilize sensitive proteins. Further research on this issue is needed to optimize the stability and performance of these nanocomposites in clinical settings. Specific examples of nanocomposite application in drug delivery are depicted in Table 10.10.

10.6.1.2 Tissue Engineering Scaffolds

A significant area of regenerative medicine is tissue engineering, which creates replacement tissues and/or encourages tissue regeneration by combining scaffolds and cell transplantation. By improving mechanical stability, biocompatibility, and biological responses, including neovascularization and antibacterial activity, nanocomposites—particularly scaffolds based on silk fibroin—are revolutionizing tissue

Table 10.10 Application of nanocomposite in drug delivery system

Application	Nanocomposite type	Description	Key applications	Reference
Cancer treatment and drug delivery	Polymer/metal/ ceramic nanoparticles	Nanocomposites enable targeted drug delivery, increasing efficacy while reducing side effects. They can be engineered for controlled release	Drug targeting, anticancer drug internalization, imaging (CT, MRI)	Nalone et al. (2018)
Transdermal drug delivery (TDD)	Hydrogels, membranes, films, microneedles	Nanocomposites improve skin penetration and bioavailability, facilitating sustained drug release	Hormones, pain relievers, vaccines, therapeutic agents through the skin	Parhi (2018)
Neurodegenerative diseases	Nanoliposomes, stimuli-responsive polymers, inorganic/organic nanoparticles	Nanocomposites overcome the blood-brain barrier (BBB) to deliver therapeutic agents to the CNS, improving neuroprotection and anti-inflammatory effects	Parkinson's (dopamine), Alzheimer's (amyloid inhibitors), MS treatment	Fatima and Siddique (2024)
Chitosan-based oral colon drug delivery for cancer	Chitosan Derivatives (trimethyl, PEGylated)	Chitosan-based nanocomposites deliver drugs specifically to targeted areas, such as the colon, for cancer therapy	Oral colon-specific drug delivery targeting colon cancer cells	Bose and Wong (2018)
Nanocomposites in genotoxicity assessment	Biocompatible, biodegradable nanoparticles (gold, silver, and plant-derived)	Made from natural, non-toxic materials, green nanocomposites are safer and more sustainable for clinical applications in drug delivery	Metal nanoparticles (gold, silver), plant-based nanocomposites for drug delivery	Cardoso et al. (2021)
Nanocomposites in diagnostic systems	Nanoparticles (polymer/metal)	Nanocomposites enhance diagnostic imaging, improving the detection and resolution of cancerous tissues	Early-stage cancer detection, improved imaging (MRI and CT)	Cardoso et al. (2021)
Nanocomposites in MS treatment	Nanoliposomes, stimuli-responsive polymers, inorganic nanoparticles	Nanocomposites improve drug delivery to areas of inflammation in the CNS, reducing systemic side effects in MS treatment	Anti-inflammatory agents for MS, localized treatment targeting the brain/spinal cord	Kondiah et al. (2018)

engineering. Typical scaffold types are hydrogels, mats, films, and sponge-like materials; each is designed for a particular use, such as bone, cartilage, cardiac, or neural regeneration. Functional characteristics, including enhanced strength, regulated release of bioactive molecules, and immunomodulation, are conferred by incorporating inorganic phases such as cerium oxide, hydroxyapatite, and gold nanoparticles. Even with significant advancements, there are still issues with maximizing scaffold lifetime, comprehending organic-inorganic interactions, and scaling to clinical applications. In order to highlight the promise of nanocomposites in regenerative medicine, future research will focus on improving scaffold design through covalent bonding, exact porosity control, and extensive in vivo testing (Nandhini et al. 2024).

By enhancing mechanical characteristics, increasing osteoconductivity and osteoinductivity, and simulating the natural bone environment at the nanoscale, nanocomposites have completely changed bone tissue engineering (BTE). They allow for the regulated administration of medications, growth hormones, and therapeutic agents while promoting improved cell adhesion, proliferation, and differentiation (Zhao et al. 2020). By recruiting osteoprogenitor cells and depositing matrix, biomimetic nanocomposites that closely mimic the extracellular matrix encourage bone remodeling and regeneration. Metal-based nanocomposites give strength and antibacterial qualities, ceramic-based ones offer structural support, and polymer-based ones offer flexibility. With encouraging preclinical findings, developments in nanotechnology, such as nanofibers, nanoparticles, and 3D-printed scaffolds, have further enhanced their effectiveness in treating severe bone abnormalities (Christy et al. 2020).

10.6.2 Packaging Industry

Packaging has revolutionized because of developments in polymeric nanocomposites, which have improved important characteristics, including mechanical robustness, solvent resistance, thermal stability, and oxygen and moisture barrier effectiveness. Packaging performance is enhanced by adding nanofillers such as carbon nanotubes, graphene, nanoclay, and inorganic nanoparticles to synthetic and biodegradable polymers (Kumar et al. 2020). Nanofillers must be properly dispersed throughout the polymer matrix to achieve optimal functioning. The usage of modified nanofillers in food, beverage, and electronic packaging has increased due to their antibacterial, non-flammable, and physiologically friendly properties. These substances have excellent stability and barrier qualities in a range of circumstances (Kaur et al. 2021).Innovative, high-performance packaging materials suited for certain industrial applications may be produced by choosing the correct polymers, nanofillers, and manufacturing techniques. This development has much promise for meeting performance and environmental requirements in contemporary packaging solutions (Basavegowda and Baek 2021).

10.6.3 *Environmental and Agricultural Applications*

Because of their distinct size, shape, and chemical reactivity, multifunctional nanoparticles, in particular, have drawn much attention for creating nanostructured materials. Numerous industries, including agriculture, healthcare, energy, sensing, and environmental cleanup, have found use for these NPs. Heavy metal ions, insecticides, medications, and dyes are among the pollutants that copper nanoparticles (NPs) effectively remove from wastewater and agroecosystems. Cu-NPs are micronutrients that increase agricultural yields by fostering plant development and protecting against disease. Metal NPs are difficult to remove from water sources and can harm humans and plants when applied directly. Metal nanoparticles in polymer-based nanocomposites provide a promising solution to these problems. These composites preserve the NPs' efficacy while lowering their toxicity. Nanocomposite-based fertilizers offer a promising solution to the environmental and health issues caused by the overuse of conventional fertilizers. These advanced materials, known for their enhanced mechanical strength, stability, and barrier properties, provide an eco-friendly alternative for slow or controlled-release fertilizer formulations (Sen et al. 2024; Menossi et al. 2022).

Nanotechnology using carbon-based nanocomposites is an emerging solution for water pollutant removal due to their cost-effectiveness, high surface area, and environmental stability. Magnetic carbon nanocomposites, derived from agro-waste such as walnut shells, rice husk, peanut shells, and sugarcane bagasse, demonstrate excellent adsorption capabilities for contaminants like heavy metals, dyes, pharmaceuticals, and organic compounds. For instance, nanocomposites from walnut shells and rice husks can remove 78% of Cd(II) from water. Hydrothermal carbonization, chemical precipitation, and fast pyrolysis enhance pollutant adsorption efficiency. These materials offer advantages like reusability, magnetic separation, and biocompatibility. Magnetic nanocomposites have been highly effective in wastewater treatment, particularly in adsorbing pollutants like rhodamine 6G, phenol, and ciprofloxacin, presenting a sustainable approach to tackling water pollution challenges (Chakraborty et al. 2023).

10.7 Conclusion

Biodegradable nanocomposites offer a promising approach in drug delivery systems due to their flexibility, sustainability, and biocompatibility. Oral, nasal, and even cancer-targeted medication administration is made possible by these nanocomposites containing a variety of nanofillers. Unlike traditional delivery methods, using biodegradable polymers guarantees safer breakdown within the body and reduces environmental impact. Nanofillers improve these composites' mechanical strength, drug loading capacity, and controlled release profiles. Prospects in this area suggest that more adaptable nanocomposites, improved control over degradation rates, and improved targeting capabilities will be developed with more research. These developments will result in more effective, individualized, and targeted treatment options for various illnesses, including cancer.

References

Abdollahi M, Alboofetileh M, Rezaei M, Behrooz R (2013) Comparing physico-mechanical and thermal properties of alginate nanocomposite films reinforced with organic and/or inorganic nanofillers. Food Hydrocoll 32(2):416–424

Alaswad SO, Mahmoud AS, Arunachalam P (2022) Recent advances in biodegradable polymers and their biological applications: a brief review. Polymers 14(22):4924

Aremu OH, Akintayo CO, Naidoo EB, Nelana SM, Ayanda OS (2021) Synthesis and applications of nanosized zinc oxide in wastewater treatment: a review. Int J Environ Sci Technol 18(10):3237–3256

Armentano I, Puglia D, Luzi F, Arciola CR, Morena F, Martino S et al (2018) Nanocomposites based on biodegradable polymers. Materials (Basel) 11(5)

Basavegowda N, Baek KH (2021) Advances in functional biopolymer-based nanocomposites for active food packaging applications. Polymers 13(23):4198

Benetti EM, Causin V, Marega C, Marigo A, Ferrara G, Ferraro A et al (2005) Morphological and structural characterization of polypropylene based nanocomposites. Polymer (Guildf) 46(19 SPEC. ISS):8275–8285

Bose A, Wong TW (2018) Oral colon cancer targeting by chitosan nanocomposites. Appl Nanocomp Mater Drug Deliv:409–429

Brun P, Zamuner A, Battocchio C, Cassari L, Todesco M, Graziani V et al (2021) Bio-functionalized chitosan for bone tissue engineering. Int J Mol Sci 22(11):5916

Cantarella M, Impellizzeri G, Di Mauro A, Privitera V, Carroccio SC (2021) Innovative polymeric hybrid nanocomposites for application in photocatalysis. Polymers (Basel) 13(8)

Cao M, Gu F, Rao C, Fu J, Zhao P (2019) Improving the electrospinning process of fabricating nanofibrous membranes to filter PM2.5. Sci Total Environ 666:1011–1021

Cardoso S, da Silva CF, Severino P, Silva AM, Souto SB, Zielińska A et al (2021) Genotoxicity assessment of metal-based nanocomposites applied in drug delivery. Materials 14(21):6551

Castillo BE, Prokhorov E, Luna-Bárcenas G, Kovalenko Y (2022) Potential use of chitosan-TiO2 nanocomposites for the electroanalytical detection of imidacloprid. Polymers 14(9):1686

Chakraborty R, Mukhopadhyay A, Paul S, Sarkar S, Mukhopadhyay R (2023) Nanocomposite-based smart fertilizers: a boon to agricultural and environmental sustainability. Sci Total Environ 863:160859

Chandra R, Rustgi R (1998) Biodegradable polymers review. Pol Sci 13(2):179–196

Christy PN, Basha SK, Kumari VS, Bashir AKH, Maaza M, Kaviyarasu K et al (2020) Biopolymeric nanocomposite scaffolds for bone tissue engineering applications—a review. J Drug Deliv Sci Technol 55:101452

Chung YL, Ansari S, Estevez L, Hayrapetyan S, Giannelis EP, Lai HM (2010) Preparation and properties of biodegradable starch-clay nanocomposites. Carbohydr Polym 79(2):391–396

Correa E, Moncada ME, Gutiérrez OD, Vargas CA, Zapata VH (2019) Characterization of polycaprolactone/rGO nanocomposite scaffolds obtained by electrospinning. Mater Sci Eng C 103:109773

Dai H, Luo Y, Huang Y, Ma L, Chen H, Fu Y et al (2023) Recent advances in protein-based emulsions: the key role of cellulose. Food Hydrocoll 136:136

Dananjaya SAV, Chevali VS, Dear JP, Potluri P, Abeykoon C (2024) 3D printing of biodegradable polymers and their composites—current state-of-the-art, properties, applications, and machine learning for potential future applications. Prog Mater Sci 146:101336

Dantas de Oliveira A, Augusto Gonçalves Beatrice C (2019) Polymer nanocomposites with different types of nanofiller. Nanocomp Recent Evol
Deshmukh K, Pandey M (2023) Biodegradable and biocompatible polymer nanocomposites: processing, characterization, and applications. Elsevier
Eichhorn SJ, Etale A, Wang J, Berglund LA, Li Y, Cai Y et al (2022) Current international research into cellulose as a functional nanomaterial for advanced applications. J Mater Sci 57(10):5697–5767
Fatima J, Siddique YH (2024) Application of nanocomposites and nanoparticles in treating neurodegenerative disorders. CNS Neurol Disord Drug Targets 23(10):1217–1233
Gobena ST, Woldeyonnes AD (2024) A review of synthesis methods, and characterization techniques of polymer nanocomposites for diverse applications. Discov Mater 4(1):1–18
Hiremath A, Thipperudrappa S, Bhat R (2022) Surface morphology analysis using atomic force microscopy and statistical method for glass fiber reinforced epoxy-zinc oxide nanocomposites. Eng Sci 18:308–319
Hou Y, Baig MM, Lu J, Zhang H, Liu P, Zhu G et al (2024) Direct ink writing 3D printing of low-dimensional nanomaterials for micro-supercapacitors. Nanoscale 16(26):12380–12396
Huang Y, Wang T, Zhao X, Wang X, Zhou L, Yang Y et al (2015) Poly(lactic acid)/graphene oxide–ZnO nanocomposite films with good mechanical, dynamic mechanical, anti-UV and antibacterial properties. J Chem Technol Biotechnol 90(9):1677–1684
Ibrahim IAA, Alzahrani AR, Alanazi IM, Shahzad N, Shahid I, Falemban AH et al (2023) Carbohydrate polymers-based surface modified nano delivery systems for enhanced target delivery to colon cancer - a review. Int J Biol Macromol 253(Pt 2):126581
Iñiguez-Franco F et al (2017) Effect of nanoparticles on the hydrolytic degradation of PLA-nanocomposites by water-ethanol solutions. Poly Degrad Stab 146:287–297
Ippolito F, Hübner G, Claypole T, Gane P (2020) Influence of the surface modification of calcium carbonate on polyamide 12 composites. Polymers 12(6):1295
Jagadeesh P, Puttegowda M, Mavinkere Rangappa S, Siengchin S (2021) Influence of nanofillers on biodegradable composites: a comprehensive review. Polym Compos 42(11):5691–5711
Jamróz E, Kulawik P, Kopel P (2019) The effect of nanofillers on the functional properties of biopolymer-based films: a review. Polymers 11(4):675
Jing Q, Liu W, Pan Y, Silberschmidt VV, Li L, Dong ZL (2015) Chemical functionalization of graphene oxide for improving mechanical and thermal properties of polyurethane composites. Mater Des 85:808–814
Kaur G, Sharma S, Mir SA, Dar BN (2021) Nanobiocomposite Films: a 'Greener Alternate' for food packaging. Food Bioprocess Technol 14(6):1013–1027
Kausar A (2020) Thermally conducting polymer/nanocarbon and polymer/inorganic nanoparticle nanocomposite: a review. Polym Technol Mater 59(8):895–909
Kausar A, Ahmad I (2023) Conducting polymer nanocomposites for electromagnetic interference shielding—radical developments. J Compos Sci 7(6):240
Khichar KK, Dangi SB, Dhayal V, Kumar U, Hashmi SZ, Sadhu V et al (2020) Structural, optical, and surface morphological studies of ethyl cellulose/graphene oxide nanocomposites. Polym Compos 41(7):2792–2802
Khosravi M, Seyfi J, Saeidi A, Khonakdar HA (2020) Spin-coated polyvinylidene fluoride/graphene nanocomposite thin films with improved β-phase content and electrical conductivity. J Mater Sci 55(15):6696–6707
Kondiah PPD, Choonara YE, Kondiah PJ, Marimuthu T, Kumar P, du Toit LC et al (2018) Nanocomposites for therapeutic application in multiple sclerosis. Appl Nanocomp Mater Drug Deliv:391–408
Kumar S, Mukherjee A, Dutta J (2020) Chitosan based nanocomposite films and coatings: Emerging antimicrobial food packaging alternatives. Trends Food Sci Technol 97:196–209
Kumbhakar P et al (2024) Optical Properties of Nanoparticles and Nanocomposites. J Nanomater. https://doi.org/10.1155/9182.si.840169

Le B et al (2020) A review on nanocomposites. Part 1: Mechanical properties. J Manuf Sci Eng 142(10):100801

Leja K, Lewandowicz G (2010) Polymer biodegradation and biodegradable polymers - a review. Polish J Environ Stud 19(2):255–266

Liberman A, Mendez N, Trogler WC, Kummel AC (2014) Synthesis and surface functionalization of silica nanoparticles for nanomedicine. Surf Sci Rep 69(2–3):132–158

Loukelis K, Helal ZA, Mikos AG, Chatzinikolaidou M (2023) Nanocomposite bioprinting for tissue engineering applications. Gels 9(2):103

Ma ZL, Tsou CH, Cui X, Wu J, Lin L, Wen H et al (2022) Barrier properties of nanocomposites from high-density polyethylene reinforced with natural attapulgite. Curr Res Green Sustain Chem 5:100314

Machado AV, Araújo A, Oliveira M (2015) Assessment of polymer-based nanocomposites biodegradability. In: Biodegradable polymers: new development. NOVA Publishers, pp 169–195

Menossi M, Casalongué C, Alvarez VA (2022) Bio-nanocomposites for modern agricultural applications. Handb Consum Nanoproducts:1201–1237

Mukherjee A, Panda S, Jeelani PG, Mossa AT, Chidambaram R (2022) Biodegradable polymers/silica nanocomposites: applications in food packaging. Nanotechnol Appl Food Saf Qual Monit:395–414

Mukherjee C, Varghese D, Krishna JS, Boominathan T, Rakeshkumar R, Dineshkumar S et al (2023) Recent advances in biodegradable polymers—properties, applications and future prospects. Eur Polym J 192:112068

Nalone LA, Amaral RG, de Lima Oliveira DM, Andrade LR, de Hollanda LM, da Silva CF et al (2018) Applications of nanocomposite materials in the delivery of anticancer drugs. Appl Nanocomp Mater Drug Deliv:339–352

Nandhini J, Karthikeyan E, Rajeshkumar S (2024) Eco-friendly bio-nanocomposites: pioneering sustainable biomedical advancements in engineering. Discov Nano 19(1):1–31

Nguyen C, Yang F (2021) Applications of electrospraying in biosensing, diagnostics, and beyond. Biomed Appl Electrospin Electrospray:277–329

Nguyen QL, Le DV, Phan AN, Nguyen VD (2023) Synthesis of biodegradable and antimicrobial nanocomposite films reinforced for coffee and agri-food product preservation. ACS Omega 8(45):42177–42185

Pandey JK, Raghunatha Reddy K, Pratheep Kumar A, Singh RP (2005) An overview on the degradability of polymer nanocomposites. Polym Degrad Stab 88(2):234–250

Parhi R (2018) Nanocomposite for transdermal drug delivery. Appl Nanocomp Mater Drug Deliv:353–389

Patwary MAS, Surid SM, Gafur MA (2020) Properties and applications of biodegradable polymers. J Res Updat Polym Sci 9:32–41

Piekarska K, Sowinski P, Piorkowska E, Haque MMU, Pracella M (2016) Structure and properties of hybrid PLA nanocomposites with inorganic nanofillers and cellulose fibers. Compos Part A Appl Sci Manuf 82:34–41

Saberi A, Baltatu MS, Vizureanu P (2024) Recent advances in magnesium–magnesium oxide nanoparticle composites for biomedical applications. Bioengineering (Basel) 11(5):508

Saharudin MS, Ilyas RA, Awang N, Hasbi S, Shyha I, Inam F (2023) Advances in sustainable nanocomposites. Sustainability 15(6):5125

Sen A, Kurian MS, Narayanan DP, Abraham A, Thomas TS, Jayalakshmi PS et al (2024) Bio-nanocomposites: Innovative solutions for addressing issues in health, agriculture, energy and environmental domains. Nanostruct Nano Obj 39:101270

Solomon NO, Simpa P, Adenekan OA, Obasi SC (2024) Sustainable nanomaterials' role in green supply chains and environmental sustainability. Eng Sci Technol J 5(5):1678–1694

Stanisz M, Klapiszewski L, Collins MN, Jesionowski T (2022) Recent progress in biomedical and biotechnological applications of lignin-based spherical nano- and microstructures: a comprehensive review. Mater Today Chem 26:101198

Ulery BD, Nair LS, Laurencin CT (2011) Biomedical applications of biodegradable polymers. J Polym Sci Part B Polym Phys 49(12):832–864

Yesaswi CS, Satya SK, Sahu SK, Badgayan ND, Sri Ram Murthy P, Kumar VMR et al (2022) Thermal properties of polymer nanocomposites. Adv Polym Nanocomposites Sci Technol Appl:99–143

Yu X, Zhang T, Li Y (2020) 3D printing and bioprinting nerve conduits for neural tissue engineering. Polymers 12(8):1637

Yu S, Pu X, Ahmed MU, Yu HH, Mutukuri TT, Li J et al (2021) Spray-freeze-dried inhalable composite microparticles containing nanoparticles of combinational drugs for potential treatment of lung infections caused by Pseudomonas aeruginosa. Int J Pharm 610:121160

Zapata P, Quijada R, Retuer J, Moncada E (2008) Preparation of nanocomposites by in situ polimerization. J Chil Chem Soc 53(1)

Zhao H, Liu M, Zhang Y, Yin J, Pei R (2020) Nanocomposite hydrogels for tissue engineering applications. Nanoscale 12(28):14976–14995

Chapter 11
Biocompatible Polymer Nanocomposites

Prashant Patil, Ganesh Sonawane, Kajal Pansare, Deepak Sonawane, and Sunil Mahajan

Abstract Biocompatible polymer nanocomposites represent a rapidly advancing class of materials with significant potential in biomedical and healthcare applications. By integrating biocompatible polymers with functional nanofillers, these materials offer unique combinations of mechanical strength, bioactivity, and controlled degradation tailored to specific physiological environments. This chapter provides a comprehensive overview of biocompatible polymer nanocomposites, covering the fundamentals of biocompatibility, selection of suitable natural and synthetic polymers, and various types of nanofillers such as carbon-based materials, metal oxides, and bioceramics. Emphasis is placed on fabrication techniques—including electrospinning, solvent casting, and 3D printing and how they influence material structure, performance, and biological interactions. The chapter further explores the wide range of biomedical applications, including tissue engineering scaffolds, drug delivery systems, wound-healing materials, biosensors, and implantable devices. Key aspects of toxicology, immunogenicity, and biological safety are critically reviewed, along with current strategies for evaluating in vitro and in vivo biocompatibility. Additionally, the chapter addresses regulatory considerations, clinical translation challenges, and ethical implications surrounding the use of nanocomposites in healthcare, by highlighting recent advancements and emerging trends such as smart, responsive, and bioresorbable systems. This chapter under-

P. Patil (✉)
Department of Pharmaceutics, N N Sattha College of Pharmacy, Ahmednagar, Maharashtra, India

G. Sonawane · S. Mahajan
Department of Pharmaceutical Chemistry, Divine College of Pharmacy, Satana, Nashik, Maharashtra, India

K. Pansare
Department of Pharmacology, Divine College of Pharmacy, Satana, Nashik, Maharashtra, India

D. Sonawane
Department of Pharmaceutics, Divine College of Pharmacy, Satana, Nashik, Maharashtra, India

S. N. Kothawade et al. (eds.), *Polymer Nanocomposites*,
https://doi.org/10.1007/978-3-032-00365-2_11

scores the transformative potential of biocompatible polymer nanocomposites in next-generation medical technologies. It concludes by outlining current limitations and future research directions aimed at overcoming challenges in scalability, reproducibility, and long-term safety.

Keywords Biocompatibility · Polymer nanocomposites · Biomedical applications · Natural and synthetic polymers · Nanofillers · Tissue engineering · Drug delivery systems · Cytotoxicity

11.1 Introduction

Polymer nanocomposites are a type of hybrid materials where nanometer-scale fillers are uniformly or semi-uniformly dispersed within a polymeric matrix (Winey and Vaia 2007). These nanofillers, which typically range in size from 1 to 100 nanometers, can include nanoparticles (e.g., silica, titanium dioxide), nanoclays, carbon nanotubes (CNTs), graphene, nanofibers, and nanocrystals. The incorporation of these nanostructures into polymer matrices results in materials that exhibit superior and often tunable physical, chemical, mechanical, thermal, electrical, and biological properties, surpassing those of conventional polymer composites (Fu et al. 2019; Krishnamoorti and Vaia 2007). The enhanced performance of polymer nanocomposites is largely attributed to the high surface-area-to-volume ratio of the nanofillers and their interaction with the polymer chains. These interactions affect the crystallinity, viscosity, degradation profile, and barrier properties of the final nanocomposite. Polymer nanocomposites have garnered great interest in recent years because of their versatility and functionality, particularly in high-performance applications in industries like aerospace, automotive, electronics, energy storage, packaging, and most importantly, biomedicine (Jordan et al. 2005; Fischer 2003; Schadler et al. 2007).

11.1.1 Importance of Biocompatibility in Biomedical Applications

In the context of biomedical applications, the integration of polymer nanocomposites presents a unique set of opportunities and challenges. While nanocomposites can offer tailored properties essential for specific clinical needs—such as controlled degradation, mechanical support, and targeted therapeutic delivery—their successful use in biological systems hinges on a critical attribute: biocompatibility (Arif et al. 2019).

Biocompatibility is the capacity of a material to fulfill its desired function in a biological environment without inducing any unwanted local or systemic effects. It encompasses several parameters, including cytotoxicity, genotoxicity, immunogenicity, and inflammation potential. A biocompatible material should support cellular adhesion, proliferation, and differentiation (where needed), while also ensuring minimal toxic or inflammatory responses from the host (Maghimaa et al. 2024).

For polymer nanocomposites used in drug delivery, tissue engineering scaffolds, implants, or wound dressings, the biocompatibility of both the polymer matrix and the incorporated nanofillers must be carefully evaluated. The introduction of nanomaterials may change the surface chemistry, degradation behavior, and interaction with cells or proteins, thereby influencing their safety and efficacy. For example, while carbon-based nanomaterials like CNTs and graphene exhibit exceptional mechanical and electrical properties, their poor biodegradability and potential toxicity raise concerns for biomedical use unless carefully modified or coated (Zare and Shabani 2016).

Thus, designing biocompatible polymer nanocomposites requires a deep understanding of material science, toxicology, cellular biology, and clinical requirements. Surface functionalization of nanofillers, selection of FDA-approved polymers (e.g., PLA, PLGA, PCL, PEG), and the use of bioactive components such as hydroxyapatite or silver nanoparticles are among the strategies employed to enhance biocompatibility.

11.1.2 Scope and Objectives of the Chapter

This chapter is intended to give an in-depth overview of biocompatible polymer nanocomposites with respect to their design principles, material choice, fabrication methods, and biomedical applications. It delves into the application of both natural and synthetic polymers in conjunction with an array of nanofillers—such as metal and metal oxide nanoparticles, carbon-based nanomaterials, and bioceramics—to develop sophisticated materials with improved mechanical, biological, and functional properties. Emphasis is placed on the importance of biocompatibility and how the structural and surface characteristics of nanocomposites influence their interaction with biological systems. The chapter also highlights commonly used synthesis methods including solvent casting, melt blending, electrospinning, and 3D printing, discussing how each technique impacts material performance. Applications across drug delivery, tissue engineering, wound healing, biosensing, and implantable devices are thoroughly reviewed, along with current strategies for assessing cytotoxicity, biodegradability, and immunogenicity. Furthermore, the chapter discusses regulatory and ethical considerations essential for the clinical translation of these materials. By identifying present challenges and emerging trends—such as smart, stimuli-responsive, and bioresorbable systems—the chapter provides a forward-looking perspective on the development of next-generation biomedical nanocomposites. The content is intended to serve as a valuable resource for researchers, academicians, clinicians, and industry professionals engaged in the fields of nanotechnology, biomaterials, and biomedical engineering.

11.2 Fundamentals of Biocompatibility

In the context of biomedical materials, particularly those used for implants, drug delivery systems, scaffolds, or wound dressings, biocompatibility is a critical determinant of clinical success. It not only ensures that the material does not trigger cytotoxic, immunogenic, mutagenic, or carcinogenic responses but also that it supports healing, tissue integration, or therapeutic action (Khane et al. 2024; Saini et al. 2017).

The principles of biocompatibility are rooted in the understanding that any foreign material introduced into the body will interact with the biological milieu. This interaction occurs at multiple levels—molecular, cellular, and tissue—and must be finely tuned to the intended function of the device or system. Biocompatibility is not a fixed property of a material; rather, it is application-specific and time-dependent. A material that is biocompatible in one context (e.g., external contact devices) may be unsuitable in another (e.g., vascular implants or neural interfaces). Thus, the assessment of biocompatibility must consider the material's composition, surface properties, degradation profile, and the physiological site of application (Jurak et al. 2021).

11.2.1 Biological Responses to Materials (Cellular and Tissue Interactions)

When a material is introduced into a biological system, a cascade of events ensues, starting from immediate contact with fluids and progressing to longer-term cellular and tissue interactions. These responses can be broadly categorized into the following stages:

- **Protein adsorption:** Immediately after implantation, host proteins like albumin, fibrinogen, and immunoglobulins adsorb onto the material surface. This adsorbed protein layer plays a key role in dictating subsequent cellular responses.
- **Cell adhesion and activation:** Depending on the material's surface chemistry and topography, cells like fibroblasts, macrophages, or endothelial cells will adhere to the surface. The quality of these interactions determines whether the material is integrated within surrounding tissue or sequestered by fibrous encapsulation.
- **Inflammatory response:** The innate immune system may recognize the material as foreign, triggering inflammation mediated by neutrophils and macrophages. In some cases, this can lead to chronic inflammation or the formation of foreign body giant cells (FBGCs).
- **Tissue remodeling and integration:** In favorable scenarios, the material supports tissue regeneration, vascularization, and integration into the surrounding

host tissue. In contrast, poor biocompatibility can result in necrosis, fibrosis, or rejection.

These biological responses are influenced by several factors, including material composition, surface roughness, porosity, hydrophilicity/hydrophobicity, and degradation products. Therefore, engineering the material interface to promote favorable cellular interactions is a key strategy in the development of biocompatible nanocomposites (Feldman 2016).

11.2.2 Standards and Testing Protocols (ISO 10993, ASTM Guidelines)

To ensure the safety and efficacy of biomaterials and medical devices, international regulatory bodies have established comprehensive testing standards. Two of the most widely recognized frameworks are:

- **ISO 10993 Series:** Developed by the International Organization for Standardization (ISO), the ISO 10993 series outlines a framework for the biological evaluation of medical devices. It includes guidelines for cytotoxicity testing (ISO 10993-5), sensitization (ISO 10993-10), systemic toxicity (ISO 10993-11), and implantation studies (ISO 10993-6), among others. The standard recommends a risk-based approach, tailoring the test methods to the nature of the device, its duration of contact with tissues, and its intended function.
- **ASTM Standards:** The American Society for Testing and Materials (ASTM) provides additional methods for evaluating biocompatibility, especially for polymers and nanomaterials. ASTM F748, for example, covers terminology related to biocompatibility, while ASTM F763 outlines procedures for assessing tissue reactions.

Preclinical testing typically includes in vitro assays (e.g., MTT, Live/Dead staining, hemocompatibility, oxidative stress assays) and in vivo animal studies to evaluate systemic and local tissue responses. Advanced imaging techniques, histological staining, and biomarker analysis are used to assess biocompatibility at the cellular and tissue levels (Harun et al. 2023; Bernard et al. 2018; Al-Zyoud et al. 2023).

11.2.3 Factors Affecting Biocompatibility

The biocompatibility of a polymer nanocomposite is influenced by a multitude of interrelated factors, which can be broadly classified as follows:

- **Material Composition:** The choice of polymer (e.g., PLA, PLGA, PCL, PEG, chitosan) and nanofillers (e.g., silver nanoparticles, carbon nanotubes, hydroxyapatite) plays a crucial role. The purity, toxicity, and degradability of each component must be thoroughly evaluated.

- **Surface Properties:** Surface chemistry, roughness, energy, and charge affect protein adsorption and cell behavior. Functionalization with peptides, growth factors, or hydrophilic groups can enhance compatibility.
- **Size and Morphology of Nanofillers:** The dimensions, shape (spherical, rod-like, sheet), and aspect ratio of nanofillers influence their distribution, cellular uptake, and potential toxicity.
- **Degradation Products:** Biodegradable polymers release byproducts over time, which can either support tissue regeneration (e.g., lactic acid from PLA) or cause adverse effects if toxic.
- **Mechanical Properties:** The stiffness, elasticity, and tensile strength must match the target tissue to avoid mechanical mismatch, which can lead to inflammation or rejection.
- **Sterilization Methods:** The tensile strength, elasticity, and stiffness should be equal to the target tissue to prevent mechanical mismatch, which can result in rejection or inflammation.
- **Host Factors:** The site of implantation, patient age, immune status, and existing comorbidities can all influence the biological response to a material.

Understanding and optimizing these factors is critical to the rational design of polymer nanocomposites that are not only functional but also safe and effective for long-term biomedical use (Armentano et al. 2010).

11.3 Types of Biocompatible Polymers Used in Nanocomposites

Biocompatible polymers play a fundamental role in the formulation of nanocomposites, particularly in the biomedical field. These polymers provide the required interface between biological systems and nanomaterials, ensuring safety, effectiveness, and appropriate biological response. Biocompatible polymers used in nanocomposites are broadly categorized into two classes: natural polymers and synthetic polymers (Table 11.1).

11.3.1 Natural Biocompatible Polymers

Natural polymers are developed from natural origins and are preferred due to their inherent biocompatibility, biodegradability, and conformational likeness to constituents of the extracellular matrix (ECM). They enhance cell adhesion, proliferation, and differentiation, rendering them extremely applicable in biomedical fields.

Table 11.1 Comparison of natural vs synthetic biocompatible polymers

Parameters	Natural Polymers	Synthetic Polymers
Examples	Chitosan, gelatin, collagen, alginate, cellulose, silk fibroin	PLA, PLGA, PCL, PEG, PU, PMMA
Biocompatibility	Excellent, mimics ECM	Good, but depends on synthesis and purity
Biodegradability	High	Tunable
Mechanical properties	Moderate, often require reinforcement	Customizable, generally stronger
Immunogenicity	May elicit immune response (source-dependent)	Generally lower if well-designed
Processability	Limited by natural variability	High, suitable for large-scale processing
Cost and availability	May vary; renewable sources (Asghari et al. 2017)	Consistent, scalable production (Armentano et al. 2018; Liu et al. 2019)

- **Chitosan:** Chitosan is a naturally occurring cationic polysaccharide obtained through the deacetylation of chitin, which is a common component found in the crustacean exoskeleton of shrimp and crab. It structurally contains β-(1 → 4)-linked D-glucosamine and N-acetyl-D-glucosamine units. Its unique characteristics are biocompatibility, biodegradability, and mucoadhesive properties. Due to its positive charge, chitosan can interact with negatively charged biological membranes, enhancing cellular uptake and drug permeability. It also exhibits antimicrobial, antifungal, and wound-healing activities. Chitosan is widely used in the preparation of nanoparticles for controlled drug delivery, gene therapy, wound dressings, and as a scaffold in tissue engineering.
- **Gelatin:** Gelatin is a protein yielded through partial hydrolysis of collagen, which constitutes the chief structural protein in animal connective tissues. It forms transparent, thermoreversible gels in aqueous solutions. Gelatin is highly biocompatible and biodegradable and supports cell attachment and proliferation. Its excellent film-forming ability and hydrophilic nature make it suitable for drug encapsulation and sustained-release formulations. Gelatin-based nanocomposites are employed in wound healing, soft tissue engineering, and as delivery systems for various pharmaceutical agents, including proteins and nucleic acids.
- **Collagen:** Collagen is the most abundant structural protein in the extracellular matrix (ECM) of mammalian connective tissues, comprising triple-helical polypeptide chains. It is naturally recognized by cells, promoting adhesion, growth, and differentiation. Collagen has high tensile strength and excellent biocompatibility. However, it is rapidly degraded in the body, necessitating crosslinking or blending with other materials for stability. Collagen-based nanocomposites are used in reconstructive surgery, bone grafts, skin regeneration, corneal implants, and as a scaffold material in regenerative medicine due to its excellent biological performance.
- **Alginate:** Alginate is a naturally found anionic polysaccharide derived from brown seaweed. It consists of β-D-mannuronic acid (M) and α-L-guluronic acid (G) residues. Alginate forms hydrogels by undergoing ionic gelation in the

presence of divalent cations like calcium. This is a mild process and hence suitable for encapsulating cells or sensitive bioactives. Alginate is biocompatible, non-toxic, and allows nutrient and oxygen diffusion. Alginate-based nanocomposites are extensively utilized in wound healing, cell encapsulation, controlled drug release systems, and tissue engineering.
- **Cellulose:** Cellulose is the most widely occurring natural polymer, made up of linear β-(1 → 4)-linked D-glucose units. It is obtained from plant material like cotton, wood, and crop waste. Cellulose has excellent mechanical strength, is non-toxic, and is highly biocompatible. It is insoluble in water but can be chemically modified to form derivatives like methylcellulose, carboxymethyl cellulose (CMC), and hydroxypropyl cellulose, which improve solubility and functionality. Cellulose-based nanocomposites are widely used in wound dressings, scaffolds, and drug delivery platforms due to their biodegradability and structural integrity.
- **Silk Fibroin:** Silk fibroin is a protein taken from the cocoons of the silkworm Bombyx mori. It consists primarily of glycine, alanine, and serine residues in crystalline β-sheet structures that are responsible for the high tensile strength. Silk fibroin is biodegradable, biocompatible, and has minimal inflammatory responses. Its unique mechanical properties and ability to be processed into various formats (films, fibers, gels, sponges) make it ideal for biomedical applications. Silk-based nanocomposites are explored in tissue engineering, bone and cartilage repair, sutures, and drug delivery due to their excellent cellular interactions and sustained degradation profiles (Asghari et al. 2017).

11.3.2 Synthetic Biocompatible Polymers

Synthetic polymers are man-made macromolecules with tailor-made properties. They offer excellent control over physical, chemical, and biological characteristics such as degradation rate, mechanical strength, and hydrophilicity. They are often used in applications requiring precision, consistency, and structural integrity.

- **Poly(lactic acid) (PLA):** PLA is a synthetic aliphatic polyester obtained from the fermentation of renewable materials like corn starch or sugarcane to obtain lactic acid, which is then polymerized. PLA is thermoplastic, biodegradable, and biocompatible. It breaks down in the body into lactic acid, which is naturally metabolized. PLA has good mechanical properties and is applied in load-bearing applications. Its rate of degradation is quite low, and hence, it is appropriate for long-term drug delivery and implantable devices. PLA-based nanocomposites have applications in orthopedic implants, sutures, tissue scaffolds, and drug delivery systems.
- **Poly(lactic-co-glycolic acid) (PLGA):** PLGA is a glycolic acid and lactic acid copolymer whose properties can be tuned based on the monomer ratio. It is among the most researched biodegradable polymers because of its superior biocompatibility and FDA regulatory approval. The degradation rate of PLGA can be adjusted by changing the lactic/glycolic acid ratio, molecular weight, and

crystallinity. PLGA biodegrades to non-toxic degradation products (glycolic acid and lactic acid), which are naturally excreted by the body. PLGA nanocomposites are extensively used in controlled drug delivery systems, vaccine carriers, cancer treatment, and biodegradable sutures.

- **Polycaprolactone (PCL):** PCL is a semi-crystalline aliphatic polyester produced by ring-opening polymerization of ε-caprolactone. It is very flexible, with a low melting temperature and an extremely slow rate of degradation, ranging from several months to years. PCL has good biocompatibility and is usually mixed with other polymers in order to improve its properties. It finds applications in long-term implants and controlled drug release systems. PCL-based nanocomposites are broadly utilized in tissue engineering, orthopedic implants, regeneration of nerves, and as a drug and gene delivery matrix.
- **Polyethylene glycol (PEG):** PEG is a hydrophilic polyether with remarkable water solubility, flexibility, and non-immunogenic behavior. It is widely used in pharmaceutical and biomedical applications to enhance the solubility and circulation time of therapeutic agents. PEGylation—the process of attaching PEG chains to molecules or nanoparticles—improves biocompatibility, reduces protein adsorption, and evades recognition by the immune system, providing "stealth" properties to nanoparticles. PEG-based nanocomposites are widely used in drug delivery, surface coatings, and hydrogel formulations.
- **Polyurethane (PU):** Polyurethanes are a broad group of polymers prepared by the reaction of diisocyanates with polyols. By adjusting the hard and soft segments, many mechanical and degradation properties can be obtained. PU can be made biodegradable or biostable. Polyurethanes used in medicine have high toughness, elasticity, and hemocompatibility. PU-based nanocomposites are also employed in cardiovascular devices like vascular grafts and heart valves and in wound dressings, drug delivery, and tissue engineering scaffolds.
- **Polymethyl methacrylate (PMMA):** PMMA is a transparent, rigid thermoplastic polymer synthesized from methyl methacrylate monomers. Unlike other biodegradable polymers, PMMA is non-biodegradable and remains stable in the biological environment. It is biocompatible and widely used in hard tissue replacement due to its mechanical strength and clarity. PMMA is commonly used in dental prosthetics, bone cement for orthopedic surgeries, intraocular lenses, and cosmetic implants. Although PMMA is not resorbable, it serves as a durable component in long-term biomedical implants and structural prosthetics (Armentano et al. 2018; Liu et al. 2019).

11.4 Types of Nanofillers Used in Biocompatible Nanocomposites

Nanofillers are nanoscale additives incorporated into polymer matrices to enhance their mechanical strength, thermal stability, biological activity, and functionality. When used in biocompatible nanocomposites, these nanofillers not only improve the physicochemical properties but also contribute to biofunctionality like

Table 11.2 Types of nanofillers used in biocompatible nanocomposites and their functional roles

Nanofiller Type	Examples	Physicochemical Properties	Functional Role in Nanocomposites
Carbon-based Nanofillers	Graphene, carbon nanotubes (CNTs), Nanodiamonds	High mechanical strength, large surface area, electrical conductivity biocompatible when functionalized	Improve mechanical strength, electrical/thermal conductivity; suitable for biosensors, neural interfaces, and drug delivery (Dhayal et al. 2021; Fauzi et al. 2021)
Inorganic nanoparticles	Hydroxyapatite (HAp)	Bioactive, osteoconductive, similar to natural bone mineral	Widely used in bone tissue engineering, dental applications, and orthopedic coatings
	Silica (SiO_2) nanoparticles	High surface area, porous structure, modifiable surface chemistry	Excellent for drug delivery (loading and release), imaging, and scaffold reinforcement
	Zinc oxide (ZnO) nanoparticles	Antimicrobial, UV-absorbing, photocatalytic properties	Used in wound dressings, antibacterial coatings, and dental composites
	Titanium dioxide (TiO_2) nanoparticles	Photocatalytic, high mechanical reinforcement potential	Antibacterial use, improved biocompatibility of implants, dental and orthopedic materials
	Gold (au) and silver (ag) nanoparticles	Antimicrobial, conductive, easy surface modification	Antibacterial agents, drug delivery, biosensors, cancer therapy (Feldman 2016)
Layered silicates	Montmorillonite (MMT), Laponite	High aspect ratio, ion-exchange capacity, swelling ability, biodegradable, bio-inert, improves polymer crystallinity	Enhance thermal/mechanical stability, gas barrier properties; useful in wound healing, drug release, facilitate cell adhesion, sustained-release profiles, and reduce cytotoxicity (Ray and Bousmina 2005)

osteoconductivity, antibacterial activity, or cellular interactions. Based on their origin and composition, nanofillers can be classified as carbon-based nanofillers, inorganic nanoparticles, and layered silicates. Types of nanofillers used in biocompatible nanocomposites and their functional roles are shown in Table 11.2.

11.4.1 Carbon-Based Nanofillers

These include materials like graphene, carbon nanotubes (CNTs), and nanodiamonds, which are known for their exceptional mechanical strength, high surface area, and electrical conductivity. These properties make carbon-based nanofillers ideal for developing conductive scaffolds, biosensors, and neural interfaces. However, their biocompatibility can vary and is often enhanced through surface modifications like PEGylation or functional group attachment to reduce toxicity and improve dispersibility in aqueous environments.

- **Graphene and Graphene Oxide (GO):** Graphene is a single sheet of carbon atoms in a two-dimensional honeycomb structure. It possesses high electrical conductivity, mechanical strength, and surface area. Graphene oxide (GO), a derivative of graphene bearing oxygen-containing functional groups (hydroxyl, carboxyl), is more hydrophilic and water-dispersible and thus more useful for biomedical purposes. GO may enhance cell adhesion, growth, and differentiation. In nanocomposites, tensile strength, conductivity, and antibacterial activities are improved by graphene. Applications of graphene involve biosensors, drug delivery, tissue scaffolds, and neural interfaces.
- **Carbon Nanotubes (CNTs):** Carbon nanotubes are rolled-up graphene sheets in the form of cylindrical nanostructures. They may be single-walled (SWCNTs) or multi-walled (MWCNTs) and possess exceptional mechanical strength, high electrical conductivity, and thermal stability. CNTs have the ability to replicate the fibrous nature of extracellular matrix, allowing cell adhesion and signaling. Pristine CNTs are hydrophobic and may pose cytotoxicity concerns, but functionalization with carboxyl, hydroxyl, or amine groups enhances their dispersion, reduces toxicity, and improves biocompatibility. CNT-based nanocomposites are studied for bone tissue engineering, nerve regeneration, biosensors, and drug delivery.
- **Nanodiamonds (NDs):** Nanodiamonds are carbon nanoparticles with a diamond-like crystalline core and functionalized surface groups. They exhibit excellent hardness, chemical stability, and biocompatibility. Due to their high surface area and tunable surface chemistry, NDs can be used to load and release drugs or genes. They are non-toxic to various cell lines and can promote osteogenic differentiation. Nanodiamond-reinforced nanocomposites have applications in drug delivery, orthopedic implants, dental composites, and wound healing (Dhayal et al. 2021; Fauzi et al. 2021).

11.4.2 *Inorganic Nanoparticles*

This group includes a variety of metal and metal oxide nanoparticles that possess desirable properties for biomedical uses:

- **Hydroxyapatite (HAp):** Hydroxyapatite [$Ca_{10}(PO_4)_6(OH)_2$] is a calcium phosphate-based mineral that closely resembles the inorganic component of bone and teeth. It is highly biocompatible, bioactive, and osteoconductive. When incorporated into polymer matrices, HAp improves the mechanical properties and promotes bone cell adhesion and mineralization. Nanoscale HAp is used in orthopedic implants, bone tissue engineering scaffolds, and dental materials. It also serves as a carrier for local drug delivery in bone-related diseases.
- **Silica Nanoparticle:** Silica (SiO_2) nanoparticles are widely used due to their high surface area, tunable pore size, biocompatibility, and ease of functionalization. Mesoporous silica nanoparticles (MSNs) are particularly valuable for controlled drug delivery due to their large pore volume and surface functionality. Silica can reinforce the mechanical and thermal properties of nanocomposites

while offering sites for drug loading, imaging agents, or targeting ligands. Silica-based nanocomposites are applied in cancer therapy, imaging, biosensing, and regenerative medicine.

- **Zinc Oxide (ZnO) Nanoparticles:** ZnO nanoparticles possess strong antibacterial and UV-blocking properties, making them suitable for biomedical coatings, wound dressings, and packaging. ZnO is biocompatible and exhibits piezoelectric properties, which can be useful in cell stimulation. ZnO also shows pro-osteogenic behavior and is used in bone regeneration. ZnO nanofillers enhance the mechanical, thermal, and barrier properties of polymer nanocomposites while conferring antimicrobial activity.
- **Titanium Dioxide (TiO_2) Nanoparticles:** TiO_2 nanoparticles are widely recognized for their photocatalytic activity, high stability, and biocompatibility. They are used in dental implants, bone prostheses, and coatings for their antibacterial and osseointegrative properties. TiO_2 enhances hardness, wear resistance, and bioactivity of nanocomposites. It also supports cell adhesion and proliferation, especially in bone and soft tissue engineering.
- **Gold and Silver Nanoparticles:** Gold nanoparticles (AuNPs) are chemically stable, non-toxic, and exhibit excellent plasmonic and imaging properties. They are used in photothermal therapy, drug delivery, and diagnostics. Silver nanoparticles (AgNPs), on the other hand, are well known for their broad-spectrum antimicrobial activity. Both Au and Ag nanoparticles can be incorporated into nanocomposites to impart therapeutic or diagnostic capabilities, particularly in wound healing, coatings, and antimicrobial dressings (Feldman 2016).

11.4.3 Layered Silicates

Layered silicates such as Montmorillonite (MMT) and Laponite are clay-based nanomaterials that consist of stacked plate-like layers. They possess a high aspect ratio, large surface area, and cation-exchange capabilities.

- **Montmorillonite (MMT):** Montmorillonite is a layered clay mineral from the smectite group, composed of hydrated aluminum silicates with a high aspect ratio and large surface area. It possesses excellent cation-exchange capacity, making it suitable for drug intercalation. When dispersed in polymers, MMT enhances thermal stability, barrier properties, and mechanical strength. Surface-modified MMT improves biocompatibility and dispersibility. It is used in wound healing, controlled drug delivery, and bone tissue regeneration.
- **Laponite:** Laponite is a synthetic layered silicate composed of magnesium lithium silicate, forming disc-shaped nanoparticles. It forms clear, thixotropic gels in aqueous solutions and is often used to mimic the extracellular matrix in hydrogels. Laponite enhances the rheological and mechanical properties of nanocomposites and supports cell adhesion and differentiation. Due to its high surface area and ion-exchange capabilities, Laponite is used in injectable scaffolds, drug delivery, and tissue regeneration, particularly in bone and cartilage applications (Ray and Bousmina 2005).

11.4.4 *Surface Functionalization and Its Role in Enhancing Compatibility*

Surface functionalization involves the modification of nanofiller surfaces with specific chemical groups, biomolecules, or polymers to improve compatibility with the polymer matrix and biological systems. This process plays a pivotal role in optimizing the dispersion, interfacial adhesion, stability, and biocompatibility of nanocomposites.

Functional groups such as –COOH, –OH, $-NH_2$, or PEG chains can be introduced to the surface of nanofillers through covalent or non-covalent methods. This improves the hydrophilicity of hydrophobic fillers like CNTs or graphene and prevents agglomeration in aqueous environments. For example, carboxylated CNTs exhibit better dispersion in polymeric matrices and reduced cytotoxicity. In the case of inorganic nanoparticles like silica or gold, surface modification with targeting ligands or antibodies enhances cellular uptake and site-specific delivery.

Functionalization also allows for bioconjugation with peptides, proteins, or drugs, making the nanofillers suitable for targeted therapies, tissue engineering, or imaging. Moreover, surface-functionalized nanofillers can influence cell-material interactions, promoting cell attachment, proliferation, and differentiation (Chen et al. 2008).

11.5 Methods of Fabrication and Processing

The choice of fabrication and processing methods plays a critical role in determining the final properties of biocompatible nanocomposites. These methods can significantly influence the nanocomposite's morphology, mechanical characteristics, and biological response, thus tailoring them for various biomedical applications. Methods of fabrication of nanocomposite are shown in Fig. 11.1.

Solvent Casting Solvent casting is a widely used technique to fabricate thin films and membranes, which involves dissolving a polymer and nanofillers in an appropriate solvent. The solution is then poured into a mold or onto a flat substrate, where the solvent is allowed to evaporate, leaving behind a solid film. This method is ideal for creating biodegradable films and drug delivery patches. Solvent casting allows for good dispersion of nanofillers within the polymer matrix, especially when the solution is sonicated or surfactants are added. However, a major challenge is the potential toxicity of residual solvents, which may leach into the surrounding environment, affecting biocompatibility. The structure-property relationship is heavily influenced by solvent evaporation rate, which affects the porosity and uniformity of the resulting film. These films are commonly used in applications such as wound healing, controlled drug delivery systems, and biodegradable coatings (Shimpi 2017).

Electrospinning Electrospinning is a method of producing nanofibers by subjecting a polymer solution or melt to a high-voltage electric field. When the polymer

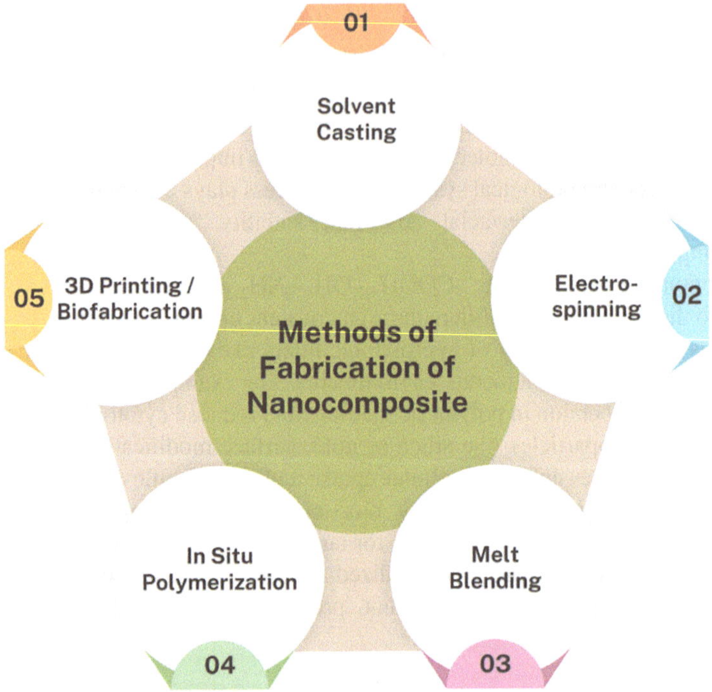

Fig. 11.1 Methods of Fabrication of Nanocomposites

solution is placed in a syringe and exposed to the electric field, it is drawn into thin fibers, which are collected on a grounded collector. This process is especially ideal for the fabrication of nanofibers with a high surface area and porosity similar to the extracellular matrix (ECM). Such nanofibers can then be processed further into scaffolds for tissue engineering, wound healing, or as drug delivery carriers. Electrospinning has benefits like the fiber diameter, orientation, and porosity can be controlled, directly affecting cell attachment, proliferation, and differentiation. Nonetheless, the technique is limited to solvable or processable-molten-state polymers and may be difficult to scale-up for industrial processes. The effect of the method on structure-property-biocompatibility is clear since fiber orientation, porosity, and surface area influence cell infiltration and nutrient diffusion in tissue scaffolds (Treushnikov and Viktorova 2015).

Melt Blending Melt blending involves the mixing of a polymer and nanofillers at high temperatures, where the polymer is brought above its melting point, and the fillers are dispersed in the molten polymer. The mixture is then cooled and solidified. This technique is commonly used for thermoplastic polymers like PLA, PCL, and polyurethane (PU). Melt blending is an environmentally friendly method, as it does not require solvents, making it suitable for large-scale industrial production, such as for implants or orthopedic devices. One of the main advantages of melt blending is its ability to integrate nanofillers into the polymer matrix effectively, enhancing the mechanical strength, thermal stability, and biodegradability of the composite. However, it may not be suitable for heat-sensitive drugs or thermally unstable fillers, as high temperatures may degrade these components. The structure-property relationship is highly influenced by the degree of dispersion of nanofillers, which determines the mechanical and thermophysical properties of the composite.

In Situ Polymerization In situ polymerization involves the dispersion of nanofillers in a monomer solution, followed by polymerization induced by heat, light, or chemical catalysts. This method results in the polymerization of monomers in the presence of nanofillers, creating strong filler-matrix interactions. In situ polymerization is widely used in the fabrication of bone cements, dental resins, and composite scaffolds for tissue engineering. The method offers several advantages, such as the ability to create highly integrated nanocomposites, where the filler and matrix are chemically bonded, improving the composite's overall performance. One key benefit is the ability to functionalize nanofillers during the polymerization process, which can enhance biological activity or cellular compatibility. However, the method requires precise control over polymerization conditions, as residual monomers or initiators can be toxic if not completely polymerized. The structure-property relationship is tightly controlled during this process, with the dispersion and bonding of nanofillers directly influencing mechanical properties and biocompatibility. This method is especially useful when designing materials for high-strength applications that require tight control over structural integrity (Han et al. 2022).

Three-Dimensional Printing and Biofabrication Techniques Three-dimensional printing or additive manufacturing refers to a cutting-edge method whereby material is built layer upon layer to create complex, three-dimensional shapes. Biocompatible materials or cell-laden hydrogels are utilized to print tissue-like constructs in biofabrication. Some of the available 3D printing methods include Fused Deposition Modeling (FDM), Stereolithography (SLA), and Inkjet-based bioprinting. It makes it possible to print customized medical devices, prostheses, and patient-specific implants, and it is transforming tissue engineering. The greatest strength of 3D printing is its capability to fabricate complex structures with high degrees of customization, with the ability to have control over geometry, porosity, and mechanical properties, which can be optimized for particular patient requirements. The use of biocompatible polymers, such as hydrogels, PLA, PCL, and bioceramics, ensures that the printed scaffolds are compatible with human tissues. However, challenges remain with bioprinting, such as the limited availability of printable bioinks and the need to create complex structures that promote tissue growth and integration. The structure–property relationship in 3D printing depends on the layer orientation, material composition, and interlayer bonding, all of which directly influence the scaffold's mechanical performance, biodegradation, and cellular response.

Each of these fabrication techniques provides distinct advantages and limitations depending on the intended biomedical application. The method chosen will directly influence the nanocomposite's morphology, mechanical performance, drug release profiles, and biocompatibility, all of which are critical factors in ensuring the material's efficacy in clinical settings (Patra and Young 2016; Puertas-Bartolomé et al. 2021).

11.5.1 Influence of Fabrication Methods on Structure–Property–Biocompatibility Relationship

The fabrication methods used to create biocompatible nanocomposites significantly influence the structure–property–biocompatibility relationship, as they directly affect the morphology, mechanical properties, and biological interactions of the material. Different fabrication techniques result in varying nanofiller dispersion, polymer matrix structure, and surface characteristics, which in turn determine how the composite will interact with biological systems. For instance, in solvent casting, the rate of solvent evaporation and polymer crystallization can influence the porosity and density of the final product, which directly impacts cell adhesion, infiltration, and growth. Similarly, the fiber diameter and alignment in electrospinning affect the surface area and inter-fiber spacing, crucial factors for tissue scaffolding applications, influencing cell migration and nutrient transport.

In melt blending, the thermal processing conditions can alter the degree of filler dispersion and interface bonding between the polymer and nanofillers, affecting the mechanical strength and degradation rate of the composite. In in situ polymerization, the polymerization kinetics and filler–matrix bonding influence the crosslink density, which governs the stiffness and biodegradability of the composite. Moreover, methods like 3D printing allow for precise control over the material's geometry and porosity, which is particularly important in tissue engineering to mimic natural tissue architecture and enable vascularization and cellular infiltration.

The biocompatibility of the final composite material is closely linked to its structural properties, such as porosity, surface roughness, and mechanical integrity, all of which are shaped by the fabrication method. For example, surface functionalization techniques, such as those used in electrospinning or 3D printing, can be employed to enhance cellular interactions and reduce immunogenic responses. The choice of fabrication method also determines how easily the material can incorporate bioactive molecules, such as drugs or growth factors, which further impact its biological performance (Kumar et al. 2016).

11.6 Structure–Property Relationships

The performance and functionality of biocompatible nanocomposites are profoundly influenced by the intricate relationship between their structure and properties. This structure–property relationship forms the cornerstone in the design and development of materials intended for biomedical applications. Factors such as morphology, nanofiller dispersion, and interfacial interactions govern the mechanical integrity, thermal stability, biodegradability, and ultimately, the biocompatibility of the nanocomposite.

11.6.1 Morphology, Dispersion, and Interfacial Interactions

The morphology of a nanocomposite refers to the internal arrangement and organization of the polymer matrix and embedded nanofillers, including fiber orientation, porosity, and surface characteristics. A homogenous and well-controlled morphology contributes to consistent physical and biological properties, while an irregular structure may lead to variability in performance.

Nanofiller dispersion plays a crucial role in defining the composite's properties. Uniform dispersion ensures that the nanofillers reinforce the matrix effectively, enhancing mechanical strength, thermal resistance, and functional properties such as conductivity or antimicrobial activity. Poor dispersion, on the other hand, leads to agglomeration, stress concentration points, and inconsistent behavior across the material.

The interfacial interactions between nanofillers and the polymer matrix—such as hydrogen bonding, van der Waals forces, or covalent bonding—are vital in determining load transfer efficiency, stress distribution, and matrix cohesion. Strong interfacial bonding enhances mechanical properties and reduces nanofiller leaching, which is essential for maintaining long-term biocompatibility and biostability. Surface functionalization of nanofillers can further improve these interactions, enabling better integration and controlled release capabilities in biomedical applications (Patil and Misra 2018).

11.6.2 Mechanical, Thermal, and Degradation Behavior

The mechanical behavior of nanocomposites—including tensile strength, modulus, toughness, and elasticity—is directly linked to their structural parameters. The inclusion of well-dispersed nanofillers like carbon nanotubes, nanoclays, or hydroxyapatite typically results in significant improvements in mechanical properties. This is particularly critical in load-bearing applications such as bone scaffolds, orthopedic implants, and dental composites.

Thermal stability is another key factor influenced by the structure. Nanofillers can act as barriers to thermal decomposition and improve the composite's resistance to temperature fluctuations, making them suitable for sterilization processes and in vivo environments where thermal stresses are encountered. The thermal conductivity and heat resistance of the nanocomposite are determined by the type, shape, and dispersion of the nanofillers as well as the interfacial thermal resistance between filler and matrix.

The degradation behavior of biocompatible nanocomposites must be finely tuned to meet specific clinical needs. Biodegradable polymers such as PLA, PLGA, and PCL break down in physiological environments over time. The degradation rate is influenced by molecular weight, crystallinity, filler content, and interfacial interaction. For instance, nanofillers that catalyze hydrolysis or alter water uptake can accelerate or decelerate degradation. Predictable and uniform degradation is essential for applications such as drug delivery, tissue engineering, and temporary implants where material resorption is required (Cui et al. 2011).

11.6.3 Impact on Biocompatibility and Performance

The structure–property relationship ultimately dictates the biocompatibility and functional performance of nanocomposites. A material's surface morphology, roughness, porosity, and chemistry have significant impacts on protein adsorption, cell adhesion, proliferation, and immune response. For instance, nanofibrous scaffolds with large surface area and tunable pore size more closely replicate the extracellular matrix (ECM) and enhance tissue regeneration. Mechanical properties such as elastic modulus must match the properties of the host tissue to avoid stress shielding and ensure functional integration. For instance, too rigid a scaffold may impair cellular infiltration, while a structure that degrades too quickly may fail to support tissue regeneration.

Moreover, the rate and mechanism of degradation should not trigger adverse immune reactions. Degradation byproducts must be non-toxic, pH-neutral, and easily metabolizable or excretable by the body. The release of bioactive molecules from the composite—whether therapeutic agents, growth factors, or antimicrobial agents—must be controlled structurally to achieve sustained, targeted activity without compromising biocompatibility (Hezi-Yamit et al. 2009).

11.7 Biomedical Applications of Biocompatible Nanocomposites

Biocompatible nanocomposites have revolutionized biomedical engineering by offering materials that combine mechanical strength, functional versatility, and biological safety. These materials, consisting of polymers reinforced with nanoscale fillers, mimic the natural extracellular environment and can be tailored for specific applications through surface modification, filler selection, and controlled fabrication methods. Their unique structure–property–biocompatibility relationships enable a wide range of medical applications, including tissue regeneration, controlled drug delivery, and bio-integrated devices. The biomedical applications of biocompatible nanocomposites are shown in Fig. 11.2.

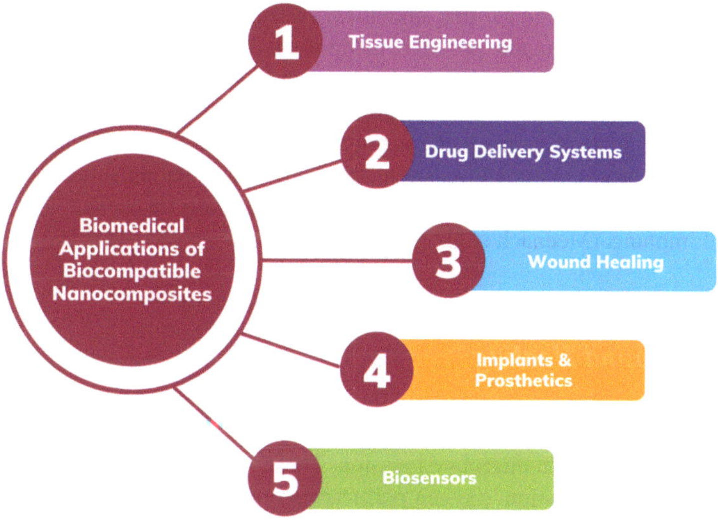

Fig. 11.2 Biomedical applications of biocompatible nanocomposites

11.7.1 Tissue Engineering

One of the most transformative uses of biocompatible nanocomposites is in tissue engineering, where they serve as scaffolds to support cell adhesion, proliferation, and differentiation, ultimately guiding tissue regeneration. These scaffolds can be designed with specific porosity, degradation rates, and mechanical properties to match the target tissue. Bone tissue engineering utilizes nanocomposites made of polymers like PLA or PCL reinforced with hydroxyapatite (HA) or carbon-based nanomaterials, mimicking the natural bone matrix and enhancing osteointegration. Cartilage scaffolds benefit from elastomeric polymers like PU or collagen-based composites, supporting load-bearing properties and chondrocyte activity. Skin regeneration leverages nanofibrous scaffolds from electrospun chitosan, gelatin, or silk fibroin, offering antimicrobial activity and a high surface area for keratinocyte growth. Nerve regeneration uses aligned nanofibers or conductive nanocomposites (e.g., polypyrrole with CNTs) to direct axon growth and restore neuronal connections. These scaffolds not only support physical regeneration but also facilitate angiogenesis and neurogenesis by incorporating bioactive cues, such as growth factors or ECM-derived peptides (Pathak et al. 2023).

11.7.2 Drug Delivery Systems

Biocompatible nanocomposites are increasingly used as platforms for controlled and targeted drug delivery, overcoming limitations such as poor solubility, rapid clearance, and off-target effects of traditional drug formulations. Controlled release is achieved through nanocomposites that degrade slowly in response to physiological triggers (e.g., pH, enzymes), releasing drugs at a sustained rate. Targeted delivery is made possible using surface-modified nanoparticles (e.g., PEGylated nanocomposites) that can home in on specific tissues or disease markers like tumor cells or inflamed tissues. Multifunctional systems can co-deliver drugs and imaging agents (e.g., magnetic nanoparticles for MRI) to monitor treatment in real time. Examples include PLGA nanoparticles encapsulating anticancer drugs or chitosan-based hydrogels for localized antibiotic delivery in wound sites or post-surgical implants (Meenach et al. 2008).

11.7.3 Wound Healing

Nanocomposite-based wound dressings are designed to accelerate healing while protecting wounds from microbial invasion. These materials combine antimicrobial activity, moisture retention, oxygen permeability, and bioactivity to create an optimal healing environment. Nanofillers such as silver nanoparticles, zinc oxide, or graphene oxide impart strong antibacterial properties, preventing infections and biofilm formation. Polymers like alginate, collagen, or gelatin maintain a moist wound interface and are often crosslinked to regulate degradation rates. Advanced nanocomposites can also release anti-inflammatory or growth-promoting agents, like epidermal growth factor (EGF), to enhance tissue regeneration. These dressings are particularly beneficial for chronic wounds, such as diabetic ulcers and burn injuries, where conventional healing is impaired (Kalia et al. 2014).

11.7.4 Implants and Prosthetics

Long-term medical devices and prosthetics require materials that can maintain mechanical integrity, biostability, and immune compatibility over extended periods. Biocompatible nanocomposites fulfill these needs while offering design flexibility and functional enhancements. Orthopedic implants reinforced with nano-hydroxyapatite or titanium dioxide nanoparticles mimic the mineral composition of bone, enhancing osteoconductivity and load-bearing strength. Dental implants and coatings use composites of PMMA or PEEK with antimicrobial fillers to reduce peri-implant infections and improve integration with gingival tissue. Cardiovascular devices, such as vascular grafts and stents, use PU- or PCL-based composites with smooth surface finishes and hemocompatible coatings to minimize thrombosis and inflammation. Prosthetic limb interfaces made from flexible nanocomposites can incorporate biosensors for real-time feedback and adaptive control. These applications highlight the importance of material–tissue interaction, where surface chemistry and mechanical matching play critical roles in device acceptance and functionality (Dheyab et al. 2020).

11.7.5 Biosensors

Biocompatible nanocomposites are increasingly used in the development of biosensors for real-time monitoring of physiological signals, disease markers, and drug levels. These sensors integrate biological recognition elements with nanostructured materials to convert biochemical information into measurable electrical, optical, or mechanical signals. Enzyme-based biosensors use polymer matrices like PEG or PVA embedded with carbon nanotubes or graphene to improve signal transduction and maintain enzyme stability. Wearable biosensors for glucose, lactate, or hydration levels employ flexible nanocomposite films that conform to skin and transmit data wirelessly. Implantable sensors must be both biocompatible and minimally invasive, using nanocomposites that resist biofouling and enable continuous function within bodily fluids. These devices are vital in personalized medicine, allowing dynamic assessment of a patient's condition and real-time feedback for therapeutic adjustments (Feng and Wang 2022).

11.8 Toxicological and Immunological Considerations in Biocompatible Nanocomposites

The integration of nanomaterials into biomedical applications necessitates a comprehensive understanding of their toxicological and immunological impacts. Despite being designed for compatibility with biological systems, biocompatible nanocomposites may still provoke cytotoxic, genotoxic, or inflammatory responses, depending on their composition, size, surface properties, degradation products, and interaction with biological tissues. Ensuring safety involves a robust set of in vitro and in vivo evaluation strategies, along with the study of past failures to inform the design of safer materials. Toxicological and immunological considerations for nanocomposites are explained in Table 11.3.

11.8.1 Cytotoxicity, Genotoxicity, and Inflammatory Responses

Toxicological assessment of nanocomposites primarily begins with evaluating cytotoxicity, which refers to the material's ability to cause damage or death to living cells. Cytotoxic effects may result from the surface characteristics, shape, size, and chemical reactivity of the nanomaterials. For instance, nanoparticles with a high surface area may produce excessive reactive oxygen species (ROS), leading to oxidative stress, mitochondrial damage, and apoptosis. Nanoparticles such as silver (AgNPs), zinc oxide (ZnO), and titanium dioxide (TiO_2) are known to release ions or catalyze ROS generation, compromising cellular function. Genotoxicity is another critical concern where nanoparticles can cause mutations by interacting

Table 11.3 Toxicological and Immunological Considerations for Nanocomposites

Parameter	Explanation	Evaluation Strategy	Importance
Cytotoxicity	Potential of material to damage or kill living cells	MTT assay, LDH assay, live/dead staining	Ensures materials do not harm surrounding tissues or cells
Genotoxicity	Potential to cause mutations or DNA damage	Comet assay, micronucleus test	Critical for long-term applications like implants and drug delivery
Inflammatory response	Activation of immune cells and cytokine release	ELISA, qPCR, histological analysis	Helps predict risk of chronic inflammation or rejection (Akturk 2023)
In vitro evaluation	Laboratory-based tests using cultured cells	Human or animal cell lines, short-term exposure	Initial screening tool before animal testing
In vivo evaluation	Testing in live animal models	Subcutaneous implantation, histopathology	Evaluates systemic effects, biodegradation, immune response in biological systems (Li et al. 2012)

with the DNA of cells either directly or indirectly through oxidative stress. This can lead to chromosomal aberrations, gene mutations, or even cancerous transformations over prolonged exposure. Additionally, inflammatory responses arise when the immune system identifies the nanomaterial or its degradation products as foreign. This results in the activation of immune cells such as macrophages and dendritic cells, release of pro-inflammatory cytokines (like IL-6, TNF-α), and recruitment of leukocytes. Chronic inflammation, in turn, can lead to fibrosis, tissue remodeling, or impaired healing, thereby severely affecting the biocompatibility of the nanocomposite (Akturk 2023).

11.8.2 In Vitro and In Vivo Evaluation Strategies

To understand and predict toxicological and immunological behaviors, a combination of in vitro and in vivo testing strategies is employed. In vitro tests are conducted in controlled laboratory environments using cultured cell lines such as fibroblasts, keratinocytes, or macrophages. These tests provide a fast, cost-effective preliminary screening of potential cytotoxicity and genotoxicity. For example, the MTT assay is used to assess cell viability by measuring mitochondrial activity, whereas LDH release assays measure cell membrane damage. ROS assays help detect oxidative stress, while the comet assay evaluates DNA strand breaks in individual cells, indicating genotoxic effects. For inflammatory studies, ELISA and qPCR are commonly used to quantify pro-inflammatory cytokines. However, these in vitro assays cannot fully replicate the complexity of whole-body biological responses.

Therefore, in vivo testing in animal models (e.g., mice, rats, rabbits) is indispensable for evaluating systemic toxicity, immunogenicity, biodistribution, metabolism, and clearance of nanocomposites. Histological examination of major organs like the liver, lungs, spleen, and kidneys helps detect structural damage, accumulation of nanomaterials, and tissue inflammation. Blood analysis provides data on liver/kidney function and immune markers. Additionally, imaging techniques like MRI or fluorescence imaging, along with radioactive labeling, allow for the real-time tracking of nanocomposites. Regulatory bodies such as ISO (especially ISO 10993 series) and OECD have established standardized testing guidelines for safety evaluation, ensuring reproducibility, ethical compliance, and relevance to human health (Li et al. 2012).

11.8.3 Case Studies on Biocompatibility Failures and Lessons Learned

Historical and recent case studies provide critical insight into the importance of thorough toxicological and immunological evaluations. A prominent example is that of silver nanoparticles used in wound dressings for their antimicrobial properties. Although initially effective, their high concentration caused cytotoxicity to keratinocytes and fibroblasts, delaying the wound healing process. This led to the re-engineering of formulations that control the rate of silver ion release to mitigate toxicity (Hebeish et al. 2014).

Another notable case is multi-walled carbon nanotubes (MWCNTs), which were once considered for various biomedical and structural applications due to their exceptional mechanical properties. However, studies revealed that certain forms of MWCNTs resembled asbestos fibers in their shape and persistence in lung tissue, leading to granuloma formation, inflammation, and fibrosis, especially after inhalation. This raised significant concerns and highlighted the importance of morphology and biodegradability (Arunkumar et al. 2020).

Furthermore, polyethylene glycol (PEG), a widely used polymer for improving solubility and circulation time of drugs, was once considered inert. However, repeated administration led to the formation of anti-PEG antibodies in some patients, resulting in hypersensitivity reactions and reduced therapeutic efficacy of PEGylated drugs. These cases demonstrate that even well-established biocompatible materials can have unforeseen immunological effects under certain conditions. The lessons learned emphasize the importance of understanding the material's long-term fate in the body, dose-dependent effects, degradation behavior, and patient-specific immune responses. As a result, future designs must integrate safety evaluation early in the development process, along with strategies like surface functionalization, particle size optimization, and controlled degradation to enhance both performance and safety (D'souza and Shegokar 2016).

11.9 Regulatory and Ethical Perspectives

The advancement of polymer-based nanocomposites in the biomedical field has opened new avenues for diagnosis, therapy, and regenerative medicine. However, the introduction of these complex materials into clinical practice is governed by stringent regulatory and ethical considerations. Regulatory frameworks play a crucial role in ensuring the safety, efficacy, and quality of nanocomposite-based products. These guidelines help manage the risks associated with the unique physicochemical properties of nanomaterials, which may not be fully addressed by conventional testing protocols. In parallel, ethical perspectives address broader societal implications, including informed consent, equitable access to healthcare innovations, and environmental responsibility. Together, regulatory and ethical frameworks shape the pathway from research and development to successful clinical application and public trust (Karim et al. 2018).

11.9.1 Overview of Global Regulatory Frameworks (FDA, EMA, CDSCO)

Globally, the approval and monitoring of polymer-based nanocomposites fall under the jurisdiction of specialized regulatory bodies. In the United States, the Food and Drug Administration (FDA) evaluates nanomedicine products through its various centers—CDRH for devices, CDER for drugs, and CBER for biologics. The FDA adopts a product-based evaluation approach, requiring comprehensive data on the physicochemical characterization, pharmacokinetics, toxicity, and performance of nanomaterials. In the European Union, the European Medicines Agency (EMA) provides regulatory oversight through the CHMP (Committee for Medicinal Products for Human Use). It ensures compliance with safety, efficacy, and environmental impact guidelines, and frequently publishes reflection papers on nanotechnology-based medicinal products. In India, the Central Drugs Standard Control Organization (CDSCO) oversees the approval of medical products and has begun issuing specific guidance documents for nanoformulations. The CDSCO, along with the Indian Pharmacopoeia Commission, is working toward harmonizing standards with international regulatory norms. These global frameworks are essential in ensuring that polymer-based nanocomposites meet rigorous safety and quality standards before being introduced into the market (Mehta et al. 2024).

11.9.2 Clinical Translation Barriers

Despite their promising potential in laboratory studies, the clinical translation of polymer-based nanocomposites faces several critical challenges. One of the foremost issues is the lack of standardized testing protocols, which can lead to inconsistent evaluation of safety, stability, and therapeutic efficacy. The complexity of nanocomposite systems—due to their nanoscale features, surface modifications, and multifunctional behavior—makes it difficult to predict their biological interactions and long-term effects. Scale-up and manufacturing consistency pose another significant barrier. Materials that perform well in small batches may exhibit altered characteristics when produced at industrial scale, impacting reproducibility and quality. Moreover, fulfilling regulatory requirements demands extensive preclinical and clinical data, including toxicological, immunological, and pharmacokinetic assessments. These evaluations are time-consuming and costly. Additional challenges include navigating GMP compliance, ethical approvals, and market readiness. Overcoming these barriers requires a multidisciplinary approach involving researchers, clinicians, industry stakeholders, and regulatory experts to streamline the transition from bench to bedside (Heidari et al. 2024).

11.9.3 Ethical Issues in the Development of Polymer-Based Nanocomposites

Ethical considerations are fundamental in the responsible development and application of polymer-based nanocomposites. A key concern is informed consent during clinical research, as the novel nature of nanocomposites may carry unknown risks. Participants must be clearly informed about potential adverse effects, mechanisms of action, and alternative treatment options. Another critical issue is equity and accessibility. High costs associated with nanotechnology-based therapies can limit access for underprivileged populations, raising concerns about health disparities. Ethical oversight must ensure that new technologies are not only safe and effective but also socially just. The use of animals in toxicological studies is another sensitive area, with ethical frameworks encouraging the adoption of alternative models wherever possible. The environmental impact of manufacturing and disposal of nanomaterials must also be considered, particularly the release of potentially toxic degradation products. Additionally, transparency in research reporting, absence of conflicts of interest, and responsible marketing are ethical imperatives. Regulatory bodies and institutional review boards must work hand-in-hand to develop comprehensive ethical guidelines that foster innovation while safeguarding human and environmental health (Palit and Hussain 2020).

11.10 Future Perspectives and Challenges

The field of biocompatible polymer-based nanocomposites is evolving rapidly, offering transformative potential in medicine and biomedical engineering. Future innovations aim to create materials that are not only biocompatible and biodegradable but also intelligent and adaptive to physiological cues. Despite the exciting progress, the translation of these materials from bench to bedside continues to face numerous scientific, technical, and regulatory challenges. Understanding and addressing these issues is critical for the sustainable development and clinical success of next-generation nanocomposites (Roy and Singha 2017).

11.10.1 Emerging Trends in Smart and Stimuli-Responsive Biocomposites

One of the major areas of development in biocompatible nanocomposites is in the development of smart or stimuli-responsive materials. These systems are engineered to change their physical or chemical properties as a response to external stimuli like pH, temperature, light, magnetic fields, or biological signals (e.g., enzymes or glucose concentration). For example, pH-sensitive nanocomposites can release drugs preferentially in acidic tumor microenvironments, enhancing targeted therapy. Similarly, temperature-sensitive hydrogels can form scaffolds at body temperature, making them ideal for minimally invasive tissue engineering applications. Magnetic or light-responsive composites are being explored for applications in hyperthermia therapy and remote-controlled drug release. These innovations provide better control over drug delivery and tissue interaction, improving therapeutic outcomes while minimizing side effects. The future will likely see the integration of multiple responsive features into a single system, paving the way for "smart implants" and self-regulating therapeutic platforms (Ramakoti et al. 2023).

11.10.2 Personalized Medicine and Bioresorbable Nanocomposites

Another promising direction involves personalized medicine, where nanocomposites are tailored to an individual's unique biological profile. This approach includes the customization of scaffold architecture, mechanical properties, degradation rate, and drug release kinetics based on a patient's specific disease state, genetic makeup, and physiological condition. Advances in 3D printing and bioprinting are enabling the fabrication of patient-specific implants using bioresorbable nanocomposites that degrade naturally within the body after fulfilling their therapeutic function, eliminating the need for secondary surgeries. Such materials can be loaded with autologous cells, growth factors, and signaling molecules to promote optimal regeneration. In cancer therapy, nanocomposites can be engineered to match the molecular signature of a patient's tumor, enhancing the precision of drug targeting. This convergence of material science, genomics, and biomedical engineering is expected to revolutionize healthcare by enabling treatments that are more effective, safer, and tailored to the individual (Silvestre et al. 2016).

11.10.3 Challenges in Scalability, Reproducibility, and Regulatory Approval

Despite these innovations, several challenges continue to hinder the large-scale deployment of nanocomposite technologies. Scalability remains a major issue, as the complex and multi-step processes used in laboratory synthesis may not be feasible for mass production. Achieving batch-to-batch consistency is particularly difficult due to the sensitivity of nanocomposite properties to small variations in composition or processing conditions. These reproducibility issues can directly affect product quality and clinical performance. Moreover, regulatory approval for nanocomposite-based products is often a lengthy and uncertain process, partly due to the lack of harmonized international guidelines specifically tailored for nanomedicine. Regulatory agencies require comprehensive data on material characterization, toxicology, pharmacokinetics, and long-term effects, which can be expensive and time-consuming to generate. Addressing these challenges will require the development of robust manufacturing protocols, advanced characterization tools, and clear regulatory frameworks. Close collaboration between scientists, industry partners, and regulatory authorities will be essential to ensure that these innovative materials can reach and benefit patients worldwide (Al Harby et al. 2022).

11.11 Conclusion

Biocompatible polymer-based nanocomposites represent a dynamic and interdisciplinary frontier in biomedical science, merging the advantages of polymers with the unique functionalities of nanoscale fillers. Throughout this chapter, we have explored the diverse categories of natural and synthetic polymers, their comparative biocompatibility, the types of nanofillers employed, and the sophisticated methods used for fabrication. These materials offer tailored properties such as enhanced mechanical strength, targeted drug release, improved degradation rates, and responsive behavior, all while maintaining a high degree of compatibility with biological systems. A deep understanding of structure–property relationships, including morphology, dispersion, and interfacial interactions, is crucial for optimizing the performance and safety of these composites. In addition, the selection of appropriate fabrication techniques—ranging from electrospinning and solvent casting to 3D bioprinting—plays a key role in defining the final behavior of the composite in biological environments. Applications in tissue engineering, wound healing, drug delivery, biosensing, and implantable devices underscore their clinical significance and versatility. From a regulatory and ethical standpoint, global frameworks such as those from the FDA, EMA, and CDSCO emphasize the need for stringent testing and careful evaluation. As the field advances, it must navigate challenges related to toxicity, immune responses, scalability, and reproducibility. Simultaneously, ethical concerns around safety, patient consent, and equitable access to emerging technologies must be addressed with transparency and responsibility. Looking ahead, biocompatible nanocomposites are expected to drive next-generation innovations in healthcare. Smart, stimuli-responsive, and bioresorbable systems are poised to transform personalized medicine, enabling more precise, effective, and minimally invasive treatments. However, realizing this potential will require collaborative efforts across research, industry, clinical, and regulatory domains to overcome current limitations and translate laboratory breakthroughs into accessible clinical solutions.

References

Akturk O (2023) Biocompatibility, toxicity, and immunological effects of functionalized carbon nanostructures. In: Handbook of functionalized carbon nanostructures: from synthesis methods to applications. Springer, Cham, pp 1–43

Al Harby NF, El-Batouti M, Elewa MM (2022) Prospects of polymeric nanocomposite membranes for water purification and scalability and their health and environmental impacts: a review. Nano 12(20):3637

Al-Zyoud W, Haddadin D, Hasan SA, Jaradat H, Kanoun O (2023) Biocompatibility testing for implants: a novel tool for selection and characterization. Materials 16(21):6881

Arif U, Haider S, Haider A, Khan N, Alghyamah AA, Jamila N, Khan MI, Almasry WA, Kang IK (2019) Biocompatible polymers and their potential biomedical applications: a review. Curr Pharm Des 25(34):3608–3619

Armentano I, Dottori M, Fortunati E, Mattioli S, Kenny JM (2010) Biodegradable polymer matrix nanocomposites for tissue engineering: a review. Polym Degrad Stab 95(11):2126–2146

Armentano I, Puglia D, Luzi F, Arciola CR, Morena F, Martino S, Torre L (2018) Nanocomposites based on biodegradable polymers. Materials 11(5):795

Arunkumar T, Karthikeyan R, Ram Subramani R, Viswanathan K, Anish M (2020) Synthesis and characterisation of multi-walled carbon nanotubes (MWCNTs). Int J Amb Ener 41(4):452–456

Asghari F, Samiei M, Adibkia K, Akbarzadeh A, Davaran S (2017) Biodegradable and biocompatible polymers for tissue engineering application: a review. Artif Cells Nanomed Biotechnol 45(2):185–192

Bernard M, Jubeli E, Pungente MD, Yagoubi N (2018) Biocompatibility of polymer-based biomaterials and medical devices–regulations, in vitro screening and risk-management. Biomater Sci 6(8):2025–2053

Chen H, Yuan L, Song W, Wu Z, Li D (2008) Biocompatible polymer materials: role of protein–surface interactions. Prog Polym Sci 33(11):1059–1087

Cui Z, Lee BH, Pauken C, Vernon BL (2011) Degradation, cytotoxicity, and biocompatibility of NIPAAm-based thermosensitive, injectable, and bioresorbable polymer hydrogels. J Biomed Mater Res A 98(2):159–166

D'souza AA, Shegokar R (2016) Polyethylene glycol (PEG): a versatile polymer for pharmaceutical applications. Expert Opin Drug Deliv 13(9):1257–1275

Dhayal V, Hashmi SZ, Kumar U, Choudhary BL, Dalela S, Dolia SN, Alvi PA (2021) Optical and electrical properties of biocompatible and novel (CS–GO) polymer nanocomposites. Opt Quant Electron 53:1–3

Dheyab MA, Aziz AA, Jameel MS, Noqta OA, Mehrdel B (2020) Synthesis and coating methods of biocompatible iron oxide/gold nanoparticle and nanocomposite for biomedical applications. Chin J Phys 64:305–325

Fauzi F, Musawwa MM, Hidayat H, Kusumaatmaja A, Dwandaru WS (2021) Nanocomposites based on biocompatible polymers and graphene oxide for antibacterial coatings. Polym Polym Compos 29(9_suppl):S1609–S1620

Feldman D (2016) Polymer nanocomposites in medicine. J Macromol Sci A 53(1):55–62

Feng W, Wang Z (2022) Biomedical applications of chitosan-graphene oxide nanocomposites. Iscience 25(1):103629

Fischer H (2003) Polymer nanocomposites: from fundamental research to specific applications. Mater Sci Eng C 23(6–8):763–772

Fu S, Sun Z, Huang P, Li Y, Hu N (2019) Some basic aspects of polymer nanocomposites: a critical review. Nano Mater Sci 1(1):2–30

Han WB, Yang SM, Rajaram K, Hwang SW (2022) Materials and fabrication strategies for biocompatible and biodegradable conductive polymer composites toward bio-integrated electronic systems. Adv Sustain Syst 6(2):2100075

Harun NH et al (2023) Hemocompatibility profiles of LLDPE with TiO2/ZnO nanocomposites for biomedical application according to ISO 10993-4 and ASTM 756-00 (2000) guidelines. BioNanoScience 13(4):2195–2204

Hebeish A, El-Rafie MH, El-Sheikh MA, Seleem AA, El-Naggar ME (2014) Antimicrobial wound dressing and anti-inflammatory efficacy of silver nanoparticles. Int J Biol Macromol 65:509–515

Heidari M, Al Hasani S, Thangavel S, Kumar A (2024) Challenges and perspectives of polymeric nanocomposites for biomedical applications. Nanomanuf Tech Sustain Healthcare Appl 29:317–340

Hezi-Yamit A, Sullivan C, Wong J, David L, Chen M, Cheng P, Shumaker D, Wilcox JN, Udipi K (2009) Impact of polymer hydrophilicity on biocompatibility: implication for DES polymer design. J Biomed Mater Res A Off J Soc Biomater Japan Soc Biomater Aust Soc Biomater Kor Soc Biomater 90(1):133–141

Jordan J, Jacob KI, Tannenbaum R, Sharaf MA, Jasiuk I (2005) Experimental trends in polymer nanocomposites—a review. Mater Sci Eng A 393(1–2):1–1

Jurak M, Wiącek AE, Ładniak A, Przykaza K, Szafran K (2021) What affects the biocompatibility of polymers? Adv Colloid Interf Sci 294:102451

Kalia S, Kango S, Kumar A, Haldorai Y, Kumari B, Kumar R (2014) Magnetic polymer nanocomposites for environmental and biomedical applications. Colloid Polym Sci 292:2025–2052

Karim ME, Munir AB, Muhammad-Sukki F, Hasanuzzaman M (2018) Polymer nanocomposites and related legal issues: an overview. In: New Polymer Nanocomposites for Environmental Remediation. Elsevier, pp 679–698

Khane Y, Albukhaty S, Sulaiman GM, Fennich F, Bensalah B, Hafsi Z, Aouf M, Amar ZH, Aouf D, Al-kuraishy HM, Saadoun H (2024) Fabrication, characterization and application of biocompatible nanocomposites: a review. Eur Polym J 31:113187

Krishnamoorti R, Vaia RA (2007) Polymer nanocomposites. J Polym Sci B Polym Phys 45(24):3252–3256

Kumar A, Nune KC, Murr LE, Misra RD (2016) Biocompatibility and mechanical behaviour of three-dimensional scaffolds for biomedical devices: process–structure–property paradigm. Int Mater Rev 61(1):20–45

Li X, Wang L, Fan Y, Feng Q, Cui FZ (2012) Biocompatibility and toxicity of nanoparticles and nanotubes. J Nanomater 2012(1):548389

Liu H, Jian R, Chen H, Tian X, Sun C, Zhu J, Yang Z, Sun J, Wang C (2019) Application of biodegradable and biocompatible nanocomposites in electronics: current status and future directions. Nanomaterials 9(7):950

Maghimaa M, Sagadevan S, Boojhana E, Fatimah I, Lett JA, Moharana S, Garg S, Al-Anber MA (2024) Enhancing biocompatibility and functionality: carbon nanotube-polymer nanocomposites for improved biomedical applications. J Drug Deliv Sci Technol 9:105958

Meenach SA, Anderson KW, Hilt JZ (2008) Hydrogel nanocomposites: biomedical applications, biocompatibility, and toxicity analysis. In: Safety of nanoparticles: from manufacturing to medical applications. Springer, New York, pp 131–157

Mehta N, Shetty S, Prajapati BG, Shetty S (2024) Regulatory and ethical concerns in the use of nanomaterials. In: Alzheimer's disease and advanced drug delivery strategies. Academic Press, pp 197–212

Palit S, Hussain CM (2020) Carbon-based polymer nanocomposite and environmental perspective. In: Emerging carbon-based nanocomposites for environmental applications. Wiley, pp 121–145

Pathak R, Punetha VD, Bhatt S, Punetha M (2023) Carbon nanotube-based biocompatible polymer nanocomposites as an emerging tool for biomedical applications. Eur Polym J 196:112257

Patil SS, Misra RD (2018) The significance of macromolecular architecture in governing structure-property relationship for biomaterial applications: an overview. Mater Technol 33(5):364–386

Patra S, Young V (2016) A review of 3D printing techniques and the future in biofabrication of bioprinted tissue. Cell Biochem Biophys 74:93–98

Puertas-Bartolomé M, Mora-Boza A, García-Fernández L (2021) Emerging biofabrication techniques: a review on natural polymers for biomedical applications. Polymers 13(8):1209

Ramakoti IS, Panda AK, Gouda N (2023) A brief review on polymer nanocomposites: current trends and prospects. J Polym Eng 43(8):651–679

Ray SS, Bousmina M (2005) Biodegradable polymers and their layered silicate nanocomposites: in greening the 21st century materials world. Prog Mater Sci 50(8):962–1079

Roy S, Singha NR (2017) Polymeric nanocomposite membranes for next generation pervaporation process: strategies, challenges and future prospects. Membranes 7(3):53

Saini RK, Bajpai AK, Jain E (2017) Fundamentals of bionanocomposites. In: Biodegradable and biocompatible polymer composites: processing, properties and applications. Woodhead Publishing, pp 351–377

Schadler LS, Brinson LC, Sawyer WG (2007) Polymer nanocomposites: a small part of the story. JOM 59:53–60

Shimpi NG (ed) (2017) Biodegradable and biocompatible polymer composites: processing, properties and applications. Woodhead Publishing

Silvestre J, Silvestre N, De Brito J (2016) Polymer nanocomposites for structural applications: recent trends and new perspectives. Mech Adv Mater Struct 23(11):1263–1277

Treushnikov VM, Viktorova EA (2015) Principles of manufacturing biocompatible and biostable polymer implants. Современные технологии в медицине 7(3 (eng)):149–164

Winey KI, Vaia RA (2007) Polymer nanocomposites. MRS Bull 32(4):314–322

Zare Y, Shabani I (2016) Polymer/metal nanocomposites for biomedical applications. Mater Sci Eng C 60:195–203

Chapter 12
Applications of Polymer Nanocomposites in Wastewater Treatment

Dipak S. Thorat, Yogesh V. Ushir, Anil Kumar Singh, Sudarshan Singh, and Sachin N. Kothawade

Abstract Wastewater serves as a significant issue today, and wastewater treatment technology continuously advances to address the growing need for potable water. The lack of pure water around the world poses major threats to the survival of all living creatures. Advanced water/wastewater treatment techniques such as filtration and ion exchange separation are required to remove biodegradable or non-biodegradable harmful and hazardous contaminants from water. Polymer nanocomposites (PNCs) can help overcome significant issues in cleaning wastewater. Thanks to their exceptional features, such as large surface and increased reactivity, in addition to this customisable functionality, PNCs are developing as efficient and adaptable materials for eliminating a wide variety of contaminants. This chapter investigates the many forms of PNCs, inorganic nanoparticle-based, carbon-based and biopolymer-based composites and their distinct roles in wastewater purification. Adsorption, catalytic degradation, membrane filtration and antimicrobial activity are all thoroughly examined to demonstrate how these materials successfully target contaminants such as heavy metals, industrial colours and biological pollutants. The chapter also discusses advanced PNC applications, such as oil/water separation and desalination. Advantages, challenges and environmental concerns are also discussed, providing a holistic view of the scalability, economic aspects and sustainability of PNCs in wastewater treatment. Finally, new research directions are suggested to aid in the development of more environmentally acceptable and commercially viable polymer nanocomposite solutions for environmental applications.

D. S. Thorat (✉) · Y. V. Ushir
SMBT College of Pharmacy, Nashik, Maharashtra, India

A. K. Singh
United Institute of Pharmacy, Naini Prayagraj, Uttar Pradesh, India

S. Singh
Office of Research Administrations, Chiang Mai University, Chiang Mai, Thailand

Faculty of Pharmacy, Chiang Mai University, Chiang Mai, Thailand

S. N. Kothawade
Department of Pharmaceutics, SCSSS's Sitabai Thite College of Pharmacy, Shirur, Pune, Maharashtra, India

S. N. Kothawade et al. (eds.), *Polymer Nanocomposites*,
https://doi.org/10.1007/978-3-032-00365-2_12

Keywords Polymer nanocomposites · Wastewater treatment · Adsorption catalysis · Heavy metal removal · Membrane filtration

12.1 Introduction

Water is a vital element for every life on Earth. Water occupies 70% of our planet, leading to the thought that it will be plentiful. Still, clean water is essential for consumption, drinking, personal hygiene and farming, and this water is too less, 3% of the water on the Earth is fresh water, with 2/3 in the form of frozen ice and just 0.06% of the natural freshwater (Musie and Gonfa 2023). According to the latest WHO reports, 14 M people die every year just because of inadequate drinking water, sanitation and hygiene (Sanitation n.d.). Contaminated water has emerged as an international environmental problem in the past few decades. The principal cause is that growing populations, industrialisation and global warming impact water for drinking resources, including streams and lakes (Divya and Oh 2022). Wastewater streams from factories are characterised by an accumulation of contaminants that include (lead, mercury, cadmium, copper, chromium, nickel and arsenic), a high biological material content, artificial colours and chemical compounds, suspended components and pathogenic microbes that are unsafe for humans, animals and aquatic life (Kumar et al. 2024). The biosphere is under threat from pollutants (Khan et al. 2022). Boundaries of present treatment technologies and argues for creative thinking to boost the elimination efficacy, high running costs and the creation of secondary contaminants (Matesun et al. 2024). Effective sewage handling is important for saving reservoirs of clean water while decreasing harmful emissions (Ulusoy et al. 2024). Therefore, there's an urgent requirement for the improvement of treatment methods. Additionally, working together is required to create an eco-friendly technology that could stop the additional release of many different contaminants into water bodies and solve all of the barriers of sewage treatment facilities. For this, polymer nanocomposites are the best option for wastewater treatment.

The word polymer is derived from the Greek words 'poly' means 'many' and 'mer' means 'parts' (Tajeddin and Arabkhedri 2020). Polymers have various applications due to their important properties like light mass, toughness and ease of use. Polymers have lower thermal, machine-driven and electric qualities than metals and ceramics. Modified polymers, including homopolymers, mixed polymers and copolymers, preserve these properties. Reinforcing fibres, platelets, nanoparticles and whiskers can accomplish the essential properties (Khan et al. 2023a). Nanotechnology is a promising technology that converts and manipulates organic and inorganic materials on an atomic and molecular size to create unique biological, chemical and physical features. Reducing the size to 1–100 nm may significantly change the physicochemical features of a material, even if the base materials remain the same. Novel approaches to wastewater treatment have received substantial research in recent years. Many projects aim to improve the efficiency and limitations of wastewater treatment materials. Scientists from diverse fields, including environmental science, physical science, chemistry, the environment, biological sciences, materials and the field of engineering, are collaborating to create innovative materials and methods for separating and transforming toxins in wastewater. Various materials have been developed for water purification, including membrane filters, adsorbents, ion exchangers and catalytic surfaces. Polymer is the best way for effective wastewater treatment. They can be customised to ensure the needs of various sewage treatment systems because of their distinct functional properties. Polymer nanocomposites are high-performance constituents that have the benefits of polymer and nanoparticles when they combine. Developing usable materials demands a thorough consideration of their fundamental features and science (Mohammad et al. 2022; Khodakarami and Bagheri 2021; Sen n.d.). From this polymer, one of the best polymers is nanocomposites.

Nanocomposites having single phases possess nanoscale physical structures like nanoparticles and nanotubes (Sen n.d.). Nanocomposites are novel composites that integrate invisible nanoparticles scattered in a biological matrix of polymers to increase efficiency. This material has multiple phases that must have a single component (fillers) at their nanoscale. Nanocomposites offer an innovative and alternative solution for substituting commonly filled polymers. These nanocomposites with fillers that are nanoscale have distinct features and characteristics that distinguish them from typical polymer materials (Khan et al. 2023a). There are presently two kinds of polymers: natural and synthetic. They are significant substances for industrial and scientific uses. They are frequently used in wrapping, electricity, safe delivery and also in military equipment (Darwish et al. 2022). Several nano-based methods have been used to manage and treat wastewater for centuries. Nano-sized particles provide significant potential for sewage treatment outstanding to their unique surface area and ability to remove microbiological, inorganic and organic pollutants. Nanomaterials suitable for water and wastewater treatment include carbon-based, polymeric and metal-containing nanoparticles (Darwish et al. 2022; Yaqoob et al. 2020). Researchers have employed several materials like adsorbents to remove harmful contaminants from sewage and water, like as adsorption, electrolysis, reverse osmosis, traditional coagulation and chemical precipitation. Many of these methods are costly and cannot be implemented in low-income countries (Ayalew 2022).

This chapter investigates the many forms of PNCs, inorganic nanoparticle-based, carbon-based and biopolymer-based composites and their distinct roles in wastewater purification. Adsorption, catalytic degradation, membrane filtration and antimicrobial activity are all thoroughly examined to demonstrate how these materials successfully target contaminants like metals, industrial colours and biological pollutants (Mohammad et al. 2022; Darwish et al. 2022; Al-Hazmi et al. 2024). The chapter also discusses advanced PNC applications, such as oil/water separation and desalination. Advantages, challenges and environmental concerns are also discussed, providing a holistic view of the scalability, economic aspects and sustainability of PNCs in wastewater treatment. Finally, new research directions are suggested to aid in the development of more environmentally acceptable and commercially viable polymer nanocomposite solutions for environmental applications.

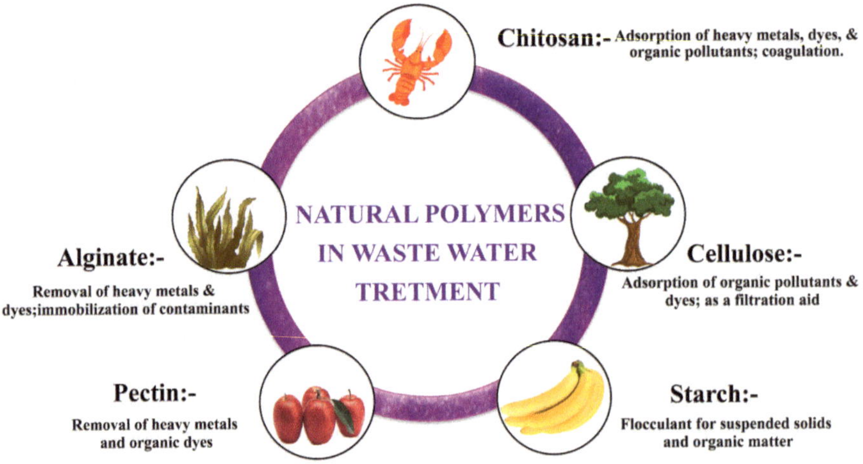

Fig. 12.1 Sources of natural polymers for wastewater treatment

12.2 Sources of Naturally Derived Polymers

Naturally derived polymers are obtained from nature and also extracted from natural sources like trees and shrubs. They include proteins, wool, cellulose, starch, DNA, etc., and rubber, hair, silk, keratin, etc. are obtained from polypeptides (Saleh 2021; Steinbüchel 2003; Feroz et al. 2020). The Polymers made from nature could vary in terms of their source. Among the most plentiful naturally occurring polysaccharides is Cellulose. The walls of plant cells are the most popular places that contain cellulose (Seddiqi et al. 2021; Kaur et al. 2024; Khan et al. 2023b). Collagen serves as a naturally found matrix polymer that is very consistent throughout species. It is normally found as aggregate cells in human skin (Khan et al. 2023b; Cheema et al. 2011). Latex, starch and pectin are all other natural polymers found in different parts of plants (Saleh 2021). Furthermore, chitin is found in the polymer waste of the fishing industry (Santos et al. 2020; Thakur et al. 2022). Polymers of gelatine have multipurpose components derived from the restricted hydrolysis/heat-denatured state of bones, skin, along with connective tissue components of the protein collagen (Rather et al. 2022). Whereas brown seaweed contains alginate, and plant cell walls contain xyloglucan (Stratilová et al. 2020; Awanthi et al. 2023) (Fig. 12.1).

12.3 Classification of Nanocomposites

There are two types of nanocomposites: the first one is the polymer, and the second one is non-polymer. They are further divided into two first nanophase particles and second nanostructured components. Nanoparticles are characterised as huge elements composed of nanoparticles. PNCs are further classified into many categories, which are listed below (Khan et al. 2022).

12.3.1 Polymeric Nanocomposite

Polymer nanocomposites are hardened polymers by adding nanoparticles as nanofillers. These nanocomposites contain a matrix of polymer, embedded with nanomaterials such as nanotubes, nanofibres and nanoparticles. Polymer nanocomposites, which serve as the foundation of nanotechnology, caught scientists' interest and found numerous applications. The nanoscale fillers increase the polymer's physical, mechanical, thermal and chemical properties. The most important factor to address in nanocomposites made of polymer is the spreading of particles with nanofillers throughout the entire matrix of polymers. The consistent spreading of nanofillers improves the characteristics of the polymer nanocomposites. Nanomaterials may be surface altered and functionalised. Surface alteration and functionalisation of nanoparticles enhance the surface interaction or compatibility amongst the matrix as well as the filler material, ensuing in higher dispersion also makes nanocomposites stronger for improved usages. Researchers disclosed the first nanocomposite material incorporating carbon nanotubes, such as a nanofiller. Further research in the applications of nanocomposites is being developed each decade. In the past several decades, carbon nanotubes have been selected as the best nanofillers for the manufacturing of potent nanocomposites just because of their superior physicochemical property, along with distinctive physical characteristics. The physical characteristics of nanocomposites can be improved by adding materials under the following conditions (Khan et al. 2022; Tofighy and Mohammadi 2020):

- The matrix must interact strongly with the nanoparticles.
- The matrix atoms must be well distributed. A greater aspect ratio would be length to height >300, with 5% of superior materials added.
- Used in water filtration, drug delivery, packaging and advanced coatings.

12.3.2 Metal Nanocomposite

Metal nanocomposites combine alloy or metal nanoparticles (such as gold, silver or titanium) into a matrix and the product shape is also referred to as a metal matrix nanocomposite. They have both properties of ceramics as well as metals. These metal nanocomposites are formed to make use of their specific electrical, catalytic and optical capabilities. The above compounds have good electrical conductivity, antibacterial activity and catalytic efficiency. These nanocomposites are composed of metals. Like other nanocomposites, nanometal nanocomposites are made by either grafting an effective monomer onto the outermost layer of the nanometal or by integrating the nanometal into the matrix of a polymer blend. The metallic filler and the biological matrix of polymers should have a high interfacial contact and a homogeneous distribution of the nanometal absent any aggregation.

- Metallic nanoparticles offer a larger surface area.
- Highly effective in photocatalysis and pollutant degradation.
- Photocatalysis, sensors, antibacterial agents for wastewater treatment, furthermore, aerospace and automotive industries.

They can be characterised (Khan et al. 2022; Rasool et al. 2023; Sen 2020) by:

- Excellent plasticity.
- Low MP.
- Improved hardness and strength.
- Enhanced magnetic features.

12.3.3 Polymer/Ceramic Nanocomposite

Ceramic-based nanocomposites are described as ceramic composites having several solid phases, in which no less than one of the phases has sizes in the nanoscale range (<50–100 nm). In these sorts of composites, both phases have integrated mechanical capabilities. These nanocomposites mix the beneficial properties of polymers (elasticity and lightweight). Ceramics are fragile materials vulnerable to breakage. Ceramics are difficult, inert and durable at greater temperature ranges. However, the low hardness of ceramic materials is still challenging in a wide range of sectors. Ceramic nanoparticles, including silica or alumina, are disseminated inside the polymer matrix. Significant effort has been invested in imparting hardness to the ceramic materials. To avoid ceramics against fracture, a ductile phase is included, which increases strength and toughness against fracture. Properties of the nanocomposite relate mostly to the ratio of surface area/volume, with the relation of distinct phases of matrix and reinforcements. To solve this problem, ceramic composites are becoming renowned because of their increased durability and their diverse uses. For instance, the fracture resistance of whiskers, fibres and various other materials can be improved by this technology. Furthermore, excellent-strength fibres to ceramic materials are beneficial in the production of materials having high thermal stability and mechanical qualities substantially higher than those of normal ceramic materials (Khan et al. 2022; Sen 2020; Rathod et al. 2017).

- Improved mechanical toughness and resistance towards heat and chemicals, and increased ductility.
- High ability to adsorb contaminants.
- Used for water filtration through membrane technology and adsorbent materials.

12.3.4 Polymer/Layer Silicate Nanocomposite

Over 50 years, layered silicate has been integrated into polymer. Indeed, one of the earliest organised investigations into the interaction of macromolecule and mineral clay was Bower's observation of DNA adsorption by montmorillonite. Although the absence of XRD data, the result showed that the macromolecule was absorbed by the silicate's lamellar framework. Organoclays were developed to improve latex-based elastomers by utilising various organic onium bases. These technologies have gained significance in academia and industry. There are two types of layer silicate, first Hectorite and second Montmorillonite. The structure of the layered silicate and matrix all have an impact on the production process. They are utilised in the production of PNC with extremely slim layers, often interconnected by counterions (Khan et al. 2022; Pavlidou and Papaspyrides 2008).

12.4 Applications of Polymer-Based Nanocomposites in Wastewater Treatment

Water treatment provides exclusive, effective and unique ways to remove a broad range of contaminants. Nowadays, PNCs have gained consideration in the area of water cleaning. These materials take advantage of polymers' inherent adaptability and chemical modification, as well as nanomaterials' increased features like large surface areas, reactivity and specificity. Nanocomposites are commonly used for water purification due to their distinctive features. Various technologies for nanocomposites involve oxidation, membrane separations, adsorption, microbial treatment, ion exchange, coagulation and flocculation and complex procedures for oxidation. The availability of freshwater is showing an ongoing decrease in quality as well as quantity. Water is necessary for plant and human life and serves as an important resource in many manufacturing processes. Due to the rapid industrial revolution, the spillage of wastewater has significantly increased, resulting in various contaminants getting released within the ecosystem, and so harming both ecological systems and the health of humans. In this case, nanocomposites can be used as an efficient and economical material for sewage cleaning. Nanocomposites consisting of polymers combined with nanoparticles show superior characteristics, including increased resistance to contamination, better heat stability, better permeability of the membrane, beneficial mechanical features and better adsorption capacity. Nanocomposites may be used in the treatment of water to reach many objectives, such as the elimination of pigments, metals and bacteria from wastewater (Khan et al. 2022; Darwish et al. 2022).

12.4.1 Metal Ion Elimination

Contamination with heavy metals in waterways is a major issue that affects all life and nature. They are entering water bodies from the company and extracting metal-contaminated soil, etc. They are commonly found in the water environment as contaminants include arsenic (V), cadmium (II), copper (II), etc. These different hazardous metals can be very harmful if found in water. These metals can enter the human body via the alimentary canal. Absorption of such metals can result in major health problems such as neurological abnormalities, nausea, lack of desire to eat, muscle weakness, diabetic complications and cancer (Khan et al. 2022). Due to a worldwide lack of fresh water, researchers explored and suggested multiple options for the treatment of wastewater for human use. This is because of the large surface hydroxyl bonds paired at the surface layer of graphene nanocellulose oxide PNcs. Researchers found that applying a thin film of GO to the CNF films to create layered-composite layers significantly improved their hardness. The study evaluated the absorbent elimination of metal from sewage using a poly(MMA-MA)Fe_3O_4 magnetic composite. Sorption capacities were determined to be 90.09 mg/g for Co^{2+}, 111.11 mg/g, 90.91 mg/g for Cr^{3+} for Cd^{2+} and 109.89 mg/g for Zn^{2+}. The researchers improved many operating factors, including contact time and pH (Soetaredjo et al. 2018). The researcher created the composites utilising the direct addition approach via chitosan and PEG as the monomers. They showed that metal absorption from the composite of nanohybrid microparticles highly depended on the proportion of rectorite and chitosan. The ability of adsorption of composites minimises with the rise content of chitosan (Soetaredjo et al. 2018). Throughout the pharmaceutical industry, the results showed that small particles of TIO_2 accelerator could significantly increase the removal of the specified chemical. Ozonation of catalytic, single and photocatalytic all achieved elimination percentages of 88.3%, 51.8% and 99.0%, respectively (Zhao et al. 2018). There is a vital need to develop and manufacture nanocomposites for the elimination of different heavy metals from water (Fig. 12.2) and contamination control in the coming years (Zhao et al. 2018).

12.4.2 Organic Contaminants Elimination

Organic contaminants, such as colours, medicines and pesticides, are common in wastewater and current problem for conventional treatment procedures. Polymer-based nanocomposites provide solutions via adsorption, photocatalysis and enhanced oxidation processes. Hydro-tolerant metals such as nickel, zinc, arsenic, selenium, copper, cadmium, etc., nuclear nuclides, petroleum-based products, pesticide residues and radioactive nuclides are among the contaminants under

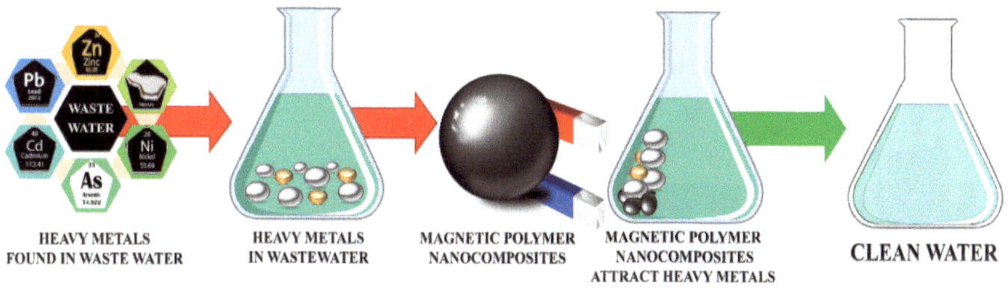

Fig. 12.2 Heavy metals removal with the help of magnetic polymer nanocomposites

consideration. The degree of hydrophobicity of the pollutant also influences its efficiency. If the pollutant strongly likes organic matter, it becomes exceedingly difficult to remove the contaminants from the wastewater (Begum et al. 2022). Photocatalytic or photochemical degradation techniques are becoming increasingly important in the cleaning of water due to their capacity to entirely mineralise organics at low temperatures and pressures. Conducting polymer nanocomposites with a lower band gap and altering chemistry have been used as photocatalysts for the last 15 years. Conduction polymers used in sewage treatment include polyethylene, poly(vinyl pyrrolidone), polypyrrole and polyaniline (Khader et al. 2024; Ahmad et al. 2021).Too much concentration of MB dye in water might lead to cardiac and skin problems. These may cause cancer, and they are difficult to degrade when exposed over an extended period. Many studies focused on removing MB in wastewater via adsorption therapy. Adding chitosan at 50–300% of clay's capacity for cation exchange increases Nc adsorption capacity by 98–282 mg/g for reacting violets 5, an anionic in nature dye (Soetaredjo et al. 2018). Researchers found that a polymer of carbon nanotubes (CNTs@poly(S-MA-DA) was made using a ring reaction. The developed nanocomposite successfully eliminated the MB colour (Fig. 12.3). The working conditions, including temperature and pH, were optimised. The investigation indicated that the removal of MB dye was rapid (Gan et al. 2020). Therefore, the present framework is competent in removing synthetic colours from fabric manufacturing discharges and Pharma company organic pollutants degradation significantly.

12.4.3 *Water and Oil Separation*

Spoilage of oil in water contaminates aquatic bodies, which is a major risk to the ecosystem. Polymer-based nanocomposites provide enhanced separation solutions due to their extremely hydrophobic and oleophilic features, which allow for a particular elimination of oil from water. Oil pollution is a concern in companies such

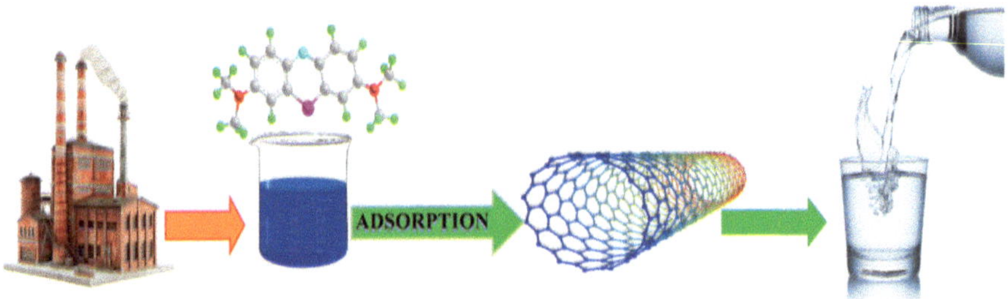

Fig. 12.3 Methylene blue removal with the help of carbon-based nanotubes polymer nanocomposites

as petroleum products, pharmaceuticals and food production (Khan et al. 2022). Grafted-lignin nanotubes of carbon (L-CNTs) are described as a new form of adsorbent in wastewater treatment. Combining the tridimensionality features, containing oxygen, and the L-CNTs, when ready, this nanocomposite additionally has the best eco-friendly and dispersibility of water. It also has an outstanding adsorption capacity for oil droplets and lead. It has an elevated lead ionic coefficient distribution (3.6×10^5 mL/g) (Li et al. 2017). The addition of a slight quantity of graphene oxide in the polybenzimidazole matrix increased oil removal efficiency to 99.9% while retaining excellent permeability (91.3, 3.4 L m^2/h/bar). The addition of graphene oxide enhanced the membrane's overall durability. These nanocomposite membranes performed significantly better in the biofouling tests over 6 months than the clean ones. Polybenzimidazole membrane (Alammar et al. 2020). The resulting nanohybrid efficiently eliminated oil from water sources. The study findings showed 96% elimination of oil as well as a sorption efficiency of 1135 mg/g for engine oil elimination. The pseudo-first-order model kinetics resulted in quick equilibrium for oil removal. The nanohybrid's magnetic, superhydrophobic and superoleophobic features were responsible for its remarkable sorption capabilities (Fig. 12.4) (Sharma et al. 2019). According to the findings, the Bulk-fill restorative materials polyethersulphone nanomembrane of SiO_2 has a 25% greater flux permeate in comparison to nanohybrid polyethersulphone-nanozinc oxide. Considering the overall performance of dual steps to one step revealed that the last flux increased by 200%, inorganic rejection capability increased by 16–18% and foulant deposition on the membrane decreased (Kusworo et al. 2017).

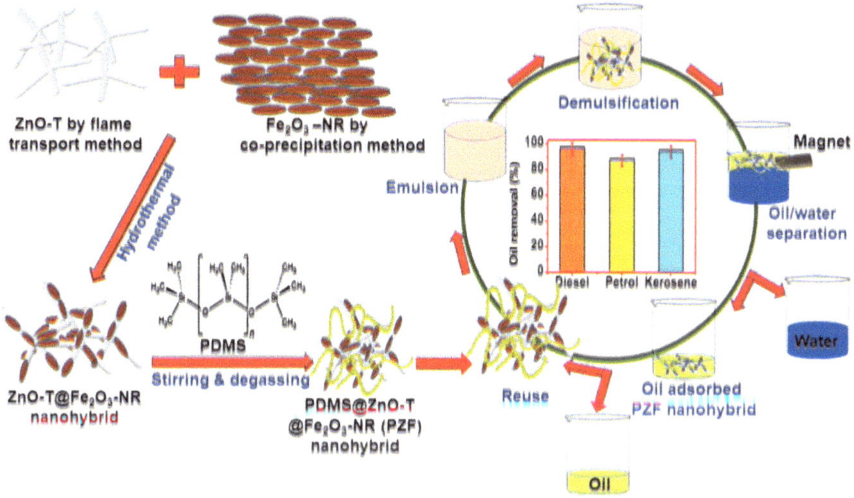

Fig. 12.4 The image of the synthesis of a superhydrophobic polymer-coated tetrapodal magnetic nanocomposite as a powerful sorbent for the purification of oil through wastewater. Reproduce with permission from (Sharma et al. 2019) under Copyright Clearance RightLinks

12.4.4 Elimination of Other Pollutants

Polymer-based nanocomposites effectively eliminate pathogens, nitrates, chemicals (such as NH_4 and NO_3) and microplastics (Al-Hazmi et al. 2024). In addition, with the above-mentioned contaminants, various other contaminants must be eliminated. The contaminants include diseases, bacteria, insecticides and organic substances. They can produce waste products, posing a significant risk. Polymer nanocomposites eliminate germs by eliminating hazardous compounds from them. While this strategy has proven beneficial over time, it does have some health-related risks. It can destroy hydrogen bonds in genetic material (DNA). Metals can be beneficial or dangerous to individuals; thus, it is important to choose the best one for our needs (Table 12.1) (Khan et al. 2022).

Table 12.1 Polymers and nanocomposites for the elimination of contaminants from wastewater

Polymer	Nanocomposite	Contaminants	Method	Adsorbent material	Adsorption capability	Reference
Chitosan	Nanochitosan with oligosaccharides and bentonite	Mycotoxins	Phase separation	Chitosan, mannan oligosaccharides, bentonite	74.77% removal efficiency	Ghobish et al. (2025)
Polymer membranes	Polymer-based nanocomposite membranes	Industrial pollutants	Electrospinning	Embedded nanoparticles in polymer matrices	High removal of heavy metals and organics	Gayatri et al. (2024)
Polymer matrix	Nanocomposites for remediation and desalination	Saline water, hazardous pollutants	Sol-gel method	Polymer matrix with nanomaterials	Effective desalination and pollutant removal	Sharaf and Kloczkowski (2024)
Nanomaterial-polymer	Nanoparticle-embedded polymeric membranes	Heavy metals, dyes and antibiotics	Solution casting	Graphene oxide, carbon nanotubes, metal oxides	Enhanced hydrophilicity and fouling resistance	Nath et al. (2024)
Chitosan biopolymer	Chitosan-based nanocomposites	Heavy metals, dyes and pharmaceuticals	Phase inversion	Chitosan reinforced with nanoparticles.	High adsorption of metal ions and organics	Suyambulingam et al. (2023)
Biopolymers	Biopolymer nanocomposites with nanoparticles	Organic pollutants, antibiotics	Green synthesis	Biopolymer nanocomposites with clay and metal oxides	High adsorption capacity for pharmaceuticals	Annu et al. (2024)
Polysaccharides	Polysaccharide nanocomposites	Pathogens, heavy metals and pharmaceutical residues	Cross-linking	Cellulose, alginate, graphene oxide	90–95% removal efficiency	Al-Hazmi et al. (2024)
Advanced polymers	Mixed polymeric nanocomposite membranes	Pesticides, dyes and antibiotics	Interfacial polymerisation	Polyvinyl alcohol (PVA) with silver nanoparticles	High flux and contaminant removal rates	Hazarika et al. (2023)
Polymeric membranes	Advanced nanocomposite membranes	Heavy metals, industrial waste	Melt blending	Zeolite-loaded polymers	Highly selective adsorption	Ahmad et al. (2023)

Polymer	Nanocomposite	Contaminants	Method	Adsorbent material	Adsorption capability	Reference
Biopolymer reinforced composites	Biopolymer reinforced composites	Heavy metals, organic pollutants	Co-precipitation	Attached are biopolymers with clay and nanoparticles	Significant adsorption of cationic contaminants	Naga Jyothi et al. (2023)
Catalytic polymers	Catalytic polymer nanocomposites	Industrial pollutants, organic contaminants	Solvent evaporation	Metal oxide nanofillers	Enhanced catalytic degradation efficiency	Das et al. (2023)
Composite membranes	Engineered polymeric nanocomposites	Saline water, effluent waste	Dip coating	Composite membranes with carbon and metal fillers	Enhanced permeability and antifouling	Hazarika et al. (2023)
Mixed hydrogels	Hybrid polymeric hydrogels	Pharmaceuticals, dyes and oil contaminants	Chemical cross-linking	Hydrogels with functional nanoparticles	Rapid adsorption capability	Sharma et al. (2024); Cheng et al. (2023)

12.5 Mechanisms of Pollutant Removal Using Polymer Nanocomposites

Polymer nanocomposites have evolved as a comprehensive and successful method for eliminating contaminants from a variety of environments. These substances take advantage of the distinctive properties of polymers reinforced together with nanoparticles to tackle a variety of contamination issues. PNCs are generally used for the purification of water just because of their unique feature. Various methods used in PNCs for water treatment consist of advanced oxidation process, membrane separations, adsorption, ion exchange, biodegradation, microbial treatment, coagulation and flocculation and degradation oxidation (Khan et al. 2022).

12.5.1 Polymer Nanocomposites Adsorbent

PNCs attract researchers, scientists and water treatment specialists (Khan et al. 2023a). Due to their special features like big surface area and customisable chemical features, polymer nanocomposites are excellent adsorbed materials. It efficiently captures dyes, other contaminants and also heavy metals by the exchange of ions, electrostatic pull, or hydrophobic reactions (Khan et al. 2023a). Adsorption is more often used for removing dyes from water (Gupta et al. 2023). By incorporating nanoparticles like graphene or metal oxides, these materials can efficiently trap dyes, other contaminants and also heavy metals through processes such as electrostatic attraction or hydrophobic interactions (Khan et al. 2023a). This adsorption process has the involvement of metal framework, nanocomposites, biomass from industry, microbes, cellulose and lignin (Idris et al. 2023). Constructing magnetic PNCs with remarkable adsorption ability and high selectivity delivers good continuity, process capacity and additional fascinating benefits related to the polymer nanoparticle interface. Furthermore, clay-polymer is superior in the elimination of different contamination, and many other polymers can eliminate heavy metals, dyes, etc. (Khan et al. 2023a). (Fig. 12.5). Adsorption has proven to be one of the

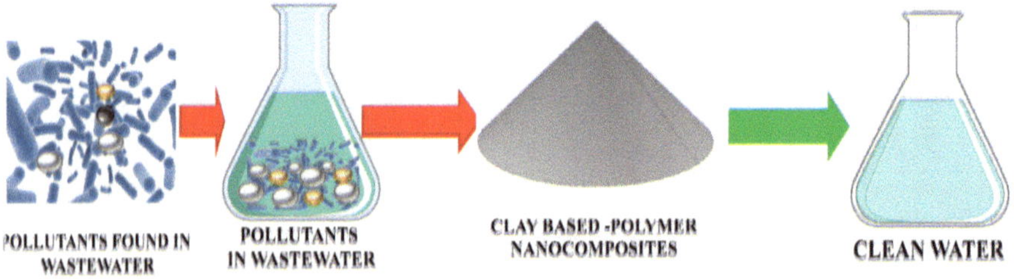

Fig. 12.5 Pollutant removal with the help of clay-based polymer nanocomposites

best methods for eliminating toxic metals from discharged industrial wastewater (Idris et al. 2023). Polymer nanocomposites increased with Fe-Mn binary oxides showed remarkable absorption for arsenic removal, with up to 96% effectiveness. These nanocomposites primarily function by chemisorption, with a capacity approaching 46.2 mg/g for arsenate (As(V)), highlighting their applications in heavy metal treatment (Nikić et al. 2024).

12.5.2 Catalytic Degradation

PNC-based catalytic degradation has emerged as a feasible wastewater treatment technique. Heterogeneous catalytic oxidation is an efficient, cost-effective and eco-friendly way to remove organic pollutants from wastewater. The most interesting properties of polymer nanocomposites are their capacity to catalyse the breakdown of contaminants. Nanoparticles, such as TiO_2 or ZVI, may start photocatalytic or redox reactions when exposed to sunlight or their surroundings. This process converts hazardous chemicals, such as organic colours or medicinal waste products, into safe by-products like H_2O and CO_2 (Singh and Susan 2018). Because of its powerful catalytic properties, biological and chemical inertness, elevated photochemical harmony and low cost, titanium dioxide, which comes in a variety of shapes from clusters as well as colloids to particles and huge single crystals, is one of the most widely used and promising materials for breaking down organic substances (Szczepanik 2017). The interaction between the polymer and the catalytic nanoparticles improves sustainability and further use. TiO_2 serves as one of the best-performing catalysts used for environmental purification for many kinds of organic matter, bacteria, cancer cells, fungi, viruses and algae (Singh and Susan 2018). The photodegradation process was carried out under direct sunlight by introducing the (MB) solution of dye with the nanocomposite. The results of the experiment demonstrate the increased photocatalytic efficiency of the NCs to facilitate the degradation of MB by 93% compared to plain NiO nanocomposites (65%) (Sadhukhan et al. 2020). Because of its inexpensive cost, stable chemical composition and excellent electrical and visual properties, it is the most commonly used photocatalyst. Arsenic-contaminated samples of water in contact with TiO_2 are treated to UV radiation (from solar or lamp sources), followed by iron additions to achieve levels of arsenic under a suitable limit (Matabola et al. 2022). Cobalt molybdate nanofibres incorporated with O_2 vacancies trigger peroxy-mono-sulphate (PMS), which produces reactive oxygen species (ROS) that destroy organic dyes like Rhodamine B. This displays outstanding catalytic activity, with reaction rates much greater than typical catalysts, which makes it perfect for treating organic contaminants (Ren et al. 2024). The recycling capacity of the NC/GO//ZnO/ adsorbate over five subsequent cycles indicates that it has the capability to remove and degrade CF from water (Orasugh and Ray 2022).

12.5.3 Membrane Filtration

In recent years, membrane filters have been a more efficient treatment than traditional procedures since they use fewer chemicals. In addition to the water industry, membrane filtration additionally serves an essential role in the treatment of wastewater, chemical industries, pharmacology, biotechnology and food industries (Dustgeer et al. 2023). Nanocomposite membranes incorporate polymer filtration effectiveness with nanoparticles' improved characteristics. They can filter out ultrafiltration (UF) and microfiltration (MF) scale pollutants such as suspended particulates, germs and pathogens (Furuya et al. 2017). Water purification methods currently make significant utilisation of membranes. They separate solids and liquids according to pore size. Due to this, smaller pore-sized screening membranes require extreme pressure for optimal functioning. Pressure-driven separation (filter/membrane) systems (Fig. 12.6), that include MF to RO, remain the most efficient in terms of energy options for water purification. The mechanism of operation of these methods of separation is simple: contaminants like molecules/metal ions travel from high-concentration regions to low-concentration regions. However, membranes made of MF operated by gravity have a smaller than typical size of pores for water, as it remains the principal energy-effective solution to water cleanup; observing this wastewater filtration method is easy (Orasugh and Ray 2022). TiO_2 nanoparticle-modified polyether sulphone membranes effectively cleanse industrial effluent. These membranes provide improved antifouling qualities, with an 89% efficiency in eliminating chemical oxygen demand (COD) and full suspended substance removal, demonstrating their adaptability in industrial use (Sathya et al. 2024). Furthermore, nanoparticles increase permeability and specificity, enabling effective contamination elimination while maintaining endurance and scalability (Ince and Kaplan Ince 2022).

12.5.4 Antimicrobial Properties

Nanocomposites containing silver, copper, or zinc oxide nanoparticles show antimicrobial capability, specifically targeting microbial and viral pollutants. These nanoparticles emit metal ions that damage bacteria's cell walls and interact with the processes of metabolism, assuring simultaneous elimination of contaminants and the disinfection process (Olawade et al. 2024). They're distinguished by a large area of surface, reactivity and antibacterial characteristics. Uses include dye destruction, hazardous metals, including organic pollutant removal and heterogeneous catalysts in natural processes. Because of its broad surface area, it can initiate numerous interactions against contaminants and remove them more quickly than traditional methods. The presence of harmful microbes in drinking water has a significant impact on human health. As a result, decontamination of water is required, either by deactivation, total elimination, or killing harmful microbes. In this case, PNCs with

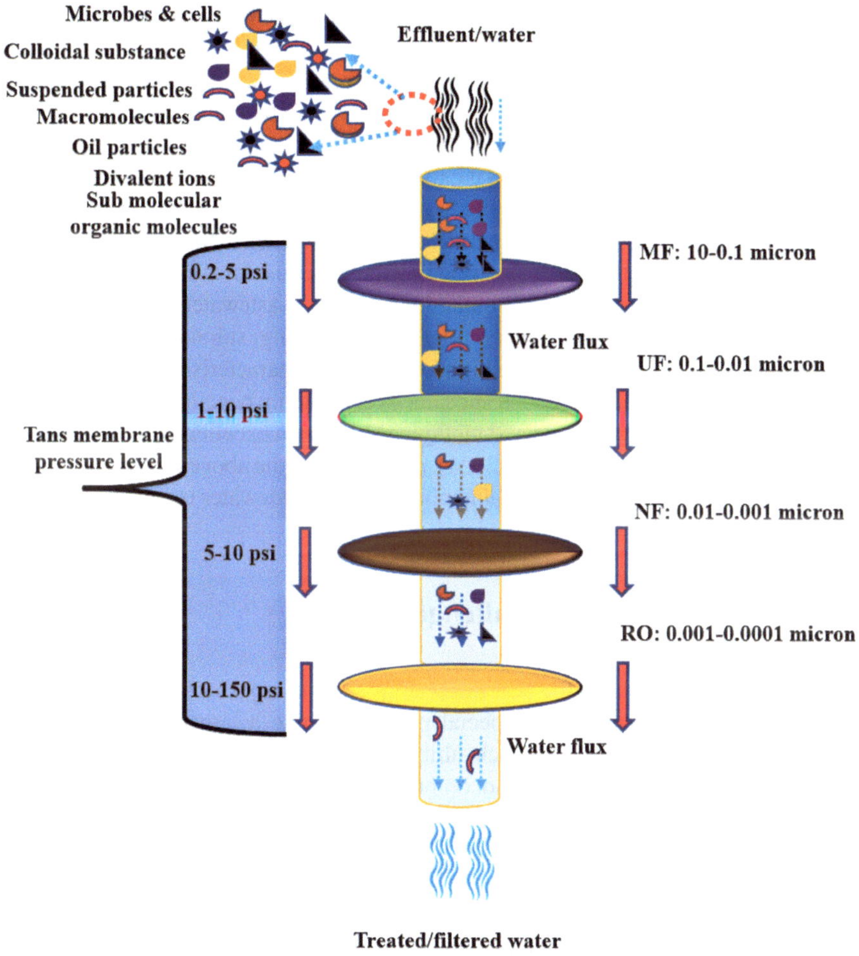

Fig. 12.6 This illustration is of a pressure-driven nanocomposite membrane for water and wastewater treatment. Reproduce with permission from (Sharaf and Kloczkowski 2024) under CCBY 3.0

antimicrobial capabilities could be used to prevent the spread of microorganisms found in drinking water. PNCs have strong antibacterial action due to the synergistic effect of their polymer and nanomaterial components. Several studies have used PNCs as antimicrobial substances for the decontamination of water. Silver cellulose nanocomposite shows enhanced antibacterial activity along with elevated water permeability, making it appropriate for the candidate as a disinfecting agent in the context of the treatment of water (Darwish et al. 2022). The in situ oxidative polymerisation of pyrrole using $AgNO_3$ yielded a nanocomposite including single-walled nanotubes of carbon. It was then utilised to kill bacteria in drinking water, with an E. coli elimination percentage ranging from 87.5 to 95. While the remarkable antibacterial behaviour of PNCs when utilised in water treatment has made them useful against a variety of pathogens, more research is needed in this area. For example, attempts to produce massive PNCs for less money must be improved, as must the PNCs' practical potential for use in real-world circumstances. Several additional problems must also be addressed, such as (i) the manufacturing of PNCs with outstanding antimicrobial capabilities against a larger number of organisms, and (ii) the study of the antimicrobial capacity of PNCs using advanced methods to gain a greater understanding of the workings of their antimicrobial properties (Darwish et al. 2022; Hassan et al. 2023). Biomolecules that cover the green nanoparticles prevent them from aggregating and offer useful sites for action, which greatly helps in the removal of contaminants from wastewater (DeLe6n-Condés et al. 2023). Furthermore, nanocomposites containing sulphonated conjugated microporous polymer (CMP) have high antibacterial characteristics and can absorb colours and pharmaceuticals. Such substances are notably beneficial for eliminating pollutants such as ciprofloxacin and methylene blue with excellent stability throughout numerous cycles (Sathya et al. 2024). According to the above information, polymer nanocomposite shows potential applications in wastewater treatment.

12.6 Advantages and Challenges

Polymer nanocomposite is the best material in various aspects, especially in wastewater treatment, because of its special properties, like a larger surface area and higher adsorption properties, which ultimately improve the removal of contaminants from water. This makes nanocomposites the choice of researchers in recent days for an effective and eco-friendly option. But they have their own advantages and challenges, as given below.

Advantages

(a) Polymer nanocomposites have heavy metal and dye removal potential from wastewater (Gayatri et al. 2024).
(b) This Polymer nanocomposite may be helpful in large-scale heavy metal removal because of its durability (Ravichandran and Santhosh 2024).
(c) Polymer nanocomposites offer a variety of properties (adsorption, filtration) that enhance heavy metal removal and water treatment (Garg et al. 2023).
(d) Polymer nanocomposite antibacterial characteristics prevent contamination and increase membrane shelf life (Gayatri et al. 2024).
(e) Polymer nanocomposites are easy to establish and used in a wide range of applications (Ravichandran and Santhosh 2024).
(f) Polymer nanocomposite uses biopolymers that are environmentally friendly, harmless and biodegradable (Al-Hazmi et al. 2024).
(g) Best option in heavy metal and pharmaceutical wastewater treatment (Garg et al. 2023).
(h) Polymer nanocomposites advance characteristics for advanced wastewater removal (Ravichandran and Santhosh 2024).

Challenges

(a) Polymer nanocomposite cannot be used for a high wastewater removal task (Garg et al. 2023).
(b) Over time, insufficient wastewater removal from membranes (Gayatri et al. 2024).
(c) Polymer nanocomposites have fewer degradation problems in the case of synthetic polymers (Al-Hazmi et al. 2024).
(d) Polymer nanocomposites have complexity while preparing the nanocomposite, and this process requires an expert (Garg et al. 2023).
(e) Polymer nanocomposite loses the potential of wastewater elimination over time (Gayatri et al. 2024).
(f) In some cases of polymer nanocomposite, it is difficult to get consistent same quality (Al-Hazmi et al. 2024).

Anyway, there are advantages and challenges associated with the polymer nanocomposite process that require a solution.

12.7 Economical Aspects

Polymer nanocomposites have various aspects in wastewater treatment, and one of the major aspects is the economic aspects. They have their sort of challenges and advantages. They have higher synthesis costs for nanoparticles, but these nanocomposites are durable and easy to use for long-term uses (Gayatri et al. 2024). This durability reduces the use cost and increases the long-term uses, ultimately decreasing the cost of running (Garg et al. 2023). This reduces the cost of the filtration process and conserves the environment (Ravichandran and Santhosh 2024). Furthermore, nanocomposite membranes are economically feasible in the industry for heavy metal elimination (Al-Hazmi et al. 2024; Garg et al. 2023). The most important advantage of reusing polymer nanocomposites is that it could decrease overall expenses by making them more environmentally friendly and easy to use for several years (Ravichandran and Santhosh 2024). Nanocomposites are economical, cost-effective and offer benefits in polymer nanocomposites in wastewater treatment.

12.8 Future Prospects and Research Directions

The ongoing growth of the worldwide water issue is directly related to world population increase and climate change. The need for clean drinking water increases tremendously (Pandey et al. 2017). Water serves as one of the earth's most important resources, and researchers must focus their efforts on cleaning up polluted water. Today, the globe is more orientated to greener alternatives for better futures (Sirach and Dave 2023). In the last decade, investigations have revealed that nanocomposites of polymers can be used for practical applications in adsorbents and membranes for wastewater treatment (Agboola et al. 2021). The total cost of cleaning water using nanocomposites of polymers would be less than that for traditional treatment systems (Panigrahy et al. 2024). Researchers use a variety of absorbent materials for the treatment of wastewater, and the toxic nature of the adsorbent when utilised in water filtering is a significant concern. Cellulose, along with other carbonaceous adsorbents, such as graphene and carbon nanotubes, has garnered significant interest in the purification of water due to their huge surface area and microbial activity during the treatment of wastewater. Carbon-based compounds may affect human health, and just because of this reason, there is a vital need to discover a polymer nanocomposite for wastewater treatment (Adeola and Nomngongo 2022). In recent days, researchers have continuously worked on the improvement of nanocomposites by combining them with carbon nanotubes and magnetic nanoparticles to improve the overall strength of nanocomposites (Adeola and Nomngongo 2022). Furthermore, polysaccharide hydrogel products used in wastewater treatment have the potential to remove wastewater (Allouss et al. 2022). Future research on the reliability of these technologies and their cost and flexibility is expected to contribute to overcoming current limits and enable the use of nanocomposites to change the purification of wastewater (Singh et al. 2019).

12.9 Case Studies

This study presents RGO-CuS-PPy, a sandwich-structured nanocomposite that can remove 94.9% of nickel ions from wastewater within 2 h and 10 min. The Langmuir isotherm method defines adsorption as having a maximal effectiveness of 166.67 mg/g. The study focuses on reuse as a sustainable method for heavy metal wastewater elimination (Mashkoor et al. 2024). Researchers combined TiO_2 and graphene nanocomposites within polyvinylidene fluoride membranes to increase MB dye removal. Results demonstrated 95% dye elimination from water within a 15-min timeframe, validated by antibiofouling activity and a Langmuir isotherm with an effectiveness of 84.5 mg/g (Sueraya et al. 2024). The study investigated the elimination of acidic medicines from aqueous medium. The best adsorption parameters were 50 mg of MIP, a 10 min extraction duration and an overall pH of 4.6. The absorption data matched the Langmuir isotherm model. Ibuprofen, diclofenac and Naproxen retrieved 38%, 69% and 87% of their sewage sources, respectively (Madikizela and Chimuka 2016). This work studied PES membranes enhanced with ZnO and CuO nanoparticles for Akçay Dam water filtration. The study indicated that eliminating heavy metals using a flow of 427.14 $L/m^2{\cdot}h$ resulted in more than 95 per cent removal efficiency, and enhanced fouling tolerance and stiffness (Güneş-Durak et al. 2024). Researchers synthesised carbon dot nanocomposites for the elimination of dyes from wastewater study showed that the adsorption efficiency was found to be 462.96 mg/g. The performance of the composite was studied in terms of its economic feasibility (Momina et al. 2024). In this study, using TiO_2 and polypyrrole, this PVDF membrane obtained 99% oil rejection. The self-cleaning ability is under visible light-maintained performance over numerous cycles, decreasing fouling and maintenance (Baig and Waheed 2023). Municipal wastewater treatment plants (WWTPs) play a critical role in collecting and redistributing plastic from both households and industries. This study assessed nanoplastics in Australian wastewater treatment plants and found a 91–96% elimination rate. The study emphasised the role of complex filtering technology in combating rising contaminants (Okoffo and Thomas 2024). In this research, cobalt-doped lignite fly ash was used to degrade Acid Red 1 in wastewater exposed to UV light. At a pH of 2.5, the catalyst was successful in removing 99% of the dye in 60 min. According to the kinetics research, the reaction is a pseudo-first-order reaction (Panigrahy et al. 2024). In this study, PES blended with an iron-based organic metal framework was used for ultrafiltration membranes, which were appropriate for industrial water treatment. This hybrid membrane showed outstanding rejection and high permeation flux above 98.5% and 165.68 L/m^2 h (Divya and Oh 2022). In this study, permeable MCM-41 silica polymer nanocomposites and non-porous silica were integrated with a PA thin-film layer using an IP technique with PSF encouraged on the lowermost. Greater surface hydrophilicity and water flux/permeability in comparison with a non-porous structure, improved NaCl (98.1%) and Na_2SO_4 (98.6%) (Yin et al. 2012).

12.10 Conclusion

The exploration of polymer nanocomposites in wastewater treatment has unveiled significant advancements and potential applications that address the pressing challenges of water pollution. This chapter has highlighted several key aspects of polymer nanocomposites, emphasising their efficiency, cost-effectiveness and environmental sustainability in purifying water. The integration of nanoparticles into polymer matrices has been shown to improve the mechanical and chemical properties of the membrane, leading to enhanced filtration capabilities. Studies indicate that polymer nanocomposites exhibit superior adsorption properties, allowing for the effective removal of various contaminants, including heavy metals and organic dyes, from wastewater. The tuneable nature of these materials facilitates their adaptation to specific treatment needs, making them versatile tools in water purification strategies. As traditional wastewater treatment methods often fall short in addressing complex pollutants, polymer nanocomposites present a promising alternative. Their ability to function effectively at lower operational costs and with reduced environmental impact aligns with global efforts to develop sustainable water treatment technologies. The chapter has underscored the importance of these materials in developing eco-friendly solutions that contribute to cleaner water sources. Looking ahead, ongoing research is essential to further optimise the synthesis and application processes of polymer nanocomposites. Innovations in material design and fabrication techniques could lead to even more efficient systems capable of tackling emerging contaminants. Additionally, addressing challenges such as agglomeration and secondary pollutant generation will be crucial for the practical implementation of polymer nanocomposites in real-world scenarios.

References

Adeola AO, Nomngongo PN (2022) Advanced polymeric nanocomposites for water treatment applications: a holistic perspective. Polymers (Basel) 14:2462. https://doi.org/10.3390/polym14122462

Agboola O, Fayomi OSI, Ayodeji A, Ayeni AO, Alagbe EE, Sanni SE, Okoro EE, Moropeng L, Sadiku R, Kupolati KW, Oni BA (2021) A review on polymer nanocomposites and their effective applications in membranes and adsorbents for water treatment and gas separation. Membranes (Basel) 11:139. https://doi.org/10.3390/membranes11020139

Ahmad N, Anae J, Khan MZ, Sabir S, Yang XJ, Thakur VK, Campo P, Coulon F (2021) Visible light-conducting polymer nanocomposites as efficient photocatalysts for the treatment of organic pollutants in wastewater. J Environ Manag 295:113362. https://doi.org/10.1016/j.jenvman.2021.113362

Ahmad A, Kumar R, Jawaid M (2023) Emerging techniques for treatment of toxic metals from wastewater. Elsevier. https://doi.org/10.1016/C2019-0-04578-5

Alammar A, Park S-H, Williams CJ, Derby B, Szekely G (2020) Oil-in-water separation with graphene-based nanocomposite membranes for produced water treatment. J Memb Sci 603:118007. https://doi.org/10.1016/j.memsci.2020.118007

Al-Hazmi HE, Łuczak J, Habibzadeh S, Hasanin MS, Mohammadi A, Esmaeili A, Kim S-J, Khodadadi Yazdi M, Rabiee N, Badawi M, Saeb MR (2024) Polysaccharide nanocomposites in wastewater treatment: a review. Chemosphere 347:140578. https://doi.org/10.1016/j.chemosphere.2023.140578

Allouss D, Makhado E, Zahouily M (2022) Recent Progress in polysaccharide-based hydrogel beads as adsorbent for water pollution remediation. In: Functional polymer nanocomposites for wastewater treatment, Springer series in materials science. Springer, Cham, pp 55–88. https://doi.org/10.1007/978-3-030-94995-2_2

Annu A, Mittal M, Tripathi S, Shin DK (2024) Biopolymeric nanocomposites for wastewater remediation: an overview on recent Progress and challenges. Polymers (Basel) 16:294. https://doi.org/10.3390/polym16020294

Awanthi MGG, Umosa M, Yuguchi Y, Oku H, Kitahara K, Ito M, Tanaka A, Konishi T (2023) Fractionation and characterization of cell wall polysaccharides from the brown alga Cladosiphon okamuranus. Carbohydr Res 523:108722. https://doi.org/10.1016/j.carres.2022.108722

Ayalew AA (2022) A critical review on clay-based nanocomposite particles for application of wastewater treatment. Water Sci Technol 85:3002–3022. https://doi.org/10.2166/wst.2022.150

Baig U, Waheed A (2023) A facile strategy for fabrication of nanocomposite ultrafiltration membrane: oily wastewater treatment and photocatalytic self-cleaning. NPJ Clean Water 6:68. https://doi.org/10.1038/s41545-023-00279-1

Begum M, Sarmah B, Goswami Kandali G, Kalita S, Ojha I, Bhagawati R, Talukdar L (2022) Persistant organic pollutants in soil and its phytoremediation. In: Biodegradation technology of organic and inorganic pollutants. IntechOpen. https://doi.org/10.5772/intechopen.99835

Cheema U, Ananta M, Muder V (2011) Collagen: applications of a natural polymer in regenerative medicine. In: Regenerative medicine and tissue engineering - cells and biomaterials. InTech. https://doi.org/10.5772/24165

Cheng Y, Xia C, Garalleh HAL, Garaleh M, Lan Chi NT, Brindhadevi K (2023) A review on optimistic development of polymeric nanocomposite membrane on environmental remediation. Chemosphere 315:137706. https://doi.org/10.1016/j.chemosphere.2022.137706

Darwish MSA, Mostafa MH, Al-Harbi LM (2022) Polymeric nanocomposites for environmental and industrial applications. Int J Mol Sci 23:1023. https://doi.org/10.3390/ijms23031023

Das TK, Jesionek M, Çelik Y, Poater A (2023) Catalytic polymer nanocomposites for environmental remediation of wastewater. Sci Total Environ 901:165772. https://doi.org/10.1016/j.scitotenv.2023.165772

DeLeón-Condés CA, Martínez-Barrera G, Roa-Morales G, Balderas-Hernández P, Ureña-Núñez F (2023) Green synthesis of metal oxide nanoparticles and gamma rays for water remediation. In: Handbook of green and sustainable nanotechnology. Springer, Cham, pp 203–219. https://doi.org/10.1007/978-3-031-16101-8_54

Divya S, Oh TH (2022) Polymer nanocomposite membrane for wastewater treatment: a critical review. Polymers (Basel) 14:1732. https://doi.org/10.3390/polym14091732

Dustgeer MR, Zakria HS, Jilani A, Ahmad SR, Shakoor MB, Othman MHD, Ansari SP (2023) Use of carbon based photocatalyst for metal removal. In: Emerging techniques for treatment of toxic metals from wastewater. Elsevier, pp 475–494. https://doi.org/10.1016/B978-0-12-822880-7.00011-X

Feroz S, Muhammad N, Ratnayake J, Dias G (2020) Keratin - based materials for biomedical applications. Bioact Mater 5:496–509. https://doi.org/10.1016/j.bioactmat.2020.04.007

Furuya K, Hafuka A, Kuroiwa M, Satoh H, Watanabe Y, Yamamura H (2017) Development of novel polysulfone membranes with embedded zirconium sulfate-surfactant micelle mesostructure for phosphate recovery from water through membrane filtration. Water Res 124:521–526. https://doi.org/10.1016/j.watres.2017.08.005

Gan D, Dou J, Huang Q, Huang H, Chen J, Liu M, Qi H, Yang Z, Zhang X, Wei Y (2020) Carbon nanotubes-based polymer nanocomposites: bio-mimic preparation and methylene blue adsorption. J Environ Chem Eng 8:103525. https://doi.org/10.1016/j.jece.2019.103525

Garg R, Sabouni R, Alaamer A, Alali A, Al-Muqbel D, Alqassem H, Almazrooei K (2023) Recent development in metal-organic framework-based hybrid nanocomposites for pollutants remediation from wastewater: challenges and opportunities. Environ Technol Innov 32:103446. https://doi.org/10.1016/j.eti.2023.103446

Gayatri R, Yuliwati E, Jaafar J, Fizal ANS, Hossain MS, Zulkifli M, Yahaya ANA, Taweepreda W (2024) Polymer-based nanocomposite membranes for industrial wastewater treatment: a review. J Environ Chem Eng 12:113276. https://doi.org/10.1016/j.jece.2024.113276

Ghobish AM, El-Nokrashy AM, Elshafey AE, Tayel AA (2025) Investigating the potential of chitosan-nanocomposites as a bio-based adsorbent for sustainable aflatoxin eradication from fish feed. Egypt J Veterin Sci 56:1719–1730. https://doi.org/10.21608/ejvs.2024.293119.2124

Güneş-Durak S, Acarer-Arat S, Tüfekci M, Pir İ, Üstkaya Z, Öz N, Tüfekci N (2024) Mechanical enhancement and water treatment efficiency of nanocomposite PES membranes: a study on Akçay dam water filtration application. ACS Omega 9:31556–31568. https://doi.org/10.1021/acsomega.4c01410

Gupta A, Singh S, Sharma AL, Deep A (2023) Green synthesis and applications of metal-organic frameworks. In: Handbook of green and sustainable nanotechnology. Springer, Cham, pp 91–110. https://doi.org/10.1007/978-3-031-16101-8_27

Hassan AA, Sayed-ElAhl RMH, El Hamaky AM, Mansour MK, Oraby NH, Barakat MH (2023) Green synthesis of metallic nanoparticles and their biomedical applications. In: Handbook of green and sustainable nanotechnology. Springer, Cham, pp 47–71. https://doi.org/10.1007/978-3-031-16101-8_4

Hazarika B, Ahmaruzzaman M, Santosh MS, Barceló D, Rtimi S (2023) Advances in polymer-based nanocomposite membranes for water remediation: preparation methods, critical issues and mechanisms. J Environ Chem Eng 11:111401. https://doi.org/10.1016/j.jece.2023.111401

Idris MO, Yaqoob AA, Ibrahim MNM, Ahmad A, Alshammari MB (2023) Introduction of adsorption techniques for heavy metals remediation. In: Emerging techniques for treatment of toxic metals from wastewater. Elsevier, pp 1–18. https://doi.org/10.1016/B978-0-12-822880-7.00024-8

Ince M, Kaplan Ince O (2022) Preparation and applications of nanocomposite membranes for water/wastewater treatment. In: Osmotically driven membrane processes. IntechOpen. https://doi.org/10.5772/intechopen.101905

Kaur R, Pathak L, Vyas P (2024) Biobased polymers of plant and microbial origin and their applications - a review. Biotechnol Sustain Mater 1:13. https://doi.org/10.1186/s44316-024-00014-x

Khader EH, Muslim SA, Saady NMC, Ali NS, Salih IK, Mohammed TJ, Albayati TM, Zendehboudi S (2024) Recent advances in photocatalytic advanced oxidation processes for organic compound degradation: a review. Desalin Water Treat 318:100384. https://doi.org/10.1016/j.dwt.2024.100384

Khan A, Malik S, Ali N, Bilal M, Yang Y, Akhter MS, Kumar V, Iqbal HMN (2022) Polymer nanocomposites for wastewater treatment. In: Environmental management technologies. CRC Press, New York, pp 295–306. https://doi.org/10.1201/9781003239956-18

Khan I, Khan I, Saeed K, Ali N, Zada N, Khan A, Ali F, Bilal M, Akhter MS (2023a) Polymer nanocomposites: an overview. In: Smart polymer nanocomposites. Elsevier, pp 167–184. https://doi.org/10.1016/B978-0-323-91611-0.00017-7

Khan A, Malik S, Ali N, Khurshid M, Zubair M, Gao X, Ni L, Bilal M (2023b) Polymers-based nanocomposites for decontamination of water matrices. In: Emerging techniques for treatment of toxic metals from wastewater. Elsevier, pp 427–457. https://doi.org/10.1016/B978-0-12-822880-7.00015-7

Khodakarami M, Bagheri M (2021) Recent advances in synthesis and application of polymer nanocomposites for water and wastewater treatment. J Clean Prod 296:126404. https://doi.org/10.1016/j.jclepro.2021.126404

Kumar R, Neha Y, Ravishankar GA, Srivatsan V (2024) Algae-based bioremediation bioproducts and biofuels for biobusiness. In: Bioremediation and bioeconomy. Elsevier, pp 583–635. https://doi.org/10.1016/B978-0-443-16120-9.00022-4

Kusworo TD, Qudratun, Utomo DP (2017) Performance evaluation of double stage process using nano hybrid PES/SiO2-PES membrane and PES/ZnO-PES membranes for oily waste water treatment to clean water. J Environ Chem Eng 5:6077–6086. https://doi.org/10.1016/j.jece.2017.11.044

Li Z, Chen J, Ge Y (2017) Removal of lead ion and oil droplet from aqueous solution by lignin-grafted carbon nanotubes. Chem Eng J 308:809–817. https://doi.org/10.1016/j.cej.2016.09.126

Madikizela LM, Chimuka L (2016) Synthesis, adsorption and selectivity studies of a polymer imprinted with naproxen, ibuprofen and diclofenac. J Environ Chem Eng 4:4029–4037. https://doi.org/10.1016/j.jece.2016.09.012

Mashkoor F, Shoeb M, Khan MN, Jeong C (2024) Facile synthesis of Polypyrrole-decorated RGO-CuS nanocomposite for efficient nickel removal from wastewater. Polymers (Basel) 16:3138. https://doi.org/10.3390/polym16223138

Matabola P, Sikhwivhilu K, Mapazi O (2022) Electrospun nanofiber-based composites for arsenic removal in water and wastewater. In: Functional polymer nanocomposites for wastewater treatment, Springer series in materials science. Springer, Cham, pp 145–174. https://doi.org/10.1007/978-3-030-94995-2_5

Matesun J, Petrik L, Musvoto E, Ayinde W, Ikumi D (2024) Limitations of wastewater treatment plants in removing trace anthropogenic biomarkers and future directions: a review. Ecotoxicol Environ Saf 281:116610. https://doi.org/10.1016/j.ecoenv.2024.116610

Mohammad ZH, Ahmad F, Ibrahim SA, Zaidi S (2022) Application of nanotechnology in different aspects of the food industry. Discov Food 2:12. https://doi.org/10.1007/s44187-022-00013-9

Momina M, Ahmad K, Khan ZA (2024) Facile synthesis of biomass derived economically viable carbon dot polymer nanocomposite: a perspective towards sustainable removal of dyes from synthetic wastewater. J Water Process Eng 58:104748. https://doi.org/10.1016/j.jwpe.2023.104748

Musie W, Gonfa G (2023) Fresh water resource, scarcity, water salinity challenges and possible remedies: a review. Heliyon 9:e18685. https://doi.org/10.1016/j.heliyon.2023.e18685

Naga Jyothi MSV, Amesh P, Maliyekkal SM (2023) Bio-polymer-reinforced nanocomposites for water and wastewater treatment: applications and future prospects. In: Technological solutions for water sustainability: challenges and prospects. IWA Publishing, pp 97–107. https://doi.org/10.2166/9781789063714_0097
Nath MR, Pal K, Bin Imran A, Chowdhury A-N (2024) Innovations in nanomaterial-embedded polymer membranes for water and wastewater treatment: a comprehensive review. Nano Express 5:022001. https://doi.org/10.1088/2632-959X/ad3e1b
Nikić J, Watson M, Jokić Govedarica J, Vujić M, Pešić J, Rončević S, Agbaba J (2024) Adsorption performance of Fe–Mn polymer nanocomposites for arsenic removal: insights from kinetic and isotherm models. Materials 17:5089. https://doi.org/10.3390/ma17205089
Okoffo ED, Thomas KV (2024) Mass quantification of nanoplastics at wastewater treatment plants by pyrolysis–gas chromatography–mass spectrometry. Water Res 254:121397. https://doi.org/10.1016/j.watres.2024.121397
Olawade DB, Wada OZ, Fapohunda O, Egbewole BI, Ajisafe O, Ige AO (2024) Nanoparticles for microbial control in water: mechanisms, applications, and ecological implications. Front Nanotechnol 6. https://doi.org/10.3389/fnano.2024.1427843
Orasugh JT, Ray SS (2022) Nanocellulose-graphene oxide-based nanocomposite for adsorptive water treatment. In: Functional polymer nanocomposites for wastewater treatment, Springer series in materials science. Springer, Cham, pp 1–53. https://doi.org/10.1007/978-3-030-94995-2_1
Pandey N, Shukla SK, Singh NB (2017) Water purification by polymer nanocomposites: an overview. Nano 3:47–66. https://doi.org/10.1080/20550324.2017.1329983
Panigrahy SK, Nandha A, Chaturvedi M, Mishra PK (2024) Novel nanocomposites with advanced materials and their role in waste water treatment. Next Sustain 4:100042. https://doi.org/10.1016/j.nxsust.2024.100042
Pavlidou S, Papaspyrides CD (2008) A review on polymer–layered silicate nanocomposites. Prog Polym Sci 33:1119–1198. https://doi.org/10.1016/j.progpolymsci.2008.07.008
Rasool A et al (2023) Thermo-responsive functionalized polymeric nanocomposites. In: Smart polymer nanocomposites. Elsevier, pp 219–240. https://doi.org/10.1016/B978-0-323-91611-0.00013-X
Rather JA, Akhter N, Ashraf QS, Mir SA, Makroo HA, Majid D, Barba FJ, Khaneghah AM, Dar BN (2022) A comprehensive review on gelatin: understanding impact of the sources, extraction methods, and modifications on potential packaging applications. Food Packag Shelf Life 34:100945. https://doi.org/10.1016/j.fpsl.2022.100945
Rathod VT, Kumar JS, Jain A (2017) Polymer and ceramic nanocomposites for aerospace applications. Appl Nanosci 7:519–548. https://doi.org/10.1007/s13204-017-0592-9
Ravichandran G, Santhosh N (2024) Synthesis, properties and applications of polymer nanocomposites - a review. J Mines Metals Fuels:611–619. https://doi.org/10.18311/jmmf/2024/44637
Ren S, Song N, Zhang Y, Wang C, Lu X (2024) Controllable fabrication of cobalt molybdate nanofibers with oxygen vacancies induced by calcinable polymer for peroxymonosulfate activation towards water treatment. Appl Surf Sci 674:160965. https://doi.org/10.1016/j.apsusc.2024.160965
Sadhukhan S, Bhattacharyya A, Rana D, Ghosh TK, Orasugh JT, Khatua S, Acharya K, Chattopadhyay D (2020) Synthesis of RGO/NiO nanocomposites adopting a green approach and its photocatalytic and antibacterial properties. Mater Chem Phys 247:122906. https://doi.org/10.1016/j.matchemphys.2020.122906
Saleh TA (2021) Polymer science and polymerization methods toward hybrid materials. In: Polymer hybrid materials and nanocomposites. Elsevier, pp 59–103. https://doi.org/10.1016/B978-0-12-813294-4.00004-2
Sanitation (n.d.). https://www.who.int/news-room/fact-sheets/detail/sanitation (accessed December 6, 2024)
Santos VP et al (2020) Seafood waste as attractive source of chitin and chitosan production and their applications. Int J Mol Sci 21:4290. https://doi.org/10.3390/ijms21124290

Sathya U, Janapriya S, Kumar VS (2024) Performance studies of Fine-tuned Poly Ether Sulphone Composite Membranes using Azadirachta indica for Antifouling Properties in Industrial wastewater treatment. https://doi.org/10.21203/rs.3.rs-5286631/v1

Seddiqi H, Oliaei E, Honarkar H, Jin J, Geonzon LC, Bacabac RG, Klein-Nulend J (2021) Cellulose and its derivatives: towards biomedical applications. Cellulose 28:1893–1931. https://doi.org/10.1007/s10570-020-03674-w

Sen M (2020) Nanocomposite materials. In: Nanotechnology and the environment. IntechOpen. https://doi.org/10.5772/intechopen.93047

Sen M (n.d.), Nanocomposite materials. www.intechopen.com

Sharaf MA, Kloczkowski A (2024) Nanocomposites-based membranes for wastewater remediation and desalination: a mini review. In: Nanocomposites - properties, preparations and applications. IntechOpen. https://doi.org/10.5772/intechopen.115166

Sharma M, Joshi M, Nigam S, Avasthi DK, Adelung R, Srivastava SK, Mishra YK (2019) Efficient oil removal from wastewater based on polymer coated superhydrophobic tetrapodal magnetic nanocomposite adsorbent. Appl Mater Today 17:130–141. https://doi.org/10.1016/j.apmt.2019.07.007

Sharma G, Verma A, García-Peñas A, Kumar A, Dhiman P, Wang T, Amirian J (2024) Polysaccharide-based biopolymeric magnetic hydrogels for remediation of antibiotics from aqueous solution. Int J Biol Macromol 283:137555. https://doi.org/10.1016/j.ijbiomac.2024.137555

Singh NB, Susan ABH (2018) Polymer nanocomposites for water treatments. In: Polymer-based nanocomposites for energy and environmental applications. Elsevier, pp 569–595. https://doi.org/10.1016/B978-0-08-102262-7.00021-0

Singh S, Kumar V, Romero R, Sharma K, Singh J (2019) Applications of nanoparticles in wastewater treatment. In: Nanobiotechnology in bioformulations. Nanotechnology in the life sciences. Springer, Cham, pp 395–418. https://doi.org/10.1007/978-3-030-17061-5_17

Sirach R, Dave PN (2023) Polymer nanocomposites in wastewater treatment. In: Handbook of green and sustainable nanotechnology. Springer, Cham, pp 1323–1353. https://doi.org/10.1007/978-3-031-16101-8_118

Soetaredjo FE, Ismadji S, Foe K, Yi-Hsu J (2018) Recent advances in the application of polymer-based nanocomposites for removal of hazardous substances from water and wastewater. In: New polymer nanocomposites for environmental remediation. Elsevier, pp 499–540. https://doi.org/10.1016/B978-0-12-811033-1.00021-4

Steinbüchel A (2003) Production of rubber-like polymers by microorganisms. Curr Opin Microbiol 6:261–270. https://doi.org/10.1016/S1369-5274(03)00061-4

Stratilová B, Kozmon S, Stratilová E, Hrmova M (2020) Plant xyloglucan xyloglucosyl transferases and the cell wall structure: subtle but significant. Molecules 25:5619. https://doi.org/10.3390/molecules25235619

Sueraya AZ, Rahman MR, Said KABM, Namakka M, Kanakaraju D, Al-Humaidi JY, Al-Baqami SM, Rahman MM, Khandaker MU (2024) Impact of titanium dioxide/graphene in polyvinylidene fluoride nanocomposite membrane to intensify methylene blue dye removal, antifouling performance, and reusability. J Appl Polym Sci 141. https://doi.org/10.1002/app.56257

Suyambulingam I, Gangadhar L, Sana SS, Divakaran D, Siengchin S, Kurup LA, Iyyadurai J, Albert Bernad Noble KE (2023) Chitosan biopolymer and its nanocomposites: emerging material as adsorbent in wastewater treatment. Adv Mater Sci Eng 2023:1–20. https://doi.org/10.1155/2023/9387016

Szczepanik B (2017) Photocatalytic degradation of organic contaminants over clay-TiO 2 nanocomposites: a review. Appl Clay Sci 141:227–239. https://doi.org/10.1016/j.clay.2017.02.029

Tajeddin B, Arabkhedri M (2020) Polymers and food packaging. In: Polymer science and innovative applications. Elsevier, pp 525–543. https://doi.org/10.1016/B978-0-12-816808-0.00016-0

Thakur M, Sharma A, Chandel M, Pathania D (2022) Modern applications and current status of green nanotechnology in environmental industry. In: Green functionalized nanomaterials for environmental applications. Elsevier, pp 259–281. https://doi.org/10.1016/B978-0-12-823137-1.00010-5

Tofighy MA, Mohammadi T (2020) Carbon nanotubes-polymer nanocomposite membranes for pervaporation. In: Polymer nanocomposite membranes for pervaporation. Elsevier, pp 105–133. https://doi.org/10.1016/B978-0-12-816785-4.00005-7

Ulusoy A, Atılgan A, Rolbiecki R, Jagosz B, Rolbiecki S (2024) Innovative approaches for sustainable wastewater resource management. Agriculture 14:2111. https://doi.org/10.3390/agriculture14122111

Yaqoob AA, Parveen T, Umar K, Mohamad Ibrahim MN (2020) Role of nanomaterials in the treatment of wastewater: a review. Water (Basel) 12:495. https://doi.org/10.3390/w12020495

Yin J, Kim E-S, Yang J, Deng B (2012) Fabrication of a novel thin-film nanocomposite (TFN) membrane containing MCM-41 silica nanoparticles (NPs) for water purification. J Memb Sci 423–424:238–246. https://doi.org/10.1016/j.memsci.2012.08.020

Zhao G, Huang X, Tang Z, Huang Q, Niu F, Wang X (2018) Polymer-based nanocomposites for heavy metal ions removal from aqueous solution: a review. Polym Chem 9:3562–3582. https://doi.org/10.1039/C8PY00484F

Chapter 13
Biomedical Applications of Polymer Nanocomposites

Upendra Chandrakant Galgatte, Parth Ramesh Gharat, Thirupataiah Boini, Lakshminarayana Mirso, and Rahul Maurya

Abstract Nanocomposites, characterized by the incorporation of nanoscale components into hybrid materials, are transforming the biomedical field by offering solutions with improved mechanical, thermal, electrical, and biological properties. These materials have diverse applications, including drug delivery, tissue engineering, diagnostics, antimicrobial coatings, biosensors, wound healing, and cancer therapy. Their adjustable characteristics enable tailored functions, thereby enhancing their ability to address complex biomedical challenges. Nanocomposites have a notable application in targeted drug delivery systems, wherein nanoparticles serve as carriers for controlled and site-specific drug release, thus reducing side effects and improving therapeutic efficacy. Nanocomposites exhibit significant potential in tissue engineering by forming scaffolds that mimic the extracellular matrices of natural tissues. This ability facilitates cell proliferation and regeneration, especially in the areas of bone and cartilage repair. Nanocomposites improve imaging contrast in diagnostics, enabling highly sensitive detection methods that support early disease diagnosis. The antimicrobial properties of metal-based nanocomposites, especially silver nanoparticles, make them beneficial for use as coatings on medical devices, aiding in infection prevention and improving patient outcomes. Biosensors utilizing nanocomposites demonstrate significant sensitivity and specificity, making them effective for detecting biomarkers linked to various diseases. Their effective-

U. Chandrakant Galgatte
Department of Pharmaceutics, Modern College of Pharmacy, Nigdi, Pune, Maharashtra, India

P. R. Gharat (✉)
Department of Pharmacognosy, Regional Ayurveda Research Institute, CCRAS, Ministry of AYUSH, Pune, Maharashtra, India

T. Boini · L. Mirso
Department of Chemistry, National Ayurveda Research Institute for Panchakarma CCRAS Ministry of AYUSH, Cheruthuruthi, Kerala, India

R. Maurya
Department of Pharmacy, Central Ayurveda Research CCRAS Ministry of AYUSH, Jhansi, UP, India

S. N. Kothawade et al. (eds.), *Polymer Nanocomposites*,
https://doi.org/10.1007/978-3-032-00365-2_13

ness in wound healing is enhanced by their ability to provide antibacterial protection while promoting tissue regeneration. The use of nanocomposites in cancer therapy enables the integration of various therapeutic strategies, such as photothermal techniques and drug delivery systems, which improves treatment efficacy. Nanocomposites exhibit considerable potential; however, their application in clinical settings faces various challenges. Key challenges to address encompass stability during formulation and long-term storage, biocompatibility, biodegradability, scalability of production, and potential toxicity. The variability in nanocomposite formulations poses challenges for consistent reproducibility and quality control. Obtaining regulatory approval is a significant challenge, as agencies require extensive data on the safety, efficacy, and environmental impacts of nanocomposites. Establishing streamlined and comprehensive regulatory guidelines for nanomaterials is essential to effectively bridge laboratory research and clinical practice.

Keywords Nanocomposites · Drug delivery · Tissue engineering · Diagnostics · Biosensors · Regulatory approval

13.1 Introduction

Composites have two materials: the reinforcing phase, which includes fibers, sheets, or particles, and the matrix phase, in which the reinforcing phase is embedded. Composite materials have greater strength than their basic materials. Wood exemplifies a prevalent natural composite. Snail and oyster shells exemplify natural composite materials. The use of nanocomposites has a long history. Egyptians used straw with clay to create bricks. The Mongols used composites in battle, and during World War II, composite-based materials were employed in military equipment. In contemporary times, a significant quantity of composites is utilized across several sectors (Gholampour and Ozbakkaloglu 2020).

The nanocomposite is a solid compound made up of several layers where at least one of the layers has one, two, or three dimensions with a nanometer size (Omanović-Mikličanin et al. 2019).

Nanocomposites are materials that combine nanoscale particles with a matrix, improving their properties for different uses, especially in biomedicine. These materials have unique features like better mechanical strength, biocompatibility, and biodegradability, which make them suitable for various biomedical uses. Scientists are currently dealing with data loss caused by the overheating of imaging spectrographs in the Hubble Space Telescope. A possible solution is using a carbon nanotube to dissipate excess heat. Nanocomposites are lighter, and it is estimated that steel and aluminum products in various engineering applications may be replaced by nanocomposites in the near future. A nanocomposite (NC) has multiple stages, with at least one, two, or three dimensions in the nanometric range. The size of objects at the nanometer level leads to the creation of optical connectors, which are crucial for developing building material structures. Nanocomposites (NCs) offer new solutions for challenges in various fields, including medicine, pharmaceuticals, food packaging, electronics, and energy production. Nanocomposites consist of materials with atoms arranged in clusters or small grains. A solid multiphase with one of the phases measuring less than 100 nm is termed nanocomposites. Nanocomposites are used in medical and engineering fields. Examples of nanocomposites include nanoparticles, nanoclays, and nanofibers. Nanocomposites are used in medical and engineering fields. This section describes the types of nanocomposites in biomedicine, their properties, and their advantages. Nanocomposites in biomedicine offer a new way to improve healthcare solutions using innovative materials. Composites that combine nanoparticles and polymers show unique properties useful for drug delivery, tissue engineering, and biosensing. This section describes the main types of nanocomposites and their uses in biomedicine. Recently, there has been a greater focus on nanocomposites compared to nanoparticles. This is why nanocomposites exhibit enhanced properties and efficiencies compared to individual nanoparticles. Nanocomposites address certain drawbacks associated with nanoparticles, including elevated band gap energy and the recombination of electron-hole pairs, which can enhance photocatalytic activity, thereby improving reaction kinetics and ultimately boosting system efficiency (Li et al. 2014; Rokesh etal. 2021).

Nanocomposites present an excellent alternative to conventional composite materials due to their distinctive characteristics, and they find extensive applications across numerous industries. Since their discovery in 1991, systems based on carbon nanotubes have captivated the attention of modern research and development. There has been a consistent and ongoing rise in the volume of publications on this topic, including occasional reviews and patents. When looking at traditional composites, the interface area between the matrix and nano-reinforcement is frequently significantly larger, often by an order of magnitude. Clays serve as a type of nanofiller material that has found extensive application in the creation of polymer matrix nanocomposites. Recently, there has been significant interest in polymer/clay nanocomposites within both academic and industrial circles, attributed to their enhanced properties (Nguyen & Baird 2007). The synthesized nanocomposites are examined by several techniques to get data on their morphology, particle size, phase, composition, thermal stability, and optical, magnetic, electrical, and thermal properties. Prevalent techniques include ultraviolet-visible light spectroscopy (UV-Vis), Fourier transform infrared spectroscopy (FTIR), X-ray diffraction (XRD), scanning electron microscopy (SEM), transmission electron microscopy (TEM), and thermal analysis (TA).

Nanocomposites can be categorized based on the types of reinforcement materials and matrix materials utilized in their fabrication. Furthermore, Parameswaranpillai, J classifies nanocomposites into three distinct categories according to the type of matrix material employed. Therefore, nanocomposites are typically categorized into the following three classes.

13.1.1 Types of Nanocomposites in Biomedicine

1. Polymer matrix nanocomposites.
2. Ceramic matrix nanocomposites.
3. Metal matrix nanocomposites.

13.1.1.1 Polymer Matrix Nanocomposites

Polymer nanocomposites are materials consisting of a polymer matrix and nanoadditives serving as reinforcement. Additives may be classified as one-dimensional (nanotubes and fibers), two-dimensional (layered materials such as clay), or three-dimensional (spherical particles). The interaction between the polymer matrix and the nanofiller at the molecular level results in observable attraction effects inside the nanocomposites (Khan et al. 2016). **As** a result, incorporating a small quantity of nanofiller with dimensions under 100 nm into the matrix alters the properties of the composite material. Among the various nanomaterials, polymer nanocomposites (PNCs) stand out as the most recognized and have been extensively studied for numerous applications. In contrast to other materials like metals and ceramics, polymers stand out due to their low manufacturing costs and impressive specific strength, resulting in reduced energy requirements for both production and recycling. Nanocomposites demonstrate exceptional performance relative to their composite counterparts, opening up a range of possibilities for various industrial applications for PNCs (Singh and Agarwal 2016). Biopolymer-based Nanocomposites—This category encompasses materials such as chitosan, alginate, and polylactic acid (PLA), utilized in drug delivery systems and scaffolds for tissue engineering (Shaikh et al. 2024). Nanocomposite Hydrogels—These hydrogels combine nanoparticles with polymer matrices, enhancing drug delivery and wound healing capabilities (Shahid et al. 2024). Polymer Nanocomposites for Implants: Utilized in various implants, these materials demonstrate significant biocompatibility and biodegradability, essential for medical applications (Li et al. 2024).

13.1.1.2 Ceramic Matrix Nanocomposites

Nihara and Gurnani's groundbreaking research showcased the potential of ceramic matrix nanocomposites, especially the combination of aluminum oxide and silicon carbide (SiC). Thorough investigations have demonstrated that incorporating a modest volume fraction (specifically 10%) of appropriately sized silicon carbide particles into the aluminum oxide matrix, followed by hot-pressing the blend, markedly enhances the strength of the matrix. Several investigations have elucidated the process of toughening by showcasing the crack-bridging effect of nanoscale reinforcements. The integration of high-strength nanofibers within ceramic matrices has led to the creation of exceptional nanocomposites, demonstrating enhanced toughness and notable failure characteristics, unlike the abrupt failures typically seen in ceramic materials (Mohamed & Bansal 2024).

Among the most common ceramic matrix nanocomposites are AL2O3/SiO2, SiO2/Ni, Al2O3/TiO2, and Al2O3/SiC. Since their discovery, carbon nanotubes (CNT) have found extensive application in the manufacturing of nanocomposites. Examples of CNT-based ceramic matrix nanocomposites include Al2O3/CNT, MgAl2O4/CNT, and MgO/CNT (Niihara 1999; Gurnani and Mukhopadhyay 2020).

13.1.1.3 Metal Matrix Nanocomposites

Nanoparticle-reinforced materials consist of metal matrix nanocomposites, formed by incorporating nanoparticle reinforcement into a ductile metal or alloy matrix. The composites consist of a metal or alloy matrix infused with nanoparticles, resulting in distinct physical, chemical, and mechanical properties that differ significantly from the matrix material itself. The incorporation of nanoparticles enhances wear resistance, mechanical properties, and damping characteristics. Recent investigations have focused on metal matrix nanocomposites, which, owing to the enhanced properties conferred by the incorporation of nanoparticles, are increasingly being utilized in various structural applications (Yu et al. 2019).

The nanoparticles serve as a hindrance to dislocation movement, consequently enhancing the mechanical properties. The prevalent methods employed in the processing of metal matrix nanocomposites encompass spray pyrolysis, liquid metal infiltration, vapor techniques, rapid solidification, electrodeposition, and various chemical approaches, such as colloidal and sol-gel methods (Mohanty et al. 2020).

13.2 Nano Synergy: Advancing Biomedical Frontiers with Nanocomposite Chemistry

13.2.1 Synthesis of Nanocomposites

Nanocomposites are special materials made by mixing tiny bits with a base material to enhance their strength, electrical performance, and light qualities. Different ways to create nanocomposites have been created to manage their size, shape, and makeup. These methods include sol-gel, coprecipitation, chemical vapor deposition (CVD), and electrospinning. This study looks closely at these synthesis methods, explaining how they work, their benefits and drawbacks, and where they can be used, along with examples of materials and important research studies. Nanocomposites have become important materials in medicine because of their special mechanical, chemical, and biological qualities.

13.2.1.1 Methods of Preparation

The synthesis of nanocomposites has different methods like sol-gel, coprecipitation, chemical vapor deposition (CVD), and electrospinning which have been used to modify these materials for specific uses, such as delivering drugs, engineering tissues, and tracking biological changes.

Sol-Gel

The sol-gel method is a widely used wet-chemical synthesis process that involves the hydrolysis and condensation of metal alkoxides or inorganic salts to make a colloidal suspension (sol). The sol turns into a gel, creating a spongy three-dimensional structure. After cooling and heating, the final structure of the nanocomposite is ready.

The sol-gel process is a flexible method for making high-quality and uniform nanocomposites. This method changes a system from a liquid state, called "sol," to a solid state, known as "gel." Silica-based nanocomposites made using the sol-gel method have great promise for drug delivery systems because they have predictable pore sizes and useful surface features (Smith et al. 2020).

Advantages

- It is a low-temperature synthesis, and the prepared material cannot withstand high heat.
- High purity and homogeneity in composition and the ability to create different tiny structures, like thin layers, small particles, and materials with holes.

Limitations

- It takes a long time to process because it gels and dries slowly.
- High shrinking and cracking while drying.

Examples

- TiO_2 and SiO_2 nanocomposites are used for the photocatalytic degradation of pollutants (Zhou et al. 2020).
- ZrO_2-Al_2O_3 nanocomposites are used in films that prevent wear.
- Silica-based bioactive nanocomposites are used in bone tissue engineering (Chen et al. 2021).

Coprecipitation

- Coprecipitation is an easy and inexpensive method where different metal solutions are combined and turned into solid form at the same time while keeping the pH, temperature, and shaking steady. A precipitating agent (e.g., NaOH, NH_4OH) is added to make solid hydroxides, which are then calcined to obtain the final nanocomposite.

Coprecipitation is a common method for making metal oxide nanocomposites. This method lets you remove several parts from a solution at the same time, which helps you control what's in it. Iron oxide nanoparticles, made using a method called coprecipitation, are widely used in magnetic resonance imaging (MRI) as contrast agents.

Advantages

- The coprecipitation method is a simple and scalable synthesis process.
- This method is rapid production with high yield.
- In this method, good control over nanoparticle size and composition is achieved.

Limitations

- This method requires precise pH and reaction control to avoid impurity formation.
- In this method, nanoparticle agglomeration occurs due to high surface energy.

Examples

- Fe_3O_4-Ag nanocomposites are used for antimicrobial and magnetic hyperthermia applications (Gupta et al. 2022).
- ZnO-CuO nanocomposites are utilized in gas sensors for detecting hazardous gases.

Chemical Vapor Deposition (CVD)

VD is a gas-phase synthesis method in which precursor gases undergo chemical reactions, resulting in the creation of nanocomposite films on a substrate. There are different types of CVD methods, such as plasma-enhanced CVD (PECVD) and atomic layer deposition (ALD). These methods allow for better control over the thin-film hybrid structures.

CVD is a method where volatile ingredients are chemically combined in the vapor phase to form a thin film or coating on a medium. Nanocomposites made with CVD, like diamond-like carbon coatings, are very safe for the body and withstand wear, making them good for orthopedic devices.

Advantages

- CVD method produces high-purity and uniform coatings.
- This method allows deposition on complex geometries.
- The CVD method enables precise control over film thickness and composition.

Limitations

- It is a high-temperature process and is limited its use for organic materials.
- In this method, expensive precursor gases and equipment are used.

Examples and Applications

- SiC/Graphene nanocomposites are employed in high-performance electronics.
- Graphene-TiO2 nanocomposites are used for a solar cell electrode and photocatalysis (Zhang et al. 2021).

- Diamond-like carbon (DLC) nanocomposite coatings are applied in wear-resistant automotive coatings.

Electrospinning

Electrospinning is a fiber production technique that uses a high-voltage electric field to pull polymer or composite solutions into ultrafine fibers. Nanoparticles can be integrated into fibers to create nanocomposites with increased functionality.

Electrospinning is a useful technology for creating nanofibers with large surface area-to-volume ratios. This approach is very beneficial for tissue engineering scaffolds and wound treatments. For example, polycaprolactone (PCL) nanofibers coated with silver nanoparticles generated using electrospinning have good antibacterial characteristics (Kim et al. 2022).

Advantages
- It is a cost-effective and simple fabrication technique.
- It produces high-surface-area nanofibers with controlled porosity.
- Electrospinning technique is versatile in processing various polymeric and hybrid materials.

Limitations
- Limited to polymer-based nanocomposites.
- This method requires careful optimization of polymer solution properties (e.g., conductivity and viscosity).

Examples and Applications
- PVDF/ZnO nanocomposite fibers are applied in piezoelectric energy harvesting devices (Sharma et al. 2022).
- Chitosan-Silver nanofibers are employed in wound dressings with antibacterial properties.
- Polymer/Carbon nanotubes (CNT), nanofibers are used in flexible supercapacitors (Kim et al. 2020).This chapter highlights the versatility of various synthesis methods in tailoring nanocomposites for specific biomedical applications emphasizing their unique properties and potential in advancing healthcare technologies are mentioned in Table 13.1.

The synthesis technique affects nanocomposites' structure, stability, and functionality. Sol-gel and coprecipitation are employed for oxide-based nanocomposites, CVD for thin-film applications, and electrospinning for fibrous ones. Improve these methods for scalability, sustainability, and nanocomposite morphology control in future studies. Advanced hybrid synthesis methods may produce nanomaterials with new characteristics for energy, health, and electronics.

Table 13.1 List of Nanocomposites preparation methods for Biomedical applications

S. No.	Method	Key features	Examples of nanocomposites	Applications	Reference
1	Sol-gel	High purity, controlled porosity	Silica-based nanocomposites	Drug delivery	Smith et al. (2020)
2	Coprecipitation	Precise composition control	Iron oxide nanoparticles	MRI contrast agents	
3	Chemical vapor deposition (CVD)	Biocompatible, wear-resistant coatings	Diamond-like carbon coatings	Orthopedic implants	
4	Electrospinning	High surface area, flexible structures	PCL nanofibers with silver nanoparticles	Tissue engineering, wound dressings	Kim et al. (2022)

13.2.1.2 Chemical Process Involved in the Synthesis of Nanocomposites

Nanocomposites are improved materials made by combining tiny particles with other materials to improve their strength, electrical, light, and heat qualities. Chemical processing is important for making them because it allows for accurate control over their makeup, structure, and shape. This study covers important methods for chemical synthesis, such as sol-gel processing, coprecipitation, hydrothermal/solvothermal synthesis, chemical vapor deposition (CVD), and in situ polymerization. The text includes examples of nanocomposites, how they are used, and links to important study results.

Sol-Gel Processing

Sol-gel processing is a flexible wet-chemical method that uses the reaction of metal compounds or artificial materials in a solution to create a mixture called a colloidal sol. This mixture then turns into a gel. The gel is dried and heated to create a nanocomposite.

Chemical Process

1. Hydrolysis:
   $M(OR)_4 + H_2O \rightarrow (RO)_3M\text{-}OH + ROH$
2. Condensation:
   $M(OH)x \rightarrow M{-}O{-}M + H_2O$

Examples and Applications

- TiO_2-SiO_2 Nanocomposites: Used in self-cleaning coatings and photocatalysis (Zhou et al. 2020).
- ZrO_2-Al_2O_3 Nanocomposites: Applied in high-temperature ceramics.
- Bioactive Glass Nanocomposites: Utilized in bone tissue engineering (Chen et al. 2021).

Advantages and Limitations

- High purity and homogeneity.
- Low-temperature processing.
- Long processing time and shrinkage during drying.

Coprecipitation Method

Coprecipitation is the simultaneous precipitation of various components from a solution while regulating the pH, temperature, and ionic strength. It is often utilized to create oxide and magnetic nanocomposites.

Chemical Process

1. Metal salt dissolution:
   Mn+ + Precipitant → M(OH)n↓ M^{n+} + Precipitant → M(OH)_n ↓Mn++Precipitant → M(OH)n↓
2. Calcination to obtain oxides:
   M(OH)n→MOn+H_2OM(OH)_n → MO_n + H_2OM(OH)n→MOn+H_2

Examples and Applications

- Fe_3O_4-Ag Nanocomposites: Used for magnetic hyperthermia and antimicrobial coatings (Gupta et al. 2022).
- Ni-Co Oxide Nanocomposites: Applied in supercapacitors and batteries.
- ZnO-CuO Nanocomposites: Utilized in gas sensors.

Advantages and Limitations

- Rapid synthesis with high yield.
- Good control over nanoparticle composition.
- Agglomeration of nanoparticles.

Hydrothermal/Solvothermal Synthesis

Hydrothermal and solvothermal synthesis are chemical processes that take place in aqueous or organic solvents at high pressures and temperatures. This approach improves crystallinity and phase purity.

Chemical Process

1. Precursors dissolution in water/solvent.
2. Reaction occurs under autoclave conditions (150–300 °C).
3. Nanoparticle formation via nucleation and growth.

Examples and Applications

- Hydroxyapatite Nanocomposites are utilized in bone regeneration.
- Graphene/ZnO Nanocomposites are used in photocatalysis and energy storage (Wang et al. 2020).
- TiO_2-CdS Nanocomposites are applied in solar cells.

Advantages and Limitations

- High crystallinity and controlled morphology.
- Suitable for complex oxides and hybrid structures.
- High energy consumption and long reaction time.

Chemical Vapor Deposition (CVD)

CVD is a gas-phase chemical process where precursor gases react and deposit nanomaterials onto a substrate. Variants include plasma-enhanced CVD (PECVD) and atomic layer deposition (ALD).

Chemical Process

1. Precursor decomposition:
   $X(g)\rightarrow M(s)+Byproducts(g)X(g) \rightarrow M(s) + Byproducts(g)$
   $X(g)\rightarrow M(s)+Byproducts(g)$
2. Controlled layer deposition on a substrate.

Examples and Applications

- Diamond-like Carbon (DLC) Coatings: Utilized in wear-resistant coatings.
- SiC/Graphene Nanocomposites: Applied in high-performance electronics.
- Graphene-TiO_2 Nanocomposites: Used in sensors and energy storage (Zhang et al. 2021).

Advantages and Limitations

- This method produces high-purity and uniform coatings.
- Precise control over film thickness.
- High processing temperatures and cost.

In Situ Polymerization

In situ polymerization involves scattering nanoparticles in a monomer solution, which is then polymerized to produce a nanocomposite.

Chemical Process

1. Nanoparticles dispersed in monomers.
2. Polymerization initiated by heat, light, or catalyst.
3. Nanocomposite formation via crosslinking.

Examples and Applications
- Chitosan-Silver Nanofibers: Applied in wound dressings.
- Polymer/Carbon Nanotube (CNT) Nanocomposites: Used in flexible electronics (Kim et al. 2020).
- Polymer-ZnO Nanocomposites: Utilized in UV-protective coatings (Sharma et al. 2022).

Advantages and Limitations
- Good dispersion of nanoparticles.
- Versatile for different polymers.
- Requires surfactants to prevent aggregation.

Chemical processing techniques allow for fine control over nanocomposite formation, which improves their usefulness in a variety of applications. The sol-gel approach is suitable for oxide nanocomposites; coprecipitation is extensively used for magnetic materials; hydrothermal synthesis improves crystallinity; CVD generates high-quality thin films; and in situ polymerization incorporates nanoparticles into polymer matrices. Future research should focus on developing environmentally friendly and scalable synthesis methods to advance nanocomposite applications in energy, medicine, and electronics.

13.3 Nanoparticle Functionalization

13.3.1 Surface Modifications for Enhancing Biocompatibility

Nanoparticles (NPs) increase medication solubility, bioavailability, and targeted administration, making them popular in drug delivery systems. Drug release, stability, and toxicity depend on nanoparticle functionalization. Functionalization modifies nanoparticle surfaces for biological uses.

13.3.1.1 Functionalization Controls Drug Release

Drug release is affected by functionalization in several ways:

1. *Surface Coating/Ligand Attachment.*

   Polymers like polyethylene glycol (PEG) can inhibit premature medication release and improve circulation time by providing steric hindrance.

   Ligand functionalization with antibodies, peptides, or folic acid allows for targeted medication delivery to specific tissues or cells.
2. *pH-responsive* functionalization of drug release in acidic tumor microenvironments or endosomes reduces systemic toxicity. For instance, PLGA nanoparticles containing pH-sensitive hydrazone bonds break down under acidic conditions, enabling site-specific drug release.

3. *Temperature-responsive functionality.*
   PNPAM, a thermosensitive polymer, changes solubility at particular temperatures, allowing for controlled medication release.
4. *Functionalization based on enzymes.*
   Specific enzymes overexpressed in sick tissues cause degradation of enzyme-sensitive linkers. Hyaluronic acid-functionalized nanoparticles release medications in response to increasing hyaluronidase levels in tumor tissues.
5. *Light-Responsive and Magnetic Functionalization.*
   Remote drug release control is possible with magnetic nanoparticles (e.g., Fe3O4) in external magnetic fields.

Near-infrared (NIR) light activates gold or silver nanoparticles, causing localized drug release by heat. Gold nanorods treated with doxorubicin release medicines during photothermal treatment when exposed to NIR irradiation.

13.3.2 Role of Functionalization in Controlling Drug Release

13.3.2.1 Drug Delivery Using Functionalized Nanoparticles

1. *Cancer Treatment.*

Targeted chemotherapeutic administration reduces adverse effects and improves medication accumulation in tumor tissues.

Example: HER2-targeted gold nanoparticles for breast cancer therapy.

2. *Neurological Disorders.*

Neurodegenerative illnesses can be treated with functionalized nanoparticles that penetrate the blood-brain barrier.

Chitosan-coated nanoparticles can deliver anti-Alzheimer's medicines.

3. *Antimicrobial Therapy.*

Antimicrobial peptides combined with silver nanoparticles enhance bacterial targeting and medication stability.

Consider functionalized silver NPs for multidrug-resistant bacterial infections.

4. *Cardiovascular Treatments.*

Functionalized liposomes effectively transport cardiovascular medicines to damaged arteries.

Examples include PEGylated liposomes containing statins for managing atherosclerosis.

5. *Delivery of Ophthalmic Drugs.*

Functionalized nanoparticles enhance medication retention and penetration in ocular tissues. Examples include chitosan-coated nanoparticles for prolonged medication release in glaucoma therapy.

13.4 Types of Nanocomposites Used in Biomedicines

13.4.1 Polymer-Based Nanocomposites

Polymer-based nanocomposites (PNCs) are hybrid materials that incorporate nanoscale fillers within polymer matrices to enhance their properties. These materials have garnered significant interest in biomedical applications due to their improved mechanical strength, biocompatibility, tunable degradation rates, and controlled drug delivery properties. This document explores the applications of PNCs in biomedicine along with examples.

- PLGA-Silica Nanocomposites—Used for controlled drug delivery of anticancer drugs.
- Chitosan-Ag Nanocomposites—Employed in wound healing due to antimicrobial activity.
- Hydrogel-GO Nanocomposites—Applied in bone tissue regeneration.
- PANI-Carbon Nanotube Nanocomposites—Used in biosensors for early disease detection.
- Gold Nanoparticle-Polymer Nanocomposites—Utilized in photothermal therapy for cancer treatment.

13.4.2 Metal-Based Nanocomposites

Metal-based nanocomposites are hybrid materials composed of metal nanoparticles dispersed within a matrix, such as carbon nanotubes (CNTs) or graphene. These composites exhibit unique properties, including high electrical conductivity, mechanical strength, and catalytic activity, making them valuable in various fields.

- Catalysis: Metal-based nanocomposites enhance catalytic efficiency in industrial and environmental applications, such as water purification and CO_2 reduction (Zhang et al. 2021).
- Energy Storage: They improve the performance of batteries and supercapacitors due to their high surface area and conductivity (Wang et al. 2020).
- Biomedical Applications: Metal nanocomposites are used in drug delivery, biosensors, and antimicrobial agents (Liu et al. 2019).
- Electronics: Used in flexible and wearable electronic devices due to their high conductivity and mechanical stability (Kumar and Singh 2022).

13.4.3 Carbon-Based Nanocomposites

A few examples of metal-based nanocomposites are mentioned in Table 13.2.

Table 13.2 Metal based nanocomposites and applications

Nanocomposite	Metal component	Matrix	Application	Reference
Pt–Graphene	Platinum (Pt)	Graphene	Fuel cells, catalysis	Shao et al. (2007)
Ag–CNT	Silver (Ag)	Carbon nanotubes	Antimicrobial coatings, sensors	Rai et al. (2016)
MnO_2–CNT	Manganese Dioxide (MnO_2)	Carbon nanotubes	Supercapacitors	Dubal et al. (2015)
Fe_3O_4–Graphene	Iron Oxide (Fe_3O_4)	Graphene	Lithium-ion batteries	Lu et al. (2007)
Au–CNT	Gold (Au)	Carbon nanotubes	Cancer therapy, imaging	Dykman and Khlebtsov (2012)
TiO_2–Graphene	Titanium Dioxide (TiO_2)	Graphene	Photocatalysis, water purification	Pelaez et al. (2012)
Ni–Graphene	Nickel (Ni)	Graphene	Gas sensors	Varghese et al. (2003)
ZnO–CNT	Zinc Oxide (ZnO)	Carbon nanotubes	Heavy metal removal	Behera et al. (2011)

Table 13.3 Ceramic-based nanocomposites for biomedical applications

Nanocomposite	Application	Key Advantages	Reference
HA-Polymer (PLA, PLGA)	Bone scaffolds, tissue engineering	Enhanced biodegradability, osteoconductivity	
HA-CNT (Carbon Nanotube)	Bone regeneration, load-bearing implants	Increased fracture toughness, osteoblast adhesion	
HA-Zirconia (ZrO_2)	Dental and orthopedic implants	Improved mechanical strength, fracture resistance	
HA-Silver (Ag)	Antibacterial bone grafts, coatings	Antimicrobial properties, infection prevention	
HA-Graphene Oxide (GO)	Bone scaffolds, osteogenic applications	Enhanced mechanical stability, osteogenic potential	
HA-Titanium (Ti)	Orthopedic and dental implants	Improved load-bearing capacity, biointegration	Kumar et al. (2019)
HA-Magnesium (Mg)	Biodegradable bone implants	Controlled degradation, bone healing support	

13.4.4 Ceramic-Based Nanocomposites

Ceramic-based nanocomposites have gained significant attention in biomedical applications due to their excellent biocompatibility, bioactivity, and mechanical properties. Among them, hydroxyapatite (HA)-based nanocomposites are widely used in bone tissue engineering, dental applications, and orthopedic implants because of their structural similarity to natural bone and ability to support osteointegration. Various modifications and reinforcements have been explored to enhance the mechanical strength, degradation rate, and antibacterial properties of HA for biomedical use.

Further, below is a concise table summarizing the various hydroxyapatite (HA)-based nanocomposites, their applications, and key advantages listed in Table 13.3.

13.5 Characterization of Nanocomposite Techniques in Biomedical Applications

Nanocomposites are materials that integrate nanoparticles within a matrix to enhance physical, chemical, and biological properties. The characterization of nanocomposites is essential to assess the size, shape, and dispersion of nanoparticles, which directly influence their performance, especially in biomedical applications.

Characterization Techniques

1. *Transmission Electron Microscopy (TEM):* TEM provides high-resolution images, allowing precise determination of nanoparticle size, shape, and distribution within the matrix. It is particularly useful for observing internal structures at the nanoscale.
2. *Scanning Electron Microscopy (SEM):* SEM offers detailed surface morphology and topography information. It helps assess nanoparticle dispersion and aggregation on the matrix surface.
3. *Dynamic Light Scattering (DLS):* DLS measures the size distribution of nanoparticles in suspension by analyzing light scattering patterns, which is critical for understanding particle uniformity.
4. *X-ray Diffraction (XRD):* XRD identifies the crystalline structure and phase composition of nanoparticles. It provides insights into particle size through peak broadening analysis.
5. *Fourier Transform Infrared Spectroscopy (FTIR):* FTIR helps in understanding the chemical interactions between nanoparticles and the matrix, indicating successful incorporation and functionalization.

Applications in Biomedical Field

Nanocomposites are extensively used in drug delivery systems, tissue engineering, medical imaging, and biosensors. For instance, polymer-based nanocomposites loaded with silver nanoparticles exhibit excellent antimicrobial properties, making them suitable for wound dressings. Magnetic nanocomposites are employed in targeted drug delivery and MRI contrast enhancement. Further techniques and applications are mentioned in Table 13.4.

The effective characterization of nanocomposites ensures their optimal design and functionality in biomedical applications. By selecting appropriate techniques, researchers can tailor nanocomposites to meet specific medical requirements, enhancing therapeutic efficacy and diagnostic accuracy.

Polymer nanocomposites have attracted to pharmaceutical and medical scientists attributable to their potential wide range of biomedical applications. The potential in biomedical applications imitates the high requirements of biomaterials in the pharmaceutical and medical field. Practically, broadly speaking, nanocomposites are the materials made up of polymers or copolymers in which the nano-sized fillers are embedded in the polymer matrix. The nanocomposites usually range in size range of 1–100 nm. This chapter overviews biomedical applications of polymeric nanocomposites.

Table 13.4 Characterization techniques and application

S. no.	Technique	Parameter assessed	Examples in biomedical applications	Reference
1	TEM	Size, shape, dispersion	Imaging of drug-loaded nanoparticles	Williams and Carter (2009)
2	SEM	Surface morphology	Analysis of scaffold surfaces in tissue engineering	Goldstein et al. (2017)
3	DLS	Size distribution	Characterization of nanoparticles in drug delivery systems	Bhattacharjee (2016)
4	XRD	Crystallinity, phase composition	Assessment of magnetic nanoparticles for MRI contrast agents	Cullity and Stock (2001)
5	FTIR	Chemical interactions	Verification of functional groups in biosensors	Stuart (2004)

13.6 Bridging Innovation: Biomaterials at the Heart of Pharmaceutical and Medical Breakthroughs

13.6.1 Tissue Engineering

Polymer nanocomposites for bone tissue engineering have been used to support bone regeneration by mimicking the natural extracellular matrix. Nanocomposites utilize a combination of biodegradable polymer matrices with bioactive nanofillers, such as nano-hydroxyapatite, graphene oxide, and titanium dioxide to enhance mechanical strength, stability, and cell compatibility. Natural polymers such as chitosan, collagen, silk fibroin-based nanocomposites, and synthetic polymer-based nanocomposites such as polylactide glycolic acid, polycaprolactone, and polyethylene glycol are used. These nanocomposites are prepared by using fabrication methods like freeze-drying, electrospinning, and 3D printing. On top of their biodegradable properties, these scaffolds promote cell adhesion, vascularization, and bone growth.

Pathak et al. (2023) described multifunctional roles of carbon dot-based polymer nanocomposites in biomedical applications. Carbon dots are quasi-zero-dimensional carbon-based nanomaterials with sizes <10 nm. They exhibit quantum confinement effects, high photoluminescence, and tunable bandgap, making them excellent for biomedical applications. CDs are integrated into polymers to enhance mechanical, thermal, and optical properties, benefiting biosensors, bioimaging, tissue engineering, and drug delivery. Carbon dot-based polymeric nanocomposites like chitosan/nanohydroxyapatite scaffolds show improved osteoinductivity and osteogenesis (Pathak et al. 2023).

Biomedical applications of polymer nanocomposites reinforced with carbon nanofillers such as graphene, carbon nanotubes (CNTs), and fullerenes have been discussed by Alosaimi et al. (2023). Synthetic polymers like polycaprolactone (PCL), polyaniline (PANI), and polyurethane (PU) are highlighted for their versatility and customizable properties in biomedical use. Carbon-based nanofillers improve the physical, mechanical, and chemical properties of polymers. Nanocomposites enhance scaffold biocompatibility, cell adhesion, and proliferation. Functionalized nanofillers such as graphene oxide (GO) improved polymer characteristics, enhancing electrical conductivity and mechanical strength. Hybrid scaffolds combining nanofillers with polymers were used for tissue regeneration and vascular applications (Alosaimi et al. 2023). Chitosan, poly(vinyl alcohol), and graphene oxide CS/PVA/GO films possess desirable mechanical properties, effective antibacterial activity, and suitable degradation rates, making them promising candidates for use in tissue engineering and other biomedical applications. Their biocompatibility further supports their potential for clinical use. Biodegradable antimicrobial films using a composite of CS/PVA/GO are intended for biomedical applications, particularly in tissue engineering, where biocompatibility and antimicrobial properties are crucial. The films were tested on numerous bacterial strains, counting both Gram-positive and Gram-negative bacteria. Results indicated that the CS/PVA/GO films exhibited significant antibacterial activity, effectively inhibiting bacterial growth. The biocompatibility of the films was assessed through in vivo tests on Wistar rats. The results demonstrated that the films were well-tolerated, with no adverse reactions observed, indicating their potential for safe use in medical applications (Ruiz et al. 2019).

13.6.2 Wound Healing

Carbon dot (CD)-based hydrogels accelerate tissue regeneration and antibacterial activity. This benefits in wound healing (Lu et al., 2025). Cellulose is a naturally abundant, biocompatible, and biodegradable polymer. Its semi-crystalline structure formed by repeated D-glucose units offers high mechanical strength and chemical stability. Nanocellulose includes not only bacterial cellulose (BC), cellulose nanofibrils (CNFs) but also cellulose nanocrystals (CNCs). These materials display distinctive belongings like an excessive surface area, of the highest standard mechanical strength besides biocompatibility. Nanocellulose holds immense promise for various biomedical applications due to its renewable nature, exceptional mechanical properties, and biocompatibility. Nanocellulose demonstrates high Young's modulus (up to 200 GPa) and tensile strength, making it ideal for load-bearing applications. The presence of hydroxyl groups facilitates chemical modifications, enhancing their functionality for specific applications. Variants like sulfate and carboxylated CNCs exhibit different charge properties, useful in biomedical formulations. Nanocellulose composites are functionalized with antimicrobial agents like silver nanoparticles for wound dressing. They exhibit excellent moisture retention, antimicrobial activity, and biocompatibility aiding tissue regeneration (Joseph et al. 2020).

13.6.3 Cancer Treatment, Bioimaging and Theranostics

Carbon dot-based polymeric nanocomposites enable high fluorescence stability and biocompatibility for cellular and deep-tissue imaging. CD-PNCs are used in theranostics including tumor ablation through photothermal treatment (Pathak et al. 2023). There is certain research that focuses on examining hierarchically organized spherical-like hematite nanoparticles (α-Fe_2O_3) synthesized through a low-temperature solution combustion method. They were uniformly distributed α-Fe_2O_3/C nanocomposites. The cytotoxic effects of the materials are assessed via MTT assay in addition to nuclear imaging, focusing on cellular morphological alterations in both the human lung cancer cell line A549 and Chang liver cells as a non-malignant control. These findings indicate both the pure as well as composite materials maintain viability more than 70% in the non-malignant cell line while demonstrating approximately 60% suppression in the A549 lung cancer cell line, suggesting that the α- α-Fe_2O_3/C nanocomposite is suitable for biological claims plus anticancer treatments. Evidence of cell death confirmed using DAPI staining (AlSalhi et al. 2020). The study of polymer/metal-organic framework (MOF) nanocomposites for varying biomedical executions has gained significant attention due to the unique characteristics of MOFs and their hybrid composite systems. Different kinds of polymer/MOF nanocomposites specifically employed not only in drug delivery but also in imaging have been explored. The synthesis approach affects the performance of these nanocomposites. Thus, a serious evaluation and comparison of the fruitfulness of these nanocomposites, emphasized their synthesis approach (Giliopoulos et al. 2020). Sunitinib malate loaded lipid polymer hybrid NPs (LPHNPs) improved cell viability reduction in breast cancer cell lines (MCF7) compared to free drugs (Andrade et al. 2024). Apigenin-loaded PLGA NPs enhanced apoptotic activity and modulation of carcinogenic pathways. Copper oxide NPs functionalized with PEG and chitosan. Folate-targeted NPs effectively expressed transgenes in cancer cells. Hydroxyapatite (HAp) nanorods were synthesized from cuttlefish bones (an eco-friendly source of calcium carbonate). Carbon nanomaterials (GO, CNTs, and GONRs) were incorporated in varying weights (1%, 3%, 5%) alongside Ag NPs using an oil bath-mediated synthesis method. Increasing the nanomaterials increased the crystallite size and particle size, improving certain properties. A 3% weight composite showed enhanced cell proliferation with MG63 osteoblast-like cells after 48 h, while 5% weight composite exhibited reduced viability due to agglomeration and toxicity. Composites up to 3% weight were hemocompatible (less than 4% hemolysis). However, 5% weight composites caused significant hemolysis (Balu et al. 2021). The emergence of multidrug-resistant microorganisms has led to a significant increase in death rates, necessitating the development of new antimicrobial agents. The study utilized Aspergillus fumigatus KIBGE-IB33 for the biosynthesis of AgNPs. Silver nitrate (AgNO3) was reduced to silver nanoparticles using the fungal biomass in an eco-friendly manner avoiding hazardous chemicals. The biosynthesized AgNPs were combined with chitosan to form silver-chitosan nanocomposites (AgCNCs) using a microwave irradiation technique. This method aimed to enhance the stability and antibacterial properties of the nanoparticles. The cytotoxic effects of the nanocomposites were evaluated on NIH/3 T3 cell lines, revealing no significant cytotoxicity, which supports their potential use in biomedical applications (Raza et al. 2021). Microfluidic biosensors offer advantages over traditional methods, including improved mass transport, low reagent consumption, and fast response times. This study presents the development of a novel microfluidics biosensor platform using a bi-enzyme functionalized nanocomposite for the rapid detection of cholesterol. The biosensor is based on a composite of nickel oxide nanoparticles and multiwalled carbon nanotubes. The nanocomposite was integrated with polydimethylsiloxane. The integrated microfluidics biosensor platform demonstrated low-cost, rapid detection capabilities, and high sensitivity, making it suitable for clinical applications in cholesterol monitoring.

13.6.4 Drug Delivery

Doxorubicin is a widely used chemotherapy drug for osteosarcoma but suffers from high toxicity and poor drug delivery efficiency. Zeolite imidazolate frameworks (ZIFs) offer excellent biocompatibility, high porosity, and pH responsiveness, making them ideal drug carriers. A Fe_3O_4-ZIF-8 nanocomposite was synthesized with ZIF-8 as the shell and Fe_3O_4 as the core. Doxorubicin was loaded into the composite, forming Doxorubicin-Fe_3O_4-ZIF-8 nanocomposites. This nanocomposite showed significantly better cytostatic and apoptosis-inducing effects on K7M2 osteosarcoma cells compared to Doxorubicin alone ($p < 0.01$). Tumor-bearing mice treated with Doxorubicin-Fe_3O_4-ZIF-8 nanocomposites exhibited a tumor inhibition rate of 67%, outperforming Doxorubicin alone. Under an external magnetic field, Doxorubicin-Fe_3O_4-ZIF-8 nanocomposites effectively accumulated in tumor tissues, enhancing anticancer performance. The Doxorubicin-Fe_3O_4-ZIF-8 nanocomposites system demonstrated superior tumor-targeting ability and antitumor efficacy compared to Doxorubicin alone. Doxorubicin-Fe_3O_4-ZIF-8 nanocomposites showed strong potential for chemotherapeutic applications in osteosarcoma treatment (Wu et al. 2023). Au/TMC/Fe_3O_4 nanocomposites possess promising characteristics for applications in biosensors and drug delivery systems. Their nontoxic nature combined with effective physical properties makes them a valuable candidate for future biomedical research and applications. Au/TMC/Fe_3O_4 nanocomposites synthesis and characterization of gold (Au) nanoparticles supported on N-trimethyl chitosan (TMC) and iron oxide (Fe_3O_4) nanocomposites. The TMC coating contributed to better dispersion and stability of the nanoparticles, making them suitable for biomedical applications (Shirazi et al. 2015). Magnetoelectric materials have ability to couple magnetic and electric properties, exhibiting a magnetoelectric (ME) effect at room temperature. This effect allows for the manipulation of electric polarization through a magnetic field. The research highlights the prospective uses of magnetoelectric nanoparticles (MENPs) in fields such as spintronics and biomedicine, particularly in targeted drug delivery for cancer treatment. Cobalt ferrite nanoparticles

(CFO) were coated with a barium titanate (BTO) shell using a sol-gel method at a 1:1 ratio of CFO to BTO. The process involved mixing titanium isopropoxide and barium hydroxide in ethanol and water, followed by thermal treatment to form the BTO shell. The crystal size of CFO was determined to be around 10 nm, indicating that the nanoparticles were primarily single crystallites. The high saturation magnetization and low coercivity at room temperature suggest that CFO and CFO@BTO nanocomposites are suitable for biomedical applications, particularly in cancer treatment (Vinţeler et al. 2024).

13.6.5 Antimicrobial Applications

Silver nanoparticles release Ag+ ions. The surface of silver nanoparticles can be modified to improve their dispersion in polymer matrices, leading to better distribution and effectiveness in the composite material. When integrated into polymer nanocomposites, silver nanoparticles can work synergistically with the polymer matrix to enhance overall antimicrobial properties. The polymer can provide a sustained release mechanism for Ag+ ions, prolonging the antimicrobial effect over time. The polymer matrix can be engineered to control the release rate of silver ions, ensuring a consistent antimicrobial effect while minimizing potential toxicity to surrounding tissues. The use of silver nanoparticles is associated with a reduced chance of bacterial resistance compared to conventional antibiotics, making them a promising alternative in the fight against infections. Silver nanoparticle-infused polymer nanocomposites are increasingly used in medical devices, wound dressings, and coatings for implants to reduce the risk of infection and improve patient outcomes. These composites can also be utilized in water treatment and filtration systems to eliminate pathogens, showcasing their versatility beyond medical applications (Dwivedi et al. 2013). A dental composite incorporating amorphous calcium phosphate nanoparticles (NACP) and silver nanoparticles (NAg) was developed to evaluate its mechanical and antibacterial properties. NACP nanoparticles (116 nm) were synthesized through spray drying, while NAg nanoparticles (2.7 nm) were incorporated into the resin. Dental plaque biofilm tests showed proportional bacterial death with Nag. The incorporation of NAg enhanced antibacterial properties without compromising mechanical strength. The study demonstrated that NACP–NAg nanocomposites have encouraging potential for dental rehabilitation, offering re-mineralizing and antibacterial effects (Cheng et al. 2012).

Cellulose nanofibers (CNFs), due to their biodegradability and low cost, are suitable for biomedical applications, including medical textiles. Modifications using nanoparticles like cerium oxide (CeO_2NPs) can address limitations in cellulose-based materials, such as low biocompatibility and mechanical strength. CeO_2NPs were bioengineered using the leaf extract of Uncaria gambir Roxb as a natural reducing agent and incorporated into CNFs. The composites showed significant antibacterial effects, especially at higher CeO_2NPs concentrations. CeO_2NPs function as antimicrobial agents through electrostatic interactions with bacterial cells.

The modified CNF/CeO_2NPs composites are promising for medical textiles due to their enhanced antibacterial activity and mechanical properties (Putri et al. 2024). Another study investigates the antibacterial properties of metal oxide nanoparticles (specifically in nanocomposites with sodium alginate) as a potential therapeutic approach. Eight types of metal oxide nanoparticles were synthesized (ZnO, Ag_2O, CuO, NiO, CoO, FeO, Fe_2O_3, Cr_2O_3) using a precipitation method. The process involved preparing metal precursor solutions, heating, and adding NaOH to form precipitates, which were then washed and calcined. Metal oxide nanoparticles were incorporated into a sodium alginate matrix to create metal oxide/alginate (MO-Alg) nanocomposites. The mixture was subjected to ultrasonic treatment to ensure uniform dispersion. The antibacterial activity was evaluated against Gram-negative (*P. aeruginosa, E. coli*) besides Gram-positive bacteria (*S. aureus, E. faecalis*). Minimum inhibitory concentrations (MICs) were determined to assess the effectiveness of the nanocomposites. The study found varying antibacterial efficacy among the different nanocomposites (Belhachem et al. 2024). Nature employs biohalogenation, a process mediated by haloperoxidases, to prevent bacterial colonization by interfering with bacterial communication (quorum sensing). This study was aimed to synthesize cerium dioxide (CeO_2) nanoparticles that mimic haloperoxidase activity and can be incorporated into transparent polycarbonate (PC) nanocomposites to repel biofilms of *P. aeruginosa*. The synthesis involved water/benzyl alcohol mixtures and oleylamine as a capping agent, significantly reducing reaction time compared to conventional methods. Plasma-treated polycarbonate panels were dip-coated in CeO_2 dispersions to create transparent nanocomposites. The resulting nanocomposites exhibited low surface roughness and enhanced mechanical properties. The CeO_2/PC nanocomposites demonstrated a lessening in bacterial biofilm creation of up to 85% against *P. aeruginosa* PA14. Additionally, the production of the virulence factor pyocyanin was reduced by up to 55%, indicating that the nanoparticles affect quorum sensing without being biocidal. The CeO_2/PC nanocomposites could be applied in various industries, including healthcare, electronics, and food packaging, where biofilm formation is a concern (Sarif et al. 2022).

13.6.6 Protein Adsorption

Hydroxyapatite (HA) is a key mineral component of vertebrate bones and teeth, recognized for its excellent bioactivity and biocompatibility, making it a significant material in biomedical applications. This study presents a novel and environmentally friendly sonochemical method for synthesizing cellulose/HA nanocomposites, aiming to enhance their mechanical properties and biological performance for applications in tissue engineering and drug delivery. The nanocomposites were synthesized using a sonochemical route in a NaOH/urea aqueous solution. Cellulose was dissolved in the solution, and calcium chloride ($CaCl_2$) and sodium dihydrogen phosphate (NaH_2PO_4) were added to form HA. The mixture was subjected to ultrasonic treatment to facilitate the creation of HA crystals on the cellulose matrix. The biological response of the cellulose/HA nanocomposites simulated body fluid (SBF) was tested for up to 28 days. The mass of the nanocomposites increased significantly, indicating the formation of carbonate-containing apatite on their surface, which is beneficial for bone regeneration. The maximum mass increase observed was 1.82 times the initial mass after 28 days of soaking. Cytotoxicity tests demonstrated that the cellulose/HA nanocomposites exhibited high biocompatibility, with cell viability exceeding 98% in MGC-803 human gastric carcinoma cells. Hemoglobin (Hb) model was used to check the protein adsorption capacity of the nanocomposites. This showed a high adsorption ability of 321.5 mg/g, indicating their potential for drug delivery applications (Fu et al. 2018).

13.7 Challenges in Biomedical Applications of Nanocomposites

13.7.1 Toxicity and Biocompatibility Issues

13.7.1.1 Potential Cytotoxicity, Inflammation, and Immune Response

Nanomaterials are present in a variety of products, including pharmaceuticals, electronics, clothing, and cosmetics. Due to their extremely small size (less than 100 nm), nanoparticles can enter the body through the airways and skin, potentially leading to health risks. Once inside the body, they may affect multiple organs such as the brain, liver, and kidneys, resulting in immune system disruptions, inflammation, and cytotoxicity. The health effects of nanomaterials depend on their specific characteristics, any underlying health conditions, and individual genetic predispositions. For instance, cobalt and nickel nanoparticles can provoke inflammation, whereas silver nanoparticles have been shown to reduce allergic inflammation. Additionally, carbon nanotubes are associated with lung damage similar to that caused by asbestos fibers. Some studies also indicate that certain nanomaterials can cause cell damage and lead to pre-cancerous lesions (Ali et al. 2016). Metallic nanoparticles (mNPs) and nanotechnology offer various advantages, but their extensive use can pose health risks due to their accumulation in biological systems and the complex migratory pathways they follow. Once inside the body, mNPs can disrupt biological processes, lead to inflammation, and affect the immune system. Their potential effects include cytotoxicity, oxidative stress, genetic damage, and mitochondrial dysfunction. Numerous studies have explored the mechanisms behind the toxicity of nanoparticles and their interactions with the environment (see Fig. 13.1). Unlike their bulk counterparts, the reactivity and toxicity of nanoparticles are not yet fully understood. While cells readily take up nanoparticles, the intracellular processes involved in this uptake remain unclear. Currently, nanotoxicology lacks standardized protocols that are present in traditional toxicology. Various nanoparticles, including silver and titanium dioxide, have shown a range of toxic effects in both in vivo and in vitro models, often through the induction of reactive

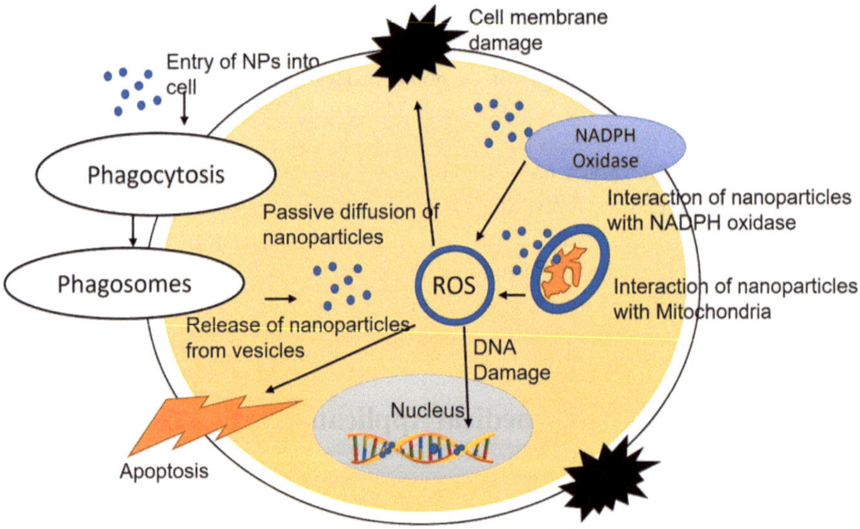

Fig. 13.1 Cellular toxicity mechanism of nanocomposite. Image reprint from Jane et al. (2021) with taking permission from Elsevier

oxygen species (ROS). This process can lead to DNA damage, protein modification, and ultimately cell death. The mechanisms that produce ROS can vary and may be unpredictable, influenced by different environmental factors (Jane et al. 2021).

Yuan et al. summarize current research on the cytotoxic effects of nanocomposites on various cell types, including macrophages, epithelial cells, and lymphocytes. It highlights three key aspects: the structural diversity of nanocomposites, the mechanisms behind their cytotoxic effects, and potential influencing factors (Table 13.5).

Yuan et al. reported that unpurified multiwalled carbon nanotubes (MWCNTs) enter cells through both active and passive mechanisms by penetrating the plasma membrane, leading to cell necrosis rather than apoptosis. Interestingly, cells treated with purified MWCNTs (with the iron catalyst removed) showed similar levels of cell death compared to those exposed to unpurified MWCNTs (Cheng et al. 2009). In another study, purified single-walled carbon nanotubes (SWCNTs) induced a low percentage of necrotic cell death in human monocyte-derived macrophages (HMMs). TEM images revealed the morphology and structure of necrotic cells, with large bundles of SWCNTs engulfed by the macrophages, either accumulating in lysosomes or freely dispersed in the cytoplasm (Porter et al. 2009). The mechanisms underlying carbon nanomaterial-induced cytotoxicity are illustrated in Fig. 13.2. Exposure of macrophages to carbon nanomaterials (CNMs) initiates a cascade of cellular and molecular processes, including reactive oxygen species (ROS) generation and lysosomal damage, which act as key mechanisms driving CNM-induced cell death, such as necrosis, apoptosis, and pyroptosis (Yuan et al. 2020). CNMs induce mitochondrial-dependent apoptotic pathways through the ROS-activated MAPK signaling cascade. ROS also activates transcription factors like NF-κB, which modulate inflammatory responses. Additionally, CNMs trigger lysosomal membrane permeabilization (LMP), leading to the release of cathepsins

Table 13.5 Cellular toxicity and immunological effects of nanocomposite (Jane et al. 2021)

Types	Length	Diameter	Dose	Target cells	Cytotoxicity
Carbon nanotubes; Nanographite; Carbon black	1.5 μm; 4.5 μm; —	9.5 nm; 12 nm; —	15 ~ 120 μg/ml	RAW 264.7	LDH release, TNF-α production, ROS production
Pristine graphene	500–1000 nm	2 ~ 3 nm	5, 10, 20, 40, 80 and 100 μg/ ml	RAW 264.7	ROS increase, apoptosis by activation of the mitochondrial pathway, activation of the MAPKs (JNK, ERK and p38) and the TGF-beta-related signaling pathways
SWCNTs	–	–	0.78 ~ 200 μg/ml	HEK293 cell	Apoptosis and cell cycle arrest in G1.
Water-soluble fullerene	–	–	–	Human dermal broblasts, HepG2, neuronal human astrocytes	Lactate dehydrogenase release, cellular membrane disruption, and lipid peroxidation
SWCNTs; MWCNTs; MWCNTs	–	1–2 nm; 10–20 nm; 30–50 nm	5 ~ 100 μg/ml	NR8383 cell	ROS generation and reduced cell viability.
Graphene oxides (GOs); Acid-functionalized SWCNTs	–	500 nm; 355 nm	10 ~ 50 μg/ml	Peritoneal macrophages	LDH release, decreased autophagic degradation, lysosomal membrane destabilization
SWCNTs	150 nm	1 ~ 2 nm	0 ~ 50 μg/ml	Mouse peritoneal macrophages	Mitochondrial damage
MWCNT1; MWCNT2	13 μm; 5 μm	40 ~ 100 nm; 30 nm	0.625 ~ 10 μg/cm^2	RAW 264.7	Mitochondrial activity reduction, LDH release
Aci- and tau-MWCNTs	5 ~ 10 μm	10 ~ 20 nm	0, 5, 20, 40, and 80 μg/ml	RAW 264.7	Apoptosis via mitochondrial pathway and scavenger receptor
SWCNT; MWCNT	1 ~ 5 μm; 1 ~ 2 μm	< 2 nm; 10 ~ 30 nm	30, 100 and 300 μg/ml	RAW 264.7	Cell death induced by SWCNT; No cell death induced by MWCNT

(continued)

Table 13.5 (continued)

Types	Length	Diameter	Dose	Target cells	Cytotoxicity
SWNTs; MWNT10; Fullerene	1 μm; 0.5 ~ 40 μm; —	1.4 nm; 10–20 nm; —	1.41 ~ 226.0 μg/cm²; 1.41 ~ 22.60 μg/cm²; 1.41 ~ 226.0 μg/cm²	Alveolar macrophage	Reduced cell viability
C-SWNTs; C60-fullerenes; Graphite particles	–	–	–	Human monocyte-derived macrophages	Apoptosis/necrosis
Carbon black nanoparticles	175 ± 80 nm	20 ± 6 nm	30 μg/cm²	RAW264.7, human alveolar macrophages	Caspase 1 and IL-1β release, LDH release, plasma membrane disruption, pyroptosis
Fe@CNPs	–	–	50 and 400 μg/ ml	HEK293 and C33A cell	ROS generation and apoptosis
SWCNHs	400 nm	~100 nm	0.01 ~ 0.3 mg/ml	RAW 264.7	Apoptosis and necrosis associated with lysosomal membrane destabilization, ROS generation, inflammatory cytokines (TNF-α, IL-1β, and IL-6) release
MWCNTs	0.5–2 μm	< 8 nm; 20 30 nm; > 50 nm	100 μg/ml	3 T3, RAW 264.7 and bronchiolar epithelial cells.	Cytotoxicity differing with particle sizes and cell types, reactive oxygen species generation, lysosomal membrane destabilization and mitochondrial permeability
SWCNTs	–	0.8 ~ 2.0 nm	25 or 50 μg/cm²	Normal and malignant human mesothelial cells	ROS generation, increased cell death, enhanced DNA damage and H2AX phosphorylation, and activated PARP, AP-1, NF-κB, p38, and Akt
MWCNTs	< 1 μm	9.5 nm	2.5 ~ 100 μg/ml	RAW264.7, A549	LDH release and oxidative stress
Carbon nanohorns	–	–	1 ~ 100 μg/ml	RAW 264.7	Reactive oxygen species generation and apoptosis lysosomal membrane permeabilization

Types	Length	Diameter	Dose	Target cells	Cytotoxicity
Pristine-SWCNTs	–	–	1 μg/cm^2	RAW264.7	Decreased cell viability and ATP production, increased ROS and NO production, activation of the MAP kinase pathway, increased levels of apoptosis- and autophagy-related proteins, and ER stress-related proteins
Functionalized MWCNTs (tau-MWCNTs); Pristine MWCNTs (raw-MWCNTs)	300 ~ 600 nm; —	10 ~ 20 nm; —	0 ~ 80 μg/ml	RAW 264.7	Apoptosis related to mitochondrial injury, less toxicity induced by tau-MWCNTs
MWCNTs	–	–	20 μg/ml	Mature human monocyte-derived macrophage cells	Apoptosis and necrosis
Short MWCNTs; Long MWCNTs	0.6 μm; 20 μm	30.6 nm; 27.8 nm	10 μg/ml	Primary human alveolar macrophage	Reduced cell viability, ROS generation and inflammatory mediator release induced by long MWCNTs
MWCNTs; Onion-like shell-shaped carbon nanoparticles	~2 μm; —	10 ~ 15 nm; 50 ~ 100 nm	0 ~ 500 μg/ml	16HBE14o-	ROS generation, reduced cell viability
MWCNTs-COOH; MWCNTs-PEG	0.9 μm; 0.8 μm	24.6 nm; 27.3 nm	0 ~ 100 μg/ml	RAW 264.7 cells, primary rat peritoneal macrophages	Activation of oxidative stress-responsive pathways, such as p38 mitogen-activated protein kinases (MAPK) and nuclear factor (NF)-κB
Purified-MWCNT; COOH-MWCNT	1122 nm; 652 nm	–	1 ~ 50 μg/ml	Human alveolar macrophage	Reduced cell viability and increased inflammatory mediator (IL-1β and IL-8) release
Two types of functionalized carbon nanotubes (1,3-dipolar cycloaddition reaction and the oxidation/amidation treatment)	–	–	1 ~ 10 μg/ml	Primary B lymphocytes, T lymphocytes, and peritoneal macrophages	Intake by B and T lymphocytes as well as macrophages in vitro without affecting cell viability

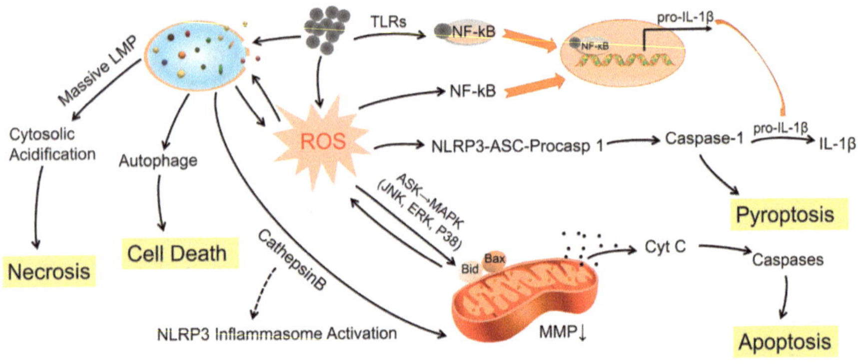

Fig. 13.2 Schematic representation of carbon-based nanocomposite inducing cytotoxicity. Image reprint with Journal permission

into the cytoplasm. ROS generation and LMP are mutually reinforcing, creating an amplification loop (Arezki et al. 2022). LMP may also impair autophagic function. Furthermore, CNM exposure activates inflammasome-dependent pyroptosis, characterized by caspase-1 cleavage and subsequent IL-1β release.

13.7.1.2 Methods for Testing and Minimizing Toxicity in Preclinical and Clinical Stages

The methods for testing and minimizing the toxicity of nanocomposites during preclinical and clinical stages involve a comprehensive, multidisciplinary approach that integrates in vitro, in vivo, and in silico techniques. There are several aspects that can be evaluated: cell, cytotoxicity, proliferation, apoptosis/necrosis, cellular morphology alteration, oxidative stress, inflammatory response, and hemotoxicity (Sanità et al. 2020). These strategies aim to ensure the safety, efficacy, and biocompatibility of nanocomposites before they are introduced into clinical practice. Computational in silico models, such as molecular modeling, QSAR, and PBPK models, further enhance the prediction of nanocomposite behavior and toxicity, potentially reducing reliance on animal studies. Kus et al. performed biosafety assessments of Gold nanoparticles (AuNPs). The key to biocompatibility lies in ensuring that the interaction between AuNPs and biological systems does not trigger adverse effects, such as oxidative stress, DNA damage, or apoptosis. Cytotoxicity, which focuses on the harmful impact on specific cell types, is usually assessed in vitro before advancing to in vivo studies. Interestingly, the results of AuNP toxicity studies have been inconsistent. Factors like the nanoparticles' size, shape, and method of administration (oral, transdermal, etc.) play a critical role in determining their safety, influencing cellular outcomes like apoptosis, ROS production, or behavioral abnormalities. Assessing AuNPs' toxicity and biocompatibility is complex and involves both in vitro and in vivo testing, with techniques like enhanced permeability and retention (EPR) effect analysis, transmission electron microscopy (TEM), and inductively coupled plasma mass spectrometry (ICP-MS) helping to

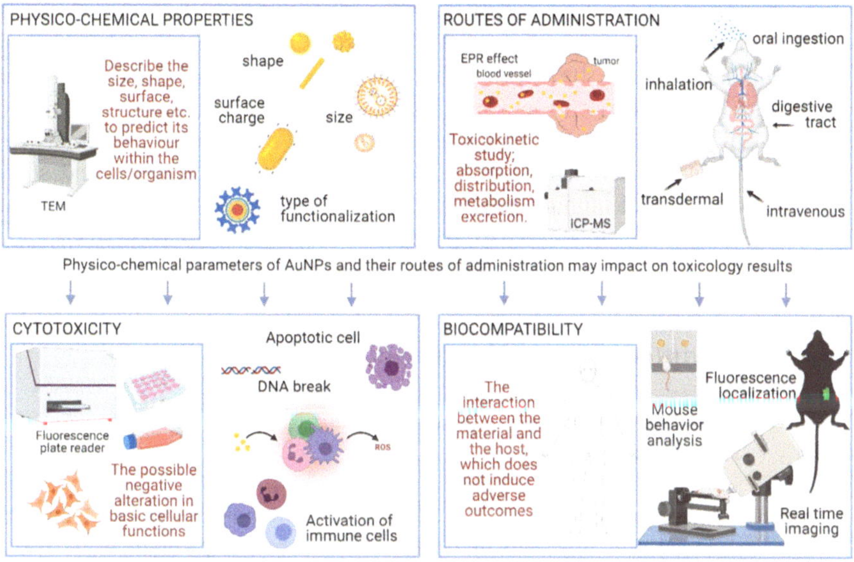

Fig. 13.3 Impact of nanocomposite on cytotoxicity and biocompatibility. Image reprint from Kus-Liskiewicz et al. (2021) open-access article distributed under CC BY license 4.0

measure their impact at the cellular and organism levels (Fig. 13.3). Understanding these variables is essential to unlocking the full potential of AuNPs in safe, effective medical applications (Kus-Liskiewicz et al. 2021).

13.7.1.3 Strategies for Improving the Biocompatibility of Nanocomposites

Organic-inorganic hybrid materials have shown great promise in the creation of biocompatible nanocomposites. These nanocomposites (NCs) can generally be divided into two main categories based on their dispersion matrix: polymeric and non-polymeric. Non-polymeric nanocomposites encompass those based on ceramics, metals, or clay, each offering distinct advantages and limitations. Nanocomposites are formed by combining two or more elements, such as carbon nanotubes (CNTs), gold (Au), silver (Ag), platinum (Pt), zinc oxide (ZnO), iron oxide (Fe_3O_4), cerium oxide (CeO_2), tin oxide (SnO_2), titanium dioxide (TiO_2), zirconium dioxide (ZrO_2), and others (Lu et al. 2025). To improve the biomedical practicality of composites, three key strategies have been explored: (i) surface modification (ii) surface engineering, and (iii) controlling biodegradation to enable the controlled release of drug. Chong et al. reported polylactic acid (PLA) biomaterial for its biocompatibility and biodegradability. The addition of zinc oxide (ZnO) nanofillers to PLA can enhance its antibacterial and tissue regenerative properties but may also compromise its biocompatibility (Chong et al. 2023). Shi et al. fabricated Injectable nanocomposites composed of biodegradable poly(propylene fumarate) and the crosslinking agent propylene fumarate–diacrylate, along with three types of single-walled carbon nanotubes (SWNTs), which were evaluated for in vitro cytotoxicity. The findings showed no cytotoxicity for purified SWNTs, SWNTs functionalized with 4-tert-butylphenylene, and ultra-short SWNTs at concentrations of 1–100 μg/mL. Remarkably, nearly 100% cell viability was observed on all crosslinked nanocomposites, with cell attachment comparable to that on standard tissue culture polystyrene (Shi et al. 2008). To minimize toxicity, several strategies are employed, including surface functionalization to improve biocompatibility, size and shape optimization to reduce immune activation, the development of biodegradable nanocomposites, and targeted delivery to specific tissues to limit off-target effects.

13.7.2 Manufacturing and Scalability

13.7.2.1 Challenges in Scaling Up the Synthesis of Nanocomposites for Commercial Use

Nanoparticles sized under 100 nm have diverse applications in the pharmaceutical industry, produced from materials like gold, silver, carbon, and polymers. They can act as drug carriers for both hydrophilic and lipophilic drugs, enhance targeting, improve bioavailability, and facilitate imaging and controlled release. While modifications with capping agents help in targeting and aggregation prevention, achieving uniform size and shape in scalable production remains a challenge. Production methods are classified as top-down, which reduces bulk materials, and bottom-up, which assembles atoms. Despite their benefits, mass-producing nanoparticles is complicated due to undefined optimization parameters, leading to issues like aggregation, contamination, and low yields. These challenges impact the economic feasibility and industry acceptance of nanoparticle production. This discussion will cover the latest advancements and challenges in processing nanocrystalline materials and coatings, including their properties and microstructure. The chapter explores the pros and cons of different processing techniques and the equipment needed for research. Figure 13.4 illustrates the consolidation techniques to be addressed.

Scaling up the synthesis of nanocomposites for commercial use faces several challenges that impede their application in biomedicine and electronics. A primary concern is achieving cost-effectiveness and reproducibility in large-scale production, where maintaining uniform product quality can be difficult. Quality control is crucial, as minor variations in processing can significantly affect performance and safety. Environmental sustainability is another major issue, as many nanocomposites are not biodegradable, raising concerns about their long-term impact. The potential toxicity and persistence of nanomaterials require thorough assessment due to their interactions with biological systems. Additionally, the regulation and standardization of nanocomposites are underdeveloped, particularly in biomedicine, where safety is critical. The lack of universal standards and testing protocols complicates regulatory approval processes, emphasizing the need for interdisciplinary approaches to develop scalable and safe nanocomposite technologies. Some of the techniques addressed to overcome the challenges associated with synthesis of nanocomposites (Hachhach 2025). Figure 13.4 illustrates the consolidation techniques to be addressed.

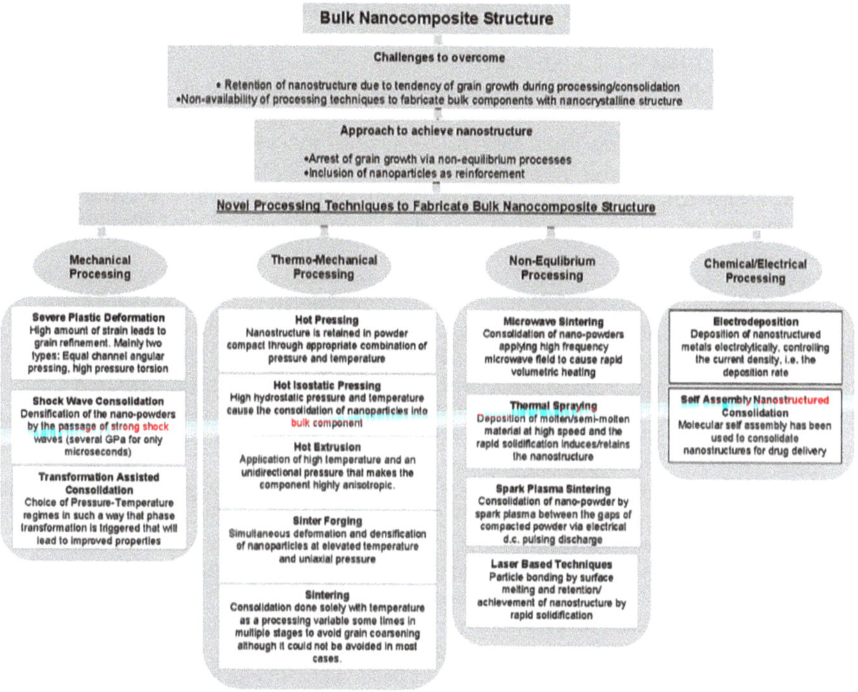

Fig. 13.4 Different manufacturing techniques. Image reprint with permission of journal Elsevier license no. 5961310732835

References

Ali A, Suhail M, Mathew S, Shah MA, Harakeh SM, Ahmad S, Kazmi Z, Rahman Alhamdan MA, Chaudhary A, Damanhouri GA, Qadri I (2016) Nanomaterial induced immune responses and cytotoxicity. J Nanosci Nanotechnol 16(1):40–57

Alosaimi AM, Alorabi RO, Katowah DF et al (2023) Recent biomedical applications of coupling nanocomposite polymeric materials reinforced with variable carbon nanofillers. Biomedicines 11(3):967. https://doi.org/10.3390/biomedicines11030967

AlSalhi MS, Devanesan S, Shanmugam P et al (2020) Synthesis and biocompatible role of hierarchical structured carbon nanoplates incorporated α-Fe_2O_3 nanocomposites for biomedical applications with respect to cancer treatment. Saudi J Biol Sci 27(2):588–593. https://doi.org/10.1016/j.sjbs.2019.11.028

Andrade S, Ramalho MJ, Loureiro JA (2024) Polymeric nanoparticles for biomedical applications. Polymers 16(2):249. https://doi.org/10.3390/polym16020249

Arezki Y, Rapp M, Lebeau L, Ronzani C, Pons F (2022) Cationic carbon nanoparticles induce inflammasome-dependent pyroptosis in macrophages via lysosomal dysfunction. Front Toxicol 4:925399

Balu SK, Sampath V, Andra S et al (2021) Fabrication of carbon and silver nanomaterials incorporated hydroxyapatite nanocomposites: enhanced biological and mechanical performances for biomedical applications. Mater Sci Eng C 128:112296. https://doi.org/10.1016/j.msec.2021.112296

Behera M, Das S, Sarkar S (2011) Zinc oxide nanostructures: novel and multifunctional applications. Int J Nanotechnol Appl 5(1):27–38

Belhachem A, Douahi O, Yahia Y et al (2024) Study of the antibacterial activity of hybrid nanocomposites "metal oxide/alginate" synthesized for therapeutic purposes. Discov Med. https://doi.org/10.1007/s44337-024-00147-y

Bhattacharjee S (2016) DLS and zeta potential—what they are and what they are not? J Control Release 235:337–351

Chen Y et al (2021) Graphene-based metal nanocomposites for energy storage applications. Adv Mater 33(12):2100053

Cheng C, Müller KH, Koziol KK, Skepper JN, Midgley PA, Welland ME et al (2009) Toxicity and imaging of multi-walled carbon nanotubes in human macrophage cells. Biomaterials 30(25):4152–4160

Cheng L, Weir MD, Xu HH et al (2012) Antibacterial nanocomposites containing amorphous calcium phosphate and silver nanoparticles with different filler levels. J Biomed Mater Resear Part B: Appl Biomater 100(5):1378–1386. https://doi.org/10.1016/j.dental.2012.01.005

Chong WJ, Shen S, Li Y, Trinchi A, Simunec DP, Kyratzis IL, Sola A, Wen C (2023) Biodegradable PLA-ZnO nanocomposite biomaterials with antibacterial properties, tissue engineering viability, and enhanced biocompatibility. Smart Mater Manufact 1:100004

Cullity BD, Stock SR (2001) Elements of X-ray diffraction, 3rd edn. Prentice Hall, Upper Saddle River, NJ

Dubal DP, Ayyad O, Ruiz V, Gómez-Romero P (2015) Hybrid energy storage: the merging of battery and supercapacitor chemistries. Chem Soc Rev 44(7):1777–1790

Dwivedi P, Narvi SS, Tewari RP (2013) Application of polymer nanocomposites in the nanomedicine landscape: envisaging strategies to combat implant-associated infections. JAppl Biomater Funct Mater 11(3):e129–e142. https://doi.org/10.5301/JABFM.2013.11544

Dykman LA, Khlebtsov NG (2012) Gold nanoparticles in biomedical applications: recent advances and perspectives. Chem Soc Rev 41(6):2256–2282

Fu LH, Qi C, Liu YJ et al (2018) Sonochemical synthesis of cellulose/hydroxyapatite nanocomposites and their application in protein adsorption. Sci Rep 8(1):8292. https://doi.org/10.1038/s41598-018-25566-7

Gholampour A, Ozbakkaloglu T (2020) A review of natural fiber composites: properties, modification and processing techniques, characterization, applications. J Mater Sci 55(3):829–892. https://doi.org/10.1007/s10853-019-03990-y

Giliopoulos D, Zamboulis A, Giannakoudakis D et al (2020) Polymer/metal organic framework (MOF) nanocomposites for biomedical applications. Molecules 25(1):185. https://doi.org/10.3390/molecules25010185

Goldstein JI, Newbury DE, Michael JR, Ritchie NWM, Scott JHJ, Joy DC (2017) Scanning electron microscopy and X-ray microanalysis, 4th edn. Springer, New York

Gupta R et al (2022) Magnetic Fe_3O_4-Ag nanocomposites for antimicrobial applications. ACS Appl Mater Interfaces 14(5):5423–5431

Guo L, Jing D, Liu M, Chen Y, Shen S, Shi J, Zhang K (2014) Functionalized nanostructures for enhanced photocatalytic performance under solar light. Beilstein J Nanotechnol 5:994–1004. https://doi.org/10.3762/bjnano.5.113

Gurnani S, Mukhopadhyay S (2020) Ceramic matrix nanocomposites: properties, processing and applications. In: Ceramic materials. Springer, pp 167–189. https://doi.org/10.1007/978-3-030-27835-6_7

Hachhach M, Bayou S, El Kasmi A, Saidi MZ, Akram H, Hanafi M, Achak O, El Moujahid C, Chafik T (2025) Towards sustainable scaling-up of nanomaterials fabrication: current situation, challenges, and future perspectives. Eng 6(7):149

Jane Cypriyana PJ, Saigeetha S, Samrot AV, Ponniah P, Chakravarthi S (2021) Overview on toxicity of nanoparticles, it's mechanism, models used in toxicity studies and disposal methods—a review. Biocatal Agric Biotechnol 36:102117

Joseph B, Sagarika VK, Chinnu S et al (2020) Cellulose nanocomposites: fabrication and biomedical applications. J Bioresour Bioprod 5(4):223–237. https://doi.org/10.1016/j.jobab.2020.10.001

Khan ZM, Saleem MR (2016) Multifunctional polymer nanocomposites in next generation smart structures. In: Development and prospective applications of nanoscience and nanotechnology nanomaterials and their fascinating attributes, vol 1, p 51. https://doi.org/10.2174/9781681081779116010005

Kim HK, Jang SJ, Cho YS, Park HH (2022) Fabrication of nanostructured polycaprolactone (PCL) film using a thermal imprinting technique and assessment of antibacterial function for its application. Polymers 14(24):5527

Kim J et al (2020) Electrospun CNT-polymer nanocomposites for flexible supercapacitors. Adv Funct Mater 30(12):2002176

Kus-Liśkiewicz M, Fickers P, Ben Tahar I (2021) Biocompatibility and cytotoxicity of gold nanoparticles: recent advances in methodologies and regulations. Int J Mol Sci 22:10952. https://doi.org/10.3390/ijms222010952

Kumar A, Singh R (2022) Metal nanocomposites in flexible electronics. Nano Today 42:101412

Kumar V, Patel R, Mehta N, Das S (2019) Dynamic light scattering techniques in nanoparticle characterization. Pharm Res 36(5):78

Lu AH, Salabas EL, Schüth F (2007) Magnetic nanoparticles: synthesis, protection, functionalization, and application. Angew Chem Int Ed 46(8):1222–1244

Li F, Chen C, Chen X (2024) Tremendous advances, multifaceted challenges and feasible future prospects of biodegradable medical polymer materials. RSC Adv 14(44):32267–32283. https://doi.org/10.1039/D4RA00075G

Liu X et al (2019) Biomedical applications of metal-based nanocomposites. ACS Nano 13(5):4563–4575

Lu T, Chen Y, Sun M, Chen Y, Tu W, Zhou Y, Li X, Hu T (2025) Multifunctional carbon-based nanocomposite hydrogels for wound healing and health management. Gels 11(5):345. https://doi.org/10.3390/gels11050345

Mohamed A, Bansal NP (2024) Structural and mechanical characterization of aluminum matrix nanocomposites reinforced with SiC and Al_2O_3 nanoparticles. Res Mater Appl 34(6):109–115. https://doi.org/10.18280/rcma.340609

Mohanty P, Mahapatra R, Padhi P, Ramana CHVV, Mishra DK (2020) Ultrasonic cavitation: an approach to synthesize uniformly dispersed metal matrix nanocomposites—a review. Nano-Struct Nano-Obj 23:100475. https://doi.org/10.1016/j.nanoso.2020.100475

Nguyen QT, Baird DG (2007) Preparation of polymer–clay nanocomposites and their properties. Adv Polym Technol 25(4):270–285

Niihara K (1999) Mechanism of toughening and strengthening in ceramic composites: transformation-toughening and microcrack toughening. J Eur Ceram Soc 19(4–5):983–991. https://doi.org/10.1016/S0955-2219(98)00212-2

Omanović-Mikličanin E, Badnjević A, Kazlagić A, Hajlovac M (2019) Nanocomposites: a brief review. Heal Technol 10(1):51–59. https://doi.org/10.1007/s12553-019-00380-x

Pathak R, Punetha VD, Bhatt S et al (2023) Multifunctional role of carbon dot-based polymer nanocomposites in biomedical applications: a review. J Mater Sci 58(15):6419–6443. https://doi.org/10.1007/s10853-023-08408-4

Pelaez M, Nolan NT, Pillai SC, Seery MK, Falaras P, Kontos AG, Dunlop PSM, Hamilton JWJ, Byrne JA, O'Shea K, Entezari MH, Dionysiou DD (2012) A review on the visible light active titanium dioxide photocatalysts for environmental applications. Appl Catal B Environ 125:331–349

Porter AE, Gass M, Bendall JS, Muller K, Goode A, Skepper JN et al (2009) Uptake of noncytotoxic acid-treated single-walled carbon nanotubes into the cytoplasm of human macrophage cells. ACS Nano 3(6):1485

Putri GE, Arifani N, Wendari TP et al (2024) Nanocomposites of cellulose-modified cerium oxide nanoparticles and their potential biomedical applications. Case Stud Chem Environ Eng 10:101013. https://doi.org/10.1016/j.cscee.2024.101013

Rai M, Deshmukh SD, Ingle AP, Gupta IR, Galdiero M, Galdiero S (2016) Metal nanoparticles: the protective nanoshield against virus infection. Crit Rev Microbiol 42(1):46–56

Raza S, Ansari A, Siddiqui NN et al (2021) Biosynthesis of silver nanoparticles for the fabrication of non-cytotoxic and antibacterial metallic polymer based nanocomposite system. Sci Rep 11(1):10500. https://doi.org/10.1038/s41598-021-90016

Rokesh K, Sakar M, Do TO (2021) Emerging hybrid nanocomposite photocatalysts for the degradation of antibiotics: insights into their designs and mechanisms. Nanomaterials (Basel, Switzerland) 11(3):572. https://doi.org/10.3390/nano11030572

Ruiz S, Tamayo JA, Delgado Ospina J et al (2019) Antimicrobial films based on nanocomposites of chitosan/poly (vinyl alcohol)/graphene oxide for biomedical applications. Biomolecules 9(3):109. https://doi.org/10.3390/biom9030109

Sanità G, Carrese B, Lamberti A (2020) Nanoparticle surface functionalization: how to improve biocompatibility and cellular internalization. Front Mol Biosci 7:587012

Sarif M, Jegel O, Gazanis A et al (2022) High-throughput synthesis of CeO_2 nanoparticles for transparent nanocomposites repelling Pseudomonas aeruginosa biofilms. Sci Rep 12(1):3935. https://doi.org/10.1038/s41598-022-07833-w

Shahid N, Erum A, Hanif S, Malik NS, Tulain UR, Syed MA (2024) Nanocomposite hydrogels—a promising approach towards enhanced bioavailability and controlled drug delivery. Curr Pharm Des 30(1):48–62. https://doi.org/10.2174/0113816128283466231219071151

Shaikh AA, Datta P, Dastidar P, Majumder A, Das MD, Manna P, Roy S (2024) Biopolymer-based nanocomposites for application in biomedicine: a review. J Polym Eng 44(2):83–116. https://doi.org/10.1515/polyeng-2023-0166

Shao Y, Yin G, Gao Y (2007) Understanding and approaches for the durability issues of Pt-based catalysts for PEM fuel cells. J Power Sources 171(2):558–566

Sharma S, Mishra SS, Kumar R, Yadav RM (2022) Recent progress on polyvinylidene difluoride-based nanocomposites: applications in energy harvesting and sensing. New J Chem 46(39):18613–18646

Shi X, Sitharaman B, Pham QP, Spicer PP, Hudson JL, Wilson LJ, Tour JM, Raphael RM, Mikos AG (2008) In vitro cytotoxicity of single-walled carbon nanotube/biodegradable polymer nanocomposites. J Biomed Mater Res A 86(3):813–823

Shirazi H, Daneshpour M, Kashanian S et al (2015) Synthesis, characterization and in vitro biocompatibility study of Au/TMC/Fe_3O_4 nanocomposites as a promising, nontoxic system for biomedical applications. Beilstein J Nanotech 6(1):1677–1689. https://doi.org/10.3762/bjnano.6.170

Singh NB, Agarwal S (2016) Nanocomposites: an overview. Emerg Mater Res 5(1):1. https://doi.org/10.1680/jemmr.15.00025

Smith A, Jones B (2020) Surface morphology analysis in tissue engineering scaffolds. Biomed Mater 15(4):445–453

Stuart B (2004) Infrared spectroscopy: fundamentals and applications. Wiley, Chichester

Varghese OK, Gong D, Paulose M, Ong KG, Grimes CA (2003) Hydrogen sensing using titania nanotubes. Sens Actuators B Chem 93(1–3):338–344

Vinţeler V, Dudric R, Bortnic R et al (2024) Synthesis and characterization of $CoFe_2O_4$ nanoparticles and $CoFe_2O_4$@$BaTiO_3$ nanocomposites for biomedical applications. Rom J Phys 69:611. https://doi.org/10.59277/RomJPhys.2024.69.611

Wang P et al (2020) Nanocomposites in energy storage: a review. J Power Sources 450:227622

Williams DB, Carter CB (2009) Transmission electron microscopy: a textbook for materials science, 2nd edn. Springer, New York

Wu W, Yu X, Sun J et al (2023) Zeolitic imidazolate framework (ZIF-8) decorated iron oxide nanoparticles loaded doxorubicin hydrochloride for osteosarcoma treatment: in vitro and in vivo preclinical studies. Int J Nanomed 18:7985–7999. https://doi.org/10.2147/IJN.S438771

Yu WH, Sing SL, Chua CK, Kuo CN, Tian XL (2019) Particle-reinforced metal matrix nanocomposites fabricated by selective laser melting: a state of the art review. Prog Mater Sci 104:330–380. https://doi.org/10.1016/j.pmatsci.2019.04.00

Yuan X, Nie W, He Z, Yang J, Shao B, Ma X, Zhang X, Bi Z, Sun L, Liang X, Tie Y (2020) Carbon black nanoparticles induce cell necrosis through lysosomal membrane permeabilization and cause subsequent inflammatory response. Theranostics 10(10):4589

Zhang H et al (2021) Metal-based nanocomposites for environmental catalysis. Chem Rev 121(7):4567–4590

Zhou H et al (2020) Sol-gel synthesis of TiO_2/SiO_2 nanocomposites for photocatalysis. J Mater Sci 55(10):3205–3214

Chapter 14
Pharmaceutical Formulations Using Polymer Nanocomposites

Kagimu Braize Ben, Anyole Arthur Amule, Delafrique Deogratias, Odongo Denis, Gershom Muganga, Kyemanta C. K. Desire, Mwandah Daniel Chans, Namutebi Josephine, Crispin Duncan Sessazi, Ntale Paulo, Okello Aaron, Mukasa Benjamin Kabuuza, Hannington Gumisiriza, Mutyaba Ronald, and Sachin N. Kothawade

Abstract Polymer nanocomposites (PNCs) are advanced materials combining polymer matrices with nanoscale fillers such as clays, metal nanoparticles, and biocompatible ceramics, resulting in composites with enhanced mechanical, thermal, and functional properties. These materials are transformative in pharmaceutical applications, offering solutions for drug formulation and delivery challenges. PNCs enable precise control over drug release kinetics, improve bioavailability, and allow targeted delivery to specific tissues or cells. Encapsulating active pharmaceutical ingredients (APIs) enhances solubility, stability, and therapeutic efficacy. Drawing from material science, nanotechnology, and biology, PNCs have revolutionized drug delivery systems, prosthetics, and sustained release technologies, bridging tra-

K. B. Ben (✉)
Department of Pharmaceutical Sciences, Faculty of Medicine, Mbarara University of Science and Technology, Mbarara, Uganda

Department of Pharmacy, Faculty of Medicine, Mbarara University of Science and Technology, Mbarara, Uganda
e-mail: kagimubraizeben@akolifesciences.com

AKO Life Sciences Uganda Limited, Mbarara, Uganda

A. A. Amule · O. Denis
Department of Pharmaceutical Sciences, Faculty of Medicine, Mbarara University of Science and Technology, Mbarara, Uganda

AKO Life Sciences Uganda Limited, Mbarara, Uganda
e-mail: arthuranyole@akolifesciences.com; denisodongo@akolifesciences.com

D. Deogratias · C. D. Sessazi · O. Aaron · M. B. Kabuuza · M. Ronald
Department of Pharmaceutical Sciences, Faculty of Medicine, Mbarara University of Science and Technology, Mbarara, Uganda
e-mail: dsesaazi@must.ac.ug

S. N. Kothawade et al. (eds.), *Polymer Nanocomposites*,
https://doi.org/10.1007/978-3-032-00365-2_14

G. Muganga
Department of Pharmaceutical Sciences, Faculty of Medicine, Mbarara University of Science and Technology, Mbarara, Uganda

Department of Pharmacy, Faculty of Medicine, Mbarara University of Science and Technology, Mbarara, Uganda

Pharm-Biotechnology and Traditional Medicine Centre of Excellence, Mbarara University of Science and Technology, Mbarara, Uganda

K. C. K. Desire · N. Josephine · N. Paulo
Department of Pharmaceutical Sciences, Faculty of Medicine, Mbarara University of Science and Technology, Mbarara, Uganda

Department of Pharmacy, Faculty of Medicine, Mbarara University of Science and Technology, Mbarara, Uganda

M. D. Chans
Department of Pharmacology and Toxicology, Kampala International University, Kampala, Uganda

Department of Pharmacology and Therapeutics, Faculty of Medicine, Mbarara University of Science and Technology, Mbarara, Uganda
e-mail: danielchans@must.ac.ug

H. Gumisiriza
Department of Chemistry, Faculty of Science, Mbarara University of Science and Technology, Mbarara, Uganda
e-mail: hgumisiriza@must.ac.ug

S. N. Kothawade
Department of Pharmaceutics, SCSSS's Sitabai Thite College of Pharmacy, Shirur, Pune, Maharashtra, India

ditional polymers with cutting-edge nanotechnology for improved medical outcomes. Polymer nanocomposites have emerged as a transformative innovation in pharmaceutical formulations, providing unprecedented versatility in drug delivery systems. This chapter will explore the design and synthesis of polymer nanocomposites tailored for pharmaceutical applications, emphasizing their physicochemical properties, biocompatibility, and enhanced functionality. Key topics include the role of polymer matrices and various nanofillers (e.g., metal oxides, carbon-based materials, and biocompatible ceramics) in modulating drug release profiles, improving bioavailability, and achieving targeted delivery. Furthermore, the chapter will review advances in surface functionalization and encapsulation strategies that enhance the stability and solubility of active pharmaceutical ingredients (APIs) within polymer nanocomposite matrices. Case studies on applications in chronic disease management, oncology, and infectious disease treatment will illustrate the therapeutic potential of polymer nanocomposites. The discussion will also highlight critical characterization techniques essential for optimizing pharmaceutical nanocomposites, including SEM, TEM, and FTIR. Finally, the chapter will address regulatory considerations, biocompatibility, and safety profiles to ensure effective translation from lab to clinical application. The insights presented in this chapter on polymer nanocomposites (PNCs) aim to bridge gaps in knowledge and practice for diverse stakeholders in the pharmaceutical field. Researchers and formulators will find valuable strategies for developing innovative drug delivery systems, while analysts will gain a deeper understanding of techniques to evaluate and optimize nanocomposite formulations. Policymakers and regulatory professionals can use the discussion on safety and compliance to inform guidelines and frameworks for integrating these materials into mainstream healthcare. By advancing scientific knowledge in material science and nanotechnology, this chapter contributes to the development of safer, more effective pharmaceutical solutions, ultimately improving patient care and inspiring future research.

Keywords Polymer nanocomposites (PNCs) · Controlled drug release · Targeted drug delivery · Surface functionalization · Biocompatibility · Stimuli-responsive systems

14.1 Introduction to Polymer Nanocomposites in Pharmaceuticals

Polymer nanocomposites (PNCs) have emerged as a transformative technology in pharmaceutical formulations. These composites offer unique properties that significantly enhance drug delivery systems by integrating polymers with nanoscale materials. This section provides a comprehensive overview of polymer nanocomposites, emphasizing their value in improving drug formulations, the types of polymers utilized, and the various nanofillers employed within a pharmaceutical context. Most of these will be discussed in later subsections of this chapter.

14.1.1 The Significance of Polymer Nanocomposites in Drug Formulations

The pharmaceutical industry faces numerous challenges, particularly concerning the solubility and bioavailability of new drug compounds. Many promising therapeutic agents exhibit poor solubility in aqueous environments, hindering their effectiveness. Polymer nanocomposites address this issue through several key mechanisms:

Enhanced Drug Solubility One of the primary advantages of using PNCs is their ability to improve the solubility of poorly water-soluble drugs. Incorporating nanoparticles into a polymer matrix can significantly increase the surface area for dissolution. For instance, studies have shown that drug-loaded polymeric nanoparticles can enhance the dissolution rates of hydrophobic drugs by providing a larger surface area for interaction with solvents (Zhang et al. 2020). This improvement in solubility can lead to higher plasma concentrations and improved therapeutic outcomes.

Controlled Release Mechanisms Controlled drug release is critical for maintaining effective therapeutic levels over extended periods. PNCs facilitate this through their unique structural properties. By carefully selecting both the polymer matrix and the type of nanoparticles used, researchers can tailor the release profiles of drugs. For example, hydrophilic polymers can create a swelling mechanism that allows for gradual drug release, while hydrophobic polymers can slow down the diffusion of drugs (Khan et al. 2021). This controlled release minimizes side effects and enhances patient compliance by reducing dosing frequency.

Targeted Delivery Targeted drug delivery is another significant advantage offered by PNCs. The ability to direct drugs to specific tissues or cells enhances treatment efficacy and reduces systemic side effects. Nanoparticles can be functionalized with targeting ligands, such as antibodies or peptides that bind specifically to receptors overexpressed on certain cancer cells (Mansoori et al. 2019). This targeted approach ensures that higher drug concentrations are delivered directly to the site of action, improving therapeutic outcomes while minimizing toxicity.

Improved Biocompatibility The biocompatibility and biodegradability of many polymers used in PNCs are crucial for their pharmaceutical applications. Polymers such as polylactic acid (PLA) and chitosan are safe for use in biological systems and break down into non-toxic byproducts that can be easily eliminated from the body (Kumar et al. 2022). This characteristic is fundamental in long-term treatments where the accumulation of materials could pose health risks.

14.1.2 Types of Polymers Used in Pharmaceutical Nanocomposites

In formulating PNCs for pharmaceutical applications, various types of polymers are utilized, each offering distinct advantages:

14.1.2.1 Synthetic Polymers

Polylactic Acid (PLA) PLA is a biodegradable thermoplastic derived from renewable resources like corn starch or sugarcane. Its excellent mechanical properties and biocompatibility make it suitable for drug delivery applications, particularly for sustained release formulations (Davis et al. 2020).

Polyglycolic Acid (PGA) Often used in combination with PLA, PGA enhances degradation rates and mechanical strength. The copolymerization of PLA and PGA allows for fine-tuning of release kinetics based on the desired therapeutic profile.

Polyvinyl Pyrrolidone (PVP) PVP is known for its solubilizing properties and ability to form stable complexes with various drugs. Its hydrophilic nature makes it an excellent choice for enhancing drug solubility and bioavailability.

14.1.2.2 Natural Polymers

Chitosan Derived from chitin, chitosan is a natural polysaccharide that exhibits favorable biocompatibility and biodegradability. It has been extensively studied for its ability to enhance drug absorption and provide mucoadhesive properties, making it ideal for oral delivery systems (Nafeeza et al. 2021).

Alginate Alginate is a naturally occurring polysaccharide extracted from brown seaweed. Its gel-forming ability allows it to be used in hydrogels for controlled drug release applications. Alginate-based systems can encapsulate drugs effectively and provide sustained release profiles due to their swelling behavior in aqueous environments.

14.1.3 Types of Nanofillers Employed in PNCs

The choice of nanofillers plays a crucial role in determining the properties and performance of polymer nanocomposites:

14.1.3.1 Inorganic Nanoparticles

Silica Nanoparticles Silica nanoparticles are commonly used in PNCs due to their high surface area and tunable porosity. They enhance mechanical strength and thermal stability while also improving drug loading capacities (Li et al. 2020).

Metal Nanoparticles Gold and silver nanoparticles are notable for their antimicrobial properties and potential use in targeted therapy. These metal nanoparticles can be modified with targeting ligands to facilitate specific interactions with diseased cells (Saha et al. 2019).

14.1.3.2 Layered Silicates

Nanoclays Nanoclays such as montmorillonite are utilized to create intercalated or exfoliated structures within polymer matrices. These structures enhance barrier properties against moisture and gases while controlling drug release rates through diffusion mechanisms.

Polymer nanocomposites represent a cutting-edge approach to addressing some of the most pressing challenges faced by the pharmaceutical industry today. Their ability to enhance drug solubility, provide controlled release mechanisms, enable targeted delivery, and ensure biocompatibility makes them invaluable in developing more effective therapeutic strategies. The careful selection of both polymers and nanofillers is essential for optimizing these systems for specific applications, paving the way for innovative solutions in modern medicine.

14.2 Role in Drug Delivery Systems

Polymer nanocomposites represent a highly advanced class of materials, distinguished by incorporating nanoscale fillers into a polymer matrix. These fillers, which may include organic or inorganic clays, metal nanoparticles, or hydroxyapatite, act as the dispersed phase, while the polymer matrix serves as the continuous phase, providing structural support. This integration significantly enhances the properties of the resulting material, making polymer nanocomposites an innovative platform with widespread applications, particularly in pharmaceutical drug delivery systems.

The multidisciplinary nature of polymer nanocomposites arises from their ability to integrate principles of materials science, nanotechnology, and biological science. This convergence has revolutionized the field of medical science by offering solutions to longstanding challenges in drug delivery, such as improving drug stability, controlled release, and targeted delivery to specific tissues or cells. The nanoscale fillers in these composites not only enhance the mechanical, thermal, and barrier properties of the polymers but also confer unique functionalities, such as bioactivity, responsiveness to external stimuli, and the ability to interact with biological systems at the molecular level.

Recent advancements underscore the potential of polymer nanocomposites in addressing critical healthcare needs. For instance, their application in fabricating nanocarriers for drug delivery leverages the high surface area and functional versatility of the fillers to achieve precise drug loading and controlled release kinetics. Moreover, the tunable nature of polymer nanocomposites enables the development of highly biocompatible and biodegradable systems tailored for specific therapeutic applications.

As highlighted by Muhammad Rahim (2020), the field of nanocomposites is not merely a fusion of different scientific disciplines but a transformative technology poised to redefine medical innovation. By harnessing the unique properties of nanomaterials and polymer matrices, polymer nanocomposites hold immense promise in advancing drug delivery systems and improving patient outcomes in the pharmaceutical sector.

Drug nanoparticles, often formulated as nanosuspensions, can be manufactured using top-down, bottom-up, or combinative approaches. Each of these methods offers distinct advantages depending on the physicochemical properties of the drug and the desired nanoparticle characteristics.

In top-down approaches, coarse, as-received drug crystals are reduced to the micro- or nanoscale using high shear-impact forces. Techniques such as ball milling, stirred media milling, and high-pressure homogenization (HPH) are widely employed for this purpose. These methods rely on mechanical energy to break down larger particles into smaller ones, resulting in a uniform particle size distribution and enhanced dissolution properties.

Conversely, bottom-up approaches focus on the generation of nanoparticles from dissolved drug molecules. This is typically achieved through precipitation techniques, such as liquid anti-solvent precipitation and supercritical fluid precipitation. These methods involve dissolving the drug in a suitable solvent and then precipitating it by introducing an anti-solvent, leading to the formation of nanoparticles. Another notable bottom-up method, melt emulsification, is particularly effective for drugs with low melting points. This technique involves dispersing the drug in an aqueous stabilizer solution, heating the mixture to melt the drug crystals, and then rapidly cooling it to form nanoparticles.

Combining these two approaches, referred to as the combinative strategy, leverages the advantages of both top-down and bottom-up techniques. By integrating mechanical size reduction with molecular-level precipitation, this approach enhances process efficiency and achieves superior control over nanoparticle properties. As described by Anagha Bhakay (2018), these methodologies provide a versatile toolkit for producing drug nanosuspensions, addressing challenges such as poor solubility, limited bioavailability, and variability in drug absorption. Each approach can be tailored to specific drug formulations, making them invaluable in the development of advanced pharmaceutical delivery systems.

This approach in drug development has enhanced drug bioavailability and enabled targeted delivery and controlled release rates, significantly improving therapeutic outcomes. Polymer nanocomposites play the following roles in achieving improved therapeutic outcomes.

14.2.1 Improving Drug Bioavailability

Polymer nanocomposites are crucial in enhancing drug bioavailability through several mechanisms, including improving drug solubility, stability, and absorption. This is particularly significant for poorly water-soluble drugs, which often exhibit limited bioavailability in conventional formulations. By incorporating these drugs into nanocomposite systems, their dissolution rate is enhanced, leading to better therapeutic outcomes. For instance, formulations utilizing Soluplus-based nanocomposites have demonstrated improved dissolution rates of drugs like aceclofenac, resulting in significantly higher bioavailability compared to traditional formulations (Begines et al. 2020).

One of the key mechanisms through which polymer nanocomposites enhance bioavailability is by creating drug nanoparticles. Reducing the particle size to the nanoscale increases the surface area available for dissolution, thereby accelerating drug solubilization and absorption. This facilitates a more efficient uptake of the drug into the systemic circulation, leading to improved bioavailability (Begines et al. 2020; Liechty et al. 2010).

In addition to improving solubility and dissolution, polymeric nanoparticles can be engineered to optimize interactions with cell membranes. By tailoring their surface properties, these nanoparticles promote faster and more efficient transport of the drug across cellular barriers, further enhancing absorption and bioavailability (Begines et al. 2020).

Another critical advantage of polymer nanocomposites is their ability to provide a stable environment for drug molecules. This stability protects drugs from degradation during storage and as they transit through physiological environments, such as the acidic conditions of the gastrointestinal tract. Such protection is vital to ensuring that the drug remains effective until it reaches its target site. Notably, studies have shown that polymeric nanoparticles exhibit greater stability than traditional lipid-based carriers like liposomes, which are often susceptible to leakage and degradation (Begines et al. 2020).

14.2.2 Drug Delivery

Polymer nanocomposites can be designed to enable site-specific drug delivery, ensuring that therapeutic agents are selectively transported to target cells or tissues while minimizing off-target effects. This precision is achieved by conjugating targeting ligands to the surface of the nanoparticles. These ligands, such as antibodies, peptides, or small molecules, are engineered to bind specifically to receptors that are overexpressed on the surface of target cells, such as cancer cells. By leveraging this specificity, polymer nanocomposites reduce drug exposure to healthy tissues, thereby minimizing side effects and improving therapeutic efficacy (Liechty et al. 2010).

A notable example of this technology is the use of polymer-graphene nanocomposites for cancer therapy. These nanocomposites are functionalized with ligands like folic acid, which bind selectively to folate receptors commonly overexpressed on cancer cells. This targeted approach ensures that the drug-loaded nanocomposites accumulate preferentially in tumor tissues, enhancing the drug's therapeutic index while sparing healthy tissues from adverse effects (Begines et al. 2020).

The ability of polymer nanocomposites to achieve site-specific targeting represents a significant advancement in drug delivery systems. By combining the properties of polymers with nanoscale engineering, these systems provide a platform for precision medicine, particularly in the treatment of complex diseases such as cancer. This approach not only enhances the efficacy of treatments but also reduces the systemic toxicity associated with conventional therapies, offering a more patient-centric solution to modern healthcare challenges.

14.2.3 Controlled Drug Release

Polymer nanocomposites play a pivotal role in enabling controlled and sustained drug release, which reduces the frequency of drug administration while improving patient compliance (Liechty et al. 2010; De Leon et al. 2012). By maintaining consistent drug concentrations in the bloodstream, these systems enhance bioavailability and minimize side effects associated with plasma concentration spikes that can lead to toxicity (Liechty et al. 2010).

One example of controlled drug release involves nanoclays embedded within polymer matrices, which are utilized in oral multiparticulate systems. These systems provide slow and controlled drug release profiles, ensuring prolonged therapeutic effects and improved efficacy (Paravastu Venkata Kamala Kumari 2019; Liechty et al. 2010).

Soluplus®-based nanocomposites are another effective controlled release system, particularly for poorly soluble drugs like aceclofenac, a nonsteroidal anti-inflammatory drug. Studies have demonstrated that Soluplus® nanocomposites significantly enhance the dissolution rate and bioavailability of aceclofenac compared to its pure form, offering a practical solution for solubility challenges in drug delivery (Begines et al. 2020).

Additionally, polymeric liposome complexes have shown promise in controlled drug delivery systems. These complexes enhance the solubility and stability of hydrophobic drugs and provide pH-responsive release mechanisms. Such systems are especially valuable for targeting specific tissues or disease sites, improving therapeutic outcomes while reducing off-target effects (Fernandes and Jozala 2022).

Dendritic polymer nanocomposites represent another advanced controlled release technology. These hybrid systems, which combine dendritic structures with traditional polymers, offer improved drug loading capacity, release control, and delivery efficiency. Their unique architecture allows for precise modulation of drug release profiles, making them suitable for a wide range of pharmaceutical applications (Fernandes 2023).

14.3 Methods for Synthesizing Polymer Nanocomposites with Precise Drug Release Properties

Polymer nanocomposites have emerged as a pivotal class of materials in drug delivery systems, primarily due to their tunable properties, biocompatibility, and ability to achieve controlled drug release. These innovative platforms incorporate nanoparticles within a polymer matrix, leveraging their synergistic properties to enhance drug release profiles. The synthesis of polymer nanocomposites with precise drug release characteristics necessitates meticulous design and integration of polymers and nanomaterials.

Selecting appropriate polymers and nanoparticles is critical for designing effective nanocomposites tailored to specific drug release properties. Biodegradable and biocompatible polymers such as polylactic-co-glycolic acid (PLGA), polyethylene glycol (PEG), and chitosan are commonly used due to their favorable degradation rates and interactions with drugs, which can be fine-tuned for various applications. Starch-based and hydrocolloid-based biopolymers are particularly noted for their biodegradability and biocompatibility, making them suitable for pharmaceutical applications.

Common nanoparticles utilized in these systems include silica, titanium dioxide, and graphene oxide. These nanoparticles can be functionalized to enhance drug loading capabilities, compatibility with the polymer matrix, and responsiveness to external stimuli. The strategic selection and modification of these nanoparticles play a crucial role in optimizing the performance of the resulting nanocomposites.

Various methods exist for synthesizing polymer nanocomposites, each offering unique advantages and challenges. Techniques such as in situ polymerization allow for the uniform distribution of nanoparticles within the polymer matrix while ensuring strong interfacial bonding. Other methods, including solution blending and melt intercalation, provide alternative routes for achieving desired properties in the final product.

By incorporating nanoparticles into the polymer matrix through these synthesis methods, researchers can create a new material class that exhibits improved mechanical integrity, enhanced drug encapsulation efficiency, and targeted release profiles. This section outlines the key strategies used in the synthesis of polymer nanocomposites with an emphasis on achieving optimal drug release characteristics.

14.3.1 Supercritical Fluid Technology

Supercritical fluid technology, particularly utilizing supercritical CO_2, has emerged as a solvent-free method for synthesizing polymer nanocomposites. This innovative approach is recognized as a green technology due to its ability to replace traditional chemical solvents with environmentally friendly alternatives. Supercritical CO_2 possesses both liquid-like density and gas-like diffusivity above its critical temperature and pressure, making it a versatile solvent for various pharmaceutical applications.

In this synthesis method, polymers, drugs, and nanoparticles are exposed to supercritical CO_2, which acts as a solvent to swell the polymer matrix. Upon depressurization, the drug and nanoparticles become encapsulated within the matrix. Smith et al. (2022) reported the successful synthesis of polylactide (PLA) nanocomposites using supercritical CO_2, demonstrating the extended release of hydrophobic drugs.

The process capitalizes on the unique properties of CO_2 in its supercritical state, where the liquid phase becomes a poor solvent for solutes upon sufficient gas expansion. This phenomenon leads to controlled precipitation of particles. The technology employs vigorous mixing techniques at a nanoliter scale, allowing for continuous operation, enhanced reproducibility, and reduced preparation steps. In a typical setup, the polymer, nanoparticles, and drug are dissolved or dispersed in a liquid solvent under high-pressure CO_2. Upon decompression, CO_2 rapidly transitions out of the supercritical state, precipitating the solute as nanoparticles embedded within the polymer matrix. This rapid phase change ensures uniform dispersion of nanoparticles and drugs throughout the composite.

Supercritical fluid extraction using carbon dioxide has been recognized as a clean and versatile method that avoids the use of harmful organic solvents. The dual properties of supercritical fluids enable various pharmaceutical applications; for instance, their liquid-like characteristics benefit extraction processes for organic solvents or impurities, drug solubilization, and polymer plasticization, while their gas-like features facilitate mass transfer processes.

This technology has proven to be versatile and environmentally attractive, effectively addressing complex challenges in pharmaceuticals. It is particularly suitable for temperature-sensitive drugs and materials and has been applied in creating porous scaffolds for tissue engineering.

The advantages of supercritical fluid technology include its environmentally friendly nature due to the avoidance of harmful solvents, precise control over particle size, morphology, and dispersion, efficient encapsulation of thermolabile drugs at relatively low temperatures, facilitation of porous structure design, and scalability for industrial applications with improved safety profiles.

Applications of this technology encompass the fabrication of porous scaffolds for tissue engineering, the development of sustained-release drug delivery systems, and the production of polymer nanocomposites with stimuli-responsive behavior for targeted drug delivery.

However, there are limitations and challenges associated with this method. These include high initial costs for equipment and operation, the necessity for careful optimization of parameters such as temperature, pressure, and polymer-solvent compatibility, as well as limited solubility of some drugs and polymers in supercritical CO_2 that may require co-solvents or additional processing steps.

14.3.2 Microfluidic Synthesis

Microfluidic synthesis represents a cutting-edge method that employs microfluidic devices to produce polymer nanocomposites with high precision and reproducibility. This technique utilizes microchannels, typically on the micron scale, to control the mixing of polymers, nanoparticles, and drugs under laminar flow conditions. By enabling precise control over reaction parameters such as flow rate, mixing time, and temperature, microfluidic devices facilitate the effective synthesis of polymer nanocomposites.

In a typical setup, polymer and nanoparticle solutions are introduced through separate inlets into a microchannel, where controlled mixing occurs. Drug molecules can either be encapsulated during synthesis or adsorbed onto the nanocomposites post-synthesis. Various designs, including T-junctions and flow-focusing geometries, are employed to fine-tune nanoparticle size and distribution. The ability to manipulate fluid dynamics at this scale allows for exceptional control over nanoparticle morphology and size due to uniform mixing.

The advantages of microfluidic synthesis include high reproducibility and scalability, reduced consumption of reagents, and minimal waste generation, making it an environmentally friendly option. Furthermore, this method allows for the incorporation of multiple functionalities into a single nanocomposite by enabling sequential or parallel mixing of components. These capabilities support high-throughput production suitable for industrial-scale applications.

Microfluidic synthesis has found numerous applications in the fabrication of polymer nanocomposites for cancer therapy, where tunable drug release profiles are crucial. It also facilitates the development of nanocarriers for hydrophobic drugs, enhancing their bioavailability. Additionally, stimuli-responsive nanocomposites produced through this method enable on-demand drug release.

Despite its many advantages, microfluidic synthesis has limitations. The complex fabrication of microfluidic devices can lead to increased initial setup costs. Moreover, scaling up microfluidic processes while maintaining precision poses significant challenges. Nonetheless, ongoing advancements in microfluidics technology continue to enhance the efficiency and applicability of this innovative synthesis method.

14.3.3 Click Chemistry

Click chemistry is a versatile and highly efficient approach for synthesizing polymer nanocomposites with precise drug release properties. This method, developed as a modular and reliable technique for forming covalent bonds, has gained prominence in nanomedicine due to its bioorthogonality, high yield, and minimal byproduct formation. At the core of click chemistry is the copper(I)-catalyzed azide-alkyne cycloaddition (CuAAC), which involves the rapid and specific coupling of an azide and an alkyne to form a stable triazole linkage under mild conditions.

The synthesis process begins with the functionalization of polymers with azide or alkyne groups, which can occur during or after their synthesis. Similarly, nanoparticles or drugs are modified to include complementary functional groups. These functionalized polymers and nanoparticles are then combined in the presence of a copper catalyst, typically under aqueous conditions or in mild organic solvents. The click reaction ensures uniform dispersion and stable integration of nanoparticles within the polymer matrix. Additionally, post-modification can introduce further functional groups, such as targeting ligands or responsive moieties, through subsequent click reactions.

The advantages of click chemistry in polymer nanocomposite synthesis are numerous. The regioselectivity and specificity of click reactions allow for precise control over the spatial arrangement of nanoparticles and drugs within the polymer matrix. Reactions proceed efficiently at ambient temperature and physiological pH, preserving the structural integrity of sensitive drugs and polymers. This method is versatile and applicable to a wide range of substrates, enabling the synthesis of hybrid nanocomposites with diverse properties. Furthermore, near-quantitative yields with minimal purification steps make click chemistry suitable for scalable production. Importantly, click reactions proceed without interference from biological molecules, making them ideal for drug delivery systems designed for in vivo applications.

Applications of click chemistry in drug delivery include the development of stimuli-responsive systems. Functional groups introduced via click chemistry can respond to pH, temperature, or light, enabling on-demand drug release. For instance, pH-sensitive linkages in triazoles can break under acidic conditions, releasing drugs specifically in tumor microenvironments (Wang et al. 2019). Additionally, click chemistry facilitates targeted drug delivery by allowing the attachment of ligands such as antibodies or peptides to polymer nanocomposites, improving specificity and reducing off-target effects. The modularity of this approach also allows for the incorporation of multiple drugs within a single nanocomposite, offering synergistic therapeutic effects.

Despite its many advantages, there are challenges associated with click chemistry. The copper catalysts used in CuAAC reactions may exhibit cytotoxicity; however, strategies such as copper-free click chemistry-like strain-promoted azide-alkyne cycloaddition (SPAAC) have been developed to mitigate this issue. Additionally, functionalizing polymers and nanoparticles with azide or alkyne groups requires additional steps that may increase complexity and cost. While highly efficient at the laboratory scale, translating click chemistry-based approaches to industrial production necessitates careful optimization of reaction conditions and catalyst recycling.

Emerging trends in click chemistry include the development of bioinspired nanocomposites that mimic extracellular matrices to enhance biocompatibility and drug retention. Furthermore, incorporating click chemistry into additive manufacturing technologies enables the precise spatial distribution of drugs within 3D-printed scaffolds for tissue engineering applications. Integrating click chemistry with other techniques such as microfluidic synthesis or layer-by-layer assembly is enhancing the functionality and complexity of polymer nanocomposites.

Several case studies illustrate the effectiveness of click chemistry in drug delivery applications. Wang et al. (2019) demonstrated the encapsulation of doxorubicin in polymer nanocomposites using click chemistry, achieving pH-responsive release profiles. Xi et al. synthesized dual-responsive polymer nanocomposites that showed enhanced therapeutic efficacy in cancer models. Additionally, Worrell et al. reported the fabrication of polymer nanocomposites with covalently attached antimicrobial peptides via click chemistry, resulting in prolonged and targeted antimicrobial activity.

By leveraging the precision, versatility, and bioorthogonality of click chemistry, researchers can develop advanced polymer nanocomposites tailored for precise drug delivery. Future innovations aim to address current limitations while paving the way for broader clinical applications.

14.3.4 Stimuli-Responsive Nanocomposites

Stimuli-responsive nanocomposites are engineered to release drugs in response to specific physiological or environmental triggers, such as pH, temperature, light, or enzymes. This targeted approach enhances therapeutic efficacy by delivering drugs precisely when and where they are needed.

The process begins with the design of polymer matrices. Polymers with functional groups sensitive to environmental stimuli are selected; for example, pH-sensitive polymers like polyacrylic acid (PAA) or poly-lactic-co-glycolic acid (PLGA) degrade under the acidic conditions typically found in tumor microenvironments. Additionally, the incorporation of nanoparticles, such as gold or quantum dots, allows for photothermal or photodynamic activation.

Next, nanoparticles that are functionalized with responsive ligands are embedded within the polymer matrix. The drug loading process can involve either covalently bonding drugs to responsive polymers or physically encapsulating them within the nanocomposites. This dual approach ensures that the drugs are effectively delivered in response to specific stimuli.

The advantages of stimuli-responsive nanocomposites include targeted and controlled drug release, which minimizes side effects and enhances drug stability and bioavailability. Furthermore, their multifunctionality allows for simultaneous drug delivery alongside imaging or diagnostic capabilities. Applications of these systems are diverse and include cancer therapy with on-demand drug release triggered by tumor acidity or external light, treatment of infections using heat-responsive nanocomposites that release antibiotics when exposed to fever-like temperatures, and the delivery of biologics such as insulin in response to glucose levels.

However, there are challenges associated with developing stimuli-responsive systems. One significant challenge is balancing responsiveness and stability under physiological conditions. Additionally, scalability for clinical use remains a concern, as does the high cost associated with advanced materials and synthesis methods.

Recent advancements in stimuli-responsive materials have led to the development of nanocomposites that can alter their properties in response to external stimuli, enabling controlled drug release. Various types of stimuli have been identified:

pH-Responsive: Nanocomposites that release drugs in response to the acidic environment characteristic of tumor tissues.

Temperature-Responsive: Systems that react to temperature changes, which can be particularly useful in hyperthermia treatments.

Overall, stimuli-responsive nanocomposites represent a promising frontier in drug delivery systems, enhancing therapeutic outcomes while minimizing adverse effects through precise control over drug release mechanisms.

14.3.5 In Situ Polymerization

One widely used technique for synthesizing polymer nanocomposites is in situ polymerization, where monomers are polymerized in the presence of nanomaterials. This method results in a uniform distribution of nanoparticles within the polymer matrix, facilitating a strong interface between the polymer and nanoparticles, which is critical for controlling drug release. For instance, Zhao et al. (2021) demonstrated the synthesis of poly(lactic-co-glycolic acid) (PLGA) nanocomposites embedded with magnetic nanoparticles via in situ polymerization, achieving precise drug release under magnetic field stimuli.

In situ polymerization is a versatile technique that ensures nanoparticles are evenly distributed within the polymer matrix and interact effectively with the polymer network. The process begins with mixing the monomer, initiator, and nanoparticles to form a homogeneous mixture. Polymerization is then initiated through thermal, photochemical, or chemical means, leading to the formation of the polymer matrix around the nanoparticles. Common monomers used in this process include methyl methacrylate (MMA), styrene, and caprolactone, which facilitate the creation of robust polymer structures.

The advantages of in situ polymerization are significant. It ensures strong interfacial bonding between the polymer and nanoparticles, resulting in enhanced mechanical and thermal properties. Additionally, it provides homogeneous nanoparticle dispersion, which is crucial for achieving consistent drug release profiles. This technique also allows for the incorporation of functionalized nanoparticles, enabling tailored interactions with specific drugs.

In situ polymerization has various applications, particularly for temperature-sensitive drugs, as the polymerization conditions can be optimized to avoid drug degradation. It is frequently utilized in developing nanocomposites for controlled release of anticancer drugs and in synthesizing hybrid nanomaterials for targeted drug delivery and tissue engineering applications (Wang et al. 2020).

However, there are challenges associated with this technique. Controlling the polymerization rate is essential to prevent nanoparticle agglomeration. Additionally, removing residual monomers or initiators is crucial since they may affect drug stability and biocompatibility. Optimization of reaction conditions is necessary to achieve the desired properties in the resulting nanocomposites.

Recent advances have further enhanced the capabilities of in situ polymerization techniques. For example, incorporating living polymerization methods such as atom transfer radical polymerization (ATRP) allows for precise control over both polymer architecture and nanoparticle distribution. Additionally, bioinspired approaches have been developed to synthesize nanocomposites with improved biocompatibility and drug release performance.

Overall, in situ polymerization remains a powerful tool in creating advanced polymer nanocomposites tailored for specific drug delivery applications.

14.3.6 Solvent Casting and Evaporation

Solvent casting and evaporation represent a prevalent methodology for the fabrication of polymer nanocomposite films and coatings, particularly within the realm of drug delivery applications. The initial step involves the preparation of a polymer solution by dissolving the polymer in a volatile organic solvent. Subsequently, nanoparticles that have been functionalized to ensure compatibility are dispersed within this solution. This dispersion is achieved through techniques such as sonication or mechanical stirring, which promote a uniform distribution of the nanoparticles throughout the polymer matrix. The resultant mixture is then cast onto a flat substrate, such as a glass plate, where the solvent is allowed to evaporate under controlled conditions, which may include ambient or vacuum drying.

The solvent casting technique offers several advantages that make it appealing for industrial applications. It is straightforward and economically viable, making it suitable for large-scale production. Additionally, this method produces nanocomposite films characterized by uniform thickness and an extensive surface area. Researchers can achieve precise control over nanoparticle loading and distribution by adjusting the composition of the solution and the parameters of the casting process. Importantly, solvent casting avoids high-temperature processing, rendering it particularly advantageous for thermosensitive drugs.

The applications of solvent casting are diverse and impactful. It is utilized in the production of thin-film drug-eluting systems essential for transdermal patches, wound dressings, and stents. Furthermore, controlled release nanocomposites are developed for oral or implantable drug delivery systems. The fabrication of multilayered nanocomposites also facilitates sequential or dual-drug release (Gupta et al. 2023).

Despite its numerous advantages, several challenges are associated with solvent casting that warrant consideration. One significant concern is the potential presence of residual solvents in the final product, which may adversely affect biocompatibility and drug stability. Achieving a homogeneous dispersion of nanoparticles necessitates careful optimization of dispersion techniques as well as solvent-polymer compatibility. Additionally, the mechanical properties of solvent-cast films may be suboptimal, potentially requiring post-processing or reinforcement to meet application demands.

In summary, while solvent casting and evaporation present significant benefits for fabricating polymer nanocomposite films, attention must be given to optimizing processes to effectively address inherent challenges.

14.3.7 Electrospinning

Electrospinning is a versatile and efficient technique for producing nanofibrous mats characterized by high surface area-to-volume ratios, making them particularly advantageous for drug delivery applications. In this process, a polymer solution containing the drug and nanoparticles is electrospun to form ultrafine fibers. Kim et al. (2023) reported that polycaprolactone (PCL) nanofibers embedded with titanium dioxide nanoparticles synthesized via electrospinning demonstrated controlled release of anti-inflammatory drugs. The method relies on the application of a high-voltage electric field to a polymer solution or melt, resulting in the formation of fibers that can incorporate both nanoparticles and drugs.

During the electrospinning process, the polymer solution or melt is pumped through a nozzle while a high voltage is applied to create an electric field. This electric force overcomes the surface tension of the solution, forming a charged jet that elongates and solidifies into nanofibers as the solvent evaporates or the melt cools. To ensure uniform incorporation of nanoparticles and drugs into the nanofibers, these components are mixed with the polymer solution prior to electrospinning. Precise control over parameters such as polymer concentration, voltage, flow rate, and the distance between the nozzle and collector is essential for achieving desired fiber diameters and properties.

The advantages of electrospinning are significant. It produces nanofibers with high porosity and interconnected pore structures, which facilitate drug diffusion and release. Additionally, this technique enables the incorporation of a wide range of drugs, including both hydrophilic and hydrophobic molecules. Furthermore, electrospinning allows for co-axial configurations that can produce multilayered fibers for sequential or dual-drug release.

Despite its numerous benefits, challenges remain in optimizing electrospinning for drug delivery systems. The design of nanofiber composites must address issues related to drug release kinetics to ensure that therapeutic effects are achieved effectively. Moreover, achieving consistent fiber morphology and mechanical properties requires careful tuning of electrospinning parameters. As research continues to advance in this area, electrospun nanofibers are increasingly recognized as promising biomaterials for applications in regenerative medicine, tissue engineering, and targeted drug delivery systems.

14.3.8 Recent Advances

The incorporation of a wide range of nanoparticles, including graphene, carbon nanotubes, and clay particles, significantly enhances the mechanical, thermal, and barrier properties of polymer nanocomposites. This versatility enables various applications such as the fabrication of drug-loaded implants for sustained release, the development of packaging materials with antimicrobial properties, and the production of structural nanocomposites that exhibit enhanced durability and flexibility for biomedical applications.

However, several challenges must be addressed to optimize the performance of these nanocomposites. Achieving uniform dispersion of nanoparticles can be difficult due to their tendency to agglomerate. Additionally, high processing temperatures may degrade sensitive drugs or polymers, limiting the applicability of certain formulations. The process also requires precise control over shear forces and processing conditions to ensure consistent quality in the final nanocomposite products.

Recent advances in this field have focused on improving nanoparticle dispersion and interfacial adhesion. For instance, the incorporation of compatibilizers and surface-modified nanoparticles has shown promise in enhancing dispersion. Furthermore, twin-screw extrusion systems have been employed to optimize mixing efficiency and scalability. The integration of in-line monitoring techniques, such as rheometry and spectroscopy, facilitates real-time quality control during melt mixing (Wang et al. 2022).

Layer-by-layer (LbL) assembly offers high precision in creating multilayered nanocomposites by alternately depositing layers of polymers and nanoparticles. This method allows for the incorporation of stimuli-responsive features. For example, Wang et al. demonstrated that multilayered polyelectrolyte films containing gold nanoparticles provided thermo-responsive drug release for localized therapy. Another approach is the sol-gel process, a hybrid synthesis technique that integrates polymers with inorganic networks, enabling precise control over the composite's architecture. In this method, hydrolysis and condensation of precursors such as metal alkoxides occur in the presence of the polymer and drug. Ahmed et al. (2020a) synthesized alginate/silica nanocomposites via the sol-gel process, which exhibited tunable drug release suitable for wound healing applications.

Melt intercalation is another effective strategy where nanoparticles are blended with a polymer melt to create nanocomposites without solvents. This environmentally friendly method is scalable for industrial applications; for instance, polyethylene glycol (PEG)-based nanocomposites incorporating layered double hydroxides demonstrated sustained drug release profiles suitable for cancer therapy (Li et al. 2022a). Additionally, solution blending involves dissolving the polymer, drug, and nanomaterials in a common solvent followed by solvent evaporation to form the composite material. This method is advantageous due to its simplicity and suitability for thermally sensitive drugs; chitosan/silica nanocomposites prepared using solution blending exhibited pH-sensitive drug release, making them ideal for treating gastrointestinal disorders (Hosseinzadeh et al. 2020).

In summary, while there are significant benefits associated with incorporating nanoparticles into polymer matrices for various applications, ongoing research is essential to overcome challenges related to dispersion uniformity and processing conditions to fully exploit their potential in advanced materials science.

14.4 Surface Modification Strategies for Improving Drug Loading, Release Kinetics, and Biocompatibility in Nanoparticle Drug Dosage Formulation

Surface modification of polymer nanocomposites is a critical strategy for enhancing their performance in drug delivery applications. By tailoring the surface properties of nanoparticles (NPs), researchers aim to improve drug loading efficiency, optimize release kinetics, and ensure biocompatibility. Nanoparticles have revolutionized drug delivery systems by enabling precise targeting, controlled release, and enhanced bioavailability of therapeutic agents. However, challenges such as low drug loading capacity, premature drug release, and inadequate biocompatibility must be addressed for their clinical success. Surface modification strategies play an essential role in overcoming these limitations by enhancing nanoparticle performance through improved stability, reduced agglomeration, and prolonged circulation time in biological environments.

These modification strategies are intricately linked to the synthesis and functionalization of NPs, where surface characteristics such as charge, hydrophobicity, and the presence of functional groups significantly influence their drug loading efficiency, release profiles, and interactions with biological systems. For instance, surface modifications can mitigate toxicity by employing biocompatible coatings that shield nanoparticles from eliciting adverse immune responses while minimizing interactions with non-target cells. Furthermore, the use of specific ligands for surface modification enhances targeting specificity, allowing nanoparticles to bind to specific receptors on cell membranes and facilitating uptake by target cells. This targeted approach not only improves the accumulation of nanoparticles in diseased tissues but also reduces off-target effects and systemic toxicity.

Recent advances in surface decoration techniques have focused on enhancing the physicochemical and biological stability of nanoparticles. These advancements include the incorporation of compatibilizers and surface-modified nanoparticles to improve dispersion and interfacial adhesion. Additionally, innovative methods such as layer-by-layer (LbL) assembly enable the creation of multilayered nanocomposites with stimuli-responsive features, further enhancing their therapeutic potential. Overall, the ongoing development of surface modification strategies is crucial for maximizing the efficacy and safety of nanoparticle-based drug delivery systems in clinical applications.

Functionalization of polymer nanocomposites is a critical strategy for enhancing their performance in drug delivery applications. By modifying the surface properties, researchers aim to improve biocompatibility, increase drug loading efficiency, and optimize release kinetics. One of the most common approaches involves the functionalization of nanocomposite surfaces with hydrophilic polymers such as polyethylene glycol (PEG) or polyvinyl alcohol (PVA). PEGylation, for instance, creates a hydration layer on the surface, reducing protein adsorption and immune recognition while enhancing drug loading by providing functional groups for drug conjugation. Wang et al. (2022) demonstrated that PEG-functionalized PLGA nanoparticles significantly prolonged drug release and reduced cytotoxicity in cancer therapy applications.

Another effective method is the *introduction of charged functional groups on the nanoparticle surfaces*. Surface functionalization with positively charged amine groups or negatively charged carboxyl groups can improve drug loading through electrostatic interactions. Hosseinzadeh et al. (2021) reported that chitosan-coated silica nanoparticles modified with amine groups exhibited enhanced loading of negatively charged anti-inflammatory drugs, along with sustained release profiles. Additionally, grafting stimuli-responsive polymers onto nanocomposite surfaces allows for precise control over drug release kinetics in response to environmental triggers such as pH, temperature, or enzymes. For example, Ahmed et al. (2020b) developed nanocomposites coated with pH-sensitive polyacrylic acid (PAA), facilitating drug release in acidic tumor environments while maintaining stability at physiological pH.

Coating nanoparticles with biomimetic materials, such as cell membranes or extracellular matrix components, can significantly enhance biocompatibility and reduce immunogenicity. Zhang et al. fabricated PLGA nanoparticles coated with red blood cell membranes, achieving high drug loading and prolonged circulation times. The layer-by-layer (LbL) assembly technique further allows for the deposition of alternating layers of polymers and other functional materials onto nanocomposite surfaces, providing precise control over coating thickness and composition for tunable drug release. Li et al. (2022b) utilized LbL assembly to create multilayered nanocomposites with pH-sensitive polyelectrolyte layers, ensuring controlled drug release under acidic conditions.

Surface engineering with targeting ligands, such as antibodies, peptides, or small molecules, can enhance site-specific drug delivery and reduce off-target effects by binding to overexpressed receptors on target cells. For instance, Smith et al. (2021) engineered nanocomposites with folic acid ligands to target cancer cells, resulting in improved drug accumulation at tumor sites and reduced systemic toxicity. Moreover, plasma treatment and chemical etching are methods used to introduce reactive functional groups onto nanocomposite surfaces, enhancing surface area and hydrophilicity to promote better drug adsorption. Kim et al. (2020a) demonstrated that plasma-treated PLGA nanoparticles exhibited increased drug loading capacity and improved release profiles for hydrophilic drugs.

Incorporating biodegradable polymers such as polylactic acid (PLA) or polycaprolactone (PCL) into surface modifications ensures gradual degradation of the nanocomposite in synchrony with drug release, which is particularly useful for long-term drug delivery applications. Ahmed et al. showed that PCL-coated nanocomposites provided a steady release of anti-inflammatory drugs over several weeks, aligning with tissue regeneration timelines.

Overall, surface modification strategies are indispensable for tailoring the properties of polymer nanocomposites to meet the demands of modern drug delivery systems. These approaches not only enhance drug loading and release kinetics but also improve biocompatibility and targeting capabilities of the nanocomposites, paving the way for advanced therapeutic applications. The synthesis of nanoparticles often involves methods such as sol-gel techniques, hydrothermal synthesis, and chemical reduction, each producing particles with distinct surface properties that can be further tailored through various functionalization techniques.

Functionalization encompasses modifying nanoparticle surfaces to introduce specific chemical or biological groups that enhance drug loading, targeting, and release kinetics. Techniques such as ligand exchange replace native ligands on nanoparticle surfaces with more functional groups like targeting peptides or antibodies to improve cell recognition and drug delivery efficacy. Covalent methods involve grafting polymers or ligands onto the surface, while non-covalent functionalization employs electrostatic interactions or hydrophobic forces to enhance stability and control over drug interactions.

In conclusion, ongoing research into surface modification strategies is crucial for maximizing the efficacy and safety of nanoparticle-based drug delivery systems in clinical applications. By tailoring these surfaces effectively, it is possible to improve therapeutic performance while addressing challenges associated with stability, biocompatibility, and targeted delivery in complex biological environments.

14.5 Characterization Techniques

14.5.1 Scanning Electron Microscopy (SEM)

Scanning electron microscopy is an imaging technique that utilizes a focused beam of electrons to scan the surface of a specimen, producing high-resolution images. The process involves several key components and interactions contributing to the formation of detailed microstructural images. An electron gun generates a fine beam of electrons, which are accelerated by high voltage (0–30 kV) and focused onto the sample surface (Buckner et al. 2016). As the electron beam scans the surface, it interacts with the atoms of the specimen, resulting in the emission of secondary and backscattered electrons, as well as X-rays (Singh and Singh 2022a; Vijayalakshmi et al. 2024). Detectors collect these emitted signals, which are then converted into digital images displayed on a computer screen, providing insights into the sample's microstructure (Buckner et al. 2016).

SEM enables the examination of the surface morphology of polymer nanocomposites, revealing the dispersion of nanoparticles within the polymer matrix. For instance, in a study on basalt-reinforced PEEK composites, SEM images indicated that optimal bonding and minimal voids were present in composites with 20% basalt, enhancing structural integrity (Arul et al. 2024). The morphology of drug-loaded polymer nanoparticles, such as erythromycin-loaded PLA and PLGA, showed spherical and rigid shapes, indicating effective encapsulation (Forna et al. 2022).

SEM assists in evaluating the structural stability of nanocomposites by identifying defects or inconsistencies in the material. As observed in SEM analyses, the presence of uniform particle distribution and strong interfacial bonding correlates with improved mechanical properties and stability (Baslak and Yildirim 2024). By visualizing the drug distribution within the polymer matrix, SEM can assess the efficacy of drug encapsulation. Studies have shown that the uniform dispersion of drugs at the nanoscale leads to enhanced release profiles and bioavailability, as evidenced by the consistent morphology of drug-loaded nanoparticles (Forna et al. 2022).

14.5.2 Transmission Electron Microscopy (TEM)

Transmission electron microscopy is a high-resolution imaging technique that utilizes electron beams to achieve high-resolution images of materials at the atomic level. It transmits electrons through a thin specimen, allowing detailed analysis of microstructures, and elemental compositions. TEM employs a focused beam of electrons that interact with the sample, producing images based on electron scattering while others are transmitted (Pattison et al. 2024; Pennycook 2024). The transmitted electrons are projected onto a screen or detector, forming an image that reveals details about the sample's microstructure (Dukes et al. 2023).

TEM allows for the visualization of polymeric nanoparticles, revealing their size, shape, and internal structure. For instance, studies have shown spherical nanoparticles with a multilayer structure, crucial for understanding drug release mechanisms (Koniuch et al. 2023). With TEM, the distribution of drug molecules and their interaction with the polymer can be visualized and aid in confirming the encapsulation of drugs within polymer matrices. This aids in assessing the stability of drug-loaded nanoparticles, ensuring that the encapsulated drugs remain effective over time (Petrushevska et al. 2019).

14.5.3 Differential Scanning Calorimetry (DSC)

Differential scanning calorimetry (DSC) is a thermal analysis technique that assesses the thermal behavior of materials, like polymer nanocomposites. DSC measures the difference in heat flow between a sample and a reference material as they are subjected to controlled temperature changes. The energy absorbed or released is recorded, allowing for the detection of endothermic and exothermic processes (Fortunato 2013). It identifies key thermal events such as glass transition temperature (Tg), melting temperature (Tm), and crystallization temperature (Tc) (Singh and Singh 2022b). The resulting thermograms provide information on specific heat capacity changes and enthalpy, essential for characterizing polymer behavior under thermal stress (Saxena et al. 2021).

By analyzing the thermal transitions, DSC can reveal the degree of crystallization and the stability of polymer matrices, which is essential for ensuring the longevity of drug delivery systems. The technique can confirm the thermal stability of encapsulated bioactive ingredients, ensuring they remain effective during processing and storage (Gharanjig et al. 2020; Shiravand et al. 2014). DSC can be coupled with other methods to validate drug encapsulation efficiency, providing a comprehensive understanding of how well drugs are retained within polymer matrices (Marques et al. 2023). The thermal properties assessed by DSC can inform the release profiles of drugs, indicating how modifications in polymer structure can enhance or hinder drug release (Shafique et al. 2023).

14.5.4 Fourier Transform Infrared Spectroscopy (FTIR)

Fourier transform infrared spectroscopy is an analytical technique used to identify molecular structures based on the vibrational transitions of chemical bonds. FTIR works by passing infrared radiation through a sample, causing molecular vibrations that result in absorption at specific wavelengths. It provides a spectrum that acts as a unique fingerprint for different materials, making it particularly useful in analyzing polymer nanocomposites. The resulting spectrum displays peaks corresponding to different functional groups, allowing for the identification of molecular structures and interactions within the material (Eid 2022; Nandiyanto et al. 2023).

FTIR can detect interactions between nanoparticles and polymer matrices, as demonstrated in PVC-Pb3O4 nanocomposites, where shifts in peak positions indicated successful integration of Pb3O4 with PVC (Taha et al. 2021). FTIR is also effective in identifying contaminants, such as DNA by measuring the absorption of infrared light by molecular bonds in a sample, providing a spectrum that identifies chemical structures revealing characteristic peaks and structural changes (Dodi et al. 2022). Changes in the FTIR spectra can indicate alterations in the polymer structure or drug interactions, which may affect the overall stability of the formulation (Song et al. 2020).

14.5.5 Dynamic Light Scattering (DLS)

Dynamic light scattering (DLS) is a powerful technique used to analyze the size and dynamics of particles in suspension by measuring the fluctuations in scattered light intensity due to Brownian motion. DLS measures the time-dependent fluctuations in scattered light intensity from particles undergoing Brownian motion. The rate of these fluctuations is inversely related to the size of the particles; smaller particles diffuse faster, leading to more rapid intensity changes (Erbe and Sigel 2013).

This technique is beneficial in understanding interactions between components, such as the aggregation behavior of nanoparticles influenced by functionalization, which is critical for applications in drug delivery systems (Anaraki et al. 2020; Cukalevski et al. 2015). DLS also effectively estimates the size distribution of nanoparticles, which is crucial for ensuring the uniformity of these particles in drug systems (Knysh et al. 2023).

14.5.6 X-Ray Diffraction (XRD)

X-ray diffraction (XRD) operates on the principle of the interaction between X-rays and the crystalline structure of materials. When X-rays, which have wavelengths comparable to the spacing between atoms, are directed at a crystal, they are scattered by the atoms within the crystal lattice. This scattering produces a diffraction pattern that is characteristic of the crystal structure, allowing for detailed analysis of the material's properties (Ghatage and Kanitkar 2019; Knysh et al. 2023). X-ray diffraction (XRD) data is collected as a series of peaks, each corresponding to specific crystal planes responsible for the diffraction. These peaks arise from the constructive interference of X-rays scattered by the crystal lattice, allowing the material's atomic structure to be deduced (Hadian-Jazi et al. 2021).

Bragg's Law explains how X-rays are diffracted by the periodic arrangement of atoms in a crystal, leading to distinct diffraction patterns that can be analyzed to reveal atomic structures. It states that $n\lambda = 2d \sin(\theta)$, where n is an integer (order of diffraction), λ is the wavelength of the incident X-rays, d is the distance between crystal planes, and θ is the angle of incidence. This relationship is crucial for determining the crystal structure through X-ray diffraction techniques (Crowley 2020).

XRD helps determine the degree of crystallinity in polymer nanocomposites, which influences mechanical properties. For instance, the incorporation of Bi_2O_3 nanoparticles in PMMA was characterized using XRD, revealing changes in crystallinity that correlate with enhanced radiation shielding properties (Alsaab and Zeghib 2023). XRD can identify different phases within the nanocomposite, aiding in the understanding of how the polymer matrix interacts with inorganic fillers (Yadav et al. 2016).

14.5.7 Nuclear Magnetic Resonance (NMR) Spectroscopy

Nuclear magnetic resonance (NMR) operates on the principle that nuclei of certain atoms possess intrinsic magnetic moments and angular momentum, known as nuclear spin. When placed in an external magnetic field, these nuclei exhibit quantized energy states, corresponding to their alignment with or against the field (Vijay Adnani 2024). When a radiofrequency (RF) pulse is applied, nuclei transition between energy levels, absorbing and re-emitting RF energy. This interaction produces an NMR signal, which decays over time and is detected as a free induction decay (FID). Fourier transformation of the FID yields an NMR spectrum, revealing the chemical environment of nuclei through chemical shifts, spin-spin coupling, and relaxation times (Hirakawa et al. 2019).

Nuclear magnetic resonance (NMR) spectroscopy can provide insights into molecular structure and interactions of polymer Nanocomposites. Key data obtained includes chemical shifts, which reveal the electronic environment of nuclei and help identify functional groups; spin-spin coupling constants, which indicate the connectivity between atoms through bonds; and relaxation times, which shed light on molecular dynamics and interactions (Jaźwiński 2022).

Chemical shifts in NMR spectroscopy reflect the electronic environment of nuclei, influenced by electronegative atoms, bond types, and molecular geometry. These shifts are unique for specific functional groups due to shielding or de-shielding effects. Electronegative atoms or groups attract electron density toward themselves, reducing the electron cloud around nearby nuclei. This results in an amplified external magnetic field experienced by the nucleus due to diminished local electron density (de-shielding). Conversely, less electronegative atoms or electron-donating groups increase electron density around the nucleus, enhancing shielding. Functional groups and molecular frameworks can be reliably identified by comparing these shifts to established databases or using correlation techniques.

Spin-spin coupling constants in NMR spectroscopy indicate atomic connectivity. When two or more nuclei are chemically bonded, their spins influence each other, splitting their resonance signals into multiplets depending on the number and type of bonds separating the nuclei and the bond angles involved. This data is crucial for mapping out structures and verifying chemical linkages.

Relaxation times in NMR spectroscopy, specifically T1 (longitudinal relaxation) and T2 (transverse relaxation), reflect how nuclear spins return to equilibrium after excitation. T1 measures the time it takes for spins to realign with the external magnetic field, a process sensitive to molecular motion and environmental interactions. Smaller molecules in solution typically have faster motion, hence shorter T1 times. T2 reflects the spin-spin relaxation due to dephasing caused by local magnetic field inhomogeneities and intermolecular interactions.

Short T2 times indicate restricted molecular motion or strong intermolecular interactions, as seen in solids or aggregated systems, while long T2 times are associated with highly mobile, non-interacting molecules. In polymer nanocomposites, relaxation times indicate nanoparticle dispersion and interaction with the polymer matrix, shedding light on material properties.

NMR is non-destructive and can probe both the bulk material and the interfaces between polymer matrices and nanoparticles. Chemical shifts reveal detailed information about the polymer's chemical structure, while spin-spin coupling constants highlight connectivity and bonding patterns. Relaxation times (T1 and T2) and diffusion measurements enable the assessment of molecular mobility, phase separation, and nanoparticle dispersion within the polymer matrix. Solid-state NMR further extends these capabilities to non-soluble or amorphous materials, capturing details of the interfacial interactions between polymers and nanoparticles, which are critical for properties such as mechanical strength and thermal stability. Additionally, advanced multidimensional NMR techniques can elucidate complex structural relationships and dynamic behaviors, making it a versatile tool for optimizing polymer nanocomposite formulations in applications like drug delivery, coatings, and packaging.

One significant drawback is its sensitivity; NMR requires high sample concentrations or isotopic enrichment (e.g., 13C) to detect signals from low-abundance nuclei, which can be costly and time-consuming. Additionally, relaxation times in highly rigid or aggregated nanocomposites can lead to broad or weak signals, complicating data interpretation. Solid-state NMR, although powerful for studying amorphous and non-soluble materials, demands long acquisition times and advanced equipment, increasing the cost and complexity of analysis. The technique also struggles with samples exhibiting significant heterogeneity, as averaging effects may obscure localized interactions. Furthermore, quantitative analysis can be challenging due to overlapping signals and variations in relaxation behavior. Despite these challenges, advances in NMR technology, such as cryoprobes and hyperpolarization, continue to mitigate many of these limitations, broadening its applicability to complex systems like polymer nanocomposites.

14.6 How Characterization of Polymer Nanoparticles Informs the Optimization of Pharmaceutical Formulations

The development of polymer nanoparticles has emerged as a transformative approach in pharmaceutical formulations, particularly for drug delivery systems. These nanoparticles can enhance solubility, bioavailability, and targeting of therapeutic agents, making them a vital focus in pharmaceutical research. Characterization of polymer nanoparticles is essential to understand their physicochemical properties, which directly influence the performance of the drug formulation (Devasier and Sang-Hyo 2014).

The particle size and distribution of polymer nanoparticles significantly influence drug solubility, bioavailability, and therapeutic efficacy. Smaller nanoparticles often enhance drug delivery by improving permeation and retention in target tissues, which is essential for effective treatment outcomes. Formulations like Piroxicam nanosuspension demonstrated a size reduction from 443 nm to 228 nm, leading to a 14- to 15-fold increase in solubility (Alhamhoom et al. 2023).

The biodistribution of nanoparticles is influenced by their size and shape, with smaller particles showing improved accumulation in specific organs, such as the brain and lungs, which can enhance therapeutic effects (Åslund et al. 2022).

The surface properties of polymer nanoparticles play a crucial role in optimizing pharmaceutical formulations, particularly in enhancing drug delivery and bioavailability. Drug encapsulation, release kinetics, and targeting capabilities, which are essential for effective therapeutic outcomes can be improved by tailoring the surface properties. For instance, spherical nanoparticles with a multilayer structure have been shown to facilitate controlled release through diffusion and degradation mechanisms (Koniuch et al. 2023).

Nanoparticles can be functionalized to enhance targeting capabilities, allowing for selective delivery to cancer cells while reducing systemic toxicity. This approach has been highlighted in the context of cancer therapies, where nanoparticles serve as carriers for chemotherapeutics (Foglizzo and Marchiò 2022).

The characterization of polymer nanoparticles is vital for optimizing pharmaceutical formulations. By providing invaluable data on size, surface properties, encapsulation efficiency, and stability, characterization guides formulators in making informed decisions which not only enhances the therapeutic efficacy of drug delivery systems but also ensures patient safety and compliance.

14.7 Current Formulations and Dosage Forms Using Polymer Nanocomposites

Nanocomposite formulations, particularly those involving polymer-based systems, have emerged as a promising solution in pharmaceutical development. These advanced materials are recognized for their ability to enhance bioavailability, enable controlled drug release, and improve patient compliance.

Polymer nanocomposite formulations have demonstrated significant value in pharmaceutical applications, particularly in achieving controlled drug delivery and improving bioavailability. By integrating advanced materials such as nanoparticles and biopolymers, these systems provide unique advantages that enhance therapeutic efficacy and patient adherence. Future research and development in this field hold immense potential for addressing ongoing challenges in drug delivery and formulation.

14.7.1 Polymer Nanocomposites in Pharmaceutics

The concept of polymer nanocomposites began gaining attention with the pioneering work of a Toyota researcher in 1987, eventually making their way into pharmaceutical formulations (Patel et al. 2006). These formulations have since gained prominence in both academia and industry, owing to their enhanced properties, particularly polymer/clay nanocomposites (Khan et al. n.d.).

Polymer nanocomposite-based drug delivery systems typically involve the association of the active pharmaceutical ingredient (API) with the polymer matrix. The matrix-forming ability of intercalated polymer composites plays a critical role in controlling drug release from formulations, especially for sustained release systems due to their unique physicochemical properties, such as mechanical, thermal, and optical properties, as well as high swelling rates, which also makes nanoparticles ideal for use in natural and synthetic polymer hydrogels (Gholamali and Yadollahi 2021).

Polymer nanocomposite structural association can occur either through encapsulation within the polymeric core or by uniformly dispersing the API throughout the polymeric matrix (Madawi et al. 2023; Patel et al. 2006). Nanoparticles play a crucial role in controlled drug delivery systems, offering cost-effective, stable, and high-efficiency solutions for drug delivery where they provide formulations that act as reservoirs that hold the drug within the core and release it into the surrounding medium in response to changes in the concentration gradient (Madawi et al. 2023) Some polymer nanocomposite materials also possess inherent therapeutic properties, contributing to a synergistic effect when combined with the API. Biopolymers, particularly natural polymers like cellulose derivatives, offer several advantages for drug formulation. These include biocompatibility, biodegradability, hydrophilicity, high mechanical strength, odorless and colorless properties, and non-toxicity, making them suitable for creating polymer nanocomposites with improved drug delivery properties (Gholamali and Yadollahi 2021).

Earlier studies have shown that polymer nanocomposites, such as calcium montmorillonite, are effective in treating various ailments, including pain, open wounds, colitis, diarrhea, hemorrhoids, stomach ulcers, and acne (Madawi et al. 2023). For certain dosage forms, such as intrathecal administration in spinal cord injury, sustained release, and a favorable risk/benefit ratio are essential. Studies have documented the stability and effectiveness of some polymer nanocomposite formulations for these types of applications (Baumann et al. 2010).

Polymer nanocomposites exhibit several desirable properties, including thermal stability, hardness, and gas barrier properties. However, some of these properties are not sufficient to meet the diverse range of applications (Thakur et al. 2016). Strong interfacial interactions, such as electrostatic and hydrogen bonding, have been shown to improve the mechanical properties, swelling behavior, water uptake, drug loading efficiency, and controlled release characteristics of biopolymer matrices (Thakur et al. 2016).

Earlier in vitro studies have demonstrated the effectiveness of polymer nanocomposites in controlled release formulations where API release happens when the polymer slowly degrades in vitro and eventually releases the nanoparticulate API as opposed to passive diffusion from the polymer nanocomposite matrix (Konishi et al. 2003). For example, an earlier study on paclitaxel, an anticancer drug, in polymer nanocomposite technology showed an initial high drug release of 22% on the first day, followed by a sustained release of up to 36% over 3 weeks. This slow release is attributed to the desorption and adsorption of drug molecules on the polymer surface (Patel et al. 2006).

Similarly, the release of buformin from a buformin/montmorillonite complex was compared to pure buformin hydrochloride in artificial intestinal juice. The results showed that the buformin/montmorillonite complex released 70% of the drug at a slower rate than the pure compound, highlighting the importance of polymer nanocomposites in controlled release systems (Patel et al. 2006).

Polymer nanocomposites have also shown promise in enhancing the bioavailability of biopharmaceuticals. For example, novel layered double hydroxides have been used to form nanohybrids by intercalating with biomolecules such as mononucleotides and DNA. This approach has been shown to enhance the bioavailability of antisense oligonucleotide molecules by facilitating their entry into cells (Patel et al. 2006).

14.7.2 Polymer Nanocomposite Formulations of Dosage Forms

14.7.2.1 Oral Dosage Forms

Polymer nanocomposite technology has significantly advanced oral dosage formulations by improving the solubility and bioavailability of poorly water-soluble drugs. Many new drug candidates face challenges of poor solubility and bioavailability with many failures in new drug products in market approval being due to poor solubility and poor bioavailability (Kesisoglou et al. 2007). Capsules have been essential for the storage and delivery of active pharmaceutical ingredients (API) into biologically active species (Baumann et al. 2010). Encapsulation of drugs into the inner core or adsorption onto the shell of polymer nanocomposites significantly improves API physicochemical properties, preventing premature degradation and offering controlled release kinetics (Madawi et al. 2023).

Formulations such as polymer nanocomposites of S-nitrosoglutathione, a drug with high environmental sensitivity and limited oral bioavailability, demonstrate improved bioavailability through efficient API loading and sustained release, coupled with cytocompatibility (Wu et al. 2015). Advances in polymer nanocomposite formulations, such as using nanocrystals, have shown improved dissolution velocity and increased passive diffusion across the gut lumen (Bhakay et al. 2018; Junghanns and Müller 2008). For instance, acyclovir formulated with semisynthetic biopolymers, including hyaluronic acid and nanoparticles, showed a 30% improvement in solubility and a 10% increase in permeation compared to conventional formulations (Sithole et al. 2018).

14.7.2.2 Topical Formulation

Topical polymer nanocomposite formulations are widely utilized in skin applications, enabling drugs to act locally on the surface or penetrate deeper skin structures. These formulations enhance drug retention in skin layers and permeation through the stratum corneum, making them effective for treating conditions such as psoriasis, dermatitis, skin infections, cancer, and wounds (Goyal et al. 2016; Madawi et al. 2023). Nanosized APIs accumulate in the stratum corneum, providing targeted delivery to inflamed tissues or pain sites (Yuniarsih et al. 2024).

Polymer nanocomposites have improved sunscreen formulations using ingredients like titanium dioxide and zinc oxide for better spreadability, transparency, and UV protection. However, concerns about nanoparticle safety in cosmetics persist (Goyal et al. 2016). Nanocomposite formulations offer enhanced physicochemical properties and targeted biodistribution for the effective treatment of skin cancers, reducing side effects compared to conventional treatments (Madawi et al. 2023).

14.7.2.3 Parenteral Dosage Forms

Parenteral drug delivery offers promising prospects such as an improved absorption profile of the drug, leading to enhanced bioavailability (Chavda et al. 2023). Because of challenges associated with parenteral drug administration, there have been endeavors to design dosage forms that offer advantages like targeting drugs to specific tissues (Shi et al. 2024). Injectable polymer nanocomposites hold promise for targeted and sustained drug delivery (Shi et al. 2024). Studies demonstrate their clinical potential, such as in pain management and oncology. Polymer nanocomposite hydrogels loaded with ropivacaine and dexamethasone showed prolonged pain relief in chronic constriction injury models (Khanal et al. 2018). Chitosan-montmorillonite nanohydrogel enabled the pulsatile release of vitamin B12 under electrostimulation, paving the way for use in implantable devices (Liu et al. 2008). Paclitaxel-loaded chitosan-polylactide composites exhibited pH-sensitive drug release, suggesting potential for improved chemotherapeutic formulations (Nanda et al. 2011; Yuniarsih et al. 2024).

Magnetic polymeric nanocomposites, combining magnetic nanoparticles within a polymeric matrix, exhibit exceptional versatility due to their magnetic responsiveness and tunable properties. These advanced materials have significant applications in drug delivery, including injectables, where they serve as innovative interfaces for precision and controlled drug administration. Their use spans magnetic micelles, liposomes, hydrogels, sponges, mesoporous silica nanoparticles, microrobots, elastomers, and scaffolds, highlighting their potential in establishing a new generation of smart, injectable drug delivery systems (Shi et al. 2024). Nguyen et al. (2016) demonstrated the potential of self-assembled magnetic nanoclusters (SAMNs) as multifunctional theranostic systems, further supporting the use of magnetic polymeric nanocomposites in advanced injectable drug delivery. By immobilizing beta-cyclodextrin (β-CD) on SPIO nanoparticles (NPs) and hosting RGD-conjugated PEG and paclitaxel (PTX) within β-CD cavities, the SAMNs achieved controlled size, enhanced magnetic resonance imaging, improved cellular uptake, efficient drug loading, and sustained release. These findings highlight the capability of magnetic nanocomposites to integrate therapeutic and diagnostic functionalities, offering a promising platform for image-guided cancer chemotherapy.

14.7.2.4 Pulmonary Drug Delivery

Pulmonary drug administration enables to target drug delivery directly to the lungs both for local and systemic treatment (Patil and Sarasija 2012). Because of the many advantages that inhaled formulations provide, for example, non-invasives, and ease of administration, inhaled polymer nanocomposite formulations have not been left unexplored in pharmaceutical dosage form designing in the field of pulmonary, and non-pulmonary diseases (Ozeki and Tagami 2014; Patil and Sarasija 2012).

The challenges of non-selectivity, side effects, and poor patient compliance in the management of some diseases like idiopathic pulmonary fibrosis (IPF) where treatments underscore the need for innovative pulmonary drug delivery systems. Further development is required to address translational and efficacy challenges. Advanced strategies like nano-delivery, hydrogels, and biological carriers show promise in enhancing drug targeting, bioavailability, and patient outcomes (Zhang et al. 2025).

Conventional nanotechnology has significantly advanced lung cancer management but its limitations in bioavailability, drug stability, and side effects remain challenges. However, the emergence of smart pulmonary nanomaterials offers a breakthrough, providing precise, stimulus-responsive drug delivery tailored to the lung cancer microenvironment. Innovative platforms are effectively a pivotal step toward more efficient and targeted lung cancer therapies (Gamal et al. 2024).

Cisplatin is one of the effective drugs for cancer, which as a polymer nanocomposite has been used by many scientists (Konishi et al. 2003; Tseng et al. 2009). The in vitro and in vivo characteristics of a polymer nanocomposite inhaled dosage form composed of gelatin nanoparticles as carriers of cisplatin the formulation exhibited high cisplatin dosage in cancerous lungs where simple aerosol delivery of the targeted drug carriers was more potent than free cisplatin and may prove useful for the clinical treatment of lung cancer patients (Tseng et al. 2009).

Ungaro et al. (2012), explored the potential of poly (lactic-co-glycolic acid) (PLGA) nanoparticles (NPs) for pulmonary antibiotic delivery, specifically tobramycin (Tb), in the form of nano-embedded micro-particles (NEM) with lactose as a carrier. The PLGA NPs were tailored using hydrophilic polymers, such as poly (vinyl alcohol) (PVA), chitosan (CS), and alginate (Alg), to optimize size, surface properties, and drug release. PVA and CS improved size and surface properties, while Alg enhanced drug entrapment and sustained release for up to 1 month. The NEM produced via spray-drying retained NP characteristics and exhibited favorable aerosolization properties. Biodistribution studies revealed that PVA-modified Alg/PLGA NPs targeted deep lung tissues, while CS-modified NPs accumulated in the upper airways. The study highlights the pivotal role of NP composition in influencing both technological properties and deposition patterns in vitro and in vivo.

Polymer nanocomposites have revolutionized pharmaceutical formulations by enhancing bioavailability, controlled release, and patient compliance across various dosage forms. In oral tablets and capsules, they improve the solubility and bioavailability of poorly water-soluble drugs, as seen in acyclovir formulations that enhance intestinal permeation. Topical gels leverage polymer nanocomposites to improve drug retention and penetration, benefiting conditions like psoriasis, infections, and cancer, while sunscreens with titanium dioxide nanocomposites offer superior UV protection and spreadability. Injectable formulations enable targeted and sustained drug delivery, exemplified by pH-sensitive hydrogels and magnetic nanocomposites used in oncology and pain management. Additionally, pulmonary delivery systems, such as inhalable cisplatin and nano-embedded tobramycin, enhance drug deposition in lung tissues, improving treatment outcomes for diseases like lung cancer and fibrosis while minimizing side effects. These advancements underscore the versatility and effectiveness of polymer nanocomposites in modern drug delivery systems.

14.8 Emerging Trends and Market Innovations

Drug delivery systems encompass technologies designed to transport drugs into the body to achieve therapeutic effects. These systems play a crucial role in improving patient outcomes, ensuring the safe and efficient administration of drugs, and enabling targeted therapy (Vargason et al. 2021). They can be customized to various routes of administration, such as oral, injectable, ocular, pulmonary, topical, implantable, nasal, or transmucosal (Tewabe et al. 2021).

According to Vargason et al. (2021), there has been a dynamic process in the evolution of Drug delivery systems marked by significant advancements in science and technology. From the early days of basic drug administration to the development of sophisticated delivery paradigms, the field has witnessed remarkable progress. The evolution can be traced through various stages, including the introduction of sustained-release capsule technology in the 1950s, the inception of controlled drug delivery and time-release systems in the 1960s, and the subsequent development of long-acting injectables and transdermal formulations. Over the years, drug delivery systems have evolved alongside the generations of therapeutics, addressing delivery challenges for small molecules, proteins, nucleic acids, and live-cell therapies. This evolution has been catalyzed by the application of established delivery approaches to improve the translation of emerging therapeutic classes.

The Pharmaceutical Drug Delivery market report from *Future Marketing Insights* has witnessed a significant growth trajectory, marked by a high compound annual growth rate (CAGR) of 9.6% from 2017 to 2022. Although the pace of growth is expected to moderate slightly, the market is still projected to exhibit a robust CAGR of 6.4% from 2023 to 2033. In 2023, the market size for Pharmaceutical Drug Delivery is estimated to be an impressive US$ 1.9 trillion, underscoring the substantial demand and economic value associated with drug delivery systems. Looking ahead, the market is poised for remarkable expansion, with the forecasted market size reaching an impressive US$ 3.5 trillion by the year 2033. This anticipated surge in market size indicates the increasing significance and widespread adoption of pharmaceutical drug delivery technologies, reflecting the dynamic landscape of the healthcare industry (Vikal et al. 2024).

Personalized medicine which can also be called precision medicine is a groundbreaking approach to healthcare that customizes medical treatments to the unique characteristics of each individual. At its core, it acknowledges the inherent diversity among patients, recognizing that genetic, environmental, and lifestyle factors can significantly influence how individuals respond to treatments. The advent of genomics and genetic research has been instrumental in this paradigm shift, allowing clinicians to analyze a patient's genetic makeup to better understand disease susceptibility, predict treatment responses, and identify potential adverse reactions to medications (Wang et al. 2023).

A key aspect of personalized medicine is pharmacogenomics, which explores how an individual's genetic profile influences their response to drugs. By leveraging this information, clinicians can optimize drug selection, dosage, and treatment plans, aiming to maximize therapeutic efficacy while minimizing side effects. Biomarkers, whether genetic, molecular, or cellular, play a crucial role in personalized medicine by aiding in disease diagnosis, predicting disease progression, and guiding treatment decisions (Oates and Lopez 2018).

Targeted therapies take center stage in personalized medicine. These therapies are designed to specifically target molecules or pathways involved in the growth and progression of diseases, offering a more precise and effective approach to treatment. Treatment plans are no longer one-size-fits-all; instead, they are tailored to the individual, considering their genetic profile, lifestyle, and environmental factors. Advanced diagnostic tools, imaging techniques, and molecular imaging contribute to early disease detection and accurate diagnosis, further informing personalized treatment strategies.

Also, personalized medicine empowers patients by providing them with information about their genetic predispositions, treatment options, and lifestyle modifications. Patients become active participants in their healthcare decisions, fostering a collaborative and informed approach to treatment (Goetz and Schork 2018).

As the field progresses, the merging of diverse datasets, even genetic information, electronic health records, and lifestyle data, becomes increasingly important. Health informatics plays a pivotal role in analyzing and interpreting this wealth of information to guide personalized treatment decisions (Husn and Kenny 2019).

Personalized medicine has demonstrated significant promise, particularly in fields like oncology, cardiovascular medicine, and psychiatry (Ozomaro et al. 2013). As technology advances and our understanding of genomics continues to deepen, personalized medicine stands poised to revolutionize healthcare, optimizing outcomes and improving patient well-being in a more tailored and precise manner (Brittain et al. 2017).

Personalized drug delivery systems (PDDS) involve customizing the dose, dosage form, frequency of administration, and drug release kinetics to individual patient characteristics (Raijada et al. 2021). This approach aims to achieve the right drug at the right dose for the right patient based on their specific conditions and needs. Nanotechnology plays a significant role in personalized medicine by enabling the customization of drug delivery systems to individual patients, considering factors such as genetic information, age, sex, and disease markers. Excipients are also being used to tailor drug products to individual patients based on their unique characteristics (Alghamdi et al. 2022). The development of stimuli-responsive drug delivery systems further enhances the ability to tune drug delivery for controlled dosing and targeted specificity based on material response to specific stimuli, both exogenous and endogenous (Raza et al. 2019). These advancements represent a shift toward more individualized and precise approaches to drug delivery, with the potential to improve treatment outcomes and reduce adverse effects.

Biodegradable nanocomposites are nanostructured materials formed by combining a *biodegradable* polymer matrix with fillers, and so biodegradable polymers are polymers derived from renewable resources such as plants, bacteria, or fungi. Examples include polyesters, poly(ester amide)s, polyurethanes, poly(lactic acid), epoxy, and poly(butyl succinate), polyhydroxyalkanoates (PHA), and starch-based bioplastics. Some common nanoparticles used in biodegradable nanocomposites include clay nanoparticles, cellulose nanocrystals, chitin nanowhiskers, and metal oxide nanoparticles among others.

14.8.1 Applications of Biodegradable Nanocomposites

Biodegradable nanocomposites are increasingly recognized for their diverse applications across various fields due to their enhanced properties. These materials are formed by incorporating nanomaterials into polymer matrices, leading to significant performance improvements that address specific challenges in therapeutic areas.

In the *packaging industry*, biodegradable nanocomposites create strong and durable materials that enhance the shelf life of food products. Their improved barrier properties against moisture and oxygen help maintain product freshness, thereby reducing food waste (Glaskova-Kuzmina et al. 2021). These advancements contribute to more sustainable packaging solutions that minimize environmental impact.

The *biomedical field* presents numerous opportunities for utilizing biodegradable nanocomposites due to their superior mechanical strength, reduced toxicity, biocompatibility, and biodegradability. These materials can be designed for slow drug delivery systems, tissue engineering scaffolds, and devices such as pacemakers that degrade after their useful lifespan. Additionally, they are effective in wound healing applications (Sonseca et al. 2022). The ability of these nanocomposites to provide a controlled release of therapeutic agents enhances treatment efficacy while minimizing side effects.

In *agriculture*, biodegradable nanocomposites have been employed to develop mulch films that improve soil quality and stabilization. They are also used for seed coating, creating biodegradable nanopesticides, and formulating controlled release fertilizers (Zafar et al. 2021). These applications not only enhance agricultural productivity but also contribute to environmental sustainability by reducing reliance on traditional plastic materials.

The *automobile industry* benefits from biodegradable nanocomposites by producing lightweight, strong, and environmentally friendly automotive parts. These materials can replace traditional petroleum-based plastics, thereby reducing the carbon footprint of the automotive sector. The shift toward sustainable materials aligns with global efforts to mitigate climate change and promote eco-friendly manufacturing practices.

The use of biodegradable nanocomposites in various industries can reduce the environmental impact of these industries and improve the quality of life for individuals. Biodegradable nanocomposites are promising materials that provide a sustainable alternative to traditional materials. The uses of nanocomposites have been limited by challenges of long-term environmental impact and high cost of production. Further research is needed to optimize the cost and sustainable methods for their production. However, studies are needed to understand the properties, applications, and potential impact on the environment (Vikal et al. 2024).

"Bioinspired" is a term used to describe the development of novel materials by imitating biological structures, functions, biological processes, or phenomena, to serve their particular purposes in various scientific areas, specifically in biomedicine. The bioinspired element can differ in features and imitate the chemical composition, nano/microstructure, and architecture of an organism. Bioinspired formulations draw inspiration from natural systems to enhance drug delivery efficiency. By mimicking biological structures and mechanisms, these formulations improve drug targeting, stability, and release kinetics (Farokhi et al. 2018; Cai et al. 2011; Wang et al. 2019; Clark et al. 2021).

To overcome the shortcomings of traditional formulations, many new, nature inspired, adhesive formulations with excellent biocompatibility and high adhesion performance have been proposed, including hydrocolloids, films, foams, hydrogels, and micro/nanofibers. The organisms living underwater have been studied to imitate their reversible wet adhesion, such as mussel-inspired polymers, barnacle-glue-inspired paste, insect-inspired secretions, and artificial octopus-inspired suckers. Great efforts have been devoted to bioinspired adhesive formulations with good biocompatibility and adhesion properties for wound care and drug delivery. Such formulations may replace conventional surgical tissue adhesives and wound dressings. Examples include liposomes and micelles designed to mimic cellular structures for effective transport across physiological barriers like the blood-brain barrier (BBB) (Chen et al. 2020a; Rouwkema et al. 2016; Mitchell et al. 2016).

Liposomes are a type of nanoparticle that can be used for targeted drug delivery (Nsairat et al. 2022). They are sphere-shaped vesicles consisting of one or more phospholipid bilayers that can encapsulate a wide range of therapeutic compounds such as drug molecules, genetic materials, and bioactive agents (Akbarzadeh et al. 2013). Liposomes can be modified to improve their properties, such as their stability, biodistribution, and accumulation in the targeted diseased area. They can be used for both passive and active targeting, with the latter using bio-affinity ligands such as small antibodies, peptides, and aptamers. Liposomes have several advantages, including biocompatibility, low toxicity, stability against chemical degradation, and site-specific targeting. Liposomes have been used in several clinical applications, including cancer therapy, and several liposomal drug products have been successfully commercialized (Liu et al. 2022; Daraee et al. 2016).

Polymeric micelles, a type of nanoparticle modified for targeted drug delivery, exhibit notable advantages in reducing toxicity, enhancing precision in organ and tissue targeting, and elevating the therapeutic effectiveness of active pharmaceutical ingredients (Negut and Bita 2023). These micelles have demonstrated the potential for high drug accumulation at specific sites (Rana and Sharma 2019). Polymeric micelles have garnered significant attention for their ability to effectively deliver poorly water-soluble drugs. They offer advantages such as prolonged circulation in the bloodstream and increased accumulation in tumor tissues. Current research attempts are concentrated on enhancing micelle stability in physiological settings. This is being pursued through various strategies, including crosslinking and non-crosslinking methods. Overall, polymeric micelles are emerging as a versatile and efficient system for targeted drug delivery. They hold immense promise for revolutionizing therapeutic interventions, particularly in the realm of delivering drugs to specific targets within the body (Lu et al. 2018).

Kazunori et al. developed pH-sensitive polymer micelles using PEO-b-poly (aspartate hydrazone doxorubicin), linking doxorubicin with acid-sensitive hydrazone linkers. These micelles released doxorubicin in acidic environments, effectively eliminating solid tumors in mice. The antitumor activity was attributed to physically encapsulated doxorubicin within the micellar core, highlighting the potential of these nanocontainers for drug delivery. Polymeric micelles have been investigated for their potential to enhance the passage of drugs across the blood-brain barrier for the treatment of brain diseases. Various block copolymers, such as poly (styrene)-poly (acrylic acid), poly (ethylene glycol)-b-poly (lactic acid), and distearyl-sn-glycero-3-phosphoethanolamine-N-methoxy poly (ethylene glycol), have been studied for this purpose. Among these, the latter showed improved cellular uptake, highlighting the capability of polymeric micelles to deliver encapsulated drugs to specific sites within the brain. This research suggests promising prospects for utilizing polymeric micelles in targeted drug delivery for neurological disorders (Bayindir et al. 2016). The development of lipopeptides derived from apolipoprotein E, which form micelles and rapidly integrate into liposomes, has underscored the importance of carrier size and cationic peptide surface density in designing target-specific drug delivery systems for the brain (Sydow et al. 2016). Recent research has concentrated on using micelles as nanocarriers for delivering curcumin to target glioma and for treating Alzheimer's disease (Tian et al. 2018; Desai and Patravale 2018). Moreover, micelles loaded with contrast agents show potential for magnetic resonance imaging of neuroinflammation and ischemic stroke injuries. These advancements highlight the versatility of micelles in neurological drug delivery and diagnostic imaging applications (Vikal et al. 2024).

Nanospheres refer to tiny, spherical particle with a diameter of 1–100 nm, and so they are a type of nanoparticles small sized, spherically shaped, uniform in size and shape, and have a high surface area. Nanoparticles can be used in targeted drug delivery to improve the bioavailability, biodistribution, and accumulation of therapeutics, preferentially in the targeted diseased area, acting as stability enhancers (Stefano 2023). These colloidal systems can deliver drugs to target sites to improve therapeutic efficiency, reduce toxicity, and reduce side effects, protecting the drug from biological degradation and achieving temporal and spatial control of therapeutics in the specific location of disease (Stefano 2023). Nanoparticles can be made of natural, synthetic/semisynthetic polymers, lipids, or metallic materials (Dmour and Taha 2018). However, the clinical application of targeted nanoparticles still faces limitations and challenges, including the variability between animal models, scaling up, and lack of reproducibility (Desai 2012). In a study by Gao et al. (2004), semiconductor quantum dots (QDs) were encapsulated in an amphiphilic copolymer, with the polymer surface modified using targeting ligands. The QDs, composed of CdSe capped with ZnS, were protected from aggregation by a coordinating ligand (tri-n-octylphosphine oxide (TOPO) (Wang et al. 2009; Gao et al. 2004). The copolymer, a triblock with hydrophobic segments (butyl acrylate and ethyl acrylate) and a hydrophilic segment (methacrylic acid), formed a structure where the carboxylic acid groups of the hydrophilic segment protruded. These acid groups served as attachment points for molecules like poly (ethylene glycol) (PEG) or affinity ligands. The resulting composite particles, 20–30 nm in diameter, were used for tumor targeting by attaching an antibody against prostate-specific membrane antigen (PSMA) to the particle surface. The study demonstrated selective binding to tumors in mice upon intravenous injection, with no observed accumulation in the brain, kidney, or lung (Gao et al. 2004). In a study by Huh et al. (2005), they detailed the application of superparamagnetic iron oxide (SPIO) nanoparticles for in vivo cancer detection using a mouse xenograft model. The nanoparticles were linked to Herceptin, an antibody specifically designed to target cancer cells.

NPs are often delivered through intranasal, intraventricular, and intraparenchymal routes, allowing them to traverse the BBB owing to their small dimensions. Upon reaching the BBB, various mechanisms come into play, including receptor-mediated pathways, active transport, and passive diffusion, facilitating the entry of nanoparticles into the brain. Due to their small size, nanoparticles can passively diffuse across the endothelial cells of the BBB and exhibit favorable interactions with brain receptors and ligands, aiding in their efficient transport and targeted delivery within the brain (Zorkina et al. 2020).

Polymeric nanoparticles effectively deliver drugs to the brain. Tacrine administered intravenously via polymeric nanoparticles increased brain concentration while reducing overall dosage (Chen et al. 2020b). Rivastigmine-loaded nanoparticles improved learning and memory. Solid Lipid Nanoparticles (SLNPs) enhanced drug retention in the brain. Piperine on SLNPs decreased brain plaques and boosted AChE enzyme activity. Huperzine A on SLNPs improved cognitive functions without skin irritation. These systems show promise for treating brain diseases (Kreuter 2001; Venkateswarlu and Manjunath 2004).

14.9 Future Directions

The integration of polymer nanocomposites with digital health technologies represents a significant advancement in medical science, with traditional pharmaceutical formulations having limitations, such as poor bioavailability, inadequate targeting, and insufficient controlled release (Xiao et al. 2023). Polymer nanocomposites-based formulations present enhanced mechanical strength, thermal stability, and biocompatibility (Khan et al. 2022). This section explores the potential growth areas, particularly focusing on smart drug delivery systems and their implications for clinical practice and patient outcomes.

14.9.1 Smart Drug Delivery Systems

Smart drug delivery systems apply polymer nanocomposites to enable targeted and controlled release of therapeutic agents. These systems can respond dynamically to patients' physiological stimuli, enhancing treatment precision. The mechanisms of action are targeted delivery and controlled release. With targeted delivery, nanoparticles are engineered for specificity hence allowing drug molecules to bind selectively to diseased receptor cells, thus minimizing off-target effects of the drug. With the controlled release mechanism, smart polymers are designed to release drugs in response to environmental changes, e.g., pH, temperature, and blood pressure facilitating sustained therapeutic effects (Sanchez-Rexach et al. 2017).

Smart drug delivery systems (SDDS) represent a significant advancement in the field of therapeutics, offering targeted and controlled release of medications to enhance treatment efficacy and improve patient outcomes. These systems encompass a variety of applications, including implantable devices, ingestible smart devices, and innovative monitoring technologies.

Implantable Devices: SDDS can be integrated into various implantable devices, such as drug-eluting coronary stents, ocular micro-stents, hydrogel depots or scaffolds containing drugs or cells, magnetic pumps with drug reservoirs, injectable hydrogel-based drug depots, and intrauterine devices. These devices facilitate localized drug delivery directly to the site of action, thereby minimizing systemic side effects and enhancing therapeutic effectiveness.

Ingestible Smart Devices: Ingestible smart devices have emerged as valuable tools for gastrointestinal monitoring and disease management. Examples include imaging capsules (e.g., PillCam SB capsules), temperature-sensing capsules (e.g., CorTemp), and pressure-sensing capsules (Thwaites et al. 2024). These devices not only assist in the diagnosis of conditions such as cancer and infections but also provide real-time data on gastrointestinal health. The ability to monitor gut disorders and assess the impact of dietary supplements through smart pills can lead to personalized dietary recommendations (Rehan et al. 2024).

The innovations in SDDS significantly enhance treatment efficacy by improving targeting precision. By localizing treatment to affected areas, these systems reduce the side effects commonly associated with systemic therapies. Furthermore, real-time monitoring capabilities foster better patient engagement and adherence to treatment regimens. Patients can receive immediate feedback on their health status, which encourages compliance with prescribed therapies.

14.9.2 Wearable Technologies

Wearable devices incorporating polymer nanocomposites monitor physiological parameters and deliver medications on demand, and with digital health technologies, such as wearable sensors and mobile health applications, for example, smart insulin delivery systems, can monitor blood glucose levels and release insulin in response, improving diabetes management (Cheng et al. 2021). Wearable devices include drug-eluting contact lenses, ocular rings, pharmaceutical jewelry, drug-delivering mouth guards, microneedles, wound healing patches, and drug-loaded self-care textiles (Lou et al. 2020). Wearable devices are believed to improve medication compliance by patients (Rubião et al. 2021), for example, human volunteers topically wearing bimatoprost eluting ocular ring exhibited continuous release of bimatoprost up to 6-month duration (Rubião et al. 2021).

One of the primary advantages is real-time monitoring, which enables healthcare providers to track various physiological parameters continuously. This capability allows for timely interventions based on individual patient needs, ensuring that adjustments to medication regimens can be made promptly. By providing immediate feedback on a patient's health status, SDDS helps prevent complications and facilitate proactive management of chronic conditions.

Additionally, SDDS supports personalized treatment by tailoring drug delivery according to real-time health metrics. Wearable devices can adjust medication dosages based on an individual's physiological responses, thereby enhancing treatment adherence and overall effectiveness. This level of customization is particularly beneficial in managing complex conditions, such as cancer or chronic pain, where optimal dosing can vary widely among patients. By aligning drug delivery with the specific needs of each, smart drug delivery systems improve therapeutic outcomes while minimizing the risk of adverse effects associated with standardized dosing regimens.

14.9.3 Tissue Engineering

The tissue engineering approach is a promising strategy added to the field of bone regenerative medicine, which aims to generate new, cell-driven, functional tissues, rather than just to implant non-living scaffolds (Salgado et al. 2004). This alternative treatment of conditions requiring bone regeneration could overcome the limitations of current therapies by combining the principles of orthopedic surgery with knowledge from biology, physics, materials science, and engineering, and its clinical application offers great potential (Salgado et al. 2004). In essence, bone-tissue engineering combines progenitor cells, such as MSCs (native or expanded) or mature cells (for osteogenesis) seeded in biocompatible scaffolds and ideally in three-dimensional tissue-like structures (for osteoconduction and vascular ingrowth), with appropriate growth factors (for osteoinduction), to generate and maintain bone (Salgado et al. 2004).

Polymer nanocomposites play a vital role in tissue engineering by creating scaffolds that support cell growth and facilitate tissue regeneration. These advanced materials have significant clinical implications, particularly in regenerative medicine, where they enhance healing processes and lead to quicker recovery times for patients undergoing surgeries or dealing with injuries. The use of polymer nanocomposites in reconstructive procedures can improve overall patient outcomes, reducing complications and increasing patient satisfaction.

In tissue engineering applications, the incorporation of nanoparticles into polymer matrices has been shown to enhance the mechanical properties of scaffolds, making them more suitable for supporting cellular activities. Nanoparticles increase surface area and improve cell adhesion, viability, proliferation, and migration, which are critical factors for successful tissue regeneration. For example, scaffolds made from poly-ε-caprolactone (PCL) infused with gold nanoparticles (AuNPs) have demonstrated increased porosity and mechanical strength, mimicking the extracellular matrix necessary for effective tissue integration.

Moreover, polymer nanocomposites can be engineered to possess specific properties that enhance their functionality in biomedical applications. These materials can be designed to degrade at controlled rates, ensuring that drugs retain their potency over time while minimizing toxicity concerns. The ability to tailor the physicochemical properties of these composites allows for sustained or controlled drug release, which is particularly beneficial in chronic wound healing and other regenerative therapies.

The versatility of polymer nanocomposites extends beyond mechanical support; they also provide opportunities for incorporating growth factors and bioactive molecules that promote tissue regeneration. This multifaceted approach not only improves the structural integrity of the scaffolds but also enhances biological interactions essential for effective healing.

14.10 Case Studies Demonstrating the Use of Polymer Nanocomposites

Polymeric nanocomposites consist of a polymer matrix reinforced with nanofillers, serving as effective agents for enhancing therapeutic applications. The advantages of these composites in drug delivery include their low cost, ease of shaping into desired forms, high surface area, low molecular weight, biocompatibility, rapid degradation, and water solubility (Bharadwaz and Jayasuriya 2020; Saravanan et al.

2023). Nanospheres and nanoparticles are the most common polymeric nanoparticles utilized as drug delivery systems (Karahmet Sher et al. 2024). These nanostructures have been shown to improve the pharmacokinetic properties of drugs, specifically absorption, distribution, metabolism, and excretion, leading to enhanced patient adherence and improved clinical outcomes (Alghamdi et al. 2022).

Biodegradable polymers are effective drug carriers, facilitating tailored and targeted drug delivery systems for various therapeutic agents. This approach offers significant advantages such as improved bioavailability, enhanced biocompatibility, and reduced toxicity. Conversely, through injection methods, non-biodegradable polymers can be employed as delivery systems for antibodies (Mondal et al. 2023). The versatility of polymeric nanocomposites allows for the encapsulation of a wide range of therapeutic agents, thereby addressing challenges associated with conventional drug delivery systems.

Incorporating nanofillers into polymer matrices enhances the mechanical integrity and functional properties of the composites. This modification enables controlled release profiles and improved stability of the encapsulated drugs. Furthermore, the nanoscale dimensions of these fillers contribute to increased surface area and reactivity, which are beneficial for enhancing drug solubility and bioavailability.

Overall, polymeric nanocomposites represent a promising platform for advancing drug delivery technologies. Their ability to combine favorable properties such as biocompatibility and tailored release kinetics makes them invaluable in modern pharmaceutical development. As research continues to evolve in this field, the potential applications of polymeric nanocomposites in therapeutics are expected to expand significantly.

Polymer nanocomposites have several applications in *antimicrobial therapeutics*, primarily due to their ability to incorporate materials such as graphene or copper, which possess inherent antibacterial properties. The addition of carbon-based nanostructures not only enhances the antibacterial characteristics of the polymer but also improves the overall properties of the composite material (Saravanan et al. 2023). For example, in *bandages and wound dressings*, the reinforcement of polymer nanocomposites with graphene and its derivatives such as graphene oxide and reduced graphene oxide plays a significant role in wound healing. These materials help reduce moisture on the wound, expedite the closure process, minimize infections, and prevent scratch formation during healing (Díez-Pascual and Luceño-Sánchez 2021).

Graphene oxide nanoparticles, particularly at a concentration of 5.0 wt%, have demonstrated improved compatibility with fibroblast cells and enhanced wound healing by promoting collagen deposition, vascularization, and re-epithelization (Shams et al. 2017). Furthermore, polylactic acid (PLA) composites reinforced with graphene oxide exhibit notable antibacterial properties against common pathogens such as *Staphylococcus aureus* and *Escherichia coli*. These composites show superior activity against bacterial proliferation and adhesion compared to PLA without graphene oxide (Arriagada et al. 2018).

The integration of these nanofillers into polymer matrices not only provides antimicrobial benefits but also enhances mechanical strength and stability. Polymer nanocomposites' unique physicochemical properties enable them to effectively hinder microbial vitality, adhesion, and biofilm formation through various mechanisms. This capability is particularly relevant in addressing the growing challenge of multidrug-resistant bacteria, which necessitates innovative solutions beyond traditional antimicrobial agents.

Drug delivery systems utilizing graphene-based polymer nanoparticles have shown significant promise in antimicrobial agents. For instance, incorporating chlorhexidine, an antiseptic agent, into graphene-based polymer nanoparticles has been reported to reduce post-surgical infections effectively (Scaffaro et al. 2019). Additionally, when ciprofloxacin, an antibiotic, was combined with poly(lactic acid) via melt-compounding, the resulting graphene-based polymer nanoparticles exhibited improved release kinetics and enhanced physical properties of ciprofloxacin, including increased matrix stiffness and reduced burst release effects (Scaffaro et al. 2017).

Graphene oxide (GO) has also been used with polysebacic anhydride as a drug carrier for levofloxacin, a fluoroquinolone antibiotic. This formulation demonstrated improved linear release characteristics, achieving over 95% drug release while maintaining the bactericidal effects of levofloxacin (Gao et al. 2011). Such findings highlight polymer nanocomposites' capacity to enhance both the pharmacokinetic and physical properties of antimicrobial drugs.

Moreover, GO-based polynanocomposites have been explored as nanocarriers for *anticancer drugs*. The hydrogels formed from graphene oxide offer a large surface area and high drug loading capacity, alongside improved physicochemical properties such as chemical stability and enhanced lipid solubility for cell membrane penetration. For example, a graphene-based hydrogel combined with polyethyleneimine (PEI), polyethylene glycol (PEG), and folic acid has demonstrated effective drug delivery for hepatocellular carcinoma (Ghawanmeh et al. 2019).

Despite these advancements, challenges regarding biocompatibility, particularly stability and blood compatibility have been noted with GO-based polynanocomposites. However, when combined with hyperbranched polyglycerol, GO showed improved biocompatibility as a carrier for doxorubicin hydrochloride. This combination retained the cytotoxicity of doxorubicin while exhibiting negligible effects on hemolysis and blood coagulation (Mu et al. 2017). The same carrier also demonstrated biocompatibility in delivering amphetamines.

In addition to graphene-based systems, poly(lactide-co-glycolide) nanoparticles have been studied for the delivery of docetaxel, an anticancer drug. This delivery system exhibited improved pharmacokinetic and biodistribution properties attributed to the favorable physicochemical characteristics of the poly(lactide-co-glycolide) polymer (Rafiei and Haddadi 2017).

However, graphene polymeric nanocomposites face limitations such as safety concerns, potential toxicity, inadequate techniques for maximizing mass production, and challenges related to the reproducibility of cost-effective nanomaterials. Furthermore, the mechanisms of action for these composites are not yet fully defined (Díez-Pascual and Luceño-Sánchez 2021), indicating a need for further exploration in this research area.

Natural polymeric nanocomposites have also been employed in delivering anticancer drugs. For instance, chitosan-based nanocomposites have demonstrated over 80% loading efficiency for cisplatin and more than 90% release rates when combined with magnetite (Abdel-Bary et al. 2020). Chitosan nanocomposites have also shown promise in delivering doxorubicin by providing controlled or delayed release mechanisms (Hosseinzadeh et al. 2019). Additionally, chitosan composites with copper oxide using rutin have exhibited enhanced anti-proliferative activity against lung carcinoma cell lines A549 in a concentration-dependent manner and induced apoptosis through fluorescent staining methods (Bharathi et al. 2019).

Furthermore, chitosan-alginate-montmorillonite composites were utilized to load 5-fluorouracil. The results indicated improved controlled drug release characteristics in the small intestine, enhancing the half-life of this anticancer drug (Farshi Azhar and Olad 2014). Chitosan-palladium nanocomposites have also demonstrated improved drug release profiles for curcumin and 5-fluorouracil to zero-order kinetics indicating a constant drug release rate, which reduces the number of doses required while enhancing cytotoxic effectiveness against colon cancer cells (Dhanavel et al. 2017).

An alginate clay-based nanocomposite has been studied as a dual delivery system for methotrexate and ciprofloxacin. The findings revealed increased release rates of both drugs within tumor cells, presenting opportunities for co-drug delivery with ciprofloxacin demonstrating improved inhibitory effects against *Escherichia coli* offering a promising combinational therapy approach for cancer patients (Mahkam et al. 2021).

Cellulose-based nanobiocomposites have also shown potential in improving delivery characteristics for anticancer drugs such as quercetin (Mahdi Eshaghi et al. 2022; Javanbakht and Namazi 2018). Additionally, copolymerized cellulose grafted hydrogels doped with calcium oxide nanocomposites have exhibited enhanced antioxidant activity for anti-arthritic applications while allowing controlled release of doxorubicin for cancer treatment (Iram Shahzadi). Starch-based nanobiocomposites have demonstrated potential in clinical applications as targeted drug delivery systems for doxorubicin (Massoumi et al. 2018) and as effective adsorbents for removing doxorubicin from aqueous solutions in magnetic targeted drug delivery systems aimed at synergistic tumor treatments (Fang et al. 2021).

Pectin-based systems have also been utilized to deliver anticancer drugs such as 5-fluorouracil via targeted oral delivery routes to minimize toxicity associated with intravenous administration (Wang et al. 2020; Rao et al. 2014). Collagen-based nanobiocomposites serve not only as drug delivery systems but also play significant roles in wound healing and tissue engineering due to collagen's natural abundance in soft and hard tissues (Chattopadhyay and Raines 2014; Wang et al. 2023).

Nanocomposites are further employed to enhance cancer diagnostics by improving specificity, sensitivity, and image resolution critical for accurate diagnosis (Jin et al. 2020; Zuo et al. 2024). These materials can function as imaging agents or contrast media during imaging techniques such as positron emission tomography (PET), magnetic resonance imaging (MRI), or computed tomography (CT), facilitating targeted delivery of diagnostic agents or therapeutic drugs directly to cancer cells.

Additionally, nanoparticle technology is being explored for developing biosensors capable of detecting pathogens while improving imaging studies related to inflammatory diseases. For instance, intranasal delivery of interferon-β using chitosan nanoparticles has shown promise in treating inflammation associated with encephalomyelitis (González et al. 2021).

Experts suggest that solid lipid nanoparticles may serve as effective drug delivery systems targeting the central nervous system by bypassing the blood-brain barrier in treating neurodegenerative diseases such as Alzheimer's and Parkinson's disease (Cacciatore et al. 2016).

Nanopolymers have significant applications in *cardiovascular medicine*, particularly in the therapeutic management of myocardial infarction and atherosclerosis (Omidian et al. 2023). These nanosystems are instrumental in molecular imaging for diagnosing atherosclerosis, as they aid in identifying atherosclerotic plaques during magnetic resonance imaging (Bonnet et al. 2021). By improving drug delivery to these plaques, nanopolymers enhance treatment precision and minimize side effects by specifically targeting inflamed endothelium and activated cardiac fibroblasts (Bonnet et al. 2021; Kim et al. 2020b; de Castro Leão et al. 2021). Various nanosystems, including nanoparticles, nanocapsules, and nanobubbles, are employed for the delivery of cardiovascular drugs (Wijaya et al. 2020; Mog et al. 2019).

In addition to facilitating non-invasive diagnostic techniques for atherosclerosis, these nanosystems support patient follow-up by monitoring disease progression and therapeutic responses (Florea et al. 2021). They also enhance thrombus treatment by improving the half-life of thrombolytic drugs, reducing side effects, and increasing targeting precision (Shen et al. 2021; Wu et al. 2018). Furthermore, there is potential for these systems to mitigate muscle destruction and oxidative stress in cardiac tissues (Hardy et al. 2015).

In stroke management, nanosystems have improved drug delivery to the central nervous system by effectively crossing the blood-brain barrier. For example, liposomes have been utilized to formulate amiodarone for controlled release, thereby reducing side effects and enhancing drug efficacy (Mahmood et al. 2020). Targeted drug delivery systems also promote angiogenesis and reduce infarction size, facilitating cardiac healing (Mahmood et al. 2020; Tan et al. 2021).

These targeted systems are also applied in treating arrhythmias following atrial fibrillation surgery and in the removal of abnormal heart muscle cells (Avula et al. 2015; Nadimi et al. 2018; Wang et al. 2022; Robinson et al. 2016; Xiong et al. 2018). Sirolimus-based nanoparticles have demonstrated the ability to reduce the release of inflammatory mediators and lower restenosis rates in lower extremity arteries after balloon injury (Xiong et al. 2018; Ang et al. 2020). Additionally, poly(lactic-co-glycolic acid) (PLGA)-coated stents loaded with antiproliferative agents enhance endothelialization and reduce restenosis risk post-stent deployment (Liang et al. 2022).

Nanoparticle delivery systems are being explored for delivering antihypertensive drugs and vasodilators into the pulmonary system to treat conditions such as pulmonary hypertension and pulmonary edema (Ferguson et al. 2019). The application of nanosystem technology is thus crucial in cardiovascular medicine, as it significantly improves patient outcomes by enhancing drug delivery mechanisms and therapeutic efficacy.

While polymer nanocomposites offer promising opportunities for therapeutic applications in various diseases, several challenges must be addressed to facilitate their clinical translation and widespread use. Key issues include concerns regarding nanotoxicity, the poor correlation between in vitro and in vivo studies, and the limited number of formulations that progress to clinical trials (Navya et al. 2019). Additionally, regulatory approval processes often lack clear guidelines, which can hinder the development and application of these innovative materials. Biocompatibility remains a concern for some polymer nanocomposites, alongside challenges related to large-scale production and the optimization of physicochemical properties such as drug loading capacity, release profiles, and surface area.

Despite these hurdles, it is crucial to address these challenges as polymer nanocomposites represent a paradigm shift in the management of diseases such as *cancer* and *antimicrobial resistance*. The unique properties of these materials enable targeted drug delivery, enhanced bioavailability, and controlled release mechanisms that can significantly improve therapeutic outcomes. For instance, recent advances in nanocomposite technology have demonstrated their potential to overcome traditional limitations associated with drug delivery systems, including reduced systemic toxicity and improved efficacy (Bonnet et al. 2021; Omidian et al. 2023).

Furthermore, increased research focus on polymer nanocomposites is essential to facilitate their clinical translation and application. By prioritizing studies that explore their safety profiles, scalability, and long-term effects, researchers can help establish a clearer path for regulatory approval and integration into standard medical practices. Ultimately, enhancing our understanding of polymer nanocomposites will contribute to their successful implementation in therapeutic settings, paving the way for innovative solutions to pressing health challenges.

14.11 Focus on How Nanocomposites Can Address Challenges in These Therapeutic Areas

Polymer nanocomposites are gaining traction in various pharmaceutical formulations due to their enhanced properties, which address several challenges in therapeutic applications. The incorporation of nanoscopic materials into polymer matrices significantly improves the performance of these composites, leading to advancements in drug delivery systems (Winey and Vaia 2011). This section focuses on how polymer nanocomposites can effectively tackle issues such as targeting tumor cells, reducing side effects associated with drug therapy, and improving the bioavailability of poorly soluble drugs.

The formation of polymer nanocomposites involves the addition of nanomaterials to polymer matrices. Natural polymers like chitin, collagen, and starch are commonly used due to their excellent biocompatibility and biodegradability. These materials degrade into non-toxic compounds, enhancing the stability of pharmaceutical products and minimizing toxicity concerns (Yasser Zare 2016). Furthermore, they can degrade at controlled rates, ensuring that the drug retains its potency and extends its shelf life.

Polymer nanocomposites provide improved thermal stability and solubility, along with other biological functionalities that make them suitable for drug delivery applications (Maghimaa 2024). Their high surface area allows for significant interactions between the drug and the target site, which is crucial for effective treatment. The structural characteristics of these composites can influence the overall properties of the drug and its release rate. For instance, variations in polyethylene glycol (PEG) density can lead to rapid or slow drug release profiles (Panangattukara Prabhakaran Praveen Kumar and Lim 2022). This capability enables manufacturers to achieve sustained or controlled release, thereby reducing the frequency of drug administration (Thakre et al. 2024).

Moreover, polymer nanocomposites can reduce first-pass metabolism and protect sensitive drugs such as peptides and nucleic acids from degradation before reaching their target sites through encapsulation or entrapment (Thakre et al. 2024). For example, carbon nanotube polymer nanocomposites serve as effective nanocarriers that target-specific sites to minimize side effects. Their hydrophobic nature allows them to remain in circulation longer, facilitating sustained drug release (Anzara et al. 2020). Other polymers like PLGA and PEGylated polymers are also utilized to achieve sustained drug release (Anzara et al. 2020).

Additionally, polymer nanocomposites can be engineered to respond to specific environmental triggers such as pH, temperature, or enzymatic activity. This feature is particularly advantageous for targeting tumor cells, which often exhibit unique microenvironments characterized by lower pH levels compared to normal cells (Lim et al. 2013).

A significant challenge in developing oral solid dosage forms is ensuring adequate aqueous solubility of drugs within the gastrointestinal tract. Poor water solubility leads to slow dissolution rates and low bioavailability (Anagha Bhakay 2018). Polymer nanocomposites have been employed in various formulation strategies to enhance the solubility and dissolution rates of poorly soluble drugs. Techniques such as the melting method are commonly used due to their simplicity and cost-effectiveness. In this method, a water-soluble polymer matrix is mixed with poorly soluble drug nanoparticles under heat to create a composite that improves solubility and bioavailability (Bhat 2015).

Some polymer nanocomposites also exhibit antibacterial properties. For instance, a composite made from cellulose acetate combined with carbon nanotubes and silver nanoparticles has shown effectiveness against both gram-negative and gram-positive bacteria by generating reactive oxygen species that damage bacterial DNA and disrupt cell walls (Alosaimi et al. 2023). Other examples include natural polymers combined with silver nanoparticles (AgNPs), which are utilized for applications such as water purification and wound disinfection while combating antibiotic resistance (Yasser Zare 2016).

Despite their many advantages, polymer nanocomposites face limitations that hinder their use in pharmaceutical formulations. Issues such as inadequate stability where some materials are prone to oxidation, and incompatibility in certain formulations can pose challenges. Additionally, aggregation due to high surface energy can lead to a loss of nanoscale properties in metal nanoparticles and similar materials (Yasser Zare 2016).

Polymer nanocomposites present innovative solutions for addressing critical challenges in therapeutic areas by enhancing drug delivery systems through improved targeting capabilities and reduced side effects. Continued research and development in this field are essential for maximizing their potential in clinical applications.

14.12 Biocompatibility, Potential Toxicity, and Regulatory Standards of Nanoparticles

The evolution of nanotechnology in the world has led to the manufacture of nanoparticles (NPs). These are submicron-sized particles ranging from 100–1000 nm (Naahidi et al. 2013). These particles have indeed demonstrated different properties when compared to their bulk counterparts, most probably due to their small size and extremely high surface-to-volume ratio. It is scientifically known that with a decrease in particle size of any substance comes a significant increase in surface area and in turn an increase in the active sites present in that particular substance (Kus-liśkiewicz et al. 2021). Owing to this fact, nanoparticles mostly from carbon, silica, and metals in different shapes (i.e., spheres, tubes, and rods) (Adabi et al. 2011; Ketabchi et al. 2016; Shakoori et al. 2015; Tavakol et al. 2014) are used widely in many sectors such as the biomedical sector for drug delivery, human organ development, and diagnosis (Adabi et al. 2017). The vast use of nanoparticles in the biomedical sector has come a great demand for knowledge expansion on their toxicity and biocompatibility since their extensive use depends on the assessment of their biosafety (Kus-liśkiewicz et al. 2021).

Biocompatibility is defined by the standards of ISO 10993–1 (International Standardization Organisation) as "The capacity of a device material to function harmoniously with an appropriate host response in a specific context" (Paqué and Özcan 2024). To simplify it, biocompatibility is the ability of a biomaterial which could be an implant, or medical device, to perform its function without causing any local or systemic effects such as oxidative stress, apoptosis, and DNA damage at both the site of application and its surrounding environment. In other words, a nanoparticle may be declared biocompatible when it performs as intended and doesn't produce any adverse effects. Nanotoxicology is a vital and interdisciplinary field that concentrates on the study of the potential toxicity of nanomaterials (Adabi et al. 2017).

Scientists have established that several factors significantly influence the biocompatibility and toxicity of these materials, including particle size, dosage regimen, route of exposure, surface chemistry, degree of aggregation, transmembrane diffusivity, excretion pathway, and immunogenicity. By controlling these factors, it is possible to manage how nanomaterials interact with biological tissues, including their ability to penetrate, diffuse, be absorbed, distributed, recognized by the immune system, and eventually cleared from the body, thereby avoiding unintended nanotoxicity. For example, modifying the surface of nanomaterials with various substances such as sodium citrate, polyvinylpyrrolidone (PVP), and surfactants can substantially reduce their potential for nanotoxicity and enhance their safety profile by minimizing harmful interactions with biological tissues (Thu et al. 2023).

The regulation of nanoparticles is a complex and evolving field that is aimed at not only ensuring the safety of human health but also the environment while promoting innovation. Depending on the country/continent, different regulations regulate nanoparticles. For instance, in Europe, nanomaterials are regulated under the Registration, Evaluation, Authorisation, and Restriction of Chemicals (REACH) regulation, the REACH regulation has specific legal requirements that have been applied since January 2020 for the characterization, safety assessment, and registration of nanoforms (European Commission 2011). Similarly, in the United States, the Food and Drug Administration (FDA) uses existing statutory and regulatory authorities to oversee nanotechnology products. The Global regulatory frameworks vary but share the common goal of managing the risks associated with nanoparticles through stringent protocols and government-sponsored research.

14.12.1 Factors Affecting Biocompatibility and Toxicity of Nanoparticles

As mentioned earlier, different factors affect the biocompatibility and toxicity of nanoparticles, and these are further explained below:

14.12.1.1 The Size of Nanoparticles

The size of nanoparticles enables them to easily penetrate cell membranes and access areas of biological systems that are normally inaccessible to larger particles (Gatti and Rivasi 2002). As mentioned earlier, when the size of a bulk material decreases, its surface area increases. As a result, the number of chemical molecules bound to the surface increases. Even though there is enhanced activity, this may also amplify the potentially toxic effects of nanoparticles compared to larger particles

(Ai et al. 2011; Linkov et al. 2008; Suh et al. 2009). Different studies have been carried out to investigate the effect of particle size on nanoparticle biocompatibility and toxicity. For instance, a study conducted on titanium dioxide (TiO_2) particles discovered that titanium dioxide (TiO_2) particles of 20 nm induced significantly more inflammation than those of 250 nm in short-term pulmonary toxicity experiments in rats (Hallock et al. 2009). Park et al. (2011), also discovered that silver nanoparticles of 20 nm were more toxic than larger ones. In addition, Shi et al. (2013), observed a negative correlation between particle size and toxicity, with smaller silver nanoparticles (5–10 nm) showing higher toxicity than larger ones (15–25 nm) in a model organism. In summary, smaller nanoparticles are more biocompatible but also more toxic than large nanoparticles.

14.12.1.2 The Shape

Studies have shown that the shape of nanoparticles can significantly impact their biocompatibility and toxicity. For instance, changing the shape of a material from equiaxed (having equal dimensions in all directions) to acicular can enhance the toxic response. Wang et al. (2008), found out that gold nanorods coated with hexadecyl cetyl trimethyl ammonium bromide (CTAB) are highly toxic to human skin cells, whereas spherical gold nanoparticles are not inherently toxic. The cytotoxicity of gold nanorods was major because CTAB, which is cytotoxic is used in their synthesis and is not present in spherical gold nanoparticles. Similarly, Hsiao and Huang (2011), reported that nanorod-shaped zinc oxide (ZnO) particles are more toxic than spherical ones to human lung epithelial cells at a fixed size. The toxicity may be attributed to the enhanced interaction forces, such as van der Waals forces, which are greater in rod-shaped nanomaterials due to their lengthwise orientation.

The rate at which nanoparticles enter cells is also influenced by their shape. Spherical particles easily penetrate cell membranes compared to elongated particles with a large length-to-radius ratio, which tend to lie parallel to the cell membrane (Decuzzi et al. 2009). For example, spherical gold nanoparticles are taken up by cells more efficiently than their rod-shaped counterparts (Chithrani et al. 2006). In summary, one ought to choose an appropriate shape when innovating a nanoparticle.

14.12.1.3 Surface Charge

The surface charge of nanoparticles significantly influences their interactions with biological systems such as cell membrane interactions, penetration, protein adsorption, and stability in biological fluids (Maffre et al. 2011). Nanoparticles with a positive surface charge are more readily taken up by cells due to the attractive electrostatic interaction between the positively charged nanoparticles and the negatively charged cell membranes (Chen et al. 2011; Zhu et al. 2010). In contrast, neutral particles exhibit slower opsonization rates compared to charged particles, impacting their cellular uptake and distribution (Owens Iii and Peppas 2006; Roser et al. 1998).

Nanoparticles with a slight negative charge tend to accumulate more efficiently in tumor tissues, making them suitable for targeted therapies (He et al. 2010; Patil et al. 2007). On the other hand, cationic nanoparticles are more potent in activating the immune response compared to neutral or anionic nanoparticles, which makes them effective in applications such as vaccine delivery (Kedmi et al. 2010; Zolnik et al. 2010). In conclusion, the surface charge of nanoparticles plays a critical role in determining their behavior in biological environments, influencing their cellular uptake, protein adsorption, stability, accumulation in tumor tissues, and ability to activate the immune response (Adabi et al. 2017).

14.12.1.4 Other Factors

The properties of nanoparticles, including hydrophilicity, hydrophobicity, surface topography, roughness, and tension, significantly influence their interactions with biological systems. These properties affect protein adsorption, platelet activation, cellular growth, and ultimately the biocompatibility of the nanoparticles (Hsu and Lin 2004).

Surface modification techniques can be employed to reduce or eliminate the toxicity of nanoparticles. For example, coating superparamagnetic iron oxide nanoparticles with pullulan has been shown to significantly reduce their toxicity. Uncoated magnetic nanoparticles (MNP) can increase the acidity of cell media leading to greater cytotoxicity compared to MNPs coated with polyethylene glycol-co-fumarate or polyvinyl alcohol (Arsianti et al. 2010, 2011). An added advantage to modifying the surface of nanoparticles is that it enhances their circulation half-life by avoiding phagocytosis and rapid clearance from the blood. Coating nanoparticles with poly (ethylene glycol) (PEG) through physical adsorption or chemical grafting also increases their hydrophilicity and blood compatibility (Jokerst et al. 2011).

Hydrophilic nanoparticles are preferred over hydrophobic nanoparticles because the hydrophobic nanoparticles tend to adsorb more biological components, such as opsonin proteins, which facilitates macrophage-mediated phagocytosis, whereas hydrophilic nanoparticles are generally safer and less toxic in biological systems due to reduced opsonization (Cedervall et al. 2007; Goppert and Muller 2005; Sadaf et al. 2012; Yan et al. 2006; Zhu et al. 2010). However, for certain cases, hydrophobic surfaces may be preferred. For instance, surface modification with hydrophobic polymers can increase the adsorption of nanoparticles in intestinal mucus (Ai et al. 2011). Modifying polymeric membranes with amphiphilic triblock copolymers, such as poly (vinyl pyrrolidone)–b-poly(methyl methacrylate)–b-poly(vinyl pyrrolidone), can enhance biocompatibility (Ran et al. 2011).

14.12.2 Biocompatibility of Nanoparticles

Below are some of the subdivisions of the biocompatibility of nanoparticles as discussed:

14.12.2.1 Immunocompatibility

Immunocompatibility is the study of the immune response to a biomaterial, prosthesis, or medical device, as a category of biocompatibility, and represents a major area of interest. Nanoparticles can either stimulate or suppress the immune system, a property that may not only affect the function of a particle for particular applications (Naahidi et al. 2013). But also the amount of adsorbed biological components, mainly proteins (opsonins) positively and negatively.

14.12.2.2 Immunostimulation

A key component of immunostimulation is inflammation which is triggered by the recognition of foreign materials, including nanoparticles leading to the secretion of cytokines to recruit immune cells. The surface properties of nanoparticles, particularly their surface charge, play a critical role in this process. Positively charged particles are more likely to cause inflammation compared to neutral or negatively charged ones, as they interact more strongly with cell membranes, leading to increased cytokine secretion and immune cell activation. Tan et al. (1999), showed that an anionic particle did not cause the secretion of cytokines while a cationic particle did. Diwan et al. (2004), also provided evidence of nanoparticle-induced inflammation. In their study, oligonucleotides bound to PLGA-based nanoparticles caused a larger amount of cytokine production and induced more T-cell proliferation than the naked oligonucleotide. Studies have shown that cationic particles induce greater cytokine production and T-cell proliferation than anionic or neutral particles.

It is common knowledge that the immune system has mechanisms for dealing with foreign bodies such as pathogens in the body. Nanoparticles can similarly be engineered to mimic pathogens by modifying their surfaces with components like Toll-like receptor (TLR) ligands or mannose, which are recognized by dendritic cells and mannose receptors, respectively (Cui et al. 2003). This approach can stimulate the immune system to respond to nanoparticles like how it responds to pathogens. However, significant research is needed to tailor these approaches for specific biomedical applications, such as targeting cancer cells without causing excessive immune reactions (Naahidi et al. 2013).

14.12.2.3 Immunosuppression

Immunosuppression is defined as the down-regulation or prevention of the immune system's activation (Naahidi et al. 2013). This process, while useful in certain medical contexts, has significant drawbacks, including increased susceptibility to infections caused by bacteria, viruses, fungi, and yeast, as well as a higher risk of developing neoplasms, such as skin cancers and lymphomas (Garibaldi 1998). Despite these risks, immunosuppression is crucial in the treatment of autoimmune diseases and in facilitating the acceptance of foreign tissues in organ transplant patients (Naahidi et al. 2013). The effectiveness of immunosuppression is influenced by several factors, including the dose of the immunosuppressant, the pathway into the body, the time of administration, the mechanism of action, and the site of activity (Říhová 2007).

Nanoparticles have been shown to exhibit immunosuppressant properties. For example, Shaunak et al. (2004), presented that Generation-3.5 polyamidoamine (PAMAM) dendrimer-glucosamine conjugates can significantly inhibit cytokine- and chemokine-induced inflammation when exposed to human macrophages and dendritic cells stimulated by bacterial endotoxin. These conjugates demonstrated novel immunomodulatory and antiangiogenic properties without apparent hematological toxicity, suggesting their potential to treat and prevent scar tissue formation.

In summary, immunosuppression is a complex therapeutic approach that, while risky, is essential for managing autoimmune diseases and ensuring the success of organ transplants. Nanotechnology, particularly through the use of nanoparticles, offers promising avenues for achieving targeted and safer immunosuppression.

14.12.3 Interaction with Blood

The surface properties of nanoparticles significantly impact their compatibility and interactions within the bloodstream. Blood constituents can react immunologically to nanoparticles and their associated drug complexes inactive (Naahidi et al. 2013). When nanoparticles enter the blood, they can be rapidly cleared by macrophages if they lack surface modifications to prevent the adsorption of opsonins, as reported by Gref et al. (1994). This highlights the importance of preclinical examinations that include studies on hemolysis, platelet aggregation, coagulation time, complement activation, leukocyte proliferation, and uptake by macrophages to assess nanoparticle biocompatibility (Dobrovolskaia and McNeil 2007). For example, positively charged nanoparticles with primary amines can damage erythrocyte membranes and cause hemolysis (Naahidi et al. 2013).

14.12.3.1 Complement Activation

The complement system, a key component of the immune response, can be activated by nanoparticles through three pathways: the classical pathway (triggered by specific antibodies), the alternate pathway (triggered by microbial surface structures), and the lectin pathway (triggered by mannose residues on microbial glycoproteins) (Říhová 2007). Understanding and reducing a surface's tendency for complement activation is essential to minimize the immunological response and ensure the biocompatibility of nanoparticles.

Modifying nanoparticle surfaces to avoid complement activation is crucial for understanding the immunological response they may trigger (Naahidi et al. 2013). For instance, nanoparticles with heterogeneous surface functional groups have been designed to minimize complement activation and enhance their immunocompatibility (Salvador-Morales et al. 2009).

In summary, the surface properties of nanoparticles, including their charge and ligand modification, are critical in determining their interactions with blood components and overall biocompatibility. Proper surface modification can help avoid immune clearance, reduce hemolysis, and minimize complement activation, ensuring the safe and effective use of nanoparticles in biomedical applications.

14.12.4 PEGlyation

The biocompatibility of a material within the body is significantly influenced by its surface characteristics, a concept first explored in the 1970s by Abuchowski et al. (1977a). They introduced the process of covalently bonding poly (ethylene glycol) (PEG) to drugs or therapeutic proteins, now known as PEGylation to enhance the biocompatibility and pharmacokinetics of therapeutic agents. In the early stages, PEGylation involved the covalent attachment of methoxy-PEGs (mPEGs) to proteins such as bovine serum albumin and liver catalase, as reported by Abuchowski et al. (1977b). This approach demonstrated that PEGylation could increase the half-life of drugs in the body, reduce dosing frequency, and prolong their activity (Abuchowski et al. 1977b; Molineux 2003).

Later, in 1980, de Gennes described two main regimes or conformations that PEG chains can adopt: the "mushroom" and "brush" conformations, depending on the grafting density. At low grafting densities, PEG polymers assume a "mushroom" configuration, while high densities result in a "brush" configuration (de Gennes 1980). The degree of surface coverage and the distance between graft sites are critical and depend on the molecular weight and graft density of the PEG polymer. PEGylation has been shown to reduce or prevent protein adsorption, thereby prolonging the circulation time of nanocarriers. This is primarily due to the physical properties of PEG, which create a steric barrier that prevents protein adsorption (Elbert and Hubbell 1996).

Studies by Allen et al. (2002) and others have highlighted that PEG molecules of different molecular weights can protect liposomes and other carriers, reducing their uptake by the mononuclear phagocyte system (MPS) and enhancing their stability in vivo. PEGylation can also lead to "dysopsonization," where specific proteins bind to the PEG-coated surface, masking the carrier and preventing its recognition by the immune system (Ahl et al. 1997). Today, it is well understood that PEGylation improves the biophysical and chemical properties of nanoparticles, enhancing their dispersibility, stability, and targeting capabilities (Naahidi et al. 2013). The flexibility and molecular weight of PEG chains are crucial factors in determining the effectiveness of PEGylation. Higher molecular weight PEGs and increased surface density can create a thicker protective layer, minimizing interactions with biomolecules and improving the overall performance of the nanoparticles (Guerrero et al. 2012).

In summary, the evolution of PEGylation from its initial introduction to its current applications has significantly advanced the field of drug delivery and nanomedicine, offering a versatile and effective method to enhance the biocompatibility and pharmacokinetics of therapeutic agents.

14.12.5 Biodegradability

Biodegradable nanoparticles are increasingly preferred for targeted drug delivery, vaccines, and the delivery of various biomolecules due to their ability to be digested internally and cleared from the body, thus avoiding the need for future removal (Naahidi et al. 2013). These nanoparticles can be prepared from a variety of materials, including proteins, polysaccharides, and synthetic biodegradable polymers such as poly-D-L-lactide-co-glycolide (PLGA), polylactic acid (PLA), poly-ε-caprolactone (PCL) (Mahapatro and Singh 2011). The selection of the base polymer depends on several factors, including the desired nanoparticle size, the properties of the drug to be encapsulated (such as aqueous solubility and stability), surface characteristics and functionality, degree of biodegradability and biocompatibility, and the drug release profile of the final product) (Mahapatro and Singh 2011).

Biodegradable nanoparticles effectively encapsulate therapeutic agents, protecting them from degradation, enhancing their solubility, and prolonging their circulation time. They enable targeted delivery to specific tissues or cells through surface modifications. Unlike non-biodegradable nanoparticles, which accumulate in the mononuclear phagocytic system (MPS) such as the liver and spleen, and can lead to potentially toxic side effects, biodegradable nanoparticles minimize waste accumulation and systemic toxicity (Naahidi et al. 2013). Further research is necessary to fully understand how the body, particularly the immune system, interacts with non-biodegradable nanoparticles (Dobrovolskaia and McNeil 2007). It is also important to consider the use of non-biodegradable nanoparticles cautiously, especially for non-terminal diseases where alternative treatment methods are available, due to the potential for lifelong side effects from irreversible accumulation in the MPS (Dobrovolskaia et al. 2009). Overall, biodegradable nanoparticles offer several advantages, including controlled and sustained drug release, targeted delivery, enhanced biocompatibility, and environmental friendliness.

14.13 Potent Toxicity of Nanoparticles to Organ Systems

Nanoparticles have proved toxic to humans once in the bloodstream, nanoparticles, spleen, bone marrow and nervous system can be transported around the body and be taken up by organ tissues and cell cultures, resulting in increased oxidative stress, inflammatory cytokine production, and cell death (Shinde et al. 2012). Research on the toxicological mechanism of NPs began in the early 2000s. The development of new analytical techniques, such as transmission electron microscopy and atomic force microscopy, has enabled researchers to study the interactions between NPs and cells in greater detail. Studies have shown that NPs can cause oxidative stress, inflammation, genotoxicity, and cytotoxicity in various cell types (Zhou et al. 2022). The toxicity of NPs can be roughly divided into acute toxicity and chronic toxicity (Xuan et al. 2023). Acute toxicity refers to adverse effects such as inflammation and tissue damage immediately after exposure to NP. Chronic toxicity refers to the long-term effects of exposure to NP, such as cancer and organ damage (Elsaesser and Howard 2012). The major types of toxicity caused by NPs include oxidative stress, inflammation, genotoxicity, and cytotoxicity (Zhou et al. 2022).

14.13.1 Oxidative Stress of Nanoparticles

Exposure to nanoparticles (NPs) can lead to significant oxidative stress due to the generation of reactive oxygen species (ROS). For instance, copper oxide (CuO) NPs have been shown to induce oxidative changes and cell death in human alveolar epithelial cells by producing ROS. This toxicity is largely attributed to the intracellular presence of CuO NPs, which interact with cellular enzymes and promote an oxidative cascade, resulting in the production of radical species like superoxide anion and hydroxyl radical. Studies have demonstrated that CuO NPs with more crystalline defects and higher ROS production capacity cause greater oxidative stress and cell viability decrease compared to fully crystalline NPs (Moschini et al. 2023).

The surface reactivity of NPs, including factors such as porosity, surface functionalization, and crystallinity, plays a crucial role in ROS generation. Smaller NPs tend to be more reactive and generate more ROS, leading to greater oxidative stress and cellular damage. For example, nanosized CuO particles were found to be more effective than micron-sized ones in producing intracellular ROS and genotoxic effects in A549 cells (Moschini et al. 2023). Understanding these mechanisms is essential for developing safer nanomaterials and for their potential therapeutic applications. The generation of ROS by NPs can lead to protein carbonylation, reduced-protein-thiol oxidation, and ultimately apoptotic or autophagic cell death, highlighting the need for a deeper understanding of the underlying toxicity pathways (Qiu et al. 2016).

Copper oxide nanoparticles (CuO NPs) have been shown to induce significant oxidative stress in various biological systems. When exposed to CuO NPs, cells experience an increase in reactive oxygen species (ROS) production, leading to oxidative stress and subsequent cytotoxicity. For instance, studies on human hepatocellular carcinoma cells (HepG2) have demonstrated that CuO NPs induce mitochondria-mediated apoptosis through the generation of ROS (Abdel-Latif et al. 2023). The mechanism of toxicity involves the intracellular presence of CuO NPs, which interact with cellular enzymes and promote an oxidative cascade, resulting in the production of radical species such as superoxide anion and hydroxyl radical.

This oxidative stress can lead to protein carbonylation, reduced-protein-thiol oxidation, and ultimately to apoptotic or autophagic cell death (Siddiqui et al. 2013).

14.13.2 Genotoxicity and Cytotoxicity of Nanoparticles

The genotoxicity and cytotoxicity of nanoparticles are significant concerns due to their potential to cause cellular damage and genetic mutations. Nanoparticles, particularly metal oxide nanoparticles such as zinc oxide (ZnO), titanium dioxide (TiO_2), and silver (Ag) nanoparticles, have been shown to exhibit both genotoxic and cytotoxic effects. For instance, ZnO nanoparticles have been found to induce cytotoxicity in human alveolar epithelial cells at concentrations as low as 5–10 μg/mL, causing approximately 50% cell damage within 24 h of exposure. This toxicity is attributed to the generation of reactive oxygen species (ROS) and the disruption of cell membrane integrity, as evidenced by lactate dehydrogenase (LDH) release assays. Genotoxicity assays have also revealed that ZnO and TiO_2 nanoparticles can cause DNA damage and genomic instability. For example, ZnO nanoparticles at concentrations of 20 μg/mL have shown significant genotoxic effects, including nuclear morphology changes and the presence of pH2AX signals, indicating DNA double-strand breaks (Handral et al. 2021).

Silver nanoparticles (AgNPs) also exhibit cytotoxic and genotoxic effects in human cells. The size and coating of AgNPs can significantly alter their toxicological potential, with smaller AgNPs generally being more toxic due to their higher surface area and reactivity. AgNPs have been shown to cause DNA damage, inhibit cell proliferation, and induce apoptosis in various cell lines (Asharani et al. 2008).

14.14 Regulatory Standards of Nanoparticles

Regulatory standards for nanotoxicity are crucial to ensure the safe use and handling of nanomaterials, given their potential toxic properties and environmental impacts. Regulatory frameworks for nanomaterials have been developed and implemented globally to address the unique challenges posed by their use. The Organisation for Economic Co-operation and Development (OECD) plays a significant role in establishing these standards. For instance, the OECD has developed test guidelines and safety testing protocols for manufactured nanomaterials, including those related to environmental sustainability, ecotoxicology, and human health safety (Journal and Ondokuzmay 2020).

14.15 Overview of Preclinical and Clinical Requirements for Translating Polymer Nanocomposites to Market

14.15.1 Preclinical Requirements

Material characterization; Physicochemical Properties involving comprehensive characterization of PNCs are crucial. This includes assessing mechanical strength, thermal stability, and chemical composition to ensure consistency and reliability in performance due to the addition of nanoparticles in nanosheets and nanotubes among others that affect the chemical and mechanical abilities of the PNCs (López-Lugo et al. 2021; Panicker and Mohamed 2023; Ramakoti et al. 2023; Wang et al. 2004). Nanofiller interactions between the polymer matrix and nanofillers are essential for forecasting the composite's behavior under physiological conditions. Utilizing the synergy with another biomolecule, the polymer and polymer dynamics, and the ability to control the performance of these new PNC materials can further improve these properties (Bustamante-Torres et al. 2021; Darwish et al. 2022; Phatak et al. 2015; Rodrigues et al. 2021).

Biocompatibility: The effects of PNCs on pertinent cell lines must be assessed using tests including cytotoxicity, hemolysis, and thrombogenicity assays that involve in vitro investigations. This helps determine their safety for use in biological systems (Andrade et al. 2019; Jamil et al. 2024; Singh et al. 2018). According to (Wang et al. 2004) wettability, chemistry, charge, roughness, and rigidity are some of the surface properties of polymeric materials that have been found to influence how certain cell types adhere and proliferate (Marchant et al. 1984; Panicker and Mohamed 2023; Singh et al. 2018).

Animal studies are required to evaluate the biocompatibility and possible systemic effects of PNCs in the event of in vivo investigations. These investigations ought to concentrate on biodistribution, pharmacokinetics, and any negative effects. PNCs' systemic effects must be evaluated using animal models. These investigations ought to assess the pharmacokinetics (absorption, distribution, metabolism, and excretion) as well as any toxicological consequences (Jamil et al. 2024; Vera et al. 2020).

Drug Discharge Investigations using Kinetic Profiling to forecast the behavior of PNCs in diverse biological contexts, preclinical assessments should incorporate drug release kinetics under a range of circumstances (such as pH and temperature) (Herdiana et al. 2022). Drug release kinetics must be understood to anticipate therapeutic effects. Research should investigate the effects of erosion, drug diffusion, and polymer swelling on release profiles. Mechanisms of diffusion-controlled release can be described by models such as Fick's law (Liang et al. 2014; Panicker and Mohamed 2023; Paravastu et al. 2019).

Stability testing is essential to guarantee the effectiveness of PNCs after delivery. It evaluates the stability of the encapsulated active pharmaceutical ingredients (APIs) over time (Bustamante-Torres et al. 2021; Panicker and Mohamed 2023).

14.15.2 Clinical Requirements

To ensure the quality, efficacy, and safety of PNCs for human use, clinical trials are mandatory. Clinical trials are the only way to determine whether the PNCs are safe and effective for human use. The use of PNCs in clinical trials depends on the approval of the production methods and assurance of the quality of the final product by implementation and verification of good manufacturing practices (GMP) (Furberg and Friedman 2012; Souto et al. 2020).

According to Souto et al. (2020), in the preclinical stage, which could take 3–6 years, the chosen API is examined in vivo in animal models to gather pertinent in vivo data for the upcoming clinical trials. A clinical trial is a research study that compares the values and effects of a particular drug formulation with a control procedure that is also administered to humans. Clinical trials differ from retrospective research in that they are mostly prospective (Furberg and Friedman 2012).

In addition to evaluating a treatment's efficacy and safety, clinical trials are used to estimate the risk of adverse reactions, determine pharmacokinetic parameters (such as absorption, distribution, metabolism, and excretion), and evaluate a product's efficacy and safety compared to an existing treatment (Friedman et al. 1980; McNally et al. 2017; Thanh 2018).

Clinical trials are categorized into four phases: I, II, III, and IV. The classification of each of these stages is contingent upon the drug's stage of development and post-marketing authorization, and thus, it is impossible to define them precisely. However, they might be arranged based on their primary goals and clinical advancements (Bustamante-Torres et al. 2021; Jamil et al. 2024; Souto et al. 2020; Vera et al. 2020).

Using human volunteers who are usually healthy, *phase I* studies include clinical pharmacology and toxicity as well as an initial assessment of medication safety, pharmacokinetics, and pharmacodynamics. To prevent serious side effects, a continual increase in the applied dose is administered to identify the appropriate drug dosage for the assessed health status. Several volunteers are included in this phase; the number of participants may range from 10 to 100 during the course of the 1.5-year trial.

In *Phase II*, studies of Preliminary Clinical Study for Medical Impact are conducted to evaluate the safety and effectiveness of an intervention for therapy on volunteers who have a medical condition for which the medication is prescribed. The comparison of the controlled approach and placebo forms the foundation of these investigations. The number of participants in this phase might range from up to 200, and each one would be registered for 2 years.

The Complete Assessment of Treatment, which will be examined in Phase III, leverages more expansive volunteer clusters to compare the medication employed in the clinical trial to the gold standard in that therapeutic area. The effectiveness, adverse effects, and even interactions of the treatment being tested must all be confirmed by *phase III* trials. Typically, this phase spans a minimum of 3 years. This phase's trials are assigned randomly and double-blinded in a bid to replicate the actual circumstances.

The objective of *phase IV*, or post-marketing surveillance, is to maintain track of the treatment post-permission commercial marketing. Here, the research concentrates on the summary of drug product (SDP), to continuously monitor the medication's side effects and any rare adverse events associated with PNC use in patients (Bustamante-Torres et al. 2021; Panicker and Mohamed 2023).

GMP is a collection of regulations designed to guarantee the quality of pharmaceuticals (PNCs). GMP mandates that PNCs be produced in a manner that guarantees their quality and avoids contamination. Participants' safety must always come first in clinical trials, which must be carried out ethically.

The approval of the production processes and the final product's quality assurance through the application and verification of GMP are prerequisites for the continued use of PNCs in clinical trials. To guarantee that PNCs are consistently produced per the requirements and with the necessary quality, GMP is a component of the quality assurance (QA) process that should be put into place. To guarantee that the PNC's product to be launched has the desired quality for the intended usage, quality control (QC) makes sure that the required quality check tests are conducted (Bustamante-Torres et al. 2021; Jamil et al. 2024; Karim et al. 2018; Souto et al. 2020).

The Nanotechnology Regulatory Science Research Plan, introduced by the Food and Drug Administration (FDA) in 2013, outlines the management and implementation of all nanotechnologies and their derivative nano products. To guarantee the safety, effectiveness, and quality of nanopharmaceuticals, such a technique sought to alleviate the breach that now exists about them. The "reflection papers" and other periodical instructions were also produced by the European Medicines Agency (EMA) (Allan et al. 2021; Bobo et al. 2016; FDA's Approach to Regulation of Nanotechnology Products _ FDA n.d.; Guo et al. 2014; Havel et al. 2016; Souto et al. 2019).

References

Abdel-Bary AS, Tolan DA, Nassar MY, Taketsugu T, El-Nahas AM (2020) Chitosan, magnetite, silicon dioxide, and graphene oxide nanocomposites: synthesis, characterization, efficiency as cisplatin drug delivery, and DFT calculations. Int J Biol Macromol 154:621–633. https://doi.org/10.1016/j.ijbiomac.2020.03.106

Abdel-Latif HMR, Dawood MAO, Mahmoud SF, Shukry M, Noreldin AE, Ghetas HA, Khallaf MA (2023) Copper oxide nanoparticles alter serum biochemical indices, induce histopathological alterations, and modulate transcription of cytokines, *HSP70*, and oxidative stress genes in *Oreochromis niloticus*. Animals 11:652

Abuchowski A, van Es T, Palczuk NC, Davis FF (1977a) Alteration of immunological properties of bovine serum albumin by covalent attachment of polyethylene glycol. J Biol Chem 252:3578–3581

Abuchowski A, McCoy JR, Palczuk NC, van Es T, Davis FF (1977b) Effect of covalent attachment of polyethylene glycol on immunogenicity and circulating life of bovine liver catalase. J Biol Chem 252:3582–3586

Adabi M, Saber R, Adabi M, Sarkar S (2011) Examination of incubation time of bare gold electrode inside cysteamine solution for immobilization of multi-walled carbon nanotubes on a gold electrode modified with cysteamine. Microchim Acta 172:83–88

Adabi M et al (2017) Biocompatibility and nanostructured materials: applications in nanomedicine. Artif Cells Nanomed Biotechnol 45(4):833–842. https://doi.org/10.1080/21691401.2016.1178134

Ahl PL, Bhatia SK, Meers P, Roberts P, Stevens R, Dause R, Perkins WR, Janoff AS (1997) Enhancement of the in vivo circulation lifetime of L-alpha- distearoylphosphatidylcholine liposomes: Importance of liposomal aggregation versus complement opsonization. Biochim Biophys Acta 1329:370–382

Ahmed S et al (2020a) Alginate/silica hybrid nanocomposites for wound healing applications. Int J Biol Macromol 152:763–773. https://doi.org/10.1016/j.ijbiomac.2020.02.110

Ahmed S et al (2020b) pH-responsive polyacrylic acid-coated nanocomposites for tumor-specific drug delivery. J Control Release 328:102–112. https://doi.org/10.1016/j.jconrel.2020.09.015

Ai J, Biazar E, Jafarpour M, Montazeri M, Majdi A, Aminifard S et al (2011) Nanotoxicology and nanoparticle safety in biomedical designs. Int J Nanomedicine 6:1117–1127

Akbarzadeh A, Sadabady RR, Davaran S, Joo SW, Zarghami N, Hanifehpour Y, Samiei M, Kouhi M, Koshki NK (2013) Liposome: classification, preparation, and applications. Nanoscale Res Lett 8(1):102. https://doi.org/10.1186/1556-276X-8-102

Alghamdi MA, Fallica AN, Virzì N, Kesharwani P, Pittalà V, Greish K (2022) The promise of nanotechnology in personalized medicine. J Pers Med 12(5):673. PMID: 35629095. https://doi.org/10.3390/jpm12050673

Alhamhoom Y, Honmane SM, Hani U, Osmani RAM, Kandasamy G, Vasudevan R, Paramshetti SR, Dudhal RK, Kengar N, Charde MS (2023) Study of formulation and process variables for optimization of Piroxicam Nanosuspension using 32 factorial design to improve solubility and in vitro bioavailability. Polymers 15(3):483. https://doi.org/10.3390/polym15030483

Allan J, Belz S, Hoeveler A, Hugas M, Okuda H, Patri A, Rauscher H, Silva P, Slikker W, Sokull-Kluettgen B, Tong W, Anklam E (2021) Regulatory landscape of nanotechnology and nanoplastics from a global perspective. Regul Toxicol Pharmacol 122(104):885. https://doi.org/10.1016/j.yrtph.2021.104885

Allen C, Dos Santos N, Gallagher R, Chiu GN, Shu Y, Li WM, Johnstone SA, Janoff AS, Mayer LD, Webb MS, Bally MB (2002) Controlling the physical behavior and biological performance of liposome formulations through use of surface grafted poly(ethylene glycol). Biosci Rep 22:225–250

Alosaimi AM, Alorabi RO et al (2023) Review on biomedical advances of hybrid nanocomposite biopolymeric materials. Bioengineering 10(3):279

Alsaab AH, Zeghib S (2023) Study of prepared Lead-free polymer nanocomposites for X- and gamma-ray shielding in healthcare applications. Polymers 15(9):2142. https://doi.org/10.3390/polym15092142

Anagha Bhakay MR (2018) Bioavailability enhancement of poorly water-soluble drugs via nanocomposites: formulation–processing aspects and challenges. Pharmaceutics 10:86

Anaraki NI, Sadeghpour A, Iranshahi K, Toncelli C, Cendrowska U, Stellacci F, Dommann A, Wick P, Neels A (2020) New approach for time-resolved and dynamic investigations on nanoparticles agglomeration. Nano Res 13(10):2847–2856. https://doi.org/10.1007/s12274-020-2940-4

Andrade LN, Oliveira DML, Chaud MV, Alves TFR, Nery M, da Silva CF, Gonsalves JKC, Nunes RS, Corrêa CB, Amaral RG, Sanchez-Lopez E, Souto EB, Severino P (2019) Praziquantel-solid lipid nanoparticles produced by supercritical carbon dioxide extraction: physicochemical characterization, release profile, and cytotoxicity. Molecules 24(21):3881. https://doi.org/10.3390/molecules24213881

Ang HY, Xiong GM, Chaw SY, Phua JL, Ng JCK, Wong PEH, Venkatraman S, Chong TT, Huang Y (2020) Adventitial injection delivery of nano-encapsulated sirolimus (Nanolimus) to injury-induced porcine femoral vessels to reduce luminal restenosis. J Control Release 319:15–24. https://doi.org/10.1016/j.jconrel.2019.12.031

Anzara N, Hasan R, Tyagi M, Yadav N (2020) Carbon nanotube—A review on synthesis, properties and plethora of applications in the field of biomedical science. Sensors Int 1:100003

Arriagada P, Palza H, Palma P, Flores M, Caviedes P (2018) Poly(lactic acid) composites based on graphene oxide particles with antibacterial behavior enhanced by electrical stimulus and biocompatibility. J Biomed Mater Res A 106:1051–1060. https://doi.org/10.1002/jbm.a.36307

Arsianti M, Lim M, Marquis CP, Amal R (2010) Polyethylenimine based magnetic iron-oxide vector: the effect of vector component assembly on cellular entry mechanism, intracellular localization, and cellular viability. Biomacromolecules 11:2521–2531

Arsianti M, Lim M, Lou SN, Goon IY, Marquis CP, Amal R (2011) Bi-functional gold-coated magnetite composites with improved biocompatibility. J Colloid Interface Sci 354:536–545

Arul QA, Arputham AJ, Mahalingam P, Debnath D, Kumaran NK (2024) Morphological characterization of an eco friendly Medi grade bio-polymer composites using SEM analysis. J Pharm Bioallied Sci 16:S3979. https://doi.org/10.4103/jpbs.jpbs_907_24

Asharani PV, Low Kah Mun G, Hande MP, Valiyaveettil S (2008) cytotoxicity and genotoxicity of silver nanoparticles in human cells. ACS Nano 3:279

Åslund AKO, Vandebriel RJ, Caputo F, de Jong WH, Delmaar C, Hyldbakk A, Rustique E, Schmid R, Snipstad S, Texier I, Vernstad K, Borgos SEF (2022) A comparative biodistribution study of polymeric and lipid-based nanoparticles. Drug Deliv Transl Res 12(9):2114–2131. https://doi.org/10.1007/s13346-022-01157-y

Avula UMR, Yoon HK, Lee CH, Kaur K, Ramirez RJ, Takemoto Y, Ennis SR, Morady F, Herron T, Berenfeld O, Kopelman R, Kalifa J (2015) Cell-selective arrhythmia ablation for photomodulation of heart rhythm. Sci Transl Med 7:311ra172. https://doi.org/10.1126/scitranslmed.aab3665

Baslak C, Yildirim M (2024) Characterization of polymer nanocomposites in biomedical implants. In: Polymeric materials for biomedical implants: characterization, properties, and applications, pp 31–66. https://doi.org/10.1016/B978-0-323-99690-7.00005-4

Baumann MD, Kang CE, Tator CH, Shoichet MS (2010) Intrathecal delivery of a polymeric nanocomposite hydrogel after spinal cord injury. Biomaterials 31(30):7631–7639. https://doi.org/10.1016/j.biomaterials.2010.07.004

Bayindir SZ, Ergin AD, Parmaksiz M, Elcin AE, Elcin YM, Yuksel N (2016) Evaluation of various block copolymers for micelle formation and brain drug delivery: in vitro characterization and cellular uptake studies. J Drug Deliv Sci Technol 36:120–129. https://doi.org/10.1016/j.jddst.2016.10.003

Begines B, Ortiz T, Pérez-Aranda M, Martínez G, Merinero M, Argüelles-Arias F, Alcudia A (2020) Polymeric nanoparticles for drug delivery: recent developments and future prospects. Nanomaterials 10(7):1403. https://doi.org/10.3390/nano10071403

Bhakay A, Rahman M, Dave RN, Bilgili E (2018) Bioavailability enhancement of poorly water-soluble drugs via nanocomposites: formulation–processing aspects and challenges. Pharmaceutics 10(3):86. https://doi.org/10.3390/pharmaceutics10030086

Bharadwaz A, Jayasuriya AC (2020) Recent trends in the application of widely used natural and synthetic polymer nanocomposites in bone tissue regeneration. Mater Sci Eng C Mater Biol Appl 110:110,698. https://doi.org/10.1016/j.msec.2020.110698

Bharathi D, Ranjithkumar R, Chandarshekar B, Bhuvaneshwari V (2019) Bio-inspired synthesis of chitosan/copper oxide nanocomposite using rutin and their anti-proliferative activity in human lung cancer cells. Int J Biol Macromol 141:476–483. https://doi.org/10.1016/j.ijbiomac.2019.08.235

Bhat MR (2015) Bionanocomposites: technique towards enhancement of solubility, drug release and bioavailability. J Med Pharmaceut Innov 2(8):6

Bobo D, Robinson KJ, Islam J, Thurecht KJ, Corrie SR (2016) Nanoparticle-based medicines: a review of FDA-approved materials and clinical trials to date. Pharm Res 33(10):2373–2387. https://doi.org/10.1007/s11095-016-1958-5

Bonnet S, Prévot G, Mornet S, Jacobin-Valat M-J, Mousli Y, Hemadou A, Duttine M, Trotier A, Sanchez S, Duonor-Cérutti M, Crauste-Manciet S, Clofent-Sanchez G (2021) A nano-emulsion platform functionalized with a fully human scFv-Fc antibody for atheroma targeting: towards a theranostic approach to atherosclerosis. Int J Mol Sci 22:5188. https://doi.org/10.3390/ijms22105188

Brittain HK, Scott R, Thomas E (2017) The rise of the genome and personalised medicine. Clin Med 17(6):545–551. PMID: 29196356. https://doi.org/10.7861/clinmedicine.17-6-545

Buckner CA, Lafrenie RM, Dénommée JA, Caswell JM, Want DA, Gan GG, Leong YC, Bee PC, Chin E, Teh AKH, Picco S, Villegas L, Tonelli F, Merlo M, Rigau J, Diaz D, Masuelli M, Korrapati S, Kurra P et al (2016) We are IntechOpen, the world ' s leading publisher of Open Access books Built by scientists, for scientists TOP 1%. INTECH 11(tourism):13

Bustamante-Torres M, Romero-Fierro D, Arcentales-Vera B, Pardo S, Bucio E (2021) Interaction between filler and polymeric matrix in nanocomposites: magnetic approach and applications. Polymers 13(17):2998. https://doi.org/10.3390/polym13172998

Cacciatore I, Ciulla M, Fornasari E, Marinelli L, Di Stefano A (2016) Solid lipid nanoparticles as a drug delivery system for the treatment of neurodegenerative diseases. Expert Opin Drug Deliv 13:1121–1131. https://doi.org/10.1080/17425247.2016.1178237

Cai S et al (2011) Mechanics and chemical thermodynamics of phase transition in temperature-sensitive hydrogels. J Mech Phys Solids 59:2259

Cedervall T, Lynch I, Lindman S, Berggård T, Thulin E, Nilsson H, Dawson KA, Linse S (2007) Understanding the nanoparticle-protein corona using methods to quantify exchange rates and affinities of proteins for nanoparticles. Proc Natl Acad Sci U S A 104:2050–2055

Chattopadhyay S, Raines RT (2014) Collagen-based biomaterials for wound healing. Biopolymers 101:821–833. https://doi.org/10.1002/bip.22486

Chavda VP, Gogoi NR, Vaghela DA, Balar PC, Dawre S, Dave DJ (2023) Parenteral microemulsions for drug delivery: advances and update. J Drug Deliv Sci Technol 89:104991. https://doi.org/10.1016/J.JDDST.2023.104991

Chen L, McCrate JM, Lee JC, Li H (2011) The role of surface charge on the uptake and biocompatibility of hydroxyapatite nanoparticles with osteoblast cells. Nanotechnology 22:105708. https://doi.org/10.1088/0957-4484/22/10/105708

Chen Y, Wei C, Lyu Y, Chen H, Jiang G, Gao X (2020a) Biomimetic drug-delivery systems for the management of brain diseases. Biomater Sci 8(4):1073–1088. https://doi.org/10.1039/C9BM01395D

Chen SX et al (2020b) Construction of microneedle-assisted co-delivery platform and its combining photodynamic/immunotherapy. J Control Release 324:218

Cheng Y, Wang K, Xu H, Li T, Jin Q, Cui D (2021) Recent developments in sensors for wearable device applications. Anal Bioanal Chem 413(24):6037–6057

Chithrani BD, Ghazani AA, Chan WC (2006) Determining the size and shape dependence of gold nanoparticle uptake into mammalian cells. Nano Lett 6:662–668

Clark RAF et al (2021) To scar or not to scar. N Engl J Med 385:469

Crowley TE (2020) Chapter 8. X-ray crystallography. In: Purification and characterization of secondary metabolites. Academic Press, New York, pp 79–84. https://doi.org/10.1016/B978-0-12-813942-4.00008-5

Cui Z, Hsu CH, Mumper RJ (2003) Physical characterization and macrophage cell uptake of mannan-coated nanoparticles. Drug Dev Ind Pharm 29:689–700

Cukalevski R, Ferreira SA, Dunning CJ, Berggård T, Cedervall T (2015) IgG and fibrinogen driven nanoparticle aggregation. Nano Res 8(8):2733–2743. https://doi.org/10.1007/s12274-015-0780-4

Daraee H, Etemadi A, Kouhi M, Alimirzalu S, Akbarzadeh A (2016) Application of liposomes in medicine and drug delivery. Artif Cells Nanomed Biotechnol 44(1):381–391. https://doi.org/10.3109/21691401.2014.953633

Darwish MSA, Mostafa MH, Al-Harbi LM (2022) Polymeric nanocomposites for environmental and industrial applications. Int J Mol Sci 23(3):1023. https://doi.org/10.3390/ijms23031023

Davis ME et al (2020) Biodegradable polymers for drug delivery. Adv Drug Deliv Rev 158:123–135

de Castro Leão M, Raffin Pohlmann A, de Cristo Soares Alves A, Helena Poliselli Farsky S, Klimuk Uchiyama M, Araki K, Sandri S, Stanisçuaski Guterres S, Castro A (2021) Docosahexaenoic acid nanoencapsulated with anti-PECAM-1 as co-therapy for atherosclerosis regression. Eur J Pharm Biopharm 159:99–107. https://doi.org/10.1016/j.ejpb.2020.12.016

de Gennes PG (1980) Conformation of polymers attached to an interface. Macromolecules 131:069–1075

De Leon VH et al (2012) Polymer nanocomposites for improved drug delivery efficiency. Mater Chem Phys 132:409–415. https://www.researchgate.net/publication/257079701_Polymer_nanocomposites_for_improved_drug_delivery_efficiency. Accessed 21 Dec 2024

Decuzzi P, Pasqualini R, Arap W, Ferrari M (2009) Intravascular delivery of particulate systems: does geometry really matter? Pharm Res 26:235–243

Desai N (2012) Challenges in development of nanoparticle-based therapeutics. AAPS J 14(2):282–295

Desai PP, Patravale VB (2018) Curcumin cocrystal micelles—multifunctional nanocomposites for management of neurodegenerative ailments. J Pharm Sci 107(4):1143

Devasier B, Sang-Hyo K (2014) Polymer nanoparticles for smart drug delivery. IntechOpen, London. https://doi.org/10.5772/58422

Dhanavel S, Nivethaa EAK, Narayanan V, Stephen A (2017) In vitro cytotoxicity study of dual drug loaded chitosan/palladium nanocomposite towards HT-29 cancer cells. Mater Sci Eng C Mater Biol Appl 75:1399–1410. https://doi.org/10.1016/j.msec.2017.03.058

Díez-Pascual AM, Luceño-Sánchez JA (2021) Antibacterial activity of polymer nanocomposites incorporating graphene and its derivatives: a state of art. Polymers 13:2105. https://doi.org/10.3390/polym13132105

Diwan M, Elamanchili P, Cao M, Samuel J (2004) Dose sparing of CpG oligodeoxynucleotide vaccine adjuvants by nanoparticle delivery. Curr Drug Deliv 1:405–412

Dmour I, Taha MO (2018) Natural and semisynthetic polymers in pharmaceutical nanotechnology. In: Organic materials as smart nanocarriers for drug delivery. Applied Science Publishers, Basel, pp 35–100. https://doi.org/10.1016/B978-0-12-813663-8.00002-6

Dobrovolskaia MA, McNeil SE (2007) Immunological properties of engineered nanomaterials. Nat Nanotechnol 2:469–478

Dobrovolskaia MA, Germolec DR, Weaver JL (2009) Evaluation of nanoparticle immunotoxicity. Nat Nanotechnol 4:411–414

Dodi G, Popescu D, Cojocaru FD, Aradoaei M, Ciobanu RC, Mihai CT (2022) Use of Fourier-transform infrared spectroscopy for DNA identification on recycled PET composite substrate. Appl Sci 12(9):4371. https://doi.org/10.3390/app12094371

Dukes MD, Krans NA, Marusak K, Walden S, Eldred T, Franks A, Larson B, Guo Y, Nackashi D, Damiano J (2023) A machine-vision approach to transmission electron microscopy workflows, results analysis and data management. J Vis Exp 2023(196):1–22. https://doi.org/10.3791/65446

Eid MM (2022) Characterization of nanoparticles by FTIR and FTIR-microscopy BT. In: Mallakpour S, Hussain CM (eds) Handbook of consumer nanoproducts. Springer Nature, Singapore, pp 645–673. https://doi.org/10.1007/978-981-16-8698-6_89

Elbert DL, Hubbell JA (1996) Surface treatments of polymers for biocompatibility. Ann Rev Mater Sci 26:365–394

Elsaesser A, Howard CV (2012) Toxicology of nanoparticles. Adv Drug Deliv Rev 64(2):129–137

Erbe A, Sigel R (2013) Incoherent dynamic light scattering by dilute dispersions of spherical particles: wavelength-dependent dynamics. Phys Chem Chem Phys 15(44):19143–19146. https://doi.org/10.1039/c3cp53220h

European Commission (2011) Commission Recommendation of 18 October 2011 on the definition of nanomaterial. Off J Eur Union 275/38

Fang K, Li K, Yang T, Li J, He W (2021) Starch-based magnetic nanocomposite as an efficient absorbent for anticancer drug removal from aqueous solution. Int J Biol Macromol 184:509–521. https://doi.org/10.1016/j.ijbiomac.2021.06.103

Farokhi M et al (2018) Overview of silk fibroin use in wound dressings. Trends Biotechnol 36:907

Farshi Azhar F, Olad A (2014) A study on sustained release formulations for oral delivery of 5-fluorouracil based on alginate–chitosan/montmorillonite nanocomposite systems. Appl Clay Sci 101:288–296. https://doi.org/10.1016/j.clay.2014.09.004

FDA's Approach to Regulation of Nanotechnology Products _ FDA (n.d.)

Ferguson SK, Pak DI, Hopkins JL, Harral JW, Redinius KM, Loomis Z, Stenmark KR, Borden MA, Schroeder T, Irwin DC (2019) Pre-clinical assessment of a water-in-fluorocarbon emulsion for the treatment of pulmonary vascular diseases. Drug Deliv 26:147–157. https://doi.org/10.1080/10717544.2019.1568621

Fernandes AI (2023) Polymers enhancing bioavailability in drug delivery, 2nd edition. Pharmaceutics 15(11):2604. https://doi.org/10.3390/pharmaceutics15112604

Fernandes AI, Jozala AF (2022) Polymers enhancing bioavailability in drug delivery. Pharmaceutics 14(10):10. https://doi.org/10.3390/pharmaceutics14102199

Florea A, Sigl JP, Morgenroth A, Vogg A, Sahnoun S, Winz OH, Bucerius J, Schurgers LJ, Mottaghy FM (2021) Sodium [18F]Fluoride PET can efficiently monitor in vivo atherosclerotic plaque calcification progression and treatment. Cells 10:275. https://doi.org/10.3390/cells10020275

Foglizzo V, Marchiò S (2022) Nanoparticles as physically-and biochemically-tuned drug formulations for cancers therapy. Cancer 14(10):2473. https://doi.org/10.3390/cancers14102473

Forna N, Damir D, Duceac LD, Dabija MG, Calin G, Ichim DL, Gutu C, Grierosu C, Eva L, Ciuhodaru MI, Goroftei ERB, Banu EA, Stafie L, Gabriela C, Mitrea G, Marcu C (2022) Nanoarchitectonics of antibiotic-loaded polymer particles as vehicles for active molecules. Appl Sci 12(4):1998. https://doi.org/10.3390/app12041998

Fortunato A (2013) DSC: history, instruments and devices. In: Drug-biomembrane interaction studies: the application of calorimetric techniques. Woodhead Publishing, Sawston. https://doi.org/10.1533/9781908818348.169

Friedman L, Passamani E, May G, Furberg C, Friedewald W (1980) Placebo controlled double blind trials. Lancet 316(8186):147–148. https://doi.org/10.1016/S0140-6736(80)90030-6

Furberg CD, Friedman LM (2012) Approaches to data analyses of clinical trials. Prog Cardiovasc Dis 54(4):330–334. https://doi.org/10.1016/J.PCAD.2011.07.002

Gamal O, Agamey T, Amr Z, El-Sherbiny IM (2024) Revolutionizing lung cancer treatment with smart pulmonary drug delivery nanosystems. J Drug Deliv Sci Technol 101:106158. https://doi.org/10.1016/J.JDDST.2024.106158

Gao X, Cui Y, Levenson RM, Chung LWK, Nie S (2004) In vivo cancer targeting and imaging with semiconductor quantum dots. Nat Biotechnol 22(8):969–976. https://doi.org/10.1038/nbt994

Gao J, Bao F, Feng L, Shen K, Zhu Q, Wang D, Chen T, Ma R, Yan C (2011) Functionalized graphene oxide modified polysebacic anhydride as drug carrier for levofloxacin controlled release. RSC Adv 1:1737–1744. https://doi.org/10.1039/C1RA00029B

Garibaldi RA (1998) Infections in organ transplant recipients. Infect Control 4:460

Gatti AM, Rivasi F (2002) Biocompatibility of micro- and nanoparticles. Part I: in liver and kidney. Biomaterials 23:2381–2387

Gharanjig H, Gharanjig K, Hosseinnezhad M, Jafari SM (2020) Differential scanning calorimetry (DSC) of nanoencapsulated food ingredients. In: Characterization of nanoencapsulated food ingredients. Elsevier, Amsterdam. https://doi.org/10.1016/B978-0-12-815667-4.00010-9

Ghatage DAK, Kanitkar PSG (2019) Characterization of materials by X-ray diffraction. Int J Eng Res Technol 8(11):156–159

Ghawanmeh AA, Ali GAM, Algarni H, Sarkar SM, Chong KF (2019) Graphene oxide-based hydrogels as a nanocarrier for anticancer drug delivery. Nano Res 12:973–990. https://doi.org/10.1007/s12274-019-2300-4

Gholamali I, Yadollahi M (2021) Bio-nanocomposite polymer hydrogels containing nanoparticles for drug delivery: a review. Regener Eng Transl Med 7(2):129–146. https://doi.org/10.1007/s40883-021-00207-0

Glaskova-Kuzmina T, Starkova O, Gaidukovs S, Platnieks O, Gaidukova G (2021) Durability of biodegradable polymer nanocomposites. Polymers 13(19):3375. https://doi.org/10.3390/polym13193375

Goetz LH, Schork NJ (2018) Personalized medicine: motivation, challenges, and progress. Fertil Steril 109(6):952–963. PMID: 29935653. https://doi.org/10.1016/j.fertnstert.2018.05.006

González LF, Acuña E, Arellano G, Morales P, Sotomayor P, Oyarzun-Ampuero F, Naves R (2021) Intranasal delivery of interferon-β-loaded nanoparticles induces control of neuroinflammation in a preclinical model of multiple sclerosis: a promising simple, effective, non-invasive, and low-cost therapy. J Control Release 331:443–459. https://doi.org/10.1016/j.jconrel.2020.11.019

Goppert TM, Muller RH (2005) Protein adsorption patterns on poloxamer- and poloxamine-stabilized solid lipid nanoparticles (SLN). Eur J Pharm Biopharm 60:361–372

Goyal R, Macri LK, Kaplan HM, Kohn J (2016) Nanoparticles and nanofibers for topical drug delivery. J Control Release 240:77–92. https://doi.org/10.1016/J.JCONREL.2015.10.049

Gref R, Minamitake Y, Peracchia MT, Trubetskoy V, Torchilin V, Langer R (1994) Biodegradable long-circulating polymeric nanospheres. Science 263:1600–1603

Guerrero S, Herance R, Rojas S, Mena J, Gispert J, Acosta G, Albericio P, Kogan MJ (2012) Synthesis and in vivo evaluation of the biodistribution of a 18 F labelled conjugate gold-nanoparticle-peptide with potential biomedical application. Bioconjug Chem 23:399–408

Guo JW, Lee YH, Huang HW, Tzou MC, Wang YJ, Tsai JC (2014) Development of Taiwan's strategies for regulating nanotechnology-based pharmaceuticals harmonized with international considerations. Int J Nanomedicine 9:4773. https://doi.org/10.2147/IJN.S68134

Gupta A et al (2023) Responsive nanoparticles for targeted drug delivery. J Drug Target 31(5):480–495

Hadian-Jazi M, Sadri A, Barty A, Yefanov O, Galchenkova M, Oberthuer D, Komadina D, Brehm W, Kirkwood H, Mills G, de Wijn R, Letrun R, Kloos M, Vakili M, Gelisio L, Darmanin C, Mancuso AP, Chapman HN, Abbey B (2021) Data reduction for serial crystallography using a robust peak finder. J Appl Cryst 54(5):1360–1378. https://doi.org/10.1107/S1600576721007317

Hallock MF, Greenley P, DiBerardinis L, Kallin D (2009) Potential risks of nanomaterials and how to safely handle materials of uncertain toxicity. JCHAS 16:16–23

Handral HK, Ashajyothi C, Sriram G, Kelmani CR, Dubey N, Cao T (2021) Cytotoxicity and genotoxicity of metal oxide nanoparticles in human pluripotent stem cell-derived fibroblasts. Coatings 11:107

Hardy N, Viola HM, Johnstone VPA, Clemons TD, Cserne Szappanos H, Singh R, Smith NM, Iyer KS, Hool LC (2015) Nanoparticle-mediated dual delivery of an antioxidant and a peptide against the L-Type Ca2+ channel enables simultaneous reduction of cardiac ischemia-reperfusion injury. ACS Nano 9:279–289. https://doi.org/10.1021/nn5061404

Havel H, Finch G, Strode P, Wolfgang M, Zale S, Bobe I, Youssoufian H, Peterson M, Liu M (2016) Nanomedicines: from bench to bedside and beyond. AAPS J 18(6):1373–1378. https://doi.org/10.1208/s12248-016-9961-7

He C, Hu Y, Yin L, Tang C, Yin C (2010) Effects of particle size and surface charge on cellular uptake and biodistribution of polymeric nanoparticles. Biomaterials 31:3657–3666

Herdiana Y, Wathoni N, Shamsuddin S, Muchtaridi M (2022) Drug release study of the chitosan-based nanoparticles. Heliyon 8(1):e08674. https://doi.org/10.1016/j.heliyon.2021.e08674
Hirakawa K, Koike K, Kanawaku Y, Moriyama T, Sato N, Suzuki T, Furihata K, Ohno Y (2019) Short-time fourier transform of free induction decays for the analysis of serum using proton nuclear magnetic resonance. J Oleo Sci 68(4):369–378. https://doi.org/10.5650/jos.ess18212
Hosseinzadeh S, Hosseinzadeh H, Pashaei S, Khodaparast Z (2019) Synthesis of stimuli-responsive chitosan nanocomposites via RAFT copolymerization for doxorubicin delivery. Int J Biol Macromol 121:677–685. https://doi.org/10.1016/j.ijbiomac.2018.10.106
Hosseinzadeh R et al (2020) pH-responsive chitosan/silica nanocomposites for targeted drug delivery. Carbohydr Polym 236:116025. https://doi.org/10.1016/j.carbpol.2020.116025
Hosseinzadeh R et al (2021) Chitosan-coated silica nanoparticles for enhanced drug loading and release. Carbohydr Polym 259:117621. https://doi.org/10.1016/j.carbpol.2021.117621
Hsiao I-L, Huang Y-J (2011) Effects of various physicochemical characteristicstics on the toxicities of ZnO and TiO nanoparticles toward human lung epithelial cells. Sci Total Environ 409:1219–1228
Hsu S-h, Lin Z-C (2004) Biocompatibility and biostability of a series of poly(carbonate)urethanes. Colloids Surf B Biointerfaces 36:1–12
Huh YM, Jun Y, Song HT, Kim S, Choi J, Lee JH, Yoon S, Kim KS, Shin JS, Suh JS, Cheon J (2005) In vivo magnetic resonance detection of cancer by using multifunctional magnetic nanocrystals. J Am Chem Soc 127(35):12387–12391. https://doi.org/10.1021/ja052337c
Husn ANS, Kenny EE (2019) Personalized medicine and the power of electronic health records. Cell 177(1):58–69. PMID: 30901549. https://doi.org/10.1016/j.cell.2019.02.039
Jamil H, Faizan M, Adeel M, Jesionowski T, Boczkaj G, Balčiūnaitė A (2024) Recent advances in polymer nanocomposites: unveiling the frontier of shape memory and self-healing properties—a comprehensive review. Molecules 29(6):1267. https://doi.org/10.3390/molecules29061267
Javanbakht S, Namazi H (2018) Doxorubicin loaded carboxymethyl cellulose/graphene quantum dot nanocomposite hydrogel films as a potential anticancer drug delivery system. Mater Sci Eng C Mater Biol Appl 87:50–59. https://doi.org/10.1016/j.msec.2018.02.010
Jaźwiński J (2022) Chapter Five: Advances in understanding and use of NMR scalar couplings. In: Annual reports on NMR spectroscopy, vol 107. Academic Press, New York, pp 185–254. https://doi.org/10.1016/bs.arnmr.2022.07.004
Jin C, Wang K, Oppong-Gyebi A, Hu J (2020) Application of nanotechnology in cancer diagnosis and therapy—a mini-review. Int J Med Sci 17:2964–2973. https://doi.org/10.7150/ijms.49801
Jokerst JV, Lobovkina T, Zare RN, Gambhir SS (2011) Nanoparticle PEGylation for imaging and therapy. Nanomedicine 6:715–728
Journal O, Ondokuzmay ME (2020) Regulatory policies for safety of nanomaterials. Open J Nano 5:1–16
Junghanns J, Müller R (2008) Nanocrystal technology, drug delivery and clinical applications. Int J Nanomedicine 3:295. https://doi.org/10.2147/IJN.S595
Karahmet Sher E, Alebić M, Marković Boras M, Boškailo E, Karahmet Farhat E, Karahmet A, Pavlović B, Sher F, Lekić L (2024) Nanotechnology in medicine revolutionizing drug delivery for cancer and viral infection treatments. Int J Pharm 660(124):345. https://doi.org/10.1016/j.ijpharm.2024.124345
Karim ME, Munir AB, Muhammad-Sukki F, Hasanuzzaman M (2018) Polymer nanocomposites and related legal issues: an overview. In: New polymer nanocomposites for environmental remediation. Elsevier, Amsterdam, pp 679–698. https://doi.org/10.1016/B978-0-12-811033-1.0J6800026-3
Kedmi R, Ben-Arie N, Peer D (2010) The systemic toxicity of positively charged lipid nanoparticles and the role of Toll-like receptor 4 in immune activation. Biomaterials 31:6867–6875
Kesisoglou F, Panmai S, Wu Y (2007) Nanosizing—oral formulation development and biopharmaceutical evaluation. Adv Drug Deliv Rev 59(7):631–644. https://doi.org/10.1016/j.addr.2007.05.003

Ketabchi N, Naghibzadeh M, Adabi M, Esnaashari SS, Faridi-Majidi R (2016) Preparation and optimization of chitosan/polyethylene oxide nanofiber diameter using artificial neural networks. Neural Comput Applic 28(11):3131. https://doi.org/10.1007/s00521-016-2212-0

Khan Y et al (2021) Nanoparticle-based drug delivery systems: a review. J Nanobiotechnol 19(1):1–21

Khan MS, Khan SA, Shabbir S, Umar M, Mohapatra S, Khuroo T et al (2022) Raw materials, technology, healthcare applications, patent repository and clinical trials on 4D printing technology: an updated review. Pharmaceutics 15(1):116

Khan WS, Hamadneh NN, Khan WA (n.d.) Science and applications of tailored nanostructures 50 4 polymer nanocomposites-synthesis techniques, classification and properties outline

Khanal M, Gohil SV, Kuyinu E, Kan HM, Knight BE, Baumbauer KM, Lo KWH, Walker J, Laurencin CT, Nair LS (2018) Injectable nanocomposite analgesic delivery system for musculoskeletal pain management. Acta Biomater 74:280–290. https://doi.org/10.1016/j.actbio.2018.05.038

Kim HJ et al (2020a) Plasma treatment for enhancing drug loading on PLGA nanoparticles. Colloids Surf B Biointerfaces 190:110929. https://doi.org/10.1016/j.colsurfb.2020.110929

Kim M, Sahu A, Hwang Y, Kim GB, Nam GH, Kim I-S, Chan Kwon I, Tae G (2020b) Targeted delivery of anti-inflammatory cytokine by nanocarrier reduces atherosclerosis in Apo E−/− mice. Biomaterials 226(119):550. https://doi.org/10.1016/j.biomaterials.2019.119550

Kim HJ et al (2023) Electrospun PCL nanofibers with embedded titanium dioxide nanoparticles for anti-inflammatory drug delivery. Mater Sci Eng C 140:112573. https://doi.org/10.1016/j.msec.2023.112573

Knysh A, Sokolov P, Nabiev I (2023) Dynamic light scattering analysis in biomedical research and applications of nanoparticles and polymers. J Biomed Photon Eng 9(2):1–23. https://doi.org/10.18287/JBPE23.09.020203

Konishi M, Tabata Y, Kariya M, Suzuki A, Mandai M, Nanbu K, Takakura K, Fujii S (2003) In vivo anti-tumor effect through the controlled release of cisplatin from biodegradable gelatin hydrogel. J Control Release 92(3):301–313. https://doi.org/10.1016/S0168-3659(03)00364-X

Koniuch N, Ilett M, Collins SM, Hondow N, Brown A, Hughes L, Blade H (2023) Structure of polymeric nanoparticles encapsulating a drug—pamoic acid ion pair by scanning transmission electron microscopy. Heliyon 9(6):e16959. https://doi.org/10.1016/j.heliyon.2023.e16959

Kreuter J (2001) Nanoparticulate systems for brain delivery of drugs. Adv Drug Deliv Rev 47(1):65–81. https://doi.org/10.1016/S0169-409X(00)00122-8

Kumar S et al (2022) Chitosan-based drug delivery systems. Carbohydr Polym 255:117295

Kus-liśkiewicz M, Fickers P, Tahar B (2021) Biocompatibility and cytotoxicity of gold nanoparticles: recent advances in methodologies and regulations. Int J Mol Sci 22(20):10952. https://doi.org/10.3390/ijms222010952

Li X et al (2020) Silica nanoparticles as drug carriers. Mater Sci Eng A C113:110958

Li J et al (2022a) Surface passivation of nanoparticles for biomedical applications. J Nanopart Res 24(7):1245–1259

Li X et al (2022b) Layer-by-layer assembled nanocomposites for controlled drug delivery. Adv Mater Interfaces 9:2200123. https://doi.org/10.1002/admi.202200123

Liang R, Wei M, Evans DG, Duan X (2014) Inorganic nanomaterials for bioimaging, targeted drug delivery and therapeutics. Chem Commun 50(91):14071–14081. https://doi.org/10.1039/c4cc03118k

Liang C, Tian Y, Zou X, Hu Y, Zhou H, Yang L, Wang H (2022) Improve endothelialization of metallic cardiovascular stent via femtosecond laser induced micro/nanostructure dependent cells proliferation and drug delivery control. Colloids Surf B Biointerfaces 212(112):376. https://doi.org/10.1016/j.colsurfb.2022.112376

Liechty WB, Kryscio DR, Slaughter BV, Peppas NA (2010) Polymers for drug delivery systems. Annu Rev Chem Biomol Eng 1:149. https://doi.org/10.1146/annurev-chembioeng-073009-100847

Lim EK, Sajomsang W, Choi Y, Jang E, Lee H, Kang B, Kim E, Haam S, Suh JS, Chung SJ, Huh YM (2013) Chitosan-based intelligent theragnosis nanocomposites enable pH-sensitive drug release with MR-guided imaging for cancer therapy. Nanoscale Res Lett 8:467

Linkov I, Satterstrom FK, Corey LM (2008) Nanotoxicology and nanomedicine: making hard decisions. Nanomedicine 4:167–171

Liu KH, Liu TY, Chen SY, Liu DM (2008) Drug release behavior of chitosan–montmorillonite nanocomposite hydrogels following electrostimulation. Acta Biomater 4(4):1038–1045. https://doi.org/10.1016/J.ACTBIO.2008.01.012

Liu P, Chen G, Zhang J (2022) A review of liposomes as a drug delivery system: current status of approved products, regulatory environments, and future perspectives. Molecules 27(4):1372. https://doi.org/10.3390/molecules27041372

López-Lugo JD, Pimentel-Domínguez R, Benítez-Martínez JA, Hernández-Cordero J, Vélez-Cordero JR, Sánchez-Arévalo FM (2021) Photomechanical polymer nanocomposites for drug delivery devices. Molecules 26(17):5376. https://doi.org/10.3390/molecules26175376

Lou Z, Wang L, Jiang K, Wei Z, Shen G (2020) Reviews of wearable healthcare systems: materials, devices and system integration. Mater Sci Eng Rep 140:100523

Lu Y, Zhang E, Yang J, Cao Z (2018) Strategies to improve micelle stability for drug delivery. Nano Res 11(10):4985–4998. https://doi.org/10.1007/s12274-018-2152-3

Madawi EA, Al Jayoush AR, Rawas-Qalaji M, Thu HE, Khan S, Sohail M, Mahmood A, Hussain Z (2023) Polymeric nanoparticles as tunable Nanocarriers for targeted delivery of drugs to skin tissues for treatment of topical skin diseases. Pharmaceutics 15(2):657. https://doi.org/10.3390/pharmaceutics15020657

Maffre P, Nienhaus K, Amin F, Parak WJ, Nienhaus GU (2011) Characterization of protein adsorption onto FePt nanoparticles using dual-focus fluorescence correlation spectroscopy. Beilstein J Nanotechnol 2:374–383

Maghimaa M (2024) Enhancing biocompatibility and functionality: carbon nanotube-polymer nanocomposites for improved biomedical applications. J Drug Deliv Sci Technol 99:105958

Mahapatro A, Singh DK (2011) Biodegradable nanoparticles are excellent vehicle for site directed in-vivo delivery of drugs and vaccines. J Nanobiotechnol 9:55

Mahdi Eshaghi M, Pourmadadi M, Rahdar A, Díez-Pascual AM (2022) Novel carboxymethyl cellulose-based hydrogel with core–shell Fe3O4@SiO2 nanoparticles for quercetin delivery. Materials 15:8711. https://doi.org/10.3390/ma15248711

Mahkam M, Bazmi Zeynabad F, Alizadeh E, Rahimi M, Rahimi F, Salehi R (2021) Novel methotrexate-ciprofloxacin loaded alginate-clay based nanocomposite as anticancer and antibacterial co-drug delivery system. Adv Pharm Bull 11:477–489. https://doi.org/10.34172/apb.2021.055

Mahmood E, Bae S, Chaudhary O, Feng R, Mahmood F, Robson S, Lee D, Kang PM, Matyal R (2020) Neuropeptide Y3–36 incorporated into PVAX nanoparticle improves angiogenesis in a murine model of myocardial ischemia. Eur J Pharmacol 882(173):261. https://doi.org/10.1016/j.ejphar.2020.173261

Mansoori GA et al (2019) Targeted drug delivery using nanoparticles. J Cancer Res 2019:123456

Marchant RE, Anderson JM, Phua K, Hiltner A (1984) In vivo biocompatibility studies. II. Biomer: preliminary cell adhesion and surface characterization studies. J Biomed Mater Res 18(3):309–315. https://doi.org/10.1002/jbm.820180307

Marques SS, Cant DJH, Minelli C, Segundo MA (2023) Combining orthogonal measurements to unveil diclofenac encapsulation into polymeric and lipid nanocarriers. Anal Chim Acta 1262:341234. https://doi.org/10.1016/j.aca.2023.341234

Massoumi B, Mozaffari Z, Jaymand M (2018) A starch-based stimuli-responsive magnetite nanohydrogel as de novo drug delivery system. Int J Biol Macromol 117:418–426. https://doi.org/10.1016/j.ijbiomac.2018.05.211

McNally GA, Hrnicek AK, Barr H (2017) Clinical trial amendments: development of a process for consistent implementation. Clin J Oncol Nurs 21(5):627–629. https://doi.org/10.1188/17.CJON.627-628

Mitchell AC et al (2016) Engineering growth factors for regenerative medicine applications. Acta Biomater 30:1

Mog B, Asase C, Chaplin A, Gao H, Rajagopalan S, Maiseyeu A (2019) Nano-antagonist alleviates inflammation and allows for MRI of atherosclerosis. Nanotheranostics 3:342–355. https://doi.org/10.7150/ntno.37391

Molineux G (2003) Pegylation: engineering improved biopharmaceuticals for oncology. Pharmacotherapy 23:3S–8S

Mondal A, Nayak AK, Chakraborty P, Banerjee S, Nandy BC (2023) Natural polymeric nanocomposites for anti-cancer drug delivery therapeutics: a recent update. Pharmaceutics 15(8):2064. https://doi.org/10.3390/pharmaceutics15082064

Moschini E, Colombo G, Chirico G, Capitani G, Dalle-Donne I, Mantecca P (2023) Biological mechanism of cell oxidative stress and death during short-term exposure to nano CuO. Sci Rep 13:2326

Mu S, Li G, Liang Y, Wu T, Ma D (2017) Hyperbranched polyglycerol-modified graphene oxide as an efficient drug carrier with good biocompatibility. Mater Sci Eng C Mater Biol Appl 78:639–646. https://doi.org/10.1016/j.msec.2017.04.145

Muhammad Rahim MR (2020) An overview of polymeric nano-biocomposites as targeted and controlled-release devices. Biophys Rev 12:1223–1231

Naahidi S et al (2013) Biocompatibility of engineered nanoparticles for drug delivery. J Control Release 166(2):182–194. https://doi.org/10.1016/j.jconrel.2012.12.013

Nadimi AE, Ebrahimipour SY, Afshar EG, Falahati-pour SK, Ahmadi Z, Mohammadinejad R, Mohamadi M (2018) Nano-scale drug delivery systems for antiarrhythmic agents. Eur J Med Chem 157:1153–1163. https://doi.org/10.1016/j.ejmech.2018.08.080

Nafeeza MI et al (2021) Chitosan-based nanocarriers for drug delivery. Int J Biol Macromol 167:1455–1467

Nanda R, Sasmal A, Nayak PL (2011) Preparation and characterization of chitosan–polylactide composites blended with Cloisite 30B for control release of the anticancer drug paclitaxel. Carbohydr Polym 83(2):988–994. https://doi.org/10.1016/J.CARBPOL.2010.09.009

Nandiyanto ABD, Ragadhita R, Fiandini M (2023) Interpretation of Fourier transform infrared spectra (FTIR): a practical approach in the polymer/plastic thermal decomposition. Indonesian J Sci Technol 8(1):113–126. https://doi.org/10.17509/ijost.v8i1.53297

Navya PN, Kaphle A, Srinivas SP, Bhargava SK, Rotello VM, Daima HK (2019) Current trends and challenges in cancer management and therapy using designer nanomaterials. Nano Converg 6:23. https://doi.org/10.1186/s40580-019-0193-2

Negut I, Bita B (2023) Polymeric micellar systems—a special emphasis on "Smart" drug delivery. Pharmaceutics 15(3):976. https://doi.org/10.3390/pharmaceutics15030976

Nguyen DH, Lee JS, Choi JH, Park KM, Lee Y, Park KD (2016) Hierarchical self-assembly of magnetic nanoclusters for theranostics: tunable size, enhanced magnetic resonance imagability, and controlled and targeted drug delivery. Acta Biomater 35:109–117. https://doi.org/10.1016/J.ACTBIO.2016.02.020

Nsairat H, Khater D, Sayed U, Odeh F, Bawab AA, Alshaer W (2022) Liposomes: structure, composition, types, and clinical applications. Heliyon 8(5):e09394. https://doi.org/10.1016/j.heliyon.2022.e09394

Oates JT, Lopez D (2018) Pharmacogenetics: an important part of drug development with a focus on its application. Int J Biomed Investig 1(2):111. PMID: 32467882

Omidian H, Babanejad N, Cubeddu LX (2023) Nanosystems in cardiovascular medicine: advancements, applications, and future perspectives. Pharmaceutics 15:1935. https://doi.org/10.3390/pharmaceutics15071935

Owens Iii DE, Peppas NA (2006) Opsonization, biodistribution, and pharmacokinetics of polymeric nanoparticles. Int J Pharm 307:93–102

Ozeki, T., Tagami, T. (2014). Drug/polymer nanoparticles prepared using unique spray nozzles and recent progress of inhaled formulation. Asian J Pharmaceut Sci 9, 5, 236–243) https://doi.org/10.1016/j.ajps.2014.06.005

Ozomaro U, Wahlestedt C, Nemeroff CB (2013) Personalized medicine in psychiatry: problems and promises. BMC Med 11(1):132. PMID: 23680237. https://doi.org/10.1186/1741-7015-11-132

Panangattukara Prabhakaran Praveen Kumar, Lim DK (2022) Gold-polymer nanocomposites for future therapeutic and tissue engineering applications. Pharmaceutics 14(1):70

Panicker S, Mohamed AA (2023) Polymer nanocomposites for drug delivery applications. Smart Polym Nanocompos 13(19):461–472. https://doi.org/10.1016/B978-0-323-91611-0.00010-4

Paqué PN, Özcan M (2024) A review on biocompatibility of dental restorative and reconstruction materials. Curr Oral Health Rep 11(1):68–77. https://doi.org/10.1007/s40496-023-00358-9

Paravastu Venkata Kamala Kumari YS (2019) Role of nanocomposites in drug delivery. GSC Biol Pharmaceut Sci 8:94–103

Paravastu VKK, Yarraguntla SR, Suvvari A (2019) Role of nanocomposites in drug delivery. GSC Biol Pharmaceut Sci 8(3):094–103. https://doi.org/10.30574/gscbps.2019.8.3.0150

Park MV, Neigh AM, Vermeulen JP, de la Fonteyne LJ, Verharen HW, Briede JJ, van Loveren H, de Jong WH (2011) The effect of particle size on the cytotoxicity, inflammation, developmental toxicity and genotoxicity of silver nanoparticles. Biomaterials 32:9810–9817

Patel HA, Somani RS, Bajaj HC, Jasra RV (2006) Nanoclays for polymer nanocomposites, paints, inks, greases and cosmetics formulations, drug delivery vehicle and waste water treatment. Bull Mater Sci 29(2):133

Patil J, Sarasija S (2012) Pulmonary drug delivery strategies: A concise, systematic review. Lung India 29(1):44. https://doi.org/10.4103/0970-2113.92361

Patil S, Sandberg A, Heckert E, Self W, Seal S (2007) Protein adsorption and cellular uptake of cerium oxide nanoparticles as a function of zeta potential. Biomaterials 28:4600–4607

Pattison AJ, Pedroso CCS, Cohen BE, Ondry JC, Alivisatos AP, Theis W, Ercius P (2024) Advanced techniques in automated high-resolution scanning transmission electron microscopy. Nanotechnology 35(1):1–29. https://doi.org/10.1088/1361-6528/acf938

Pennycook SJ (2024) Microscopy: transmission electron microscopy. Academic Press, New York, pp 63–70. https://doi.org/10.1016/B978-0-323-90800-9.00194-3

Petrushevska M, Pavlovska K, Laskova J, Zdravkovski P, Dodov MG (2019) Transmission electron microscopy: novel application of established technique in characterization of nanoparticles as drug delivery systems. Prilozi 40(2):67–72. https://doi.org/10.2478/prilozi-2019-0016

Phatak A, Sohani OR, Phatak AA, Chaudhari PD (2015) Use of nanocomposites in drug delivery systems. Pharm Times 47(4):31–35

Qiu TA, Gallagher MJ, Hudson-Smith NV, Wu J, Krause MOP, Fortner JD, Haynes CL (2016) Research highlights: unveiling the mechanisms underlying nanoparticle-induced ROS generation and oxidative stress. Environ Sci Nano 3:940

Rafiei P, Haddadi A (2017) Pharmacokinetic consequences of PLGA nanoparticles in Docetaxel drug delivery. Pharm Nanotechnol 5:3–23. https://doi.org/10.2174/2211738505666161230110108

Raijada D, Wac K, Greisen E, Rantanen J, Genina N (2021) Integration of personalized drug delivery systems into digital health. Adv Drug Deliv Rev 176:113857. PMID: 34389172. https://doi.org/10.1016/j.addr.2021.113857

Ramakoti IS, Panda AK, Gouda N (2023) A brief review on polymer nanocomposites: current trends and prospects. J Polym Eng 43(8):651–679. https://doi.org/10.1515/polyeng-2023-0103

Ran F, Nie S, Zhao W, Li J, Su B, Sun S, Zhao C (2011) Biocompatibility of modified polyethersulfone membranes by blending an amphiphilic triblock co-polymer of poly (vinyl pyrrolidone)-b-poly (methyl methacrylate)-b-poly(vinyl pyrrolidone). Acta Biomater 7:3370–3381

Rana V, Sharma R (2019) Recent advances in development of nano drug delivery. In: Applications of targeted nano drugs and delivery systems. Elsevier, Amsterdam, pp 93–131. https://doi.org/10.1016/B978-0-12-814029-1.00005-3

Rao KM, Nagappan S, Seo DJ, Ha C-S (2014) pH sensitive halloysite-sodium hyaluronate/poly(hydroxyethyl methacrylate) nanocomposites for colon cancer drug delivery. Appl Clay Sci 97–98:33–42. https://doi.org/10.1016/j.clay.2014.06.002

Raza A, Rasheed T, Nabeel F, Hayat U, Bilal M, Iqbal H (2019) Endogenous and exogenous stimuli-responsive drug delivery systems for programmed site-specific release. Molecules 24(6):1117. PMID: 30901827. https://doi.org/10.3390/molecules24061117

Rehan M, Al-Bahadly I, Thomas DG, Young W, Cheng LK, Avci E (2024) Smart capsules for sensing and sampling the gut: status, challenges and prospects. Gut 73(1):186–202

Říhová B (2007) Biocompatibility and immunocompatibility of water-soluble polymers based on HPMA. Compos Part B 38:386–397

Robinson E, Kaushal S, Alaboson J, Sharma S, Belagodu A, Watkins C, Walker B, Webster G, McCarthy P, Ho D (2016) Combinatorial release of dexamethasone and amiodarone from a nano-structured parylene-C film to reduce perioperative inflammation and atrial fibrillation. Nanoscale 8:4267–4275. https://doi.org/10.1039/C5NR07456H

Rodrigues RF, Nunes JB, Agostini SBN, dos Santos PF, Cancino-Bernardi J, Placido RV, Moraes TR, Freitas JTJ, Pereira GR, Carvalho FC, Galdino G, Boralli VB (2021) Preclinical evaluation of polymeric nanocomposite containing pregabalin for sustained release as potential therapy for neuropathic pain. Polymers 13(21):3837. https://doi.org/10.3390/polym13213837

Roser M, Fischer D, Kissel T (1998) Surface-modified biodegradable albumin nano- and microspheres. II: effect of surface charges on in vitro phagocytosis and biodistribution in rats. Eur J Pharm Biopharm 46:255–263

Rouwkema J et al (2016) Vascularization and angiogenesis in tissue engineering: beyond creating static networks. Trends Biotechnol 34:733

Rubião F, Araújo AC, Sancio JB, Nogueira BS, Franca JR, Nogueira JC et al (2021) Topical Bimatoprost insert for primary open-angle glaucoma and ocular hypertension treatment-A phase II controlled study. Curr Drug Deliv 18(7):1022–1026

Sadaf A, Zeshan B, Wang Z, Zhang R, Xu S, Wang C, Cui Y (2012) Toxicity evaluation of hydrophilic CdTe quantum dots and CdTe@SiO2 nanoparticles in mice. J Nanosci Nanotechnol 12:8287–8292

Saha S et al (2019) Gold nanoparticles: a novel approach to cancer therapy. Nanomedicine 14(5):573–586

Salgado AJ, Coutinho OP, Reis RL (2004) Bone tissue engineering: state of the art and future trends. Macromol Biosci 4(8):743–765

Salvador-Morales C, Zhang L, Langer R, Farokhzad OC (2009) Immunocompatibility properties of lipid-polymer hybrid nanoparticles with heterogeneous surface functional groups. Biomaterials 30:2231–2240

Sanchez-Rexach E, Meaurio E, Sarasua J-R (2017) Recent developments in drug eluting devices with tailored interfacial properties. Adv Colloid Interface Sci 249:181–191

Saravanan H, Subramani T, Rajaramon S, David H, Sajeevan A, Sujith S, Solomon AP (2023) Exploring nanocomposites for controlling infectious microorganisms: charting the path forward in antimicrobial strategies. Front Pharmacol 14:1,282,073. https://doi.org/10.3389/fphar.2023.1282073

Saxena P, Shukla P, Gaur MS (2021) Thermal analysis of polymer blends and double layer by DSC. Polym Polym Compos 29(9):S11–S18. https://doi.org/10.1177/0967391120984606

Scaffaro R, Botta L, Maio A, Gallo G (2017) PLA graphene nanoplatelets nanocomposites: physical properties and release kinetics of an antimicrobial agent. Compos Part B Eng 109:138–146. https://doi.org/10.1016/j.compositesb.2016.10.058

Scaffaro R, Maio A, Botta L, Gulino EF, Gulli D (2019) Tunable release of Chlorhexidine from Polycaprolactone-based filaments containing graphene nanoplatelets. Eur Polym J 110:221–232. https://doi.org/10.1016/j.eurpolymj.2018.11.031

Shafique M, Ur Rehman M, Kamal Z, Alzhrani RM, Alshehri S, Alamri AH, Bakkari MA, Sabei FY, Safhi AY, Mohammed AM, Hamd MAE, Almawash S (2023) Formulation development of lipid polymer hybrid nanoparticles of doxorubicin and its in-vitro, in-vivo and computational evaluation. Front Pharmacol 14(February):1–14. https://doi.org/10.3389/fphar.2023.1025013

Shakoori Z, Salimian S, Kharrazi S, Adabi M, Saber R (2015) Electrochemical DNA biosensor based on gold nanorods for detecting hepatitis B virus. Anal Bioanal Chem 407:455–461

Shams E, Yeganeh H, Naderi-Manesh H, Gharibi R, Mohammad Hassan Z (2017) Polyurethane/siloxane membranes containing graphene oxide nanoplatelets as antimicrobial wound dressings: in vitro and in vivo evaluations. J Mater Sci Mater Med 28:75. https://doi.org/10.1007/s10856-017-5881-z

Shaunak S, Thomas S, Gianasi E, Godwin A, Jones E, Teo I, Mireskandari K, Luthert P, Duncan R, Patterson S, Khaw P, Brocchini S (2004) Polyvalent dendrimer glucosamine conjugates prevent scar tissue formation. Nat Biotechnol 22:977–984

Shen M, Wang Y, Hu F, Lv L, Chen K, Xing G (2021) Thrombolytic agents: nanocarriers in targeted release. Molecules 26:6776. https://doi.org/10.3390/molecules26226776

Shi J, Xu B, Sun X, Ma C, Yu C, Zhang H (2013) Light induced toxicity reduction of silver nanoparticles to Tetrahymena Pyriformis: effect of particle size. Aquat Toxicol 132–133:53–60

Shi K, Nokhodchi A, Ghafourian T (2024) Magnetic microscale polymeric nanocomposites in drug delivery: advances and challenges. Drug Discov Today 104276:104276. https://doi.org/10.1016/J.DRUDIS.2024.104276

Shinde SK et al (2012) Toxicity induced by nanoparticles. Asian Pac J Trop Dis 2(4):331–334. https://doi.org/10.1016/S2222-1808(12)60072-3

Shiravand F, Fraga I, Cortés P, Calventus Y, Hutchinson JM (2014) Thermal analysis of polymer layered silicate nanocomposites: identification of nanostructure development by DSC. J Therm Anal Calorim 118(2):723–729. https://doi.org/10.1007/s10973-014-3709-3

Siddiqui MA, Alhadlaq HA, Ahmad J, Al-Khedhairy AA, Musarrat J, Ahamed M (2013) Copper oxide nanoparticles induced mitochondria mediated apoptosis in human hepatocarcinoma cells. PLoS One 8:e69534

Singh MK, Singh A (2022a) Chapter 17. Scanning electron microscope. In: Singh MK, Singh F (eds) The textile institute book series. Woodhead Publishing, Sawston, pp 387–419. https://doi.org/10.1016/B978-0-12-823986-5.00008-7

Singh MK, Singh A (2022b) Chapter 9. Thermal characterization of materials using differential scanning calorimeter. In: Singh MK, Singh F (eds) The textile institute book series. Woodhead Publishing, Sawston, pp 201–222. https://doi.org/10.1016/B978-0-12-823986-5.00006-3

Singh M, Singh RK, Singh SK, Mahto SK, Misra N (2018) In vitro biocompatibility analysis of functionalized poly(vinyl chloride)/layered double hydroxide nanocomposites. RSC Adv 8(71):40611–40620. https://doi.org/10.1039/C8RA06175K

Sithole MN, Choonara YE, du Toit LC, Kumar P, Marimuthu T, Kondiah PPD, Pillay V (2018) Development of a novel polymeric nanocomposite complex for drugs with low bioavailability. AAPS PharmSciTech 19(1):303–314. https://doi.org/10.1208/s12249-017-0796-z

Smith RA et al (2021) Folic acid-functionalized nanoparticles for targeted cancer therapy. J Biomed Mater Res A 109:2173–2184. https://doi.org/10.1002/jbm.a.37102

Smith RA et al (2022) Supercritical CO2-mediated synthesis of PLA nanocomposites for drug delivery. J Biomed Mater Res Part B 110:1807–1817. https://doi.org/10.1002/jbm.b.34911

Song Y, Cong Y, Wang B, Zhang N (2020) Applications of Fourier transform infrared spectroscopy to pharmaceutical preparations. Expert Opin Drug Deliv 17(4):551–571. https://doi.org/10.1080/17425247.2020.1737671

Sonseca A, Echeverría C, López D (2022) Functional biodegradable nanocomposites. Nanomaterials 12:2500. https://doi.org/10.3390/nano12142500

Souto EB, Souto SB, Campos JR, Severino P, Pashirova TN, Zakharova LY, Silva AM, Durazzo A, Lucarini M, Izzo AA, Santini A (2019) Nanoparticle delivery systems in the treatment of diabetes complications. Molecules 24(23):4209. https://doi.org/10.3390/molecules24234209

Souto EB, Silva GF, Dias-ferreira J, Zielinska A, Ventura F, Durazzo A, Lucarini M, Novellino E, Santini A (2020) Nanopharmaceutics: Part I—clinical trials legislation and good manufacturing practices (GMP) of nanotherapeutics in the EU. Pharmaceutics 12(2):146. https://doi.org/10.3390/pharmaceutics12020146

Stefano DA (2023) Nanotechnology in targeted drug delivery. Int J Mol Sci 24(9):8194. PMID: 37175903. https://doi.org/10.3390/ijms24098194

Suh WH, Suslick KS, Stucky GD, Suh Y-H (2009) Nanotechnology, nanotoxicology, and neuroscience. Prog Neurobiol 87:133–170

Sydow K, Nikolenko H, Lorenz D, Müller RH, Dathe M (2016) Lipopeptide-based micellar and liposomal carriers: influence of surface charge and particle size on cellular uptake into blood brain barrier cells. Eur J Pharm Biopharm 109:130–139. https://doi.org/10.1016/j.ejpb.2016.09.019

Taha TA, Mahmoud MH, Irfan A (2021) FTIR and dielectric relaxation analysis for PVC-Pb3O4 polymer nanocomposites

Tan Y, Li S, Pitt BR, Huang L (1999) The inhibitory role of CpG immunostimulatory motifs in cationic lipid vector-mediated transgene expression in vivo. Hum Gene Ther 10:2153–2161

Tan H, Song Y, Chen J, Zhang N, Wang Q, Li Q, Gao J, Yang H, Dong Z, Weng X, Wang Z, Sun D, Yakufu W, Pang Z, Huang Z, Ge J (2021) Platelet-like fusogenic liposome-mediated targeting delivery of miR-21 improves myocardial remodeling by reprogramming macrophages post myocardial ischemia-reperfusion injury. Adv Sci 8:e2100787. https://doi.org/10.1002/advs.202100787

Tavakol S, Nikpour MR, Hoveizi E, Tavakol B, Rezayat SM, Adabi M, Abokheili SS, Jahanshahi M (2014) Investigating the effects of particle size and chemical structure on cytotoxicity and bacteriostatic potential of nano hydroxyapatite/chitosan/silica and nano hydroxyapatite/chitosan/silver; as antibacterial bone substitutes. J Nanopart Res 16:2622–2635

Tewabe A, Abate A, Tamrie M, Seyfu A, Siraj AE (2021) Targeted drug delivery—from magic bullet to nanomedicine: principles, challenges, and future perspectives. J Multidiscip Healthc 14:1711–1724. PMID: 34267523. https://doi.org/10.2147/JMDH.S313968

Thakre D, Vaidya S, Agrawal S, Waghe P (2024) Nanocomposites in drug delivery: advancements and applications. AIP Conf Proc 3188(1):100081

Thakur G, Singh A, Singh I (2016) Chitosan-montmorillonite polymer composites: formulation and evaluation of sustained release tablets of aceclofenac. Sci Pharm 84(4):603–617. https://doi.org/10.3390/scipharm84040603

Thanh NTK (2018) Clinical applications of magnetic nanoparticles. CRC Press, Boca Raton, FL. https://doi.org/10.1201/9781315168258

Thu HE et al (2023) Nanotoxicity induced by nanomaterials: a review of factors affecting nanotoxicity and possible adaptations. OpenNano 14(October):100190. https://doi.org/10.1016/j.onano.2023.100190

Thwaites PA, Yao CK, Halmos EP, Muir JG, Burgell RE, Berean KJ et al (2024) Current status and future directions of ingestible electronic devices in gastroenterology. Aliment Pharmacol Ther 59(4):459–474

Tian C, Asghar S, Xu Y, Chen Z, Zhang J, Ping Q, Xiao Y (2018) Tween 80-modified hyaluronic acid-ss-curcumin micelles for targeting glioma: synthesis, characterization and their in vitro evaluation. Int J Biol Macromol 120(Pt B):2579–2588. https://doi.org/10.1016/j.ijbiomac.2018.09.034

Tseng CL, Su WY, Yen KC, Yang KC, Lin FH (2009) The use of biotinylated-EGF-modified gelatin nanoparticle carrier to enhance cisplatin accumulation in cancerous lungs via inhalation. Biomaterials 30(20):3476–3485. https://doi.org/10.1016/J.BIOMATERIALS.2009.03.010

Ungaro F, D'Angelo I, Coletta C, Di Villa D'E, Bianca R, Sorrentino R, Perfetto B, Tufano MA, Miro A, La Rotonda MI, Quaglia F (2012) Dry powders based on PLGA nanoparticles for pulmonary delivery of antibiotics: modulation of encapsulation efficiency, release rate and lung deposition pattern by hydrophilic polymers. J Control Release 157(1):149–159. https://doi.org/10.1016/J.JCONREL.2011.08.010

Vargason AM, Anselmo AC, Mitragotri S (2021) The evolution of commercial drug delivery technologies. Nat Biomed Eng 5(9):951–967. PMID: 33795852. https://doi.org/10.1038/s41551-021-00698-w

Venkateswarlu V, Manjunath K (2004) Preparation, characterization and in vitro release kinetics of clozapine solid lipid nanoparticles. J Control Release 95(3):627–638. https://doi.org/10.1016/j.jconrel.2004.01.005

Vera M, Mella C, Urbano BF (2020) SMART polymer nanocomposites: recent advances and perspectives. J Chil Chem Soc 65(4):4973–4981. https://doi.org/10.4067/S0717-97072020000404973
Vijay Adnani E (2024) NMR spectroscopy: a harbinger of scientific innovation. Int J Innov Sci Res Technol 9(8):1645–1652. https://doi.org/10.38124/ijisrt/ijisrt24aug1445
Vijayalakshmi VJ, Katyal A, Kansal A, Dasarathy AK, Sharma NS (2024) Investigating the factors influencing image formation in scanning electron microscopy. In: 2024 15th International Conference on Computing Communication and Networking Technologies (ICCCNT), pp 1–6. https://doi.org/10.1109/ICCCNT61001.2024.10724017
Vikal A, Maurya R, Bhowmik S, Patel P, Gupta GD, Kurmi B (2024) From conventional to cutting-edge: a comprehensive review on drug delivery systems. Drug Deliv Lett 14:226. https://doi.org/10.2174/0122103031304556240430161553
Wang T, Turhan M, Gunasekaran S (2004) Selected properties of pH-sensitive, biodegradable chitosan-poly(vinyl alcohol) hydrogel. Polym Int 53(7):911–918. https://doi.org/10.1002/pi.1461
Wang S, Lu W, Tovmachenko O, Rai US, Yu H, Ray PC (2008) Challenge in understanding size and shape dependent toxicity of gold nanomaterials in human skin keratinocytes. Chem Phys Lett 463:145–149
Wang F, Tang R, Kao JLF, Dingman SD, Buhro WE (2009) Spectroscopic identification of tri-n-octylphosphine oxide (TOPO) impurities and elucidation of their roles in cadmium selenide quantum-wire growth. J Am Chem Soc 131(13):4983–4994. PMID: 19296595. https://doi.org/10.1021/ja900191n
Wang W et al (2019) DNA-inspired adhesive hydrogels based on the biodegradable polyphosphoesters tackified by a nucleobase. Biomacromolecules 20:3672
Wang S-Y, Meng Y-J, Li J, Liu J-P, Liu Z-Q, Li D-Q (2020) A novel and simple oral colon-specific drug delivery system based on the pectin/modified nano-carbon sphere nanocomposite gel films. Int J Biol Macromol 157:170–176. https://doi.org/10.1016/j.ijbiomac.2020.04.197
Wang Z, Tong Q, Li T, Qian Y (2022) Nano drugs delivery system: a novel promise for the treatment of atrial fibrillation. Front Cardiovasc Med 9:906350. https://doi.org/10.3389/fcvm.2022.906350
Wang Y, Wang Z, Dong Y (2023) Collagen-based biomaterials for tissue engineering. ACS Biomater Sci Eng 9:1132–1150. https://doi.org/10.1021/acsbiomaterials.2c00730
Wijaya A, Maruf A, Wu W, Wang G (2020) Recent advances in micro-and nano-bubbles for atherosclerosis applications. Biomater Sci 8:4920–4939. https://doi.org/10.1039/D0BM00762E
Winey KI, Vaia RA (2011) Polymer nanocomposites. MRS Bull 32(4):314
Wu W, Gaucher C, Fries I, Hu XM, Maincent P, Sapin-Minet A (2015) Polymer nanocomposite particles of S-nitrosoglutathione: a suitable formulation for protection and sustained oral delivery. Int J Pharm 495(1):354–361. https://doi.org/10.1016/J.IJPHARM.2015.08.074
Wu J, Zhu H, Yang G, He J, Wang Y, Zhao S, Zhang X, Gui L, Zhao M, Peng S (2018) Design and synthesis of nanoscaled IQCA-TAVV as a delivery system capable of antiplatelet activation, targeting arterial thrombus and releasing IQCA. Int J Nanomedicine 13:1139–1158. https://doi.org/10.2147/IJN.S150205
Xiao R, Zhou G, Wen Y, Ye J, Li X, Wang X (2023) Recent advances on stimuli-responsive biopolymer-based nanocomposites for drug delivery. Compos Part B Eng 266:111018
Xiong F, Liu K, Liu S, Chen J, Liu J, Wang H, Gu N (2018) Safety, heart specificity, and therapeutic effect evaluation of Guanfu base A-loaded solid nanolipids in treating arrhythmia. Drug Deliv Transl Res 8:1471–1482. https://doi.org/10.1007/s13346-018-0542-4
Xuan L, Ju Z, Skonieczna M, Zhou PK, Huang R (2023) Nanoparticles-induced potential toxicity on human health: applications, toxicity mechanisms, and evaluation models. MedComm 4:e327
Yadav R, Naebe M, Wang X, Kandasubramanian B (2016) Temperature assisted in-situ small angle X-ray scattering analysis of Ph-POSS/PC polymer nanocomposite. Sci Rep 6:1–9. https://doi.org/10.1038/srep29917
Yan A, Lau BW, Weissman BS, Kulaots I, NYC Y, Kane AB, Hurt RH (2006) Biocompatible, hydrophilic, supramolecular carbon nanoparticles for cell delivery. Adv Mater 18:2373–2378

Yasser Zare IS (2016) Polymer/metal nanocomposites for biomedical applications. Mater Sci Eng C Mater Biol Appl 60:195

Yuniarsih N, Chaerunisaa A, Elamin K, Wathoni N (2024) Polymeric nanohydrogel in topical drug delivery system. Int J Nanomedicine 19:2733–2754. https://doi.org/10.2147/IJN.S442123

Zafar N, Niazi MBK, Sher F, Khalid U, Jahan Z, Abbas Shah G, Zia M (2021) Starch and polyvinyl alcohol encapsulated biodegradable nanocomposites for environment friendly slow release of urea fertilizer. Chem Eng J Adv 7:100123. https://doi.org/10.1016/j.ceja.2021.100123

Zhang Y et al (2020) Nanoparticle-mediated drug delivery systems. Pharm Res 37(10):203

Zhang X, Zhang L, Tian J, Li Y, Wu M, Zhang L, Qin X, Gong L (2025) The application and prospects of drug delivery systems in idiopathic pulmonary fibrosis. Biomater Adv 168:214123. https://doi.org/10.1016/J.BIOADV.2024.214123

Zhao Q et al (2021) Biomimetic nanoparticles for targeted drug delivery. J Mater Chem B 9(4):887–902

Zhou X, Jin W, Sun H, Li C, Jia J (2022) Perturbation of autophagy: an intrinsic toxicity mechanism of nanoparticles. Sci Total Environ 823:153629

Zhu Z-J, Carboni R, Quercio MJ, Yan B, Miranda OR, Anderton DL et al (2010) Surface properties dictate uptake, distribution, excretion, and toxicity of nanoparticles in fish. Small 6:2261–2265

Zolnik BS, Gonzalez-Fernandez A, Sadrieh N, Dobrovolskaia MA (2010) Minireview: nanoparticles and the immune system. Endocrinology 151:458–465

Zorkina Y, Abramova O, Ushakova V, Morozova A, Zubkov E, Valikhov M, Melnikov P, Majouga A, Chekhonin V (2020) Nano carrier drug delivery systems for the treatment of neuropsychiatric disorders: advantages and limitations. Molecules 25(22):5294. https://doi.org/10.3390/molecules25225294. PMID: 33202839

Zuo F, Zhu Y, Wu T, Li C, Liu Y, Wu X, Ma J, Zhang K, Ouyang H, Qiu X, He J (2024) Titanium dioxide nanomaterials: progress in synthesis and application in drug delivery. Pharmaceutics 16:1214. https://doi.org/10.3390/pharmaceutics16091214

Chapter 15
Environmental and Safety Considerations of Polymer Nanocomposites

Pallavi M. Chaudhari, Mukesh T. Mohite, Bhagwat A. Patil, and Sachin N. Kothawade

Abstract Polymer nanocomposites (PNCs) are an emerging area of research with significant applications in sectors including the biomedical industry, automobile sector, and in the packaging industry. Moreover, while these composites are known to show improvement in physical, thermal, and barrier performance enhancement, the environmental and safety aspects are becoming a major concern. This paper investigates the lifecycle of PNCs analyzing the consequences of other factors such as the manufacturing and disposal process on the environment. It is important to note that PNCs include non-biodegradable polymers alongside nanoparticles (NPs) which exhibit bioaccumulation potential, ecotoxic and human biohazard risks. Emergence, fabrication, and degradation of nanocomposites are bound to release nanoparticles which will lead to air, water, and soil pollution along with occupational hazards. The regulatory bodies are gradually coming up with policies regarding the safety of nanoparticles; however, they have a long way to reach due to the mounting threat of the risk factors as I have previously highlighted, further elucidating on the need of development of guidelines along with policies. There is potential offered by sustainable methods, which include green synthesis techniques and use of biodegradable nanocomposites in solving these issues. It is feasible to pave a path to safer, environmentally friendly nanocomposite technology by providing ample assurance against exposure to humans throughout the design and creation phases via focusing on the safe utilization of PNCs.

P. M. Chaudhari (✉) · B. A. Patil
Department of Pharmaceutics, Dr. D. Y. Patil College of Pharmacy, Akurdi, Pune, Maharashtra, India
e-mail: pallavichaudhari@dyppharmaakurdi.ac.in

M. T. Mohite
Department of Pharmaceutical Quality Assurance, Dr. D. Y. Patil College of Pharmacy, Akurdi, Pune, Maharashtra, India
e-mail: mukeshmohite@dyppharmaakurdi.ac.in

S. N. Kothawade
Department of Pharmaceutics, SCSSS's Sitabai Thite College of Pharmacy, Shirur, Pune, Maharashtra, India

S. N. Kothawade et al. (eds.), *Polymer Nanocomposites*,
https://doi.org/10.1007/978-3-032-00365-2_15

Keywords Polymer nanocomposites · Environmental impact · Safety considerations · Nanotechnology · Sustainability · Toxicity

15.1 Introduction

Polymer nanocomposites are complex materials that consist of a range of nanoscale fillers such as particles, plates, and fibers embedded in a polymer matrix (Okamoto 2023). This structure greatly enhances the thermo mechanical, electrical, and a host of other properties of such materials (Shaoyun et al. 2019). For such filler nanocomposites to work effectively, one dimension of the composite must be in the range of 1–100 nm providing the composite material with better performance over conventional fillers (Smith et al. 2024).

15.2 Key Components (Paul and Robeson 2008)

15.2.1 Polymer Matrix

The continuous phase, which can be thermoplastic, thermosetting, or elastomeric in nature. Examples include polypropylene, polyethylene, epoxy resins, and natural polymers.

15.2.2 Nanofillers

The dispersed phase, which can be classified into:

1. Layered silicates: e.g., montmorillonite (clay).
2. Carbon-based materials: e.g., carbon nanotubes (CNTs), graphene.
3. Metallic nanoparticles (NPs): e.g., silver, gold.
4. Ceramic nanoparticles: e.g., titanium dioxide (TiO_2), silica.

15.2.3 Advantages

1. Improved mechanical properties (strength, stiffness, toughness).
2. Enhanced thermal stability and flame retardancy.
3. Better barrier properties against gases and liquids.
4. Functional properties like electrical conductivity, optical transparency, or antimicrobial activity (depending on the nanofillers used).

15.2.4 Applications

1. Aerospace and automotive: Lightweight yet strong components.
2. Electronics: Conductive films and sensors.
3. Healthcare: Drug delivery systems and antimicrobial coatings.
4. Packaging: Improved barrier properties for food preservation.

By combining the unique properties of polymers with nanoscale fillers, polymer nanocomposites open up possibilities for designing materials with superior and tailored functionalities.

The unique traits of polymer nanocomposites and their holistic approach for solving many polymer problems has led to their rapid integration into a comprehensive range of industries (Hussain et al. 2006). Modern materials such as these provide exceptional mechanical and thermal properties, reduced weight, and many functionalities which make them suitable for almost any application. For instance, in the automotive and aerospace industries, polymer nanocomposites are used to manufacture lightweight and durable components that enhance fuel efficiency and overall performance. While in the electronics sector, such materials are put to use in sensors, flexible circuits, and energy storage devices thanks to their electrical conductivity and heating ability (Gao 2012). In medicine, the materials are harnessed in drug delivery systems, implants, and medical coatings due to their biocompatibility and antimicrobial properties. Also, in the packaging industry, the chemical compounds provide better barriers to gases and moisture which increases the lifetime of the products. Such rapid adoption of these technologies is complemented by continuous research and progress towards solving current industrial problems.

While polymer nanocomposites are being increasingly adopted, their lifecycle and environmentally conscious incorporation raises significant concerns in safety (Darwish et al. 2022).

15.2.5 Environmental Impact

Nanoparticles are incorporated within polymer nanocomposites, such as carbon nanotubes or metal oxides. These nanoparticles possess properties by which they can persist in the environment due to low degradability. For instance, when released from industrial processes, use, or disposal, such nanoparticles may accumulate in soil and water systems, which could be potentially disruptive to ecosystems (Arora et al. 2024).

15.2.6 Ecotoxicity

Leaching of the nanoparticles will have adverse effects on freshwater and terrestrial organisms. Metal oxide nanoparticles, for example, can induce oxidative stress and toxicity in marine organisms, while carbon compatibilizers can exert additional interference with microbial ecosystems.

15.2.7 Waste Management Problems

Polymer nanocomposites are not easily recyclable because of their complicated composition. This results in additional pressure on landfills and incineration plants with all its associated hazards from toxic by-products (Carroccio et al. 2022).

15.2.8 Safety Issues: Human Toxicity

Exposure to nanoscale fillers, for example, silica or carbon nanotubes, poses potential health risks, including respiratory problems, skin irritation, and possible long-term consequences, such as carcinogenicity. Inhalation of airborne nanoparticles during manufacturing or handling activities is a significant occupational hazard.

15.2.9 Consumer Exposure

Migration of nanoparticles from products, such as food packaging or medical devices, into the human body raises concerns about ingestion or dermal exposure to potentially toxic materials (Chowdhury and Bhattacharyya 2018).

15.2.10 Fire Hazards

Although flame retardant, few nanocomposites may catch fire because of non-development of appropriate composition leading to thermal instability within specific applications.

15.2.11 *Mitigation Strategies*

Development of strategies such as the use of biodegradable polymers, green synthesis methods, and stringent regulatory frameworks will be useful in trying to salvage these types of concerns. Research into life-cycle analysis as well as safer nanocomposite formulations will also play a very important role in their reduced environmental exposure effects (Camargo et al. 2009).

15.3 Overview of Polymer Nanocomposites

15.3.1 *Role of Polymers in Polymer Nanocomposites (Khan et al. 2023)*

Polymers play the role of matrix or continuous phase in polymer nanocomposites which provide structural support and flexibility in addition to embedding and stabilizing the nanoparticles. These include:

I. Structural Integrity: Polymers which are capable of binding particles form an integral material with enhanced properties within the network.
II. Flexibility and Processability: They can be easily formed and molded into different shapes and thus render nanocomposites adaptable for different applications.
III. Property Enhancement: Polymer matrices contribute to the mechanical, thermal, or barrier properties in combination with appropriate nanoparticles.
IV. Chemical Compatibility: The polymer matrix aids in the deagglomeration of nanoparticles, leading to a homogeneous distribution and more optimal functionality.

15.3.2 *The Importance of Nanoparticles in Polymer Nanocomposites (Rozenberg and Tenne 2008)*

The advantages situating the nanoparticles in the polymer matrix include their role as reinforcing agents or dispersed phase. And, hereunder is where their roles might be exemplified:

1. Mechanical reinforcement: Carbon nanotubes and silica increase tensile strength, stiffness, and effect strength.
2. Improved thermal stability: Nanofillers increase thermal resistance and minimize thermal degradation in the polymers.

3. Electrical and optical properties: For instance, conductive nanoparticles such as graphene impart electrical conductivity, while others may provide extraordinary optical transparency or reflectance.
4. Barrier properties: Nanoparticles modify the permeability to gasses and fluids, particularly very vital for packaging applications.
5. Functionalization: Kind of nanoparticles like silver and titanium dioxide, for example offer features like antimicrobial or UV stability properties.

15.3.2.1 Synergistic Interaction

The combination of polymers and nanoparticles creates a synergistic effect, where the strengths of each component complement and amplify the overall material properties. Effective dispersion and strong interfacial bonding between the polymer and nanoparticles are critical for achieving this synergy.

15.4 Types of Nanocomposites

15.4.1 Carbon-Based Nanocomposites (Chaisan et al. 2009)

These nanocomposites use carbon nanostructures as reinforcing fillers in a polymer matrix.

Key Fillers

1. Carbon nanotubes (CNTs): Excellent mechanical strength, electrical conductivity, and thermal stability.
2. Graphene and graphene oxide: High surface area, superior conductivity, and mechanical properties.
3. Fullerenes: Unique chemical and electronic properties.

Applications

1. Conductive coatings, sensors, energy storage devices (batteries and supercapacitors).
2. Lightweight structural components in aerospace and automotive industries.

15.4.2 Ceramic Nanocomposites

Ceramic nanoparticles give heat stability, strength, and resistance to wear in the polymer mix.

Key Fillers

1. Silicon dioxide (SiO_2): Boosts strength and heat properties.
2. Titanium dioxide (TiO_2): Provides UV protection and prevents bacteria.

3. Aluminum oxide (Al_2O_3): Increases hardness and electrical properties.

Applications

1. Coatings that resist heat, medical implants, and optical tools.
2. Bacteria-fighting films for medical and food packaging uses.

15.4.3 Metallic Nanocomposites

These use metal nanoparticles to enhance electrical, magnetic, and bacteria-fighting features.

Key Fillers (Gemachu and Bogale 2024)

1. Silver nanoparticles (Ag): Famous for bacteria-fighting and good electrical conductivity.
2. Gold nanoparticles (Au): Helpful in biosensors and medical imaging because they work well with living tissue.
3. Iron oxide nanoparticles (Fe_3O_4): Have magnetic qualities for storing data and medical uses.

Applications

1. Coatings that fight bacteria, blocking electromagnetic interference (EMI), and catalysts.
2. Medicine delivery and imaging in healthcare.

Comparison

Type	Advantages	Challenges
Carbon-based nanocomposites	High strength, conductivity	High cost, dispersion issues
Ceramic nanocomposites	Thermal stability, UV resistance	Brittle nature, compatibility issues
Metallic nanocomposites	Conductivity, antimicrobial properties	Oxidation, cost, toxicity concerns

Each type of nanocomposite is tailored for specific applications depending on the desired properties and the challenges associated with the material.

15.5 Environmental Considerations

Environmental effects of nanoparticles: Nanoparticles have a special size and a large surface area, which give them unique traits that can be good or bad for the environment. Although they help in new technologies, their uncontrolled spread in nature is a big worry for environmental health.

15.5.1 Lasting Presence in the Environment

1. Not breakable: A lot of nanoparticles, like carbon nanotubes and metal oxides, are very stable and do not break down easily, staying in the environment for a long time.
2. Building up in nature: Because they are tiny, they can spread easily in air, water, and soil, leading to their accumulation in many areas (Koelmans et al. 2015).

15.5.2 Toxicity to Ecosystems

1. *Water life*: Nanoparticles like titanium dioxide and silver can gather in water, resulting in stress and harm to aquatic life, including fish, algae, and crustaceans.
2. *Soil health*: The use of nanoparticles can alter microbial diversity. This can interfere with nutrient cycling and soil fertility.
3. *Bioaccumulation*: Some nanoparticles can enter the food chain. accumulated in higher living beings and cause long-term damage to the ecosystem (Shah et al. 2017).

15.5.3 Interaction with Pollutants Chemical Reaction

Nanoparticles can react with existing pollutants. And it can form more harmful compounds. For example, in sunlight, titanium dioxide nanoparticles can create reactive oxygen species (ROS), which increases toxicity. Contaminant transport: Nanoparticles can absorb and transport pollutants, which spreads pollution in the ecosystem (Shah et al. 2017).

15.5.4 Air Pollution

1. Air quality: Nanoparticles released during production or combustion can cause air pollution. Affects human health and the environment.
2. Climate impacts: Some nanoparticles, such as carbon black, can affect atmospheric processes, which contributes to climate change.

15.5.5 Challenges in Waste Management

1. Difficulty in recycling: Nanoparticles embedded in materials are difficult to separate and recycle.
2. Landfill leachate: Improper disposal causes leaching of nanoparticles, which contaminates the soil and groundwater.

15.5.6 Mitigation Strategies

1. Green synthesis: Developing biodegradable nanoparticles using sustainable methods.
2. Life cycle analysis: A comprehensive assessment of nanoparticle production, use and disposal to reduce environmental impact.
3. Regulations: Set guidelines for the release and use of nanoparticles to protect ecosystems. Addressing these effects through innovation and regulation can reduce the environmental impact of nanoparticles and responsibly exploit technological benefits (Mahlambi et al. 2015).

15.6 Impact on the Ecosystem

15.6.1 Impact of Nanoparticles on Ecosystems

The release of nanoparticles into the environment raises concerns because they can damage ecosystems. Their small size, high reactivity, and persistence can cause a number of negative effects on aquatic and terrestrial microbial ecosystems.

15.6.2 Disruption of Aquatic Ecosystems

1. Toxicity to aquatic life: Nanoparticles such as silver (Ag) and titanium dioxide (TiO_2) can be toxic to fish, algae, and crustaceans. It causes oxidative stress and damages cell membranes.
2. Bioaccumulation: Nanoparticles can accumulate in aquatic organisms. It enters the food chain and affects higher nutritional levels, including humans.
3. Inhibiting algal growth: Some nanoparticles inhibit photosynthesis by destroying chloroplasts. This results in a decrease in primary production in the aquatic environment (Patel et al. 2023a).

15.6.3 Impact on Terrestrial Ecosystems

1. Soil toxicity: Nanoparticles such as zinc oxide (ZnO) and copper oxide (CuO) can change soil pH and reduce nutrient availability, which has a negative effect on plant growth.
2. Effects on soil microorganisms: The introduction of nanoparticles can destroy microbial diversity. Affects important processes such as nitrogen fixation, and the decomposition of organic substances (Sánchez-García et al. 2021).

15.6.4 Disruption of Microbial Communities

1. Antimicrobial effects: Although silver and other nanoparticles are used for their antimicrobial properties, releasing them into the environment can kill beneficial microorganisms that are essential for nutrient cycling and ecosystem balance.
2. Development of resistance: Prolonged exposure to nanoparticles can lead to the development of nanoparticle-resistant microbial strains. This may disturb the natural microbial population (Qian et al. 2018).

15.6.5 Ecotoxicological Effects

1. Chemical interactions: Nanoparticles can absorb other pollutants such as pesticides and heavy metals. This increases toxicity and mobility in the ecosystem.
2. Tropical transfer: Nanoparticles can move to different trophic levels. Affects predators and prey, and cause long-term disturbances to the ecological balance (Zhang et al. 2022).

15.6.6 Change of Residence

1. Sediment pollution: Nanoparticles living in sediment can change the chemical and physical properties of aquatic habitats. Affecting living things that depend on shallow water ecosystems.
2. Airborne nanoparticles: In terrestrial ecosystems, nanoparticles can accumulate on plants. This affects the processes of photosynthesis and transpiration (Kumar et al. 2021).

Nanoparticles pose a significant risk to ecosystems by affecting biodiversity, nutrient cycling, and ecological balance. Addressing these challenges requires a robust risk assessment framework. Sustainable synthesis methods and strict environmental regulations to reduce the ecological footprint.

15.7 Toxicity of Nanoparticles to Aquatic and Terrestrial Organisms

The unique physicochemical properties of nanoparticles such as their small size and high surface area and reaction makes these particles highly effective in a variety of applications. But these same properties also contribute to their potential toxicity to aquatic and terrestrial life (Kumar et al. 2016).

15.7.1 Toxicity to Aquatic Organisms (Liu et al. 2010)

Nanoparticles that enter water bodies through industrial discharges, runoff, or improper waste management can have a negative effect on aquatic ecosystems.

(a) Oxidative stress: Nanoparticles such as titanium dioxide (TiO_2) and zinc oxide (ZnO) create reactive oxygen species (ROS) that damage cell structures, lipids, proteins, and DNA in aquatic organisms.
(b) Membrane disruption: Nanoparticles such as silver (Ag) can bind to cell membranes. This causes structural damage and interferes with ion exchange. Resulting in cell death.
(c) Bioaccumulation and bio measurement: Persistent nanoparticles accumulate in aquatic organisms such as fish and crustaceans. This causes transfer up the food chain and poses a risk to higher organisms, including humans (Miller et al. 2010).
(d) Growth inhibition: Algae and phytoplankton, which are important in aquatic food webs, are sensitive to nanoparticles. This can reduce photosynthesis and nutrient absorption, disturbing the productivity of the ecosystem (Keller et al. 2010).

15.7.2 Toxicity to Terrestrial Life

Nanoparticles released into the soil from industrial waste, fertilizers, or airborne deposits can have adverse effects on plants, animals, and soil organisms (Garner and Keller 2014).

(a) Phytotoxicity: Nanoparticles such as copper oxide (CuO) and zinc oxide (ZnO) can inhibit seed germination, root elongation, and photosynthesis by interfering with cellular processes and causing oxidative stress in plants (Zhou et al. 2012).
(b) Impact on soil microorganisms: Beneficial microorganisms involved in nutrient cycling and organic matter decomposition are particularly sensitive to nanoparticles. Reduced activity can affect soil fertility and plant growth.
(c) Toxicity to invertebrates: Glassworms and insects exposed to nanoparticles may undergo bioaccumulation, reduced reproduction, and mortality due to the oxidative damage caused by nanoparticles and changes in enzyme activity.
(d) Risk to wildlife: Large land animals exposed to nanoparticles through contaminated water, soil, or feed may suffer organ damage, immune deficiency, and reproductive system problems.

15.7.3 Factors Affecting Toxicity (Zhang et al. 2024)

1. Nanoparticle type: Metal-based nanoparticles (e.g., Ag, ZnO) are generally more toxic than carbon-based nanoparticles (e.g., graphene).
2. Concentration: Higher nanoparticle concentration increases toxicity.

3. Size and shape: Nanoparticles that are smaller in size and have sharp edges can more easily penetrate biological barriers. Increases the potential for toxicity.
4. Surface properties: Coated or functionalized nanoparticles may interact differently with living organisms. This may reduce or increase toxicity.

Nanoparticles pose a significant risk to aquatic and terrestrial life. This is because it is highly reactive and persists in the environment. Developing safer nanomaterials Establishing regulations for the release of nanoparticles and comprehensive eco-toxicological studies. It is important to reduce these impacts (Handy et al. 2008).

15.8 Strategies for Reducing Environmental Impacts

The following strategies can be used to reduce the environmental impact of biodegradable polymers. Recycling and reuse of nanocomposites and the use of environmentally friendly synthesis techniques (Kalia et al. 2021).

15.8.1 Strategies for Biodegradable Polymers (Raza et al. 2020)

1. Sustainable design and use of increased biodegradability: Use natural polymers (such as cellulose, starch, or chitosan) or blends to ensure complete biodegradation.
2. Target use: Deploy biodegradable polymers in applications with potential exposure to the environment (e.g., packaging, agriculture).
3. Life cycle analysis: Assess the environmental impact of biodegradable polymers throughout their life cycle.
4. Control decomposition and optimize the degradation rate: Design the polymer to decompose within a specified time frame according to the application needs.
5. Environmental compatibility: Ensure that the decomposition by-products are non-toxic, environmentally friendly, and production efficiency.
6. Renewable raw materials: Use renewable resources, such as monomers derived from plants or agricultural residues.
7. Energy efficiency optimization: Use energy-efficient production processes with minimal waste production.

15.8.2 Recycling and Recovery of Nanocomposites Separation Techniques

1. Mechanical separation: Develop processes to isolate nanocomposite components (e.g., polymers and fillers) for reuse.

2. Chemical recycling: Use solvents or catalysts to break down nanocomposites into their base materials.
3. Thermal methods: Employ pyrolysis or gasification to recover valuable elements.

15.8.3 Closed-Loop Systems

1. Circular economy: Design nanocomposites for easy recycling and reusability, reducing the need for virgin materials.
2. Collection programs: Establish systems for collecting and recycling end-of-life nanocomposite products (Liu et al. 2024).

15.8.4 Material Recovery Innovations

1. Nano-additive extraction: Develop methods to recover nanoparticles from composite waste for reuse.
2. Matrix decomposition: Research biodegradable matrices for nanocomposites to minimize persistent waste.

15.8.5 Adoption of Green Synthesis Techniques

Environmentally friendly methods

(a) Biosynthesis: Using microorganisms, enzymes, or plant extracts to produce nanoparticles and polymers.
(b) Solvent-free process: Replace toxic solvents with water or other environmentally friendly alternatives.
(c) Low energy techniques: Use synthesis methods that require the least amount of energy, such as reactions at room temperature.

To reduce waste and emissions

(a) Nuclear economics: Optimizing reactions to ensure maximum incorporation of reactants into the final product.
(b) Waste assessment: Converting synthesis by-products into useful secondary products.

Scalable and secure protocol

(a) Scalability: Develop green synthesis methods suitable for industrial applications.
(b) Toxicity testing: Ensure that all reagents and procedures pose minimal health and environmental risks.

Strategy integration

(a) Combination method: Using biodegradable polymers in nanocomposites to produce environmentally friendly materials.
(b) Research and development investment: Fund research on advanced regeneration methods and green synthesis innovations.
(c) Policy support: Support regulations that promote sustainable materials and technologies.

By using these strategies industry and researchers can effectively reduce their environmental impact, while increasing the sustainability of materials and processes (Zhang et al. 2023).

15.9 Safety Considerations

It is important to consider safety when working with biodegradable polymers, nanocomposites, and green synthesis techniques. Below is a comprehensive guide for managing toxicological concerns, occupational health hazards, and consumer safety:

15.9.1 Toxicology Concerns of Biodegradable Polymers

(a) Toxic decomposition byproducts: Ensure that decomposition products (e.g., microplastics, monomers) are non-toxic and environmentally friendly.
(b) Allergic reactions: Evaluation of polymers for possible allergic reactions. This is especially true in biomedical applications.
(c) Ecotoxicity: Assess the impact of biodegradable materials on aquatic and terrestrial ecosystems (Bischoff 1972).

Nanocomposites

(a) Nanotoxicity: Monitor the size, shape, and reactivity of nanoparticles to prevent unintended biological reactions.
(b) Bioaccumulation: Assess the risk of nanoparticles accumulating in living organisms and entering the food chain.
(c) Reaction with pollutants: Investigate how nanocomposites interact with environmental pollutants. This may increase the toxicity of those substances.

Green Synthesis

(a) Chemical residue: Reduces potentially hazardous solvents, reagents, or by-products.
(b) Biological substances: Tests for pathogenicity or allergic properties of biological substances used in synthesis (Shanker et al. 2016).

15.9.2 Occupational Health Hazards (Donoghue 2004)

Risk of exposure

(a) Breathing: Use appropriate ventilation and respirators to minimize exposure to nanoparticles, fine powders, or volatile compounds.
(b) Skin contact: Use gloves and protective clothing to prevent skin contact with this potentially reactive and toxic material.
(c) Consumption: Establish strict hygiene protocols to avoid accidental ingestion of hazardous substances.

Safety measures in the workplace

(a) Training program: Educate employees about the risks associated with nanoparticles, polymers, and green synthesis techniques.
(b) Engineering controls: Install a fume hood, closed system, and nanoparticle filtration unit.
(c) Monitoring system: Uses sensors to detect and monitor nanoparticle and chemical vapor levels in the workplace. Long-term health tracking.

Long-Term Exposure Studies

(a) Chronic exposure studies: Monitor employees for long-term effects from exposure to small amounts of nanoparticles or synthetic by-products.
(b) Data collection: Maintain employee exposure records to ensure appropriate risk assessment and management.

15.9.3 Consumer Safety

15.9.3.1 Product Safety

Verify by migration testing that no food, water, or other consumer goods are contaminated by nanoparticles or degradation byproducts. In order to prevent negative reactions, biocompatibility should be tested for safety in medical applications (such as implants and medication delivery systems).

Labeling: Indicate the safety characteristics and disposal procedures of products that incorporate nanocomposites or biodegradable polymers.

15.9.3.2 Environmental Considerations

End-of-life safety: Make sure that items break down without leaving dangerous residues in water or land.

To ensure the safety of children, make sure that the materials used in toys or childcare goods are non-toxic and devoid of small parts that could be consumed (Viscusi 1985).

15.9.3.3 Regulatory Compliance

Consumer Protection Standards: Comply with safety regulations for chemicals and materials established by agencies such as the FDA, EPA, and REACH. Acquire eco-labels or certifications to reassure customers about the sustainability and safety of the product.

15.9.3.4 Integrated Risk Management

(a) Holistic risk assessment: At every stage of the product lifecycle, carry out comprehensive risk studies taking toxicological, worker safety, and end-user effect into account.
(b) Stakeholder collaboration: To harmonize safety procedures, collaborate with industry experts, regulatory authorities, and consumer advocates.
(c) Constant monitoring: Evaluate and revise safety procedures on a regular basis in light of new scientific discoveries.
(d) By addressing these safety considerations, the risks associated with biodegradable polymers, nanocomposites, and green synthesis can be effectively minimized, promoting sustainable development without compromising health or safety (Meulbroek 2008).

15.9.4 Toxicological Concerns

Because of their special qualities, nanoparticles (NPs) are used extensively, but they also present serious toxicological risks, especially to people in commercial and residential settings. A thorough examination of these issues is provided below:

15.10 Toxicity of Nanoparticles to Humans

15.10.1 Size-Dependent Effects

(a) Cellular penetration: Nanoparticles can get collected in delicate tissues like the brain, lungs, and liver because of their small size, which allows them to pass through biological barriers like the placenta and blood-brain barrier.

(b) Enhanced reactivity: Nanoparticles have a greater surface area to volume ratio, which raises their oxidative stress potential and chemical reactivity.

15.10.2 Toxicological Mechanisms

(a) *Oxidative stress*: Reactive oxygen species (ROS) produced by nanoparticles can cause inflammation, apoptosis, and cellular damage.
(b) *Genotoxicity*: Certain nanoparticles have the potential to cause cancer by causing chromosomal abnormalities, DNA damage, or mutations.
(c) *Immunotoxicity*: Nanoparticles have the potential to cause aberrant immunological reactions, such as immunosuppression or hypersensitivity.
(d) *Cytotoxicity*: Direct exposure may cause damage to organelles, disrupted metabolic processes, and cell death (Sahu et al. 2014).

Examples of Toxic Nanoparticles

1. At high quantities, metallic nanoparticles (such as gold and silver) are known to have cytotoxic and genotoxic effects.
2. Carbon-based nanoparticles, such as carbon nanotubes, have been connected to inflammation and respiratory toxicity.
3. Because they leak heavy metals like cadmium, quantum dots have the potential to be harmful (Kumah et al. 2023).

15.10.3 Long-Term Exposure Risks in Industrial Settings

Occupational Health Concerns

1. Risks of Inhalation: Nanoparticles can enter the lungs deeply and cause cancer, fibrosis, or chronic respiratory disorders.
2. Long-term contact to the skin might cause nanoparticles to enter through hair follicles or micro abrasions.
3. Systemic Distribution: Following absorption, nanoparticles have the ability to move through the bloodstream, where they may build up in organs and result in systemic toxicity (Peckham et al. 2017).

Chronic Conditions

1. Respiratory Conditions: Asthma, bronchitis, and pulmonary fibrosis have all been linked to prolonged exposure to airborne nanoparticles.
2. Neurotoxicity: Because of the buildup of nanoparticles in the brain, prolonged exposure might cause neurodegenerative disorders.
3. Carcinogenicity: Regulatory agencies such as the IARC have identified certain nanoparticles (such as carbon nanotubes and titanium dioxide) as possible carcinogens.

Exposure Pathways in Industries

1. Manufacturing and processing: Employees who work with nanoparticles have a serious danger of ingesting, inhalation, or skin exposure.
2. Accidental releases: Localized contamination may result from spills or leaks that occur during the manufacture of nanoparticles (Cheng et al. 2022).

15.10.4 Long-Term Exposure Risks in Consumer Settings

Chronic Low-Dose Exposure

1. Cosmetics and personal care items: Extended exposure to nanoparticles through the skin or respiratory system (such as titanium dioxide in sunscreens) can result in cumulative toxicity.
2. Food Packaging: Over time, nanoparticles used in packaging may seep into food and cause ingestion.
3. Household Products: Nanoparticles in paints, cleaning products, and electronics might accidentally expose people through their skin or through their breath (Hurlbert et al. 2022).

Bioaccumulation

1. Persistence in the Body: Long-term health impacts such organ damage or dysfunction might result from nanoparticle accumulation in the liver, kidneys, and spleen.
2. Environmental Reintroduction: By bioaccumulating in food chains, nanoparticles from consumer trash may re-enter ecosystems and have an indirect effect on human health (Wang 2016).

15.10.5 Risk Mitigation Strategies (Shamsuzzaman et al. 2024)

In Industrial Settings

1. Engineering Controls: To reduce exposure, use fume hoods, closed systems, and nanoparticle filters.
2. Personal Protective Equipment (PPE): Make sure that gloves, respirators, and protective clothes are used correctly.
3. Monitoring and Surveillance: Keep a close eye on surface and air pollution and do worker health surveillance on a regular basis.

In Consumer Settings

1. Safe product design: prevent the leaching of nanoparticles in consumer goods, particularly in food and cosmetics.

2. Regulatory compliance: comply with EPA and REACH safety regulations for the use of nanoparticles.
3. Public awareness: inform customers about how to use and dispose of products that include nanoparticles

Long-Term Research

1. Toxicological studies: Conduct in-depth studies on chronic exposure to nanoparticles and their long-term effects.
2. Alternative materials: Explore safer nanomaterials with similar functional properties but lower toxicity.

By addressing these concerns proactively, industries can ensure the safe use of nanoparticles while minimizing health risks to workers and consumers.

15.11 Current Guidelines and Regulatory Framework

Global efforts are being made to regulate nanocomposites in order to solve the particular difficulties presented by their nanoscale characteristics and guarantee the safety of customers, employees, and the environment (Torens et al. 2022).

15.11.1 Synopsis of International Nanocomposites Regulations

15.11.1.1 United States

1. The Environmental Protection Agency (EPA) mandates that producers comply with Significant New Use Rules (SNURs) and submit pre-manufacture notices for nanomaterials under the Toxic Substances Control Act (TSCA).
2. To guarantee product safety, the Food and Drug Administration (FDA) regulates nanomaterials in medications, cosmetics, and materials that come into contact with food.
3. The Occupational Safety and Health Administration (OSHA) is responsible for enforcing current safety regulations, which include workplace controls and hazard communication to reduce exposure to nanoparticles (Garcia et al. 2018).

15.11.1.2 European Union (EU)

1. REACH (Registration, Evaluation, Authorization, and Restriction of Chemicals): Requires that nanomaterials, including nanocomposites, that are produced or imported in amounts greater than one ton annually be registered and have their risks evaluated.

2. The Classification, Labeling, and Packaging (CLP) Regulation mandates that nanomaterials be labeled in order to communicate hazards.
3. The Cosmetic Products Regulation makes sure that certain safety tests are performed on nanocomposites used in cosmetics.

15.11.1.3 The Asia-Pacific Area

1. China is in charge of ensuring the safety of nanomaterials through its chemical laws and new rules tailored to nanotechnology.
2. Japan: Uses the Chemical Substances Control Law to promote adherence to voluntary industry standards for nanomaterials.
3. India: The Ministry of Environment, Forests, and Climate Change is currently creating frameworks to address the concerns associated with nanomaterials.

15.11.1.4 Global Institutions

1. The Organization for Economic Co-operation and Development, or OECD, offers guidelines for evaluating the safety of nanomaterials and encourages uniformity across its member nations.
2. International Organization for Standardization (ISO) Develops international standards for terminology, characterization, and testing of nanomaterials (Gressler et al. 2020).

15.12 Challenges in Implementing Regulations for Nanocomposites (Heidari et al. 2024)

15.12.1 Nanocomposites complexity

1. Multifaceted Structures: Because nanocomposites frequently include bulk and nanoscale elements, it can be difficult to forecast their dangers, interactions, and behavior.
2. Diverse uses: From electronics to biomedical devices, the broad range of uses for nanocomposite materials is difficult for regulatory frameworks to handle.

15.12.2 Uncertainty in Science

1. Lack of Standardized Testing: Regulatory choices are made more difficult by the lack of widely recognized techniques for exposure assessment and toxicity testing.

2. Long-Term Risks: Risk assessment is hampered by a lack of information regarding the long-term effects of nanocomposites on the environment and human health.

15.12.3 International Harmonization Deficits

1. Inconsistent Standards: International trade and compliance are hampered by differences in definitions and regulatory thresholds among nations
2. Limited Cooperation: The establishment of unified international standards is delayed by regulatory organizations' inadequate cooperation.

15.12.4 Observation and Implementation

1. Resource Limitations: Regulatory bodies frequently lack the tools and know-how necessary to properly monitor compliance with nanomaterials.
2. Dynamic Industry: Nanotechnology is developing at a faster rate than rules are being developed and put into place.

15.12.5 Opposition from Industry

1. Economic Concerns: Because of the possible costs associated with compliance and innovation, industries may oppose strict rules.
2. Businesses may conceal important information about nanoparticles due to proprietary concerns, which is known as Confidential Business Information (CBI).

15.12.6 Recommendations for Improving Regulatory Frameworks

1. Standardization efforts: Create globally accepted procedures for nanocomposites testing and risk evaluation.
2. Capacity building: Give regulatory agencies the tools and instruction they need to effectively monitor and enforce laws.
3. Public-private collaboration: To harmonize safety and innovation goals, promote communication among industries, researchers, and regulators.
4. Global harmonization: Encourage global accords to create uniform legal structures.
5. Transparency: While preserving intellectual property, require the publication of safety information and the composition of nanomaterials.

Global initiatives can guarantee the safe and sustainable use of nanocomposites by tackling these issues and fortifying regulatory structures.

15.13 Sustainable Strategies

The sustainable use of the environment to access eco-friendly nanocomposites are a diverse material of huge applications. Nevertheless, both during production and disposal, these materials have environmental implications. A sustainable strategy will consider the use of biodegradable resources, which could be cow-pea. Purposely designed masses of these substances minimalize their environmental impacts (El-Sayed and Youssef 2023).

1. Renewable and natural resources possibilities in agropolymer matrices: PLA, PHA, starch, cellulose, and chitosan are very familiar renewable biopolymers.
2. Advantage: Derived from renewable sources such as plants, algae, and microbes, these gains mark a sustainable beginning. Meanwhile, decay at the end helps restore the land that was once burdened by waste.

15.13.1 Natural Nanofillers (Kausar 2021)

Examples:

1. Cellulose nanocrystals (CNCs): High mechanical strength and renewable.
2. Chitin and chitosan nanofibers: Biocompatible and biodegradable.
3. Clay nanoparticles: Naturally occurring and widely available.

Benefits:

1. Non-toxic and renewable.
2. Enhance properties like strength, thermal stability, and barrier performance.

15.13.2 Green Synthesis Techniques

1. **Biological methods:** Steps must be taken to harness the incredible power of mixed microbial cultures, isolated microbes, their extracellular enzymatic activities, and cell-free extract from microorganisms, yeast, and plant sources. In words, according to conventional strategies, selective sucrose or glucose fermentation into bioethanol is obtained. Biotechnological platforms and infrastructural development are ideas for a green technology (Patel et al. 2023b).
2. **Solid waste management:** The environment should be an essential aspect to consider as technology progresses. Paris has held conferences. Both strategies recognize the same importance related to the natural environment (Aswathi et al. 2023).

3. **Evidence-based science:** One of the crucial aims of green chemistry is the screening, evaluation, and validation of various technical solutions pioneered by green chemists. Creation of a new area, designing a global target, and promoting educational instruction are some of the approaching paths.

15.13.3 Utilization of Waste in Agriculture

Applications: The application of nanocomposite materials for the extraction of lignin, cellulose, or silica nanoparticles from agricultural waste (e.g., rice husk, wheat straw).

Advantages: Waste reduction, resource recovery, and reduction in dependency of fossil-based materials (Saratale et al. 2018).

15.13.4 Observed Circular Economy by the Polymer Nanocomposite Industry

Designing for Recycling

1. *Modular composition*: Design a nanocomposite that has components that can be separated easily to aid in their recycling.
2. *Reprocessable polymers*: Use thermoplastics to develop systems with reversible cross-linkage to ensure a repetition in processing.
3. *Biodegradable system*: Develop nanocomposites that get biodegraded in nature after use (Torres-Giner 2023).

Waste Management and Recycling

1. *Development of the recovery of nanofillers*: Develop a process for recovering and reusing nanofillers from disposed-of nanocomposites.
2. *Recycling stream development*: Create a designated pathway for the dismantling of polymer nanocomposites so as to stop contamination of the environment.
3. *Upcycling*: Translate waste into nanocomposites with higher reserves or resources (pyrolysis and gasification provide alternatives) (Sharma et al. 2024).

Closed-Loop Production System

1. Circular supply chains incorporating post-consumer and post-industrial wastes into production cycles.
2. Industrial symbiosis uses waste materials from one industry to play into another as raw inputs.
3. Life-cycle assessment (LCA) was used for production, distribution, use, and disposal, and market forces are assembled toward better design.

Closed-Loop Production System: All nanocomposite manufacturing facilities incorporate renewable energy sources such as solar, wind, or bioenergy.

15.13.5 Innovation in Sustainable Nanocomposites

Hybrid systems: Blend biodegradable matrices with natural nanofillers to create fully environmentally sustainable systems. e.g., PLA reinforced with cellulose nanofibers for packaging.

Smart materials: Create nanocomposites that decompose under specific conditions, e.g., UV light, water, or enzymes, in order to bypass their accumulation in the environment.

Bioinspired approaches: Realizing what is found in nature (nacre, wood) that help in the creation of high-performance materials with less generations of pollution (Andersen et al. 2023).

Benefits of Sustainable Nanocomposites

1. *Environmental impact reduction*: Lesser carbon footprint and consequent pollution.
2. *Resource efficiency*: Maximize the use of sustainable and recycled materials.
3. *Economic value*: To support an industrial sector that subscribes to green economies and adheres to consumer preferences regarding eco-initiatives.
4. *Regulatory compliance*: Comply with tougher environmental regulations and standards.

The industry in polymer nanocomposites would advance toward a sustainable future by harnessing renewable resources and conformation to the principles of the circular economy.

Eco-friendly nanocomposites are developing quickly, but in order to maximize sustainability, a number of research gaps still need to be filled. Investigating novel approaches to augment the environmental, economic, and social advantages of nanocomposites is crucial as the need for nanoparticles increases. Some important research gaps and future directions are listed below (Shaik and Franchetti 2022).

15.14 Research Deficits in Eco-Friendly Nanomaterials

15.14.1 Limited Knowledge on the Environmental Impact of Nanomaterials

1. *Toxicity and ecotoxicity data*: Information about the toxicity and ecotoxicity of nanoparticles is available, but there are not many thorough research works on how they affect ecosystems, wildlife, and people over the long run.
2. *Biodegradation and bioaccumulation*: Further study is required to comprehend the processes by which nanomaterials degrade and accumulate in biological systems.

3. *Nano-bio-interactions*: Investigating the interactions between nanomaterials and living organisms to assess potential environmental risks is crucial for developing safer nanocomposites (Balusamy et al. 2023).

15.14.2 Sustainable Methods' Scalability and Industrialization

1. *Green synthesis techniques*: A lot of environmentally friendly synthesis techniques, such as solvent-free or biological processes, are still in the lab stage. It is still difficult to scale these methods for industrial manufacturing.
2. *Incorporation of natural nanofillers*: Although natural fillers such as chitin and cellulose exhibit considerable potential, little is known about how to economically include them into the manufacturing of large-scale nanocomposite materials.
3. *Energy consumption in production*: To be cost-effective on an industrial scale, the energy requirements of contemporary sustainable production techniques (such as low-temperature synthesis) must be optimized (Oshilalu 2024).

15.14.3 Nanocomposites: Recycling and Reuse

Recycling difficulties: Because of their small size and dispersion, nanomaterials are difficult to recover and reuse from end-of-life items.

End-of-life management: To create techniques for recycling and disposing of nanocomposites that avoid contaminating the environment, research is required (Zhang et al. 2011).

Effective closed-loop solutions: As recycling nanocomposite materials are scarce, improved waste management techniques and recyclability-focused design are required (Patel et al. 2021).

15.15 Absence of Regulatory Frameworks and Standardization

Developing industry-wide best practices is hampered by the lack of a unified standard for testing the safety and environmental impact of nanocomposites. Global regulatory alignment: Disparate national regulatory regimes for nanocomposites make international trade and compliance difficult. For cross-border uniformity and safety, it is essential to harmonize laws governing sustainable nanocomposites.

To create life-cycle assessment (LCA) models for nanocomposites that take into consideration all environmental effects throughout the manufacture, use, and disposal stages, more study is needed (Ingvarson and Hassel 2023).

15.16 Prospects for Sustainable Nanocomposites in the Future

15.16.1 *Promoting Biodegradable Materials and Green Chemistry*

Bio-based polymers: Future studies should concentrate on creating biopolymers that are completely biodegradable, use renewable feedstocks, and require less processing energy.

Green synthesis pathways: To lessen the ecological impact of nanocomposites, it will be essential to expand sustainable synthesis approaches that employ low-energy, ecologically friendly solvents, reagents, and procedures.

Enzyme-driven synthesis: By eliminating the need for hazardous chemicals, enzymatic methods may provide a low-energy, eco-friendly method of creating nanomaterials (Iroegbu and Ray 2022).

15.16.2 *Developments in the Circular Economy and Nanocomposite Recycling*

Effective recovery techniques: New developments in recovery procedures that enable the effective removal of nanomaterials from waste streams (for example, by chemical, magnetic, or electrostatic means) will make them easier to reuse and lessen their negative effects on the environment.

Recycling-compatible nanocomposites: By creating nanocomposites with recyclability in mind (like reversible polymer networks), a circular economy will be promoted by their ability to be reused in various product lifecycles.

Circular supply chains: By including recycling into the design and manufacturing processes, closed-loop systems for nanocomposite materials can be established, improving industrial sustainability (Khosravi et al. 2024).

15.16.3 *Improved Environmental Risk Evaluation and Regulations*

Nanomaterial safety protocols: To evaluate the safety and long-term impacts of nanocomposite materials, standardized procedures for testing their toxicity, persistence, and eco-compatibility will be crucial.

Regulation of emerging nanomaterials: Creating strong regulatory frameworks that take into consideration the unique properties and possible hazards of innovative nanocomposites will aid in directing their safe application across a range of industries.

Models for risk-benefit analysis: To ensure sustainable innovation, future studies should investigate models for weighing the possible ecological dangers of nanocomposites against their environmental advantages (such as their durability and light weight) (Nielsen et al. 2021).

15.16.4 Innovation and Collaboration Across Disciplines

Interdisciplinary cooperation between material scientists, environmental scientists, engineers, and policymakers will be essential for addressing the intricate problems related to sustainable nanocomposites.

Innovation in consumer products: In order to meet global sustainability targets, researchers should look into novel uses of nanocomposites in green technology, such as biodegradable packaging or renewable energy devices (solar cells, energy storage).

Public awareness and adoption: Educating customers about the advantages and environmental effects of sustainable nanocomposites can increase demand for green products and motivate producers to use more environmentally friendly processes (Wang and Su 2022).

15.17 Conclusion

Resolving the present research gaps in environmental effect, scalability, recycling, and standardization is crucial for the future of sustainable nanocomposites. Nanocomposites have the potential to play a significant role in a sustainable, circular economy if concentrated efforts are made to create environmentally acceptable materials, enhance recycling infrastructure, and create thorough regulatory frameworks. The nanocomposite sector may make a substantial contribution to lessening the environmental impact of contemporary materials through ethical design and cross-disciplinary innovation.

References

Andersen AL, Brunoe TD, Bockholt MT, Napoleone A, Hemdrup Kristensen J, Colli M, Vejrum Wæhrens B, Nielsen K (2023) Changeable closed-loop manufacturing systems: challenges in product take-back and evaluation of reconfigurable solutions. Int J Prod Res 61(3):839–858

Arora R, Roy T, Adak P (2024) A review of the impact of nanoparticles on environmental processes. In: BIO web of conferences, vol 86. EDP Sciences, Les Ulis, p 01001

Aswathi VP, Meera S, Maria CA, Nidhin M (2023) Green synthesis of nanoparticles from biodegradable waste extracts and their applications: a critical review. Nanotechnol Environ Eng 8(2):377–397

Balusamy SR, Joshi AS, Perumalsamy H, Mijakovic I, Singh P (2023) Advancing sustainable agriculture: a critical review of smart and eco-friendly nanomaterial applications. J Nanobiotechnol 21(1):372
Bischoff F (1972) Organic polymer biocompatibility and toxicology. Clin Chem 18(9):869–894
Camargo PHC, Satyanarayana KG, Wypych F (2009) Nanocomposites: synthesis, structure, properties, and new application opportunities. Mater Res 12(1):1–39
Carroccio SC, Scarfato P, Bruno E, Aprea P, Dintcheva NT, Filippone G (2022) Impact of nanoparticles on the environmental sustainability of polymer nanocomposites based on bioplastics or recycled plastics—a review of life-cycle assessment studies. J Clean Prod 335:130322
Chaisan W, Yimnirun R, Ananta S (2009) Preparation and characterization of ceramic: nanocomposites in the PZT-BT system. Ceram Int 35(1):121–124
Cheng L, Wei W, Guo A, Zhang C, Sha K, Wang R, Wang K, Cheng S (2022) Health risk assessment of hazardous VOCs and its associations with exposure duration and protection measures for coking industry workers. J Clean Prod 379:134919
Chowdhury PR, Bhattacharyya KG (2018) 26—Toxicology and environmental fate of polymer nanocomposites. In: Hussain CM, Mishra AK (eds) New polymer nanocomposites for environmental remediation. Elsevier, Amsterdam, pp 649–677
Darwish MSA, Mostafa MH, Al-Harbi LM (2022) Polymeric nanocomposites for environmental and industrial applications. Int J Mol Sci 23(3):1023
Donoghue AM (2004) Occupational health hazards in mining: an overview. Occup Med 54(5):283–289
El-Sayed SM, Youssef AM (2023) Eco-friendly biodegradable nanocomposite materials and their recent use in food packaging applications: a review. Sustain Food Technol 1(2):215–227
Gao F (ed) (2012) Advances in polymer nanocomposites: types and application. Woodhead Publishing, Cambridge
Garcia CV, Shin GH, Kim JT (2018) Metal oxide-based nanocomposites in food packaging: applications, migration, and regulations. Trends Food Sci Technol 82:21–31
Garner KL, Keller AA (2014) Emerging patterns for engineered nanomaterials in the environment: a review of fate and toxicity studies. J Nanopart Res 16(10):1–14
Gemachu LY, Bogale RF (2024) A review on the three types of nanocomposites synthesis, characterization and their applications in different areas. Preprints 2024:011201
Gressler S, Prenner S, Kurz A, Resch S, Pavlicek A, Part F (2020) Polymer nanocomposites—additives, properties, applications, environmental aspects. In: NanoTrust-Dossier No. 052 en
Handy RD, Owen R, Valsami-Jones E (2008) The ecotoxicology of nanoparticles and nanomaterials: current status, knowledge gaps, challenges, and future needs. Ecotoxicology 17(5):315–325
Heidari M, Al Hasani S, Thangavel S, Kumar A (2024) Challenges and perspectives of polymeric nanocomposites for biomedical applications. In: Nanomanufacturing techniques in sustainable healthcare applications, pp 317–340
Hurlbert M, Shasko L, Neetz M (2022) Addressing risk perceptions of low-dose radiation exposure. Dose Response 20(2):15593258221088428
Hussain F, Hojjati M, Okamoto M, Gorga RE (2006) Review article: polymer-matrix nanocomposites, processing, manufacturing, and application: an overview. J Compos Mater 40:1511–1575
Ingvarson J, Hassel H (2023) On the strength of arguments related to standardization in risk management regulations. Saf Sci 158:105998
Iroegbu AO, Ray SS (2022) Recent developments and future perspectives of biorenewable nanocomposites for advanced applications. Nanotechnol Rev 11(1):1696–1721
Kalia S, Thakur VK, Kaur R et al (2021) Recent advances in biodegradable polymers and their nanocomposites for sustainable packaging applications. J Polym Environ 29(1):1–15
Kausar A (2021) Progress in green nanocomposites for high-performance applications. Mater Res Innov 25(1):53–65
Keller AA, Wang H, Zhou D, Lenihan HS, Cherr G (2010) Stability and aggregation of metal oxide nanoparticles in natural aqueous matrices. Environ Sci Technol 44(6):1962–1967

Khan I, Khan I, Saeed K, Ali N, Zada N, Khan A, Ali F, Bilal M, Akhter MS (2023) Polymer nanocomposites: an overview. In: Smart polymer nanocomposites, pp 167–184
Khosravi A, Zarepour A, Iravani S, Varma RS, Zarrabi A (2024) Sustainable synthesis: natural processes shaping the nanocircular economy. Environ Sci Nano 11(3):688–707
Koelmans AA, Besseling E, Shim WJ (2015) Nanoplastics in the aquatic environment. Critical review. In: Bergmann M, Gutow L, Klages M (eds) Marine anthropogenic litter. Springer, Cham, pp 325–340
Kumah EA, Fopa RD, Harati S, Boadu P, Zohoori FV, Pak T (2023) Human and environmental impacts of nanoparticles: a scoping review of the current literature. BMC Public Health 23:1059
Kumar S, Krishnan S, Mohanty S, Nayak SK (2016) Toxicology and environmental fate of polymer nanocomposites. In: Thomas S (ed) Spectroscopy of polymer nanocomposites. Elsevier, Amsterdam, pp 462–488
Kumar R, Verma A, Shome A, Sinha R, Sinha S, Jha PK et al (2021) Impacts of plastic pollution on ecosystem services, sustainable development goals, and need to focus on circular economy and policy interventions. Sustainability 13(17):9963
Liu J, Wang Z, Li Y et al (2010) Fate and risks of nanomaterials in aquatic and terrestrial environments. Environ Sci Technol 44(6):2169–2175
Liu Y, Wang H, Li Z et al (2024) Recycling and sustainability of nanocomposites. In: Shaik S, Franchetti M (eds) Advances in nanocomposites. CRC Press, Boca Raton, FL, pp 235–250
Mahlambi MM, Ngila CJ, Mamba BB (2015) Recent developments in environmental photocatalytic degradation of organic pollutants: the case of titanium dioxide nanoparticles—a review. J Nanomater 2015(1):790173
Meulbroek LK (2008) A senior manager's guide to integrated risk management. In: Corporate risk management. Columbia University Press, New York, pp 63–86
Miller RJ, Lenihan HS, Muller EB, Tseng RJ, Hanna SK (2010) Impacts of metal oxide nanoparticles on marine phytoplankton. Environ Sci Technol 44(19):7329–7334
Nielsen MB, Baun A, Mackevica A, Thit A, Wallinder IO, Gallego JA, Clausen LP, Rissler J, Skjolding L, Nilsson AC, Cedervall T (2021) Nanomaterials in the European chemicals legislation–methodological challenges for registration and environmental safety assessment. Environ Sci Nano 8(3):731–747
Okamoto M (2023) Polymer nanocomposites. Eng 4:457–479
Oshilalu AZ (2024) Sustainability meets scalability: transforming energy infrastructure projects into economic catalysts through supply chain innovation. Int J Res Publ Rev 5(12):762–779
Patel S, Frost K, Smith JA et al (2021) Recycled carbon nanofiber-polypropylene nanocomposite. Materials 14(11):332
Patel M, Patel V, Patel K et al (2023a) Investigating natural polymer nanocomposites in relation with eco-friendly potential. Int J Res Appl Sci Eng Technol 11(1):123–130
Patel RR, Singh SK, Singh M (2023b) Green synthesis of silver nanoparticles: methods, biological applications, delivery and toxicity. Mater Adv 4(8):1831–1849
Paul DR, Robeson LM (2008) Polymer nanotechnology: nanocomposites. Polymer 49:3187–3204
Peckham TK, Baker MG, Camp JE, Kaufman JD, Seixas NS (2017) Creating a future for occupational health. Ann Work Exposures Health 61(1):3–15
Qian H, Zhang M, Liu G et al (2018) Effects of soil residual plastic film on soil microbial community structure and fertility. Water Air Soil Pollut 229(8):266
Raza ZA, Noreen A, Shahnaz G et al (2020) Enzymatic synthesis of biodegradable polymers: a green approach towards sustainable materials. J Polym Environ 28(3):654–669
Rozenberg BA, Tenne R (2008) Polymer—assisted fabrication of nanoparticles and nanocomposites. Prog Polym Sci 33(1):40–112
Sahu D, Kannan GM, Vijayaraghavan R (2014) Size-dependent effect of zinc oxide on toxicity and inflammatory potential of human monocytes. J Toxic Environ Health A 77(4):177–191
Sánchez-García MD, López-Martínez JE, González-Carrillo S et al (2021) Impact of nanoparticles on the environmental sustainability of polymer nanocomposites. J Environ Manage 287:112–123

Saratale RG, Saratale GD, Shin HS, Jacob JM, Pugazhendhi A, Bhaisare M, Kumar G (2018) New insights on the green synthesis of metallic nanoparticles using plant and waste biomaterials: current knowledge, their agricultural and environmental applications. Environ Sci Pollut Res 25:10164–10183
Shah SN, Shah Z, Hussain M, Khan M (2017) Hazardous effects of titanium dioxide nanoparticles in ecosystem. Bioinorg Chem Appl 2017(1):4101735
Shaik S, Franchetti MJ (2022) Recycling and sustainability of nanocomposites. In: Applications of nanocomposites. CRC Press, Boca Raton, FL, pp 236–259
Shamsuzzaman HM, Mosleuzzaman MD, Mia A, Nandi A (2024) Cybersecurity risk mitigation in industrial control systems analyzing physical hybrid and virtual test bed applications. Acad J Artif Intell Mach Learn Data Sci Manag Inf Syst 1(01):19–39
Shanker U, Jassal V, Rani M, Kaith BS (2016) Towards green synthesis of nanoparticles: from bio-assisted sources to benign solvents. A review. Int J Environ Anal Chem 96(9):801–835
Shaoyun F, Sun Z, Huang P, Li Y, Ning H (2019) Some basic aspects of polymer nanocomposites: a critical review. Nano Mater Sci 1(1):2–30
Sharma V, Negi AS, Sharma N, Parmar Y, Prashanthi B, Sharma P (2024) Reuse and recycling of waste materials for green nanocomposite fabrication. In: E3S web of conferences, vol 537. EDP Sciences, Les Ulis, p 07014
Smith R, Jones M, Taylor S et al (2024) Some basic aspects of polymer nanocomposites: a critical review. J Polym Sci Part B Polym Phys 62(5):345–360
Torens C, Durak U, Dauer JC (2022) Guidelines and regulatory framework for machine learning in aviation. In: AIAA Scitech 2022 Forum, p 1132
Torres-Giner S (2023) Sustainable polymer technologies for a circular economy. Appl Sci 13(10):5864
Viscusi WK (1985) Consumer behavior and the safety effects of product safety regulation. J Law Econ 28(3):527–553
Wang WX (2016) Bioaccumulation and biomonitoring. In: Marine ecotoxicology, pp 99–119
Wang S, Su D (2022) Sustainable product innovation and consumer communication. Sustainability 14(14):8395
Zhang Y, Li J, Li Y et al (2011) Comprehensive and sustainable recycling of polymer nanocomposites. J Mater Chem 21(43):17363–17369
Zhang Y, Li J, Li Y et al (2022) The effects on the function and diversity of environmental microbial communities by nanomaterials. Microorganisms 10(10):2080
Zhang Y, Li J, Li Y et al (2023) A review of plastic waste nanocomposites: assessment of features and recycling potential. Int J Environ Res Public Health 20(12):3050
Zhang Y, Li J, Li Y et al (2024) Toxicity mechanisms of plastic nanoparticles in three terrestrial species: antioxidant system imbalance and neurotoxicity. Environ Toxicol Chem 43(3):567–578
Zhou D, Abdel-Fattah AI, Keller AA (2012) Clay particles destabilize engineered nanoparticles in aqueous environments. Environ Sci Technol 46(14):7696–7703

MIX
Papier aus verantwortungsvollen Quellen
Paper from responsible sources
FSC® C105338

If you have any concerns about our products,
you can contact us on
ProductSafety@springernature.com

In case Publisher is established outside the EU,
the EU authorized representative is:
Springer Nature Customer Service Center GmbH
Europaplatz 3, 69115 Heidelberg, Germany

Printed by Libri Plureos GmbH
in Hamburg, Germany